Ramanujan's Theta Functions

Shaun Cooper

# Ramanujan's Theta Functions

Springer

Shaun Cooper
Institute of Natural and Mathematical
Sciences
Massey University
Auckland, New Zealand

ISBN 978-3-319-56171-4 ISBN 978-3-319-56172-1 (eBook)
DOI 10.1007/978-3-319-56172-1

Library of Congress Control Number: 2017937318

Mathematics Subject Classification: 11-02 (primary); 05A30, 11A55, 11F11, 11F27, 11Y55, 33-02 (secondary)



Printed on acid-free paper

This Springer imprint is published by Springer Nature
The registered company is Springer International Publishing AG
The registered company address is: Gewerbestrasse 11, 6330 Cham, Switzerland

*To Hilary*

# Preface

A theta function is a series of the form

$$\sum_{n=-\infty}^{\infty} q^{n^2} x^n, \quad \text{where} \quad |q| < 1 \quad \text{and} \quad 0 < |x| < \infty,$$

or a generalization or specialization of such a series. Theta functions were studied extensively by Ramanujan and form the central theme of this book.

Quotients of theta functions can be used to construct elliptic functions and modular functions. One of the incredible results discovered by Jacobi is that the quotient of theta functions defined by

$$x = \left( \frac{\sum_{n=-\infty}^{\infty} q^{(n+\frac{1}{2})^2}}{\sum_{n=-\infty}^{\infty} q^{n^2}} \right)^4 \quad \text{for} \quad 0 < q < 1 \tag{1}$$

has an inverse function that is given by the formula

$$q = \exp\left(-\pi \frac{F(1-x)}{F(x)}\right), \quad \text{where} \quad F(x) = \sum_{n=0}^{\infty} \binom{2n}{n}^2 \left(\frac{x}{16}\right)^n \tag{2}$$

and $\binom{2n}{n}$ is the binomial coefficient. Moreover,

$$\sum_{n=0}^{\infty} \binom{2n}{n}^2 \left(\frac{x}{16}\right)^n = \left(\sum_{n=-\infty}^{\infty} q^{n^2}\right)^2. \tag{3}$$

Jacobi's results are fundamental to the theory, computation, and applications of Jacobian elliptic functions and elliptic integrals.

Ramanujan discovered three analogues of Jacobi's result. At the time, they barely received attention. But since the 1990s they have become the subject of intense interest.

The goal of this book is to provide a systematic development of the results of Jacobi and Ramanujan and to extend the results to a general theory. The treatment is comprehensive and provides a detailed study of theta functions and modular forms for levels up to 12. Jacobi's results in (1)–(3) above arise in the theory for level 4, while Ramanujan's alternative results occur in the corresponding theories for levels 1, 2, and 3. The organization of the material lends itself to serving as a useful reference. Additionally, the book contains a rich source of examples for theoreticians in number theory and modular forms. The topics, especially those in the second half of the book, have been the subject of much recent research; many of these also appearing for the first time in book form. Further results are summarized in the numerous exercises at the end of each chapter.

This book has been written for advanced undergraduates, graduate students, and researchers. It is user friendly and in principle can be used to learn the subject from scratch. The prerequisites are fairly straightforward and rely mainly on an understanding of undergraduate complex analysis (especially in terms of the zeros and poles of meromorphic functions, and convergence of infinite series and infinite products) and the basic theory of linear differential equations. The reader is expected to come up to speed quickly with manipulations of series and products; that is inevitable in any serious study of Ramanujan's mathematics.

Chapter 0 contains an introduction to the main identities in $q$-analysis: Jacobi's triple product identity, the quintuple product identity, Ramanujan's summation formula, and the $q$-binomial theorem. We also meet $q$-analogues of the sine and cosine functions, which are denoted by $s(\theta)$ and $c(\theta)$, respectively. The function $s(\theta)$ has zeros that are simple and lie on a lattice, and because of this fundamental structure, it plays an important role throughout the book.

Chapter 1 develops the theory of elliptic functions to deduce identities that can be used to convert infinite products to series (and vice versa). Specific examples that lead to the number of representations of an integer by the quadratic forms $x^2 + 7y^2$ and $x^2 + y^2 + 3z^2 + 3w^2$ are worked through in detail. Properties of the Weierstrass, Jacobian, and Dixon elliptic functions are developed from $q$-infinite products, the latter being the first (to the author's knowledge) in a book.

Chapter 2 contains an introduction to modular transformations with an emphasis on examples. Most books present only the simplest examples and expend a considerable amount of labor to handle conditionally convergent series. We treat Eisenstein series that arise from either the modular group $\Gamma$ or the Hecke congruence subgroup $\Gamma_0(p)$ case in a simple way by starting with absolutely convergent infinite products. The analysis also works for modular forms of weights one and two and avoids the difficulties of conditional convergence that arise in the more standard approaches.

Chapter 3 provides a comprehensive and systematic treatment of the main properties of Ramanujan's theta functions

$$\varphi(q) = \sum_{j=-\infty}^{\infty} q^{j^2} \quad \text{and} \quad \psi(q) = \sum_{j=0}^{\infty} q^{j(j+1)/2}$$

and the Borweins' cubic theta functions $a(q)$, $b(q)$, and $c(q)$. The approach emphasizes the interconnections between the two sets of functions and pays attention to how properties for one set of functions carry over to the other set of functions. The identities are classified by weight, and some have not been published before. The chapter ends with an analysis of representations of integers by quadratic forms and includes results for sums of 2, 4, 6, 8, 10, and 12 squares (and also triangular numbers) as well as a discussion of Ramanujan's general result for sums of any even number of squares and any even number of triangular numbers. Cubic analogues are also analyzed. Exercises include representations by several other quadratic forms, including $x^2 + my^2$ where $m = 2, 3, 4, 5, 6$, and 15. The cases $m = 7$ and $m = 8$ are detailed in Chapters 1 and 8, respectively.

In Chapter 4, the theory of theta functions is used to prove Jacobi's inversion theorem given by (1)–(3). The same procedure is used to prove the main results for Ramanujan's alternative theories. The most significant result is Theorem 4.30, in which all four theories occur in a unified way and where ${}_3F_2$ hypergeometric functions are parameterized by theta functions. The arithmetic-geometric mean and other iterations are analyzed as a consequence, and hypergeometric transformation formulas are obtained.

In Chapter 5 we start with the Rogers–Ramanujan continued fraction and establish its expansion as an infinite product. We prove the Rogers–Ramanujan identities and Ramanujan's partition congruence modulo 5. From there, it is shown that the Rogers–Ramanujan continued fraction is central to a level 5 theory that is analogous to Jacobi's results in (1)–(3). This gives rise to Apéry-type numbers and Zagier's sporadic sequences, and also to two stunning and recent analogues of Clausen's identity for the squares of certain power series.

Chapter 6 centers around Ramanujan's cubic continued fraction. The results in this chapter are the level 6 analogues of the results in Chapter 5. An interesting aspect is that there are three analogues in the level 6 theory of every function for level 5.

The corresponding theories for levels 7–12 are developed in Chapters 7–12, respectively. Each theory has its own idiosyncrasies and individual features. The results in Chapter 11 reach the limits of our approach, which up to this point has been by theta functions. We obtain a glimpse into the important role of the theory of modular forms in studying higher levels. The theory for level 23 can be developed with no extra effort and is also included.

In Chapter 13, the level 10 and level 12 theories are used to develop and systematize an enormous number of hypergeometric modular transformation formulas. An example of one of the beautiful special cases that arise from the identities in this chapter is

$$\begin{aligned}
&\frac{1}{(1-4t)^{5/2}}\, {}_3F_2\left( \begin{matrix} \frac{1}{2}, \frac{1}{2}, \frac{1}{2} \\ 1, 1 \end{matrix} ; -64t\left(\frac{1+t}{1-4t}\right)^5 \right) \\
&= \frac{1}{(1-4t)^{1/2}}\, {}_3F_2\left( \begin{matrix} \frac{1}{2}, \frac{1}{2}, \frac{1}{2} \\ 1, 1 \end{matrix} ; -64t^5\left(\frac{1+t}{1-4t}\right) \right) \\
&= \frac{1}{(1+4t^2)^{3/2}} \sum_{n=0}^{\infty} \left\{ \sum_{j=0}^{n} \binom{n}{j}^4 \right\} \left( \frac{t\,(1+t)\,(1-4t)}{(1+4t^2)^2} \right)^n .
\end{aligned}$$

Finally, in Chapter 14 we show how everything in the book can be brought together to explain a class of series for $1/\pi$, which are due to Ramanujan. This culminates in Tables 14.1 and 14.2, which provide a concise summary of the main results.

My teacher Richard Askey opened the door to Ramanujan's mathematics to me and I have been captivated by the subject ever since. My career has been enriched through interactions with coauthors, and I especially acknowledge Heng Huat Chan, Michael Hirschhorn, and my recent student Dongxi Ye. Chapters 0–4 have been used as the basis of a one-semester graduate course at Massey University since 2001. I am grateful to all of the students who have participated in the course and especially to those who have gone on to write theses, including Uros Abaz, Antesar Aldawoud, Jinqi Ge, Heung Yeung Lam, Lynette O'Brien, and Dongxi Ye. The manuscript was prepared over a number of years and was completed during long leave at the University of Newcastle in Australia. Michael Coons, Mumtaz Hussain, Wadim Zudilin, and the late Jonathan Borwein provided excellent hospitality and a wonderful working environment at Newcastle. Invigorating discussions with Zudilin helped shape the last two chapters. Hilary Boyd, Jesús Guillera, Michael Hirschhorn, Winston Sweatman, and Pee Choon Toh supplied feedback on parts of the manuscript. Elizabeth Loew at Springer provided enthusiastic and expert guidance to bring the manuscript to publication. To all of the people mentioned in this paragraph, I offer my heartfelt thanks.

Auckland, New Zealand
February 2017

Shaun Cooper

# Contents

# List of Tables

# Chapter 0
# Sum to Product Identities

## 0.1 Introduction

The results in this chapter are basic tools that will be used throughout the book. They include the fundamental identities for theta functions such as Jacobi's triple product identity, the quintuple product identity, Ramanujan's summation formula, and the $q$-binomial theorem. We also encounter functions $s(\theta)$ and $c(\theta)$ that are generalizations of the sine and cosine functions, respectively. A study of the coefficients in the power series expansions of $s(\theta)$, $c(\theta)$ and their logarithmic derivatives leads to a system of nonlinear differential equations, called Ramanujan's differential equations, that play an important role in Chapters 4–12.

Throughout this book, $\tau$ will always be a complex variable with positive imaginary part, and $q$ will be defined by

$$q = \exp(2\pi i \tau).$$

Observe that $|q| < 1$.

## 0.2 Euler's identity for infinite products

We will often need to simplify and rearrange infinite products. Here are three examples to illustrate some of the techniques.



S. Cooper, *Ramanujan's Theta Functions*, DOI 10.1007/978-3-319-56172-1_1

First, consider the following sequence of manipulations:

$$\begin{aligned}\prod_{n=1}^{\infty}(1+q^n) &= \prod_{n=1}^{\infty}(1+q^n)\times\frac{(1-q^n)}{(1-q^n)}\\ &= \prod_{n=1}^{\infty}\frac{(1-q^{2n})}{(1-q^n)}\\ &= \prod_{n=1}^{\infty}\frac{(1-q^{2n})}{(1-q^{2n-1})(1-q^{2n})}\\ &= \prod_{n=1}^{\infty}\frac{1}{(1-q^{2n-1})}.\end{aligned}$$

The resulting identity

$$\prod_{n=1}^{\infty}(1+q^n) = \prod_{n=1}^{\infty}\frac{1}{(1-q^{2n-1})} \tag{0.1}$$

is called Euler's identity for infinite products.

The second example is

$$\begin{aligned}\prod_{n=1}^{\infty}(1-(-q)^n) &= \prod_{n=1}^{\infty}(1-q^{2n})(1+q^{2n-1})\\ &= \prod_{n=1}^{\infty}(1-q^{2n})(1+q^{2n-1})\times\frac{(1+q^{2n})}{(1+q^{2n})}\\ &= \prod_{n=1}^{\infty}(1-q^{2n})\times(1+q^n)\times\frac{1}{(1+q^{2n})}\\ &= \prod_{n=1}^{\infty}(1-q^{2n})\times\frac{(1-q^{2n})}{(1-q^n)}\times\frac{(1-q^{2n})}{(1-q^{4n})}.\end{aligned}$$

That is,

$$\prod_{n=1}^{\infty}(1-(-q)^n) = \prod_{n=1}^{\infty}\frac{(1-q^{2n})^3}{(1-q^n)(1-q^{4n})}. \tag{0.2}$$

Euler's product $E(q)$ is defined by

$$E(q) = \prod_{n=1}^{\infty}(1-q^n). \tag{0.3}$$

The identity (0.2) gives the result of replacing $q$ with $-q$ in Euler's product, namely

$$E(-q) = \frac{E^3(q^2)}{E(q)E(q^4)}. \tag{0.4}$$

Here is the third example. Let $\omega = \exp(2\pi i/3)$ and consider the product

$$\prod_{n=1}^{\infty}(1 - q^n)(1 - (\omega q)^n)(1 - (\omega^2 q)^n).$$

Separating the terms into cases according to whether $n$ is divisible by 3 and noting

$$(1 - x)(1 - \omega x)(1 - \omega^2 x) = 1 - x^3$$

we obtain

$$\begin{aligned}\prod_{n=1}^{\infty}(1 - q^n)(1 - (\omega q)^n)(1 - (\omega^2 q)^n) &= \prod_{3|n}(1 - q^n)^3 \times \prod_{3\nmid n}(1 - q^{3n})\\ &= \prod_{n=1}^{\infty}(1 - q^{3n})^3 \times \prod_{n=1}^{\infty}\frac{(1 - q^{3n})}{(1 - q^{9n})}\\ &= \prod_{n=1}^{\infty}\frac{(1 - q^{3n})^4}{(1 - q^{9n})}. \end{aligned} \tag{0.5}$$

In terms of Euler's product, this is

$$E(q)E(\omega q)E(\omega^2 q) = \frac{E(q^3)^4}{E(q^9)}.$$

More generally, if $m \geq 2$ is any integer and $\omega = \exp(2\pi i/m)$, then similar reasoning leads to

$$\prod_{j=0}^{m-1} E(\omega^j q) = \frac{E(q^m)^{m+1}}{E(q^{m^2})}. \tag{0.6}$$

The identities (0.4) and (0.5) correspond to the special cases $m = 2$ and $m = 3$, respectively, of (0.6).

## 0.3 Jacobi's triple product identity

Jacobi's triple product identity is one of the most useful tools in analytic number theory. It provides a factorization of the series $\sum_{n=-\infty}^{\infty} q^{n^2} x^n$. The name "triple product" refers to the three factors that occur in the factorization.

**Theorem 0.1 (Jacobi's triple product identity).** *For $|q| < 1$ and $x \neq 0$, we have*

$$\sum_{n=-\infty}^{\infty} q^{n^2} x^n = \prod_{n=1}^{\infty} (1 + xq^{2n-1})(1 + x^{-1}q^{2n-1})(1 - q^{2n}). \tag{0.7}$$

*Proof.* Fix a value of $q$ and let $f(x)$ be defined by

$$f(x) = \sum_{n=-\infty}^{\infty} q^{n^2} x^n.$$

Clearly,

$$f(x^{-1}) = f(x). \tag{0.8}$$

Shifting the index of summation, we obtain

$$f(x) = \sum_{n=-\infty}^{\infty} q^{(n+1)^2} x^{n+1} = qx \sum_{n=-\infty}^{\infty} q^{n^2} (q^2x)^n = qxf(q^2x). \tag{0.9}$$

On taking $x = -q^{-1}$ in the functional equations (0.8) and (0.9) we find that

$$f(-q) = f(-q^{-1}) = -f(-q)$$

and it follows that

$$f(-q) = 0. \tag{0.10}$$

By repeated application of (0.9) to (0.10) we deduce that $f(-q^{2n-1}) = 0$ for every integer $n$. This analysis does not preclude $f$ from having other zeros. It is easy, however, to construct a function that has simple zeros at each point $x = -q^{2n-1}$ and no other zeros, for example

$$g(x) = \prod_{n=1}^{\infty} (1 + xq^{2n-1})(1 + x^{-1}q^{2n-1})$$

is such a function. Now,

$$\frac{g(x)}{g(q^2x)} = \prod_{n=1}^{\infty} \frac{(1 + xq^{2n-1})(1 + x^{-1}q^{2n-1})}{(1 + xq^{2n+1})(1 + x^{-1}q^{2n-3})} = \frac{1 + qx}{1 + q^{-1}x^{-1}} = qx,$$

so $g$ satisfies the same functional equation as $f$ in (0.9).

The quotient $h(x) = f(x)/g(x)$ has no poles in the annulus $0 < |x| < \infty$, is analytic in the annulus, and satisfies the functional equation

$$h(q^2x) = h(x). \tag{0.11}$$

If we denote the Laurent expansion of $h(x)$ by

$$h(x) = \sum_{n=-\infty}^{\infty} a_n(q)x^n$$

where the coefficients $a_n(q)$ may depend on $q$ but not on $x$, then the functional equation (0.11) implies

$$(1 - q^{2n})a_n(q) = 0.$$

It follows that $a_n(q) = 0$ unless $n = 0$. Therefore,

$$h(x) = a_0(q).$$

That is, $h(x)$ is independent of $x$ and depends only on $q$. Since $f(x) = g(x)h(x)$, we deduce that

$$\sum_{n=-\infty}^{\infty} q^{n^2}x^n = a_0(q)\prod_{n=1}^{\infty}(1 + xq^{2n-1})(1 + x^{-1}q^{2n-1}). \tag{0.12}$$

In order to determine $a_0(q)$, we first put $x = i$, where $i^2 = -1$, in (0.12) to obtain

$$\sum_{n=-\infty}^{\infty} q^{n^2}i^n = a_0(q)\prod_{n=1}^{\infty}(1 + iq^{2n-1})(1 - iq^{2n-1}).$$

On simplifying each side, we deduce that

$$\sum_{n=-\infty}^{\infty} q^{4n^2}(-1)^n = a_0(q)\prod_{n=1}^{\infty}(1 + q^{4n-2}). \tag{0.13}$$

Next, take $x = -1$ in (0.12) and then replace $q$ with $q^4$, to obtain

$$\sum_{n=-\infty}^{\infty} q^{4n^2}(-1)^n = a_0(q^4)\prod_{n=1}^{\infty}(1-q^{8n-4})^2. \tag{0.14}$$

On comparing (0.13) and (0.14) we find that

$$\begin{aligned}\frac{a_0(q)}{a_0(q^4)} &= \prod_{n=1}^{\infty}\frac{(1-q^{8n-4})^2}{(1+q^{4n-2})}\\ &= \prod_{n=1}^{\infty}(1-q^{8n-4})(1-q^{4n-2})\\ &= \prod_{n=1}^{\infty}\frac{(1-q^{2n})}{(1-q^{8n})}.\end{aligned}$$

It follows that

$$\frac{a_0(q)}{\prod_{n=1}^{\infty}(1-q^{2n})} = \frac{a_0(q^4)}{\prod_{n=1}^{\infty}(1-q^{8n})} = \frac{a_0(q^{16})}{\prod_{n=1}^{\infty}(1-q^{32n})} = \cdots = \frac{a_0(q^{4^k})}{\prod_{n=1}^{\infty}(1-q^{2\times 4^k n})}.$$

Now take the limit as $k \to \infty$, and note from (0.12) that $a_0(0) = 1$, to obtain

$$a_0(q) = \prod_{n=1}^{\infty}(1-q^{2n}).$$

On substituting the value of $a_0(q)$ back into (0.12) we complete the proof. □

For a fixed value of $q$, the series on the left-hand side of (0.7) may be regarded as the Laurent expansion in powers of $x$ valid in the annulus $0 < |x| < \infty$. The point $x = 0$ is an essential singularity.

The next result provides another useful but less symmetric form of the triple product identity.

**Theorem 0.2.** *For $|q| < 1$ and $x \neq 0$, we have*

$$\sum_{n=-\infty}^{\infty}(-1)^n q^{n(n-1)/2} x^n = \prod_{n=1}^{\infty}(1-xq^{n-1})(1-x^{-1}q^n)(1-q^n). \tag{0.15}$$

*Proof.* This can be obtained from (0.7) in two steps: first replace $q$ with $q^{1/2}$, then replace $x$ with $-xq^{-1/2}$. □

Here are two more forms of the triple product identity.

**Theorem 0.3.** *For $|q| < 1$ and $x \neq 0$, we have*

$$1+\sum_{n=1}^{\infty}(-1)^n\left(x^n+x^{n-1}+\cdots+x^{-n}\right)q^{n(n+1)/2}$$
$$=\prod_{n=1}^{\infty}(1-xq^n)(1-x^{-1}q^n)(1-q^n) \tag{0.16}$$

*and*

$$\sum_{n=-\infty}^{\infty} q^{(4n+1)^2}\left(\frac{x^{2n+\frac{1}{2}}-x^{-2n-\frac{1}{2}}}{x^{\frac{1}{2}}-x^{-\frac{1}{2}}}\right)$$
$$=q\prod_{n=1}^{\infty}(1-xq^{8n})(1-x^{-1}q^{8n})(1-q^{8n}). \tag{0.17}$$

*Proof.* Start with the identity (0.15), rearrange the series using

$$\sum_{n=-\infty}^{\infty} a_n=\sum_{n=0}^{\infty}(a_{n+1}+a_{-n})$$

and divide by $1-x$, to obtain

$$\sum_{n=0}^{\infty}(-1)^n q^{n(n+1)/2}\left(\frac{x^{n+\frac{1}{2}}-x^{-n-\frac{1}{2}}}{x^{\frac{1}{2}}-x^{-\frac{1}{2}}}\right)=\prod_{n=1}^{\infty}(1-xq^n)(1-x^{-1}q^n)(1-q^n). \tag{0.18}$$

The identity (0.16) can be obtained by expanding the term in parentheses on the left-hand side as a geometric series.

To prove (0.17), we rearrange the series on the left-hand side of (0.18) using

$$\sum_{n=0}^{\infty} a_n=\sum_{n=0}^{\infty} a_{2n}+\sum_{n=-\infty}^{-1} a_{-2n-1} \tag{0.19}$$

to obtain

$$\sum_{n=-\infty}^{\infty} q^{n(2n+1)}\left(\frac{x^{2n+\frac{1}{2}}-x^{-2n-\frac{1}{2}}}{x^{\frac{1}{2}}-x^{-\frac{1}{2}}}\right)=\prod_{n=1}^{\infty}(1-xq^n)(1-x^{-1}q^n)(1-q^n).$$

Then replace $q$ with $q^8$ and multiply by $q$. The result is (0.17). □

## 0.4 Ramanujan's theta function $f(a, b)$

Ramanujan's theta function $f(a, b)$ is defined, for $|ab| < 1$, by

$$f(a, b) = \sum_{n=-\infty}^{\infty} a^{n(n+1)/2} b^{n(n-1)/2}.$$

A trivial property is that $f$ is symmetric, that is $f(a, b) = f(b, a)$. For future reference, we record three useful identities.

**Theorem 0.4.** *Ramanujan's theta function has a representation as an infinite product given by*

$$f(a, b) = \prod_{n=1}^{\infty} (1 + a^n b^{n-1})(1 + a^{n-1} b^n)(1 - a^n b^n).$$

*Proof.* If $ab \neq 0$, the required identity follows from Jacobi's triple product identity by taking $q = (ab)^{1/2}$ and $x = (a/b)^{1/2}$ in (0.7).

If $a = 0$ or $b = 0$, then both the sum and the product for $f(a, b)$ should be interpreted as the limiting cases $a \to 0$ or $b \to 0$, respectively. Thus, either from the sum or the product for $f(a, b)$, we have

$$f(a, 0) = \lim_{b \to 0} f(a, b) = 1 + a$$

and similarly, $f(0, b) = 1 + b$. Therefore, the result reduces to a triviality if $ab = 0$. □

**Theorem 0.5.** *The following identity holds:*

$$f(a, b) = f(a^3 b, ab^3) + af\left(a^5 b^3, \frac{b}{a}\right).$$

*Proof.* By separating the terms in the series according to whether the summation index is even or odd, we find that

$$\begin{aligned} f(a, b) &= \sum_{n=-\infty}^{\infty} a^{n(n+1)/2} b^{n(n-1)/2} \\ &= \sum_{n=-\infty}^{\infty} a^{n(2n+1)} b^{n(2n-1)} + \sum_{n=-\infty}^{\infty} a^{(2n+1)(n+1)} b^{(2n+1)n}. \end{aligned}$$

If follows that

$$
\begin{aligned}
f(a,b) &= \sum_{n=-\infty}^{\infty} (a^3b)^{n(n+1)/2}\,(ab^3)^{n(n-1)/2} \\
&\quad + a \sum_{n=-\infty}^{\infty} (a^5b^3)^{n(n+1)/2} \left(\frac{b}{a}\right)^{n(n-1)/2} \\
&= f(a^3b, ab^3) + af\left(a^5b^3, \frac{b}{a}\right).
\end{aligned}
$$

□

**Theorem 0.6.** *If $ab = cd$ then*

$$f(a,b)f(c,d) = f(ac,bd)f(ad,bc) + af\left(\frac{b}{c}, \frac{c}{b}abcd\right)f\left(\frac{b}{d}, \frac{d}{b}abcd\right).$$

*Hence,*

$$f(a,b)f(c,d) + f(-a,-b)f(-c,-d) = 2f(ac,bd)f(ad,bc)$$

*and*

$$f(a,b)f(c,d) - f(-a,-b)f(-c,-d) = 2af\left(\frac{b}{c}, \frac{c}{b}abcd\right)f\left(\frac{b}{d}, \frac{d}{b}abcd\right).$$

*Proof.* Put $q = ab = cd$. Then, using the series rearrangement

$$\sum_{m=-\infty}^{\infty}\sum_{n=-\infty}^{\infty} t_{m,n} = \sum_{j=-\infty}^{\infty}\sum_{k=-\infty}^{\infty} t_{j+k,j-k} + \sum_{j=-\infty}^{\infty}\sum_{k=-\infty}^{\infty} t_{j+k+1,j-k}$$

we find that

$$
\begin{aligned}
f(a,b)f(c,d) &= \sum_{m=-\infty}^{\infty} a^{m(m+1)/2}b^{m(m-1)/2} \sum_{n=-\infty}^{\infty} c^{n(n+1)/2}d^{n(n-1)/2} \\
&= \sum_{m=-\infty}^{\infty}\sum_{n=-\infty}^{\infty} q^{(m^2+n^2)/2} \left(\frac{a}{b}\right)^{m/2} \left(\frac{c}{d}\right)^{n/2} \\
&= \sum_{j=-\infty}^{\infty}\sum_{k=-\infty}^{\infty} q^{j^2+k^2} \left(\frac{a}{b}\right)^{(j+k)/2} \left(\frac{c}{d}\right)^{(j-k)/2} \\
&\quad + \sum_{j=-\infty}^{\infty}\sum_{k=-\infty}^{\infty} q^{j^2+k^2+j+k+\frac{1}{2}} \left(\frac{a}{b}\right)^{(j+k+1)/2} \left(\frac{c}{d}\right)^{(j-k)/2}.
\end{aligned}
\tag{0.20}
$$

Since $q = (abcd)^{1/2}$, the first sum on the right-hand side of (0.20) is

$$\sum_{j=-\infty}^{\infty}\sum_{k=-\infty}^{\infty} q^{j^2+k^2}\left(\frac{a}{b}\right)^{(j+k)/2}\left(\frac{c}{d}\right)^{(j-k)/2}$$

$$= \sum_{j=-\infty}^{\infty}(abcd)^{j^2/2}\left(\frac{ac}{bd}\right)^{j/2}\sum_{k=-\infty}^{\infty}(abcd)^{k^2/2}\left(\frac{ad}{bc}\right)^{k/2}$$

$$= f(ac, bd)f(ad, bc). \tag{0.21}$$

In a similar way, since $a/d = c/b$ and $a/c = d/b$, the second sum on the right-hand side of (0.20) is

$$\sum_{j=-\infty}^{\infty}\sum_{k=-\infty}^{\infty} q^{j^2+k^2+j+k+\frac{1}{2}}\left(\frac{a}{b}\right)^{(j+k+1)/2}\left(\frac{c}{d}\right)^{(j-k)/2}$$

$$= \left(\frac{aq}{b}\right)^{1/2}\sum_{j=-\infty}^{\infty} q^{j^2+j}\left(\frac{ac}{bd}\right)^{j/2}\sum_{k=-\infty}^{\infty} q^{k^2+k}\left(\frac{ad}{bc}\right)^{k/2}$$

$$= a\sum_{j=-\infty}^{\infty}(abcd)^{(j^2+j)/2}\left(\frac{c}{b}\right)^{j}\sum_{k=-\infty}^{\infty}(abcd)^{(k^2+k)/2}\left(\frac{d}{b}\right)^{k}$$

$$= af\left(\frac{c}{b}abcd, \frac{b}{c}\right)f\left(\frac{d}{b}abcd, \frac{b}{d}\right) = af\left(\frac{b}{c}, \frac{c}{b}abcd\right)f\left(\frac{b}{d}, \frac{d}{b}abcd\right). \tag{0.22}$$

The proof of the first identity may be completed by substituting (0.21) and (0.22) into (0.20).

The remaining two identities follow from the first identity by replacing $a$, $b$, $c$, and $d$ with $-a$, $-b$, $-c$, and $-d$, respectively, and adding or subtracting. □

## 0.5 The quintuple product identity

The method in Section 0.3 can also be used to factorize the series $\sum_{n=-\infty}^{\infty} \chi_3(n)q^{n^2}x^n$, where

$$\chi_3(n) = \begin{cases} 1 \text{ if } n \equiv 1 \pmod 3, \\ -1 \text{ if } n \equiv 2 \pmod 3, \\ 0 \text{ if } n \equiv 0 \pmod 3. \end{cases} \tag{0.23}$$

The name "quintuple product" refers to the five infinite products that will occur in the factorization.

**Theorem 0.7 (Quintuple product identity).** *For $|q| < 1$ and $x \neq 0$, we have*

$$\sum_{n=-\infty}^{\infty} \chi_3(n) q^{n^2} x^n = q(x - x^{-1}) \prod_{n=1}^{\infty} (1 - xq^{6n-3})(1 - x^{-1}q^{6n-3}) \times (1 - x^2q^{12n})(1 - x^{-2}q^{12n})(1 - q^{6n}). \qquad (0.24)$$

*Proof.* Fix a value of $q$ and let $f(x)$ be defined by

$$f(x) = \sum_{n=-\infty}^{\infty} \chi_3(n) q^{n^2} x^n. \qquad (0.25)$$

Observe that

$$\sum_{n=-\infty}^{\infty} \chi_3(n) q^{n^2} x^n = \sum_{n=-\infty}^{\infty} q^{(3n+1)^2} \left(x^{3n+1} - x^{-3n-1}\right). \qquad (0.26)$$

It follows that

$$f(x^{-1}) = -f(x). \qquad (0.27)$$

Shifting the index of summation by 3 in (0.25) gives

$$f(x) = \sum_{n=-\infty}^{\infty} \chi_3(n+3) q^{(n+3)^2} x^{n+3} = q^9 x^3 f(q^6 x). \qquad (0.28)$$

On taking $x = q^{-3}$ in each of the functional equations (0.27) and (0.28) and combining the results, we find that

$$f(q^3) = -f(q^{-3}) = -f(q^3).$$

It follows that $f(q^3) = 0$. If we put $x = 1$ or $x = -1$ in (0.27) we find that

$$f(1) = 0 \qquad \text{and} \qquad f(-1) = 0.$$

Using (0.28), we deduce that $f$ has zeros at $x = q^{6n-3}$, $q^{6n}$, and $-q^{6n}$, for every integer $n$. In what follows, we will show that these are the only zeros of $f$. Let

$$g(x) = (x - x^{-1}) \prod_{n=1}^{\infty} (1 - xq^{6n-3})(1 - x^{-1}q^{6n-3})(1 - x^2q^{12n})(1 - x^{-2}q^{12n}).$$

Then $g$ has simple zeros at $x = q^{6n-3}$, $q^{6n}$, and $-q^{6n}$ for every integer $n$, and these are the only zeros of $g$. Moreover, it is easy to check that

$$q^9x^3g(q^6x) = g(x).$$

It follows that the quotient $h(x) = f(x)/g(x)$ has no poles and is analytic in the annulus $0 < |x| < \infty$, and satisfies the functional equation

$$h(q^6x) = h(x).$$

Therefore, if the Laurent expansion of $h(x)$ is given by

$$h(x) = \sum_{n=-\infty}^{\infty} a_n(q)x^n,$$

the coefficients satisfy the condition $(1 - q^{6n})a_n(q) = 0$. Consequently, $a_n(q) = 0$ unless $n = 0$, and so

$$f(x) = a_0(q)g(x),$$

that is,

$$\sum_{n=-\infty}^{\infty} q^{(3n+1)^2}\left(x^{3n+1} - x^{-3n-1}\right)$$
$$= a_0(q)(x - x^{-1})\prod_{n=1}^{\infty}(1 - xq^{6n-3})(1 - x^{-1}q^{6n-3})(1 - x^2q^{12n})(1 - x^{-2}q^{12n}). \tag{0.29}$$

In order to determine $a_0(q)$, first let $x = i$ in (0.29) and divide by $2i$ to deduce that

$$\sum_{n=-\infty}^{\infty} q^{(3n+1)^2}\sin\frac{(3n+1)\pi}{2} = a_0(q)\prod_{n=1}^{\infty}(1 + q^{12n-6})(1 + q^{12n})^2.$$

This simplifies to

$$\sum_{n=-\infty}^{\infty}(-1)^n q^{(6n+1)^2} = a_0(q)\prod_{n=1}^{\infty}\frac{(1 - q^{24n})}{(1 - q^{6n})}. \tag{0.30}$$

One way of completing the proof is to use the triple product identity (0.7) to convert the series in (0.30) to a product. Another way, that is independent of the triple product identity, is to use a different specialization, viz., put $x = -q^{-3}$ in (0.29), to give

$$\sum_{n=-\infty}^{\infty} q^{(3n+1)^2}(-1)^{n+1}\left(q^{-9n-3}-q^{9n+3}\right)$$
$$= 2a_0(q)\,q^{-3}\prod_{n=1}^{\infty}(1+q^{6n})^2(1-q^{12n-6})^2. \tag{0.31}$$

The series on the left may be simplified by noting that it is of the form

$$\sum_{n=-\infty}^{\infty}(a_n+a_{n+1}) = 2\sum_{n=-\infty}^{\infty} a_n.$$

The product on the right-hand side of (0.31) can be simplified using Euler's identity (0.1). After performing the simplifications and multiplying by $q^3/2$, the identity (0.31) reduces to

$$\sum_{n=-\infty}^{\infty}(-1)^n q^{9n^2-3n+1} = a_0(q).$$

Replace $q$ with $q^4$, and change $n$ to $-n$ in the sum, to obtain

$$q^3\sum_{n=-\infty}^{\infty}(-1)^n q^{(6n+1)^2} = a_0(q^4). \tag{0.32}$$

On combining (0.30) and (0.32) to eliminate the series we deduce that

$$\frac{a_0(q)}{q\prod_{n=1}^{\infty}(1-q^{6n})} = \frac{a_0(q^4)}{q^4\prod_{n=1}^{\infty}(1-q^{24n})}.$$

This may be iterated to give

$$\frac{a_0(q)}{q\prod_{n=1}^{\infty}(1-q^{6n})} = \frac{a_0(q^{4^k})}{q^{4^k}\prod_{n=1}^{\infty}(1-q^{6\times 4^k n})}, \tag{0.33}$$

which holds for any positive integer $k$. Note that from (0.30) we have $a_0(q)/q \to 1$ as $q \to 0$. Therefore, on taking the limit as $k \to \infty$ in (0.33), we deduce that

$$a_0(q) = q\prod_{n=1}^{\infty}(1-q^{6n}).$$

The proof may be completed by substituting this value for $a_0(q)$ back into (0.29).

□

The analogous factorization of $\sum_{n=-\infty}^{\infty} \chi_{12}(n) q^{n^2} x^n$, where

$$\chi_{12}(n) = \begin{cases} 1 \text{ if } n \equiv 1 \text{ or } 11 \pmod{12}, \\ -1 \text{ if } n \equiv 5 \text{ or } 7 \pmod{12}, \\ 0 \text{ otherwise,} \end{cases}$$

will be given in the exercises at the end of the chapter.

Here are three other forms of the quintuple product identity. They may be compared with the alternative form of Jacobi's triple product identity in (0.17).

**Theorem 0.8 (Alternative forms of the quintuple product identity).** *For $|q| < 1$ and $x \neq 0$, the following identities hold:*

$$\sum_{n=-\infty}^{\infty} q^{(3n+1)^2} \left( \frac{x^{3n+1} - x^{-3n-1}}{x - x^{-1}} \right)$$

$$= q \prod_{n=1}^{\infty} (1 - xq^{6n-3})(1 - x^{-1}q^{6n-3})(1 - x^2q^{12n})(1 - x^{-2}q^{12n})(1 - q^{6n}), \qquad (0.34)$$

$$\sum_{n=-\infty}^{\infty} q^{(6n+1)^2} \left( \frac{x^{6n+1} - x^{-6n-1}}{x - x^{-1}} \right)$$

$$= q \prod_{n=1}^{\infty} (1 - x^2q^{24n})(1 - x^{-2}q^{24n})(1 - x^4q^{48n-24})(1 - x^{-4}q^{48n-24})(1 - q^{24n}) \qquad (0.35)$$

*and*

$$\sum_{n=-\infty}^{\infty} (-1)^n q^{(6n+1)^2} \left( \frac{x^{6n+1} + x^{-6n-1}}{x + x^{-1}} \right)$$

$$= q \prod_{n=1}^{\infty} \frac{(1 - x^4q^{24n})(1 - x^{-4}q^{24n})}{(1 - x^2q^{24n})(1 - x^{-2}q^{24n})} \times (1 - q^{24n}). \qquad (0.36)$$

*Proof.* The identity (0.34) is clearly equivalent to (0.24) by the series rearrangement (0.26).

We will now deduce (0.35) from (0.34). Multiply both sides of (0.34) by $(x - x^{-1})$, replace $q$ with $q^4$, and then replace $x$ with $q^{-12}x^2$. The left-hand side of (0.34) becomes

$$\sum_{n=-\infty}^{\infty} q^{36n^2-12n-8}x^{6n+2} - \sum_{n=-\infty}^{\infty} q^{36n^2+60n+16}x^{-6n-2}$$

$$= q^{-9}x^3 \sum_{n=-\infty}^{\infty} q^{(6n+1)^2}\left(x^{-6n-1} - x^{6n+1}\right), \tag{0.37}$$

where we have replaced $n$ with $-n$ in the first sum, and $n$ with $-n-1$ in the second sum. The right-hand side of (0.34) becomes

$$q^4(x^2q^{-12} - x^{-2}q^{12})\prod_{n=1}^{\infty}(1 - x^2q^{24n-24})(1 - x^{-2}q^{24n})$$
$$\times(1 - x^4q^{48n-24})(1 - x^{-4}q^{48n+24})(1 - q^{24n})$$
$$= q^{-8}x^3(x^{-1} - x)\prod_{n=1}^{\infty}(1 - x^2q^{24n})(1 - x^{-2}q^{24n})$$
$$\times(1 - x^4q^{48n-24})(1 - x^{-4}q^{48n-24})(1 - q^{24n}). \tag{0.38}$$

On combining (0.37) and (0.38) we obtain (0.35). The identity (0.36) can be obtained from (0.35) by replacing $x$ with $ix$, where $i^2 = -1$, and rearranging the products. □

## 0.6 Examples

This section contains some results that follow readily from the triple product and quintuple product identities. The results involve infinite products of the form $\prod_{n=1}^{\infty}(1 - q^{kn})$ for various positive integers $k$. To maximize the symmetry, the identities are often best stated in terms of $\eta_k$ defined by

$$\eta_k = q^{k/24}\prod_{n=1}^{\infty}(1 - q^{kn}), \tag{0.39}$$

where $k$ is any positive integer. The reason for the factor $q^{k/24}$ will be seen later in the statement of the modular transformation in Theorem 2.7. For now, we offer the simple explanation that the factor $q^{k/24}$ completes the square in the quadratic exponent in the power of $q$ in the series, as can be seen in (0.40).

**Theorem 0.9.** *With $\eta_k$ as defined by* (0.39), *the following identities hold*:

$$\eta_1 = \sum_{n=-\infty}^{\infty} (-1)^n q^{(6n+1)^2/24}, \tag{0.40}$$

$$\frac{\eta_1^2}{\eta_2} = \sum_{n=-\infty}^{\infty} (-1)^n q^{n^2}, \tag{0.41}$$

$$\frac{\eta_2^5}{\eta_1^2 \eta_4^2} = \sum_{n=-\infty}^{\infty} q^{n^2}, \tag{0.42}$$

*and*

$$\frac{\eta_2^2}{\eta_1} = \sum_{n=0}^{\infty} q^{(2n+1)^2/8} = \sum_{n=-\infty}^{\infty} q^{(4n+1)^2/8}. \tag{0.43}$$

*The identities* (0.40) *and* (0.43) *may be restated as*

$$E(q) = \sum_{n=-\infty}^{\infty} (-1)^n q^{n(3n-1)/2} \tag{0.44}$$

*and*

$$\frac{E^2(q^2)}{E(q)} = \sum_{n=0}^{\infty} q^{n(n+1)/2}, \tag{0.45}$$

*respectively, where* $E(q)$ *is Euler's product defined by* (0.3).

*Proof.* The standard way to prove (0.40) is to replace $q$ with $q^3$ in the triple product identity (0.15) and then set $x = q$ to obtain

$$\sum_{n=-\infty}^{\infty} (-1)^n q^{n(3n-1)/2} = \prod_{n=1}^{\infty} (1 - q^{3n-2})(1 - q^{3n-1})(1 - q^{3n}) = \prod_{n=1}^{\infty} (1 - q^n).$$

In fact, this proves (0.44). The identity (0.40) may be obtained by replacing the summation index $n$ with $-n$, multiplying both sides by $q^{1/24}$ and completing the square in the power of $q$ in the series. An even easier way of proving (0.40) is to take $x = 1$ in the quintuple product identity (0.36) and then replace $q$ with $q^{1/24}$.

The identity (0.41) is obtained by setting $x = -1$ in the triple product identity (0.7) and rearranging the infinite products.

The identity (0.42) may be deduced from (0.41) by replacing $q$ with $-q$, and then using (0.2) to simplify the infinite products.

It remains to prove (0.43) and (0.45). Take $x = -1$ in Jacobi's triple product identity (0.16) to obtain

$$1 + \sum_{n=1}^{\infty} q^{n(n+1)/2} = \prod_{n=1}^{\infty} (1 + q^n)^2 (1 - q^n).$$

This is equivalent to

$$\sum_{n=0}^{\infty} q^{n(n+1)/2} = \prod_{n=1}^{\infty} \frac{(1-q^{2n})^2}{(1-q^n)}$$

which is (0.45). Multiplying by $q^{1/8}$ we obtain

$$\sum_{n=0}^{\infty} q^{(2n+1)^2/8} = q^{1/8} \prod_{n=1}^{\infty} \frac{(1-q^{2n})^2}{(1-q^n)}$$

and this is the first identity in (0.43). The remaining equality in (0.43), namely

$$\sum_{n=0}^{\infty} q^{(2n+1)^2/8} = \sum_{n=-\infty}^{\infty} q^{(4n+1)^2/8},$$

is obtained by series rearrangement using (0.19). □

The identities in the next theorem arise as limiting cases of the triple product and quintuple product identities.

**Theorem 0.10.** *With* $\eta_k$ *as defined by* (0.39), *the following identities hold*:

$$\eta_1^3 = \sum_{n=-\infty}^{\infty} (4n+1) q^{(4n+1)^2/8}, \tag{0.46}$$

$$\frac{\eta_1^2 \eta_4^2}{\eta_2} = \sum_{n=-\infty}^{\infty} (3n+1) q^{(3n+1)^2/3}, \tag{0.47}$$

$$\frac{\eta_1^5}{\eta_2^2} = \sum_{n=-\infty}^{\infty} (6n+1) q^{(6n+1)^2/24}. \tag{0.48}$$

*Proof.* The identity (0.46) follows from the triple product identity (0.17) by taking the limit as $q \to 1$ and replacing $q$ with $q^{1/8}$. The identities (0.47) and (0.48) may be obtained by letting $x \to 1$ in the two forms of the quintuple product identity (0.34) and (0.35), respectively, and then replacing $q$ with an appropriate power of $q$. □

The identity (0.46) is often stated in terms of Euler's product as

$$E^3(q) = \sum_{n=0}^{\infty} (-1)^n (2n+1) q^{n(n+1)/2}. \tag{0.49}$$

It can be obtained by rearranging the series in (0.46), or simply by taking $x = 1$ in (0.16).

The identity (0.47) may be expressed in the form

$$\frac{E^2(q)E^2(q^4)}{E(q^2)} = \sum_{n=-\infty}^{\infty} (3n+1)q^{3n^2+2n}.$$

Replacing $q$ with $-q$ and applying (0.4) to the $E(-q)$ term, we obtain

$$\frac{E^2(-q)E^2(q^4)}{E(q^2)} = \frac{E^5(q^2)}{E^2(q)} = \sum_{n=-\infty}^{\infty} (-1)^n(3n+1)q^{3n^2+2n}.$$

This may be written as

$$\frac{\eta_2^5}{\eta_1^2} = \sum_{n=-\infty}^{\infty} (-1)^n(3n+1)q^{(3n+1)^2/3},$$

where the product on the left-hand side bears analogy with (0.48).

A product of the form

$$\eta_1^{e_1}\eta_2^{e_2}\cdots\eta_N^{e_N},$$

where $e_1, e_2, \ldots, e_N$ are integers, is called an eta-quotient. If all of the integers $e_1$, $e_2, \ldots, e_N$ are nonnegative, then the product is called an eta-product. Thus, the left-hand sides of (0.40)–(0.43) and (0.46)–(0.48) are eta-quotients, while the left-hand sides of (0.40) and (0.46) are also eta-products. Some other examples of eta-quotients that can be derived from the triple product identity are given in Exercise 1 at the end of the chapter. The identities (0.40) and (0.46) provide expansions of $\eta_1$ and its cube as theta series. Analogous expansions for the 8th, 10th, 14th, and 26th powers can also be determined; see Exercise 11 at the end of this chapter, and Exercises 10–13 at the end of the next chapter.

## 0.7 Partitions

A partition of a positive integer $n$ is a finite sequence of positive integers $(\lambda_1, \lambda_2, \ldots, \lambda_r)$ such that

$$\lambda_1 + \lambda_2 + \cdots + \lambda_r = n \qquad \text{and} \qquad \lambda_1 \ge \lambda_2 \ge \cdots \ge \lambda_r.$$

The number of partitions of $n$ is denoted by $p(n)$. We define $p(0) = 1$ and note that the empty sequence is the only partition of 0. For $0 \le n \le 6$, the partitions of $n$ together with the corresponding values of $p(n)$ are listed in Table 0.1.

If $(\lambda_1, \lambda_2, \ldots, \lambda_r)$ is a partition of $n$, then the numbers $\lambda_1, \lambda_2, \ldots, \lambda_r$ are called the parts, and $r$ is the number of parts.

**Table 0.1** Partitions

| $n$ | Partitions of $n$ | $p(n)$ |
|---|---|---|
| 0 | ( ) | 1 |
| 1 | (1) | 1 |
| 2 | (2), (1, 1) | 2 |
| 3 | (3), (2, 1), (1, 1, 1) | 3 |
| 4 | (4), (3, 1), (2, 2), (2, 1, 1), (1, 1, 1, 1) | 5 |
| 5 | (5), (4, 1), (3, 2), (3, 1, 1), (2, 2, 1), (2, 1, 1, 1), (1, 1, 1, 1, 1) | 7 |
| 6 | (6), (5, 1), (4, 2), (4, 1, 1), (3, 3), (3, 2, 1), (3, 1, 1, 1), (2, 2, 2), (2, 2, 1, 1), (2, 1, 1, 1, 1), (1, 1, 1, 1, 1, 1) | 11 |

The generating function for the sequence $p(n)$ is

$$\sum_{n=0}^{\infty} p(n)\, q^n = 1 + q + 2q^2 + 3q^3 + 5q^4 + 7q^5 + 11q^6 + \cdots .$$

It has a simple representation as an infinite product.

**Theorem 0.11.** *The following identity holds:*

$$\sum_{n=0}^{\infty} p(n)\, q^n = \prod_{j=1}^{\infty} \frac{1}{1-q^j}.$$

*Proof.* Write the infinite product as

$$\begin{aligned}
\prod_{j=1}^{\infty} \frac{1}{1-q^j} &= \prod_{j=1}^{\infty} (1 + q^j + q^{2j} + q^{3j} + \cdots) \\
&= (1 + q^1 + q^{1+1} + q^{1+1+1} + q^{1+1+1+1} + \cdots) \\
&\quad \times (1 + q^2 + q^{2+2} + q^{2+2+2} + \cdots) \\
&\quad \times (1 + q^3 + q^{3+3} + q^{3+3+3} + \cdots) \\
&\quad \times (1 + q^4 + q^{4+4} + q^{4+4+4} + \cdots) \times \cdots .
\end{aligned}$$

Now multiply the series together and observe that every partition corresponds to exactly one term in the expansion, so that

$$\begin{aligned}
\prod_{j=1}^{\infty} \frac{1}{1-q^j} &= 1 + (q^1) + (q^2 + q^{1+1}) + (q^3 + q^{2+1} + q^{1+1+1}) \\
&\quad + (q^4 + q^{3+1} + q^{2+2} + q^{2+1+1} + q^{1+1+1+1}) + \cdots \\
&= \sum_{n=0}^{\infty} p(n)\, q^n.
\end{aligned}$$

□

The proof given above relies on multiplying an infinite number of infinite series. The idea can be made rigorous by considering a finite number of infinite series and passing to the limit. For details, see the books by Andrews [22, p. 5] and Hardy and Wright [175, p. 275].

The partition $(\lambda_1, \lambda_2, \ldots, \lambda_r)$ is said to be a partition into odd parts if $\lambda_1, \lambda_2, \ldots, \lambda_r$ are all odd. The partition $(\lambda_1, \lambda_2, \ldots, \lambda_r)$ is called a partition into distinct parts if

$$\lambda_1 > \lambda_2 > \cdots > \lambda_r.$$

The number of partitions of $n$ into odd parts or distinct parts is denoted by $p_o(n)$ or $p_d(n)$, respectively. For example, the partitions of 6 into odd parts are

$$(5,1),\ (3,3),\ (3,1,1,1),\ (1,1,1,1,1,1)$$

and the partitions of 6 into distinct parts are

$$(6),\ (5,1),\ (4,2),\ (3,2,1).$$

Hence, $p_o(6) = 4$ and $p_d(6) = 4$. In general, we have:

**Theorem 0.12.** *Let $p_o(n)$ and $p_d(n)$ denote the number of partitions of $n$ into odd parts and into distinct parts, respectively. Then $p_o(n) = p_d(n)$.*

*Proof.* By the same reasoning as for the proof of Theorem 0.11 we find that

$$\sum_{n=0}^{\infty} p_o(n)\, q^n = \prod_{j=1}^{\infty} \frac{1}{1-q^{2j-1}}$$

and

$$\sum_{n=0}^{\infty} p_d(n)\, q^n = \prod_{j=1}^{\infty} (1+q^j).$$

By Euler's identity (0.1) we have

$$\prod_{j=1}^{\infty} \frac{1}{1-q^{2j-1}} = \prod_{j=1}^{\infty} (1+q^j),$$

and it follows that $p_o(n) = p_d(n)$. □

The next result gives a recurrence relation for $p(n)$.

**Theorem 0.13.** *For any positive integer $n$ we have*

$$p(n) = \sum_{\substack{j\geq 1 \\ \frac{j(3j-1)}{2}\leq n}} (-1)^{j+1} p\left(n - \tfrac{j(3j-1)}{2}\right) + \sum_{\substack{j\geq 1 \\ \frac{j(3j+1)}{2}\leq n}} (-1)^{j+1} p\left(n - \tfrac{j(3j+1)}{2}\right).$$

*Proof.* By (0.44) and Theorem 0.11 we have

$$1 = \prod_{j=1}^{\infty}(1-q^j) \times \prod_{k=1}^{\infty} \frac{1}{1-q^k}$$

$$= \left(1 + \sum_{j=1}^{\infty}(-1)^j \left(q^{j(3j-1)/2} + q^{j(3j+1)/2}\right)\right) \times \sum_{k=0}^{\infty} p(k)\, q^k.$$

Equate coefficients of $q^n$ on each side to obtain

$$0 = p(n) + \sum_{\substack{j\geq 1 \\ \frac{j(3j-1)}{2}+k=n}} (-1)^j\, p(k) + \sum_{\substack{j\geq 1 \\ \frac{j(3j+1)}{2}+k=n}} (-1)^j\, p(k).$$

The proof may be completed by solving for the $p(n)$ term and expressing the summation index $k$ in each sum in terms of $j$. □

For example, using Theorem 0.13 and data from Table 0.1 we find that

$$p(7) = p(6) + p(5) - p(2) - p(0) = 11 + 7 - 2 - 1 = 15,$$
$$p(8) = p(7) + p(6) - p(3) - p(1) = 15 + 11 - 3 - 1 = 22$$

and

$$p(9) = p(8) + p(7) - p(4) - p(2) = 22 + 15 - 5 - 2 = 30.$$

## 0.8 $q$-analogues of sine and cosine

Jacobi's triple product identity (0.17) may be expressed in the form

$$\sum_{n=-\infty}^{\infty} q^{(4n+1)^2/8}(x^{2n+\frac{1}{2}} - x^{-2n-\frac{1}{2}})$$
$$= q^{1/8}\,(x^{\frac{1}{2}} - x^{-\frac{1}{2}}) \prod_{n=1}^{\infty}(1-xq^n)(1-x^{-1}q^n)(1-q^n). \qquad (0.50)$$

The product on the right has simple zeros at each point in the bilateral geometric progression $x = q^n$, $n \in \mathbb{Z}$. Put $x = e^{i\theta}$ and divide by $iq^{1/8}\prod_{n=1}^{\infty}(1-q^n)^3$, and consider the function $s$ defined by

$$s(\theta) = s(\theta, \tau) = 2\sin\frac{\theta}{2}\prod_{n=1}^{\infty}\frac{(1-e^{i\theta}q^n)(1-e^{-i\theta}q^n)}{(1-q^n)^2} \qquad (0.51)$$

where, as usual, $q = e^{2\pi i\tau}$ and $\operatorname{Im}\tau > 0$. For a fixed value of $\tau$, $s(\theta)$ is an entire function of $\theta$. The zeros of $s(\theta)$ are all simple and are located at $\theta = 2\pi m + 2\pi n\tau$, where $m, n \in \mathbb{Z}$. The function $s(\theta)$ has been normalized so that

$$\lim_{\theta\to 0} \frac{s(\theta)}{\theta} = 1. \tag{0.52}$$

In a similar way, starting with the quintuple product identity (0.36), we are led to consider the function $c$ defined by

$$c(\theta) = c(\theta, \tau) = \cos\frac{\theta}{2} \prod_{n=1}^{\infty} \frac{(1-e^{2i\theta}q^n)(1-e^{-2i\theta}q^n)}{(1-e^{i\theta}q^n)(1-e^{-i\theta}q^n)}. \tag{0.53}$$

Since

$$c(\theta) = \cos\frac{\theta}{2} \prod_{n=1}^{\infty}(1+e^{i\theta}q^n)(1+e^{-i\theta}q^n)(1-e^{2i\theta}q^{2n-1})(1-e^{-2i\theta}q^{2n-1}),$$

it follows that $c(\theta)$ is an entire function of $\theta$. The zeros of $c(\theta)$ are all simple and are located at $\theta = \pi m + \pi n\tau$ where $m$ and $n$ are integers that are not both even. The function $c(\theta)$ has been normalized so that

$$\lim_{\theta\to 0} c(\theta) = 1.$$

In both (0.51) and (0.53), $s(\theta, \tau)$ and $c(\theta, \tau)$ are functions of two variables $\theta$ and $\tau$. When the value of $\tau$ doesn't need to be emphasized, we will omit it and just write $s(\theta)$ and $c(\theta)$, respectively. From the definitions (0.51) and (0.53) we deduce that

$$s(2\theta) = 2s(\theta)c(\theta), \tag{0.54}$$

which is an analogue of the double angle formula for the sine function. When $q = 0$ (that is, when $\tau \to i\infty$) we have

$$s(\theta) = 2\sin\frac{\theta}{2} \qquad \text{and} \qquad c(\theta) = \cos\frac{\theta}{2}.$$

Thus, $s(\theta)$ and $c(\theta)$ may be viewed as analogues of the sine and cosine functions, respectively.

Fourier expansions for $s(\theta)$ and $c(\theta)$ can be derived using the triple product and quintuple product identities, respectively. From (0.50) and (0.51) we find that

$$s(\theta) = \frac{2}{\eta_1^3} \sum_{n=-\infty}^{\infty} q^{(4n+1)^2/8} \sin\frac{(4n+1)\theta}{2}. \tag{0.55}$$

Letting $x = e^{i\theta/2}$ and replacing $q$ with $q^{1/24}$ in (0.36), then dividing by $\eta_1$, gives

$$c(\theta) = \frac{1}{\eta_1} \sum_{n=-\infty}^{\infty} (-1)^n q^{(6n+1)^2/24} \cos \tfrac{(6n+1)\theta}{2}. \tag{0.56}$$

The Fourier series can be used to derive Taylor expansions for $s(\theta)$ and $c(\theta)$ in powers of $\theta$. As $s(\theta)$ and $c(\theta)$ are entire functions, the Taylor expansions converge for all complex values of $\theta$.

**Theorem 0.14.** *The following power series expansions hold for all complex values of* $\theta$:

$$s(\theta) = \sum_{k=0}^{\infty} \frac{(-1)^k U_{2k}}{2^{2k}(2k+1)!} \theta^{2k+1} \tag{0.57}$$

*and*

$$c(\theta) = \sum_{k=0}^{\infty} \frac{(-1)^k V_{2k}}{2^{2k}(2k)!} \theta^{2k}, \tag{0.58}$$

*where* $U_k$ *and* $V_k$ *are defined by*

$$U_k = \frac{\sum\limits_{n=-\infty}^{\infty} (4n+1)^{k+1} q^{(4n+1)^2/8}}{\sum\limits_{n=-\infty}^{\infty} (4n+1) q^{(4n+1)^2/8}} \tag{0.59}$$

*and*

$$V_k = \frac{\sum\limits_{n=-\infty}^{\infty} (-1)^n (6n+1)^k q^{(6n+1)^2/24}}{\sum\limits_{n=-\infty}^{\infty} (-1)^n q^{(6n+1)^2/24}}. \tag{0.60}$$

*Proof.* Expand the Fourier series (0.55) in powers of $\theta$ and interchange the order of summation to obtain

$$\begin{aligned} s(\theta) &= \frac{2}{\eta_1^3} \sum_{n=-\infty}^{\infty} q^{(4n+1)^2/8} \sum_{k=0}^{\infty} \frac{(-1)^k (4n+1)^{2k+1}}{2^{2k+1}(2k+1)!} \theta^{2k+1} \\ &= \sum_{k=0}^{\infty} \frac{(-1)^k \theta^{2k+1}}{2^{2k}(2k+1)!} \left( \sum_{n=-\infty}^{\infty} (4n+1)^{2k+1} q^{(4n+1)^2/8} \right) \bigg/ \eta_1^3. \end{aligned}$$

Now apply (0.46) to expand $\eta_1^3$ in powers of $q$. This completes the proof of (0.57). The identity (0.58) can be obtained in a similar way starting with the Fourier expansion (0.56) and making use of the expansion of $\eta_1$ in (0.40). □

The next theorem gives differential recurrence relations for $U_k$ and $V_k$.

**Theorem 0.15.** *Let $U_k$ and $V_k$ be given by* (0.59) *and* (0.60)*, respectively, and let $P$ be defined by*

$$P = q\frac{\mathrm{d}}{\mathrm{d}q}\log\left(q\prod_{n=1}^{\infty}(1-q^n)^{24}\right) = 1 - 24\sum_{n=1}^{\infty}\frac{nq^n}{1-q^n}. \tag{0.61}$$

*Then $U_k$ and $V_k$ satisfy the differential recurrence relations*

$$U_{k+2} = PU_k + 8q\frac{\mathrm{d}U_k}{\mathrm{d}q} \tag{0.62}$$

*and*

$$V_{k+2} = PV_k + 24q\frac{\mathrm{d}V_k}{\mathrm{d}q}. \tag{0.63}$$

*Proof.* By (0.46) and (0.59) we have

$$U_k = \frac{\sum_{n=-\infty}^{\infty}(4n+1)^{k+1}q^{(4n+1)^2/8}}{\sum_{n=-\infty}^{\infty}(4n+1)q^{(4n+1)^2/8}} = \frac{\sum_{n=-\infty}^{\infty}(4n+1)^{k+1}q^{(4n+1)^2/8}}{q^{1/8}\prod_{n=1}^{\infty}(1-q^n)^3}.$$

Take the logarithmic derivative with respect to $q$ and multiply by $8q$ to obtain

$$\begin{aligned} 8q\frac{U_k'}{U_k} &= \frac{\sum_{n=-\infty}^{\infty}(4n+1)^{k+3}q^{(4n+1)^2/8}}{\sum_{n=-\infty}^{\infty}(4n+1)^{k+1}q^{(4n+1)^2/8}} - \left(1 - 24\sum_{n=1}^{\infty}\frac{nq^n}{1-q^n}\right) \\ &= \frac{U_{k+2}}{U_k} - P. \end{aligned}$$

This proves (0.62). The identity (0.63) may be proved in the same way using (0.40) and (0.60). □

**Corollary 0.16.** *With $U_k$, $V_k$ and $P$ as defined by* (0.59), (0.60), *and* (0.61), *we have*

$$U_0 = V_0 = 1$$

*and*

$$U_2 = V_2 = P.$$

*Proof.* The values of $U_0$ and $V_0$ are immediate from the definitions (0.59) and (0.60), and the values of $U_2$ and $V_2$ follow from Theorem 0.15. □

*Remark 0.17.* The power series expansions in Theorem 0.14 involve $U_k$ and $V_k$ for even values of $k$ only. The coefficient $V_1$ also turns out to be useful and it will be studied in Theorem 3.18. For the value of $U_1$, see Exercise 12 at the end of the chapter.

## 0.9 Ramanujan's differential equations

In the previous section, the functions $s(\theta)$ and $c(\theta)$ were studied, and it was noted that they are analogues of the sine and cosine, respectively. In this section we shall consider their logarithmic derivatives, namely $s'(\theta)/s(\theta)$ and $c'(\theta)/c(\theta)$. They are analogues of the cotangent and tangent functions, respectively. We will expand each function in powers of $\theta$ and use the expansions to deduce a system of three nonlinear differential equations that have come to be known as Ramanujan's differential equations.

The Bernoulli numbers $B_n$ are defined by

$$\frac{x}{e^x - 1} = \sum_{n=0}^{\infty} \frac{B_n x^n}{n!}. \tag{0.64}$$

Since $\dfrac{x}{e^x - 1} + \dfrac{x}{2}$ is an even function of $x$, it follows that

$$\frac{x}{e^x - 1} = 1 - \frac{x}{2} + \sum_{n=1}^{\infty} \frac{B_{2n} x^{2n}}{(2n)!}.$$

Hence, $B_1 = -\frac{1}{2}$ and $B_{2n+1} = 0$ for all positive integers $n$. Values of the first few Bernoulli numbers $B_n$, for even values of $n$, are given in Table 0.2. The cotangent function may be expanded in powers of $\theta$ using the Bernoulli numbers, as

**Table 0.2** Bernoulli numbers

| $n$ | 0 | 2 | 4 | 6 | 8 | 10 | 12 | 14 |
|---|---|---|---|---|---|---|---|---|
| $B_n$ | 1 | $\frac{1}{6}$ | $-\frac{1}{30}$ | $\frac{1}{42}$ | $-\frac{1}{30}$ | $\frac{5}{66}$ | $-\frac{691}{2730}$ | $\frac{7}{6}$ |

$$\cot\theta = i\frac{e^{i\theta}+e^{-i\theta}}{e^{i\theta}-e^{-i\theta}} = i + \frac{2i}{e^{2i\theta}-1} = \frac{1}{\theta} + \sum_{n=1}^{\infty} \frac{2^{2n}(-1)^n B_{2n}}{(2n)!}\theta^{2n-1}. \tag{0.65}$$

For any positive integer $k$, the Eisenstein series $E_k$ is defined by

$$E_k = -\frac{B_k}{2k} + \sum_{j=1}^{\infty} \frac{j^{k-1}q^j}{1-q^j}. \tag{0.66}$$

The Eisenstein series $E_{2k}$ arise naturally as coefficients in the power series expansions of the logarithmic derivatives of $s$ and $c$, as follows.

**Lemma 0.18.** *Let $s(\theta)$ and $c(\theta)$ be defined by* (0.51) *and* (0.53). *Then*

$$\frac{s'(\theta)}{s(\theta)} = \frac{1}{\theta} + 2\sum_{k=1}^{\infty} \frac{(-1)^{k-1}E_{2k}}{(2k-1)!}\theta^{2k-1} \tag{0.67}$$

*and*

$$\frac{c'(\theta)}{c(\theta)} = 2\sum_{k=1}^{\infty} \frac{(2^{2k}-1)(-1)^{k-1}E_{2k}}{(2k-1)!}\theta^{2k-1}, \tag{0.68}$$

*where the primes denote differentiation with respect to $\theta$.*

*Proof.* On applying logarithmic differentiation to (0.51) we find that

$$\frac{s'(\theta)}{s(\theta)} = \frac{1}{2}\cot\frac{\theta}{2} + \frac{1}{i}\sum_{n=1}^{\infty}\left(\frac{e^{i\theta}q^n}{1-e^{i\theta}q^n} - \frac{e^{-i\theta}q^n}{1-e^{-i\theta}q^n}\right). \tag{0.69}$$

Expanding as a double series and interchanging the order of summation, we obtain

$$\begin{aligned}\frac{s'(\theta)}{s(\theta)} &= \frac{1}{2}\cot\frac{\theta}{2} + \frac{1}{i}\sum_{n=1}^{\infty}\sum_{j=1}^{\infty} q^{jn}\left(e^{ij\theta}-e^{-ij\theta}\right)\\ &= \frac{1}{2}\cot\frac{\theta}{2} + 2\sum_{j=1}^{\infty}\frac{q^j}{1-q^j}\sin j\theta.\end{aligned} \tag{0.70}$$

Now expand in powers of $\theta$, using (0.65), to give

$$\frac{s'(\theta)}{s(\theta)} = \frac{1}{\theta} + 2\sum_{k=1}^{\infty}\left(-\frac{B_{2k}}{4k} + \sum_{j=1}^{\infty}\frac{j^{2k-1}q^j}{1-q^j}\right)\frac{(-1)^{k-1}\theta^{2k-1}}{(2k-1)!}.$$

This proves (0.67). The identity (0.68) can be obtained from (0.67) by using the double angle formula (0.54) in the form

$$c(\theta) = \frac{s(2\theta)}{2s(\theta)}.$$

□

For future reference, the Eisenstein series $E_2$, $E_4$, $E_6$, and $E_8$ are given by

$$E_2 = -\frac{1}{24} + \sum_{j=1}^{\infty} \frac{jq^j}{1-q^j}, \qquad E_4 = \frac{1}{240} + \sum_{j=1}^{\infty} \frac{j^3q^j}{1-q^j},$$

$$E_6 = -\frac{1}{504} + \sum_{j=1}^{\infty} \frac{j^5q^j}{1-q^j} \quad \text{and} \quad E_8 = \frac{1}{480} + \sum_{j=1}^{\infty} \frac{j^7q^j}{1-q^j}.$$

It will sometimes be useful to work with a different normalization of the Eisenstein series for which the constant term is 1. The series $E_2$ is a multiple of the function $P$ defined in (0.61), that is, $P = -24E_2$. Let $Q$, $R$ and $S$ be defined by $Q = 240E_4$, $R = -504E_6$ and $S = 480E_8$, so that

$$P = P(q) = -24E_2 = 1 - 24\sum_{j=1}^{\infty} \frac{jq^j}{1-q^j}, \tag{0.71}$$

$$Q = Q(q) = 240E_4 = 1 + 240\sum_{j=1}^{\infty} \frac{j^3q^j}{1-q^j}, \tag{0.72}$$

$$R = R(q) = -504E_6 = 1 - 504\sum_{j=1}^{\infty} \frac{j^5q^j}{1-q^j} \tag{0.73}$$

and

$$S = S(q) = 480E_8 = 1 + 480\sum_{j=1}^{\infty} \frac{j^7q^j}{1-q^j}. \tag{0.74}$$

The functions $P$, $Q$, and $R$ are called Ramanujan's Eisenstein series. Occasionally, we will write

$$P_n = P(q^n) \qquad \text{and} \qquad Q_n = Q(q^n), \tag{0.75}$$

where $n$ is a positive integer.

The next result provides recurrence relations for $U_{2k}$ and $V_{2k}$. The coefficients involve the Eisenstein series $E_{2k}$.

**Lemma 0.19.** *Let $U_k$, $V_k$, and $E_k$ be defined by* (0.59), (0.60), *and* (0.66), *respectively. For any positive integer k, we have*

$$U_{2k} = -\frac{(2k+1)}{2k}\sum_{j=1}^{k}\binom{2k}{2j-1}2^{2j+1}U_{2k-2j}E_{2j} \tag{0.76}$$

*and*

$$V_{2k} = -\sum_{j=1}^{k}\binom{2k-1}{2j-1}2^{2j+1}(2^{2j}-1)V_{2k-2j}E_{2j}, \tag{0.77}$$

*where* $\binom{k}{j} = \dfrac{k!}{j!(k-j)!}$ *is the binomial coefficient.*

*Proof.* Start by writing

$$s'(\theta) = s(\theta) \times \left(\frac{s'(\theta)}{s(\theta)}\right)$$

and expand each term in powers of $\theta$ using (0.57) and (0.67) to obtain

$$\sum_{k=0}^{\infty}\frac{(-1)^k U_{2k}}{2^{2k}(2k)!}\theta^{2k}$$
$$= \left(\sum_{k=0}^{\infty}\frac{(-1)^k U_{2k}}{2^{2k}(2k+1)!}\theta^{2k+1}\right) \times \left(\frac{1}{\theta} + 2\sum_{k=1}^{\infty}\frac{(-1)^{k-1}E_{2k}}{(2k-1)!}\theta^{2k-1}\right).$$

If we equate the coefficients of $\theta^{2k}$ and solve for $U_{2k}$ we obtain (0.76). The result (0.77) can be obtained in the same way starting with

$$c'(\theta) = c(\theta) \times \left(\frac{c'(\theta)}{c(\theta)}\right).$$

□

The next result gives formulas for $U_{2k}$ and $V_{2k}$, for $1 \le k \le 4$, in terms of $P$, $Q$, $R$, and $S$.

**Lemma 0.20.** *Let $U_k$ and $V_k$ be defined by* (0.59) *and* (0.60) *and let $P$, $Q$, $R$, and $S$ be defined by* (0.71)–(0.74). *Then*

$$U_2 = P,$$
$$U_4 = \frac{1}{3}\left(5P^2 - 2Q\right),$$

$$U_6 = \frac{1}{9}\left(35P^3 - 42PQ + 16R\right),$$

$$U_8 = \frac{1}{15}\left(175P^4 - 420P^2Q + 320PR + 84Q^2 - 144S\right)$$

*and*

$$V_2 = P,$$

$$V_4 = 3P^2 - 2Q,$$

$$V_6 = 15P^3 - 30PQ + 16R,$$

$$V_8 = 105P^4 - 420P^2Q + 448PR + 140Q^2 - 272S.$$

*Proof.* By definition, $U_0 = V_0 = 1$. The required results follow by successively taking $k = 1, 2, 3, 4$ in Lemma 0.19 and using (0.71)–(0.74) to express $E_2$, $E_4$, $E_6$, and $E_8$ in terms of $P$, $Q$, $R$, and $S$. □

We are now ready to prove the main result of this section.

**Theorem 0.21 (Ramanujan's differential equations).** *Let $P$, $Q$, and $R$ be defined by* (0.71)–(0.73). *Then*

$$q\frac{\mathrm{d}P}{\mathrm{d}q} = \frac{P^2 - Q}{12}, \tag{0.78}$$

$$q\frac{\mathrm{d}Q}{\mathrm{d}q} = \frac{PQ - R}{3}, \tag{0.79}$$

*and*

$$q\frac{\mathrm{d}R}{\mathrm{d}q} = \frac{PR - Q^2}{2}. \tag{0.80}$$

*Proof.* Take $k = 2$ in the differential recurrence relation (0.62) to obtain

$$U_4 = PU_2 + 8q\frac{\mathrm{d}U_2}{\mathrm{d}q}.$$

On substituting the values of $U_2$ and $U_4$ from Lemma 0.20 we deduce that

$$\frac{1}{3}(5P^2 - 2Q) = P^2 + 8q\frac{\mathrm{d}P}{\mathrm{d}q},$$

and this can be rearranged to give (0.78).

Next, take $k = 4$ in the differential recurrence relation (0.62) to obtain

$$U_6 = PU_4 + 8q\frac{\mathrm{d}U_4}{\mathrm{d}q}$$

and then substitute the values of $U_4$ and $U_6$ from Lemma 0.20 to give

$$\frac{1}{9}\left(35P^3 - 42PQ + 16R\right) = \frac{P}{3}\left(5P^2 - 2Q\right) + 8q\frac{\mathrm{d}}{\mathrm{d}q}\left(\frac{1}{3}\left(5P^2 - 2Q\right)\right)$$

The derivative of the $P^2$ term can be computed by the chain rule and (0.78), and this leads to

$$\frac{1}{9}\left(35P^3 - 42PQ + 16R\right) = \frac{P}{3}\left(5P^2 - 2Q\right) + \frac{8}{3}\left(10P\left(\frac{P^2 - Q}{12}\right) - 2q\frac{\mathrm{d}Q}{\mathrm{d}q}\right).$$

This may be simplified to give (0.79).

Now take $k = 6$ in the differential recurrence relation (0.62) and then substitute the values of $U_6$ and $U_8$ from Lemma 0.20 to obtain

$$\begin{aligned}&\frac{1}{15}\left(175P^4 - 420P^2Q + 320PR + 84Q^2 - 144S\right)\\ &\quad = \frac{35P^4 - 42P^2Q + 16PR}{9} + 8q\frac{\mathrm{d}}{\mathrm{d}q}\left(\frac{35P^3 - 42PQ + 16R}{9}\right).\end{aligned} \tag{0.81}$$

The derivatives of the $P^3$ and $PQ$ terms can be calculated using (0.78) and (0.79). Then (0.81) simplifies to give

$$q\frac{\mathrm{d}R}{\mathrm{d}q} = \frac{1}{40}\left(20PR + 7Q^2 - 27S\right). \tag{0.82}$$

In a similar way, if we use the differential recurrence relation for $V_k$ with $k = 6$, namely

$$V_8 = PV_6 + 24q\frac{\mathrm{d}V_6}{\mathrm{d}q},$$

and substitute the values of $V_6$ and $V_8$ from Lemma 0.20, we find that

$$q\frac{\mathrm{d}R}{\mathrm{d}q} = \frac{1}{24}\left(12PR + 5Q^2 - 17S\right). \tag{0.83}$$

The differential equation (0.80) is obtained by eliminating $S$ from (0.82) and (0.83). □

The following three corollaries are worth noting.

**Corollary 0.22.** *Let $Q$ and $S$ be defined by* (0.72) *and* (0.74). *Then*

$$S = Q^2.$$

*Proof.* Eliminate the $q\dfrac{dR}{dq}$ terms from (0.82) from (0.83). □

Further results of this type will be given in the next chapter; see Corollary 1.40.

**Corollary 0.23.** *Let Q and R be defined by* (0.72) *and* (0.73). *Then*

$$Q^3 - R^2 = 1728q\prod_{j=1}^{\infty}(1-q^j)^{24}.$$

*Proof.* By a calculation using Ramanujan's differential equations in Theorem 0.21 we have

$$q\frac{d}{dq}(Q^3 - R^2) = 3Q^2\left(\frac{PQ-R}{3}\right) - 2R\left(\frac{PR-Q^2}{2}\right) = (Q^3-R^2)P.$$

So,

$$\frac{d}{dq}\log(Q^3-R^2) = \frac{1}{q}P = \frac{1}{q} - 24\sum_{j=1}^{\infty}\frac{jq^{j-1}}{1-q^j}.$$

Integrating both sides with respect to $q$, we find that

$$\log(Q^3-R^2) = \log q + 24\sum_{j=1}^{\infty}\log(1-q^j) + \log c$$

for some constant $c$, and so

$$Q^3 - R^2 = cq\prod_{j=1}^{\infty}(1-q^j)^{24}.$$

On comparing coefficients of $q$ on both sides we find that

$$c = 3\times 240 - 2\times(-504) = 1728.$$

□

**Corollary 0.24.** *Let $U_n$ and $V_n$ be defined by* (0.59) *and* (0.60) *and let P, Q, and R be Ramanujan's Eisenstein series given by* (0.71)–(0.73). *For any nonnegative integer n, $U_{2n}$ and $V_{2n}$ may be expressed as polynomials in P, Q, and R. In fact,*

$$U_{2n} = \sum_{i+2j+3k=n} a(i,j,k)P^iQ^jR^k \tag{0.84}$$

*and*

$$V_{2n} = \sum_{i+2j+3k=n} b(i,j,k)P^i Q^j R^k \tag{0.85}$$

*where the sums are over all nonnegative integers $i$, $j$, and $k$ that satisfy $i+2j+3k=n$. Moreover, the coefficients $a(i,j,k)$ are all rational numbers, and the $b(i,j,k)$ are all integers.*

*Proof.* The results for the cases $n = 0$ and $n = 1$ are true by Corollary 0.16, and the results for the cases $n = 2$ and $n = 3$ hold by Lemma 0.20. The general results may be proved by induction on $n$ using the differential recurrence relations in Theorem 0.15 and Ramanujan's differential equations in Theorem 0.21. □

The instances $n = 1, 2$, and 3 of (0.84) and (0.85) are given in Lemma 0.20. The instances $n = 4$ of (0.84) and (0.85) can be obtained from the formulas for $U_8$ and $V_8$ in Lemma 0.20 by using Corollary 0.22 to replace $S$ with $Q^2$. The results are

$$U_8 = \frac{1}{3}\left(35P^4 - 84P^2Q + 64PR - 12Q^2\right)$$

and

$$V_8 = 105P^4 - 420P^2Q + 448PR - 132Q^2.$$

There is an analogue of (0.84) and (0.85) for the Eisenstein series $E_{2k}$; see (1.82) in Corollary 1.40.

## 0.10 Ramanujan's summation formula

The triple product identity (0.7) gives the Laurent expansion, in powers of $x$, of the function

$$f(x) = \prod_{n=1}^{\infty}(1+xq^{2n-1})(1+x^{-1}q^{2n-1}), \tag{0.86}$$

namely

$$f(x) = \sum_{n=-\infty}^{\infty} c_n x^n$$

where

$$c_n = c_n(q) = q^{n^2}\prod_{j=1}^{\infty}(1-q^{2j})^{-1}.$$

The function $f$ has simple zeros at $x = -q^{2n-1}$ for every integer $n$ and an essential singularity at $x = 0$. These are the only zeros and singularities of $f$.

Ramanujan generalized the infinite product (0.86) to include a denominator that has two infinite products. Specifically, let

$$g(x, \alpha, \beta) = \prod_{n=1}^{\infty} \frac{(1 + xq^{2n-1})(1 + x^{-1}q^{2n-1})}{(1 + \alpha xq^{2n-1})(1 + \beta x^{-1}q^{2n-1})} \tag{0.87}$$

and note that

$$g(x, 0, 0) = f(x).$$

The zeros and poles of $g$ can be determined from the factors in the infinite product. The zeros of $g$ are the same as the zeros of $f$, described above. The poles of $g$ are simple and are given by

$$x \in \left\{-(\alpha q)^{-1}, -(\alpha q^3)^{-1}, -(\alpha q^5)^{-1}, \dots\right\} \cup \left\{-\beta q, -\beta q^3, -\beta q^5, \dots\right\}.$$

This description of the zeros and poles of $g$ assumes that $\alpha$ and $\beta$ are not even powers of $q$, or 0, which ensures there is no cancellation of numerator factors with denominator factors in (0.87). If $\alpha$ or $\beta$ is an even power of $q$, or 0, then some of the zeros and poles will coalesce and cancel. In this case, the zeros and poles will be subsets of the sets described above.

From now on we will assume that

$$|\beta q| < |x| < |\alpha q|^{-1}. \tag{0.88}$$

This is to ensure that $g$ is an analytic function of $x$ in an annulus that separates the two families of poles. Ramanujan's summation formula gives an explicit formula for the Laurent expansion of $g$ in this annulus.

**Theorem 0.25 (Ramanujan's summation formula).** *Suppose $|\beta q| < |x| < |\alpha q|^{-1}$ and $|q| < 1$. Then*

$$\prod_{n=1}^{\infty} \frac{(1 + xq^{2n-1})(1 + x^{-1}q^{2n-1})(1 - q^{2n})(1 - \alpha\beta q^{2n})}{(1 + \alpha xq^{2n-1})(1 + \beta x^{-1}q^{2n-1})(1 - \alpha q^{2n})(1 - \beta q^{2n})}$$

$$= 1 + \sum_{n=1}^{\infty} \frac{(1-\alpha)(q^2-\alpha)\cdots(q^{2n-2}-\alpha)}{(1-\beta q^2)(1-\beta q^4)\cdots(1-\beta q^{2n})} q^n x^n$$

$$+ \sum_{n=1}^{\infty} \frac{(1-\beta)(q^2-\beta)\cdots(q^{2n-2}-\beta)}{(1-\alpha q^2)(1-\alpha q^4)\cdots(1-\alpha q^{2n})} q^n x^{-n}. \tag{0.89}$$

*Proof.* By Laurent's theorem, the function $g$ defined by (0.87) has an expansion of the form

$$g(x, \alpha, \beta) = \sum_{n=-\infty}^{\infty} c_n(\alpha, \beta) x^n$$

in the annulus $|\beta q| < |x| < |\alpha q|^{-1}$. From the definition (0.87) we deduce that

$$\frac{g(q^2 x, \alpha, \beta)}{g(x, \alpha, \beta)} = \frac{(1 + \alpha q x)(1 + q^{-1} x^{-1})}{(1 + \beta q^{-1} x^{-1})(1 + qx)} = \frac{1 + \alpha q x}{\beta + qx},$$

so

$$(1 + \alpha q x) g(x, \alpha, \beta) = (\beta + qx) g(q^2 x, \alpha, \beta). \tag{0.90}$$

If we put

$$h(x) = (1 + \alpha q x) g(x, \alpha, \beta),$$

then $h(x)$ is analytic in the larger annulus

$$|\beta q| < |x| < |\alpha q^3|^{-1}, \tag{0.91}$$

because the pole of $g(x, \alpha, \beta)$ at $x = -(\alpha q)^{-1}$ has been eliminated. Now

$$h(x) = (1 + \alpha q x) \sum_{n=-\infty}^{\infty} c_n(\alpha, \beta) x^n = \sum_{n=-\infty}^{\infty} (c_n(\alpha, \beta) + \alpha q c_{n-1}(\alpha, \beta)) x^n, \tag{0.92}$$

and by (0.90),

$$\begin{aligned} h(x) &= (\beta + qx) \sum_{n=-\infty}^{\infty} c_n(\alpha, \beta) q^{2n} x^n \\ &= \sum_{n=-\infty}^{\infty} (\beta c_n(\alpha, \beta) q^{2n} + c_{n-1}(\alpha, \beta) q^{2n-1}) x^n. \end{aligned} \tag{0.93}$$

Both of these expansions for $h$ are valid in the larger annulus given by (0.91). Comparing coefficients of $x^n$ in (0.92) and (0.93), we find that

$$c_n(\alpha, \beta) + \alpha q c_{n-1}(\alpha, \beta) = \beta c_n(\alpha, \beta) q^{2n} + c_{n-1}(\alpha, \beta) q^{2n-1}.$$

Hence, we obtain the recurrence relation

$$c_n(\alpha, \beta) = \frac{q^{2n-2} - \alpha}{1 - \beta q^{2n}} q c_{n-1}(\alpha, \beta).$$

By iteration, it follows that for any positive integer $n$,

$$c_n(\alpha,\beta) = \frac{(1-\alpha)(q^2-\alpha)\cdots(q^{2n-2}-\alpha)}{(1-\beta q^2)(1-\beta q^4)\cdots(1-\beta q^{2n})}\, q^n c_0(\alpha,\beta). \tag{0.94}$$

To find a formula for $c_n(\alpha,\beta)$ for negative values of $n$, observe that from the definition (0.87), we have

$$g(x,\alpha,\beta) = g(x^{-1},\beta,\alpha).$$

Expanding in powers of $x$ and comparing coefficients of $x^{-n}$ on each side, we deduce that

$$c_{-n}(\alpha,\beta) = c_n(\beta,\alpha). \tag{0.95}$$

Using (0.95) in (0.94) it follows that for any positive integer $n$,

$$\begin{aligned} c_{-n}(\alpha,\beta) &= c_n(\beta,\alpha) \\ &= \frac{(1-\beta)(q^2-\beta)\cdots(q^{2n-2}-\beta)}{(1-\alpha q^2)(1-\alpha q^4)\cdots(1-\alpha q^{2n})} q^n c_0(\beta,\alpha) \\ &= \frac{(1-\beta)(q^2-\beta)\cdots(q^{2n-2}-\beta)}{(1-\alpha q^2)(1-\alpha q^4)\cdots(1-\alpha q^{2n})} q^n c_0(\alpha,\beta), \end{aligned}$$

where the last step follows by taking $n = 0$ in (0.95).

Summarizing the results so far, we have

$$\begin{aligned} g(x,\alpha,\beta) &= \prod_{n=1}^{\infty} \frac{(1+xq^{2n-1})(1+x^{-1}q^{2n-1})}{(1+\alpha xq^{2n-1})(1+\beta x^{-1}q^{2n-1})} \\ &= c_0(\alpha,\beta)\left(1+\sum_{n=1}^{\infty}\frac{(1-\alpha)(q^2-\alpha)\cdots(q^{2n-2}-\alpha)}{(1-\beta q^2)(1-\beta q^4)\cdots(1-\beta q^{2n})}\, q^n x^n\right. \\ &\qquad \left. +\sum_{n=1}^{\infty}\frac{(1-\beta)(q^2-\beta)\cdots(q^{2n-2}-\beta)}{(1-\alpha q^2)(1-\alpha q^4)\cdots(1-\alpha q^{2n})} q^n x^{-n}\right). \end{aligned} \tag{0.96}$$

It remains to find the value of $c_0(\alpha,\beta)$. This can be done by calculating the residue of $g(x,\alpha,\beta)$ at the simple pole $x = -(\alpha q)^{-1}$. We split the series by separating the negative powers of $x$ and write

$$(1+\alpha qx)g(x,\alpha,\beta) = (1+\alpha qx)\sum_{n=0}^{\infty} c_n(\alpha,\beta)x^n + (1+\alpha qx)\sum_{n=1}^{\infty} c_{-n}(\alpha,\beta)x^{-n}. \tag{0.97}$$

By the ratio test, the series in (0.97) that involves negative powers of $x$ converges for $|x| > |\beta q|$, and since $|\alpha q|^{-1} > |\beta q|$ it follows that

$$\lim_{x\to-(\alpha q)^{-1}} (1+\alpha q x)\sum_{n=1}^{\infty} c_{-n}(\alpha,\beta)x^{-n} = 0.$$

The series in (0.97) that involves nonnegative powers of $x$ is

$$(1+\alpha q x)\sum_{n=0}^{\infty} c_n(\alpha,\beta)x^n = c_0(\alpha,\beta) + \sum_{n=1}^{\infty}(\alpha q c_{n-1}(\alpha,\beta)+c_n(\alpha,\beta))x^n, \tag{0.98}$$

and by the ratio test, this converges for $|x| < |\alpha q^3|^{-1}$. Hence, the limit as $x \to -(\alpha q)^{-1}$ can be determined by substituting the value $x = -(\alpha q)^{-1}$ in the right-hand side of (0.98). Taking the above into account and letting $x \to -(\alpha q)^{-1}$ in (0.97) we find that

$$\begin{aligned}
&\prod_{n=1}^{\infty}\frac{(1-\alpha^{-1}q^{2n-2})(1-\alpha q^{2n})}{(1-q^{2n})(1-\alpha\beta q^{2n})}\\
&= \lim_{N\to\infty}\left\{c_0(\alpha,\beta)+\sum_{n=1}^{N}(\alpha q c_{n-1}(\alpha,\beta)+c_n(\alpha,\beta))(-\alpha q)^{-n}\right\}\\
&= \lim_{N\to\infty} c_N(\alpha,\beta)(-\alpha q)^{-N}\\
&= \lim_{N\to\infty}\frac{(1-\alpha^{-1})(1-\alpha^{-1}q^2)\cdots(1-\alpha^{-1}q^{2N-2})}{(1-\beta q^2)(1-\beta q^4)\cdots(1-\beta q^{2N})}\,c_0(\alpha,\beta)\\
&= c_0(\alpha,\beta)\prod_{n=1}^{\infty}\frac{(1-\alpha^{-1}q^{2n-2})}{(1-\beta q^{2n})}.
\end{aligned}$$

It follows that

$$c_0(\alpha,\beta) = \prod_{n=1}^{\infty}\frac{(1-\alpha q^{2n})(1-\beta q^{2n})}{(1-q^{2n})(1-\alpha\beta q^{2n})}.$$

On using this in (0.96), we complete the proof. □

Convergence of the series on the right-hand side of (0.89) may be analyzed as follows. By the ratio test, the series involving positive powers of $x$ in (0.89) converges for $|\alpha q x| < 1$, while the series involving negative powers of $x$ converges for $|\beta q x^{-1}| < 1$. The two series will converge simultaneously and represent an analytic function of $x$ when both conditions are satisfied, and this is precisely the annular region $|\beta q| < |x| < |\alpha q|^{-1}$ given by (0.88) that was obtained by considering the poles of $g$.

## 0.11 Special cases of Ramanujan's summation formula

Ramanujan's summation formula contains a wealth of special cases. For example, taking the limit as $\alpha,\ \beta \to 0$ in (0.89) gives

$$\prod_{n=1}^{\infty}(1+xq^{2n-1})(1+x^{-1}q^{2n-1})(1-q^{2n}) = \sum_{n=-\infty}^{\infty} q^{n^2}x^n,$$

and the annulus of convergence given by (0.88) becomes $0 < |x| < \infty$. Thus, Jacobi's triple product identity is the special case $\alpha = \beta = 0$ of Ramanujan's summation formula.

Another consequence of Ramanujan's formula is:

**Theorem 0.26.** *For* $|q| < |x| < 1$ *we have*

$$\sum_{n=-\infty}^{\infty} \frac{x^n}{1-yq^n} = \prod_{n=1}^{\infty} \frac{(1-xyq^{n-1})(1-x^{-1}y^{-1}q^n)(1-q^n)^2}{(1-xq^{n-1})(1-x^{-1}q^n)(1-yq^{n-1})(1-y^{-1}q^n)}.$$

*Proof.* In Ramanujan's summation formula (0.89), take $\beta = 1/\alpha$ and divide both sides by $1-\alpha$, to obtain

$$\prod_{n=1}^{\infty} \frac{(1+xq^{2n-1})(1+x^{-1}q^{2n-1})(1-q^{2n})^2}{(1+\alpha xq^{2n-1})(1+\alpha^{-1}x^{-1}q^{2n-1})(1-\alpha q^{2n-2})(1-\alpha^{-1}q^{2n})}$$

$$= \frac{1}{1-\alpha} + \sum_{n=1}^{\infty} \frac{(-1)^n}{q^{2n}-\alpha}(\alpha qx)^n + \sum_{n=1}^{\infty} \frac{(-1)^n}{1-\alpha q^{2n}}\left(\frac{q}{\alpha x}\right)^n$$

$$= \sum_{n=-\infty}^{\infty} \frac{1}{1-\alpha q^{2n}}\left(-\frac{q}{\alpha x}\right)^n,$$

valid in the annulus $|q| < |\alpha x| < |q|^{-1}$. Now replace $x$ with $-q/\alpha x$, then replace $q^2$ with $q$ and $\alpha$ with $y$ to obtain the required identity. □

The Jordan–Kronecker function $F(x, y)$ is defined to be the product on the right-hand side of the identity in Theorem 0.26, that is,

$$F(x,y) = \prod_{n=1}^{\infty} \frac{(1-xyq^{n-1})(1-x^{-1}y^{-1}q^n)(1-q^n)^2}{(1-xq^{n-1})(1-x^{-1}q^n)(1-yq^{n-1})(1-y^{-1}q^n)}. \tag{0.99}$$

Clearly, $F$ is symmetric in $x$ and $y$, that is, $F(x,y) = F(y,x)$. It is not obvious that the series on the left-hand side of the identity in Theorem 0.26 is symmetric in $x$ and $y$. The symmetry in the series is restored in the following result.

**Theorem 0.27.** *The identity*

$$F(x,y) = \frac{1-xy}{(1-x)(1-y)} + \sum_{m=1}^{\infty}\sum_{n=1}^{\infty} q^{mn}(x^m y^n - x^{-m}y^{-n})$$

*holds for all $x$ and $y$ that satisfy $|q| < |x|, |y| < |q|^{-1}$, $x, y \neq 1$.*

*Proof.* By Theorem 0.26 and simple manipulations of series we have

$$\begin{aligned} F(x,y) &= \frac{1}{1-y} + \sum_{m=1}^{\infty} \frac{x^m}{1-yq^m} + \sum_{m=1}^{\infty} \frac{x^{-m}}{1-yq^{-m}} \\ &= \frac{1}{1-y} + \sum_{m=1}^{\infty} \frac{x^m(1-yq^m) + x^m yq^m}{1-yq^m} - \sum_{m=1}^{\infty} \frac{x^{-m}y^{-1}q^m}{1-y^{-1}q^m} \\ &= \frac{1}{1-y} + \frac{x}{1-x} + \sum_{m=1}^{\infty} \frac{x^m yq^m}{1-yq^m} - \sum_{m=1}^{\infty} \frac{x^{-m}y^{-1}q^m}{1-y^{-1}q^m}. \end{aligned}$$

Now expand each series on the right-hand side into a double series to obtain

$$F(x,y) = \frac{1-xy}{(1-x)(1-y)} + \sum_{m=1}^{\infty}\sum_{n=1}^{\infty} q^{mn}(x^m y^n - x^{-m}y^{-n}).$$

□

Theorem 0.27 can be used to derive expansions for the reciprocal of the product in Jacobi's triple product identity and its square.

**Theorem 0.28.** *For $|q| < |x| < |q|^{-1}$, we have*

$$\prod_{n=1}^{\infty} \frac{(1-q^n)^2}{(1-xq^n)(1-x^{-1}q^n)} = 1 + \left(x^{\frac{1}{2}} - x^{-\frac{1}{2}}\right) \sum_{m=1}^{\infty}\sum_{n=1}^{\infty} q^{mn}(x^{m-\frac{n}{2}} - x^{-m+\frac{n}{2}}) \tag{0.100}$$

*and*

$$\prod_{n=1}^{\infty} \frac{(1-q^n)^4}{(1-xq^n)^2(1-x^{-1}q^n)^2} = 1 + \left(x^{\frac{1}{2}} - x^{-\frac{1}{2}}\right)^2 \sum_{m=1}^{\infty}\sum_{n=1}^{\infty} nq^{mn}(x^{m-n} + x^{n-m}). \tag{0.101}$$

*Proof.* Multiply the result of Theorem 0.27 by $(1-x)(1-y)/(1-xy)$ to obtain

$$\prod_{n=1}^{\infty} \frac{(1-xyq^n)(1-x^{-1}y^{-1}q^n)(1-q^n)^2}{(1-xq^n)(1-x^{-1}q^n)(1-yq^n)(1-y^{-1}q^n)}$$
$$= 1 + \frac{(1-x)(1-y)}{1-xy} \sum_{m=1}^{\infty}\sum_{n=1}^{\infty} q^{mn}(x^m y^n - x^{-m}y^{-n}).$$

If we put $y = x^{-1/2}$ we obtain (0.100); or, if we take the limit as $y \to 1/x$ we obtain (0.101). □

If we take $\beta = 1$ in Ramanujan's summation formula, the series in negative powers of $x$ becomes identically zero. The resulting identity is called the "$q$-binomial theorem." The reason for the name is given after Theorem 0.31.

**Theorem 0.29 ($q$-binomial theorem).** *For $|x| < 1$ we have*

$$\prod_{n=0}^{\infty} \frac{(1-axq^n)}{(1-xq^n)} = 1 + \sum_{n=1}^{\infty} \frac{(1-a)(1-aq)\cdots(1-aq^{n-1})}{(1-q)(1-q^2)\cdots(1-q^n)} x^n.$$

*Proof.* Let $\beta = 1$ in Ramanujan's summation formula to obtain

$$\prod_{n=1}^{\infty} \frac{(1+xq^{2n-1})}{(1+\alpha xq^{2n-1})} = 1 + \sum_{n=1}^{\infty} \frac{(1-\alpha)(q^2-\alpha)\cdots(q^{2n-2}-\alpha)}{(1-q^2)(1-q^4)\cdots(1-q^{2n})} q^n x^n. \tag{0.102}$$

This has been obtained under the assumption that $|q| < |x| < |\alpha q|^{-1}$. The condition $|q| < |x|$ can be dropped, because the product on the left-hand side no longer has poles at the points $x = -\beta q, -\beta q^3, -\beta q^5, \ldots$ and therefore represents a function that is analytic in the region $|x| < |\alpha q|^{-1}$. Moreover, as only nonnegative powers of $x$ remain in the series, the Laurent expansion has reduced to a Taylor expansion. It follows that the identity (0.102) holds for $|x| < |\alpha q|^{-1}$.

Finally, replace $x$ with $-x/\alpha q$, then replace $\alpha$ and $q^2$ with $1/a$ and $q$, respectively, to complete the proof. □

**Theorem 0.30.** *Suppose $|q| < 1$. The following identities hold:*

$$\prod_{n=0}^{\infty} \frac{1}{(1-xq^n)} = 1 + \sum_{n=1}^{\infty} \frac{x^n}{(1-q)(1-q^2)\cdots(1-q^n)}, \qquad \text{for } |x| < 1,$$

*and*

$$\prod_{n=0}^{\infty}(1-xq^n) = 1 + \sum_{n=1}^{\infty} \frac{(-1)^n q^{n(n-1)/2} x^n}{(1-q)(1-q^2)\cdots(1-q^n)}, \qquad \text{for all } x.$$

*Proof.* Let $a = 0$ in the $q$-binomial theorem to obtain the first result. The second result follows by replacing $x$ with $x/a$ in the $q$-binomial theorem and taking the limit as $a \to \infty$. □

In order to state the next theorem, it will be convenient to introduce the $q$-factorial and $q$-binomial coefficient. Before doing that, recall that the factorial of any nonnegative integer $n$ is defined by

$$n! = \begin{cases} 1, & \text{if } n = 0 \text{ or } n = 1, \\ 1 \times 2 \times 3 \times \cdots \times (n-1) \times n, & \text{if } n \geq 2. \end{cases}$$

The binomial coefficient is defined, for a nonnegative integer $k$ and any complex number $n$, by

$$\binom{n}{k} = \begin{cases} \prod_{j=1}^{k} \frac{n+1-j}{j} & \text{if } k \geq 1, \\ 1 & \text{if } k = 0, \end{cases} \tag{0.103}$$

e.g., see Knuth's book [195]. For all values of $n$ we have

$$\binom{n}{k} = (-1)^k \binom{k-n-1}{k}, \tag{0.104}$$

while for integer values of $n$ the definition implies

$$\binom{n}{k} = \begin{cases} \frac{n!}{k!(n-k)!} & \text{if } 0 \leq k \leq n \\ 0 & \text{if } 0 \leq n < k. \end{cases}$$

The $q$-factorial is defined for any nonnegative integer $n$ by

$$0!_q = 1!_q = 1$$

and

$$n!_q = 1 \times (1+q) \times (1+q+q^2) \times \cdots \times (1+q+q^2+\cdots+q^{n-1}) = \prod_{j=1}^{n} \frac{1-q^j}{1-q}$$

if $n \geq 2$. By analogy with (0.103), the $q$-binomial coefficient is defined, for a nonnegative integer $k$ and any complex number $n$, by

$$\begin{bmatrix} n \\ k \end{bmatrix}_q = \begin{cases} \prod_{j=1}^{k} \frac{1-q^{n+1-j}}{1-q^j} & \text{if } k \geq 1, \\ 1 & \text{if } k = 0. \end{cases} \tag{0.105}$$

For integer values of $n$, the definition implies

$$\begin{bmatrix} n \\ k \end{bmatrix}_q = \begin{cases} \dfrac{n!_q}{k!_q (n-k)!_q} & \text{if } 0 \leq k \leq n \\ 0 & \text{if } 0 \leq n < k. \end{cases} \tag{0.106}$$

The $q$-binomial coefficients satisfy two recurrence relations—see Exercise 16. When $q = 1$, the $q$-factorial and $q$-binomial coefficient reduce to the ordinary factorial and binomial coefficient, respectively. That is,

$$\lim_{q \to 1} n!_q = n! \qquad \text{and} \qquad \lim_{q \to 1} \begin{bmatrix} n \\ k \end{bmatrix}_q = \binom{n}{k}.$$

**Theorem 0.31 (Finite form of the $q$-binomial theorem).** *For any positive integer $n$,*

$$(x+y)(x+qy)(x+q^2y)\cdots(x+q^{n-1}y) = \sum_{j=0}^{n} \begin{bmatrix} n \\ j \end{bmatrix}_q q^{j(j-1)/2} x^{n-j} y^j.$$

*Proof.* Take $a = q^{-n}$ in the $q$-binomial theorem to obtain

$$\prod_{j=0}^{\infty} \frac{(1-q^{-n+j}x)}{(1-q^j x)} = 1 + \sum_{j=1}^{\infty} \frac{(1-q^{-n})(1-q^{-n+1})\cdots(1-q^{-n+j-1})}{(1-q)(1-q^2)\cdots(1-q^j)} x^n. \tag{0.107}$$

On the left-hand side, each factor in the denominator cancels with a corresponding factor from the numerator to leave a finite number of factors in the numerator. On the right-hand side, all terms with $j > n$ are zero, due to the occurrence of the factor $(1 - q^{-n+n})$. Accordingly, (0.107) simplifies to

$$\begin{aligned}
&\prod_{j=0}^{n-1} (1 - q^{-n+j}x) \\
&= 1 + \sum_{j=1}^{n} \frac{(1-q^{-n})(1-q^{-n+1})\cdots(1-q^{-n+j-1})}{(1-q)(1-q^2)\cdots(1-q^j)} x^j \\
&= 1 + \sum_{j=1}^{n} \frac{(1-q^{n})(1-q^{n-1})\cdots(1-q^{n+1-j})}{(1-q)(1-q^2)\cdots(1-q^j)} (-1)^j q^{-nj+j(j-1)/2} x^j \\
&= \sum_{j=0}^{n} \begin{bmatrix} n \\ j \end{bmatrix}_q (-1)^j q^{-nj+j(j-1)/2} x^j.
\end{aligned}$$

Replace $x$ with $-xq^n$ to obtain

$$\prod_{j=0}^{n-1}(1+xq^j) = \sum_{j=0}^{n} \begin{bmatrix} n \\ j \end{bmatrix}_q q^{j(j-1)/2} x^j .$$

Finally, replace $x$ with $y/x$ and multiply both sides by $x^n$ to complete the proof. □

We can now explain why Theorem 0.29 is called the $q$-binomial theorem. Replace $q$ with $q^a$ in Theorem 0.29 to obtain

$$\prod_{n=0}^{\infty} \frac{(1-xq^{a+n})}{(1-xq^n)} = 1 + \sum_{n=1}^{\infty} \frac{(1-q^a)(1-q^{a+1})\cdots(1-q^{a+n-1})}{(1-q)(1-q^2)\cdots(1-q^n)} x^n. \tag{0.108}$$

On taking the limit as $q \to 1$, the series on the right-hand side becomes

$$1 + \sum_{n=1}^{\infty} \frac{a(a+1)\cdots(a+n-1)}{n!} x^n.$$

This is the binomial series, and its sum is $(1-x)^{-a}$. That is, the limit $q \to 1$ of (0.108) gives the binomial series in infinite series form

$$(1-x)^{-a} = 1 + \sum_{n=1}^{\infty} \frac{a(a+1)\cdots(a+n-1)}{n!} x^n. \tag{0.109}$$

Similarly, the identity in Theorem 0.31 reduces to the binomial theorem in the finite form

$$(x+y)^n = \sum_{j=0}^{n} \binom{n}{j} x^{n-j} y^j$$

in the limiting case $q \to 1$. This is the reason for the names "$q$-binomial theorem" and "finite form of the $q$-binomial theorem" given to Theorems 0.29 and 0.31, respectively.

It is possible to prove Theorem 0.31 directly; see Exercise 17.

## 0.12 Notes

The proof of Jacobi's triple product identity in Section 0.3 is from Hardy and Wright [175, p. 282]. It is the same as a proof of Jacobi apart from the determination of the constant term $a_0(q)$. Jacobi's original proof appeared near the end of his book on elliptic functions [189, Secs. 62, 63] and a simplified version is given in Exercise 20. Jacobi's book contains a second proof [189, Section 64] and this is outlined in Exercise 21.

The triple product identity was known to Gauss, although he didn't publish it; see the book by R. Roy [259, Chapter 29] for more information.

There are many other proofs of Jacobi's triple product identity. The analytic proof in Section 0.3 was chosen because the technique can be used to prove other identities, for example, the quintuple product identity in Section 0.5 and Ramanujan's summation formula in Section 0.10.

Ramanujan's theta function $f(a, b)$ occurs in Chapter 16 of Ramanujan's second notebook [253] from Entry 18 onwards. See [46, pp. 34–86] for the analysis by B. C. Berndt.

The quintuple product identity was known to L. Kiepert in 1879. For, equating the series [193, (27)] with the product [193, (31)] gives the quintuple product identity in the form (0.36). A survey of the quintuple product identity, which includes details of 29 proofs, has been given in [106]. The proof of the quintuple product identity that we have presented is a combination of proofs of Mordell [234] (as far as (0.29)) and H.-C. Chan [73] (to determine the constant term). A bijective proof of the quintuple product identity has been given by S. Kim [194].

The identity (0.44) is due to Euler; (0.41), (0.42), and (0.45) are due to Gauss; (0.49) is due to Jacobi; and (0.47) and (0.48) are due to Ramanujan.

The coefficients $U_{2k}$ and $V_{2k}$ were introduced by Ramanujan. The differential recurrence relations in Theorem 0.15, together with a proof, are in Ramanujan's lost notebook [254, p. 369]. Corollary 0.24 is also due to Ramanujan.

The differential equations in Theorem 0.21 were known to G.-H. Halphen in 1886 [173, p. 450]. They have come to be known as Ramanujan's differential equations as they play a fundamental role in Ramanujan's paper [249]. The proof of Theorem 0.21 we have given is due to H. H. Chan [78]. References to other proofs are also given in [78]. Chan's methods have been extended by P. C. Toh [282] to derive the analogous system of differential equations for level 2.

Ramanujan's summation formula (0.89) is in Ramanujan's second notebook [253, Chapter 16, Entry 17]. For some historical background and references to other proofs, see the book by G. E. Andrews and B. C. Berndt [26, pp. 54–56] and the papers by R. Askey [30] and W. P. Johnson [190]. The identities in Theorem 0.30 are due to Euler.

Theorem 0.27 was known to Jordan [191, pp. 507–511] and Kronecker [200]. For this reason, the function $F(x, y)$ defined by (0.99) is called the Jordan–Kronecker function.

The product in (0.100) occurs in Ramanujan's lost notebook; see the analysis by Andrews and Berndt [27, Entries 3.2.1, 4.2.1 and 4.2.2] for details and further information. See also the identity (9.2) in Fine's book [151, p. 10].

An excellent introduction to the $q$-binomial theorem and to $q$-series in general is Chapter 10 of the book by R. Askey, G. E. Andrews and R. Roy [24].

## 0.13 Exercises

1. Let $\omega = \exp(i\pi/3)$. Use Jacobi's triple product identity to prove:

$$\frac{\eta_2\eta_3^2}{\eta_1\eta_6} = \sum_{n=-\infty}^{\infty} q^{(6n+1)^2/24},$$

$$\frac{\eta_1\eta_6^2}{\eta_2\eta_3} = \sum_{n=-\infty}^{\infty} (-1)^n q^{(3n+1)^2/3},$$

$$\frac{\eta_1^2\eta_6}{\eta_2\eta_3} = \sum_{n=-\infty}^{\infty} \omega^{2n+1} q^{(2n+1)^2/8}$$

and

$$\frac{\eta_2^2\eta_3}{\eta_1\eta_6} = \sum_{n=-\infty}^{\infty} \omega^n q^{n^2}.$$

See the article by D. Zagier [306, p. 30] and the book by G. Köhler [196, p. 115] for more information about these identities.

2. Let $\chi_{12} : \mathbb{Z} \to \{-1, 0, 1\}$ be defined by

$$\chi_{12}(n) = \chi_{12}(n) = \begin{cases} 1 & \text{if } n \equiv 1 \text{ or } 11 \pmod{12}, \\ -1 & \text{if } n \equiv 5 \text{ or } 7 \pmod{12}, \\ 0 & \text{otherwise.} \end{cases}$$

Observe that $\chi$ is periodic with period 12 and $\chi_{12}(n) = 0$ unless $n$ is coprime to 12.

a. Prove that $\chi_{12}$ is completely multiplicative, that is, $\chi_{12}(mn) = \chi_{12}(m)\chi_{12}(n)$, for all integers $m$ and $n$.
b. Prove the following analogue of (0.24):

$$\sum_{n=-\infty}^{\infty} \chi_{12}(n) q^{n^2} x^n$$

$$= q(x + x^{-1}) \prod_{n=1}^{\infty} \frac{(1 - x^4q^{24n})(1 - x^{-4}q^{24n})}{(1 - x^2q^{24n})(1 - x^{-2}q^{24n})} \times (1 - q^{24n}).$$

Hint: This is immediate from (0.36).

c. Prove that

$$\eta_1 = \sum_{n=1}^{\infty} \chi_{12}(n) q^{n^2/24}.$$

Hint: Either use (0.40), or take $x = 1$ in Part b and replace $q$ with $q^{1/24}$.

3. Prove that

$$\begin{aligned}
&\prod_{j=1}^{\infty}(1-q^j)\\
&= \prod_{j=1}^{\infty}(1-q^{16j-2})(1-q^{16j-14})(1-q^{16j})(1-q^{32j-12})(1-q^{32j-20})\\
&- q\prod_{j=1}^{\infty}(1-q^{16j-6})(1-q^{16j-10})(1-q^{16j})(1-q^{32j-4})(1-q^{32j-28}),
\end{aligned}$$

and observe that this gives a representation of Euler's product as the sum of an even function and an odd function.
Hint: By (0.44) we have

$$\prod_{j=1}^{\infty}(1-q^j) = \sum_{j=-\infty}^{\infty}(-1)^j q^{j(3j-1)/2}.$$

Separate the terms in the series according to whether the power is even or odd, by noting that $j(3j-1)/2$ is even if $j \equiv 0$ or $-1$ modulo 4 and odd if $j \equiv 1$ or 2 modulo 4. Each of the two resulting series can be summed by the quintuple product identity.

See the work with M. Hirschhorn [118] for further results of this type.

4. By expanding and collecting terms according to the residue class of the power of $q$ modulo 3, prove that

$$\begin{aligned}
\prod_{j=1}^{\infty}(1-q^j) &= \prod_{j=1}^{\infty}(1-q^{27j-12})(1-q^{27j-15})(1-q^{27j})\\
&- q\prod_{j=1}^{\infty}(1-q^{27j-6})(1-q^{27j-21})(1-q^{27j})\\
&- q^2\prod_{j=1}^{\infty}(1-q^{27j-3})(1-q^{27j-24})(1-q^{27j}).
\end{aligned}$$

For further results of this type and their implications, see Theorems 5.8 and 7.1.

5. Let $T = 2\tau$ so that $q = e^{i\pi T} = e^{2\pi i\tau}$. The theta functions $\theta_1, \theta_2, \theta_3$ and $\theta_4$ may be defined by the Fourier expansions

$$\theta_1(z,q) = 2\sum_{n=0}^{\infty}(-1)^n q^{(n+\frac{1}{2})^2}\sin(2n+1)z,$$

$$\theta_2(z,q) = 2\sum_{n=0}^{\infty} q^{(n+\frac{1}{2})^2}\cos(2n+1)z,$$

$$\theta_3(z,q) = 1 + 2\sum_{n=1}^{\infty} q^{n^2}\cos 2nz,$$

$$\theta_4(z,q) = 1 + 2\sum_{n=1}^{\infty}(-1)^n q^{n^2}\cos 2nz.$$

When it is not required to indicate the dependence on $q$, we will just write $\theta_1(z), \ldots, \theta_4(z)$, for $\theta_1(z,q), \ldots, \theta_4(z,q)$, respectively.

a. Use Jacobi's triple product identity to show that

$$\theta_1(z) = 2q^{1/4}\sin z\prod_{n=1}^{\infty}(1-q^{2n}e^{2iz})(1-q^{2n}e^{-2iz})(1-q^{2n}),$$

$$\theta_2(z) = 2q^{1/4}\cos z\prod_{n=1}^{\infty}(1+q^{2n}e^{2iz})(1+q^{2n}e^{-2iz})(1-q^{2n}),$$

$$\theta_3(z) = \prod_{n=1}^{\infty}(1+q^{2n-1}e^{2iz})(1+q^{2n-1}e^{-2iz})(1-q^{2n}),$$

$$\theta_4(z) = \prod_{n=1}^{\infty}(1-q^{2n-1}e^{2iz})(1-q^{2n-1}e^{-2iz})(1-q^{2n}).$$

b. Deduce the periodic and quasi-periodic properties of the theta functions that are indicated in Table 0.3.

**Table 0.3** Periodic and quasi-periodic properties of theta functions

| $j$ | 1 | 2 | 3 | 4 |
|---|---|---|---|---|
| $\dfrac{\theta_j(z+\pi)}{\theta_j(z)}$ | $-1$ | $-1$ | $1$ | $1$ |
| $\dfrac{\theta_j(z+\pi T)}{\theta_j(z)}$ | $-q^{-1}e^{-2iz}$ | $q^{-1}e^{-2iz}$ | $q^{-1}e^{-2iz}$ | $-q^{-1}e^{-2iz}$ |

c. Let us say that two complex numbers $z$ and $w$ are equivalent if $z = w + \pi m + \pi n T$ for some integers $m$ and $n$. Show that the zeros of $\theta_1(z)$, $\theta_2(z)$, $\theta_3(z)$, and $\theta_4(z)$ are simple and are located at points equivalent to 0, $\pi/2$, $\pi/2 + \pi T/2$ and $\pi T/2$, respectively. Observe that these four points are the vertices of a parallelogram, traversed counterclockwise.

d. Show that

$$\theta_1'(0) = 2\eta_2^3,$$

where the prime denotes differentiation with respect to $z$, and $\eta_k$ is defined by (0.39).

Hint: Use the definition of $\theta_1$ as a Fourier series, and (0.49).

e. Use the result of Part a to show that

$$\theta_1(z)\theta_2(z)\theta_3(z)\theta_4(z) = \eta_2^3\,\theta_1(2z).$$

f. Use Parts d and e to deduce that

$$\theta_1'(0) = \theta_2(0)\theta_3(0)\theta_4(0).$$

g. Use the results of Part a and the quintuple product identity to prove that

$$\eta_2 \times \frac{\theta_1(2z,q)}{\theta_1(z,q)} = 2\sum_{n=-\infty}^{\infty}(-1)^n q^{(6n+1)^2/12}\cos(6n+1)z,$$

$$\eta_2 \times \frac{\theta_1(2z,q)}{\theta_2(z,q)} = 2\sum_{n=-\infty}^{\infty} q^{(6n+1)^2/12}\sin(6n+1)z,$$

$$\eta_2 \times \frac{\theta_1(2z,q)}{\theta_3(z,q)} = 2\sum_{n=-\infty}^{\infty} q^{(6n+2)^2/12}\sin(6n+2)z,$$

$$\eta_2 \times \frac{\theta_1(2z,q)}{\theta_4(z,q)} = 2\sum_{n=-\infty}^{\infty}(-1)^{n+1} q^{(6n+2)^2/12}\sin(6n+2)z.$$

Parts a–f are standard results in the theory of theta functions, e.g., see the book by Whittaker and Watson [299, Chapter 21]. For Part g, see the work of L.-C. Shen [269].

6. Show that the number of partitions of $n$ into parts that do not exceed $k$ is equal to the number of partitions of $n$ into no more than $k$ parts.
7. a. Show that

$$\prod_{j=1}^{\infty}\frac{1}{(1-q^{3j-1})(1-q^{3j-2})} = \prod_{j=1}^{\infty}(1+q^j+q^{2j}).$$

b. Deduce that the number of partitions of $n$ with no part divisible by 3 is equal to the number of partitions of $n$ with no part repeated more than twice.

c. Let $m \geq 2$ be an integer. Generalize the result of Part b by considering the product

$$\prod_{j=1}^{\infty} \frac{1}{(1-q^{mj-1})(1-q^{mj-2})\cdots(1-q^{mj-(m-1)})}.$$

The result should specialize to Theorem 0.12 when $m = 2$, and to the result of Part b when $m = 3$.

8. Let $A(n)$ denote the number of partitions of $n$ into parts that are congruent to 2, 5, or 11 modulo 12. Let $B(n)$ denote the number of partitions of $n$ into distinct parts that are congruent to either 2, 4, or 5 modulo 6. Prove that $A(n) = B(n)$. See [49, p. 5] for this exercise, and others.

9. Let $D$ be the differential operator defined by $D = q\frac{\mathrm{d}}{\mathrm{d}q}$. Use Ramanujan's differential equations to show that $P$ satisfies the Chazy equation:

$$D^3P - PD^2P + \frac{3}{2}(DP)^2 = 0.$$

10. If $E_{2k}$ is defined by (0.66), show that

$$E_{2k}(-q) = -E_{2k}(q) + \left(2^{2k}+2\right)E_{2k}(q^2) - 2^{2k}E_{2k}(q^4).$$

Hence, deduce that the effect of replacing $q$ with $-q$ in Ramanujan's Eisenstein series $P$, $Q$, and $R$ is given by

$$P(-q) = -P(q) + 6P(q^2) - 4P(q^4), \tag{0.110}$$

$$Q(-q) = -Q(q) + 18Q(q^2) - 16Q(q^4), \tag{0.111}$$

$$R(-q) = -R(q) + 66R(q^2) - 64R(q^4). \tag{0.112}$$

11. a. Use Corollary 0.24 to prove Ramanujan's results [254, p. 369]:

$$\begin{aligned}
V_0 &= 1,\\
V_2 &= P,\\
V_4 &= 3P^2 - 2Q,\\
V_6 &= 15P^3 - 30PQ + 16R,\\
V_8 &= 105P^4 - 420P^2Q + 448PR - 132Q^2,\\
V_{10} &= 945P^5 - 6300P^3Q + 10080P^2R - 5940PQ^2 + 1216QR
\end{aligned}$$

and

$$V_{12} = 10395P^6 - 103950P^4Q + 221760P^3R - 196020P^2Q^2 + 80256PQR - 2712Q^3 - 9728R^2.$$

b. Use the results from Part a to show that

$$\sum_{k=0}^{6} \binom{12}{2k} (-1)^k V_{2k} V_{12-2k} = 64\,(3999\,Q^3 - 4000\,R^2)$$

and hence deduce that

$$2^6 \times \eta_1^2 \times (3999\,Q^3(q) - 4000\,R^2(q)) = \sum_{\substack{a\equiv 1 \pmod 6\\ b\equiv 1 \pmod 6}} (-1)^{(a+b-2)/6}\,\mathrm{Re}\left((a+ib)^{12}\right) q^{(a^2+b^2)/24},$$

where the sum is over all integers $a$ and $b$ that satisfy the given congruences.

c. It can be shown, e.g., see work with H. H. Chan and P. C. Toh [84, Lemma 2], that

$$3^6 \times \eta_1^2 \times (5439\,Q^3(q) - 5438\,R^2(q)) = \sum_{\substack{c\equiv 0 \pmod 6\\ d\equiv 1 \pmod 6}} (-1)^{(c+d-1)/6}\,\mathrm{Re}\left((c+id\sqrt{3})^{12}\right) q^{(c^2+3d^2)/36}.$$

Use this together with Part b and Corollary 0.23 to deduce that

$$\begin{aligned}(2^7 \times 3^4 \times 11^2 \times 13) \times \eta_1^{26} &= \frac{1}{2^6} \sum_{\substack{a\equiv 1 \pmod 6\\ b\equiv 1 \pmod 6}} (-1)^{(a+b-2)/6}\,\mathrm{Re}\left((a+ib)^{12}\right) q^{(a^2+b^2)/24}\\ &\quad + \frac{1}{3^6} \sum_{\substack{c\equiv 0 \pmod 6\\ d\equiv 1 \pmod 6}} (-1)^{(c+d-1)/6}\,\mathrm{Re}\left((c+id\sqrt{3})^{12}\right) q^{(c^2+3d^2)/36}.\end{aligned}$$

Analogous formulas for $\eta_1^8$, $\eta_1^{10}$, and $\eta_1^{14}$ will be obtained in the exercises at the end of the next chapter.

12. Let $U_k$ be as defined by (0.59). Prove that

$$U_{-1} = \frac{\eta_2^2}{\eta_1^4} \quad \text{and} \quad U_1 = \frac{4P(q^2) - P(q)}{3} \times U_{-1}.$$

Hint: Use (0.43), (0.46), and Theorem 0.15.

13. Use (0.54) to prove that

$$(2^{2k}-1)U_{2k}=\sum_{j=0}^{k-1}\binom{2k+1}{2j+1}U_{2j}V_{2k-2j}.$$

14. a. Suppose $0 < |a|,\ |b|,\ |q| < 1$, and let $\lambda$ be any complex number. Use Ramanujan's summation formula to prove that

$$\frac{(q;q)_\infty^3(ab;q)_\infty}{(a;q)_\infty^2(b;q)_\infty^2}=\sum_{n=0}^{\infty}\frac{(q/a;q)_n}{(b;q)_{n+1}}(n+\lambda+\frac{1}{2})a^n+\sum_{n=0}^{\infty}\frac{(q/b;q)_n}{(a;q)_{n+1}}(n-\lambda+\frac{1}{2})b^n$$

where we are using the compact notation for products defined by

$$(x;q)_n=\begin{cases}1 & \text{if } n=0,\\ \prod_{j=0}^{n-1}(1-q^jx) & \text{if } n=1,2,3,\dots\end{cases} \tag{0.113}$$

and

$$(x;q)_\infty=\lim_{n\to\infty}(x;q)_n=\prod_{j=0}^{\infty}(1-q^jx),\quad \text{provided } |q|<1.$$

b. Show that the result in Part a reduces to (0.49) in the limiting case $a, b \to 0$. See S. Bhargava and D. D. Somashekara [56].

15. For real values of $\theta$, let $f(\theta)$ be defined by

$$f(\theta)=\sum_{k=-\infty}^{\infty}\frac{x^{-k}q^ke^{ik\theta}}{1-x^2q^{2k}}$$

where $x$ and $q$ are constants and $|q| < |x| < |q|^{-1}$.

a. Use Theorem 0.26 to show that

$$f(\theta)=\prod_{j=1}^{\infty}\frac{(1-xe^{i\theta}q^{2j-1})(1-x^{-1}e^{-i\theta}q^{2j-1})(1-q^{2j})^2}{(1-xe^{-i\theta}q^{2j-1})(1-x^{-1}e^{i\theta}q^{2j-1})(1-x^2q^{2j-2})(1-x^{-2}q^{2j})}.$$

b. Deduce that $f(\theta)f(-\theta)=(f(0))^2$.

c. Integrate the result of Part b over the interval $0 \le \theta \le 2\pi$ and divide by $2\pi$, to deduce the "freshman's dream"

$$\sum_{k=-\infty}^{\infty} \left( \frac{x^{-k} q^k}{1 - x^2 q^{2k}} \right)^2 = \left( \sum_{k=-\infty}^{\infty} \frac{x^{-k} q^k}{1 - x^2 q^{2k}} \right)^2 .$$

See K. B. Ford [152].

16. a. Show that

$$\begin{bmatrix} n \\ k \end{bmatrix}_q = (-1)^n \, q^{nk - k(k-1)/2} \begin{bmatrix} k - n - 1 \\ k \end{bmatrix}_q ,$$

where $n$ is any complex number and $k$ is a nonnegative integer. This is an analogue of (0.104).

b. Use the definition (0.105) to prove that the $q$-binomial coefficients satisfy the recurrence relations

$$\begin{bmatrix} n+1 \\ k \end{bmatrix}_q = \begin{bmatrix} n \\ k \end{bmatrix}_q + q^{n+1-k} \begin{bmatrix} n \\ k-1 \end{bmatrix}_q$$

and

$$\begin{bmatrix} n+1 \\ k \end{bmatrix}_q = q^k \begin{bmatrix} n \\ k \end{bmatrix}_q + \begin{bmatrix} n \\ k-1 \end{bmatrix}_q ,$$

where $n$ is any complex number and $k$ is a positive integer.

c. Suppose $k$ and $n$ are integers and $0 \le k \le n$. Use induction on $n$ and either of the recurrence relations in Part b to show that $\begin{bmatrix} n \\ k \end{bmatrix}_q$ is a polynomial in $q$ of degree $k(n-k)$.

17. Give a direct proof of the finite form of the $q$-binomial theorem by completing the steps outlined below.

a. Let $f(x) = (1+x)(1+qx)(1+q^2x)\cdots(1+q^{n-1}x)$. Observe that for any fixed value of $q, f(x)$ is a polynomial of degree $n$ in $x$. That is,

$$f(x) = \sum_{j=0}^{n} c(n,j) x^j$$

where the coefficients $c(n,j)$ depend on $q$, and $c(n,0) = 1$.

b. Show that

$$f(x) \times (1 + q^n x) = (1 + x) \times f(qx). \tag{0.114}$$

c. Expand both sides of (0.114) in powers of $x$ and equate coefficients of $x^j$ to show that

$$c(n,j)+q^n c(n,j-1)=q^j c(n,j)+q^{j-1}c(n,j-1).$$

d. Deduce that

$$c(n,j)=\frac{q^{j-1}-q^n}{1-q^j}\,c(n,j-1)$$

and iterate to obtain

$$c(n,j)=\prod_{k=1}^{j}\frac{q^{k-1}-q^n}{1-q^k}=\begin{bmatrix} n \\ j \end{bmatrix}_q q^{j(j-1)/2}.$$

e. By Parts a–d we have

$$(1+x)(1+qx)(1+q^2x)\cdots(1+q^{n-1}x)=\sum_{j=0}^{n}\begin{bmatrix} n \\ j \end{bmatrix}_q q^{j(j-1)/2}x^j. \qquad (0.115)$$

Replace $x$ with $y/x$ and multiply both sides by $x^n$ to complete this proof of Theorem 0.31.

See Pólya and Alexanderson [243], written for high school students, for the above proof and also for a combinatorial interpretation of the $q$-binomial coefficients.

18. The purpose of this exercise is to use the finite form of the $q$-binomial theorem to prove the Jacobi triple product identity. This proof was given by Gauss [159] and Cauchy [71].

a. In (0.115), replace $n$ with $2n$ and then replace $x$ with $q^{-n}x$ to obtain

$$\prod_{j=1}^{n}(1+xq^{-j})(1+xq^{j-1})=\sum_{j=0}^{2n}\begin{bmatrix} 2n \\ j \end{bmatrix}_q q^{j(j-1)/2-nj}x^j.$$

b. Multiply the result of Part a by $x^{-n}q^{n(n+1)/2}$ and shift the summation index on the right-hand side by setting $j=n+k$ to obtain

$$\prod_{j=1}^{n}(1+xq^{j-1})(1+x^{-1}q^j)=\sum_{k=-n}^{n}\begin{bmatrix} 2n \\ n+k \end{bmatrix}_q q^{k(k-1)/2}x^k.$$

c. Observe that for $|q|<1$ and any fixed value of $k$,

$$\lim_{n\to\infty}\begin{bmatrix} 2n \\ n+k \end{bmatrix}_q=\prod_{j=1}^{\infty}\frac{1}{1-q^j}.$$

d. Take the limit as $n \to \infty$ in the result of Part b and make use of Part c to deduce

$$\prod_{j=1}^{n}(1+xq^{j-1})(1+x^{-1}q^{j})(1-q^{j}) = \sum_{k=-\infty}^{\infty} q^{k(k-1)/2}x^{k}.$$

On replacing $x$ with $-x$, this gives another proof of Jacobi's triple product identity (0.15).

19. Here is a proof of Jacobi's triple product identity in the form (0.15). It has been extracted from work of I. Macdonald [229] where multivariate extensions of Jacobi's triple product identity were studied. Further details of the proof, sketched below, are in [101].

Let $f(x) = \prod_{n=1}^{\infty}(1-xq^{n-1})(1-x^{-1}q^{n})$ and denote the Laurent expansion of $f$ in the annulus $0 < |x| < \infty$ by $f(x) = \sum_{n=-\infty}^{\infty} c_n(q)x^n$.

a. Show that $f(x) = -xf(qx)$.
b. Show that for any positive integer $n$, $c_n(q) = -q^{n-1}c_{n-1}(q)$ and $c_{-n}(q) = -q^n c_{1-n}(q)$. Deduce that, for any integer $n$,

$$c_n(q) = (-1)^n q^{n(n-1)/2} c_0(q).$$

c. Use Part b to conclude that

$$\prod_{n=1}^{\infty}(1-xq^{n-1})(1-x^{-1}q^{n}) = c_0(q)\sum_{n=-\infty}^{\infty}(-1)^n q^{n(n-1)/2}x^n.$$

d. Replace $q$ with $q^3$ in Part c and then put $x = q$ to obtain

$$\prod_{n=1}^{\infty}\frac{(1-q^n)}{(1-q^{3n})} = c_0(q^3)\sum_{n=-\infty}^{\infty}(-1)^n q^{n(3n-1)/2}.$$

e. Let $x = \exp(2\pi i/3)$ in Part c, and simplify the result, to show that

$$\prod_{n=1}^{\infty}\frac{(1-q^{3n})}{(1-q^n)} = c_0(q)\sum_{n=-\infty}^{\infty}(-1)^n q^{3n(3n-1)/2}.$$

f. Replace $q$ with $q^3$ in the result of Part d, and combine with Part e, to deduce

$$c_0(q)\prod_{n=1}^{\infty}(1-q^n) = c_0(q^9)\prod_{n=1}^{\infty}(1-q^{9n}).$$

g. Iterate the result of Part f $k$ times and let $k \to \infty$ to deduce

$$c_0(q)\prod_{n=1}^{\infty}(1-q^n) = \lim_{k\to\infty} c_0(q^{9^k})\prod_{n=1}^{\infty}(1-q^{9^k n}) = 1.$$

Hence complete the proof of Jacobi's triple product identity in the form (0.15).

20. This exercise outlines Jacobi's first proof of the triple product identity from [189, Secs. 62, 63]. The point to be noted is the procedure for determining the constant term $a_0(q)$.

Let $f$ and $g$ be defined by

$$f(x,q) = \prod_{n=1}^{\infty}(1+xq^{2n-1})(1+x^{-1}q^{2n-1})$$

and

$$g(x,q) = \prod_{n=1}^{\infty}(1+xq^{2n})(1+x^{-1}q^{2n-2}).$$

a. Proceed as in the proof of (0.7) in Section 0.3 to show that

$$f(x,q) = a_0(q)\sum_{n=-\infty}^{\infty} q^{n^2}x^n,$$

where $a_0(q)$ denotes a function of $q$ that is independent of $x$.

b. Observe that $g(x,q) = f(qx,q)$ and deduce that

$$g(x,q) = a_0(q)\sum_{n=-\infty}^{\infty} q^{n^2+n}x^n.$$

c. Show that $g(x,q) = f(x,q^2)g(x,q^2)$.

d. Expand both sides of the result of Part c in powers of $x$ and extract terms independent of $x$, that is, the coefficient of $x^0$, to deduce

$$a_0(q) = \left(a_0(q^2)\right)^2\sum_{n=-\infty}^{\infty} q^{2n^2+n}.$$

e. Take $x = 1$ in the result of Part b to show that

$$2\prod_{n=1}^{\infty}(1+q^{2n})^2 = a_0(q)\sum_{n=-\infty}^{\infty} q^{n^2+n}.$$

f. Use series rearrangements to show that

$$\sum_{n=-\infty}^{\infty} q^{2n^2+n} = \sum_{n=0}^{\infty} q^{n^2+n} = \frac{1}{2}\sum_{n=-\infty}^{\infty} q^{n^2+n}.$$

g. The result of Part f can be used to eliminate the series from Parts d and e. Hence, deduce that

$$(a_0(q))^2 = \left(a_0(q^2)\right)^2 \prod_{n=1}^{\infty}(1+q^{2n})^2$$

and therefore

$$a_0(q)\prod_{n=1}^{\infty}(1-q^{2n}) = a_0(q^2)\prod_{n=1}^{\infty}(1-q^{4n}).$$

h. Iterate the result of Part g and take the limit as $k \to \infty$ to obtain

$$a_0(q)\prod_{n=1}^{\infty}(1-q^{2n}) = \lim_{k\to\infty} a_0(q^{2^k})\prod_{n=1}^{\infty}(1-q^{2^{k+1}n}) = a_0(0) = 1.$$

This gives the value of $a_0(q)$.

i. Use the results of Parts a and h to complete this proof of Jacobi's triple product identity.

21. This exercise outlines Jacobi's second proof of the triple product identity from [189, Section 64].

a. Show that for all $x$ and $|q| < 1$, the following identity holds:

$$\prod_{n=1}^{\infty}(1+xq^{2n-1}) = 1 + \sum_{n=1}^{\infty}\frac{q^{n^2}x^n}{(1-q^2)(1-q^4)\cdots(1-q^{2n})}.$$

Hints: Denote the product on the left-hand side by $g(x)$. It is an entire function and hence has a Taylor expansion in powers of $x$. Compare $g(x)$ with $g(q^2x)$ to obtain a recurrence relation for the coefficients. The constant term can be obtained by letting $x = 0$.

Note: After a change of variable, this provides a direct and alternative proof of one of the identities in Theorem 0.30.

b. Consider the problem of determining the coefficients $b_n(q)$ in the expansion

$$\prod_{n=1}^{\infty}\frac{1}{(1-xq^n)} = b_0(q) + \sum_{n=1}^{\infty}\frac{b_n(q)\,x^n}{(1-xq)(1-xq^2)\cdots(1-xq^n)}.$$

Show that $b_0(q) = 1$ and

$$b_n(q) = \frac{q^{n^2}}{(1-q)(1-q^2)\cdots(1-q^n)} \qquad \text{for } n = 1, 2, 3, \ldots.$$

Hint: The value of $b_0(q)$ is obtained by setting $x = 0$. Denote the product on the left-hand side by $h(x)$. It satisfies the functional equation

$$h(x) = \frac{h(qx)}{1-qx}.$$

Therefore,

$$\begin{aligned}
&1 + \sum_{n=1}^{\infty} \frac{b_n(q)\, x^n}{(1-xq)(1-xq^2)\cdots(1-xq^n)} \\
&= \frac{1}{1-qx} + \sum_{n=1}^{\infty} \frac{b_n(q)\,(qx)^n}{(1-xq)(1-xq^2)\cdots(1-xq^{n+1})} \\
&= 1 + \sum_{n=1}^{\infty} \frac{(q^{2n-1}b_{n-1}(q) + q^n b_n(q))x^n}{(1-xq)(1-xq^2)\cdots(1-xq^n)},
\end{aligned}$$

where the last step is obtained by using

$$\frac{1}{1-xq^{n+1}} = 1 + \frac{xq^{n+1}}{1-xq^{n+1}}.$$

Use this to obtain and then solve a recurrence relation for $b_n(q)$.

c. Let $f(x) = \prod_{n=1}^{\infty}(1+xq^{2n-1})(1+x^{-1}q^{2n-1})$ and consider the Laurent expansion of $f(x)$ in powers of $x$. That is, write

$$f(x) = \sum_{m=-\infty}^{\infty} a_m(q)x^m.$$

Use the result of Part a twice to show that for any nonnegative integer $m$,

$$a_m(q) = a_{-m}(q) = \frac{q^{m^2}}{(q^2;q^2)_m} \sum_{n=0}^{\infty} \frac{q^{2n^2+2mn}}{(q^2;q^2)_n(q^{2m+2};q^2)_n},$$

where we are using the compact notation for products defined by (0.113).

d. Replace $q$ with $q^2$ in the result of Part b, then put $x = q^{2m}$, to show that the formula in Part c simplifies to

$$a_m(q) = a_{-m}(q) = q^{m^2} \times \prod_{n=1}^{\infty} \frac{1}{(1-q^{2n})}.$$

Hence, complete this proof of (0.7).

22. Prove the partial fractions expansion for the reciprocal of the function that occurs in Jacobi's triple product identity:

$$\prod_{n=1}^{\infty} \frac{(1-q^n)^2}{(1-xq^{n-1})(1-x^{-1}q^n)} = \sum_{n=-\infty}^{\infty} \frac{(-1)^n q^{n(n+1)/2}}{(1-xq^n)}.$$

See Tannery and Molk [280, Sec. 486]. Further information has been given by Andrews and Berndt [27, pp. 47, 72].

# Chapter 1
# Elliptic Functions

## 1.1 Introduction and the Weierstrass elliptic function

The entire function $s(\theta)$ introduced in (0.51), viz.,

$$s(\theta) = s(\theta, \tau) = 2\sin\frac{\theta}{2}\prod_{n=1}^{\infty}\frac{(1-e^{i\theta}q^n)(1-e^{-i\theta}q^n)}{(1-q^n)^2} \tag{1.1}$$

has simple zeros at each point in the set

$$\Lambda = \{2\pi m + 2\pi n\tau \,|m, n \in \mathbb{Z}\}.$$

Here, as always, $\operatorname{Im}\tau > 0$ and $q = \exp(2\pi i\tau)$. Since $\Lambda$ is invariant under translations by $2\pi$ or by $2\pi\tau$, it is natural to consider the behavior of $s(\theta)$ under the transformations $\theta \to \theta + 2\pi$ and $\theta \to \theta + 2\pi\tau$. Clearly,

$$s(\theta + 2\pi) = -s(\theta). \tag{1.2}$$

A short calculation gives

$$\begin{aligned}\frac{s(\theta+2\pi\tau)}{s(\theta)} &= \left(\frac{e^{i\theta/2}q^{1/2} - e^{-i\theta/2}q^{-1/2}}{e^{i\theta/2} - e^{-i\theta/2}}\right)\prod_{n=1}^{\infty}\frac{(1-e^{i\theta}q^{n+1})(1-e^{-i\theta}q^{n-1})}{(1-e^{i\theta}q^n)(1-e^{-i\theta}q^n)} \\ &= q^{-1/2}\left(\frac{1-e^{i\theta}q}{1-e^{i\theta}}\right)\left(\frac{1-e^{-i\theta}}{1-e^{i\theta}q}\right) \\ &= -e^{-i\theta}q^{-1/2},\end{aligned}$$

that is,

$$s(\theta + 2\pi\tau) = -e^{-i\theta}q^{-1/2}s(\theta). \tag{1.3}$$



S. Cooper, *Ramanujan's Theta Functions*, DOI 10.1007/978-3-319-56172-1_2

We will apply differentiation to (1.2) and (1.3) to produce a function that is invariant under translations by $2\pi$ and by $2\pi\tau$. Let $z$ and $p$ be the functions defined by

$$z(\theta) = z(\theta,\tau) = \frac{s'(\theta)}{s(\theta)} \tag{1.4}$$

and

$$p(\theta) = p(\theta,\tau) = -z'(\theta) \tag{1.5}$$

where the primes denote differentiation with respect to $\theta$. In each of (1.1), (1.4), and (1.5), $s$, $z$, and $p$ are functions of two variables $\theta$ and $\tau$. When the value of $\tau$ doesn't need to be emphasized, we will omit it and just write $s(\theta)$, $z(\theta)$, and $p(\theta)$, respectively.

Properties of $z$ and $p$ are inherited from properties of $s$. The singularities of $z$ are simple poles at each point of $\Lambda$ and the residue at each pole is 1. By (1.2) and (1.3) it follows that

$$z(\theta + 2\pi) = z(\theta) \qquad \text{and} \qquad z(\theta + 2\pi\tau) = z(\theta) - i. \tag{1.6}$$

From (1.5) and (1.6) we deduce that the singularities of $p$ are poles of order two at each point of $\Lambda$, and

$$\lim_{\theta\to 0} \theta^2 p(\theta) = 1.$$

Moreover,

$$p(\theta + 2\pi) = p(\theta) \qquad \text{and} \qquad p(\theta + 2\pi\tau) = p(\theta); \tag{1.7}$$

that is to say, $p$ is doubly periodic with periods $2\pi$ and $2\pi\tau$. These properties provide motivation for the following definition.

**Definition 1.1 (Elliptic function).**

1. A function is said to be meromorphic if it is analytic (i.e., holomorphic) on $\mathbb{C}$ except for poles.
2. Let $\omega$ be a nonzero complex number. A function $f$ is said to be periodic with period $\omega$ if

$$f(\theta + \omega) = f(\theta)$$

   for all $\theta \in \mathbb{C}$.
3. A function $f$ is doubly periodic if it has two nonzero periods $\omega_1$ and $\omega_2$, where $\omega_2/\omega_1$ is not real. That is to say,

$$f(\theta + \omega_1) = f(\theta) \qquad \text{and} \qquad f(\theta + \omega_2) = f(\theta)$$

   for all $\theta \in \mathbb{C}$.

4. A function that is meromorphic and doubly periodic is called an elliptic function.

The function $p(\theta)$ is an elliptic function of $\theta$ with periods $2\pi$ and $2\pi\tau$. Neither $s(\theta)$ nor $z(\theta)$ is an elliptic function.

Some general properties of elliptic functions will be outlined in the next section. Before that, we record some further properties of $s$, $z$, and $p$, and introduce the Weierstrass elliptic function $\wp(\theta)$. By Theorem 0.14 and Lemma 0.18, we have the following expansions in powers of $\theta$:

$$s(\theta) = \theta + \sum_{k=1}^{\infty} \frac{(-1)^k U_{2k}}{2^{2k}(2k+1)!} \theta^{2k+1}, \tag{1.8}$$

$$z(\theta) = \frac{1}{\theta} + 2\sum_{k=1}^{\infty} \frac{(-1)^{k-1} E_{2k}}{(2k-1)!} \theta^{2k-1}. \tag{1.9}$$

By differentiation of (1.9), we obtain

$$p(\theta) = \frac{1}{\theta^2} - 2E_2 - 2\sum_{k=1}^{\infty} \frac{(-1)^k E_{2k+2}}{(2k)!} \theta^{2k}. \tag{1.10}$$

As $s$ is an entire function, the Taylor series (1.8) converges for all complex values of $\theta$. The Laurent series (1.9) and (1.10) converge in the annulus $0 < |\theta| < r$ where $r$ is the modulus of the nonzero pole(s) nearest to the origin, that is,

$$r = \inf_{\substack{\lambda \in \Lambda \\ \lambda \neq 0}} |\lambda| = \inf_{\substack{m,n \in \mathbb{Z} \\ (m,n) \neq (0,0)}} |2\pi m + 2\pi n\tau|. \tag{1.11}$$

The Weierstrass elliptic function $\wp$ is defined to be the function obtained by removing the constant term from the expansion (1.10), that is,

$$\wp(\theta) = \wp(\theta, \tau) = \frac{1}{\theta^2} - 2\sum_{k=1}^{\infty} \frac{(-1)^k E_{2k+2}}{(2k)!} \theta^{2k}. \tag{1.12}$$

Thus,

$$\wp(\theta, \tau) = p(\theta, \tau) + 2E_2 = p(\theta, \tau) - \frac{P(q)}{12} = -\frac{\mathrm{d}}{\mathrm{d}\theta} z(\theta, \tau) - \frac{P(q)}{12}. \tag{1.13}$$

The Weierstrass function $\wp$ is an elliptic function with the same periods and poles as $p$. That is,

$$\wp(\theta + 2\pi) = \wp(\theta) \qquad \text{and} \qquad \wp(\theta + 2\pi\tau) = \wp(\theta), \tag{1.14}$$

and the singularities of $\wp(\theta)$ are poles of order two at points in $\Lambda$.

Alternative series expressions for $z(\theta)$ and $\wp(\theta)$ that hold for all values of $\theta$ can be obtained as follows. By the definitions of $s$ and $z$ in (1.1) and (1.4) and direct calculation, we have

$$z(\theta) = \frac{1}{2}\cot\frac{\theta}{2} + \frac{1}{i}\sum_{n=1}^{\infty}\left(\frac{e^{i\theta}q^n}{1-e^{i\theta}q^n} - \frac{e^{-i\theta}q^n}{1-e^{-i\theta}q^n}\right). \tag{1.15}$$

The series on the right-hand side has been encountered before in (0.69). Differentiating again and using the definitions (1.5) and (1.13) we find that

$$\begin{aligned}\wp(\theta) &= \frac{1}{4}\csc^2\frac{\theta}{2} - \sum_{n=1}^{\infty}\left(\frac{e^{i\theta}q^n}{(1-e^{i\theta}q^n)^2} + \frac{e^{-i\theta}q^n}{(1-e^{-i\theta}q^n)^2}\right) - \frac{P(q)}{12}\\ &= -\frac{P(q)}{12} - \sum_{n=-\infty}^{\infty}\frac{e^{i\theta}q^n}{(1-e^{i\theta}q^n)^2}.\end{aligned} \tag{1.16}$$

The series (1.15) and (1.16) converge for all values of $\theta$ except for points $\theta \in \Lambda$ where there are poles of order one and two, respectively. The double periodicity properties of the Weierstrass $\wp$ function in (1.14) are immediate from (1.16).

The Fourier series for $z$ has been derived in (0.70):

$$z(\theta) = \frac{1}{2}\cot\frac{\theta}{2} + 2\sum_{j=1}^{\infty}\frac{q^j}{1-q^j}\sin j\theta. \tag{1.17}$$

By differentiation and the definition (1.13), we obtain

$$\wp(\theta) = \frac{1}{4}\csc^2\frac{\theta}{2} - \frac{1}{12} + 2\sum_{j=1}^{\infty}\frac{jq^j}{1-q^j}(1-\cos j\theta). \tag{1.18}$$

The Fourier series (1.17) and (1.18) converge at all points in the horizontal strip $|\operatorname{Im}\theta| < 2\pi \operatorname{Im}\tau$, except for the points $\theta = 2\pi m$, $m \in \mathbb{Z}$, where there are poles.

For completeness it may be mentioned that the Weierstrass $\zeta$ and $\sigma$ functions[1] are defined in terms of the Weierstrass $\wp$ function by the formulas

$$\frac{\mathrm{d}}{\mathrm{d}\theta}\zeta(\theta) = -\wp(\theta) \qquad \text{and} \qquad \zeta(\theta) = \frac{1}{\sigma(\theta)}\frac{\mathrm{d}}{\mathrm{d}\theta}\sigma(\theta),$$

[1]The Weierstrass zeta function should not be confused with Riemann's zeta function $\zeta(s) = \sum_{n=1}^{\infty}\frac{1}{n^s}$, which is something else!

where the constants of integration are determined by the conditions

$$\lim_{\theta\to 0}\left(\zeta(\theta)-\frac{1}{\theta}\right)=0 \qquad \text{and} \qquad \lim_{\theta\to 0}\frac{\sigma(\theta)}{\theta}=1.$$

It follows that

$$\zeta(\theta)=z(\theta)+\frac{P\theta}{12} \qquad \text{and} \qquad \sigma(\theta)=\exp\left(\frac{P\theta^2}{24}\right)s(\theta)$$

where $P$ is Ramanujan's Eisenstein series defined by (0.61). Although we will make considerable use of the Weierstrass elliptic function $\wp$, we will not use the $\zeta$ or $\sigma$ functions.

For future reference, let $\rho_1$ be the function defined by

$$\rho_1(x)=\frac{1+x}{2(1-x)}+\sum_{n=1}^{\infty}\left(\frac{xq^n}{1-xq^n}-\frac{x^{-1}q^n}{1-x^{-1}q^n}\right) \tag{1.19}$$

$$=\frac{1}{2}+\sum_{n=1}^{\infty}\left(\frac{xq^{n-1}}{1-xq^{n-1}}-\frac{x^{-1}q^n}{1-x^{-1}q^n}\right). \tag{1.20}$$

By (1.15) it follows that

$$\rho_1(e^{i\theta})=iz(\theta). \tag{1.21}$$

## 1.2 General properties of elliptic functions

Before giving further examples of elliptic functions, it will be useful to present the main properties that are shared by all elliptic functions. We assume a basic knowledge of complex analysis and follow the procedure set out by Apostol [29, Chapter 1].

If $f$ is an elliptic function with periods $\omega_1$ and $\omega_2$, and $m$ and $n$ are integers, then $m\omega_1+n\omega_2$ is also a period of $f$. If every period of $f$ is of the form $m\omega_1+n\omega_2$ for some integers $m$ and $n$, then $(\omega_1,\omega_2)$ is called a fundamental pair of periods. For example, $(2\pi, 2\pi\tau)$ is a fundamental pair of periods for the elliptic function $\wp(\theta)$.

Every nonconstant elliptic function has a fundamental pair of periods; see [29, Theorem 1.3]. Given a fundamental pair of periods $(\omega_1,\omega_2)$, the set

$$\{\alpha\omega_1+\beta\omega_2 : 0\le\alpha,\beta<1\}$$

is called a period parallelogram.

A nonconstant meromorphic function can have only finitely many poles or zeros in a bounded subset of the complex plane. Therefore, a period parallelogram for a nonconstant elliptic function $f$ can be translated by some complex number $z_0$ to produce a set

$$\{z_0 + \alpha\omega_1 + \beta\omega_2 : 0 \leq \alpha, \beta < 1\}$$

for which $f$ has no zeros or poles on the boundary. Such a set is called a cell.

The next five theorems are standard in the theory of elliptic functions. The most useful tool for obtaining identities involving elliptic functions is:

**Theorem 1.2 (Liouville).** *An elliptic function without poles is a constant.*

*Proof.* Suppose $(\omega_1, \omega_2)$ is a fundamental pair of periods for the elliptic function. A meromorphic function without poles is continuous and therefore bounded on the compact set

$$\{\alpha\omega_1 + \beta\omega_2 : 0 \leq \alpha, \beta \leq 1\}.$$

By double periodicity, the elliptic function will be bounded on the complex plane. But, an entire function that is bounded is a constant, by another theorem of Liouville. □

**Theorem 1.3.** *An elliptic function $f$ without zeros is a constant.*

*Proof.* Apply Theorem 1.2 to the elliptic function $1/f$. □

**Theorem 1.4.** *The sum of the residues of an elliptic function $f$ at its poles in any cell is zero.*

*Proof.* By the residue theorem, the sum of the residues in a cell is given by

$$\frac{1}{2\pi i}\int_C f(z)\mathrm{d}z$$

where the contour of integration $C$ is the boundary of the cell. By double periodicity the integrals along parallel edges cancel, and therefore the integral is zero. □

**Theorem 1.5.** *The number of zeros of an elliptic function $f$ in any cell is equal to the number of poles, each counted according to their multiplicity.*

*Proof.* By the argument principle, e.g., see [299, Section 6.31, p. 119], the difference between the number of zeros and the number of poles in a cell is given by

$$\frac{1}{2\pi i}\int_C \frac{f'(z)}{f(z)}\mathrm{d}z$$

where $C$ is the boundary of the cell. As $f'(z)/f(z)$ is an elliptic function, the integral is zero. □

The number of poles (equivalently, the number of zeros) of an elliptic function in any cell, counted by multiplicity, is called the order of the elliptic function.

**Theorem 1.6.** *The order of a nonconstant elliptic function must be at least two.*

*Proof.* The order of a nonconstant elliptic function is at least one, by Theorem 1.2. If an elliptic function had order one, then a cell would contain just a single simple pole with nonzero residue. This contradicts Theorem 1.4. □

It follows that a nonconstant elliptic function must have at least two simple poles, or at least one pole of order two (or higher) in any cell. The Weierstrass elliptic function $\wp$ studied in the previous section has order two. It has poles of order 2 at each point $2\pi m + 2\pi\tau n$, $m, n \in \mathbb{Z}$. Examples of elliptic functions of order 2 with two simple poles in any cell will be given in Section 1.8.

## 1.3 Elliptic function identities

We will now prove five identities involving the functions $s$, $z$, and $\wp$. They are of fundamental importance because of the vast number of identities that can be derived from them by specializing the variables.

**Theorem 1.7 (Weierstrass' identity).** *Let $s(\theta)$ and $\wp(\theta)$ be defined by* (1.1) *and* (1.13). *Then*

$$\wp(\theta) - \wp(\alpha) = -\frac{s(\theta - \alpha)s(\theta + \alpha)}{s^2(\theta)s^2(\alpha)}.$$

*Proof.* Fix a value of $\alpha$ that is not a period or half-period of the elliptic function $\wp(\theta)$, so that $\alpha \neq m\pi + n\pi\tau$, $m, n \in \mathbb{Z}$. Consider the function $f$ defined by

$$f(\theta) = \wp(\theta) - \wp(\alpha)$$

as a function of $\theta$. It is an elliptic function of order two with periods $2\pi$ and $2\pi\tau$. There are poles of order two at $\theta = 2\pi m + 2\pi n\tau$ for all $m, n \in \mathbb{Z}$, and these are the only singularities. By double periodicity and since $\wp$ is an even function, we find that $f$ has zeros at $\theta = \pm\alpha + 2\pi m + 2\pi n\tau$ for all $m, n \in \mathbb{Z}$.

A function with similar properties can be created in terms of the function $s$. For example, consider the function $g$ defined by

$$g(\theta) = \frac{s(\theta - \alpha)s(\theta + \alpha)}{s^2(\theta)}.$$

Since $s$ is an entire function, the zeros and poles of $g$ are determined by the zeros in the numerator and denominator, respectively. We find that $g$ has simple zeros at $\theta = \pm\alpha + 2\pi m + 2\pi n\tau$ for all $m, n \in \mathbb{Z}$ and these are the only zeros. Moreover, $g$ has poles of order two at $\theta = 2\pi m + 2\pi n\tau$ for all $m, n \in \mathbb{Z}$, and these are the only singularities. It is easy to check, using (1.2) and (1.3), that $g(\theta)$ is a doubly periodic function of $\theta$ with periods $2\pi$ and $2\pi\tau$.

It follows that the quotient

$$h(\theta) = \frac{f(\theta)}{g(\theta)} \tag{1.22}$$

is an entire, doubly periodic function of $\theta$. By Theorem 1.2 it follows that $h(\theta)$ is a constant that is independent of $\theta$. The value of the constant may be determined by examining the behavior at $\theta = 0$. By (1.8), (1.12), and Taylor's theorem, we find that

$$f(\theta) = \frac{1}{\theta^2} + O(1) \quad \text{and} \quad g(\theta) = -\frac{s^2(\alpha)}{\theta^2} + O(1)$$

as $\theta \to 0$, and it follows that

$$h(\theta) = \lim_{\theta\to 0} h(\theta) = \lim_{\theta\to 0} \frac{f(\theta)}{g(\theta)} = \frac{-1}{s^2(\alpha)}. \tag{1.23}$$

The proof is completed by using the result of (1.23) in (1.22) and simplifying the resulting identity. □

An immediate consequence of Theorem 1.7 is the three-term identity of Weierstrass.

**Theorem 1.8 (Weierstrass' three-term identity).** *The function $s(\theta)$ defined by* (1.1) *satisfies the identity*

$$\begin{aligned} &s(\theta-\alpha)s(\theta+\alpha)s(\beta-\gamma)s(\beta+\gamma)\\ &+s(\theta-\beta)s(\theta+\beta)s(\gamma-\alpha)s(\gamma+\alpha)\\ &+s(\theta-\gamma)s(\theta+\gamma)s(\alpha-\beta)s(\alpha+\beta) \quad = \quad 0. \end{aligned}$$

*Proof.* By two applications of Theorem 1.7 we may deduce that

$$(\wp(\theta)-\wp(\alpha))\,(\wp(\beta)-\wp(\gamma)) = \frac{s(\theta-\alpha)s(\theta+\alpha)s(\beta-\gamma)s(\beta+\gamma)}{s^2(\theta)s^2(\alpha)s^2(\beta)s^2(\gamma)}.$$

Replace $(\alpha,\beta,\gamma)$ with the cyclic permutations $(\beta,\gamma,\alpha)$ and $(\gamma,\alpha,\beta)$ and add, to obtain

$$\begin{aligned}
&(\wp(\theta)-\wp(\alpha))\,(\wp(\beta)-\wp(\gamma))\\
&+(\wp(\theta)-\wp(\beta))\,(\wp(\gamma)-\wp(\alpha))\\
&+(\wp(\theta)-\wp(\gamma))\,(\wp(\alpha)-\wp(\beta)) = \frac{s(\theta-\alpha)s(\theta+\alpha)s(\beta-\gamma)s(\beta+\gamma)}{s^2(\theta)s^2(\alpha)s^2(\beta)s^2(\gamma)}\\
&\qquad\qquad +\frac{s(\theta-\beta)s(\theta+\beta)s(\gamma-\alpha)s(\gamma+\alpha)}{s^2(\theta)s^2(\alpha)s^2(\beta)s^2(\gamma)}\\
&\qquad\qquad +\frac{s(\theta-\gamma)s(\theta+\gamma)s(\alpha-\beta)s(\alpha+\beta)}{s^2(\theta)s^2(\alpha)s^2(\beta)s^2(\gamma)}.
\end{aligned}$$

The left-hand side simplifies to zero. Then clear the denominators on the right-hand side to obtain the required result. □

The next result is a consequence of Theorem 1.8.

**Theorem 1.9 (Halphen's identity).** *The functions $s(\theta)$ and $z(\theta)$ defined by* (1.1) *and* (1.4) *satisfy the identity*

$$\frac{s(2\alpha)s(\beta+\gamma)s(\beta-\gamma)}{s(\alpha+\beta)s(\alpha-\beta)s(\alpha+\gamma)s(\alpha-\gamma)} = z(\alpha+\beta)+z(\alpha-\beta)-z(\alpha+\gamma)-z(\alpha-\gamma).$$

*Proof.* As $s$ is an odd function, the result of Theorem 1.8 may be written in the form

$$\begin{aligned}
&s(\theta-\alpha)s(\theta+\alpha)s(\beta-\gamma)s(\beta+\gamma)\\
&= s(\theta-\beta)s(\theta+\beta)s(\alpha-\gamma)s(\alpha+\gamma)\\
&\quad -s(\alpha-\beta)s(\alpha+\beta)s(\theta-\gamma)s(\theta+\gamma).
\end{aligned}$$

Now divide both sides by $\theta-\alpha$ and take the limit as $\theta\to\alpha$. To compute the limits, use (0.52) on the left-hand side and use L'Hôpital's rule on the right. We find that

$$\begin{aligned}
&s(2\alpha)s(\beta-\gamma)s(\beta+\gamma)\\
&= \lim_{\theta\to\alpha}\frac{\mathrm{d}}{\mathrm{d}\theta}\{s(\theta-\beta)s(\theta+\beta)s(\alpha-\gamma)s(\alpha+\gamma)\\
&\qquad -s(\alpha-\beta)s(\alpha+\beta)s(\theta-\gamma)s(\theta+\gamma)\}\\
&= \{z(\alpha-\beta)+z(\alpha+\beta)-z(\alpha-\gamma)-z(\alpha+\gamma)\}\\
&\quad \times s(\alpha-\beta)s(\alpha+\beta)s(\alpha-\gamma)s(\alpha+\gamma).
\end{aligned}$$

On rearranging, we complete the proof. □

In the analysis above, we started with Theorem 1.7 and successively deduced Theorems 1.8 and 1.9. We will now show that Theorem 1.7 can be deduced from

Theorem 1.9. Thus, all three theorems are equivalent in the sense that any one can be deduced from any of the others. If we divide the result of Theorem 1.9 by $2\alpha$ and take the limit as $\alpha \to 0$, we find that

$$\begin{aligned}\frac{s(\beta+\gamma)s(\beta-\gamma)}{s^2(\beta)s^2(\gamma)} &= \lim_{\alpha\to 0} \frac{z(\beta+\alpha)-z(\beta-\alpha)}{2\alpha} - \frac{z(\gamma+\alpha)-z(\gamma-\alpha)}{2\alpha} \\ &= z'(\beta) - z'(\gamma) \\ &= -p(\beta) + p(\gamma) \\ &= \wp(\gamma) - \wp(\beta).\end{aligned}$$

Thus we have obtained Theorem 1.7 from Theorem 1.9, as required.

Here are two further identities that will be useful.

**Theorem 1.10.** *The functions* $s(\theta)$ *and* $\wp(\theta)$ *defined by* (1.1) *and* (1.13) *satisfy the identity*

$$\wp'(\theta) = -\frac{s(2\theta)}{s^4(\theta)}.$$

*Proof.* Divide both sides of the identity in Theorem 1.7 by $\theta - \alpha$ and take the limit as $\alpha \to \theta$. □

**Theorem 1.11 (Frobenius and Stickelberger).** *Let* $z(\theta)$ *and* $\wp(\theta)$ *be defined by* (1.4) *and* (1.13). *If* $\alpha + \beta + \gamma = 0$ *then*

$$(z(\alpha) + z(\beta) + z(\gamma))^2 = \wp(\alpha) + \wp(\beta) + \wp(\gamma).$$

*Proof.* Fix a value of $\beta \notin \Lambda$ and let $f(\alpha)$ and $g(\alpha)$ be defined by

$$f(\alpha) = (z(\alpha) + z(\beta) - z(\alpha+\beta))^2$$

and

$$g(\alpha) = \wp(\alpha) + \wp(\beta) + \wp(\alpha+\beta).$$

By (1.6) and (1.7), $f(\alpha)$ and $g(\alpha)$ are both doubly periodic functions of $\alpha$ with periods $2\pi$ and $2\pi\tau$. The poles of $f$ and $g$ occur when $\alpha \in \Lambda$ or $\alpha + \beta \in \Lambda$.

Let us consider the behavior of each function $f(\alpha)$ and $g(\alpha)$ at the pole $\alpha = 0$. By (1.9), (1.10), (1.12), and Taylor's theorem,

$$\begin{aligned}f(\alpha) &= \left(\frac{1}{\alpha} + 2E_2\alpha - \alpha z'(\beta) + O(\alpha^2)\right)^2 \\ &= \frac{1}{\alpha^2} + 2\wp(\beta) + O(\alpha)\end{aligned}$$

and

$$g(\alpha) = \frac{1}{\alpha^2} + 2\wp(\beta) + O(\alpha)$$

as $\alpha \to 0$. It follows that $f(\alpha) - g(\alpha)$ has a removable singularity at $\alpha = 0$ and in fact

$$f(\alpha) - g(\alpha) \to 0 \quad \text{as} \quad \alpha \to 0. \tag{1.24}$$

Now consider the behavior of each function at the pole $\alpha = -\beta$. Put $\gamma = -\alpha - \beta$ and expand in powers of $\gamma$ to obtain

$$f(\alpha) = \frac{1}{\gamma^2} + 2\wp(\beta) + O(\gamma)$$

and

$$g(\alpha) = \frac{1}{\gamma^2} + 2\wp(\beta) + O(\gamma)$$

as $\alpha \to -\beta$. It follows that $f(\alpha) - g(\alpha)$ has a removable singularity at $\alpha = -\beta$. By double periodicity it follows that $f(\alpha) - g(\alpha)$ is an entire function of $\alpha$. Therefore by Theorem 1.2 it is constant, and by (1.24) it is identically zero. □

## 1.4 Explicit forms of Halphen's identity

The identities in Theorems 1.7–1.11 can be specialized to produce a vast number of identities for theta functions. To gain access to those results, we will need explicit forms of the theorems in the previous section. The most powerful and useful result for converting products to series is Halphen's identity, so we begin with that and give five different forms. Each of the five forms has its own advantages. For example, the first form has interesting specializations when roots of unity are substituted for the variables. The second and third forms are well suited for substituting powers of $q$ for the variables. The fourth form is beautiful, and the fifth form will be used in the study of Jacobian elliptic functions.

Because as many as sixteen infinite products appear in some of the formulas, we make use of the compact notation for products given by

$$(x_1, \ldots, x_m; q)_\infty = \prod_{j=0}^{\infty} (1 - x_1 q^j) \cdots (1 - x_m q^j). \tag{1.25}$$

**Theorem 1.12 (First explicit form of Halphen's identity).** *For* $|q| < |x|, |y|, |z|, |xyz| < |q|^{-1}$, *we have*

$$\frac{(qxy, qxz, qyz, q/xy, q/xz, q/yz, q, q; q)_\infty}{(qx, qy, qz, qxyz, q/x, q/y, q/z, q/xyz; q)_\infty}$$
$$= 1 + A\sum_{j=1}^{\infty} \frac{q^j}{1-q^j}\left(x^j + y^j + z^j - (xyz)^j - x^{-j} - y^{-j} - z^{-j} + (xyz)^{-j}\right), \tag{1.26}$$

*where*

$$A = \frac{(1-x)(1-y)(1-z)(1-xyz)}{(1-xy)(1-xz)(1-yz)}. \tag{1.27}$$

*Proof.* Substitute the explicit formulas for $s(\theta)$ and $z(\theta)$ given by (1.1) and (1.17) into the result of Theorem 1.9 and multiply both sides by

$$B := 2\,\frac{\sin\frac{\alpha+\beta}{2}\sin\frac{\alpha-\beta}{2}\sin\frac{\alpha+\gamma}{2}\sin\frac{\alpha-\gamma}{2}}{\sin\alpha\,\sin\frac{\beta+\gamma}{2}\sin\frac{\beta-\gamma}{2}}.$$

Then make the change of variables

$$x = e^{i(\alpha+\beta)}, \quad y = e^{i(\alpha-\beta)} \quad \text{and} \quad z = e^{i(\gamma-\alpha)}$$

so that

$$e^{2i\alpha} = xy, \quad e^{i(\beta+\gamma)} = xz, \quad e^{i(\beta-\gamma)} = \frac{1}{yz} \quad \text{and} \quad e^{i(\alpha+\gamma)} = xyz,$$

to deduce

$$\frac{(qxy, qxz, qyz, q/xy, q/xz, q/yz, q, q; q)_\infty}{(qx, qy, qz, qxyz, q/x, q/y, q/z, q/xyz; q)_\infty}$$
$$= C + A\sum_{j=1}^{\infty} \frac{q^j}{1-q^j}\left(x^j + y^j + z^j - (xyz)^j - x^{-j} - y^{-j} - z^{-j} + (xyz)^{-j}\right), \tag{1.28}$$

where

$$C = \frac{B}{2} \times \left(\cot\frac{\alpha+\beta}{2} + \cot\frac{\alpha-\beta}{2} - \cot\frac{\alpha+\gamma}{2} - \cot\frac{\alpha-\gamma}{2}\right)$$

and

$$A = \frac{B}{i} = \frac{(1-x)(1-y)(1-z)(1-xyz)}{(1-xy)(1-xz)(1-yz)}.$$

By letting $q = 0$ in (1.28) we deduce that $C = 1$. This completes the proof. □

**Theorem 1.13 (Second explicit form of Halphen's identity).** *Let $\rho_1$ be the function defined by* (1.20)*, that is,*

$$\rho_1(x;q) = \frac{1}{2} + \sum_{k=1}^{\infty}\left(\frac{xq^{k-1}}{1-xq^{k-1}} - \frac{x^{-1}q^k}{1-x^{-1}q^k}\right).$$

*Then for $|q| < 1$ and all x, y, and z we have*

$$\frac{(xy, xz, yz, q/xy, q/xz, q/yz, q, q; q)_\infty}{(x, y, z, xyz, q/x, q/y, q/z, q/xyz; q)_\infty} = \rho_1(x;q) + \rho_1(y;q) + \rho_1(z;q) - \rho_1(xyz;q).$$

*Proof.* Multiply (1.26) by $1/A$, where the value of $A$ is given by (1.27), and note that

$$\frac{1}{A} = 1 + \frac{x}{1-x} + \frac{y}{1-y} + \frac{z}{1-z} - \frac{xyz}{1-xyz}$$

to obtain

$$\begin{aligned}
&\frac{(xy,\ xz,\ yz,\ q/xy,\ q/xz,\ q/yz,\ q,\ q; q)_\infty}{(x,\ y,\ z,\ xyz,\ q/x,\ q/y,\ q/z,\ q/xyz; q)_\infty} \\
&= 1 + \frac{x}{1-x} + \frac{y}{1-y} + \frac{z}{1-z} - \frac{xyz}{1-xyz} \\
&\quad + \sum_{j=1}^{\infty}\frac{q^j}{1-q^j}\left(x^j + y^j + z^j - (xyz)^j - x^{-j} - y^{-j} - z^{-j} + (xyz)^{-j}\right). \quad (1.29)
\end{aligned}$$

Since

$$\sum_{j=1}^{\infty}\frac{q^j x^j}{1-q^j} = \sum_{j=1}^{\infty}\sum_{k=1}^{\infty} q^{jk}x^j = \sum_{k=1}^{\infty}\frac{xq^k}{1-xq^k},$$

the right-hand side of (1.29) is

$$1+\frac{x}{1-x}+\frac{y}{1-y}+\frac{z}{1-z}-\frac{xyz}{1-xyz}$$
$$+\sum_{k=1}^{\infty}\left(\frac{xq^k}{1-xq^k}-\frac{x^{-1}q^k}{1-x^{-1}q^k}+\frac{yq^k}{1-yq^k}-\frac{y^{-1}q^k}{1-y^{-1}q^k}+\frac{zq^k}{1-zq^k}-\frac{z^{-1}q^k}{1-z^{-1}q^k}\right.$$
$$\left.-\frac{xyzq^k}{1-xyzq^k}+\frac{(xyz)^{-1}q^k}{1-(xyz)^{-1}q^k}\right)$$

and this is just $\rho_1(x;q)+\rho_1(y;q)+\rho_1(z;q)-\rho_1(xyz;q)$. □

**Theorem 1.14 (Third explicit form of Halphen's identity).** *Let $r$ be the function defined by*

$$r(x;q)=\sum_{k=1}^{\infty}\left(\frac{xq^{2k-1}}{1-xq^{2k-1}}-\frac{x^{-1}q^{2k-1}}{1-x^{-1}q^{2k-1}}\right). \tag{1.30}$$

*Then for $|q|<1$ and for all $a$, $b$, and $c$ we have*

$$\frac{q}{abc}\frac{(a^2,b^2,c^2,q^2/a^2,q^2/b^2,q^2/c^2,q^2,q^2;q^2)_\infty}{(qabc,qbc/a,qac/b,qab/c,qa/bc,qb/ac,qc/ab,q/abc;q^2)_\infty}$$
$$=r(ab/c;q)+r(ac/b;q)+r(bc/a;q)-r(abc;q).$$

*Proof.* In Theorem 1.13, replace $q$ with $q^2$, then replace $x$, $y$, and $z$ with $aq/bc$, $bq/ac$, and $cq/ab$, respectively, to obtain

$$\frac{\left(a^2,b^2,c^2,q^2/a^2,q^2/b^2,q^2/c^2,q^2,q^2;q^2\right)_\infty}{(abc/q,bcq/a,acq/b,abq/c,aq/bc,bq/ac,cq/ab,q^3/abc;q^2)_\infty}$$
$$=\rho_1\left(aq/bc;q^2\right)+\rho_1\left(bq/ac;q^2\right)+\rho_1\left(cq/ab;q^2\right)-\rho_1\left(q^3/abc;q^2\right). \tag{1.31}$$

Since

$$(abc/q,q^3/abc;q^2)_\infty=-\frac{abc}{q}(qabc,q/abc;q^2)_\infty,$$

$$\rho_1(q^3/abc;q^2)=1+\rho_1(q/abc;q^2)$$

and

$$\rho_1(qx;q^2)=\frac{1}{2}-r(x^{-1},q),$$

the identity (1.31) simplifies to

$$\begin{aligned}&\frac{q}{abc}\frac{(a^2,b^2,c^2,q^2/a^2,q^2/b^2,q^2/c^2,q^2,q^2;q^2)_\infty}{(qabc,qbc/a,qac/b,qab/c,qa/bc,qb/ac,qc/ab,q/abc;q^2)_\infty}\\&=-\rho_1(aq/bc;q^2)-\rho_1(bq/ac;q^2)-\rho_1(cq/ab;q^2)+\rho_1(q/abc;q^2)+1\\&=r(bc/a;q)+r(ac/b;q)+r(ab/c;q)-r(abc;q).\end{aligned}$$

□

**Theorem 1.15 (Fourth explicit form of Halphen's identity).** *For* $|qa^{\pm1}b^{\pm1}c^{\pm1}|<1$, *we have*

$$\begin{aligned}&\frac{q}{abc}\frac{(a^2,\ b^2,\ c^2,\ q^2/a^2,\ q^2/b^2,\ q^2/c^2,\ q^2,\ q^2;q^2)_\infty}{(qabc,\ qbc/a,\ qac/b,\ qab/c,\ qa/bc,\ qb/ac,\ qc/ab,\ q/abc;q^2)_\infty}\\&\qquad=\sum_{j=1}^{\infty}\frac{(a^j-a^{-j})(b^j-b^{-j})(c^j-c^{-j})}{(q^j-q^{-j})}.\end{aligned}$$

Before starting the proof, observe that without loss of generality it can be assumed that $|a|,|b|,|c|\le 1$. Then, the eight conditions $|qa^{\pm1}b^{\pm1}c^{\pm1}|<1$ reduce to the single condition $|q|<|abc|$.

*Proof.* Since

$$\begin{aligned}r(x;q)&=\sum_{k=1}^{\infty}\left(\frac{xq^{2k-1}}{1-xq^{2k-1}}-\frac{x^{-1}q^{2k-1}}{1-x^{-1}q^{2k-1}}\right)\\&=\sum_{k=1}^{\infty}\sum_{j=1}^{\infty}\left(x^jq^{j(2k-1)}-x^{-j}q^{j(2k-1)}\right)\\&=\sum_{j=1}^{\infty}\frac{q^j}{1-q^{2j}}(x^j-x^{-j}),\end{aligned}$$

the result of Theorem 1.14 may be expressed as

$$\begin{aligned}&\frac{q}{abc}\frac{\left(a^2,b^2,c^2,\frac{q^2}{a^2},\frac{q^2}{b^2},\frac{q^2}{c^2},q^2,q^2;q^2\right)_\infty}{\left(abcq,\frac{bcq}{a},\frac{acq}{b},\frac{abq}{c},\frac{aq}{bc},\frac{bq}{ac},\frac{cq}{ab},\frac{q}{abc};q^2\right)_\infty}\\&=\sum_{j=1}^{\infty}\frac{q^j}{1-q^{2j}}\left(\left(\frac{ab}{c}\right)^j+\left(\frac{ac}{b}\right)^j+\left(\frac{bc}{a}\right)^j-\left(\frac{a}{bc}\right)^j\right.\\&\qquad\left.-\left(\frac{b}{ac}\right)^j-\left(\frac{c}{ab}\right)^j-(abc)^j+\left(\frac{1}{abc}\right)^j\right)\\&=\sum_{j=1}^{\infty}\frac{(a^j-a^{-j})(b^j-b^{-j})(c^j-c^{-j})}{(q^j-q^{-j})}.\end{aligned}$$

□

**Theorem 1.16 (Fifth explicit form of Halphen's identity).** *Let $F(x,y)$ and $\rho_1(x)$ be defined by* (0.99) *and* (1.20)*, respectively, that is,*

$$F(x,y)=\prod_{n=1}^{\infty}\frac{(1-xyq^{n-1})(1-x^{-1}y^{-1}q^n)(1-q^n)^2}{(1-xq^{n-1})(1-x^{-1}q^n)(1-yq^{n-1})(1-y^{-1}q^n)}$$

*and*

$$\rho_1(x;q)=\frac{1}{2}+\sum_{k=1}^{\infty}\left(\frac{xq^{k-1}}{1-xq^{k-1}}-\frac{x^{-1}q^k}{1-x^{-1}q^k}\right).$$

*Then*

$$F(x,t)F(y,t)=t\frac{\partial}{\partial t}F(xy,t)+\left(\rho_1(x)+\rho_1(y)\right)F(xy,t).$$

*Proof.* Let us divide by $F(xy,t)$ and prove the equivalent identity

$$\frac{F(x,t)F(y,t)}{F(xy,t)}=t\frac{\partial}{\partial t}\log F(xy,t)+\rho_1(x)+\rho_1(y).$$

Since

$$\begin{aligned}t\frac{\partial}{\partial t}\log F(xy,t)&=t\frac{\partial}{\partial t}\log\prod_{n=1}^{\infty}(1-xytq^{n-1})(1-x^{-1}y^{-1}t^{-1}q^n)\\&\quad-t\frac{\partial}{\partial t}\log\prod_{n=1}^{\infty}(1-tq^{n-1})(1-t^{-1}q^n)\\&=-\rho_1(xyt)+\rho_1(t),\end{aligned}$$

the identity we are required to prove is

$$\frac{F(x,t)F(y,t)}{F(xy,t)}=\rho_1(x)+\rho_1(y)+\rho_1(t)-\rho_1(xyt).$$

On substituting infinite products for $F(x,t)$, $F(y,t)$, and $F(xy,t)$, we find that the claimed identity is equivalent to Theorem 1.13. □

## 1.5 Explicit forms of Weierstrass' identity

The results in this section provide three different explicit formulations of Weierstrass' identity Theorem 1.7.

**Theorem 1.17 (First explicit form of Weierstrass' identity).**
*For* $|q| < |x|, |y| < |q|^{-1}$, *we have*

$$\frac{(qxy, qx/y, qy/x, q/xy; q)_\infty (q;q)_\infty^4}{(qx, qy, q/x, q/y; q)_\infty^2} = 1 + \frac{(1-x)^2(1-y)^2}{(1-xy)(x-y)} \sum_{j=1}^{\infty} \frac{jq^j}{1-q^j} \left(x^j + x^{-j} - y^j - y^{-j}\right). \tag{1.32}$$

*Proof.* Take the limit as $z \to x^{-1}$ in (1.26) and note that

$$\lim_{z \to x^{-1}} \frac{1}{1-xz} \left(x^j + y^j + z^j - (xyz)^j - x^{-j} - y^{-j} - z^{-j} + (xyz)^{-j}\right) = -j(x^j + x^{-j} - y^j - y^{-j}).$$

□

**Theorem 1.18 (Second explicit form of Weierstrass' identity).** *For* $|qa^{\pm 1}b^{\pm 1}| < 1$, *we have*

$$\frac{q}{ab} \frac{(a^2, b^2, q^2/a^2, q^2/b^2; q^2)_\infty (q^2;q^2)_\infty^4}{(qab, qb/a, qa/b, q/ab; q^2)_\infty^2} = -\sum_{j=1}^{\infty} \frac{j(a^j - a^{-j})(b^j - b^{-j})}{(q^j - q^{-j})}. \tag{1.33}$$

Before starting the proof, observe that without loss of generality it can be assumed that $|a|, |b| \le 1$. Then, the four conditions $|qa^{\pm 1}b^{\pm 1}| < 1$ reduce to the single condition $|q| < |ab|$.

*Proof.* This follows from the result of Theorem 1.14 by dividing by $(c^{-1} - c)$ and taking the limit as $c \to 1$. □

**Theorem 1.19 (Third explicit form of Weierstrass' identity).** *Let* $\wp(\theta)$ *and* $F(x, y)$ *be defined by* (1.13) *and* (0.99), *respectively. Then*

$$\wp(\theta) - \wp(\alpha) = F(e^{i\theta}, e^{i\alpha}) F(e^{-i\theta}, e^{i\alpha}).$$

*Proof.* By (1.1) we have

$$s(\theta) = 2\sin\frac{\theta}{2} \prod_{n=1}^{\infty} \frac{(1 - e^{i\theta}q^n)(1 - e^{-i\theta}q^n)}{(1-q^n)^2} = ie^{-i\theta/2} \frac{(e^{i\theta}, qe^{-i\theta}; q)_\infty}{(q, q; q)_\infty}.$$

Therefore by Theorem 1.7, and since $s$ is an odd function, we have

$$\begin{aligned}\wp(\theta) - \wp(\alpha) &= -\frac{s(\theta-\alpha)s(\theta+\alpha)}{s^2(\theta)s^2(\alpha)} \\ &= -\frac{s(\alpha-\theta)s(\alpha+\theta)}{s(\theta)s(-\theta)s^2(\alpha)}\end{aligned}$$

$$
\begin{aligned}
&= \frac{(e^{i(\alpha-\theta)}, qe^{-i(\alpha-\theta)}, e^{i(\alpha+\theta)}, qe^{-i(\alpha+\theta)}, q, q, q, q; q)_\infty}{(e^{i\theta}, qe^{-i\theta}, e^{-i\theta}, qe^{i\theta}, e^{i\alpha}, qe^{-i\alpha}, e^{i\alpha}, qe^{-i\alpha}; q)_\infty} \\
&= \frac{(e^{i(\alpha+\theta)}, qe^{-i(\alpha+\theta)}, q, q; q)_\infty}{(e^{i\theta}, qe^{-i\theta}, e^{i\alpha}, qe^{-i\alpha}; q)_\infty} \times \frac{(e^{i(\alpha-\theta)}, qe^{-i(\alpha-\theta)}, q, q; q)_\infty}{(e^{-i\theta}, qe^{i\theta}, e^{i\alpha}, qe^{-i\alpha}; q)_\infty} \\
&= F(e^{i\theta}, e^{i\alpha})F(e^{-i\theta}, e^{i\alpha}),
\end{aligned}
$$

where the last step follows from the definition of $F$ as an infinite product in (0.99).

□

## 1.6 Explicit forms of Theorems 1.8, 1.10, and 1.11

We now provide explicit formulations for the remaining identities in Theorems 1.8, 1.10, and 1.11. We begin with an explicit form of Weierstrass' three-term identity.

**Theorem 1.20 (Explicit form of Weierstrass' three-term identity).** *Suppose $a^2 = bcde$ and $|q| < 1$. Then*

$$
[a/b; q]_\infty [a/c; q]_\infty [a/d; q]_\infty [a/e; q]_\infty - [b; q]_\infty [c; q]_\infty [d; q]_\infty [e; q]_\infty
$$
$$
= b\,[a; q]_\infty [a/bc; q]_\infty [a/bd; q]_\infty [a/be; q]_\infty
$$

*where*

$$
[x; q]_\infty = (x; q)_\infty (q/x; q)_\infty = \prod_{j=1}^{\infty} (1 - q^{j-1}x)(1 - q^j x^{-1}). \tag{1.34}
$$

*Proof.* First note that by (1.1) and (1.34) we have

$$
s(\theta) = 2\sin\frac{\theta}{2} \prod_{n=1}^{\infty} \frac{(1 - e^{i\theta}q^n)(1 - e^{-i\theta}q^n)}{(1 - q^n)^2} = ie^{-i\theta/2} \frac{[e^{i\theta}; q]_\infty}{(q; q)_\infty^2}. \tag{1.35}
$$

Now let $B$, $C$, $D$, and $E$ be the arguments in the first term in Weierstrass' three-term identity in Theorem 1.8, that is,

$$
B = \theta - \alpha, \quad C = \theta + \alpha, \quad D = \beta - \gamma, \quad E = \beta + \gamma,
$$

and let

$$
A = \frac{1}{2}(B + C + D + E) = \theta + \beta.
$$

Then,

$$
A - B = \alpha + \beta, \quad A - C = -(\alpha - \beta), \quad A - D = \theta + \gamma, \quad A - E = \theta - \gamma,
$$

and

$$A - B - C = -(\theta - \beta), \quad A - B - D = \alpha + \gamma, \quad A - B - E = -(\gamma - \alpha).$$

With this change of variables, Weierstrass' three-term identity in Theorem 1.8 becomes

$$\begin{aligned} s(B)\, s(C)\, s(D)\, s(E) \;&-\; s(A-B)\, s(A-C)\, s(A-D)\, s(A-E) \\ &+\; s(A)\, s(A-B-C)\, s(A-B-D)\, s(A-B-E) \;=\; 0. \end{aligned}$$

Now put $a = \exp(iA)$, $b = \exp(iB)$, $c = \exp(iC)$, $d = \exp(iD)$, $e = \exp(iE)$, and use (1.35) to obtain the required identity. □

Next, we give three explicit forms of Theorem 1.10.

**Theorem 1.21 (First explicit form of Theorem 1.10).** *For $|q| < |x| < |q|^{-1}$, we have*

$$\frac{(qx^2,\, q/x^2; q)_\infty\, (q;q)_\infty^6}{(qx,\, q/x; q)_\infty^4} = 1 + \frac{(1-x)^3}{x(1+x)} \sum_{j=1}^{\infty} \frac{j^2 q^j}{1-q^j} \left(x^j - x^{-j}\right).$$

*Proof.* Take the limit as $y \to x$ in (1.32). □

**Theorem 1.22 (Second explicit form of Theorem 1.10).** *For $|q| < |a| < |q|^{-1}$, we have*

$$\frac{q}{a} \frac{(a^2,\, q^2/a^2; q^2)_\infty\, (q^2;q^2)_\infty^6}{(qa,\, q/a; q^2)_\infty^4} = \sum_{j=1}^{\infty} \frac{j^2 (a^j - a^{-j})}{(q^j - q^{-j})}.$$

*Proof.* Divide both sides of the identity (1.33) by $(b^{-1} - b)$ and take the limit as $b \to 1$. □

**Theorem 1.23 (Third explicit form of Theorem 1.10).** *For $|q| < |x| < 1$, we have*

$$\frac{x(x^2, q/x^2; q)_\infty\, (q;q)_\infty^6}{(x,\, q/x; q)_\infty^4} = \sum_{j=1}^{\infty} \frac{j^2 (x^j - q^j x^{-j})}{1-q^j}.$$

*Proof.* In Theorem 1.22, let $a = x/q$, then replace $q^2$ with $q$ and multiply both sides by $-1$. The result follows after simplification. □

Finally, here is an explicit formulation of the identity of Frobenius and Stickelberger.

**Theorem 1.24.** *Suppose $\alpha$, $\beta$, and $\gamma$ are complex numbers that satisfy*

$$\alpha + \beta + \gamma = 0$$

*and*

$$-2\pi \text{ Im } \tau < \text{Im}\,\alpha,\ \text{Im}\,\beta,\ \text{Im}\,\gamma < 2\pi \text{ Im } \tau.$$

*As usual, let* $q = \exp(2\pi i\tau)$. *Then*

$$\left(\cot\frac{\alpha}{2} + \cot\frac{\beta}{2} + \cot\frac{\gamma}{2} + 4\sum_{j=1}^{\infty}\frac{q^j}{1-q^j}\left(\sin j\alpha + \sin j\beta + \sin j\gamma\right)\right)^2$$

$$= \csc^2\frac{\alpha}{2} + \csc^2\frac{\beta}{2} + \csc^2\frac{\gamma}{2} - 1 + 8\sum_{j=1}^{\infty}\frac{jq^j}{1-q^j}\left(3 - \cos j\alpha - \cos j\beta - \cos j\gamma\right).$$

*Proof.* Substitute the Fourier expansions (1.17) and (1.18) into the result of Theorem 1.11. □

## 1.7 Examples

In this section, some worked examples are given as applications of the explicit formulas in the previous section. The reader is encouraged to work through the details in order to become familiar with the techniques. In future chapters the computational details will normally be omitted from similar proofs. Examples 1.25 and 1.26 lead to theorems about the number of representations of integers by certain quadratic forms and by certain quadratic functions; see Theorems 3.47 and 3.48.

The first example is an application of the explicit forms of Halphen's identity in Theorems 1.12 and 1.14.

*Example 1.25.* The following identities hold:

$$\sum_{m=-\infty}^{\infty}\sum_{n=-\infty}^{\infty} q^{m^2+7n^2} = 1 + 2\sum_{j=1}^{\infty}\chi_7(j)\frac{q^j}{1-(-q)^j} \tag{1.36}$$

and

$$\sum_{\substack{m>0 \\ m \text{ odd}}}\sum_{\substack{n>0 \\ n \text{ odd}}} q^{(m^2+7n^2)/8} = \sum_{\substack{j>0 \\ j \text{ odd}}}\chi_7(j)\frac{q^j}{1-q^j} \tag{1.37}$$

$$= \sum_{j=1}^{\infty}\chi_7(j)\frac{q^j}{1-q^{2j}}, \tag{1.38}$$

where the values of $\chi_7(j)$ are given by

$$\chi_7(j) = \begin{cases} 1 \text{ if } j \equiv 1, 2 \text{ or } 4 \pmod 7, \\ -1 \text{ if } j \equiv 3, 5 \text{ or } 6 \pmod 7, \\ 0 \text{ if } j \equiv 0 \pmod 7. \end{cases} \tag{1.39}$$

*Proof.* Let $\omega = \exp(i\pi/7)$ and take $(x, y, z) = (\omega^3, \omega^5, \omega^7)$ in (1.26) to obtain

$$\frac{(q\omega^8, q\omega^{10}, q\omega^{12}, q\omega^6, q\omega^4, q\omega^2, q, q; q)_\infty}{(q\omega^3, q\omega^5, q\omega^7, q\omega, q\omega^{11}, q\omega^9, q\omega^7, q\omega^{13}; q)_\infty}$$
$$= 1 + A\sum_{j=1}^{\infty} \frac{q^j}{1-q^j}\left(\omega^{3j} + \omega^{5j} + \omega^{7j} - \omega^j - \omega^{-3j} - \omega^{-5j} - \omega^{-7j} + \omega^{-j}\right) \tag{1.40}$$

where

$$A = \frac{(1-\omega^3)(1-\omega^5)(1-\omega^7)(1-\omega)}{(1-\omega^8)(1-\omega^{10})(1-\omega^{12})}.$$

Using

$$(1-\omega^2x)(1-\omega^4x)(1-\omega^6x)(1-\omega^8x)(1-\omega^{10}x)(1-\omega^{12}x)(1-x) = 1-x^7$$

and

$$(1-\omega x)(1-\omega^3x)(1-\omega^5x)(1-\omega^7x)(1-\omega^9x)(1-\omega^{11}x)(1-\omega^{13}x) = \frac{1-x^{14}}{1-x^7},$$

the product on the left-hand side of (1.40) simplifies to

$$\prod_{j=1}^{\infty} \frac{(1-q^j)^2(1-q^{7j})^2}{(1-q^{2j})(1-q^{14j})}.$$

By (0.41) this is equal to the sum

$$\left(\sum_{j=-\infty}^{\infty} (-1)^j q^{j^2}\right)\left(\sum_{j=-\infty}^{\infty} (-1)^j q^{7j^2}\right).$$

Now we will simplify the right-hand side of (1.40). Instead of attempting to simplify the complicated terms in the sum, we make use of the fact that the coefficients are periodic with period 14. Therefore, the identity (1.40) takes the form

$$\left(\sum_{j=-\infty}^{\infty}(-1)^j q^{j^2}\right)\left(\sum_{j=-\infty}^{\infty}(-1)^j q^{7j^2}\right) = 1 + \sum_{j=1}^{\infty} h(j)\frac{q^j}{1-q^j}, \tag{1.41}$$

where $h(j)$ is periodic with period 14. In order to determine $h(j)$ it is enough to compute the first 14 values. This can be done by expanding each side of (1.41) in powers of $q$ as far as $q^{14}$. The left-hand side is

$$\begin{aligned}(1 - 2q + 2q^4 - 2q^9 + \cdots)\,&(1 - 2q^7 + \cdots) \\ &= 1 - 2q + 2q^4 - 2q^7 + 4q^8 - 2q^9 - 4q^{11} + O(q^{15}).\end{aligned}$$

The expansion of the right-hand side is given by

$$1 + \sum_{j=1}^{\infty} h(j)\frac{q^j}{1-q^j} = 1 + \sum_{j=1}^{\infty}\sum_{k=1}^{\infty} h(j)q^{jk} = 1 + \sum_{n=1}^{\infty}\left(\sum_{d|n} h(d)\right)q^n.$$

On comparing coefficients of $q, q^2, \ldots, q^{14}$ in the expansions of (1.41), we deduce that

$$\begin{aligned}
-2 &= h(1),\\
0 &= h(1) + h(2),\\
0 &= h(1) + h(3),\\
2 &= h(1) + h(2) + h(4),\\
0 &= h(1) + h(5),\\
0 &= h(1) + h(2) + h(3) + h(6),\\
-2 &= h(1) + h(7),\\
4 &= h(1) + h(2) + h(4) + h(8),\\
-2 &= h(1) + h(3) + h(9),\\
0 &= h(1) + h(2) + h(5) + h(10),\\
-4 &= h(1) + h(11),\\
0 &= h(1) + h(2) + h(3) + h(4) + h(6) + h(12),\\
0 &= h(1) + h(13),\\
0 &= h(1) + h(2) + h(7) + h(14).
\end{aligned}$$

Solving, we find

$$h(j) = \begin{cases} 2 \text{ if } j &= 2, 3, 4, 5, 8 \text{ or } 13, \\ -2 \text{ if } j &= 1, 6, 9, 10, 11 \text{ or } 12, \\ 0 \text{ if } j &= 0 \text{ or } 7, \end{cases}$$

that is,

$$h(j) = 2(-1)^j \chi_7(j) \quad \text{for } 1 \le j \le 14.$$

By periodicity, the formula for $h(j)$ holds for all positive integers $j$. It follows that (1.41) simplifies to

$$\left(\sum_{m=-\infty}^{\infty} (-1)^m q^{m^2}\right)\left(\sum_{n=-\infty}^{\infty} (-1)^n q^{7n^2}\right) = 1 + 2\sum_{j=1}^{\infty} (-1)^j \chi_7(j) \frac{q^j}{1-q^j}.$$

On replacing $q$ with $-q$ we obtain (1.36).

Now we shall prove (1.37). In Theorem 1.14, replace $q$ with $q^7$, then let $(a, b, c) = (q, q^2, q^3)$, to obtain

$$\begin{aligned} &q\frac{(q^2, q^4, q^6, q^8, q^{10}, q^{12}, q^{14}, q^{14}; q^{14})_\infty}{(q, q^3, q^5, q^7, q^7, q^9, q^{11}, q^{13}; q^{14})_\infty} \\ &= \sum_{j=0}^{\infty} \left( \frac{q^{14j+1}}{1-q^{14j+1}} - \frac{q^{14j+3}}{1-q^{14j+3}} - \frac{q^{14j+5}}{1-q^{14j+5}} \right. \\ &\qquad \left. + \frac{q^{14j+9}}{1-q^{14j+9}} + \frac{q^{14j+11}}{1-q^{14j+11}} - \frac{q^{14j+13}}{1-q^{14j+13}} \right). \end{aligned}$$

This simplifies to

$$q\prod_{j=1}^{\infty} \frac{(1-q^{2j})^2(1-q^{14j})^2}{(1-q^j)(1-q^{7j})} = \sum_{\substack{j>0 \\ j \text{ odd}}} \chi_7(j) \frac{q^j}{1-q^j}.$$

On applying (0.43) to the left-hand side we obtain (1.37).

Finally, on noting that

$$\chi_7(2j) = \chi_7(j)$$

and adding and subtracting terms that correspond to even values of $j$, we find that

$$\sum_{\substack{j>0\\ j \text{ odd}}} \chi_7(j)\frac{q^j}{1-q^j} = \sum_{j=1}^{\infty} \chi_7(j)\frac{q^j}{1-q^j} - \sum_{j=1}^{\infty} \chi_7(2j)\frac{q^{2j}}{1-q^{2j}}$$

$$= \sum_{j=1}^{\infty} \chi_7(j)\frac{q^j}{1-q^{2j}}.$$

Thus, (1.38) follows from (1.37). □

A series of the form

$$a_0 + \sum_{j=1}^{\infty} \frac{a_j\, q^j}{1-q^j}$$

is called a Lambert series. Since

$$\frac{q^j}{1+q^j} = \frac{q^j}{1-q^j} - \frac{2q^{2j}}{1-q^{2j}}$$

and

$$\frac{q^j}{1-q^{2j}} = \frac{q^j}{1-q^j} - \frac{q^{2j}}{1-q^{2j}},$$

the series in (1.36) and (1.38) may be regarded as Lambert series.

The next example is an application of Weierstrass' identity in Theorems 1.17 and 1.18.

*Example 1.26.* Let $P(q)$ be Ramanujan's Eisenstein series be defined by (0.61), that is,

$$P(q) = 1 - 24\sum_{j=1}^{\infty} \frac{j\,q^j}{1-q^j}.$$

Then

$$\sum_{k,\ell,m,n\in\mathbb{Z}} q^{k^2+\ell^2+3m^2+3n^2} = \frac{1}{6}\left(-P(q) + 2P(q^2) + 3P(q^3)\right.$$
$$\left. -4P(q^4) - 6P(q^6) + 12P(q^{12})\right) \tag{1.42}$$

and

$$\sum_{\substack{k,\ell,m,n>0\\ k,\ell,m,n \text{ odd}}} q^{(k^2+\ell^2+3m^2+3n^2)/8} = \frac{1}{24}\left(-P(q) + P(q^2) + 3P(q^3) - 3P(q^6)\right). \tag{1.43}$$

*Proof.* Let $\omega = \exp(\pi i/3)$ and replace $(x, y)$ with $(\omega, \omega^3)$ in (1.32). On noting that

$$\frac{(1-x\omega^2)^2(1-x\omega^4)^2}{(1-x\omega)^2(1-x\omega^3)^4(1-x\omega^5)^2} = \frac{(1-x^3)^4}{(1-x^2)^2(1-x^6)^2},$$

the result simplifies to

$$\prod_{j=1}^{\infty} \frac{(1-q^j)^4(1-q^{3j})^4}{(1-q^{2j})^2(1-q^{6j})^2} = 1 + \sum_{j=1}^{\infty} h(j) \frac{j q^j}{1-q^j},$$

where $h(j)$ is periodic with period 6. We expand both sides in powers of $q$ as far as $q^6$. The left-hand side is, by (0.41),

$$\begin{aligned}\prod_{j=1}^{\infty} \frac{(1-q^j)^4(1-q^{3j})^4}{(1-q^{2j})^2(1-q^{6j})^2} &= \left(\sum_{j=-\infty}^{\infty} (-1)^j q^{j^2}\right)^2 \left(\sum_{j=-\infty}^{\infty} (-1)^j q^{3j^2}\right)^2 \\ &= (1-2q+2q^4+\cdots)^2(1-2q^3+\cdots)^2 \\ &= 1-4q+4q^2-4q^3+20q^4-24q^5+4q^6+\cdots,\end{aligned}$$

and the right-hand side is

$$1 + \sum_{j=1}^{\infty} h(j) \frac{j q^j}{1-q^j} = 1 + \sum_{j=1}^{\infty}\sum_{k=1}^{\infty} j\, h(j) q^{jk} = 1 + \sum_{n=1}^{\infty} \left(\sum_{d|n} d\, h(d)\right) q^n.$$

On comparing coefficients of $q, q^2, \ldots, q^6$, we deduce that

$$\begin{aligned} -4 &= h(1), \\ 4 &= h(1) + 2h(2), \\ -4 &= h(1) + 3h(3), \\ 20 &= h(1) + 2h(2) + 4h(4), \\ -24 &= h(1) + 5h(5), \\ 4 &= h(1) + 2h(2) + 3h(3) + 6h(6). \end{aligned}$$

Solving for $h(1), \ldots, h(6)$ and using the fact that $h(j)$ is periodic with period 6, we deduce that

$$h(j) = \begin{cases} -4 \text{ if } j \equiv 1 \text{ or } 5 \pmod{6}, \\ \ \ 4 \text{ if } j \equiv 2 \text{ or } 4 \pmod{6}, \\ \ \ 0 \text{ otherwise.} \end{cases}$$

It follows that

$$\prod_{j=1}^{\infty} \frac{(1-q^j)^4(1-q^{3j})^4}{(1-q^{2j})^2(1-q^{6j})^2}$$
$$= 1 - 4\sum_{j=1}^{\infty} \frac{jq^j}{1-q^j} + 16\sum_{j=1}^{\infty} \frac{jq^{2j}}{1-q^{2j}} + 12\sum_{j=1}^{\infty} \frac{jq^{3j}}{1-q^{3j}} - 48\sum_{j=1}^{\infty} \frac{jq^{6j}}{1-q^{6j}}.$$

The left-hand side can be converted a series by using (0.41), and the right-hand side can be expressed in terms of Ramanujan's Eisenstein series $P$. We find that

$$\left(\sum_{m,n=-\infty}^{\infty} (-1)^{m+n} q^{m^2+3n^2}\right)^2 = \frac{1}{6}\left(P(q) - 4P(q^2) - 3P(q^3) + 12P(q^6)\right). \tag{1.44}$$

Now replace $q$ with $-q$ and use (0.110) to obtain (1.42).

In order to prove (1.43), replace $(q,a,b)$ with $(q^3,q,q)$ in (1.33) to obtain

$$q\prod_{j=1}^{\infty} \frac{(1-q^{6j-4})^2(1-q^{6j-2})^2(1-q^{6j})^4}{(1-q^{6j-5})^2(1-q^{6j-3})^4(1-q^{6j-1})^2} = -\sum_{j=1}^{\infty} j\,\frac{(q^j-q^{-j})^2}{(q^{3j}-q^{-3j})}.$$

If we simplify the product on the left-hand side, and apply the identity

$$-\frac{(x-x^{-1})^2}{x^3-x^{-3}} = \frac{x}{1-x} - \frac{x^2}{1-x^2} - \frac{3x^3}{1-x^3} + \frac{3x^6}{1-x^6}$$

to the right-hand side, we find that

$$q\prod_{j=1}^{\infty} \frac{(1-q^{2j})^4(1-q^{6j})^4}{(1-q^j)^2(1-q^{3j})^2} = \sum_{j=1}^{\infty}\left(\frac{jq^j}{1-q^j} - \frac{jq^{2j}}{1-q^{2j}} - \frac{3jq^{3j}}{1-q^{3j}} + \frac{3jq^{6j}}{1-q^{6j}}\right).$$

The left-hand side can be converted to a series by using (0.43), and the right-hand side can be expressed in terms of Ramanujan's Eisenstein series $P$. The resulting identity is (1.43). □

## 1.8 Jacobian elliptic functions

At the end of Section 1.2 it was shown that the order of a nonconstant elliptic function must be at least two. There are two possibilities for an elliptic function of order two:

- either, it has a single pole of order two in each cell with residue zero;
- or, it has two simple poles in each cell with opposite residues.

The Weierstrass elliptic function $\wp(\theta)$ is an example of the first case: it has a single pole of order two in each cell. In this section we will study elliptic functions that are instances of the second case. Such functions can be easily constructed using the function $F(x, y)$ defined by (0.99), namely

$$F(x, y) = \prod_{n=1}^{\infty} \frac{(1 - xyq^{n-1})(1 - x^{-1}y^{-1}q^n)(1 - q^n)^2}{(1 - xq^{n-1})(1 - x^{-1}q^n)(1 - yq^{n-1})(1 - y^{-1}q^n)}, \tag{1.45}$$

where as usual, $q = e^{2\pi i\tau}$. Put $y = e^{iv}$ and consider the four special cases

$$v \in \{0, \pi, \pi\tau, \pi + \pi\tau\}.$$

The four values of $v$ range over the vertices of a parallelogram. Corresponding to $v = \pi,\ \pi\tau$ and $\pi + \pi\tau$, let $f_1, f_2$ and $f_3$ be defined by

$$f_1(\theta) = f_1(\theta, q) = -iF(e^{i\theta}, e^{i\pi}), \tag{1.46}$$

$$f_2(\theta) = f_2(\theta, q) = -ie^{i\theta/2}F(e^{i\theta}, e^{i\pi\tau}), \tag{1.47}$$

and

$$f_3(\theta) = f_3(\theta, q) = -ie^{i\theta/2}F(e^{i\theta}, e^{i\pi+i\pi\tau}). \tag{1.48}$$

The factors $-i$ and $-ie^{i\theta/2}$ have been inserted to make the functions real valued when $\theta$ is real. For the reader who is already familiar with the Jacobian elliptic functions, $f_1, f_2$, and $f_3$ are the Jacobian functions cs, ns, and ds, respectively, up to rescaling. The precise identifications will be given in Exercises 15 and 16 at the end of the chapter.

The case $v = 0$ is less straightforward because the function $F(e^{i\theta}, e^{iv})$ has a singularity at $v = 0$. In order to obtain something meaningful, we remove the singular part. We will need the expansion of $F(e^{i\theta}, e^{iv})$ in powers of $v$.

**Lemma 1.27.** *Let $\rho_1(x)$ be defined by* (1.19)*, that is,*

$$\rho_1(x) = \frac{1 + x}{2(1 - x)} + \sum_{n=1}^{\infty} \left( \frac{xq^n}{1 - xq^n} - \frac{x^{-1}q^n}{1 - x^{-1}q^n} \right). \tag{1.49}$$

*For $m = 2, 3, 4, \ldots$, let $\rho_m(x)$ be defined by*

$$\rho_m(x) = -\frac{B_m}{m} + \sum_{n=1}^{\infty} \left( \frac{n^{m-1}xq^n}{1 - xq^n} + (-1)^m \frac{n^{m-1}x^{-1}q^n}{1 - x^{-1}q^n} \right) \tag{1.50}$$

*where $B_m$ is the mth Bernoulli number as defined by* (0.64). *Then for $|v|$ sufficiently small and $v \neq 0$, we have*

$$F(x, e^{iv}) = -\frac{1}{iv} + \sum_{m=0}^{\infty} \rho_{m+1}(x) \frac{(iv)^m}{m!}. \tag{1.51}$$

*Proof.* Since $F(x, y) = F(y, x)$, by Theorem 0.26 and series manipulations, we have

$$\begin{aligned} F(x, y) &= \sum_{n=-\infty}^{\infty} \frac{y^n}{1 - xq^n} \\ &= \frac{1}{1-x} + \sum_{n=1}^{\infty} \frac{y^n(1 - xq^n + xq^n)}{1 - xq^n} + \sum_{n=1}^{\infty} \frac{y^{-n}}{1 - xq^{-n}} \\ &= \frac{1}{1-x} + \frac{1}{1-y} - 1 + \sum_{n=1}^{\infty} \frac{xq^n y^n}{1 - xq^n} - \sum_{n=1}^{\infty} \frac{x^{-1} q^n y^{-n}}{1 - x^{-1} q^n}. \end{aligned}$$

Put $y = e^{iv}$ and expand in powers of $v$ to obtain

$$\begin{aligned} F(x, e^{iv}) &= \frac{1}{1-x} + \frac{1}{1-e^{iv}} - 1 + \sum_{n=1}^{\infty} \left( \frac{xq^n e^{inv}}{1 - xq^n} - \frac{x^{-1} q^n e^{-inv}}{1 - x^{-1} q^n} \right) \\ &= \frac{1}{1-x} - 1 - \sum_{m=0}^{\infty} \frac{B_m}{m!} (iv)^{m-1} \\ &\quad + \sum_{n=1}^{\infty} \left( \frac{xq^n}{1 - xq^n} \sum_{m=0}^{\infty} \frac{(inv)^m}{m!} - \frac{x^{-1} q^n}{1 - x^{-1} q^n} \sum_{m=0}^{\infty} \frac{(-inv)^m}{m!} \right). \end{aligned}$$

On collecting terms according to the powers of $v$, we obtain

$$\begin{aligned} F(x, e^{iv}) &= -\frac{1}{iv} + \left( \frac{1+x}{2(1-x)} + \sum_{n=1}^{\infty} \frac{xq^n}{1 - xq^n} - \frac{x^{-1} q^n}{1 - x^{-1} q^n} \right) \\ &\quad + \sum_{m=1}^{\infty} \left( -\frac{B_{m+1}}{m+1} + \sum_{n=1}^{\infty} \frac{n^m xq^n}{1 - xq^n} - \frac{(-1)^m n^m x^{-1} q^n}{1 - x^{-1} q^n} \right) \frac{(iv)^m}{m!} \\ &= -\frac{1}{iv} + \sum_{m=0}^{\infty} \rho_{m+1}(x) \frac{(iv)^m}{m!}. \end{aligned}$$

□

We are now ready to consider the function $F(e^{i\theta}, e^{iv})$ in the case $v = 0$. We subtract the singular part and define $f_0$ by

$$f_0(\theta) = f_0(\theta, q) = -i \lim_{v\to 0} \left\{ F(e^{i\theta}, e^{iv}) + \frac{1}{iv} \right\}. \tag{1.52}$$

By (1.51) and (1.21), we find that

$$f_0(\theta) = -i \lim_{v\to 0} \left\{ F(e^{i\theta}, e^{iv}) + \frac{1}{iv} \right\} = -i\rho_1(e^{i\theta}) = z(\theta), \tag{1.53}$$

where $z(\theta)$ is the function introduced at the beginning of the chapter in (1.4).

For future reference, let $g_0$ be defined by

$$g_0(\theta) = \rho_2(e^{i\theta}) \tag{1.54}$$

where $\rho_2$ is given by (1.50). Its Fourier expansion will be determined in (1.69).

We now proceed to study properties of the functions $f_0$, $f_1$, $f_2$, and $f_3$ defined by (1.46)–(1.48) and (1.52). We start by recording infinite product formulas for the functions $f_1, f_2$, and $f_3$. There is no similar, simple infinite product formula for $f_0$.

**Theorem 1.28.** *The following infinite product formulas hold:*

$$f_1(\theta) = \frac{1}{2}\cot\frac{\theta}{2}\prod_{n=1}^{\infty}\frac{(1+e^{i\theta}q^n)(1+e^{-i\theta}q^n)(1-q^n)^2}{(1-e^{i\theta}q^n)(1-e^{-i\theta}q^n)(1+q^n)^2},$$

$$f_2(\theta) = \frac{1}{2}\csc\frac{\theta}{2}\prod_{n=1}^{\infty}\frac{(1-e^{i\theta}q^{n-1/2})(1-e^{-i\theta}q^{n-1/2})(1-q^n)^2}{(1-e^{i\theta}q^n)(1-e^{-i\theta}q^n)(1-q^{n-1/2})^2}$$

*and*

$$f_3(\theta) = \frac{1}{2}\csc\frac{\theta}{2}\prod_{n=1}^{\infty}\frac{(1+e^{i\theta}q^{n-1/2})(1+e^{-i\theta}q^{n-1/2})(1-q^n)^2}{(1-e^{i\theta}q^n)(1-e^{-i\theta}q^n)(1+q^{n-1/2})^2}.$$

*Proof.* These are immediate from the product (1.45) and the definitions (1.46)–(1.48). □

The next result gives Fourier expansions for $f_0, f_1, f_2$, and $f_3$.

**Theorem 1.29.** *The following Fourier expansions hold for* $-2\pi \operatorname{Im}\tau < \operatorname{Im}\theta < 2\pi \operatorname{Im}\tau$:

$$f_0(\theta) = \frac{1}{2}\cot\frac{\theta}{2} + 2\sum_{n=1}^{\infty}\frac{q^n}{1-q^n}\sin n\theta, \tag{1.55}$$

$$f_1(\theta) = \frac{1}{2}\cot\frac{\theta}{2} - 2\sum_{n=1}^{\infty}\frac{q^n}{1+q^n}\sin n\theta, \tag{1.56}$$

$$f_2(\theta) = \frac{1}{2}\csc\frac{\theta}{2} + 2\sum_{n=1}^{\infty}\frac{q^{n-1/2}}{1-q^{n-1/2}}\sin\left(n-\frac{1}{2}\right)\theta, \tag{1.57}$$

*and*

$$f_3(\theta) = \frac{1}{2}\csc\frac{\theta}{2} - 2\sum_{n=1}^{\infty}\frac{q^{n-1/2}}{1+q^{n-1/2}}\sin\left(n-\frac{1}{2}\right)\theta. \tag{1.58}$$

*Proof.* It has already been noted in (1.53) that $f_0(\theta) = z(\theta)$. Therefore the Fourier expansion of $f_0$ is just the Fourier expansion of $z$ given by (1.17), and this is (1.55).

The Fourier expansions for $f_1, f_2$, and $f_3$ are obtained by using Theorem 0.27. For example, by the definition of $f_1$, we deduce that

$$\begin{aligned} f_1(\theta) = -iF(e^{i\theta}, -1) &= -i\left(\frac{1+e^{i\theta}}{2(1-e^{i\theta})} + \sum_{m=1}^{\infty}\sum_{n=1}^{\infty} q^{mn}(-1)^n(e^{im\theta} - e^{-im\theta})\right) \\ &= \frac{1}{2}\cot\frac{\theta}{2} - 2\sum_{m=1}^{\infty}\frac{q^m}{1+q^m}\sin m\theta. \end{aligned}$$

This proves (1.56). The Fourier expansions (1.57) and (1.58) may be proved in the same way. □

The next result gives the periodicity properties of $f_0, f_1, f_2$, and $f_3$.

**Theorem 1.30.** *For any integers m and n we have*

$$\begin{aligned} f_0(\theta + 2\pi m + 2\pi\tau n) &= -in + f_0(\theta), \\ f_1(\theta + 2\pi m + 2\pi\tau n) &= (-1)^n f_1(\theta), \\ f_2(\theta + 2\pi m + 2\pi\tau n) &= (-1)^m f_2(\theta) \end{aligned}$$

*and*

$$f_3(\theta + 2\pi m + 2\pi\tau n) = (-1)^{m+n} f_3(\theta).$$

*Proof.* Since $f_0(\theta) = z(\theta)$, the result for $f_0$ follows from the properties of $z$ given by (1.6). The results for $f_1$, $f_2$ and $f_3$ can be deduced from the infinite product formulas in Theorem 1.28. □

**Table 1.1** Periods, zeros, and poles of Jacobian elliptic functions

| $f(\theta)$ | periods | zeros | poles |
|---|---|---|---|
| $f_1(\theta)$ | $2\pi,\ 4\pi\tau$ | $\theta = (2m+1)\pi + 2n\pi\tau$ | $\theta = 2m\pi + 2n\pi\tau$ |
| $f_2(\theta)$ | $4\pi,\ 2\pi\tau$ | $\theta = 2m\pi + (2n+1)\pi\tau$ | $\theta = 2m\pi + 2n\pi\tau$ |
| $f_3(\theta)$ | $4\pi,\ 2\pi + 2\pi\tau$ | $\theta = (2m+1)\pi + (2n+1)\pi\tau$ | $\theta = 2m\pi + 2n\pi\tau$ |

The functions $f_1, f_2$, and $f_3$ are elliptic functions. By Theorem 1.30 they are doubly periodic with periods given by Table 1.1. The zeros and poles can be determined from the infinite products in Theorem 1.28 and they are listed in Table 1.1. By the Fourier expansions in Theorem 1.29, each of $f_1(\theta), f_2(\theta)$, and $f_3(\theta)$ is an odd function of $\theta$, and

$$f_j(\theta) = \frac{1}{\theta} + O(\theta) \quad \text{as } \theta \to 0. \tag{1.59}$$

It follows that each function $f_1, f_2$, and $f_3$ has a simple pole at $\theta = 0$ and the residue is 1. By (1.59) and the double periodicity properties in Theorem 1.30, it follows further that the poles of $f_1, f_2$, and $f_3$ are all simple, and the residue of each function at any pole is $\pm 1$. There are two simple poles in any cell and no other singularities. Therefore, $f_1, f_2$, and $f_3$ are elliptic functions of order two. They are called Jacobian elliptic functions.

The function $f_0$ is not doubly periodic, so it is not an elliptic function. By (1.15), the poles of $f_0$ are all simple and occur at the points $\theta = 2m\pi + 2n\pi\tau$. The zeros of $f_0$ do not lie on a lattice. The derivative of $f_0(\theta)$ is an elliptic function, and from (1.13) we have

$$-f_0'(\theta) - \frac{P(q)}{12} = \wp(\theta) \tag{1.60}$$

where $\wp(\theta)$ is the Weierstrass elliptic function.

The Weierstrass invariants $e_1$, $e_2$, and $e_3$ are defined by

$$e_1 = \wp(\pi), \tag{1.61}$$

$$e_2 = \wp(\pi\tau), \tag{1.62}$$

$$e_3 = \wp(\pi + \pi\tau). \tag{1.63}$$

The next result relates squares of $f_0, f_1, f_2$, and $f_3$ to the Weierstrass elliptic function.

**Theorem 1.31.** *The following identities hold:*

$$f_1^2(\theta) = \wp(\theta) - e_1, \tag{1.64}$$

$$f_2^2(\theta) = \wp(\theta) - e_2, \tag{1.65}$$

$$f_3^2(\theta) = \wp(\theta) - e_3. \tag{1.66}$$

*Moreover, if* $g_0$ *is defined by* (1.54), *then*

$$f_0^2(\theta) = \wp(\theta) + 2g_0(\theta). \tag{1.67}$$

*Proof.* The proofs of (1.64)–(1.66) are similar, so let us prove (1.65). Let $\alpha = \pi\tau$ in Theorem 1.19 to obtain

$$\begin{aligned}\wp(\theta) - e_2 &= \wp(\theta) - \wp(\pi\tau)\\ &= F(e^{i\theta}, e^{i\pi\tau})F(e^{-i\theta}, e^{i\pi\tau})\\ &= -\left(-ie^{i\theta/2}F(e^{i\theta}, e^{i\pi\tau})\right) \times \left(-ie^{-i\theta/2}F(e^{-i\theta}, e^{i\pi\tau})\right)\\ &= -f_2(\theta)f_2(-\theta)\\ &= f_2^2(\theta).\end{aligned}$$

The identities (1.64) and (1.66) can be proved in a similar way, by taking $\alpha = \pi$ and $\alpha = \pi + \pi\tau$, respectively, in Theorem 1.19.

It remains to prove (1.67). Interchanging the variables in Theorem 1.19 and using the symmetry property $F(x, y) = F(y, x)$ gives

$$\wp(\alpha) - \wp(\theta) = F(e^{i\theta}, e^{i\alpha})F(e^{i\theta}, e^{-i\alpha}).$$

Expand both sides in powers of $\alpha$ by (1.12) and (1.51) to obtain

$$\begin{aligned}&\frac{1}{\alpha^2} - \wp(\theta) + O(\alpha^2)\\ &= \left(-\frac{1}{i\alpha} + \sum_{m=0}^{\infty} \rho_{m+1}(e^{i\theta})\frac{(i\alpha)^m}{m!}\right)\left(\frac{1}{i\alpha} + \sum_{m=0}^{\infty} \rho_{m+1}(e^{i\theta})\frac{(-i\alpha)^m}{m!}\right).\end{aligned}$$

On equating the coefficients of $\alpha^0$ we find that

$$-\wp(\theta) = \rho_1^2(e^{i\theta}) + 2\rho_2(e^{i\theta}),$$

and by (1.53) and (1.54) this is equivalent to (1.67). □

**Corollary 1.32 (Ramanujan).** *For* $|\operatorname{Im}\theta| < 2\pi \operatorname{Im}\tau$, *we have*

$$\begin{aligned}&\left(\frac{1}{2}\cot\frac{\theta}{2} + 2\sum_{j=1}^{\infty}\frac{q^j}{1-q^j}\sin j\theta\right)^2\\ &= \frac{1}{4}\cot^2\frac{\theta}{2} + 2\sum_{j=1}^{\infty}\frac{jq^j}{1-q^j}(1-\cos j\theta) + 4\sum_{j=1}^{\infty}\frac{q^j}{(1-q^j)^2}\cos j\theta.\end{aligned} \tag{1.68}$$

*Proof.* We expand both sides of the identity (1.67) as Fourier series. The expansion of $\rho_2$ may be obtained from the definition (1.50), as follows:

$$\begin{aligned}\rho_2(x) &= -\frac{1}{12} + \sum_{n=1}^{\infty}\left(\frac{nxq^n}{1-xq^n} + \frac{nx^{-1}q^n}{1-x^{-1}q^n}\right)\\ &= -\frac{1}{12} + \sum_{n=1}^{\infty}\sum_{j=1}^{\infty} nq^{jn}(x^j + x^{-j})\\ &= -\frac{1}{12} + \sum_{j=1}^{\infty}\frac{q^j}{(1-q^j)^2}(x^j + x^{-j}).\end{aligned}$$

Since $g_0(\theta) = \rho_2(e^{i\theta})$, this gives

$$g_0(\theta) = \rho_2(e^{i\theta}) = -\frac{1}{12} + 2\sum_{j=1}^{\infty}\frac{q^j}{(1-q^j)^2}\cos j\theta. \tag{1.69}$$

The required result follows by substituting the Fourier expansions of $\wp$, $f_0$, and $g_0$ given respectively by (1.18), (1.55), and (1.69), into the identity (1.67). □

The identity (1.68) plays a significant role in Ramanujan's paper [249]. Other proofs will be given in the exercises at the end of the chapter.

The results in the next theorem may be regarded as analogues of the trigonometric identity $\csc^2\theta - \cot^2\theta = 1$.

**Theorem 1.33.** *The following identities hold*:

$$\begin{aligned} f_2^2(\theta) - f_1^2(\theta) &= e_1 - e_2,\\ f_2^2(\theta) - f_3^2(\theta) &= e_3 - e_2,\\ f_3^2(\theta) - f_1^2(\theta) &= e_1 - e_3.\end{aligned}$$

*Proof.* Combine the results in Theorem 1.31 two at a time to eliminate the $\wp(\theta)$ term. □

The right-hand sides of the identities in Theorem 1.33 have simple expressions as infinite products.

**Theorem 1.34.** *The Weierstrass invariants $e_1$, $e_2$, and $e_3$ defined by* (1.61)–(1.63) *satisfy the identities*

$$e_1 - e_2 = \frac{1}{4}\prod_{j=1}^{\infty}\frac{(1+q^{j-1/2})^4(1-q^j)^4}{(1-q^{j-1/2})^4(1+q^j)^4},$$

$$e_3 - e_2 = 4q^{1/2} \prod_{j=1}^{\infty} \frac{(1+q^j)^4(1-q^j)^4}{(1+q^{j-1/2})^4(1-q^{j-1/2})^4}$$

*and*

$$e_1 - e_3 = \frac{1}{4} \prod_{j=1}^{\infty} \frac{(1-q^{j-1/2})^4(1-q^j)^4}{(1+q^{j-1/2})^4(1+q^j)^4}.$$

*Proof.* Successively let $\theta = \pi\tau$ in (1.64), $\theta = \pi + \pi\tau$ in (1.65) and $\theta = \pi$ in (1.66) and use the infinite products in Theorem 1.28. □

As $\operatorname{Im}\tau > 0$, Theorem 1.34 implies that $e_1$, $e_2$, and $e_3$ are distinct. Moreover, if $\tau$ is purely imaginary then $0 < q < 1$, and therefore $e_1$, $e_2$, and $e_3$ are real and $e_1 > e_3 > e_2$. Further details about the signs of $e_1, e_2$, and $e_3$ are given in Exercise 8.

**Corollary 1.35.** *The Weierstrass invariants $e_1$, $e_2$, and $e_3$ defined by* (1.61)–(1.63) *satisfy the identity*

$$16(e_1 - e_2)^2(e_1 - e_3)^2(e_2 - e_3)^2 = q \prod_{j=1}^{\infty} (1-q^j)^{24}.$$

*Proof.* This follows from Theorem 1.34 and noting that

$$\prod_{j=1}^{\infty} (1-q^{j-1/2})(1+q^{j-1/2})(1+q^j) = \prod_{j=1}^{\infty} (1-q^{2j-1})(1+q^j) = 1,$$

where the last identity holds by Euler's identity for infinite products (0.1). □

## 1.9 Differential equations and addition formulas

We will now obtain differential equations and addition formulas for the Jacobian elliptic functions and for the Weierstrass elliptic function $\wp$. We begin by calculating derivatives of $f_0, f_1, f_2$, and $f_3$.

**Theorem 1.36.** *The following differentiation formulas hold:*

$$f_1'(\theta) = -f_2(\theta)f_3(\theta), \tag{1.70}$$

$$f_2'(\theta) = -f_1(\theta)f_3(\theta), \tag{1.71}$$

$$f_3'(\theta) = -f_1(\theta)f_2(\theta). \tag{1.72}$$

*And,*

$$f_0'(\theta) = -f_0^2(\theta) + 2g_0(\theta) - \frac{P(q)}{12}. \tag{1.73}$$

*Proof.* The proofs of (1.70)–(1.72) are similar, so we will prove (1.72). Let $t = e^{i\theta}$ in Theorem 1.16 to obtain

$$F(x, e^{i\theta})F(y, e^{i\theta}) = -i\frac{\partial}{\partial\theta}F(xy, e^{i\theta}) + F(xy, e^{i\theta})(\rho_1(x) + \rho_1(y)). \tag{1.74}$$

Now let $x = e^{i\pi}$ and $y = e^{i\pi\tau}$. From (1.49) we may deduce that

$$\rho_1(e^{i\pi}) = 0 \qquad \text{and} \qquad \rho_1(e^{i\pi\tau}) = \frac{1}{2}.$$

Therefore (1.74) implies

$$F(e^{i\pi}, e^{i\theta})F(e^{i\pi\tau}, e^{i\theta}) = -i\frac{\partial}{\partial\theta}F(e^{i\pi+i\pi\tau}, e^{i\theta}) + \frac{1}{2}F(e^{i\pi+i\pi\tau}, e^{i\theta}). \tag{1.75}$$

The left-hand side of (1.75) is

$$F(e^{i\pi}, e^{i\theta})F(e^{i\pi\tau}, e^{i\theta}) = \big(if_1(\theta)\big)\big(ie^{-i\theta/2}f_2(\theta)\big) = -e^{-i\theta/2}f_1(\theta)f_2(\theta).$$

The right-hand side of (1.75) is

$$\begin{aligned}
&-i\frac{\partial}{\partial\theta}\left(ie^{-i\theta/2}f_3(\theta)\right) + \frac{i}{2}e^{-i\theta/2}f_3(\theta) \\
&= e^{-i\theta/2}f_3'(\theta) - \frac{i}{2}e^{-i\theta/2}f_3(\theta) + \frac{i}{2}e^{-i\theta/2}f_3(\theta) \\
&= e^{-i\theta/2}f_3'(\theta).
\end{aligned}$$

On combining the above results we obtain (1.72).

In a similar way, if we let $\alpha = e^{i\pi\tau}$, $\beta = e^{i\pi+i\pi\tau}$ or $\alpha = e^{i\pi+i\pi\tau}$, $\beta = e^{i\pi}$ in (1.74), then we obtain (1.70) and (1.71), respectively.

Lastly, from (1.60) and (1.67) we have

$$f_0'(\theta) = -\wp(\theta) - \frac{P(q)}{12} = -f_0^2(\theta) + 2g_0(\theta) - \frac{P(q)}{12}.$$

□

**Theorem 1.37.** *$f_1, f_2$, and $f_3$ satisfy the differential equations:*

$$\left(f_1'\right)^2 = \left(f_1^2 + e_1 - e_2\right)\left(f_1^2 + e_1 - e_3\right),$$
$$\left(f_2'\right)^2 = \left(f_2^2 + e_2 - e_1\right)\left(f_2^2 + e_2 - e_3\right),$$
$$\left(f_3'\right)^2 = \left(f_3^2 + e_3 - e_1\right)\left(f_3^2 + e_3 - e_2\right).$$

*Furthermore,*

$$f_0'' = 2f_1f_2f_3.$$

*And, the Weierstrass elliptic function $\wp = \wp(\theta)$ satisfies the differential equation*

$$\left(\wp'\right)^2 = 4\left(\wp - e_1\right)\left(\wp - e_2\right)\left(\wp - e_3\right). \tag{1.76}$$

*Proof.* By Theorems 1.33 and 1.36 we have

$$\left(f_1'\right)^2 = f_2^2f_3^2 = \left(f_1^2 + e_1 - e_2\right)\left(f_1^2 + e_1 - e_3\right).$$

This proves the first result, and the corresponding results for $f_2$ and $f_3$ may be obtained similarly.

Next, by (1.60), (1.64) and (1.70) we have

$$f_0'' = -\wp' = -\left(f_1^2 + e_1\right)' = -2f_1f_1' = 2f_1f_2f_3.$$

Squaring, gives

$$(\wp')^2 = 4f_1^2f_2^2f_3^2 = 4\left(\wp - e_1\right)\left(\wp - e_2\right)\left(\wp - e_3\right),$$

where the last step follows from Theorem 1.31. □

**Theorem 1.38.** *Let $e_1$, $e_2$, and $e_3$ be the Weierstrass invariants defined by* (1.61)–(1.63) *and let $Q$, $R$, $E_4$, and $E_6$ be the Eisenstein series defined by* (0.72) *and* (0.73). *Then*

$$e_1 + e_2 + e_3 = 0,$$

$$e_1e_2 + e_2e_3 + e_3e_1 = -5E_4 = -\frac{Q}{48}$$

*and*

$$e_1e_2e_3 = -\frac{7E_6}{12} = \frac{R}{864}.$$

*Hence, the Weierstrass differential equation* (1.76) *is equivalent to*

$$(\wp'(\theta))^2 = 4\wp^3(\theta) - \frac{Q}{12}\wp(\theta) - \frac{R}{216}. \tag{1.77}$$

*Proof.* Expanding the right-hand side of (1.76) we obtain

$$\begin{aligned}(\wp'(\theta))^2 &= 4\wp^3(\theta) - 4(e_1+e_2+e_3)\wp^2(\theta)\\ &\quad + 4(e_1e_2+e_2e_3+e_3e_1)\wp(\theta) - 4e_1e_2e_3.\end{aligned} \tag{1.78}$$

On the other hand, from (1.12) we have

$$\wp(\theta) = \frac{1}{\theta^2} + E_4\theta^2 - \frac{E_6}{12}\theta^4 + O(\theta^6).$$

Hence,

$$\begin{aligned}\wp^2(\theta) &= \frac{1}{\theta^4} + 2E_4 - \frac{E_6}{6}\theta^2 + O(\theta^4),\\ \wp^3(\theta) &= \frac{1}{\theta^6} + \frac{3E_4}{\theta^2} - \frac{E_6}{4} + O(\theta^2),\\ \wp'(\theta) &= \frac{-2}{\theta^3} + 2E_4\theta - \frac{E_6}{3}\theta^3 + O(q^5)\end{aligned}$$

and

$$(\wp'(\theta))^2 = \frac{4}{\theta^6} - \frac{8E_4}{\theta^2} + \frac{4E_6}{3} + O(\theta^2).$$

Substituting the above expansions into (1.78) we deduce that

$$\begin{aligned}&\frac{4}{\theta^6} - \frac{8E_4}{\theta^2} + \frac{4E_6}{3} + O(\theta^2)\\ &= 4\left(\frac{1}{\theta^6} + \frac{3E_4}{\theta^2} - \frac{E_6}{4} + O(\theta^2)\right) - 4(e_1+e_2+e_3)\left(\frac{1}{\theta^4} + 2E_4 + O(\theta^2)\right)\\ &\qquad + 4(e_1e_2+e_2e_3+e_3e_1)\left(\frac{1}{\theta^2} + O(\theta^2)\right) - 4e_1e_2e_3.\end{aligned}$$

Equating the coefficients of $\theta^{-4}$, $\theta^{-2}$, and $\theta^0$ we find that

$$e_1 + e_2 + e_3 = 0,$$

$$4(e_1e_2 + e_2e_3 + e_3e_1) + 12E_4 = -8E_4$$

and

$$-4e_1e_2e_3 - 8(e_1 + e_2 + e_3)E_4 - E_6 = \frac{4E_6}{3},$$

respectively. The required results follow by simplification. □

The differential equation (1.77) has several interesting consequences.

**Corollary 1.39.** *The following differential equation holds*:

$$\wp''(\theta) = 6\wp^2(\theta) - \frac{Q}{24}. \tag{1.79}$$

*Hence,*

$$\left(\frac{1}{8}\cot^2\frac{\theta}{2} + \frac{1}{12} + \sum_{j=1}^{\infty}\frac{jq^j}{1-q^j}(1-\cos j\theta)\right)^2$$
$$= \left(\frac{1}{8}\cot^2\frac{\theta}{2} + \frac{1}{12}\right)^2 + \frac{1}{12}\sum_{j=1}^{\infty}\frac{j^3q^j}{1-q^j}(5+\cos j\theta). \tag{1.80}$$

*Proof.* The differential equation (1.79) is obtained by taking the derivative of (1.77) and simplifying the result. It remains to prove (1.80). The identity (1.79) may be written in the form

$$\left(\frac{\wp(\theta)}{2}\right)^2 = \frac{\wp''(\theta)}{24} + \frac{Q}{576}.$$

On substituting the Fourier expansion for $\wp(\theta)$ and given by (1.18) and making use of the definition of $Q$ from (0.72), we obtain

$$\left(\frac{1}{8}\csc^2\frac{\theta}{2} - \frac{1}{24} + \sum_{j=1}^{\infty}\frac{jq^j}{1-q^j}(1-\cos j\theta)\right)^2$$
$$= \frac{1}{96}\left(\csc^2\frac{\theta}{2}\right)'' + \frac{1}{12}\sum_{j=1}^{\infty}\frac{j^3q^j}{1-q^j}\cos j\theta + \frac{1}{576} + \frac{5}{12}\sum_{j=1}^{\infty}\frac{j^3q^j}{1-q^j}.$$

Since

$$\frac{1}{8}\csc^2\frac{\theta}{2} - \frac{1}{24} = \frac{1}{8}\cot^2\frac{\theta}{2} + \frac{1}{12}$$

and

$$\frac{1}{96}\left(\csc^2\frac{\theta}{2}\right)'' + \frac{1}{576} = \left(\frac{1}{8}\cot^2\frac{\theta}{2} + \frac{1}{12}\right)^2,$$

the required result follows. □

The next result may be compared with Corollary 0.24, where the analogous results for $U_{2n}$ and $V_{2n}$ were given.

**Corollary 1.40.** *Let $E_k$ be the Eisenstein series defined by* (0.66). *Then, for $n \geq 2$ we have*

$$E_{2n+4} = 12\frac{(n+1)(2n+1)}{(n-1)(2n+5)}\sum_{j=1}^{n-1}\binom{2n}{2j}E_{2j+2}E_{2n+2-2j}. \tag{1.81}$$

*Hence for $n \geq 2$, there exist rational numbers $c_{j,k}$ such that*

$$E_{2n} = \sum_{\substack{j,k\geq 0\\ 2j+3k=n}} c_{j,k}Q^jR^k. \tag{1.82}$$

*Proof.* Expand the result of Corollary 1.39 in powers of $\theta$ using (1.12) to obtain

$$\frac{6}{\theta^4} - 2\sum_{j=1}^{\infty}\frac{(-1)^jE_{2j+2}}{(2j-2)!}\theta^{2j-2} = 6\left(\frac{1}{\theta^2} - 2\sum_{j=1}^{\infty}\frac{(-1)^jE_{2j+2}}{(2j)!}\theta^{2j}\right)^2 - \frac{Q}{24}.$$

Equate coefficients of $q^{2n}$, for $n \geq 1$, to obtain

$$\frac{2(-1)^nE_{2n+4}}{(2n)!}$$

$$= \frac{24(-1)^nE_{2n+4}}{(2n+2)!} + 48\sum_{j=1}^{n-1}\left(\frac{(-1)^jE_{2j+2}}{(2j)!}\right)\left(\frac{(-1)^{n-j}E_{2n+2-2j}}{(2n-2j)!}\right).$$

For $n \geq 2$ this may be solved for $E_{2n+4}$ to give (1.81).

The cases $n = 2$ and $n = 3$ of (1.82) are true trivially by the definitions of $Q$ and $R$ given by (0.72) and (0.73), respectively. The identity (1.82) is true for $n \geq 4$ by mathematical induction, using (1.81). □

The rational numbers $c_{j,k}$ in (1.82) can be computed for any given value of $n$ by expanding in powers of $q$ and equating coefficients. The results in the cases $2 \leq n \leq 7$ are given by:

$$1 + 240 \sum_{j=1}^{\infty} \frac{j^3 q^j}{1 - q^j} = Q,$$

$$1 - 504 \sum_{j=1}^{\infty} \frac{j^5 q^j}{1 - q^j} = R,$$

$$1 + 480 \sum_{j=1}^{\infty} \frac{j^7 q^j}{1 - q^j} = Q^2,$$

$$1 - 264 \sum_{j=1}^{\infty} \frac{j^9 q^j}{1 - q^j} = QR,$$

$$691 + 65520 \sum_{j=1}^{\infty} \frac{j^{11} q^j}{1 - q^j} = 441Q^3 + 250R^2,$$

$$1 - 24 \sum_{j=1}^{\infty} \frac{j^{13} q^j}{1 - q^j} = Q^2 R.$$

The next result provides addition formulas for the Jacobian elliptic functions.

**Theorem 1.41.** *The following addition formulas hold*:

$$f_1(a+b) = \frac{f_1(a)f_2(b)f_3(b) - f_1(b)f_2(a)f_3(a)}{f_1^2(b) - f_1^2(a)}, \tag{1.83}$$

$$f_2(a+b) = \frac{f_2(a)f_3(b)f_1(b) - f_2(b)f_3(a)f_1(a)}{f_2^2(b) - f_2^2(a)}, \tag{1.84}$$

$$f_3(a+b) = \frac{f_3(a)f_1(b)f_2(b) - f_3(b)f_1(a)f_2(a)}{f_3^2(b) - f_3^2(a)}. \tag{1.85}$$

*Proof.* In Theorem 1.16 make the change of variables $x = e^{ia}$, $y = e^{ib}$, $t = e^{i\theta}$ to obtain

$$F(e^{ia}, e^{i\theta})F(e^{ib}, e^{i\theta}) = \frac{1}{i}\frac{\partial}{\partial\theta}F(e^{i(a+b)}, e^{i\theta}) + F(e^{i(a+b)}, e^{i\theta})(\rho_1(e^{ia}) + \rho_1(e^{ib})).$$

Apply the differential operator $\partial/\partial a - \partial/\partial b$ to both sides. The result is

$$\frac{\partial}{\partial a}F(e^{ia}, e^{i\theta})F(e^{ib}, e^{i\theta}) - F(e^{ia}, e^{i\theta})\frac{\partial}{\partial b}F(e^{ib}, e^{i\theta})$$
$$= F(e^{i(a+b)}, e^{i\theta})\left(\frac{\mathrm{d}}{\mathrm{d}a}\rho_1(e^{ia}) - \frac{\mathrm{d}}{\mathrm{d}b}\rho_1(e^{ib})\right).$$

Rearranging and using (1.60) to write the derivatives of $\rho_1$ in terms of the Weierstrass $\wp$ function, we find that

$$F(e^{i(a+b)}, e^{i\theta}) = i\left(\frac{\frac{\partial}{\partial a}F(e^{ia}, e^{i\theta})F(e^{ib}, e^{i\theta}) - F(e^{ia}, e^{i\theta})\frac{\partial}{\partial b}F(e^{ib}, e^{i\theta})}{\wp(a) - \wp(b)}\right). \tag{1.86}$$

Let $\theta = \pi$ and multiply both sides by $-i$. Referring to the definitions of $f_1, f_2$, and $f_3$ in (1.46)–(1.48), we deduce that

$$f_1(a+b) = \frac{(if_1'(a))(if_1(b)) - (if_1(a))(if_1'(b))}{\wp(a) - \wp(b)}.$$

The derivative of $f_1$ is given by Theorem 1.36, and from Theorem 1.31 we have

$$\wp(a) - \wp(b) = f_1^2(a) - f_1^2(b).$$

Hence, we obtain (1.83). In a similar way, letting $\theta = \pi\tau$ and $\theta = \pi + \pi\tau$ in (1.86), we obtain (1.84) and (1.85), respectively. □

In order to obtain the addition formula for the Weierstrass elliptic function, we first prove a result of K. Venkatachaliengar that is interesting in its own right.

**Theorem 1.42 (Venkatachaliengar's theorem).** *Suppose $x$, $y$, and $z$ are complex numbers that satisfy $xyz = 1$. Let $\rho_1$ and $\rho_2$ be defined by* (1.49) *and* (1.50). *Then*

$$\rho_1(x)\rho_1(y) + \rho_1(y)\rho_1(z) + \rho_1(z)\rho_1(x) = \rho_2(x) + \rho_2(y) + \rho_2(z).$$

*Proof.* Under the change of variable $t = e^v$, the identity in Theorem 1.16 becomes

$$F(x, e^v)F(y, e^v) = \frac{\partial}{\partial v}F(xy, e^v) + F(xy, e^v)(\rho_1(x) + \rho_1(y)).$$

We expand both sides in powers of $v$ using (1.51) to obtain

$$\left(-\frac{1}{v} + \sum_{m=0}^{\infty}\rho_{m+1}(x)\frac{v^m}{m!}\right)\left(-\frac{1}{v} + \sum_{m=0}^{\infty}\rho_{m+1}(y)\frac{v^m}{m!}\right)$$
$$= \left(\frac{1}{v^2} + \sum_{m=0}^{\infty}\rho_{m+2}(xy)\frac{v^m}{m!}\right) + \left(-\frac{1}{v} + \sum_{m=0}^{\infty}\rho_{m+1}(xy)\frac{v^m}{m!}\right)(\rho_1(x) + \rho_1(y)).$$

Equating the coefficients of $v^0$, we deduce

$$\rho_1(x)\rho_1(y) = \rho_2(x) + \rho_2(y) + \rho_2(xy) + \rho_1(xy)\left(\rho_1(x) + \rho_1(y)\right). \tag{1.87}$$

From (1.49) and (1.50) we may deduce that

$$\rho_1(xy) = -\rho_1(x^{-1}y^{-1}) \qquad \text{and} \qquad \rho_2(xy) = \rho_2(x^{-1}y^{-1}).$$

The required identity now follows on putting $z = x^{-1}y^{-1}$ in (1.87). □

The identity in Venkatachaliengar's theorem can be rephrased in terms of Fourier series.

**Corollary 1.43.** *If $\alpha + \beta + \gamma = 0$, then*

$$f_0(\alpha)f_0(\beta) + f_0(\beta)f_0(\gamma) + f_0(\gamma)f_0(\alpha) = -\left(g_0(\alpha) + g_0(\beta) + g_0(\gamma)\right),$$

*where*

$$f_0(\theta) = -i\rho_1(e^{i\theta}) = \frac{1}{2}\cot\frac{\theta}{2} + 2\sum_{n=1}^{\infty}\frac{q^n}{1-q^n}\sin n\theta$$

*and*

$$g_0(\theta) = \rho_2(e^{i\theta}) = -\frac{1}{12} + 2\sum_{j=1}^{\infty}\frac{q^j}{(1-q^j)^2}\cos j\theta.$$

*Proof.* This is just a restatement of Theorem 1.42. The functions $f_0$ and $g_0$ are defined in terms of $\rho_1$ and $\rho_2$ by (1.52) and (1.54), respectively. Their Fourier expansions are given by (1.55) and (1.69). □

**Theorem 1.44.** *The Weierstrass elliptic function satisfies the addition theorem*

$$\wp(a+b) = \frac{1}{4}\left(\frac{\wp'(a)-\wp'(b)}{\wp(a)-\wp(b)}\right)^2 - \wp(a) - \wp(b). \tag{1.88}$$

*Alternatively, if $a + b + c = 0$, then*

$$\det\begin{pmatrix} 1 & 1 & 1 \\ \wp(a) & \wp(b) & \wp(c) \\ \wp'(a) & \wp'(b) & \wp'(c) \end{pmatrix} = 0. \tag{1.89}$$

*Proof.* From (1.73) we have

$$f_0^2(\theta) + f_0'(\theta) = 2g_0(\theta) - \frac{P(q)}{12}. \tag{1.90}$$

Substitute $a$, $b$, and $c$ for $\theta$ in (1.90) and add to obtain

$$f_0^2(a)+f_0^2(b)+f_0^2(c)+f_0'(a)+f_0'(b)+f_0'(c)$$
$$=2\left(g_0(a)+g_0(b)+g_0(c)\right)-\frac{P(q)}{4}, \tag{1.91}$$

for all values of $a$, $b$, and $c$.

On the other hand, if $a+b+c=0$, then from Corollary 1.43 we have

$$f_0(a)f_0(b)+f_0(b)f_0(c)+f_0(c)f_0(a)=-\left(g_0(a)+g_0(b)+g_0(c)\right). \tag{1.92}$$

For the remainder of the proof we assume that $a+b+c=0$. Multiply (1.92) by 2 and add the result to (1.91) to obtain

$$\left(f_0(a)+f_0(b)+f_0(c)\right)^2+f_0'(a)+f_0'(b)+f_0'(c)=-\frac{P(q)}{4}. \tag{1.93}$$

Apply the differential operator $\partial/\partial a-\partial/\partial b$ to (1.93) to obtain

$$2\left(f_0(a)+f_0(b)+f_0(c)\right)\left(f_0'(a)-f_0'(b)\right)+f_0''(a)-f_0''(b)=0, \tag{1.94}$$

or equivalently,

$$f_0(a)+f_0(b)+f_0(c)=-\frac{1}{2}\left(\frac{f_0''(a)-f_0''(b)}{f_0'(a)-f_0'(b)}\right). \tag{1.95}$$

On substituting (1.95) into (1.93) we obtain

$$\frac{1}{4}\left(\frac{f_0''(a)-f_0''(b)}{f_0'(a)-f_0'(b)}\right)^2+f_0'(a)+f_0'(b)+f_0'(c)=-\frac{P(q)}{4}.$$

This may be expressed in terms of the Weierstrass elliptic function using (1.60), to give

$$\frac{1}{4}\left(\frac{\wp'(a)-\wp'(b)}{\wp(a)-\wp(b)}\right)^2-\wp(a)-\wp(b)-\wp(c)=0,$$

where $a+b+c=0$. Since the Weierstrass elliptic function is an even function, this is equivalent to (1.88).

The alternative form of the addition theorem can be obtained by applying the operator $\partial^2/\partial a\partial b$ to (1.94), to obtain

$$\det\begin{pmatrix} 1 & 1 & 1 \\ f_0'(a) & f_0'(b) & f_0'(c) \\ f_0''(a) & f_0''(b) & f_0''(c) \end{pmatrix} = 0,$$

which holds provided $a + b + c = 0$. This may be expressed in terms of the Weierstrass elliptic function using (1.60), and the result simplifies to (1.89). □

The identity (1.93) is equivalent to the identity of Frobenius and Stickelberger in Theorem 1.11. This gives another proof of Theorem 1.11.

## 1.10 Dixon's elliptic functions

We give a brief introduction to an interesting but not-so-well-known class of elliptic functions of order three, originally studied by A. C. Dixon in 1890. Let $\mathrm{sm}^*(\theta, \tau)$ and $\mathrm{cm}^*(\theta, \tau)$ be defined by

$$\mathrm{sm}^*(\theta) = \mathrm{sm}^*(\theta, \tau) = \frac{s(\theta, \tau)}{s(\theta + \frac{2\pi}{3}, \tau)} \tag{1.96}$$

and

$$\mathrm{cm}^*(\theta) = \mathrm{cm}^*(\theta, \tau) = \frac{s(\frac{2\pi}{3} - \theta, \tau)}{s(\theta + \frac{2\pi}{3}, \tau)}. \tag{1.97}$$

We follow the usual convention and abbreviate the notation to $\mathrm{sm}^*(\theta)$ and $\mathrm{cm}^*(\theta)$ when the value of $\tau$ does not need to be emphasized. The reason for the asterisks is that later it will be useful to rescale the variable and define new functions sm and cm by

$$\mathrm{sm}\,\theta = \mathrm{sm}^*(k\theta) \quad \text{and} \quad \mathrm{cm}\,\theta = \mathrm{cm}^*(k\theta)$$

where $k$ is a particular constant that depends only on $\tau$ and not on $\theta$. At this stage it is not clear that any rescaling is preferable, so we will work with $\mathrm{sm}^*$ and $\mathrm{cm}^*$ until a better option becomes apparent.

The periodicity properties of $\mathrm{sm}^*$ and $\mathrm{cm}^*$ can be obtained from the corresponding properties for $s(\theta)$.

**Theorem 1.45.** *The functions* $\mathrm{sm}^*$ *and* $\mathrm{cm}^*$ *are elliptic functions of order three with periods* $2\pi$ *and* $6\pi\tau$. *The zeros and poles of each function are all simple and are located as follows*:

$$\begin{array}{ll} \text{zeros of } \mathrm{sm}^*(\theta): & \theta = 2\pi m + 2\pi n\tau, \\ \text{zeros of } \mathrm{cm}^*(\theta): & \theta = 2\pi(m + \frac{1}{3}) + 2\pi n\tau, \\ \text{poles of } \mathrm{sm}^*(\theta) \text{ and } \mathrm{cm}^*(\theta): & \theta = 2\pi(m - \frac{1}{3}) + 2\pi n\tau \end{array}$$

*where* $m, n \in \mathbb{Z}$.

*Proof.* The locations and orders of the zeros and poles are immediate from the definitions of sm* and cm* together with the fact that the zeros of the entire function $s(\theta)$ are all simple and located at $\theta = 2\pi m + 2\pi n\tau$. By the properties for $s(\theta)$ in (1.2) and (1.3), namely

$$s(\theta + 2\pi) = -s(\theta) \qquad \text{and} \qquad s(\theta + 2\pi\tau) = -e^{-i\theta} q^{-1/2} s(\theta),$$

we deduce that

$$\mathrm{sm}^*(\theta + 2\pi) = \mathrm{sm}^*(\theta), \qquad \mathrm{sm}^*(\theta + 2\pi\tau) = \omega\, \mathrm{sm}^*(\theta)$$

and

$$\mathrm{cm}^*(\theta + 2\pi) = \mathrm{cm}^*(\theta), \qquad \mathrm{cm}^*(\theta + 2\pi\tau) = \omega^2\, \mathrm{cm}^*(\theta),$$

where $\omega = \exp(2\pi i/3)$. It follows that sm* and cm* are elliptic functions with periods $2\pi$ and $6\pi\tau$. A count of the number of poles in a cell shows that the order of each function is three. □

Since $s(\theta + 2\pi) = -s(\theta)$ and because $s(\theta)$ is an odd function of $\theta$, it follows from the definitions of sm* and cm* that

$$\mathrm{cm}^*\left(\frac{2\pi}{3} - \theta\right) = \frac{s(\theta)}{s\left(\frac{4\pi}{3} - \theta\right)} = \frac{s(\theta)}{s\left(\theta + \frac{2\pi}{3}\right)} = \mathrm{sm}^*(\theta).$$

In a similar way, we find that

$$\mathrm{sm}^*\left(\theta + \frac{2\pi}{3}\right) = \frac{1}{\mathrm{cm}^*(\theta)}, \qquad \mathrm{sm}^*\left(\theta + \frac{4\pi}{3}\right) = -\frac{\mathrm{cm}^*(\theta)}{\mathrm{sm}^*(\theta)},$$

and

$$\mathrm{cm}^*\left(\theta + \frac{2\pi}{3}\right) = -\frac{\mathrm{sm}^*(\theta)}{\mathrm{cm}^*(\theta)}, \qquad \mathrm{cm}^*\left(\theta + \frac{4\pi}{3}\right) = \frac{1}{\mathrm{sm}^*(\theta)}.$$

The functions sm* and cm* are neither even nor odd, but satisfy the properties

$$\mathrm{sm}^*(-\theta) = -\frac{\mathrm{sm}^*(\theta)}{\mathrm{cm}^*(\theta)} \qquad \text{and} \qquad \mathrm{cm}^*(-\theta) = \frac{1}{\mathrm{cm}^*(\theta)}.$$

The next result shows that the elliptic functions sm* and cm* satisfy a beautiful cubic identity.

**Theorem 1.46.** *The following identity holds*:

$$\left(\mathrm{sm}^*(\theta)\right)^3 + \left(\mathrm{cm}^*(\theta)\right)^3 + 3\gamma \left(\mathrm{sm}^*(\theta)\right)\left(\mathrm{cm}^*(\theta)\right) = 1,$$

*where $\gamma$ is independent of $\theta$ and is given by the formula*

$$\gamma = \left(1 + 6\sum_{j=1}^{\infty} \chi_3(j)\frac{q^j}{1-q^j}\right) \times \prod_{j=1}^{\infty} \frac{(1-q^{3j})}{(1-q^j)^3}. \tag{1.98}$$

Before proving the theorem we need to establish two lemmas.

**Lemma 1.47.** *The following function evaluations hold:*

$$s\left(\frac{2\pi}{3}\right) = s\left(\frac{4\pi}{3}\right) = \sqrt{3}\prod_{j=1}^{\infty} \frac{(1-q^{3j})}{(1-q^j)^3} \tag{1.99}$$

*and*

$$z\left(\frac{2\pi}{3}\right) = \frac{1}{2\sqrt{3}}\left(1 + 6\sum_{j=1}^{\infty} \chi_3(j)\frac{q^j}{1-q^j}\right). \tag{1.100}$$

*Proof.* By the definition of $s(\theta)$ as an infinite product in (1.1) we have

$$\begin{aligned} s\left(\frac{2\pi}{3}\right) = s\left(\frac{4\pi}{3}\right) &= \sqrt{3}\prod_{j=1}^{\infty} \frac{(1-e^{2\pi i/3}q^j)(1-e^{-2\pi i/3}q^j)}{(1-q^j)^2} \\ &= \sqrt{3}\prod_{j=1}^{\infty} \frac{(1-q^{3j})}{(1-q^j)^3}. \end{aligned}$$

The Fourier series (1.17) implies that

$$\begin{aligned} z\left(\frac{2\pi}{3}\right) &= \frac{1}{2}\cot\frac{\pi}{3} + 2\sum_{j=1}^{\infty} \frac{q^j}{1-q^j}\sin\frac{2\pi j}{3} \\ &= \frac{1}{2\sqrt{3}}\left(1 + 6\sum_{j=1}^{\infty} \chi_3(j)\frac{q^j}{1-q^j}\right). \end{aligned}$$

□

The reciprocal of the infinite product in (1.99) and the series in (1.100) will be studied in detail in Chapter 3 where they will be denoted by $b(q)$ and $a(q)$, respectively. Precisely, we have

$$b = b(q) = \prod_{j=1}^{\infty} \frac{(1-q^j)^3}{(1-q^{3j})} \quad \text{and} \quad a = a(q) = 1 + 6\sum_{j=1}^{\infty} \chi_3(j)\frac{q^j}{1-q^j}. \tag{1.101}$$

See (3.28) and (3.34), respectively.

Here is the other lemma that is needed.

**Lemma 1.48.** *Consider the function h defined by*

$$h(\theta) = \frac{s^2(\theta)}{s(\theta - \frac{2\pi}{3})s(\theta + \frac{2\pi}{3})}.$$

*The following identity holds:*

$$h(\theta) + h\left(\theta + \frac{2\pi}{3}\right) + h\left(\theta + \frac{4\pi}{3}\right) = 3\gamma,$$

*where $\gamma$ is the constant independent of $\theta$ given by* (1.98).

*Proof.* By the definition of $s(\theta)$ in (1.1) and the quasi-periodicity properties (1.2) and (1.3), it follows that $h$ is an elliptic function of order two, with periods $2\pi$ and $2\pi\tau$. The singularities of $h$ are simple poles at the points $\theta = 2\pi(m \pm \frac{1}{3}) \pm 2\pi\tau n$. Let $H$ be the function defined by

$$H(\theta) = h(\theta) + h\left(\theta + \frac{2\pi}{3}\right) + h\left(\theta + \frac{4\pi}{3}\right).$$

By the periodicity properties of $h$ it follows that

$$\begin{aligned} H\left(\theta + \frac{2\pi}{3}\right) &= h\left(\theta + \frac{2\pi}{3}\right) + h\left(\theta + \frac{4\pi}{3}\right) + h\left(\theta + 2\pi\right) \\ &= h\left(\theta + \frac{2\pi}{3}\right) + h\left(\theta + \frac{4\pi}{3}\right) + h\left(\theta\right) \\ &= H(\theta), \end{aligned}$$

and clearly,

$$H(\theta + 2\pi\tau) = H(\theta).$$

That is, $H$ is periodic with periods $2\pi/3$ and $2\pi\tau$. The singularities of $H$ are poles of order at most one at each point

$$\theta = \frac{2\pi m}{3} + 2\pi n\tau, \quad m, n \in \mathbb{Z}.$$

It follows that the order of $H$ is at most one. By Theorem 1.6, the singularities are removable and therefore $H$ is a constant function. To determine the value of the constant, we compute the limiting value of $H(\theta)$ as $\theta \to 0$. Since $h(0) = 0$, we have

$$\lim_{\theta\to 0} H(\theta) = \lim_{\theta\to 0} \left\{ h(\theta) + h\left(\theta + \frac{2\pi}{3}\right) + h\left(\theta + \frac{4\pi}{3}\right) \right\}$$
$$= \lim_{\theta\to 0} \left\{ \frac{s^2(\theta + \frac{2\pi}{3})}{s(\theta)s(\theta + \frac{4\pi}{3})} + \frac{s^2(\theta + \frac{4\pi}{3})}{s(\theta + \frac{2\pi}{3})s(\theta + 2\pi)} \right\}.$$

Using the facts that $s(\theta)$ is an odd function and $s(\theta + 2\pi) = -s(\theta)$, we deduce that

$$\lim_{\theta\to 0} H(\theta) = \lim_{\theta\to 0} \frac{s^3(\theta + \frac{2\pi}{3}) - s^3(\frac{2\pi}{3} - \theta)}{s(\theta)} \times \frac{1}{s(\theta + \frac{2\pi}{3})s(\theta + \frac{4\pi}{3})}$$
$$= \lim_{\theta\to 0} \frac{s(\theta + \frac{2\pi}{3}) - s(\frac{2\pi}{3} - \theta)}{s(\theta)}$$
$$\times \frac{s^2(\theta + \frac{2\pi}{3}) + s(\theta + \frac{2\pi}{3})s(\frac{2\pi}{3} - \theta) + s^2(\frac{2\pi}{3} - \theta)}{s(\theta + \frac{2\pi}{3})s(\theta + \frac{4\pi}{3})}.$$

Evaluating the limits, we obtain

$$\lim_{\theta\to 0} H(\theta) = 2s'\left(\frac{2\pi}{3}\right) \times 3\frac{s(\frac{2\pi}{3})}{s(\frac{4\pi}{3})} = 6s'\left(\frac{2\pi}{3}\right), \tag{1.102}$$

where (1.99) has been used in the last step. This confirms that the singularity of $H(\theta)$ at $\theta = 0$ is removable. The right-hand side of (1.102) can made explicit by using the definition of $z(\theta)$ in (1.4) and the evaluations in (1.99) and (1.100) to obtain

$$\lim_{\theta\to 0} H(\theta) = 6\, s\left(\frac{2\pi}{3}\right) z\left(\frac{2\pi}{3}\right) = 3\left(1 + 6\sum_{j=1}^{\infty} \chi_3(j)\frac{q^j}{1-q^j}\right) \times \prod_{j=1}^{\infty} \frac{(1-q^{3j})}{(1-q^j)^3} = 3\gamma.$$

Since we have already determined that $H(\theta)$ is a constant function, this completes the proof. □

We are now ready for the proof of Theorem 1.46.

*Proof.* The identity in Lemma 1.48 is

$$\frac{s^2(\theta)}{s(\theta - \frac{2\pi}{3})s(\theta + \frac{2\pi}{3})} + \frac{s^2(\theta + \frac{2\pi}{3})}{s(\theta)s(\theta + \frac{4\pi}{3})} + \frac{s^2(\theta + \frac{4\pi}{3})}{s(\theta + \frac{2\pi}{3})s(\theta + 2\pi)} = 3\gamma.$$

Replace $\theta$ with $-\theta$ and simplify, using the properties

$$s(\theta + 2\pi) = -s(\theta) \quad \text{and} \quad s(-\theta) = -s(\theta),$$

to obtain

$$\frac{s^2(\theta)}{s(\theta - \frac{2\pi}{3})s(\theta + \frac{2\pi}{3})} - \frac{s^2(\theta - \frac{2\pi}{3})}{s(\theta)s(\theta + \frac{2\pi}{3})} - \frac{s^2(\theta + \frac{2\pi}{3})}{s(\theta - \frac{2\pi}{3})s(\theta)} = 3\gamma.$$

Now multiply by $s(\theta)s(\theta - \frac{2\pi}{3})/s^2(\theta + \frac{2\pi}{3})$, rearrange, and again use the fact that $s$ is an odd function to obtain

$$\frac{s^3(\theta)}{s^3(\theta + \frac{2\pi}{3})} + \frac{s^3(\frac{2\pi}{3} - \theta)}{s^3(\theta + \frac{2\pi}{3})} + 3\gamma\frac{s(\theta)s(\frac{2\pi}{3} - \theta)}{s^2(\theta + \frac{2\pi}{3})} = 1.$$

This is equivalent to the required identity. □

The elliptic functions sm* and cm* can be expressed in terms of the function $F(x, y)$ defined by (0.99). This will allow us to deduce other properties of the functions, including differentiation formulas and Fourier expansions.

**Lemma 1.49.** *Let $F(x, y)$ be defined by* (0.99), *that is,*

$$F(x, y) = \prod_{n=1}^{\infty} \frac{(1 - xyq^{n-1})(1 - x^{-1}y^{-1}q^n)(1 - q^n)^2}{(1 - xq^{n-1})(1 - x^{-1}q^n)(1 - yq^{n-1})(1 - y^{-1}q^n)},$$

*and let* sm* *and* cm* *be as defined at the beginning of this section. Let* $\omega = \exp(2\pi i/3)$ *and let $b(q)$ be as in* (1.101). *Then*

$$\mathrm{sm}^*(\theta) = \frac{i\sqrt{3}}{b(q)} F(\omega^2, \omega e^{i\theta})$$

*and*

$$\mathrm{cm}^*(\theta) = -\frac{i\sqrt{3}}{b(q)} F(\omega, \omega e^{i\theta}).$$

*Proof.* This is simply an exercise in converting from one notation to another, using the definitions of sm* and cm* at the beginning of this section, the definition of $s$ in (1.1), and the definition of $F$ as given above. □

Ramanujan's summation formula can be applied to the results of Lemma 1.49 to produce Fourier expansions for sm* and cm*. See Exercise 17 at the end of the chapter. The derivatives of sm* and cm* may be computed, as follows.

**Theorem 1.50.** *Let $a$ and $b$ be as in* (1.101). *The following differentiation formulas hold*:

$$\sqrt{3}\,\frac{\mathrm{d}}{\mathrm{d}\theta}\,\mathrm{sm}^*(\theta) = b\,(\mathrm{cm}^*(\theta))^2 + a\ \mathrm{sm}^*(\theta)$$

*and*

$$\sqrt{3}\,\frac{\mathrm{d}}{\mathrm{d}\theta}\,\mathrm{cm}^*(\theta) = -b\,(\mathrm{sm}^*(\theta))^2 - a\ \mathrm{cm}^*(\theta).$$

*Proof.* Take $x = y = \omega$ and $t = \omega e^{i\theta}$, where $\omega = e^{2\pi i/3}$, in Theorem 1.16 to obtain

$$F(\omega, \omega e^{i\theta})^2 = t\frac{\partial}{\partial t}F(\omega^2, \omega e^{i\theta}) + 2\rho_1(\omega)F(\omega^2, \omega e^{i\theta}). \tag{1.103}$$

By the chain rule we have

$$t\frac{\partial}{\partial t} = -i\frac{\mathrm{d}}{\mathrm{d}\theta}$$

while by (1.21) and (1.100) we have

$$\rho_1(\omega) = \rho_1\left(e^{2\pi i/3}\right) = iz\left(\frac{2\pi}{3}\right) = \frac{i}{2\sqrt{3}}a(q).$$

Hence, Lemma 1.49 can be used to express each side of (1.103) in terms of sm* and cm*, to obtain

$$\left(\frac{ib(q)}{\sqrt{3}}\ \mathrm{cm}^*(\theta)\right)^2 = -i\frac{\mathrm{d}}{\mathrm{d}\theta}\left(\frac{-ib(q)}{\sqrt{3}}\ \mathrm{sm}^*(\theta)\right) + \left(\frac{ia(q)}{\sqrt{3}}\right)\left(\frac{-ib(q)}{\sqrt{3}}\ \mathrm{sm}^*(\theta)\right).$$

This simplifies to

$$\sqrt{3}\,\frac{\mathrm{d}}{\mathrm{d}\theta}\,\mathrm{sm}^*(\theta) = b\,(\mathrm{cm}^*(\theta))^2 + a\ \mathrm{sm}^*(\theta).$$

The other result can be proved in the same way by taking $x = y = \omega^2$ and $t = \omega e^{i\theta}$ in Theorem 1.16. □

The differentiation formulas in Theorem 1.50 can be expressed in a simpler form by scaling the variable $\theta$. Specifically, consider the functions sm and cm defined by

$$\mathrm{sm}\,\theta = \mathrm{sm}^*\left(\frac{\sqrt{3}\,\theta}{b}\right) \quad \text{and} \quad \mathrm{cm}\,\theta = \mathrm{cm}^*\left(\frac{\sqrt{3}\,\theta}{b}\right), \tag{1.104}$$

where $b$ is given by (1.101), viz.,

$$b = b(q) = \prod_{j=1}^{\infty} \frac{(1-q^j)^3}{(1-q^{3j})}.$$

The functions $\operatorname{sm}\theta$ and $\operatorname{cm}\theta$ are called Dixon's elliptic functions.

**Theorem 1.51.** *The basic properties of the Dixon elliptic functions are as follows.*

1. *The functions* $\operatorname{sm}\theta$ *and* $\operatorname{cm}\theta$ *are elliptic functions of order three with periods* $3\lambda_1 = 2\pi b/\sqrt{3}$ *and* $3\lambda_2 = 6\pi\tau b/\sqrt{3}$*, where* $b$ *is as above.*
2. *We have* $\operatorname{sm}0 = 0$, $\operatorname{cm}0 = 1$ *and* $\operatorname{cm}(\lambda_1 - \theta) = \operatorname{sm}\theta$.
3. *The following differentiation formulas hold*:

$$\frac{\mathrm{d}}{\mathrm{d}\theta}\operatorname{sm}\theta = \operatorname{cm}^2\theta + \gamma\operatorname{sm}\theta$$

*and*

$$\frac{\mathrm{d}}{\mathrm{d}\theta}\operatorname{cm}\theta = -\operatorname{sm}^2\theta - \gamma\operatorname{cm}\theta$$

*where* $\gamma$ *is given by* (1.98)*, that is,*

$$\gamma = \left(1 + 6\sum_{j=1}^{\infty}\chi_3(j)\frac{q^j}{1-q^j}\right) \times \prod_{j=1}^{\infty}\frac{(1-q^{3j})}{(1-q^j)^3}.$$

4. *The cubic identity holds*:

$$\operatorname{sm}^3\theta + \operatorname{cm}^3\theta + 3\gamma\operatorname{sm}\theta\operatorname{cm}\theta = 1.$$

*Proof.* These are immediate from the corresponding properties for sm* and cm*. □

The differentiation formulas in the previous theorem can be used to compute the first few terms of the Taylor expansions about $\theta = 0$. We find that

$$\operatorname{sm}\theta = \theta - \gamma\frac{\theta^2}{2!} + 3\gamma^2\frac{\theta^3}{3!} - (4+5\gamma^3)\frac{\theta^4}{4!} + (28\gamma + 11\gamma^4)\frac{\theta^5}{5!} + \cdots \tag{1.105}$$

and

$$\operatorname{cm}\theta = 1 - \gamma\theta + \gamma^2\frac{\theta^2}{2!} - (2+\gamma^3)\frac{\theta^3}{3!} + (8\gamma+\gamma^4)\frac{\theta^4}{4!} - (38\gamma^2+\gamma^5)\frac{\theta^5}{5!} + \cdots . \tag{1.106}$$

## 1.11 Dixon's elliptic functions: special case $\gamma = 1$

Let us consider Dixon's elliptic functions in the case $q = 0$. Then $\gamma = 1$ by (1.98), and the cubic identity in Theorem 1.51 factorizes as

$$(\operatorname{sm}\theta + \operatorname{cm}\theta - 1)(\operatorname{sm}\theta + \omega\operatorname{cm}\theta - \omega^2)(\operatorname{sm}\theta + \omega\operatorname{cm}\theta - \omega^2) = 0,$$

where $\omega = \exp(2\pi i/3)$. Since $\operatorname{sm} 0 = 0$ and $\operatorname{cm} 0 = 1$ it follows that

$$\operatorname{sm}\theta + \operatorname{cm}\theta = 1.$$

Moreover, when $q = 0$ we have $s(\theta) = 2\sin\frac{\theta}{2}$ and $a = b = 1$. Therefore, the Dixon functions reduce to the singly periodic meromorphic functions

$$\operatorname{sm}\theta = \frac{\sin\left(\frac{\sqrt{3}}{2}\theta\right)}{\sin\left(\frac{\sqrt{3}}{2}\theta + \frac{\pi}{3}\right)}, \qquad \operatorname{cm}\theta = \frac{\sin\left(\frac{\pi}{3} - \frac{\sqrt{3}}{2}\theta\right)}{\sin\left(\frac{\sqrt{3}}{2}\theta + \frac{\pi}{3}\right)}.$$

The functions $\operatorname{sm}\theta$ and $\operatorname{cm}\theta$ are no longer elliptic functions because the value $q = 0$ occurs as the limiting case $\operatorname{Im}\tau \to \infty$. This corresponds to one of the two periods of the elliptic function becoming infinite.

## 1.12 Dixon's elliptic functions: special case $\gamma = 0$

The case $\gamma = 0$ is much more interesting, as we will now demonstrate. As preparation, we will need the following lemma.

**Lemma 1.52.** *The functions $a$ and $b$ defined by* (1.101) *have the special values*

$$a\left(-e^{-\pi/\sqrt{3}}\right) = 0 \quad \textit{and} \quad b\left(-e^{-\pi/\sqrt{3}}\right) = \frac{\sqrt{3}}{2\pi} \times \frac{\Gamma(1/3)^2}{\Gamma(2/3)}.$$

*Proof.* We defer a proof of this result to Exercise 6 at the end of Chapter 4. The Dixon elliptic functions are not used elsewhere in this book, so there is no circular reasoning. □

Let

$$\tau = -\frac{1}{2} + \frac{i\sqrt{3}}{6}$$

so that the corresponding value of $q$ is given by

$$q = e^{2\pi i\tau} = -e^{-\pi/\sqrt{3}}.$$

The values of $a$ and $b$ are given by Lemma 1.52, and by (1.98) and (1.101) we have $\gamma = a/b = 0$. For the remainder of this section, let

$$\operatorname{sm}\theta = \operatorname{sm}\left(\theta, -\frac{1}{2} + \frac{i\sqrt{3}}{6}\right),$$

that is, consider the Dixon elliptic function $\operatorname{sm}\theta$ that corresponds to the parameter values

$$\tau = -\frac{1}{2} + \frac{i\sqrt{3}}{6}, \quad q = -e^{-\pi/\sqrt{3}}, \quad a = 0, \quad b = \frac{\sqrt{3}}{2\pi} \times \frac{\Gamma(1/3)^2}{\Gamma(2/3)} \quad \text{and} \quad \gamma = 0.$$

By Theorem 1.51 the periods of $\operatorname{sm}\theta$ are $3\lambda$ and $3\mu$ where

$$(3\lambda,\ 3\mu) = \left(\frac{2\pi b}{\sqrt{3}}, \frac{6\pi\tau b}{\sqrt{3}}\right) = \frac{\Gamma(1/3)^2}{\Gamma(2/3)}\left(1, -\frac{3}{2} + \frac{i\sqrt{3}}{2}\right).$$

We may also take $(3\lambda, 3\lambda + 3\mu)$ as a fundamental pair of periods. Moreover, $(3\lambda, -6\lambda - 3\mu)$ is another fundamental pair of periods. Since

$$3\lambda + 3\mu = \frac{2\pi b}{\sqrt{3}}\left(-\frac{1}{2} + \frac{i\sqrt{3}}{2}\right) = 3\omega\lambda$$

and

$$-6\lambda - 3\mu = \frac{2\pi b}{\sqrt{3}}\left(-\frac{1}{2} - \frac{i\sqrt{3}}{2}\right) = 3\omega^2\lambda$$

where $\omega = \exp(2\pi i/3)$, it follows that $3\lambda$, $3\omega\lambda$, and $3\omega^2\lambda$ are all periods of $\operatorname{sm}\theta$, where $3\lambda = \Gamma(1/3)^2/\Gamma(2/3)$.

A diagram showing three different period parallelograms is given in Figure 1.1. The figure also shows that it is possible to take a regular hexagon as a fundamental periodic region.

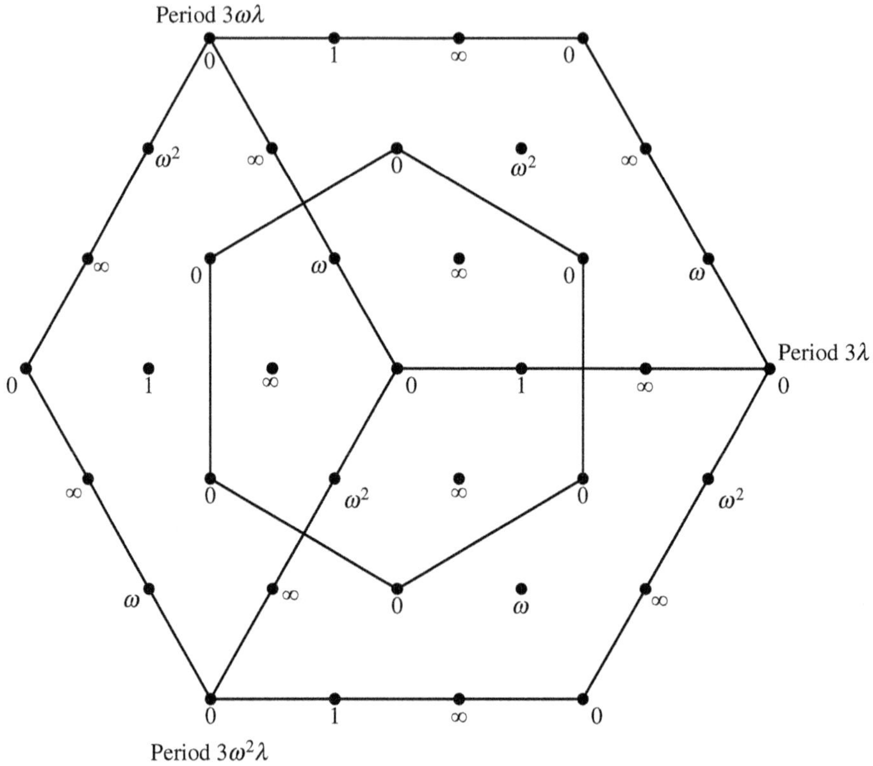


**Fig. 1.1** Values of the Dixon elliptic function $\operatorname{sm}\theta$ with $\gamma = 0$ at the one-third periods, in three period parallelograms generated by $3\lambda$, $3\omega\lambda$, and $3\omega^2\lambda$, where $3\lambda = \Gamma^2(1/3)/\Gamma(2/3)$ and $\omega = \exp(2\pi i/3)$. Also shown is a fundamental period hexagon.

Finally, it may be noted that the differentiation formulas and cubic identity in Theorem 1.51 simplify when $\gamma = 0$ to give

$$\frac{\mathrm{d}}{\mathrm{d}\theta}\operatorname{sm}\theta = \operatorname{cm}^2\theta, \qquad \frac{\mathrm{d}}{\mathrm{d}\theta}\operatorname{cm}\theta = -\operatorname{sm}^2\theta$$

and

$$\operatorname{sm}^3\theta + \operatorname{cm}^3\theta = 1,$$

respectively. Furthermore, the identities

$$\operatorname{sm}(\omega\theta) = \omega\operatorname{sm}\theta, \quad \operatorname{sm}(\omega^2\theta) = \omega^2\operatorname{sm}\theta$$

and

$$\operatorname{cm}\theta = \operatorname{cm}(\omega\theta) = \operatorname{cm}(\omega^2\theta)$$

hold when $\gamma = 0$. Since $\operatorname{sm}\theta$ and $\operatorname{cm}\theta$ are real valued when $\theta$ is real, it follows that the Taylor expansions of $\operatorname{sm}\theta$ and $\operatorname{cm}\theta$ about the origin contain only terms of the form $\theta^{3n+1}$ and $\theta^{3n}$, respectively. The first few terms in the expansions are given by taking $\gamma = 0$ in (1.105) and (1.106).

## 1.13 Notes

The identities named after Weierstrass in Theorems 1.7 and 1.8, as well as the identity in Theorem 1.10, were given in lectures by Weierstrass that have been written up by Schwarz [265, pp. 13, 14, 47]. An interesting discussion of the Weierstrass three-term identity, including its connection with a five-term identity of Jacobi as well as its applications in other branches of mathematics, has been given by T. Koornwinder in [199]. Applications of the Weierstrass three-term identity to combinatorics are discussed by M. Schlosser [262]. Halphen's identity in Theorem 1.9 appears in [173, p. 187]. Theorem 1.11 was given by Frobenius and Stickelberger, [156, Eq. (9)].

The explicit form of Halphen's identity in Theorem 1.13 was given by G. Andrews, R. Lewis and Z.-G. Liu, [28]. Another attractive explicit form of Weierstrass' identity is due to W. Bailey [36]; see Exercise 2, below. The formulation of the identity in Theorem 1.16 is due to K. Venkatachaliengar [288, p. 43]. The identity in Theorem 1.20 has been given by many authors; this particular form together with a different proof was given by W. Chu and L. Di Claudio [97, Theorem G5.2]. Theorem 1.21 was given by L. Carlitz [69].

A powerful multivariate generalization of Theorem 1.13 has been given by S. H. Chan [95]. To state the result, let

$$[x_1, \ldots, x_m; q]_\infty = (x_1, q/x_1, \ldots, x_m, q/x_m; q)_\infty$$

and define

$$\begin{aligned} &F(b_1, b_2, \ldots, b_m) + \text{idem}\,(b_1; b_2, \ldots, b_m) \\ &= F(b_1, b_2, \ldots, b_m) + F(b_2, b_1, b_3, \ldots, b_m) + \cdots + F(b_m, b_2, \ldots, b_{m-1}, b_1). \end{aligned}$$

Then, for nonnegative integers $r < s$, we have

$$\frac{[a_1, \ldots, a_r; q]_\infty (q;q)_\infty^2}{[b_1, \ldots, b_s; q]_\infty}$$

$$= \frac{[a_1/b_1, \ldots, a_r/b_1; q]_\infty}{[b_2/b_1, \ldots, b_s/b_1; q]_\infty} \sum_{k=-\infty}^{\infty} \frac{(-1)^{(s-r)k} q^{(s-r)k(k+1)/2}}{1 - b_1 q^k} \left( \frac{a_1 \cdots a_r b_1^{s-r-1}}{b_2 \cdots b_s} \right)^k$$

$$+ \text{ idem } (b_1; b_2, \ldots, b_s).$$

For $r = s$, the result holds provided that $|q| < \left|\frac{a_1 \cdots a_r}{b_1 \cdots b_r}\right| < 1$. The proof in [95] is by partial fractions and uses the Mittag-Leffler theorem. Chan showed [95, Corollary 3.1] that the case $r = s = 1$ is equivalent to Theorem 0.26. Moreover [95, Corollary 3.2], Theorem 1.13 may be obtained by specializing the case $r = s = 2$ and taking $(a_1, a_2, b_1, b_2) = (xy, xz, x, xyz)$.

Example 1.25 was recorded by Ramanujan in his notebooks [253, Chapter 19, Entry 17]; see also the analysis by Berndt [46, pp. 302–304]. Example 1.26 has been studied by Ramanujan [253, Chapter 19, Entry 3], Fine [151, (32.66), (33.2)] and Berndt [46, pp. 223–225].

The identities (1.68) and (1.80) are in Ramanujan's paper [249]. Ramanujan gave a direct proof of (1.68) using only series manipulations and noted that similar techniques may be used to prove the other identity (1.80). Hardy liked Ramanujan's proof of (1.68) so much that he reproduced it in his book on number theory with Wright [175, p. 312] and also in his lectures [174, pp. 133–135]. Ramanujan's proof is outlined below in Exercise 4. Another proof, due to van der Pol, is given in Exercise 3. Generalizations of (1.68) have been considered by V. Ramamani [246], K. Venkatachaliengar [288, pp. 4–7] and by A. Alaca, Ş. Alaca and K. S. Williams [13].

Corollary 1.40 is classical. For example, see the notes from lectures of Weierstrass written up and published by Schwarz [265, pp. 10, 11]. A different approach has been given by N.-P. Skoruppa [270].

The methods used in Sections 1.8 and 1.9 are due to K. Venkatachaliengar [288]. The functions $f_1$, $f_2$, and $f_3$ are rescalings of the Jacobian elliptic functions cs, ns and ds, respectively. The precise identifications are given in Exercises 15 and 16 below.

The Dixon elliptic functions sm and cm are named after A. C. Dixon who first studied them in [138]. Dixon's parameter $\alpha$ is the negative of our $\gamma$, i.e., $\alpha = -\gamma$. Dixon's approach uses elliptic integrals and is quite different from the development we have given. The nome $q$ is not mentioned in either Dixon's paper [138] or his book [139]. In contrast to Dixon, we have made no use of elliptic integrals and instead rely on properties of $q$-infinite products and their expansions as series. Our calculation of the derivatives of the Dixon functions in Theorem 1.50 is an adaptation of Venkatachaliengar's method for the Jacobian functions, as given in Theorem 1.36. Representations for the Dixon functions as quotients of theta functions were given by Y. Ohyama [240]. The proof of the cubic identity in Theorem 1.46 is from H. H. Chan, Z.-G. Liu and S. T. Ng [89]. Continued fractions related to Dixon's elliptic functions have been studied by E. Conrad [99] and connections with combinatorics have been given by Conrad and P. Flajolet [100].

## 1.14 Exercises

1. By expanding as a double series and interchanging the order of summation, show that

$$\sum_{k=1}^{\infty} \frac{kq^k}{1-q^k} = \sum_{j=1}^{\infty} \frac{q^j}{(1-q^j)^2}$$

and

$$\sum_{k=1}^{\infty} \frac{q^k}{1+q^k+q^{2k}} = \sum_{j=1}^{\infty} \chi_3(j)\frac{q^j}{1-q^j}.$$

2. Prove that Weierstrass' identity in Theorem 1.7 is equivalent to

$$a\frac{(ab, q/ab, b/a, qa/b, q, q, q, q; q)_\infty}{(a, a, q/a, q/a, b, b, q/b, q/b; q)_\infty} = \sum_{n=-\infty}^{\infty} \left( \frac{aq^n}{(1-aq^n)^2} - \frac{bq^n}{(1-bq^n)^2} \right).$$

See Bailey [36].

3. This exercise outlines a proof of Ramanujan's identity (1.68) that is due to B. van der Pol [284]. For $\operatorname{Re}(t) > 0$, let $q = e^{-t}$ and define $f = f(\theta, t)$ by

$$f(\theta, t) = q^{1/8} \sin\frac{\theta}{2} \prod_{j=1}^{\infty} (1 - e^{i\theta}q^j)(1 - e^{-i\theta}q^j)(1 - q^j). \tag{1.107}$$

a. Use Jacobi's triple product identity (0.17) to show that

$$f(\theta, t) = \sum_{n=-\infty}^{\infty} q^{(4n+1)^2/8} \sin\frac{(4n+1)}{2}\theta.$$

b. Use the result of Part a to show that

$$2\frac{\partial f}{\partial t} = \frac{\partial^2 f}{\partial \theta^2}.$$

c. Let $g(\theta, t) = \log f(\theta, t)$. Use the result of Part b to prove that

$$\left(\frac{\partial g}{\partial \theta}\right)^2 = 2\frac{\partial g}{\partial t} - \frac{\partial^2 g}{\partial \theta^2}.$$

d. Use (1.107) to obtain the Fourier expansions

$$\frac{\partial g}{\partial \theta} = \frac{1}{2}\cot\frac{\theta}{2} + 2\sum_{k=1}^{\infty}\frac{q^k}{1-q^k}\sin k\theta,$$

$$\frac{\partial^2 g}{\partial \theta^2} = -\frac{1}{4} - \frac{1}{4}\cot^2\frac{\theta}{2} + 2\sum_{k=1}^{\infty}\frac{kq^k}{1-q^k}\cos k\theta$$

and

$$\frac{\partial g}{\partial t} = -\frac{1}{8} + 2\sum_{k=1}^{\infty}\frac{q^k}{(1-q^k)^2}\cos k\theta + \sum_{k=1}^{\infty}\frac{kq^k}{1-q^k}.$$

e. Hence deduce Ramanujan's identity (1.68), viz.,

$$\left(\frac{1}{2}\cot\frac{\theta}{2} + 2\sum_{k=1}^{\infty}\frac{q^k}{1-q^k}\sin k\theta\right)^2$$
$$= \frac{1}{4}\cot^2\frac{\theta}{2} + 2\sum_{k=1}^{\infty}\frac{kq^k}{1-q^k}(1-\cos k\theta) + 4\sum_{k=1}^{\infty}\frac{q^k}{(1-q^k)^2}\cos k\theta.$$

4. This exercise outlines Ramanujan's proof of (1.68), from [249].

a. Show that the function

$$\left(\frac{1}{2}\cot\frac{\theta}{2} + 2\sum_{k=1}^{\infty}\frac{q^k}{1-q^k}\sin k\theta\right)^2 - \frac{1}{4}\cot^2\frac{\theta}{2}$$

is analytic in the strip $-2\pi\tau < \operatorname{Im}\theta < 2\pi\tau$. That is, show that the singularities on the real axis are removable. Further, show that the function is even and periodic with period $2\pi$. It follows that a Fourier cosine expansion of the form

$$\left(\frac{1}{2}\cot\frac{\theta}{2} + 2\sum_{k=1}^{\infty}\frac{q^k}{1-q^k}\sin k\theta\right)^2 - \frac{1}{4}\cot^2\frac{\theta}{2} = \sum_{n=0}^{\infty}C_n\cos n\theta,$$

holds.

b. Show that

$$1 + \left(2\sum_{j=1}^{k-1}\cos j\theta\right) + \cos k\theta = \cot\frac{\theta}{2}\sin k\theta.$$

c. By direct calculation, and using the result of Part b, show that

$$\begin{aligned}
&\left(\frac{1}{2}\cot\frac{\theta}{2} + 2\sum_{k=1}^{\infty}\frac{q^k}{1-q^k}\sin k\theta\right)^2 - \frac{1}{4}\cot^2\frac{\theta}{2} \\
&= 2\cot\frac{\theta}{2}\sum_{k=1}^{\infty}\frac{q^k}{1-q^k}\sin k\theta + 4\left(\sum_{k=1}^{\infty}\frac{q^k}{1-q^k}\sin k\theta\right)^2 \\
&= 2\sum_{k=1}^{\infty}\frac{q^k}{1-q^k}\left(1 + \left(2\sum_{j=1}^{k-1}\cos j\theta\right) + \cos k\theta\right) \\
&\quad + 2\sum_{j=1}^{\infty}\sum_{k=1}^{\infty}\frac{q^{j+k}}{(1-q^j)(1-q^k)}\left(\cos(j-k)\theta - \cos(j+k)\theta\right).
\end{aligned}$$

d. Use Part c and the result of Exercise 1 to deduce that

$$C_0 = 2\sum_{j=1}^{\infty}\frac{q^j}{1-q^j} + 2\sum_{j=1}^{\infty}\left(\frac{q^j}{1-q^j}\right)^2 = 2\sum_{k=1}^{\infty}\frac{kq^k}{1-q^k}.$$

e. For $n \geq 1$, show that

$$\begin{aligned}
C_n = \frac{2q^n}{1-q^n} &+ 4\sum_{k=n+1}^{\infty}\frac{q^k}{1-q^k} + 4\sum_{k=1}^{\infty}\frac{q^{n+2k}}{(1-q^{n+k})(1-q^k)} \\
&- 2\sum_{k=1}^{n-1}\frac{q^n}{(1-q^k)(1-q^{n-k})}.
\end{aligned}$$

f. Show that

$$\begin{aligned}
&\sum_{k=n+1}^{\infty}\frac{q^k}{1-q^k} + \sum_{k=1}^{\infty}\frac{q^{n+2k}}{(1-q^{n+k})(1-q^k)} \\
&= \sum_{k=1}^{\infty}\left(\frac{q^{n+k}}{1-q^{n+k}} + \frac{q^{n+2k}}{(1-q^{n+k})(1-q^k)}\right) \\
&= \frac{q^n}{1-q^n}\sum_{k=1}^{\infty}\left(\frac{q^k}{1-q^k} - \frac{q^{n+k}}{1-q^{n+k}}\right) = \frac{q^n}{1-q^n}\sum_{k=1}^{n}\frac{q^k}{1-q^k}.
\end{aligned}$$

g. Show that

$$\sum_{k=1}^{n-1} \frac{q^n}{(1-q^k)(1-q^{n-k})} = \frac{q^n}{1-q^n} \sum_{k=1}^{n-1} \left(1 + \frac{q^k}{1-q^k} + \frac{q^{n-k}}{1-q^{n-k}}\right)$$
$$= \frac{(n-1)q^n}{1-q^n} + \frac{2q^n}{1-q^n} \sum_{k=1}^{n-1} \frac{q^k}{1-q^k}.$$

h. Use Parts e, f, and g to deduce that for $n \geq 1$,

$$C_n = \frac{4q^n}{(1-q^n)^2} - \frac{2nq^n}{1-q^n}.$$

i. Use the results of (a), (d), and (h) to complete Ramanujan's proof of (1.68).

5. a. Show that

$$\left(1 + 6\sum_{n=1}^{\infty} \chi_3(n) \frac{q^n}{1-q^n}\right)^2 = \frac{1}{2}\left(3P(q^3) - P(q)\right)$$

where $P$ is Ramanujan's Eisenstein series given by (0.61).
Hint: Let $\theta = 2\pi/3$ in Ramanujan's identity in Exercise 3 e.

b. Show that

$$\left(1 + 4\sum_{n=1}^{\infty} \sin\left(\frac{n\pi}{2}\right) \frac{q^n}{1-q^n}\right)^2 = \frac{1}{3}\left(4P(q^4) - P(q)\right).$$

Hint: Let $\theta = \pi/2$ in Ramanujan's identity in Exercise 3 e.
The identities in this exercise were given by Ramanujan [249].

6. Let $g_2$ and $g_3$ be any complex numbers and consider the polynomial of degree 3 defined by

$$f(x) = 4x^3 - g_2 x - g_3. \tag{1.108}$$

Suppose the roots of $f$ are $e_1$, $e_2$, and $e_3$, so that

$$f(x) = 4(x-e_1)(x-e_2)(x-e_3). \tag{1.109}$$

The discriminant of $f$ is defined to be

$$\Delta = 16(e_1-e_2)^2(e_2-e_3)^2(e_3-e_1)^2.$$

Observe that the discriminant of $f$ is zero if and only if $f$ has a repeated root. The purpose of this exercise is to find a formula for the discriminant in terms of $g_2$ and $g_3$.

a. Show that

$$e_1 + e_2 + e_3 = 0, \quad e_1e_2 + e_2e_3 + e_3e_1 = -\frac{g_2}{4} \quad \text{and} \quad e_1e_2e_3 = \frac{g_3}{4}.$$

b. Show that

$$e_1^2 + e_2^2 + e_3^2 = \frac{g_2}{2}.$$

Hint: Write $e_1^2 = e_1(-e_2 - e_3)$, and give similar expressions for $e_2^2$ and $e_3^2$.

c. Show that

$$(e_1e_2 + e_2e_3 + e_3e_1)^2 = e_1^2e_2^2 + e_2^2e_3^2 + e_3^2e_1^2.$$

d. By differentiating (1.108), show that

$$f'(e_1)f'(e_2)f'(e_3) = (12e_1^2 - g_2)(12e_2^2 - g_2)(12e_3^2 - g_2).$$

e. By differentiating (1.109), show that

$$f'(e_1)f'(e_2)f'(e_3) = -4\Delta.$$

f. Use Parts d and e to show that

$$\Delta = \frac{g_2^3}{4} - 3g_2^2(e_1^2 + e_2^2 + e_3^2) + 36g_2(e_1^2e_2^2 + e_2^2e_3^2 + e_3^2e_1^2) - 432e_1^2e_2^2e_3^2.$$

g. Use the results of Parts a, b, c, and f to show that

$$\Delta = g_2^3 - 27g_3^2.$$

7. Consider the differential equation for the Weierstrass elliptic function given by (1.77):

$$(\wp'(\theta))^2 = 4\wp^3(\theta) - \frac{Q}{12}\wp(\theta) - \frac{R}{216}. \tag{1.110}$$

a. Use the result of the previous exercise to show that the discriminant of the cubic on the right-hand side of (1.110) is

$$\Delta = \frac{1}{1728}(Q^3 - R^2).$$

b. Use the result of Part a and Corollary 0.23 to deduce that

$$\Delta = q \prod_{j=1}^{\infty} (1 - q^j)^{24}.$$

This gives another proof of Corollary 1.35, and hence another proof that the roots of the cubic in (1.110) are distinct.

8. a. Use the definitions (1.61)–(1.63) and the Fourier expansion (1.18) to show

$$e_1 = \frac{1}{6} + 4 \sum_{n=1}^{\infty} \frac{(2n-1)q^{2n-1}}{1 - q^{2n-1}},$$

$$e_2 = -\frac{1}{12} - 2 \sum_{n=1}^{\infty} \frac{(2n-1)q^{n-1/2}}{1 - q^{n-1/2}}$$

and

$$e_3 = -\frac{1}{12} + 2 \sum_{n=1}^{\infty} \frac{(2n-1)q^{n-1/2}}{1 + q^{n-1/2}}.$$

b. As $q$ increases from 0 to 1, show that:

i. $e_1$ increases monotonically from $1/6$ to $+\infty$;
ii. $e_2$ decreases monotonically from $-1/12$ to $-\infty$;
iii. $e_3$ increases monotonically from $-1/12$ to $+\infty$.

In particular, if $0 < q < 1$, then $e_1 > 0$ and $e_2 < 0$.
In Exercise 10 at the end of the next chapter, it will be shown that $e_3$ changes sign when $q = e^{-\pi}$.

9. This exercise uses the functions $f_0$ and $g_0$ whose Fourier expansions were given in Corollary 1.43, namely

$$f_0(\theta) = \frac{1}{2} \cot \frac{\theta}{2} + 2 \sum_{n=1}^{\infty} \frac{q^n}{1 - q^n} \sin n\theta$$

and

$$g_0(\theta) = -\frac{1}{12} + 2 \sum_{n=1}^{\infty} \frac{q^n}{(1 - q^n)^2} \cos n\theta.$$

a. Show that

$$g_0''(\theta) = -q \frac{\partial}{\partial q} f_0'(\theta)$$

where the primes denote partial differentiation with respect to $\theta$.

b. Let $u_1$, $u_2$ and $u_3$ be defined by

$$u_1 = -4f_0'(\pi), \quad u_1 = -4f_0'(\pi\tau) \quad \text{and} \quad u_3 = -4f_0'(\pi + \pi\tau).$$

Show that

$$u_1 = 1 - 8\sum_{n=1}^{\infty} \frac{(-1)^n nq^n}{1-q^n}, \qquad u_2 = -8\sum_{n=1}^{\infty} \frac{nq^{n/2}}{1-q^n}$$

and

$$u_3 = -8\sum_{n=1}^{\infty} \frac{(-1)^n nq^{n/2}}{1-q^n}.$$

c. Show that $f_0''(\pi) = f_0''(\pi\tau) = f_0''(\pi + \pi\tau) = 0$.
Hint: Use the facts that $f_0''(\theta)$ is an odd elliptic function with periods $2\pi$ and $2\pi\tau$.

d. Show that

$$f_0''(\alpha)f_0(\beta) - f_0''(\alpha+\beta)f_0(\beta) - f_0''(\alpha)f_0(\alpha+\beta)$$
$$- 2f_0'(\alpha)f_0'(\alpha+\beta) - f_0(\alpha)f_0''(\alpha+\beta) = -g_0''(\alpha) - g_0''(\alpha+\beta).$$

Hint: From Corollary 1.43, if $\alpha + \beta + \gamma = 0$, then

$$f_0(\alpha)f_0(\beta) + f_0(\beta)f_0(\gamma) + f_0(\gamma)f_0(\alpha) = -\left(g_0(\alpha) + g_0(\beta) + g_0(\gamma)\right).$$

Now take the partial derivative twice with respect to $\alpha$.

e. Let $\alpha = \pi$ and $\beta = \pi\tau - \pi$ in the result of Part d, and make use of Part c, to show that

$$2f_0'(\pi)f_0'(\pi\tau) = g_0''(\pi) + g_0''(\pi\tau).$$

Hence, use the results of Parts a and b to deduce

$$q\frac{\mathrm{d}}{\mathrm{d}q}(u_1 + u_2) = \frac{1}{2}u_1u_2.$$

f. Similarly, show that

$$q\frac{\mathrm{d}}{\mathrm{d}q}(u_1 + u_3) = \frac{1}{2}u_1u_3 \quad \text{and} \quad q\frac{\mathrm{d}}{\mathrm{d}q}(u_2 + u_3) = \frac{1}{2}u_2u_3.$$

Hint: Substitute $(\alpha, \beta) = (\pi, \pi\tau)$ and $(\alpha, \beta) = (\pi\tau, \pi)$ in the result of Part d.
It follows that the functions $u_1$, $u_2$, and $u_3$ satisfy a system of differential equations, called Halphen's differential equations, given by

$$q\frac{\mathrm{d}}{\mathrm{d}q}(u_j + u_k) = \frac{1}{2}u_j u_k, \qquad 1 \le j < k \le 3.$$

This system of differential equations, together with the solution given above, was studied by Halphen [173, p. 330] in response to a questions of Darboux [134, p. 149] in mechanics and differential geometry. The proof outlined in the exercise is due to K. Venkatachaliengar [288, p. 65–67].

The next four exercises are based on work with H. H. Chan and P. C. Toh [85, Section 3].

10. Suppose $\operatorname{Im} T > 0$ and let $q = e^{i\pi T}$. Let $\theta_1$, $\theta_2$, $\theta_3$, and $\theta_4$ be the Jacobian theta functions studied in Exercise 5 in Chapter 0. Let $M$ and $N$ be defined by

$$M(x, y, q) = \frac{\theta_1(2x, q)}{\theta_2(x, q)}\,\theta_2(y, q) \quad - \quad \frac{\theta_1(2x, q)}{\theta_3(x, q)}\,\theta_3(y, q)$$

and

$$N(x, y, q) = \theta_1\left(x, q^{1/2}\right)\theta_1\left(\frac{x+y}{2}, q^{1/2}\right)\theta_1\left(\frac{x-y}{2}, q^{1/2}\right).$$

a. Use the results in Table 0.3 to show that the functions $M$ and $N$ both satisfy the periodicity and quasi-periodicity properties

$$f(x + 2\pi, y, q) = f(x, y, q), \quad f(x + \pi T, y, q) = q^{-3}e^{-6ix}f(x, y, q),$$
$$f(x, y + 2\pi, q) = f(x, y, q), \quad f(x, y + \pi T, q) = q^{-1}e^{-2iy}f(x, y, q).$$

b. Fix a value of $y$ not of the form $2\pi m + \pi n T$ where $m, n \in \mathbb{Z}$, and consider $M$ and $N$ as functions of $x$. Show that $N$ has simple zeros at $x = \pi m + \pi n T/2$ and $x = \pm y + 2\pi m + \pi n T$ for $m, n \in \mathbb{Z}$, and that these are the only zeros of $N$. Show that $M$ also has zeros at these points. (Do not attempt to show that these are the only zeros of $M$). Conclude that $M(x, y, q)/N(x, y, q)$ is an entire, doubly periodic function of $x$, and hence is independent of $x$.

c. Now fix a value of $x$ not of the form $\pi m + \pi n T/2$ where $m, n \in \mathbb{Z}$, and consider $M$ and $N$ as functions of $y$. Show that $N$ has simple zeros at $y = \pm x + 2\pi m + \pi n T$ for $m, n \in \mathbb{Z}$, and that these are the only zeros of $N$. Show that $M$ also has zeros at these points. Conclude that $M(x, y, q)/N(x, y, q)$ is an entire, doubly periodic function of $y$, and hence is independent of $y$.

d. It follows from Parts b and c that $M(x, y, q)/N(x, y, q)$ is independent of $x$ and $y$ and depends only on $q$. Let $x = \pi/2$ and $y = \pi$ and use the infinite product formulas for theta functions in Exercise 5 in Chapter 0 to conclude that

$$\eta_2 \times \left( \frac{\theta_1(2x,q)}{\theta_2(x,q)}\,\theta_2(y,q) - \frac{\theta_1(2x,q)}{\theta_3(x,q)}\,\theta_3(y,q) \right)$$
$$= \frac{1}{\eta_1} \times \theta_1\left(x,q^{1/2}\right)\theta_1\left(\frac{x+y}{2},q^{1/2}\right)\theta_1\left(\frac{x-y}{2},q^{1/2}\right). \qquad (1.111)$$

e. Apply the differential operator $\partial^3/\partial x^3$ to the identity in Part d and then let $x = y = 0$ to deduce that

$$\sum_{\substack{\alpha+\beta\equiv 0 \pmod 2 \\ \alpha\equiv 1 \pmod 3}} \alpha^3 q^{(\alpha^2+3\beta^2)/12} = -6\eta_1^8,$$

where the sum is over all integer values of $\alpha$ and $\beta$ that satisfy the given conditions.

Hint: Derivatives of the terms on the left-hand side of (1.111) can be calculated using the series in Part g of Exercise 5 of Chapter 0. The derivative of the right-hand side of (1.111), when evaluated at $x = y = 0$, will be a linear combination of terms of the form

$$\frac{1}{\eta_1} \times \theta_1^{(j)}(0,q^{1/2})\,\theta_1^{(k)}(0,q^{1/2})\,\theta_1^{(\ell)}(0,q^{1/2})$$

where $j$, $k$, and $\ell$ are nonnegative integers such that $j + k + \ell = 3$. The only nonzero term of this type is the one with $j = k = \ell = 1$.

f. Apply the operator $\partial^3/\partial x \partial y^2$ to the identity in Part d and then let $x = y = 0$, to show that

$$\sum_{\substack{\alpha+\beta\equiv 0 \pmod 2 \\ \alpha\equiv 1 \pmod 3}} \alpha\beta^2 q^{(\alpha^2+3\beta^2)/12} = 2\eta_1^8.$$

The results of applying higher order differential operators to the identity in Part d have been considered in [85].

11. a. Apply the techniques of the previous exercise and the specialization $x = \pi/2$, $y = \pi/6$ to prove that

$$\eta_2^2 \times \left( \frac{\theta_1(2x,q)}{\theta_2(x,q)}\frac{\theta_1(2y,q)}{\theta_3(y,q)} - \frac{\theta_1(2x,q)}{\theta_3(x,q)}\frac{\theta_1(2y,q)}{\theta_2(y,q)} \right)$$
$$= \frac{1}{\eta_1^2} \times \theta_1\left(x,q^{1/2}\right)\theta_1\left(y,q^{1/2}\right)\theta_1\left(\frac{x+y}{2},q^{1/2}\right)\theta_1\left(\frac{x-y}{2},q^{1/2}\right).$$

This is a form of Winquist's identity [303, Theorem 1.1] that was originally used to prove Ramanujan's partition congruence

$$p(11n+6) \equiv 0 \pmod{11}.$$

b. Apply the differential operator $\partial^4/\partial x^3 \partial y$ to the identity in Part a and then let $x = y = 0$ to deduce that

$$\sum_{\substack{\alpha \equiv 1 \pmod 6 \\ \beta \equiv 4 \pmod 6}} \alpha\beta(\alpha^2 - \beta^2) q^{(\alpha^2+\beta^2)/12} = 6\eta_1^{10}.$$

12. a. Apply the techniques of Exercise 10 and the specialization $x = -\pi/8$, $y = 7\pi/8$ to prove that

$$\begin{aligned}
\eta_2 \times \Bigg( &\frac{\theta_1(4x,q)}{\theta_4(2x,q)}\theta_1(2y) + \frac{\theta_1(2x-2y,q)}{\theta_4(x-y,q)}\theta_1(3x+y) \\
&\qquad + \frac{\theta_1(2x+2y,q)}{\theta_4(x+y,q)}\theta_1(-3x+y)\Bigg) \\
= \frac{1}{\eta_1^4} &\times \theta_1\left(x,q^{1/2}\right)\theta_1\left(y,q^{1/2}\right)\theta_1\left(\frac{x+y}{2},q^{1/2}\right)\theta_1\left(\frac{x-y}{2},q^{1/2}\right) \\
&\times \theta_1\left(\frac{3x+y}{2},q^{1/2}\right)\theta_1\left(\frac{-3x+y}{2},q^{1/2}\right).
\end{aligned}$$

b. Apply the differential operator $D_xD_y(D_x^2-D_y^2)(D_x^2-9D_y^2)$, where $D_x = \partial/\partial x$ and $D_y = \partial/\partial y$, to the identity in Part a and then let $x = y = 0$ to deduce that

$$\sum_{\substack{\alpha \equiv 4 \pmod 6 \\ \beta \equiv 1 \pmod 4}} (-1)^{(\alpha+2)/6}\alpha\beta(\alpha^2-\beta^2)(\alpha^2-9\beta^2) q^{(\alpha^2+3\beta^2)/12} = 30\eta_1^{14}.$$

13. Use the results of Exercises 10–12 to deduce that

$$\eta_1^8 = \frac{-1}{24} \sum_{\substack{\alpha+\beta \equiv 0 \pmod 2 \\ \alpha \equiv 1 \pmod 3}} \left(\alpha + i\beta\sqrt{3}\right)^3 q^{(\alpha^2+3\beta^2)/12},$$

$$\eta_1^{10} = \frac{1}{24} \sum_{\substack{\alpha \equiv 1 \pmod 6 \\ \beta \equiv 4 \pmod 6}} \operatorname{Im}\left((\alpha+i\beta)^4\right) q^{(\alpha^2+\beta^2)/12}$$

and

$$\eta_1^{14} = \frac{1}{180\sqrt{3}} \sum_{\substack{\alpha \equiv 4 \pmod 6 \\ \beta \equiv 1 \pmod 4}} (-1)^{(\alpha+2)/6} \operatorname{Im}\left((\alpha+i\beta\sqrt{3})^6\right) q^{(\alpha^2+3\beta^2)/12}.$$

Generalizations of these results have been given in [85, Section 7]. The analogous result for the 26th power was given in Exercise 11 in Chapter 0.

14. For this exercise, let $c_1 = e_2 - e_3$, $c_2 = e_3 - e_1$, and $c_3 = e_1 - e_2$.

a. Prove that

$$c_1 f_1^2(x) + c_2 f_2^2(x) + c_3 f_3^2(x) = 0.$$

Hint: Use Theorem 1.31.

b. Suppose that $x + y + z = 0$. Prove that

$$c_1 f_1(x) f_1(y) f_1(z) + c_2 f_2(x) f_2(y) f_2(z) + c_3 f_3(x) f_3(y) f_3(z) = 0.$$

Hint: Fix a value of $y$ and show that the left-hand side is a doubly periodic function of $x$. Determine the singularities (as a function of $x$), use (1.59) and the result of Part a to show that the singularities are removable, and hence deduce that the left-hand side is a constant that is independent of $x$. Then use the result of Part a again to show that the constant is zero.

c. Suppose that $w + x + y + z = 0$. Prove that

$$c_1 f_1(w) f_1(x) f_1(y) f_1(z) + c_2 f_2(w) f_2(x) f_2(y) f_2(z)$$
$$+ c_3 f_3(w) f_3(x) f_3(y) f_3(z) = c_1 c_2 c_3.$$

Hint: Fix the values of $w$ and $y$, and show that the left-hand side is an elliptic function of $x$ with removable singularities and hence is a constant independent of $x$. By symmetry in the variables, the left-hand side is a constant independent of $w$, $x$, $y$, and $z$, and its value can be found by specialization, e.g., by taking $(w, x, y, z) = (\pi, -\pi, \pi\tau, -\pi\tau)$ and using (1.61)–(1.66).

d. Suppose that $v + w + x + y + z = 0$. Prove that

$$c_1 f_1(v) f_1(w) f_1(x) f_1(y) f_1(z)$$
$$+ c_2 f_2(v) f_2(w) f_2(x) f_2(y) f_2(z)$$
$$+ c_3 f_3(v) f_3(w) f_3(x) f_3(y) f_3(z)$$
$$= c_1 c_2 c_3 (f_0(v) + f_0(w) + f_0(x) + f_0(y) + f_0(z)). \quad (1.112)$$

Hint: Fix the values of $v$, $w$, and $y$ and consider the left- and right-hand sides as functions of $x$. Use (1.59) to show that the difference is an entire elliptic function of $x$, which is therefore constant. By symmetry, the difference is independent of $v$, $w$, $x$, $y$, and $z$ and is a constant that depends only on $q$. The value of the constant may be shown to be zero by specializing the variables, e.g., by taking $(v, w, x, y, z) = (\pi, -\pi, -\pi, -\pi\tau, \pi + \pi\tau)$. The values $f_0(\pi) = 0$, $f_0(\pi\tau) = -i/2$, and $f_0(\pi + \pi\tau) = -i/2$ can be determined using (1.6) and (1.53). The values $f_1(\pi) = f_2(\pi\tau) = f_3(\pi + \pi\tau) = 0$ were obtained in the discussion following Theorem 1.30.

The results in Parts c and d are equivalent to identities in [299, pp. 458–459, Ex. 22] and [224, Theorem 3.1], respectively. An application of the identity in Part d will be given in Exercise 3 at the end of Chapter 11.

15. Let $\mathsf{s}$, $\mathsf{c}$, and $\mathsf{d}$ be defined by

$$\mathsf{s}(\theta) = \sum_{n=1}^{\infty} \frac{\sin(n-\frac{1}{2})\theta}{\sinh(n-\frac{1}{2})y}, \quad \mathsf{c}(\theta) = \sum_{n=1}^{\infty} \frac{\cos(n-\frac{1}{2})\theta}{\cosh(n-\frac{1}{2})y}$$

and

$$\mathsf{d}(\theta) = \frac{1}{2} + \sum_{n=1}^{\infty} \frac{\cos n\theta}{\cosh ny},$$

where $y = -i\tau$ so that $q = e^{2\pi i\tau} = e^{-2\pi y}$.

a. Use Theorem 0.26 to show that

$$\mathsf{s}(\theta) = 2q^{1/4} \sin\frac{\theta}{2} \prod_{n=1}^{\infty} \frac{(1-e^{i\theta}q^n)(1-e^{-i\theta}q^n)(1-q^n)^2}{(1-e^{i\theta}q^{n-1/2})(1-e^{-i\theta}q^{n-1/2})(1-q^{n-1/2})^2},$$

$$\mathsf{c}(\theta) = 2q^{1/4} \cos\frac{\theta}{2} \prod_{n=1}^{\infty} \frac{(1+e^{i\theta}q^n)(1+e^{-i\theta}q^n)(1-q^n)^2}{(1-e^{i\theta}q^{n-1/2})(1-e^{-i\theta}q^{n-1/2})(1+q^{n-1/2})^2}$$

and

$$\mathsf{d}(\theta) = \frac{1}{2} \prod_{n=1}^{\infty} \frac{(1+e^{i\theta}q^{n-1/2})(1+e^{-i\theta}q^{n-1/2})(1-q^n)^2}{(1-e^{i\theta}q^{n-1/2})(1-e^{-i\theta}q^{n-1/2})(1+q^n)^2}.$$

b. Deduce that $\mathsf{s}$, $\mathsf{c}$, and $\mathsf{d}$ are elliptic functions, with periods, poles, and zeros as indicated in Table 1.2.

**Table 1.2** Periodicity properties, zeros, and poles of Jacobian elliptic functions

| $f(\theta)$ | $\dfrac{f(\theta+2m\pi+2n\pi\tau)}{f(\theta)}$ | zeros | poles |
|---|---|---|---|
| $\mathsf{s}(\theta)$ | $(-1)^m$ | $\theta = 2m\pi + 2n\pi\tau$ | $\theta = 2m\pi + (2n+1)\pi\tau$ |
| $\mathsf{c}(\theta)$ | $(-1)^{m+n}$ | $\theta = (2m+1)\pi + 2n\pi\tau$ | $\theta = 2m\pi + (2n+1)\pi\tau$ |
| $\mathsf{d}(\theta)$ | $(-1)^n$ | $\theta = (2m+1)\pi + (2n+1)\pi\tau$ | $\theta = 2m\pi + (2n+1)\pi\tau$ |
| $f_1(\theta)$ | $(-1)^n$ | $\theta = (2m+1)\pi + 2n\pi\tau$ | $\theta = 2m\pi + 2n\pi\tau$ |
| $f_2(\theta)$ | $(-1)^m$ | $\theta = 2m\pi + (2n+1)\pi\tau$ | $\theta = 2m\pi + 2n\pi\tau$ |
| $f_3(\theta)$ | $(-1)^{m+n}$ | $\theta = (2m+1)\pi + (2n+1)\pi\tau$ | $\theta = 2m\pi + 2n\pi\tau$ |

c. By comparing infinite products, show that

$$f_1(\theta) = a_1 \frac{\mathsf{c}(\theta)}{\mathsf{s}(\theta)}, \qquad f_2(\theta) = \frac{a_2}{\mathsf{s}(\theta)} \qquad \text{and} \qquad f_3(\theta) = a_3 \frac{\mathsf{d}(\theta)}{\mathsf{s}(\theta)},$$

where $a_1$, $a_2$, and $a_3$ are independent of $\theta$ and are given by

$$a_1 = \frac{1}{2} \prod_{n=1}^{\infty} \frac{(1 + q^{n-1/2})^2 (1 - q^n)^2}{(1 - q^{n-1/2})^2 (1 + q^n)^2} = \sqrt{e_1 - e_2},$$

$$a_3 = 2q^{1/4} \prod_{n=1}^{\infty} \frac{(1 + q^n)^2 (1 - q^n)^2}{(1 + q^{n-1/2})^2 (1 - q^{n-1/2})^2} = \sqrt{e_3 - e_2},$$

and

$$a_2 = a_1 a_3.$$

d. Show that $\mathsf{s}^2(\theta) + \mathsf{c}^2(\theta) = e_3 - e_2$ and $\mathsf{s}^2(\theta) + \mathsf{d}^2(\theta) = e_1 - e_2$.

e. Show that

$$\mathsf{s}'(\theta) = \mathsf{c}(\theta)\mathsf{d}(\theta), \quad \mathsf{c}'(\theta) = -\mathsf{s}(\theta)\mathsf{d}(\theta) \quad \text{and} \quad \mathsf{d}'(\theta) = -\mathsf{s}(\theta)\mathsf{c}(\theta).$$

16. The Jacobian elliptic functions sn, cn, and dn may be defined by

$$\operatorname{sn}(\theta) = \frac{1}{\sqrt{e_3 - e_2}} \mathsf{s}\left(\frac{\theta}{\sqrt{e_1 - e_2}}\right),$$

$$\operatorname{cn}(\theta) = \frac{1}{\sqrt{e_3 - e_2}} \mathsf{c}\left(\frac{\theta}{\sqrt{e_1 - e_2}}\right)$$

and

$$\operatorname{dn}(\theta) = \frac{1}{\sqrt{e_1 - e_2}} \mathsf{d}\left(\frac{\theta}{\sqrt{e_1 - e_2}}\right),$$

where s, c, and d are the functions in Exercise 15.

a. Show that $\operatorname{sn}^2\theta + \operatorname{cn}^2\theta = 1$ and $\operatorname{dn}^2\theta + k^2 \operatorname{sn}^2\theta = 1$ where

$$k^2 = \frac{e_3 - e_2}{e_1 - e_2}.$$

b. Show that $\operatorname{sn}'(\theta) = \operatorname{cn}(\theta)\operatorname{dn}(\theta)$, $\quad \operatorname{cn}'(\theta) = -\operatorname{sn}(\theta)\operatorname{dn}(\theta)$ and $\operatorname{dn}'(\theta) = -k^2 \operatorname{sn}(\theta)\operatorname{cn}(\theta)$.

c. Use Theorem 1.41 to show that

$$\operatorname{sn}(a+b) = \frac{\operatorname{sn} a \operatorname{cn} b \operatorname{dn} b + \operatorname{sn} b \operatorname{cn} a \operatorname{dn} a}{1 - k^2 \operatorname{sn}^2 a \operatorname{sn}^2 b},$$

$$\operatorname{cn}(a+b) = \frac{\operatorname{cn} a \operatorname{cn} b - \operatorname{sn} a \operatorname{sn} b \operatorname{dn} a \operatorname{dn} b}{1 - k^2 \operatorname{sn}^2 a \operatorname{sn}^2 b}$$

and

$$\operatorname{dn}(a+b) = \frac{\operatorname{dn} a \operatorname{dn} b - k^2 \operatorname{sn} a \operatorname{sn} b \operatorname{cn} a \operatorname{cn} b}{1 - k^2 \operatorname{sn}^2 a \operatorname{sn}^2 b}.$$

17. Use Lemma 1.49 and the special case of Ramanujan's summation formula in Theorem 0.27 to show that

$$\operatorname{sm}^*\left(\theta - \frac{2\pi}{3}\right) = \frac{1}{b(q)}\left(\frac{\sin(\frac{\theta}{2} - \frac{\pi}{3})}{\sin\frac{\theta}{2}} - 2\sqrt{3}\sum_{j=1}^{\infty} \frac{q^j \sin(j\theta + \frac{4\pi}{3}) - q^{2j}\sin(j\theta)}{1 + q^j + q^{2j}}\right)$$

and

$$\operatorname{cm}^*\left(\theta - \frac{2\pi}{3}\right) = \frac{1}{b(q)}\left(\frac{\sin(\frac{\theta}{2} + \frac{\pi}{3})}{\sin\frac{\theta}{2}} + 2\sqrt{3}\sum_{j=1}^{\infty} \frac{q^j \sin(j\theta + \frac{2\pi}{3}) - q^{2j}\sin(j\theta)}{1 + q^j + q^{2j}}\right).$$

# Chapter 2
# Transformations

## 2.1 Lattices

Let $\omega_1$ and $\omega_2$ be complex numbers that satisfy $\operatorname{Im}(\omega_2/\omega_1) > 0$. The set

$$\Lambda(\omega_1, \omega_2) = \{m\omega_1 + n\omega_2 : m, n \in \mathbb{Z}\}$$

is called the lattice generated by $\omega_1$ and $\omega_2$. The set $\Lambda$ introduced at the beginning of the previous chapter is the lattice generated by $2\pi$ and $2\pi\tau$. Many of the functions studied in the previous chapter have zeros or poles that lie in a lattice, or in the translation of a lattice. The goal of this chapter is to show how properties of the lattice imply transformation properties of such functions.

Let us fix some notation that will be used throughout the chapter. Let

$$\Gamma = \left\{ \begin{pmatrix} a & b \\ c & d \end{pmatrix} : a, b, c, d \in \mathbb{Z},\ ad - bc = 1 \right\}.$$

For $\begin{pmatrix} a & b \\ c & d \end{pmatrix} \in \Gamma$, let $\omega_3$ and $\omega_4$ be defined in terms of $\omega_1$ and $\omega_2$ by

$$\begin{pmatrix} \omega_4 \\ \omega_3 \end{pmatrix} = \begin{pmatrix} a & b \\ c & d \end{pmatrix} \begin{pmatrix} \omega_2 \\ \omega_1 \end{pmatrix},$$

that is,

$$\omega_4 = a\omega_2 + b\omega_1 \quad \text{and} \quad \omega_3 = c\omega_2 + d\omega_1. \tag{2.1}$$

Let $\tau = \omega_2/\omega_1$ and note that

$$\frac{\omega_4}{\omega_3} = \frac{a\tau + b}{c\tau + d} \quad \text{and} \quad \frac{\omega_3}{\omega_1} = c\tau + d. \tag{2.2}$$



S. Cooper, *Ramanujan's Theta Functions*, DOI 10.1007/978-3-319-56172-1_3

For future reference we record the fact

$$a\omega_3 - \omega_1 = c\omega_4. \tag{2.3}$$

A simple calculation shows that if $\operatorname{Im}\tau > 0$ and $\begin{pmatrix} a & b \\ c & d \end{pmatrix} \in \Gamma$ then

$$\operatorname{Im}\left(\frac{a\tau + b}{c\tau + d}\right) = \frac{\operatorname{Im}\tau}{|c\tau + d|^2}.$$

Since $\operatorname{Im}(\omega_2/\omega_1) > 0$ it follows that $\operatorname{Im}(\omega_4/\omega_3) > 0$.

Let $q$ and $r$ be defined by

$$q = \exp\left(\frac{2\pi i\omega_2}{\omega_1}\right) = \exp(2\pi i\tau) \tag{2.4}$$

*and*

$$r = \exp\left(\frac{2\pi i\omega_4}{\omega_3}\right) = \exp\left(2\pi i\,\frac{a\tau + b}{c\tau + d}\right). \tag{2.5}$$

By the discussion above, $|q| < 1$ and $|r| < 1$.

The following result indicates that there is more than one description of a lattice.

**Lemma 2.1.** *Let* $\begin{pmatrix} a & b \\ c & d \end{pmatrix} \in \Gamma$ *and let* $\omega_1$, $\omega_2$, $\omega_3$, *and* $\omega_4$ *be as above. Then*

$$\Lambda(\omega_1, \omega_2) = \Lambda(\omega_3, \omega_4).$$

*Proof.* On the one hand,

$$\begin{aligned}
\Lambda(\omega_3, \omega_4) &= \{m\omega_3 + n\omega_4 : m, n \in \mathbb{Z}\} \\
&= \{m(c\omega_2 + d\omega_1) + n(a\omega_2 + b\omega_1) : m, n \in \mathbb{Z}\} \\
&= \{(bn + dm)\omega_1 + (an + cm)\omega_2 : m, n \in \mathbb{Z}\} \\
&\subseteq \Lambda(\omega_1, \omega_2).
\end{aligned}$$

On the other hand, since $ad - bc = 1$ the relations in (2.1) can be solved for $\omega_1$ and $\omega_2$ to give

$$\omega_2 = d\omega_4 - b\omega_3 \quad \text{and} \quad \omega_1 = -c\omega_4 + a\omega_3.$$

Therefore,

$$\begin{aligned}
\Lambda(\omega_1, \omega_2) &= \{m\omega_1 + n\omega_2 : m, n \in \mathbb{Z}\} \\
&= \{m(-c\omega_4 + a\omega_3) + n(d\omega_4 - b\omega_3) : m, n \in \mathbb{Z}\}
\end{aligned}$$

$$= \{(am - bn)\omega_3 + (dn - cm)\omega_4 : m, n \in \mathbb{Z}\}$$
$$\subseteq \Lambda(\omega_3, \omega_4).$$

Hence, $\Lambda(\omega_1, \omega_2) = \Lambda(\omega_3, \omega_4)$.

□

## 2.2 Transformation formulas

The location and nature of the poles of $z$ can be seen from the formula (1.17), viz.,

$$z(\theta, \tau) = \frac{1}{2}\cot\frac{\theta}{2} + \frac{1}{i}\sum_{n=1}^{\infty}\left(\frac{e^{i\theta}q^n}{1 - e^{i\theta}q^n} - \frac{e^{-i\theta}q^n}{1 - e^{-i\theta}q^n}\right).$$

The poles are simple and are located at each point $\theta \in \Lambda(2\pi, 2\pi\tau)$, and the residue at each pole is 1. Furthermore, by (1.6), we have

$$z(\theta + 2\pi, \tau) = z(\theta, \tau) \qquad \text{and} \qquad z(\theta + 2\pi\tau, \tau) = z(\theta, \tau) - i. \tag{2.6}$$

The function $z(\theta, \tau)$ can be rescaled to produce a function that has poles at each point in the general lattice $\Lambda(\omega_1, \omega_2)$. In fact, the function $Z$ defined by

$$Z(\theta, \omega_1, \omega_2) = \frac{2\pi}{\omega_1} z\left(\frac{2\pi\theta}{\omega_1}, \frac{\omega_2}{\omega_1}\right) \tag{2.7}$$

is such a function. The next result summarizes some of its properties.

**Lemma 2.2.** *The function $Z(\theta, \omega_1, \omega_2)$ has simple poles with residue 1 at each point $\theta \in \Lambda(\omega_1, \omega_2)$ and no other singularities. It satisfies the periodic property*

$$Z(\theta + \omega_1, \omega_1, \omega_2) = Z(\theta, \omega_1, \omega_2) \tag{2.8}$$

*and the quasi-periodic property*

$$Z(\theta + \omega_2, \omega_1, \omega_2) = Z(\theta, \omega_1, \omega_2) - \frac{2\pi i}{\omega_1}. \tag{2.9}$$

*Proof.* This is immediate from the definition (2.7) and the corresponding properties of $z(\theta, \tau)$ described above and in (2.6).

□

In view of Lemmas 2.1 and 2.2, it may be expected that the function $Z(\theta, \omega_3, \omega_4)$ is closely related to $Z(\theta, \omega_1, \omega_2)$. The next result gives the precise relationship.

**Theorem 2.3.** *Let* $\begin{pmatrix} a & b \\ c & d \end{pmatrix} \in \Gamma$ *and let* $\omega_1$, $\omega_2$, $\omega_3$ *and* $\omega_4$ *be as in* (2.1). *Then the function* $Z$ *defined by* (2.7) *satisfies the transformation formula*

$$Z(\theta, \omega_3, \omega_4) = Z(\theta, \omega_1, \omega_2) + \frac{2\pi i c\theta}{\omega_1\omega_3}.$$

*Proof.* Let $f(\theta) = Z(\theta, \omega_3, \omega_4) - Z(\theta, \omega_1, \omega_2)$. By Lemmas 2.1 and 2.2, the functions $Z(\theta, \omega_1, \omega_2)$ and $Z(\theta, \omega_3, \omega_4)$ have simple poles with residue 1 at each point in the common set $\Lambda(\omega_1, \omega_2) = \Lambda(\omega_3, \omega_4)$, and no other singularities. It follows that $f(\theta)$ is an entire function of $\theta$.

Next, by the definitions of $\omega_3$ and $\omega_4$ and several applications of (2.8) and (2.9),

$$\begin{aligned} f(\theta + \omega_3) &= Z(\theta + \omega_3, \omega_3, \omega_4) - Z(\theta + \omega_3, \omega_1, \omega_2) \\ &= Z(\theta + \omega_3, \omega_3, \omega_4) - Z(\theta + c\omega_2 + d\omega_1, \omega_1, \omega_2) \\ &= Z(\theta, \omega_3, \omega_4) - \left( Z(\theta, \omega_1, \omega_2) - \frac{2\pi i c}{\omega_1} \right) \\ &= f(\theta) + \frac{2\pi i c}{\omega_1} \end{aligned} \tag{2.10}$$

and

$$\begin{aligned} f(\theta + \omega_4) &= Z(\theta + \omega_4, \omega_3, \omega_4) - Z(\theta + \omega_4, \omega_1, \omega_2) \\ &= Z(\theta + \omega_4, \omega_3, \omega_4) - Z(\theta + a\omega_2 + b\omega_1, \omega_1, \omega_2) \\ &= \left( Z(\theta, \omega_3, \omega_4) - \frac{2\pi i}{\omega_3} \right) - \left( Z(\theta, \omega_1, \omega_2) - \frac{2\pi i a}{\omega_1} \right) \\ &= f(\theta) + \frac{2\pi i c\omega_4}{\omega_1\omega_3}, \end{aligned} \tag{2.11}$$

where the last step follows by using (2.3).

But the simple linear function $g(\theta) = 2\pi i c\theta/\omega_1\omega_3$ also satisfies the functional equations (2.10) and (2.11), that is,

$$g(\theta + \omega_3) = g(\theta) + \frac{2\pi i c}{\omega_1} \quad \text{and} \quad g(\theta + \omega_4) = g(\theta) + \frac{2\pi i c\omega_4}{\omega_1\omega_3}.$$

It follows that the difference $f(\theta) - g(\theta)$ is an entire, doubly periodic function, with periods $\omega_3$ and $\omega_4$. By Liouville's theorem, $f(\theta) - g(\theta)$ is constant, and since $f$ and $g$ are odd functions of $\theta$, the constant is zero. This completes the proof.

□

The above analysis leads to transformation formulas for Eisenstein series.

**Corollary 2.4.** *Let $E_k$ be the Eisenstein series defined by* (0.66), *that is,*

$$E_k(\tau) = -\frac{B_k}{2k} + \sum_{j=1}^{\infty} \frac{j^{k-1}q^j}{1-q^j}$$

*where* $\operatorname{Im}\tau > 0$ *and* $q = \exp(2\pi i\tau)$. *Let* $\begin{pmatrix} a & b \\ c & d \end{pmatrix} \in \Gamma$. *Then*

$$E_2\left(\frac{a\tau+b}{c\tau+d}\right) = (c\tau+d)^2 E_2(\tau) + \frac{ic(c\tau+d)}{4\pi} \tag{2.12}$$

*and for* $k = 2, 3, 4, \ldots,$

$$E_{2k}\left(\frac{a\tau+b}{c\tau+d}\right) = (c\tau+d)^{2k} E_{2k}(\tau). \tag{2.13}$$

*Proof.* Expand both sides of the identity in Theorem 2.3 in powers of $\theta$ using (1.9) and (2.7) to obtain

$$\begin{aligned}
&\frac{1}{\theta} + \frac{4\pi}{\omega_3} \sum_{k=1}^{\infty} \frac{(-1)^{k-1}}{(2k-1)!} E_{2k}\left(\frac{\omega_4}{\omega_3}\right) \left(\frac{2\pi\theta}{\omega_3}\right)^{2k-1} \\
&= \frac{1}{\theta} + \frac{2\pi i c\theta}{\omega_1\omega_3} + \frac{4\pi}{\omega_1} \sum_{k=1}^{\infty} \frac{(-1)^{k-1}}{(2k-1)!} E_{2k}\left(\frac{\omega_2}{\omega_1}\right) \left(\frac{2\pi\theta}{\omega_1}\right)^{2k-1}.
\end{aligned}$$

On equating coefficients of $\theta$ we find that

$$\frac{8\pi^2}{\omega_3^2} E_2\left(\frac{\omega_4}{\omega_3}\right) = \frac{2\pi i c}{\omega_1\omega_3} + \frac{8\pi^2}{\omega_1^2} E_2\left(\frac{\omega_2}{\omega_1}\right)$$

and this simplifies to (2.12) on using (2.2). Comparing coefficients of $\theta^{2k-1}$ for $k \geq 2$ gives (2.13).

□

The results in Corollary 2.4 for $E_2$, $E_4$, and $E_6$ may be expressed in terms of Ramanujan's Eisenstein series $P$, $Q$, and $R$.

**Corollary 2.5.** *Let* $\begin{pmatrix} a & b \\ c & d \end{pmatrix} \in \Gamma$ *and let $q$ and $r$ be given by* (2.4) *and* (2.5), *that is,*

$$q = \exp(2\pi i\tau) \qquad \text{and} \qquad r = \exp\left(2\pi i\,\frac{a\tau+b}{c\tau+d}\right).$$

*Then*

$$P(r) = (c\tau + d)^2 P(q) - \frac{6ic(c\tau + d)}{\pi}, \tag{2.14}$$

$$Q(r) = (c\tau + d)^4 Q(q) \tag{2.15}$$

*and*

$$R(r) = (c\tau + d)^6 R(q). \tag{2.16}$$

*Proof.* By the definitions (0.71)–(0.73),

$$P(q) = -24E_2(q), \quad Q(q) = 240E_4(q) \quad \text{and} \quad R(q) = -504E_6(q).$$

The required results now follow from Corollary 2.4. □

The following special case of (2.14) is noteworthy.

**Corollary 2.6.** *The following transformation formula holds for positive values of* $n$:

$$nP\left(e^{-2\pi\sqrt{n}}\right) + P\left(e^{-2\pi/\sqrt{n}}\right) = \frac{6\sqrt{n}}{\pi}. \tag{2.17}$$

*Hence,*

$$\sum_{k=1}^{\infty} \frac{k}{e^{2\pi k} - 1} = \frac{1}{24} - \frac{1}{8\pi}. \tag{2.18}$$

*Proof.* Let $\begin{pmatrix} a & b \\ c & d \end{pmatrix} = \begin{pmatrix} 0 & -1 \\ 1 & 0 \end{pmatrix}$ in (2.14) to deduce that

$$P\left(e^{-2\pi i/\tau}\right) = \tau^2 P\left(e^{2\pi i\tau}\right) - \frac{6i\tau}{\pi}. \tag{2.19}$$

Now put $\tau = i\sqrt{n}$ to obtain (2.17).

If we take $n = 1$ in (2.17), we find that $P(e^{-2\pi}) = 3/\pi$, that is,

$$1 - 24\sum_{k=1}^{\infty} \frac{ke^{-2\pi k}}{1 - e^{-2\pi k}} = \frac{3}{\pi}.$$

This implies (2.18). □

Dedekind's eta function is defined for $\operatorname{Im}\tau > 0$ by

$$\eta(\tau) = e^{i\pi\tau/12} \prod_{j=1}^{\infty} (1 - e^{2\pi i j\tau}) = q^{1/24} \prod_{j=1}^{\infty} (1 - q^j). \tag{2.20}$$

The next result gives the transformation formula for the 24th power of the eta function.

**Theorem 2.7.** *Let* $\begin{pmatrix} a & b \\ c & d \end{pmatrix} \in \Gamma$ *and suppose* $\operatorname{Im}\tau > 0$. *Then*

$$\eta^{24}\left(\frac{a\tau + b}{c\tau + d}\right) = (c\tau + d)^{12}\eta^{24}(\tau).$$

*Proof.* By Corollary 0.23,

$$\eta^{24}(\tau) = q\prod_{j=1}^{\infty}(1 - q^j)^{24} = \frac{1}{1728}\left(Q^3(q) - R^2(q)\right).$$

Therefore, on using the transformation formulas for $Q$ and $R$ given by (2.15) and (2.16) we find that

$$\begin{aligned}
\eta^{24}\left(\frac{a\tau + b}{c\tau + d}\right) &= \frac{1}{1728}\left(Q^3(r) - R^2(r)\right) \\
&= \frac{1}{1728}(c\tau + d)^{12}\left(Q^3(q) - R^2(q)\right) \\
&= (c\tau + d)^{12}\eta^{24}(\tau).
\end{aligned}$$

□

**Theorem 2.8.** *The following transformation formula holds*:

$$\eta\left(\frac{-1}{\tau}\right) = \sqrt{\frac{\tau}{i}}\,\eta(\tau),$$

*where the branch of the square root is determined by taking the positive square root when* $\tau$ *is purely imaginary.*

*Proof.* Let $\begin{pmatrix} a & b \\ c & d \end{pmatrix} = \begin{pmatrix} 0 & -1 \\ 1 & 0 \end{pmatrix}$ in the previous theorem and take the 24th root to obtain

$$\eta\left(\frac{-1}{\tau}\right) = \epsilon\,\tau^{1/2}\,\eta(\tau)$$

where $\epsilon$ is some 24th root of unity. When $\tau$ is purely imaginary, both $\eta(-1/\tau)$ and $\eta(\tau)$ are real valued and positive, therefore $\epsilon\,\tau^{1/2}$ is real and positive. It follows that

$$\epsilon\,\tau^{1/2} = \sqrt{\frac{\tau}{i}}$$

where the positive branch of the square root is selected when $\tau$ is purely imaginary. This completes the proof. □

For any $\begin{pmatrix} a & b \\ c & d \end{pmatrix} \in \Gamma$ it is possible to express the value of $\eta\left(\frac{a\tau+b}{c\tau+d}\right)$ in terms of $\eta(\tau)$, but the determination of the 24th root of unity is more complicated. We state the following result without proof.

**Theorem 2.9.** *If* $\begin{pmatrix} a & b \\ c & d \end{pmatrix} \in \Gamma$, $c > 0$ *and* $\operatorname{Im} \tau > 0$, *then*

$$\eta\left(\frac{a\tau+b}{c\tau+d}\right) = \epsilon(a,b,c,d)\sqrt{\frac{c\tau+d}{i}}\,\eta(\tau)$$

*where*

$$\epsilon(a,b,c,d) = \exp\left(\pi i\left(\frac{a+d}{12c} + s(-d,c)\right)\right),$$

$$s(h,k) = \sum_{r=1}^{k-1} \frac{r}{k}\left(\frac{hr}{k} - \left\lfloor \frac{hr}{k} \right\rfloor - \frac{1}{2}\right)$$

*and* $\lfloor x \rfloor$ *is the greatest integer that is less than or equal to* $x$. *The sum* $s(h,k)$ *is called a Dedekind sum.*

*Proof.* See [29, Ch. 3], [86], or any standard book on modular forms. □

The modular $j$-invariant is defined by

$$j = j(\tau) = \frac{1728Q^3(q)}{Q^3(q) - R^2(q)}, \qquad q = \exp(2\pi i\tau). \tag{2.21}$$

By the transformation formulas for $Q$ and $R$ in Corollary 2.5, we have that $j(\tau)$ is invariant under $\Gamma$, that is,

$$j\left(\frac{a\tau+b}{c\tau+d}\right) = j(\tau) \quad \text{for any} \begin{pmatrix} a & b \\ c & d \end{pmatrix} \in \Gamma.$$

By Corollary 0.23,

$$j(\tau) = \frac{Q^3(q)}{q\prod_{j=1}^{\infty}(1-q^j)^{24}}. \tag{2.22}$$

The denominator in (2.22) is never zero for $\operatorname{Im}\tau > 0$ and it follows that $j(\tau)$ is analytic for $\operatorname{Im}\tau > 0$. The first few terms in the expansion in powers of $q$ are given by

$$j(\tau) = \frac{1}{q} + 744 + 196884q + 21493760q^2 + 864299970q^3 + \cdots .$$

Here are the results of the transformation $\tau \to -1/\tau$ for the functions $s$, $z$, and $\wp$ defined by (0.51), (1.4), and (1.12).

**Theorem 2.10.** *The following transformation formulas hold*:

$$s(\theta,\tau) = \tau \exp\left(\frac{\theta^2}{4\pi i\tau}\right) s\left(\frac{\theta}{\tau}, \frac{-1}{\tau}\right), \tag{2.23}$$

$$z(\theta,\tau) = \frac{1}{\tau} z\left(\frac{\theta}{\tau}, \frac{-1}{\tau}\right) + \frac{\theta}{2\pi i\tau}, \tag{2.24}$$

*and*

$$\wp(\theta,\tau) = \frac{1}{\tau^2} \wp\left(\frac{\theta}{\tau}, \frac{-1}{\tau}\right). \tag{2.25}$$

*Proof.* We prove (2.24) first, then (2.23) and (2.25). Let $\omega_1 = 2\pi$, $\omega_2 = 2\pi\tau$ and take $\begin{pmatrix} a & b \\ c & d \end{pmatrix} = \begin{pmatrix} 0 & -1 \\ 1 & 0 \end{pmatrix}$ in Theorem 2.3 to obtain

$$\frac{1}{\tau} z\left(\frac{\theta}{\tau}, \frac{-1}{\tau}\right) = z(\theta,\tau) + \frac{i\theta}{2\pi\tau}.$$

This gives (2.24).

On integrating (2.24) with respect to $\theta$ and using (1.4) we find that

$$\log s(\theta,\tau) = \log s\left(\frac{\theta}{\tau}, \frac{-1}{\tau}\right) + \frac{\theta^2}{4\pi i\tau} + \log a_\tau$$

where $a_\tau$ is a constant that may depend on $\tau$ but is independent of $\theta$. Hence,

$$s(\theta,\tau) = a_\tau \exp\left(\frac{\theta^2}{4\pi i\tau}\right) s\left(\frac{\theta}{\tau}, \frac{-1}{\tau}\right).$$

Now divide by $\theta$ and take the limit as $\theta \to 0$. Since $\lim_{\theta\to 0} s(\theta)/\theta = 1$ by (0.52), it follows that $a_\tau = \tau$. This proves (2.23).

By the definitions in (1.5) and (1.13),

$$\wp(\theta,\tau) = p(\theta,\tau) + 2E_2(\tau) = -\frac{\mathrm{d}}{\mathrm{d}\theta} z(\theta,\tau) + 2E_2(\tau).$$

Hence, the transformation formula for $\wp$ can be obtained from the corresponding formulas for $E_2$ and $z$ in (2.12) and (2.24), to give

$$\begin{aligned}\wp(\theta,\tau) &= -\frac{\mathrm{d}}{\mathrm{d}\theta}\left\{\frac{1}{\tau}z\left(\frac{\theta}{\tau},\frac{-1}{\tau}\right)+\frac{\theta}{2\pi i\tau}\right\}+2\left\{\frac{1}{\tau^2}E_2\left(\frac{-1}{\tau}\right)-\frac{i}{4\pi\tau}\right\}\\ &= \frac{1}{\tau^2}\left\{p\left(\frac{\theta}{\tau},\frac{-1}{\tau}\right)+2E_2\left(\frac{-1}{\tau}\right)\right\}\\ &= \frac{1}{\tau^2}\wp\left(\frac{\theta}{\tau},\frac{-1}{\tau}\right).\end{aligned}$$

This proves (2.25). □

The next result is an elegant consequence of (2.23).

**Theorem 2.11.** *Suppose $m$ is a fixed positive integer and let $\omega = \exp(2\pi i/m)$. Suppose $\operatorname{Im}\tau > 0$, and let*

$$q = \exp(2\pi i\tau) \quad \textit{and} \quad q_1 = \exp(-2\pi i/m\tau).$$

*Let $j$ and $k$ be integers with $0 < j,k < m$. Then*

$$\frac{\sin\frac{\pi j}{m}}{\sin\frac{\pi k}{m}}\prod_{n=1}^{\infty}\frac{(1-\omega^j q^n)(1-\omega^{-j}q^n)}{(1-\omega^k q^n)(1-\omega^{-k}q^n)} = q_1^E\prod_{n=1}^{\infty}\frac{(1-q_1^{mn-j})(1-q_1^{mn-m+j})}{(1-q_1^{mn-k})(1-q_1^{mn-m+k})}$$

*where the value of the exponent $E$ is given by*

$$E = \frac{(2j-m)^2-(2k-m)^2}{8m} = \frac{(j+k-m)(j-k)}{2m}.$$

*Proof.* The left-hand side is

$$\frac{s(2\pi j/m,\tau)}{s(2\pi k/m,\tau)}.$$

The required identity follows on applying the transformation formula for $s$ given by (2.23) and simplifying the result. □

A simple explanation for the value of the exponent $E$ in Theorem 2.11 is that it completes the square in the quadratic exponents when the numerator and denominator are expanded by Jacobi's triple product identity. For, by (0.15) we have

$$q^E\prod_{n=1}^{\infty}\frac{(1-q^{mn-j})(1-q^{mn-m+j})}{(1-q^{mn-k})(1-q^{mn-m+k})}$$

$$= q^{((2j-m)^2-(2k-m)^2)/8m} \frac{\sum_{n=-\infty}^{\infty} (-1)^n q^{mn^2/2+n(2j-m)/2}}{\sum_{n=-\infty}^{\infty} (-1)^n q^{mn^2/2+n(2k-m)/2}}$$

$$= \frac{\sum_{n=-\infty}^{\infty} (-1)^n q^{(mn+j-m/2)^2/2m}}{\sum_{n=-\infty}^{\infty} (-1)^n q^{(mn+k-m/2)^2/2m}}.$$

Here are two special cases of the identity in Theorem 2.11. Taking $m = 5$, $j = 1$ and $k = 2$, we obtain

$$\frac{\sqrt{5}-1}{2} \prod_{n=1}^{\infty} \frac{(1-\omega q^n)(1-\omega^4 q^n)}{(1-\omega^2 q^n)(1-\omega^3 q^n)} = q_1^{1/5} \prod_{n=1}^{\infty} \frac{(1-q_1^{5n-1})(1-q_1^{5n-4})}{(1-q_1^{5n-2})(1-q_1^{5n-3})} \tag{2.26}$$

where $\omega = \exp(2\pi i/5)$, $q = \exp(2\pi i\tau)$ and $q_1 = \exp(-2\pi i/5\tau)$. The product on the right-hand side of (2.26) is the central topic of Chapter 5.

Taking $m = 8$, $j = 1$, and $k = 3$ in Theorem 2.11 we find that

$$(\sqrt{2}-1) \prod_{n=1}^{\infty} \frac{(1-\omega q^n)(1-\omega^7 q^n)}{(1-\omega^3 q^n)(1-\omega^5 q^n)} = q_1^{1/2} \prod_{n=1}^{\infty} \frac{(1-q_1^{8n-1})(1-q_1^{8n-7})}{(1-q_1^{8n-3})(1-q_1^{8n-5})} \tag{2.27}$$

where $\omega = \exp(\pi i/4)$, $q = \exp(2\pi i\tau)$ and $q_1 = \exp(-\pi i/4\tau)$. The identity (2.27) occurs again in Chapter 8; see (8.4) and Theorem 8.7.

## 2.3 The weight 2, level $\ell$, modular form $Z(\tau, \ell)$

In Corollary 2.4 we saw that the Eisenstein series $E_4(\tau), E_6(\tau), \ldots,$ are invariant under transformations by $\Gamma$, viz., for any $\begin{pmatrix} a & b \\ c & d \end{pmatrix} \in \Gamma$, we have

$$E_{2k}\left(\frac{a\tau+b}{c\tau+d}\right) = (c\tau+d)^{2k} E_{2k}(\tau), \quad k = 2, 3, \ldots. \tag{2.28}$$

For $E_2(\tau)$ the transformation formula is given by

$$E_2\left(\frac{a\tau+b}{c\tau+d}\right) = (c\tau+d)^2 E_2(\tau) + \frac{ic(c\tau+d)}{4\pi} \tag{2.29}$$

and there is an additional term on the right-hand side. It will now be shown how the additional term can be eliminated by scaling and taking a linear combination of two such functions.

Let us first consider the effect of scaling.

**Lemma 2.12.** *Let $k$ be an integer. Let $f$ be meromorphic on the upper half plane and suppose there is a function $h(\tau, c, d)$ such that the transformation formula*

$$f\left(\frac{a\tau + b}{c\tau + d}\right) = (c\tau + d)^k f(\tau) + h(c, d, \tau)$$

*holds for all* $\begin{pmatrix} a & b \\ c & d \end{pmatrix} \in \Gamma$. *Suppose $\ell \geq 1$ is an integer and let $g(\tau) = f(\ell\tau)$. Then, the transformation formula*

$$g\left(\frac{a\tau + b}{c\tau + d}\right) = (c\tau + d)^k g(\tau) + h\left(\frac{c}{\ell}, d, \ell\tau\right)$$

*holds for all* $\begin{pmatrix} a & b \\ c & d \end{pmatrix} \in \Gamma$ *that satisfy the additional condition* $c \equiv 0 \pmod{\ell}$.

Before starting the proof, it may be mentioned that by (2.29), the functions

$$f(\tau) = E_2(\tau) \quad \text{and} \quad h(c, d, \tau) = \frac{ic(c\tau + d)}{4\pi}$$

satisfy the hypotheses of Lemma 2.12 with $k = 2$. And by (2.28), the functions

$$f(\tau) = E_4(\tau),\ E_6(\tau), \ldots, \quad \text{with} \quad h(c, d, \tau) = 0$$

satisfy the hypotheses of the lemma with $k = 4,\ 6, \ldots$, respectively.

*Proof.* Let $e = c/\ell$ so that by assumption, $e$ is an integer. Then by definition,

$$g\left(\frac{a\tau + b}{c\tau + d}\right) = f\left(\frac{a\ell\tau + b\ell}{c\tau + d}\right) = f\left(\frac{a\tau_1 + b\ell}{e\tau_1 + d}\right)$$

where $\tau_1 = \ell\tau$. Since $\begin{pmatrix} a & b\ell \\ e & d \end{pmatrix} \in \Gamma$, the transformation property for $f$ implies

$$\begin{aligned} f\left(\frac{a\tau_1 + b\ell}{e\tau_1 + d}\right) &= (e\tau_1 + d)^k f(\tau_1) + h(e, d, \tau_1) \\ &= (c\tau + d)^k g(\tau) + h\left(\frac{c}{\ell}, d, \ell\tau\right). \end{aligned}$$

□

Because of the natural appearance of the condition $c \equiv 0 \pmod{\ell}$ in Lemma 2.12, we make the following definition. For any positive integer $\ell$, let

$$\Gamma_0(\ell) = \left\{ \begin{pmatrix} a & b \\ c & d \end{pmatrix} \in \Gamma \quad \text{such that} \quad c \equiv 0 \pmod{\ell} \right\}.$$

For $\ell \geq 2$, the set $\Gamma_0(\ell)$ is called the Hecke congruence subgroup of level $\ell$. It is easy to check that $\Gamma_0(\ell)$ is a subgroup of $\Gamma$. When $\ell = 1$, we simply have $\Gamma_0(1) = \Gamma$.

We are now ready to eliminate the additional term from the right-hand side of (2.29). For any integer $\ell \geq 2$, let[1] $Z(\tau, \ell)$ be defined by

$$Z(\tau, \ell) = \frac{\ell P(q^\ell) - P(q)}{\ell - 1} \tag{2.30}$$

where, as usual, $q = \exp(2\pi i \tau)$ and $\operatorname{Im} \tau > 0$. The division by $\ell - 1$ is used so that the first term in the $q$-expansion is 1. Since $P(q) = -24E_2(\tau)$, we have

$$Z(\tau, \ell) = \frac{-24}{\ell - 1} \left( \ell E_2(\ell\tau) - E_2(\tau) \right).$$

**Theorem 2.13.** *The function $Z(\tau, \ell)$ satisfies the transformation formulas*

$$Z\left( \frac{a\tau + b}{c\tau + d}, \ell \right) = (c\tau + d)^2 Z(\tau, \ell) \quad \textit{for any} \quad \begin{pmatrix} a & b \\ c & d \end{pmatrix} \in \Gamma_0(\ell), \tag{2.31}$$

*and*

$$Z\left( \frac{-1}{\tau\sqrt{\ell}}, \ell \right) = -\tau^2 Z\left( \frac{\tau}{\sqrt{\ell}}, \ell \right). \tag{2.32}$$

*Proof.* Let $\begin{pmatrix} a & b \\ c & d \end{pmatrix} \in \Gamma_0(\ell)$. Then by (2.29) and Lemma 2.12 we have

$$\begin{aligned} &Z\left( \frac{a\tau + b}{c\tau + d}, \ell \right) \\ &= \frac{-24}{\ell - 1} \left( \ell E_2 \left( \ell \frac{a\tau + b}{c\tau + d} \right) - E_2 \left( \frac{a\tau + b}{c\tau + d} \right) \right) \\ &= \frac{-24}{\ell - 1} \left( \ell \left[ (c\tau + d)^2 E_2(\ell\tau) + h\left( \frac{c}{\ell}, d, \ell\tau \right) \right] - \left[ (c\tau + d)^2 E_2(\tau) + h(c, d, \tau) \right] \right), \end{aligned}$$

[1]The function $Z(\tau, \ell)$ being defined in (2.30) is completely different from the function $Z(\theta, \omega_1, \omega_2)$ in (2.7). They can be distinguished from each other because one is a function of two variables whereas the other is a function of three variables. No confusion will arise, because the functions occur in different contexts and do not normally appear together.

where

$$h(c,d,\tau) = \frac{ic(c\tau + d)}{4\pi}.$$

Since

$$\ell\, h\left(\frac{c}{\ell}, d, \ell\tau\right) - h(c,d,\tau) = 0$$

identically, the above simplifies to

$$\begin{aligned} Z\left(\frac{a\tau + b}{c\tau + d}, \ell\right) &= \frac{-24}{\ell - 1}(c\tau + d)^2 \,(\ell\, E_2(\ell\tau) - E_2(\tau)) \\ &= (c\tau + d)^2\, Z(\tau, \ell). \end{aligned}$$

This gives (2.31).

Next, taking $\begin{pmatrix} a & b \\ c & d \end{pmatrix} = \begin{pmatrix} 0 & -1 \\ 1 & 0 \end{pmatrix}$ in (2.29) we obtain

$$E_2\left(\frac{-1}{\tau}\right) = \tau^2\, E_2(\tau) + \frac{i\tau}{4\pi}.$$

Therefore,

$$\begin{aligned} Z\left(\frac{-1}{\tau\sqrt{\ell}}, \ell\right) &= \frac{-24}{\ell - 1}\left(\ell\, E_2\left(-\frac{\sqrt{\ell}}{\tau}\right) - E_2\left(\frac{-1}{\tau\sqrt{\ell}}\right)\right) \\ &= \frac{-24}{\ell - 1}\left(\ell\left[\frac{\tau^2}{\ell} E_2\left(\frac{\tau}{\sqrt{\ell}}\right) + \frac{i\tau}{4\pi\sqrt{\ell}}\right] - \left[\tau^2 \ell\, E_2\left(\tau\sqrt{\ell}\right) + \frac{i\tau\sqrt{\ell}}{4\pi}\right]\right) \\ &= \frac{-24}{\ell - 1}\,\tau^2\left(E_2\left(\frac{\tau}{\sqrt{\ell}}\right) - \ell\, E_2(\tau\sqrt{\ell})\right) \\ &= -\tau^2\, Z\left(\frac{\tau}{\sqrt{\ell}}, \ell\right), \end{aligned}$$

and this proves (2.32). □

An analytic function defined for $\operatorname{Im}\tau > 0$ that satisfies the transformation formula

$$f\left(\frac{a\tau + b}{c\tau + d}\right) = (c\tau + d)^k f(\tau) \quad \text{for all} \quad \begin{pmatrix} a & b \\ c & d \end{pmatrix} \in \Gamma_0(\ell) \tag{2.33}$$

will be called a modular form of weight $k$ for the group $\Gamma_0(\ell)$. There is an additional requirement that a modular form should also be analytic at $i\infty$ and the other cusps, and some of this can be explained as follows. Since $\begin{pmatrix} 1 & 1 \\ 0 & 1 \end{pmatrix} \in \Gamma_0(\ell)$, the transformation formula (2.33) implies that $f(\tau + 1) = f(\tau)$, that is, a modular form is periodic with period 1. The periodicity and analyticity properties imply that a modular form has a Fourier expansion in powers of $q$. The analytic property at $i\infty$ simply means that the Fourier expansion of a modular form contains only nonnegative powers of $q$, that is,

$$f(\tau) = \sum_{n=0}^{\infty} c_n q^n, \quad q = \exp(2\pi i \tau).$$

The reader is referred to books that specialize in modular forms, e.g., [232] or [307], for details about other cusps and what it means to be analytic at those points.

We will often conflate the group $\Gamma_0(\ell)$ with its level, and simply say "$f$ is a modular form of weight $k$ and level $\ell$" as an abbreviation for "$f$ is a modular form of weight $k$ for the group $\Gamma_0(\ell)$, where $\Gamma_0(\ell)$ is the Hecke congruence subgroup of level $\ell$." There are other congruence subgroups of level $\ell$ besides $\Gamma_0(\ell)$, although we will not meet them in this book.

A meromorphic function defined for $\operatorname{Im} \tau > 0$ that satisfies the transformation formula

$$f\left(\frac{a\tau + b}{c\tau + d}\right) = f(\tau) \quad \text{for all} \quad \begin{pmatrix} a & b \\ c & d \end{pmatrix} \in \Gamma_0(\ell)$$

will be called a modular function of level $\ell$. There is also a requirement that a modular function be meromorphic at $i\infty$ and the other cusps, e.g., see [232] or [307]. This implies that a modular function has a Fourier expansion with at most finitely many nonnegative powers of $q$, that is,

$$f(\tau) = \sum_{n=-N}^{\infty} c_n q^n, \quad q = \exp(2\pi i \tau)$$

for some integer $N$.

*Example 2.14.*

1. The function $Z(\tau, \ell)$ is a modular form of weight 2 and level $\ell$, by (2.31).
2. By (2.13), $E_{2k}(\tau)$ is a modular form of weight $2k$ and level 1, provided $k \geq 2$.
3. By Theorem 2.7, $\eta^{24}(\tau)$ is a modular form of weight 12 and level 1.
4. If $f(\tau)$ is a modular form of weight $k$ and level $\ell$, and $N \geq 2$ is an integer, then the proof of Lemma 2.12 can be adapted to show that $f(N\tau)$ is a modular form of weight $k$ and level $N\ell$.

5. A modular function can be obtained by taking the quotient of two modular forms of the same weight. The level of a modular function obtained in this way will be the least common multiple of the levels of the numerator and denominator. For example, the modular $j$-invariant in (2.21), viz.,

$$j(\tau) = \frac{1728Q^3(q)}{Q^3(q) - R^2(q)} = \frac{Q^3(q)}{\eta^{24}(\tau)}, \qquad q = \exp(2\pi i\tau)$$

is a modular function of level 1. For further examples of modular functions, see the columns labeled $X$ in Tables 14.1 and 14.2 at the end of Chapter 14.
6. If $N$ is a positive integer and $(f(\tau))^N$ is a modular form of weight $k$ and level $\ell$, we will say that $f(\tau)$ is a modular form of weight $k/N$ and level $\ell$. For example, $\eta(\tau)$ is a modular form of weight 1/2 because $\eta^{24}(\tau)$ is a modular form of weight 12. See Theorem 2.9 where the factor $(c\tau + d)^{1/2}$ is made explicit; there is also a formula for the 24th root of unity given by $\epsilon(a, b, c, d)$, in this case.
7. If $f(\tau)$ is a modular form of weight $k$ and level $\ell$, and $N \geq 2$ is any integer, then $f(\tau)$ is also a modular form of weight $k$ and level $N\ell$. So the level of a modular form, as we have defined it, is not unique. We will often, but not always, take the minimal value of $\ell$ for the level.

## 2.4 Lattice subsets

Let $p$ be an odd prime and let $n$ be any integer. If $p \nmid n$ and the congruence

$$x^2 \equiv n \pmod{p}$$

has a solution, then $n$ is called a quadratic residue modulo $p$. The Legendre symbol $\left(\frac{n}{p}\right)$ is defined by

$$\left(\frac{n}{p}\right) = \begin{cases} 1 & \text{if } n \text{ is a quadratic residue modulo } p, \\ 0 & \text{if } p|n, \\ -1 & \text{otherwise.} \end{cases} \tag{2.34}$$

If $\left(\frac{n}{p}\right) = -1$ then $n$ is said to be a quadratic non-residue modulo $p$.

For example, when $p = 3$ or $p = 7$, we have

$$\left(\frac{n}{3}\right) = \chi_3(n) \quad \text{and} \quad \left(\frac{n}{7}\right) = \chi_7(n),$$

where $\chi_3$ and $\chi_7$ are the characters defined by (0.23) and (1.39), respectively.

The Legendre symbol is periodic modulo $p$, that is,

$$\left(\frac{n}{p}\right) = \left(\frac{k}{p}\right) \quad \text{if } n \equiv k \pmod{p}.$$

**Table 2.1** The numbers in R and NR, for small primes $p$

| | $p=3$ | $p=5$ | $p=7$ | $p=11$ | $p=13$ |
|---|---|---|---|---|---|
| R | 1 | 1, 4 | 1, 2, 4 | 1, 3, 4, 5, 9 | 1, 3, 4, 9, 10, 12 |
| NR | 2 | 2, 3 | 3, 5, 6 | 2, 6, 7, 8, 10 | 2, 5, 6, 7, 8, 11 |

A less trivial property, proved in almost all books on elementary number theory, is that the Legendre symbol is completely multiplicative, that is,

$$\left(\frac{nk}{p}\right) = \left(\frac{n}{p}\right)\left(\frac{k}{p}\right)$$

for all integers $n$ and $k$. For an odd prime $p$, let

$$\mathrm{R} = \left\{ r : 1 \le r \le p-1, \left(\frac{r}{p}\right) = 1 \right\}$$

and

$$\mathrm{NR} = \left\{ r : 1 \le r \le p-1, \left(\frac{r}{p}\right) = -1 \right\}.$$

Thus, R and NR are the sets of quadratic residues modulo $p$ and quadratic non-residues modulo $p$, respectively, among the numbers $1, 2, \ldots, p-1$. The sets R and NR for odd primes $p \le 13$ are given in Table 2.1.
Let $\omega_1$ and $\omega_2$ be as in Section 2.1 and let $p$ be an odd prime. Let $\Lambda^+(\omega_1, \omega_2)$, $\Lambda^-(\omega_1, \omega_2)$, $\Omega^+(\omega_1, \omega_2)$ and $\Omega^-(\omega_1, \omega_2)$ be the lattice subsets defined by

$$\Lambda^{\pm}(\omega_1, \omega_2) = \left\{ m\omega_1 + n\omega_2 : m, n \in \mathbb{Z}, \quad \left(\frac{n}{p}\right) = \pm 1 \right\}$$

and

$$\Omega^{\pm}(\omega_1, \omega_2) = \left\{ \frac{m}{p}\omega_1 + n\omega_2 : m, n \in \mathbb{Z}, \quad \left(\frac{m}{p}\right) = \pm 1 \right\}.$$

Here, and in what follows, symbols like $\pm$ or $\mp$ are used to denote two statements: one when the upper part of each symbol is used throughout, and another when the lower part of each symbol is used. The lattice subsets $\Lambda^+(\omega_1, \omega_2)$, $\Lambda^-(\omega_1, \omega_2)$, $\Omega^+(\omega_1, \omega_2)$, and $\Omega^-(\omega_1, \omega_2)$ depend on $p$ as well as $\omega_1$ and $\omega_2$. We will be interested in how the lattices depend on $\omega_1$ and $\omega_2$, whereas $p$ will remain fixed. Therefore $p$ will be omitted from the notation. Clearly,

$$\Lambda^+(\omega_1, \omega_2) \cap \Lambda^-(\omega_1, \omega_2) = \emptyset \quad \text{and} \quad \Omega^+(\omega_1, \omega_2) \cap \Omega^-(\omega_1, \omega_2) = \emptyset.$$

The next lemma records the fundamental relations between the lattice subsets.

**Lemma 2.15.** *Let $p$ be an odd prime, let $\begin{pmatrix} a & b \\ c & d \end{pmatrix} \in \Gamma_0(p)$, and let $\omega_1$, $\omega_2$, $\omega_3$ and $\omega_4$ be as in Section 2.1. Then*

$$\Lambda^{\pm}(\omega_1, \omega_2) = \Lambda^{\pm\left(\frac{d}{p}\right)}(\omega_3, \omega_4),$$

$$\Omega^{\pm}(\omega_1, \omega_2) = \Omega^{\pm\left(\frac{d}{p}\right)}(\omega_3, \omega_4)$$

*and*

$$\Lambda^{\pm}(\omega_1, \omega_2) = \Omega^{\pm}(p\omega_2, -\omega_1).$$

*Proof.* We have

$$\begin{aligned}
\Lambda^{\pm}(\omega_3, \omega_4) &= \left\{ m\omega_3 + n\omega_4 : \left(\frac{n}{p}\right) = \pm 1 \right\} \\
&= \left\{ m(c\omega_2 + d\omega_1) + n(a\omega_2 + b\omega_1) : \left(\frac{n}{p}\right) = \pm 1 \right\} \\
&= \left\{ (bn + dm)\omega_1 + (an + cm)\omega_2 : \left(\frac{n}{p}\right) = \pm 1 \right\}.
\end{aligned}$$

Make the change of variable $k = an + cm$, $j = bn + dm$, that is,

$$\begin{pmatrix} k & j \end{pmatrix} = \begin{pmatrix} n & m \end{pmatrix} \begin{pmatrix} a & b \\ c & d \end{pmatrix}, \quad \text{so that} \quad \begin{pmatrix} n & m \end{pmatrix} = \begin{pmatrix} k & j \end{pmatrix} \begin{pmatrix} d & -b \\ -c & a \end{pmatrix}.$$

By the periodic and multiplicative properties of the Legendre symbol along with the fact $c \equiv 0 \pmod{p}$, we have

$$\begin{aligned}
\Lambda^{\pm}(\omega_3, \omega_4) &= \left\{ j\omega_1 + k\omega_2 : \left(\frac{kd - cj}{p}\right) = \pm 1 \right\} \\
&= \left\{ j\omega_1 + k\omega_2 : \left(\frac{k}{p}\right)\left(\frac{d}{p}\right) = \pm 1 \right\} \\
&= \left\{ j\omega_1 + k\omega_2 : \left(\frac{k}{p}\right) = \pm \left(\frac{d}{p}\right) \right\} \\
&= \Lambda^{\pm\left(\frac{d}{p}\right)}(\omega_1, \omega_2).
\end{aligned}$$

This proves the first result. Next, let $e = c/p$ and observe that $e$ is an integer by hypothesis. Now

$$\Omega^{\pm}(\omega_3, \omega_4) = \left\{ \frac{m}{p}\omega_3 + n\omega_4 : \left(\frac{m}{p}\right) = \pm 1 \right\}$$

$$= \left\{ \frac{m}{p}(c\omega_2 + d\omega_1) + n(a\omega_2 + b\omega_1) : \left(\frac{m}{p}\right) = \pm 1 \right\}$$
$$= \left\{ \frac{(bnp + dm)}{p}\omega_1 + (an + em)\omega_2 : \left(\frac{m}{p}\right) = \pm 1 \right\}.$$

Make the change of variable

$$\begin{pmatrix} k & j \end{pmatrix} = \begin{pmatrix} n & m \end{pmatrix} \begin{pmatrix} a & pb \\ e & d \end{pmatrix} \quad \text{so that} \quad \begin{pmatrix} n & m \end{pmatrix} = \begin{pmatrix} k & j \end{pmatrix} \begin{pmatrix} d & -pb \\ -e & a \end{pmatrix}$$

and observe that since $ad = 1 + bc \equiv 1 \pmod{p}$, it follows that

$$\left(\frac{a}{p}\right)\left(\frac{d}{p}\right) = \left(\frac{ad}{p}\right) = \left(\frac{1}{p}\right) = 1 \quad \text{and so} \quad \left(\frac{a}{p}\right) = \left(\frac{d}{p}\right).$$

Therefore

$$\begin{aligned} \Omega^{\pm}(\omega_3, \omega_4) &= \left\{ \frac{j}{p}\omega_1 + k\omega_2 : \left(\frac{aj - pbk}{p}\right) = 1 \right\} \\ &= \left\{ \frac{j}{p}\omega_1 + k\omega_2 : \left(\frac{a}{p}\right)\left(\frac{j}{p}\right) = 1 \right\} \\ &= \left\{ \frac{j}{p}\omega_1 + k\omega_2 : \left(\frac{d}{p}\right)\left(\frac{j}{p}\right) = 1 \right\} \\ &= \Omega^{\pm\left(\frac{d}{p}\right)}(\omega_3, \omega_4). \end{aligned}$$

This proves the second result. The third result follows immediately from the definitions of $\Lambda^{\pm}$ and $\Omega^{\pm}$. □

## 2.5 Elliptic functions for level $p$

For each odd prime $p$ we will define two functions $F$ and $G$, show that they are elliptic functions and establish their transformation properties. Recall the function $s(\theta, \tau)$ defined by (0.51) and studied in Section 1.1:

$$s(\theta, \tau) = 2 \sin \frac{\theta}{2} \prod_{n=1}^{\infty} \frac{(1 - e^{i\theta}q^n)(1 - e^{-i\theta}q^n)}{(1 - q^n)^2}.$$

It is an entire function of $\theta$. Its zeros are simple and are located at points in the lattice $\Lambda(2\pi, 2\pi\tau)$, and it obeys the quasi-periodic properties

$$s(\theta + 2\pi, \tau) = -s(\theta, \tau) \tag{2.35}$$

and

$$s(\theta + 2\pi\tau, \tau) = -e^{-i\theta} q^{-1/2} s(\theta, \tau). \tag{2.36}$$

Let $p$ be an odd prime and R and NR be the sets of quadratic residues and quadratic non-residues modulo $p$, respectively. Let $F$ and $G$ be the functions defined by

$$F(\theta, \omega_1, \omega_2, p) = \frac{\prod_{r \in \mathrm{R}} \exp\left(\frac{-2\pi i r\theta}{p\omega_1}\right) s\left(\frac{2\pi}{\omega_1}(\theta - r\omega_2)\,,\, \frac{p\omega_2}{\omega_1}\right)}{\prod_{r \in \mathrm{NR}} \exp\left(\frac{-2\pi i r\theta}{p\omega_1}\right) s\left(\frac{2\pi}{\omega_1}(\theta - r\omega_2)\,,\, \frac{p\omega_2}{\omega_1}\right)}$$

and

$$G(\theta, \omega_1, \omega_2, p) = \frac{\prod_{r \in \mathrm{R}} s\left(2\pi\left(\frac{\theta}{\omega_1} - \frac{r}{p}\right)\,,\, \frac{\omega_2}{\omega_1}\right)}{\prod_{r \in \mathrm{NR}} s\left(2\pi\left(\frac{\theta}{\omega_1} - \frac{r}{p}\right)\,,\, \frac{\omega_2}{\omega_1}\right)}.$$

For example, when $p = 3$,

$$G(\theta, 2\pi, 2\pi\tau, 3) = \frac{s(\theta - \frac{2\pi}{3})}{s(\theta - \frac{4\pi}{3})} = \frac{s(\frac{2\pi}{3} - \theta)}{s(\theta + \frac{2\pi}{3})} = \mathrm{cm}^*\,\theta,$$

where $\mathrm{cm}^*\,\theta$ is one of the Dixon elliptic functions studied in Section 1.10.

**Theorem 2.16.** *The functions $F(\theta, \omega_1, \omega_2, p)$ and $G(\theta, \omega_1, \omega_2, p)$ are elliptic functions of $\theta$. The periods, zeros, poles, and orders of the elliptic functions are given in Tables 2.2 and 2.3. All of the zeros and poles have order 1. Moreover, F and G are homogeneous of degree 0; that is, for any nonzero complex number $\lambda$,*

$$F(\lambda\theta, \lambda\omega_1, \lambda\omega_2, p) = F(\theta, \omega_1, \omega_2, p)$$

**Table 2.2** Periods, zeros, poles, and orders of elliptic functions for $p > 3$

| | periods | zeros | poles | order |
|---|---|---|---|---|
| $F(\theta, \omega_1, \omega_2, p)$ | $\omega_1,\ p\omega_2$ | $\theta \in \Lambda^+(\omega_1, \omega_2)$ | $\theta \in \Lambda^-(\omega_1, \omega_2)$ | $(p-1)/2$ |
| $G(\theta, \omega_1, \omega_2, p)$ | $\omega_1,\ \omega_2$ | $\theta \in \Omega^+(\omega_1, \omega_2)$ | $\theta \in \Omega^-(\omega_1, \omega_2)$ | $(p-1)/2$ |

**Table 2.3** Periods, zeros, poles, and orders of elliptic functions when $p = 3$

| | periods | zeros | poles | order |
|---|---|---|---|---|
| $F(\theta, \omega_1, \omega_2, 3)$ | $3\omega_1,\ 3\omega_2$ | $\theta \in \Lambda^+(\omega_1, \omega_2)$ | $\theta \in \Lambda^-(\omega_1, \omega_2)$ | 3 |
| $G(\theta, \omega_1, \omega_2, 3)$ | $\omega_1,\ 3\omega_2$ | $\theta \in \Omega^+(\omega_1, \omega_2)$ | $\theta \in \Omega^-(\omega_1, \omega_2)$ | 3 |

*and*

$$G(\lambda\theta, \lambda\omega_1, \lambda\omega_2, p) = G(\theta, \omega_1, \omega_2, p).$$

*Proof.* First consider the function $F(\theta, \omega_1, \omega_2, p)$. By (2.35),

$$\frac{F(\theta + \omega_1, \omega_1, \omega_2, p)}{F(\theta, \omega_1, \omega_2, p)} = \frac{\prod_{r\in \mathrm{R}} \exp(-2\pi i r/p)}{\prod_{r\in \mathrm{NR}} \exp(-2\pi i r/p)} = \exp\left(\frac{-2\pi i}{p} \sum_{r=1}^{p-1} r\left(\frac{r}{p}\right)\right).$$

If $p > 3$, then by Exercise 1 at the end of the chapter we have

$$\sum_{r=1}^{p-1} r\left(\frac{r}{p}\right) \equiv 0 \pmod{p},$$

and it follows that $\omega_1$ is a period of $F$. If $p = 3$ then

$$F(\theta + \omega_1, \omega_1, \omega_2, 3) = e^{2\pi i/3} F(\theta, \omega_1, \omega_2, 3)$$

and it follows that $3\omega_1$ is a period of $F$. In a similar way using (2.36) we find that for any odd prime $p$,

$$\frac{F(\theta + p\omega_2, \omega_1, \omega_2, p)}{F(\theta, \omega_1, \omega_2, p)} = 1,$$

therefore $p\omega_2$ is a period of $F$. Hence $F(\theta, \omega_1, \omega_2, p)$ is doubly periodic. By definition, $F(\theta, \omega_1, \omega_2, p)$ is a quotient of entire functions, therefore it is meromorphic. Consequently, $F$ is an elliptic function.

The zeros of $F$ are located at the zeros of $s(2\pi(\theta - r\omega_2)/\omega_1,\ p\omega_2/\omega_1)$, where $r \in \mathrm{R}$. The zeros are simple and occur when

$$2\pi(\theta - r\omega_2)/\omega_1 = 2m\pi + 2np\pi\omega_2/\omega_1, \quad \text{where} \quad r \in \mathrm{R}.$$

It follows that the zeros of $F$ are simple and are located at the points

$$\theta = m\omega_1 + n\omega_2, \quad \text{where} \quad \left(\frac{n}{p}\right) = 1.$$

That is, the zeros of $F$ are simple and are located at the points $\theta \in \Lambda^+(\omega_1, \omega_2)$, and these are the only zeros. The poles of $F$ may be determined similarly.

The statements about the order of $F$ follow by counting the poles in a period parallelogram and observing that there is no smaller period parallelogram. The homogeneity property follows directly from the definition.

The results for $G$ may be proved similarly. □

The transformation properties for $F$ and $G$ are as follows.

**Theorem 2.17.** *Let* $\begin{pmatrix} a & b \\ c & d \end{pmatrix} \in \Gamma_0(p)$*, and let* $\omega_1$*,* $\omega_2$*,* $\omega_3$*, and* $\omega_4$ *be as in Section 2.1. Then*

$$\frac{F(\theta,\omega_1,\omega_2,p)}{F(0,\omega_1,\omega_2,p)} = \left(\frac{F(\theta,\omega_3,\omega_4,p)}{F(0,\omega_3,\omega_4,p)}\right)^{\left(\frac{d}{p}\right)},$$

$$\frac{G(\theta,\omega_1,\omega_2,p)}{G(0,\omega_1,\omega_2,p)} = \left(\frac{G(\theta,\omega_3,\omega_4,p)}{G(0,\omega_3,\omega_4,p)}\right)^{\left(\frac{d}{p}\right)}$$

*and*

$$\frac{F(\theta,\omega_1,\omega_2,p)}{F(0,\omega_1,\omega_2,p)} = \frac{G(\theta,p\omega_2,-\omega_1,p)}{G(0,p\omega_2,-\omega_1,p)}.$$

*Proof.* First let us consider the case $\left(\frac{d}{p}\right) = 1$. By Lemma 2.15 and Theorem 2.16, the functions $F(\theta,\omega_1,\omega_2,p)$ and $F(\theta,\omega_3,\omega_4,p)$ have the same zeros and the same poles, namely the points in $\Lambda^+(\omega_1,\omega_2) = \Lambda^+(\omega_3,\omega_4)$ and the points in $\Lambda^-(\omega_1,\omega_2) = \Lambda^-(\omega_3,\omega_4)$, respectively. Furthermore both functions are doubly periodic, with periods $\omega_1$ and $p\omega_2$ if $p > 3$, and with periods $3\omega_1$ and $3\omega_2$ if $p = 3$. To see this, observe that for $p > 3$,

$$\begin{aligned} F(\theta+\omega_1,\omega_3,\omega_4,p) &= F(\theta + a\omega_3 - c\omega_4,\omega_3,\omega_4,p) \\ &= F(\theta,\omega_3,\omega_4,p), \end{aligned}$$

where the last equality holds because $c \equiv 0 \pmod{p}$, and by the periodicity properties of $F(\theta,\omega_3,\omega_4,p)$. Similarly we find

$$F(\theta + p\omega_2,\omega_3,\omega_4,p) = F(\theta,\omega_3,\omega_4,p).$$

Thus for $p > 3$, $F(\theta,\omega_1,\omega_2,p)$ and $F(\theta,\omega_3,\omega_4,p)$ are both doubly periodic with periods $\omega_1$ and $p\omega_2$. Similar calculations show that if $p = 3$, then $F(\theta,\omega_1,\omega_2,3)$ and $F(\theta,\omega_3,\omega_4,3)$ are both doubly periodic with periods $3\omega_1$ and $3\omega_2$.

Hence, for any odd prime $p$, $F(\theta,\omega_1,\omega_2,p)/F(\theta,\omega_3,\omega_4,p)$ is an entire elliptic function, which by Liouville's theorem is constant. It follows that

$$\frac{F(\theta,\omega_1,\omega_2,p)}{F(\theta,\omega_3,\omega_4,p)} = \frac{F(0,\omega_1,\omega_2,p)}{F(0,\omega_3,\omega_4,p)},$$

and this proves the first result in the case that $\left(\frac{d}{p}\right) = 1$.

In the case that $\left(\frac{d}{p}\right) = -1$, the zeros and poles of $F(\theta, \omega_1, \omega_2, p)$ coincide with the poles and zeros, respectively, of $F(\theta, \omega_3, \omega_4, p)$. It follows that the product $F(\theta, \omega_1, \omega_2, p)F(\theta, \omega_3, \omega_4, p)$ is an entire elliptic function, which is therefore constant. The value of the constant may be determined by setting $\theta = 0$.

Combining the two cases gives the first result in the theorem. The other two results may be proved similarly. □

## 2.6 Further elliptic functions for level *p*

We shall now analyze the logarithmic derivatives of $F$ and $G$. We continue to assume that $p$ is an odd prime and let

$$f(\theta, \omega_1, \omega_2, p) = \frac{F'(\theta, \omega_1, \omega_2, p)}{F(\theta, \omega_1, \omega_2, p)} \quad \text{and} \quad g(\theta, \omega_1, \omega_2, p) = \frac{G'(\theta, \omega_1, \omega_2, p)}{G(\theta, \omega_1, \omega_2, p)}, \tag{2.37}$$

where the primes denote differentiation with respect to $\theta$. The transformation properties for $f$ and $g$ are inherited from the corresponding properties of $F$ and $G$, and the results are as follows.

**Theorem 2.18.** *Let* $\begin{pmatrix} a & b \\ c & d \end{pmatrix} \in \Gamma_0(p)$, *and let* $\omega_1$, $\omega_2$, $\omega_3$, *and* $\omega_4$ *be as in Section 2.1. Then*

$$f(\theta, \omega_1, \omega_2, p) = \left(\frac{d}{p}\right) f(\theta, \omega_3, \omega_4, p),$$

$$g(\theta, \omega_1, \omega_2, p) = \left(\frac{d}{p}\right) g(\theta, \omega_3, \omega_4, p)$$

*and*

$$f(\theta, \omega_1, \omega_2, p) = g(\theta, p\omega_2, -\omega_1, p).$$

*Proof.* Apply logarithmic differentiation to the results in Theorem 2.17. □

The next goal is to obtain Fourier expansions for $f$ and $g$. Two cases will arise according to whether $p$ is congruent to 1 or 3 modulo 4, and it will be useful to introduce some notation for this. For an odd prime $p$, let $c_p$ and $t_p$ be defined by

$$c_p = \begin{cases} 1, & \text{if } p \equiv 1 \pmod 4, \\ i, & \text{if } p \equiv 3 \pmod 4 \end{cases} \tag{2.38}$$

and

$$t_p(\theta) = \begin{cases} \sin\theta, & \text{if } p \equiv 1 \pmod 4, \\ \cos\theta, & \text{if } p \equiv 3 \pmod 4. \end{cases}$$

We require a technical lemma.

**Lemma 2.19.** *Let p be an odd prime, let n be a positive integer and let u be a complex variable. Then*

$$\sum_{r=1}^{p-1} \left(\frac{r}{p}\right) e^{2irn\pi/p} = \sum_{r=0}^{p-1} e^{2ir^2 n\pi/p} = c_p\sqrt{p}\left(\frac{n}{p}\right), \tag{2.39}$$

$$\sum_{r=1}^{p-1} \left(\frac{r}{p}\right) \cot\left(u - \frac{r\pi}{p}\right) = \frac{2i\sqrt{p}}{c_p(e^{2ipu}-1)} \sum_{k=1}^{p-1} \left(\frac{k}{p}\right) e^{2iku} \tag{2.40}$$

*and*

$$\sum_{r=1}^{p-1} \left(\frac{r}{p}\right) \sin 2n\left(u - \frac{r\pi}{p}\right) = c_p^2\sqrt{p}\left(\frac{n}{p}\right) t_p(2nu). \tag{2.41}$$

*Proof.* The first equality in (2.39) follows by simple rearrangement, considering the quadratic residues and quadratic non-residues separately. The second equality in (2.39) is a standard result for Gauss sums; see Exercise 2 at the end of the chapter for a proof.

Let us prove (2.40). To begin with, assume $\operatorname{Im} u > 0$. Using

$$\cot x = -i\left(1 + \frac{2e^{2ix}}{1-e^{2ix}}\right)$$

and writing $\overline{\theta}$ for the complex conjugate of $\theta$, we have

$$\begin{aligned} \sum_{r=1}^{p-1} \left(\frac{r}{p}\right) \cot\left(u - \frac{r\pi}{p}\right) &= -i\sum_{r=1}^{p-1}\left(\frac{r}{p}\right) - 2i\sum_{r=1}^{p-1}\left(\frac{r}{p}\right)\frac{e^{2i(u-r\pi/p)}}{1-e^{2i(u-r\pi/p)}} \\ &= 0 - 2i\sum_{r=1}^{p-1}\left(\frac{r}{p}\right)\sum_{k=1}^{\infty} e^{2ik(u-r\pi/p)} \\ &= -2i\sum_{k=1}^{\infty} e^{2iku}\sum_{r=1}^{p-1}\left(\frac{r}{p}\right)\overline{e^{2ikr\pi/p}} \end{aligned}$$

$$= -2i\sqrt{p}\,\overline{c_p}\sum_{k=1}^{\infty}\left(\frac{k}{p}\right)e^{2iku}$$

$$= \frac{-2i\sqrt{p}}{c_p}\frac{\sum_{k=1}^{p-1}\left(\frac{k}{p}\right)e^{2iku}}{(1-e^{2ipu})}.$$

The identity in (2.39) has been used in the second-to-last step. This proves (2.40) for $\operatorname{Im} u > 0$. The result holds true for other complex values of $u$ by analytic continuation, as both sides are meromorphic functions of $u$. This completes the proof of (2.40).

It remains to prove (2.41). We have

$$\sum_{r=1}^{p-1}\left(\frac{r}{p}\right)e^{2in(u-r\pi/p)} = e^{2inu}\sum_{r=1}^{p-1}\left(\frac{r}{p}\right)\overline{e^{2irn\pi/p}}$$

$$= e^{2inu}\sqrt{p}\left(\frac{n}{p}\right)\overline{c_p},$$

where (2.39) has been used to obtain the last step. Equation (2.41) now follows for real values of $u$ by equating imaginary parts of both sides. The result holds for complex values of $u$ by analytic continuation. This completes the proof of (2.41).

□

Here are the Fourier expansions of $f$ and $g$.

**Theorem 2.20.** *Let $p$ be an odd prime and let $\omega_1$, $\omega_2$ and $q$ be as in Section 2.1. Let*

$$A_p = \frac{1}{p}\sum_{r=1}^{p-1} r\left(\frac{r}{p}\right).$$

*If $p \equiv 1 \pmod 4$ then the following Fourier expansions hold:*

$$f(\theta,\omega_1,\omega_2,p) = \frac{4\pi}{\omega_1}\sum_{j=1}^{\infty}\frac{\sum_{k=1}^{p-1}\left(\frac{k}{p}\right)q^{jk}}{1-q^{pj}}\sin\left(\frac{2\pi j\theta}{\omega_1}\right)$$

*and*

$$g(\theta,\omega_1,\omega_2,p) = \frac{4\pi\sqrt{p}}{\omega_1}\left(\frac{i\sum_{k=1}^{p-1}\left(\frac{k}{p}\right)e^{2\pi ik\theta/\omega_1}}{2(e^{2\pi ip\theta/\omega_1}-1)} + \sum_{j=1}^{\infty}\left(\frac{j}{p}\right)\frac{q^j}{1-q^j}\sin\left(\frac{2\pi j\theta}{\omega_1}\right)\right).$$

*If* $p \equiv 3 \pmod 4$ *then the following Fourier expansions hold*:

$$f(\theta,\omega_1,\omega_2,p) = \frac{4\pi i}{\omega_1}\left(\frac{-A_p}{2} + \sum_{j=1}^{\infty} \frac{\sum_{k=1}^{p-1}\left(\frac{k}{p}\right) q^{jk}}{1-q^{pj}} \cos\left(\frac{2\pi j\theta}{\omega_1}\right)\right)$$

*and*

$$g(\theta,\omega_1,\omega_2,p) = \frac{4\pi\sqrt{p}}{\omega_1}\left(\frac{\sum_{k=1}^{p-1}\left(\frac{k}{p}\right) e^{2\pi i k\theta/\omega_1}}{2(e^{2\pi i p\theta/\omega_1}-1)} - \sum_{j=1}^{\infty}\left(\frac{j}{p}\right)\frac{q^j}{1-q^j}\cos\left(\frac{2\pi j\theta}{\omega_1}\right)\right).$$

*Proof.* First note that

$$\cot\theta = i + \frac{2ie^{-2i\theta}}{1-e^{-2i\theta}}.$$

By the definitions of $f$ and $F$,

$$\begin{aligned} f(\theta,\omega_1,\omega_2,p) &= \frac{\mathrm{d}}{\mathrm{d}\theta}\log F(\theta,\omega_1,\omega_2,p) \\ &= \sum_{r=1}^{p-1}\left(\frac{r}{p}\right)\left(\frac{-2\pi i r}{p\omega_1} + \frac{2\pi}{\omega_1}\frac{s'}{s}\left(\frac{2\pi}{\omega_1}(\theta - r\omega_2)\,,\ \frac{p\omega_2}{\omega_1}\right).\right) \end{aligned}$$

By (1.4), we have $s'(\theta)/s(\theta) = z(\theta)$. Using the expansion of $z(\theta)$ given by (1.15), along with the definition of $A_p$ in the statement of the theorem, we find that

$$\begin{aligned} &f(\theta,\omega_1,\omega_2,p) \\ &= \frac{-2\pi i}{\omega_1}\left(A_p + \sum_{r=1}^{p-1}\left(\frac{r}{p}\right)\sum_{k=1}^{\infty}\left[\frac{q^{pk-r}e^{2\pi i\theta/\omega_1}}{1-q^{pk-r}e^{2\pi i\theta/\omega_1}} - \frac{q^{pk-p+r}e^{-2\pi i\theta/\omega_1}}{1-q^{pk-p+r}e^{-2\pi i\theta/\omega_1}}\right]\right) \\ &= \frac{-2\pi i}{\omega_1}\left(A_p + \sum_{k=1}^{\infty}\left(\frac{k}{p}\right)\left[\left(\frac{-1}{p}\right)\frac{q^k e^{2\pi i\theta/\omega_1}}{1-q^k e^{2\pi i\theta/\omega_1}} - \frac{q^k e^{-2\pi i\theta/\omega_1}}{1-q^k e^{-2\pi i\theta/\omega_1}}\right]\right) \\ &= \frac{-2\pi i}{\omega_1}\left(A_p + \sum_{k=1}^{\infty}\sum_{j=1}^{\infty}\left(\frac{k}{p}\right)q^{jk}\left[\left(\frac{-1}{p}\right)e^{2\pi i j\theta/\omega_1} - e^{-2\pi i j\theta/\omega_1}\right]\right) \\ &= \frac{-2\pi i}{\omega_1}\left(A_p + 2i\sum_{j=1}^{\infty}\sum_{k=1}^{\infty}\left(\frac{k}{p}\right)q^{jk}c_p t_p\left(\frac{2\pi j\theta}{\omega_1}\right)\right) \\ &= \frac{4\pi c_p}{\omega_1}\left(\frac{-A_p}{2} + \sum_{j=1}^{\infty}\frac{\sum_{k=1}^{p-1}\left(\frac{k}{p}\right)q^{jk}}{1-q^{pj}}t_p\left(\frac{2\pi j\theta}{\omega_1}\right)\right). \end{aligned}$$

The fact $A_p = 0$ if $p \equiv 1 \pmod 4$ has been used to obtain the last step; see Exercise 1 at the end of the chapter. On separating the formula into cases according to the residue of $p$ modulo 4 we obtain the required results for $f$.

Next,

$$\begin{aligned}
&g(\theta, \omega_1, \omega_2, p)\\
&= \frac{\mathrm{d}}{\mathrm{d}\theta} \log G(\theta, \omega_1, \omega_2, p)\\
&= \frac{2\pi}{\omega_1} \sum_{r=1}^{p-1} \left(\frac{r}{p}\right) z\left(2\pi\left(\frac{\theta}{\omega_1} - \frac{r}{p}\right), \frac{\omega_2}{\omega_1}\right)\\
&= \frac{2\pi}{\omega_1} \sum_{r=1}^{p-1} \left(\frac{r}{p}\right) \left\{\frac{1}{2}\cot \pi\left(\frac{\theta}{\omega_1} - \frac{r}{p}\right) + 2\sum_{j=1}^{\infty} \frac{q^j}{1-q^j} \sin 2\pi j\left(\frac{\theta}{\omega_1} - \frac{r}{p}\right)\right\}.
\end{aligned}$$

Now apply (2.40) and (2.41) to obtain

$$\begin{aligned}
&g(\theta, \omega_1, \omega_2, p)\\
&= \frac{4\pi\sqrt{p}}{\omega_1} \left(\frac{i}{2c_p} \frac{\sum_{k=1}^{p-1} \left(\frac{k}{p}\right) e^{2\pi i k\theta/\omega_1}}{(e^{2\pi i p\theta/\omega_1} - 1)} + c_p^2 \sum_{j=1}^{\infty} \left(\frac{j}{p}\right) \frac{q^j}{1-q^j}\, t_p\left(\frac{2\pi j\theta}{\omega_1}\right)\right).
\end{aligned}$$

The proof may be completed by separating into cases according the value of $p$ modulo 4. □

The function $f(\theta, 2\pi, 2\pi\tau, 3)$ occurs in Exercise 17 of Chapter 3, and the functions $f(\theta, 2\pi, 2\pi\tau, 5)$ and $g(\theta, 2\pi, 2\pi\tau, 5)$ will be encountered in Exercise 7 of Chapter 5.

## 2.7 Eisenstein series for $\Gamma_0(p)$

The Fourier expansions in Theorem 2.20 can be expanded further in powers of $\theta$. We will examine the coefficients that arise.

Let $p$ be an odd prime. The generalized Bernoulli numbers $B_{n,p}$ are defined by

$$\frac{x}{e^{px} - 1} \sum_{k=1}^{p-1} \left(\frac{k}{p}\right) e^{kx} = \sum_{n=0}^{\infty} B_{n,p} \frac{x^n}{n!}. \tag{2.42}$$

Letting $x \to 0$, we deduce that $B_{0,p} = 0$. If we divide by $x$ and take the limit as $x \to 0$, we find that

$$B_{1,p} = \frac{1}{p} \sum_{r=1}^{p-1} r\left(\frac{r}{p}\right),$$

that is, $B_{1,p} = A_p$ where $A_p$ is defined in Theorem 2.20. It is easily checked that the left-hand side of (2.42) is an even or odd function of $x$, depending on whether $p \equiv 1$ or $3 \pmod 4$, respectively. Consequently, $B_{2n+1,p} = 0$ if $p \equiv 1 \pmod 4$, and $B_{2n,p} = 0$ if $p \equiv 3 \pmod 4$.

For an odd prime $p$ and any positive integer $n$ that satisfies $n \equiv (p-1)/2 \pmod 2$, let $E_n^0(\tau, p)$ and $E_n^\infty(\tau, p)$ be the Eisenstein series defined by

$$E_n^0(\tau, p) = -\frac{B_{1,p}}{2}\delta_{n,1} + \sum_{j=1}^{\infty} \frac{j^{n-1}}{1-q^{pj}} \sum_{k=1}^{p-1} \left(\frac{k}{p}\right) q^{jk} \tag{2.43}$$

and

$$E_n^\infty(\tau, p) = -\frac{B_{n,p}}{2n} + \sum_{j=1}^{\infty} \left(\frac{j}{p}\right) \frac{j^{n-1}q^j}{1-q^j}, \tag{2.44}$$

where, as usual $q = \exp(2\pi i \tau)$, and

$$\delta_{m,n} = \begin{cases} 1 \text{ if } m = n, \\ 0 \text{ if } m \neq n \end{cases}$$

is the Kronecker delta. The Eisenstein series $E_n^0(\tau, p)$ and $E_n^\infty(\tau, p)$, along with the condition $n \equiv (p-1)/2 \pmod 2$, arise naturally in the power series expansions of $f$ and $g$. For, since $\tau = \omega_2/\omega_1$ and $q = \exp(2\pi i \tau)$, expanding the results of Theorem 2.20 in powers of $\theta$ gives

$$\begin{aligned} &f(\theta, \omega_1, \omega_2, p) \\ &= \begin{cases} \dfrac{4\pi}{\omega_1} \displaystyle\sum_{n=0}^{\infty} E_{2n+2}^0(\tau, p) \frac{(-1)^n}{(2n+1)!} \left(\frac{2\pi\theta}{\omega_1}\right)^{2n+1} & \text{if } p \equiv 1 \pmod 4, \\ \dfrac{4\pi i}{\omega_1} \displaystyle\sum_{n=0}^{\infty} E_{2n+1}^0(\tau, p) \frac{(-1)^n}{(2n)!} \left(\frac{2\pi\theta}{\omega_1}\right)^{2n} & \text{if } p \equiv 3 \pmod 4 \end{cases} \end{aligned} \tag{2.45}$$

and

$$\begin{aligned} &g(\theta, \omega_1, \omega_2, p) \\ &= \begin{cases} \dfrac{4\pi\sqrt{p}}{\omega_1} \displaystyle\sum_{n=0}^{\infty} E_{2n+2}^\infty(\tau, p) \frac{(-1)^n}{(2n+1)!} \left(\frac{2\pi\theta}{\omega_1}\right)^{2n+1} & \text{if } p \equiv 1 \pmod 4, \\ \dfrac{-4\pi\sqrt{p}}{\omega_1} \displaystyle\sum_{n=0}^{\infty} E_{2n+1}^\infty(\tau, p) \frac{(-1)^n}{(2n)!} \left(\frac{2\pi\theta}{\omega_1}\right)^{2n} & \text{if } p \equiv 3 \pmod 4. \end{cases} \end{aligned} \tag{2.46}$$

The transformation formulas for $f$ and $g$ in Theorem 2.18 imply transformation formulas for Eisenstein series.

**Theorem 2.21.** *Let $p$ be an odd prime, let $\begin{pmatrix} a & b \\ c & d \end{pmatrix} \in \Gamma_0(p)$, and let $n$ be a positive integer that satisfies $n \equiv (p-1)/2 \pmod 2$. Let $E_n^0(\tau, p)$ and $E_n^\infty(\tau, p)$ be defined by* (2.43) *and* (2.44), *respectively, and let $c_p$ be defined by* (2.38). *Then*

$$
\begin{aligned}
E_n^0\left(\frac{a\tau+b}{c\tau+d}, p\right) &= \left(\frac{d}{p}\right)(c\tau+d)^n E_n^0(\tau, p), \\
E_n^\infty\left(\frac{a\tau+b}{c\tau+d}, p\right) &= \left(\frac{d}{p}\right)(c\tau+d)^n E_n^\infty(\tau, p), \\
E_n^\infty\left(\frac{-1}{p\tau}, p\right) &= \frac{1}{c_p\sqrt{p}}(p\tau)^n E_n^0(\tau, p), \\
E_n^0\left(\frac{-1}{p\tau}, p\right) &= \frac{\sqrt{p}}{c_p}\tau^n E_n^\infty(\tau, p).
\end{aligned}
$$

*Proof.* The first three results may be obtained by substituting the expansions (2.45) and (2.46) into the results of Theorem 2.18 and equating coefficients of powers of $\theta$. The fourth result follows from the third by replacing $\tau$ with $-1/p\tau$. □

Let us consider in detail the simplest case $n = 1$. The corresponding primes $p$ are of the form $p \equiv 3 \pmod 4$. Expanding as a double series and interchanging the order of summation we obtain

$$
\begin{aligned}
E_1^0(\tau, p) &= -\frac{B_{1,p}}{2} + \sum_{j=1}^\infty \frac{1}{1-q^{pj}} \sum_{k=1}^{p-1} \left(\frac{k}{p}\right) q^{jk} \\
&= -\frac{B_{1,p}}{2} + \sum_{j=1}^\infty \sum_{k=1}^\infty \left(\frac{k}{p}\right) q^{jk} \\
&= -\frac{B_{1,p}}{2} + \sum_{k=1}^\infty \left(\frac{k}{p}\right) \frac{q^k}{1-q^k} \\
&= E_1^\infty(\tau, p).
\end{aligned}
$$

It should be noted that $E_n^0(\tau, p) \neq E_n^\infty(\tau, p)$, if $n \geq 2$.

Using the fact that $E_1^\infty(\tau, p) = E_1^0(\tau, p)$, the transformation formulas in Theorem 2.21 reduce to

$$
E_1^0\left(\frac{a\tau+b}{c\tau+d}, p\right) = \left(\frac{d}{p}\right)(c\tau+d)E_1^0(\tau, p) \quad \text{for any } \begin{pmatrix} a & b \\ c & d \end{pmatrix} \in \Gamma_0(p),
$$

and

$$E_1^0\left(\frac{-1}{p\tau}, p\right) = \sqrt{p}\,\frac{\tau}{i}\,E_1^0(\tau, p),$$

where $p$ is a prime that is congruent to 3 modulo 4.

For the primes $p < 50$ and $p \equiv 3 \pmod 4$, we have

$$E_1^0(\tau, 3) = E_1^\infty(\tau, 3) = \frac{1}{6} + \sum_{k=1}^{\infty}\left(\frac{k}{3}\right)\frac{q^k}{1-q^k}, \tag{2.47}$$

$$E_1^0(\tau, p) = E_1^\infty(\tau, p) = \frac{1}{2} + \sum_{k=1}^{\infty}\left(\frac{k}{p}\right)\frac{q^k}{1-q^k}, \qquad \text{for } p = 7, 11, 19, 43,$$

$$E_1^0(\tau, p) = E_1^\infty(\tau, p) = \frac{3}{2} + \sum_{k=1}^{\infty}\left(\frac{k}{p}\right)\frac{q^k}{1-q^k}, \qquad \text{for } p = 23, 31,$$

$$E_1^0(\tau, p) = E_1^\infty(\tau, p) = \frac{5}{2} + \sum_{k=1}^{\infty}\left(\frac{k}{p}\right)\frac{q^k}{1-q^k}, \qquad \text{for } p = 47.$$

We state without proof or explanation that in general, for any prime $p > 3$ of the form $p \equiv 3 \pmod 4$,

$$E_1^0(\tau, p) = E_1^\infty(\tau, p) = \frac{h(p)}{2} + \sum_{k=1}^{\infty}\left(\frac{k}{p}\right)\frac{q^k}{1-q^k}$$

where $h(p)$ is the class number for an imaginary quadratic field of discriminant $-p$; see the books by Landau [203, Part Four] and Cox [133, p. 149].

## 2.8 Notes

The usual construction of Eisenstein series for the modular group begins with the function

$$G_k(\tau) = \sum_n \sum_m{}' \frac{1}{(m+n\tau)^k},$$

where the prime means that the summation is over all integer values of $m$ and $n$ except $(m, n) = (0, 0)$, and $k$ is a positive integer. If $k > 2$, the series is absolutely convergent. If $k$ is odd, the terms corresponding to $(m, n)$ and $(-m, -n)$ cancel and the series simplifies identically to zero. Therefore, we consider only the series

$$G_{2k}(\tau) = \sum_n \sum_m{}' \frac{1}{(m+n\tau)^{2k}}, \tag{2.48}$$

where $k$ is an integer and $k > 1$. It is easy to check that

$$G_{2k}\left(\frac{a\tau+b}{c\tau+d}\right) = (c\tau+d)^{2k} G_{2k}(\tau) \tag{2.49}$$

for any integers $a$, $b$, $c$, and $d$ with $ad - bc = 1$. The Fourier series for $G_{2k}(\tau)$ can be determined—see Exercise 7—and the result is

$$G_{2k}(\tau) = \frac{2(2\pi i)^{2k}}{(2k-1)!} E_{2k}(\tau). \tag{2.50}$$

From (2.49) and (2.50) we deduce the transformation formula (2.13). When $k = 1$, the series in (2.48) is not absolutely convergent, term by term rearrangement is not permitted, and the transformation formula (2.49) does not hold. A more delicate analysis is required; the details are outlined in Exercise 8.

The analogous approach for $\Gamma_0(p)$ would be to start with the series

$$G_k^0(\tau,p) = \sum_n \sum_m{}' \frac{\left(\frac{n}{p}\right)}{(m+n\tau)^k} \quad \text{and} \quad G_k^\infty(\tau,p) = \sum_n \sum_m{}' \frac{\left(\frac{m}{p}\right)}{(m+pn\tau)^k} \tag{2.51}$$

where $k$ is an integer and $k > 2$. If $k \equiv (p+1)/2 \pmod 2$ then the both of the series for $G_k^0(\tau,p)$ and $G_k^\infty(\tau,p)$ simplify identically to zero. Therefore, we will assume that $k \equiv (p-1)/2 \pmod 2$. For $k > 2$ the series converge absolutely, and it is easy to check that the following transformation formulas hold for any for any $\begin{pmatrix} a & b \\ c & d \end{pmatrix} \in \Gamma_0(p)$:

$$G_k^0\left(\frac{a\tau+b}{c\tau+d}, p\right) = \left(\frac{d}{p}\right)(c\tau+d)^k G_k^0(\tau,p), \tag{2.52}$$

$$G_k^\infty\left(\frac{a\tau+b}{c\tau+d}, p\right) = \left(\frac{d}{p}\right)(c\tau+d)^k G_k^\infty(\tau,p) \tag{2.53}$$

and

$$G_k^\infty\left(\frac{-1}{p\tau}, p\right) = \tau^k G_k^0(\tau,p). \tag{2.54}$$

The Fourier series can be shown to be

$$G_k^0(\tau,p) = \frac{2(-2\pi i)^k}{(k-1)!} E_k^0(\tau,p) \tag{2.55}$$

and

$$G_k^\infty(\tau,p) = \frac{2(2\pi i)^k}{(k-1)!p^{k-1/2}c_p} E_k^\infty(\tau,p), \tag{2.56}$$

where $c_p$ is as for Lemma 2.19. The results in Theorem 2.21 in the case $k > 2$ then follow from the results above. When $k = 1$ or 2, the series for $G_k^0(\tau,p)$ and $G_k^\infty(\tau,p)$ do not converge absolutely, so a different approach is needed. One method, originally due to Hecke, is to consider the series

$$G_k(\tau,s,p) = \sum_n \sum_m{}' \frac{\chi_1(m,p)\chi_2(n,p)}{(m+n\tau)^k|m+n\tau|^s}$$

as analytic functions of $s$ for $\text{Re}(s) > 2-k$. Here $\chi_1(m,p)$ and $\chi_2(n,p)$ are Dirichlet characters modulo $p$. It turns out that $G_k(\tau,s,p)$ may be analytically continued in the variable $s$ to a meromorphic function which is analytic at $s = 0$. The transformation formulas for $G_k(\tau,s,p)$ can be worked out, and the transformation formulas for $G_k^0(\tau,p)$ and $G_k^\infty(\tau,p)$ in the cases $k = 1$ or 2 can be obtained by considering the behavior at $s = 0$. The method is illustrated in detail in the books by Miyake [232, pp. 268–293] and Schoeneberg [261, pp. 63–68 and 164–169].

The methods used in this chapter are based on [108] and are similar to those used by O. Kolberg [198]. They have the significant advantage that all positive weights are handled in the same way, and the conditionally convergent cases of weights 1 and 2 present no extra difficulty.

## 2.9 Exercises

1. Let $p$ be an odd prime and define $g(p)$ by

$$g(p) = \sum_{r=1}^{p-1} r\left(\frac{r}{p}\right).$$

   a. Show that $g(p) \equiv 0 \pmod p$ if $p > 3$.
   b. Show that $g(3) = -1$.
   c. Show that $g(p) = 0$ if $p \equiv 1 \mod 4$.

2. The goal of this exercise is to evaluate the Gauss sum $\sum_{j=0}^{N-1} e^{2\pi i j^2/N}$, for any positive integer $N$.

Let $f(t) = \sum_{j=0}^{N-1} e^{2\pi i(j+t)^2/N}$ for $0 \leq t < 1$, and extend $f$ to the whole real line by defining it to be periodic with period 1.

a. Show that $f$ is continuous. This simply requires checking that

$$\sum_{j=0}^{N-1} e^{2\pi i j^2/N} = \sum_{j=0}^{N-1} e^{2\pi i(j+1)^2/N}.$$

b. Consider the Fourier expansion $f(t) = \sum_{n=-\infty}^{\infty} a_n\, e^{2\pi int}$. Show that

$$a_n = \int_0^1 f(t)e^{-2\pi int}\mathrm{d}t = e^{-\pi in^2N/2} \int_{-nN/2}^{N-nN/2} e^{2\pi iu^2/N}\mathrm{d}u.$$

c. Observe that $f(0) = \sum_{j=0}^{N-1} e^{2\pi ij^2/N} = \sum_{n=-\infty}^{\infty} a_n$. Deduce that

$$\sum_{j=0}^{N-1} e^{2\pi ij^2/N} = \sum_{n=-\infty}^{\infty} e^{-\pi in^2N/2} \int_{-nN/2}^{N-nN/2} e^{2\pi iu^2/N}\mathrm{d}u.$$

d. Split the sum on the right-hand side of Part c into two, according to whether $n$ is even or odd, to show that

$$\begin{aligned}\sum_{j=0}^{N-1} e^{2\pi ij^2/N} &= \sum_{n=-\infty}^{\infty} \int_{-nN}^{N-nN} e^{2\pi iu^2/N}\mathrm{d}u \\ &\quad + i^{-N} \sum_{n=-\infty}^{\infty} \int_{-(n+1/2)N}^{N-(n+1/2)N} e^{2\pi iu^2/N}\mathrm{d}u \\ &= \left(1 + i^{-N}\right) \int_{-\infty}^{\infty} e^{2\pi iu^2/N}\mathrm{d}u \\ &= \left(1 + i^{-N}\right) \sqrt{N} \int_{-\infty}^{\infty} e^{2\pi iu^2}\mathrm{d}u.\end{aligned}$$

e. Let $N = 1$ in Part d to deduce that

$$1 = \left(1 + i^{-1}\right) \int_{-\infty}^{\infty} e^{2\pi iu^2}\mathrm{d}u.$$

f. Combine Parts d and e to show that, for any positive integer $N$,

$$\sum_{j=0}^{N-1} e^{2\pi i j^2/N} = \frac{1+i^{-N}}{1+i^{-1}}\sqrt{N}.$$

g. Use Part f to deduce that if $p$ is an odd prime, then

$$\sum_{j=0}^{p-1} e^{2\pi i j^2/p} = c_p\sqrt{p}$$

where the value of $c_p$ is given by (2.38), that is,

$$c_p = \begin{cases} 1, & \text{if } p \equiv 1 \,(\text{mod } 4), \\ i, & \text{if } p \equiv 3 \,(\text{mod } 4). \end{cases}$$

3. This exercise follows K. Venkatachaliengar [287] to derive Euler's infinite product formulas for the sine and cosine functions. For a nonnegative integer $n$ and a complex variable $z$, let

$$I_n(z) = \int_0^{\pi/2} \cos zt \cos^n t \, dt.$$

a. Use integration by parts twice to show that for $n \geq 2$,

$$n(n-1)I_{n-2}(z) = (n^2 - z^2)I_n(z).$$

Hence, deduce that

$$\frac{I_{n-2}(z)}{I_{n-2}(0)} = \left(1 - \frac{z^2}{n^2}\right)\frac{I_n(z)}{I_n(0)}.$$

b. Show that

$$\sin\frac{\pi z}{2} = \frac{\pi z}{2}\frac{I_0(z)}{I_0(0)} \quad \text{and} \quad \cos\frac{\pi z}{2} = (1-z^2)\frac{I_1(z)}{I_1(0)}.$$

c. Assuming that $\lim_{n\to\infty} I_n(z)/I_n(0) = 1$, use the results of Parts a and b to deduce that

$$\sin\frac{\pi z}{2} = \frac{\pi z}{2}\prod_{k=1}^{\infty}\left(1 - \frac{z^2}{(2k)^2}\right) \quad \text{and} \quad \cos\frac{\pi z}{2} = \prod_{k=0}^{\infty}\left(1 - \frac{z^2}{(2k+1)^2}\right).$$

d. It remains to show that $\lim_{n\to\infty} I_n(z)/I_n(0) = 1$. For $0 \le t \le \pi/2$ and $z = x + iy$, verify that

$$|1 - \cos zt| = 2\left(\sin^2 \frac{xt}{2} + \sinh^2 \frac{yt}{2}\right) \le \frac{1}{2}(x^2 + \eta^2)t^2 = \frac{1}{2}|\zeta|^2 t^2$$

where $\pi\eta/4 = \sinh \pi y/4$ and $\zeta = x + i\eta$.

e. Deduce that

$$|I_n(0) - I_n(z)| \le \frac{1}{2}|\zeta|^2 \int_0^{\pi/2} t^2 \cos^n t \, dt.$$

f. Use the inequality $t < \tan t$ for $0 < t < \pi/2$ to show that

$$\int_0^{\pi/2} t^2 \cos^n t \, dt < \int_0^{\pi/2} t \cos^{n-1} t \sin t \, dt = \frac{1}{n} I_n(0).$$

g. Conclude that

$$\left|1 - \frac{I_n(z)}{I_n(0)}\right| \le \frac{|\zeta|^2}{2n}$$

and hence $\lim_{n\to\infty} I_n(z)/I_n(0) = 1$.

4. This exercise outlines another proof of the infinite product formula for the sine function given by W. F. Eberlein [143]. It is elementary in the sense that it relies only on simple properties of the trigonometric and the exponential functions and there is no need for sophisticated or deep theorems from real or complex analysis to justify the limiting processes. Let $p_n(z)$ be the polynomial defined by

$$p_n(z) = \frac{1}{2}\left(\left(1 + \frac{z}{n}\right)^n - \left(1 - \frac{z}{n}\right)^n\right).$$

The idea is to factorize the polynomial in the case that $n$ is odd and investigate the limit as $n \to \infty$.

a. Show that $p_n(z) = 0$ if and only if

$$\left(1 + \frac{z}{n}\right) = \omega \left(1 - \frac{z}{n}\right)$$

for some $\omega$ with $\omega^n = 1$.

b. Suppose that $n$ is odd and write $n = 2m + 1$. Use the result of Part a to show that the zeros of $p_{2m+1}(z)$ are the points $z_k$ where

$$z_k = (2m+1)\, i \tan\left(\frac{\pi k}{2m+1}\right), \quad k \in \mathbb{Z} \quad \text{and} \quad -m \le k \le m.$$

Hence deduce that

$$p_{2m+1}(z) = A\, z \prod_{k=1}^{m} \left(1 + \frac{z^2}{(2m+1)^2 \tan^2 \frac{\pi k}{2m+1}}\right)$$

for some constant $A$, and show that $A = 1$ by considering the coefficient of $z$ on each side.

c. Let $m \to \infty$ in the result of Part b to deduce that

$$\sinh z = z \prod_{k=1}^{\infty} \left(1 + \frac{z^2}{\pi^2 k^2}\right).$$

Replace $z$ with $i\pi z$ to show that

$$\sin \pi z = \pi z \prod_{j=1}^{\infty} \left(1 - \frac{z^2}{j^2}\right). \tag{2.57}$$

The limiting process in Part c relies on the property

$$\lim_{m\to\infty} \prod_{k=1}^{\infty} f(m,k) = \prod_{k=1}^{\infty} \lim_{m\to\infty} f(m,k)$$

where

$$f(m,k) = \begin{cases} 1 + \dfrac{z^2}{(2m+1)^2 \tan^2 \frac{\pi k}{2m+1}} & \text{if } 1 \le k \le m, \\ 1 & \text{if } k > m. \end{cases}$$

A justification for this interchange of limit and product goes as follows.

d. First, assume that $z$ is real and nonnegative and write $z = x \ge 0$. For any positive integers $\ell$ and $m$ with $\ell < m$, show that

$$x \prod_{k=1}^{\ell} \left(1 + \frac{x^2}{(2m+1)^2 \tan^2 \frac{\pi k}{2m+1}}\right) \le p_{2m+1}(x),$$

and let $m \to \infty$ to obtain

$$x\prod_{k=1}^{\ell}\left(1+\frac{x^2}{k^2\pi^2}\right)\leq \sinh x.$$

e. Use the result of Part b and the inequality $\theta \leq \tan\theta$ that holds for $0 \leq \theta < \pi/2$ to deduce that

$$p_{2\ell+1}(x) \leq x\prod_{k=1}^{\ell}\left(1+\frac{x^2}{k^2\pi^2}\right).$$

f. Parts d and e imply that

$$p_{2\ell+1}(x) \leq x\prod_{k=1}^{\ell}\left(1+\frac{x^2}{k^2\pi^2}\right)\leq \sinh x.$$

Let $\ell \to \infty$ to conclude that

$$x\prod_{k=1}^{\infty}\left(1+\frac{x^2}{k^2\pi^2}\right)= \sinh x.$$

This establishes the result for real and nonnegative values of $x$.

g. Now suppose $z \in \mathbb{C}$. Suppose $m, n \in \mathbb{Z}$ and $m > \ell \geq 1$. Show that

$$\begin{aligned}
&\left|p_{2m+1}(z) - z\prod_{k=1}^{\ell}\left(1+\frac{z^2}{(2m+1)^2\tan^2\frac{k\pi}{2m+1}}\right)\right| \\
&= \left|z\prod_{k=1}^{\ell}\left(1+\frac{z^2}{(2m+1)^2\tan^2\frac{k\pi}{2m+1}}\right)\right| \\
&\qquad\times\left|\prod_{k=\ell+1}^{m}\left(1+\frac{z^2}{(2m+1)^2\tan^2\frac{k\pi}{2m+1}}\right)-1\right| \\
&\leq |z|\prod_{k=1}^{\ell}\left(1+\frac{|z|^2}{(2m+1)^2\tan^2\frac{k\pi}{2m+1}}\right) \\
&\qquad\times\left(\prod_{k=\ell+1}^{m}\left(1+\frac{|z|^2}{(2m+1)^2\tan^2\frac{k\pi}{2m+1}}\right)-1\right) \\
&= p_{2m+1}(|z|) - |z|\prod_{k=1}^{\ell}\left(1+\frac{|z|^2}{(2m+1)^2\tan^2\frac{k\pi}{2m+1}}\right).
\end{aligned}$$

h. Let $m \to \infty$ in the result of Part g to obtain

$$\left| \sinh z - z \prod_{k=1}^{\ell} \left(1 + \frac{z^2}{k^2\pi^2}\right) \right| \leq \sinh |z| - |z| \prod_{k=1}^{\ell} \left(1 + \frac{|z|^2}{k^2\pi^2}\right).$$

Now let $\ell \to \infty$ and note that the right-hand side $\to 0$ by Part f to show that

$$\sinh z - z \prod_{k=1}^{\infty} \left(1 + \frac{z^2}{k^2\pi^2}\right) = 0$$

for any complex number $z$. Thus, the limiting process that led to the results in Part c has been justified.

5. a. Apply logarithmic differentiation to the identity (2.57) to deduce the partial fraction expansion of the cotangent function:

$$\pi \cot \pi z = \frac{1}{z} + \sum_{j=1}^{\infty} \frac{2z}{z^2 - j^2} = \frac{1}{z} + \sum_{j=1}^{\infty} \left(\frac{1}{z+j} + \frac{1}{z-j}\right) = \lim_{N\to\infty} \sum_{j=-N}^{N} \frac{1}{z+j}. \tag{2.58}$$

Direct proofs of (2.58), that do not rely on the infinite product formula for the sine, are also possible. The neatest is the "Herglotz trick" as described in [5, pp. 169–171].

b. Deduce that for $|z| < 1$,

$$\pi z \cot \pi z = 1 - 2 \sum_{j=1}^{\infty} \frac{z^2/j^2}{1 - z^2/j^2} = 1 - 2 \sum_{n=1}^{\infty} \left\{ \sum_{j=1}^{\infty} \frac{1}{j^{2n}} \right\} z^{2n}.$$

c. Use (0.65) to show that

$$\pi z \cot \pi z = \sum_{n=0}^{\infty} (-1)^n B_{2n} \frac{(2\pi)^{2n}}{(2n)!} z^{2n} \quad \text{for } |z| < 1,$$

where $B_n$ are the Bernoulli numbers.

d. Use Parts b and c to deduce that for any positive integer $n$,

$$\sum_{j=1}^{\infty} \frac{1}{j^{2n}} = (-1)^{n+1} \frac{(2\pi)^{2n}}{2(2n)!} B_{2n}.$$

6. a. With $q = \exp(2\pi i\tau)$ and $\operatorname{Im} \tau > 0$, show that

$$\pi \cot \pi \tau = i\pi - \frac{2\pi i}{1-q} = i\pi - 2\pi i \sum_{n=0}^{\infty} q^n.$$

b. Hence, use the result of Exercise 5, Part a, to show that

$$\frac{1}{\tau} + \sum_{j=1}^{\infty}\left(\frac{1}{\tau+j} + \frac{1}{\tau-j}\right) = i\pi - 2\pi i \sum_{j=0}^{\infty} q^j .$$

c. By differentiation, show that for any integer $n \geq 2$,

$$\sum_{j=-\infty}^{\infty} \frac{1}{(\tau+j)^n} = \frac{(-2\pi i)^n}{(n-1)!} \sum_{j=1}^{\infty} j^{n-1} q^j . \tag{2.59}$$

7. For $\operatorname{Im}\tau > 0$ and $k$ a positive integer, let $G_{2k}(\tau)$ be defined by

$$G_{2k}(\tau) = \sum_n {\sum_m}' \frac{1}{(m+n\tau)^{2k}},$$

where the prime denotes that the sum is over all $m, n \in \mathbb{Z}$ except that the term corresponding to $(m, n) = (0, 0)$ is omitted. The Fourier expansion for $G_{2k}(\tau)$ may be obtained by completing the following steps.

a. By considering the cases $n = 0$ and $n \neq 0$ separately, show that

$$G_{2k}(\tau) = 2\sum_{m=1}^{\infty} \frac{1}{m^{2k}} + 2\sum_{n=1}^{\infty} \sum_{m=-\infty}^{\infty} \frac{1}{(m+n\tau)^{2k}} .$$

b. Use the results of Exercises 5 and 6 to deduce that

$$\begin{aligned} G_{2k}(\tau) &= (-1)^{k+1}\frac{(2\pi)^{2k}}{(2k)!}B_{2k} + \frac{2(-2\pi i)^{2k}}{(2k-1)!}\sum_{n=1}^{\infty}\sum_{j=1}^{\infty} j^{2k-1} q^{nj} \\ &= \frac{2(-1)^k(2\pi)^{2k}}{(2k-1)!}\left\{-\frac{B_{2k}}{4k} + \sum_{j=1}^{\infty} \frac{j^{2k-1}q^j}{1-q^j}\right\} . \end{aligned}$$

8. This exercise shows how rearranging the terms in a conditionally convergent double series can affect the sum of the series. It also shows the attention to detail that is required in dealing with such series. We follow Serre [267, pp. 95–96] who attributes the idea to Hurwitz. Let

$$g_1(\tau) = \sum_n {\sum_m}' \frac{1}{(m+n\tau)^2},$$

$$g(\tau) = \sum_m {\sum_n}' \frac{1}{(m+n\tau)^2},$$

$$h_1(\tau) = \sum_n \sum_m{}' \frac{1}{(m-1+n\tau)(m+n\tau)}$$

and

$$h(\tau) = \sum_m \sum_n{}' \frac{1}{(m-1+n\tau)(m+n\tau)},$$

where the primes denote that the sums are over all $m, n \in \mathbb{Z}$ except that the term corresponding to $(m, n) = (0, 0)$ is omitted from the series for $g_1$ and $g$, and the terms corresponding to $(m, n) = (0, 0)$ and $(m, n) = (1, 0)$ are omitted from the series for $h_1$ and $h$.

a. Use partial fractions

$$\frac{1}{(m-1+n\tau)(m+n\tau)} = \frac{1}{m-1+n\tau} - \frac{1}{m+n\tau}$$

to show that

$$h_1(\tau) = 2 \quad \text{and} \quad h(\tau) = 2 - \frac{2\pi i}{\tau}.$$

Hint: The identity for $h_1$ is relatively straightforward by summing the series in the order indicated. For the other identity, use the partial fractions for the cotangent in Exercise 5 to write

$$\pi \cot \pi z - \pi \cot \pi y = \sum_{j=-\infty}^{\infty} \left( \frac{1}{z+j} - \frac{1}{y+j} \right),$$

and deduce that

$$h(\tau) = 2 - \frac{2\pi}{\tau} \lim_{N\to\infty} \cot \frac{N\pi}{\tau}.$$

Then use the fact that $\operatorname{Im} \tau > 0$ to evaluate the limit.

b. Show that

$$h_1(\tau) - g_1(\tau) = h(\tau) - g(\tau)$$

and deduce that

$$g_1(\tau) - g(\tau) = h_1(\tau) - h(\tau) = \frac{2\pi i}{\tau}.$$

Hint: The double series with term

$$\frac{1}{(m-1+n\tau)(m+n\tau)} - \frac{1}{(m+n\tau)^2} = \frac{1}{(m+n\tau)^2(m-1+n\tau)}$$

is absolutely convergent, so interchanging the order of summation does not affect the sum of the series.

c. Show that

$$g_1(-1/\tau) = \tau^2 g(\tau)$$

and deduce that

$$g_1(-1/\tau) = \tau^2 g_1(\tau) - 2\pi i\tau.$$

d. Use the result of (2.59) to show that

$$g_1(\tau) = \frac{\pi^2}{3}P(q), \qquad \text{where} \qquad P(q) = 1 - 24\sum_{j=1}^{\infty}\frac{jq^j}{1-q^j}.$$

e. Use Parts c and d to show that Ramanujan's Eisenstein series $P(q)$ satisfies the transformation formula, cf. (2.19):

$$P\left(e^{-2\pi i/\tau}\right) = \tau^2 P\left(e^{2\pi i\tau}\right) - \frac{6i\tau}{\pi}.$$

9. Let $p$ be an odd prime, let $k > 2$ be an integer, and let $G_k^0(\tau, p)$ and $G_k^\infty(\tau, p)$ be defined by (2.51).

a. Prove that $G_k^0(\tau, p)$ and $G_k^\infty(\tau, p)$ are identically zero if $k \equiv (p+1)/2 \pmod 2$.

For Parts b and c, assume $k \equiv (p-1)/2 \pmod 2$.

b. Prove the transformation formulas in (2.52)–(2.54).

c. Prove the Fourier expansions (2.55) and (2.56).

10. Consider the Eisenstein series $R(q) = 1 - 504\sum_{j=1}^{\infty}\frac{j^5q^j}{1-q^j}$.

a. Show that, as $q$ increases from 0 to 1, $R(q)$ decreases monotonically from 1 to $-\infty$.

b. Show that $R(e^{-\pi}) = 0$.

Hint: Take $\begin{pmatrix} a & b \\ c & d \end{pmatrix} = \begin{pmatrix} 0 & -1 \\ 1 & 0 \end{pmatrix}$ in (2.16) and then put $\tau = i$.

c. Use the result that $R(q) = 864e_1e_2e_3$ from Theorem 1.38 and the results of Exercise 8 from the previous chapter to deduce that $e_3 = 0$ when $q = e^{-\pi}$.

# Chapter 3
# Theta Functions

## 3.1 Introduction

In this chapter we study Ramanujan's theta functions as well as the cubic theta functions of J. M. Borwein and P. B. Borwein.

**Definition 3.1 (Ramanujan's theta functions).** The theta functions $\varphi(q)$ and $\psi(q)$ are defined by

$$\varphi(q) = \sum_{j=-\infty}^{\infty} q^{j^2},$$

$$\psi(q) = \sum_{j=0}^{\infty} q^{j(j+1)/2},$$

so that $\varphi(q)$ is the generating function of the squares and $\psi(q)$ is the generating function of the triangular numbers.

There are several other equivalent series for $\psi$. For example,

$$\sum_{j=-\infty}^{\infty} q^{j(j+1)/2} = 2\sum_{j=0}^{\infty} q^{j(j+1)/2} = 2\psi(q). \tag{3.1}$$

Hence, on completing the square in the exponent and replacing $q$ with $q^2$,

$$\sum_{j=-\infty}^{\infty} q^{(j+\frac{1}{2})^2} = 2q^{1/4}\psi(q^2). \tag{3.2}$$



S. Cooper, *Ramanujan's Theta Functions*, DOI 10.1007/978-3-319-56172-1_4

Similarly,

$$\sum_{j=-\infty}^{\infty} q^{(4j+1)^2} = \sum_{j=0}^{\infty} q^{(2j+1)^2} = q\psi(q^8). \tag{3.3}$$

**Definition 3.2 (Borweins' cubic theta functions).** Let $\omega = \exp(2\pi i/3)$. The cubic theta functions $a(q)$, $b(q)$, and $c(q)$ are defined by

$$a(q) = \sum_{j=-\infty}^{\infty} \sum_{k=-\infty}^{\infty} q^{j^2+jk+k^2}, \tag{3.4}$$

$$b(q) = \sum_{j=-\infty}^{\infty} \sum_{k=-\infty}^{\infty} \omega^{j-k} q^{j^2+jk+k^2}, \tag{3.5}$$

$$c(q) = \sum_{j=-\infty}^{\infty} \sum_{k=-\infty}^{\infty} q^{(j+\frac{1}{3})^2+(j+\frac{1}{3})(k+\frac{1}{3})+(k+\frac{1}{3})^2}. \tag{3.6}$$

The cubic theta functions $a(q)$, $b(q)$, and $c(q)$ are analogues of

$$\varphi^2(q) = \sum_{j=-\infty}^{\infty} \sum_{k=-\infty}^{\infty} q^{j^2+k^2},$$

$$\varphi^2(-q) = \sum_{j=-\infty}^{\infty} \sum_{k=-\infty}^{\infty} (-1)^{j+k} q^{j^2+k^2}$$

and

$$4q^{1/2}\psi^2(q^2) = \sum_{j=-\infty}^{\infty} \sum_{k=-\infty}^{\infty} q^{(j+\frac{1}{2})^2+(k+\frac{1}{2})^2},$$

respectively. The analogy will become more evident as properties of the functions are developed.

## 3.2 Ramanujan's theta functions

We shall now establish some of the basic properties of the theta functions $\varphi(q)$ and $\psi(q)$.

**Lemma 3.3.**

$$\varphi(q) + \varphi(-q) = 2\varphi(q^4), \tag{3.7}$$

$$\varphi(q) - \varphi(-q) = 4q\psi(q^8), \tag{3.8}$$

$$\varphi^2(q) + \varphi^2(-q) = 2\varphi^2(q^2), \tag{3.9}$$

$$\varphi^2(q) - \varphi^2(-q) = 8q\psi^2(q^4). \tag{3.10}$$

*Proof.* By definition, we have

$$\varphi(q) + \varphi(-q) = \sum_{j=-\infty}^{\infty} \left(q^{j^2} + (-q)^{j^2}\right) = 2\sum_{j \text{ even}} q^{j^2} = 2\varphi(q^4).$$

This proves (3.7). The proof of (3.8) is similar and involves odd powers of $q$:

$$\varphi(q) - \varphi(-q) = 2\sum_{j \text{ odd}} q^{j^2} = 2\sum_{j=-\infty}^{\infty} q^{(2j+1)^2} = 4q\psi(q^8),$$

where the last step follows from (3.2). In order to prove (3.9) and (3.10) we use the rearrangements for double series:

$$\sum_{j+k \text{ even}} f(j,k) = \sum_{m,n} f(m+n, m-n)$$

and

$$\sum_{j+k \text{ odd}} f(j,k) = \sum_{m,n} f(m+n+1, m-n),$$

where the sums are over all integers $m$ and $n$, and all integers $j$ and $k$ that satisfy the given parity conditions. Thus,

$$\begin{aligned}
\varphi^2(q) + \varphi^2(-q) &= \sum_{j,k} \left(q^{j^2+k^2} + (-q)^{j^2+k^2}\right) \\
&= 2\sum_{j+k \text{ even}} q^{j^2+k^2} \\
&= 2\sum_{m,n} q^{(m+n)^2+(m-n)^2} \\
&= 2\sum_{m,n} q^{2m^2+2n^2} \\
&= 2\varphi^2(q^2),
\end{aligned}$$

and

$$\begin{aligned}
\varphi^2(q) - \varphi^2(-q) &= \sum_{j,k} \left( q^{j^2+k^2} - (-q)^{j^2+k^2} \right) \\
&= 2 \sum_{j+k \text{ odd}} q^{j^2+k^2} \\
&= 2 \sum_{m,n} q^{(m+n+1)^2+(m-n)^2} \\
&= 2 \sum_{m,n} q^{2m^2+2m+2n^2+2n+1} \\
&= 8q\psi^2(q^4),
\end{aligned}$$

where the last step follows from (3.1).

□

An immediate consequence of Lemma 3.3 is the representation of $\varphi(q)$, and its square, as a sum of an even function and an odd function.

**Corollary 3.4.** *The following identities hold:*

$$\varphi(q) = \varphi(q^4) + 2q\psi(q^8),$$
$$\varphi^2(q) = \varphi^2(q^2) + 4q\psi^2(q^4).$$

The next result records infinite product formulas for the functions $\varphi$ and $\psi$.

**Theorem 3.5.**

$$\varphi(-q) = \prod_{j=1}^{\infty} \frac{(1-q^j)^2}{(1-q^{2j})} = \prod_{j=1}^{\infty} \frac{(1-q^j)}{(1+q^j)},$$

$$\varphi(q) = \prod_{j=1}^{\infty} \frac{(1-q^{2j})^5}{(1-q^j)^2(1-q^{4j})^2},$$

$$\psi(q) = \prod_{j=1}^{\infty} \frac{(1-q^{2j})^2}{(1-q^j)} = \prod_{j=1}^{\infty} \frac{(1-q^{2j})}{(1-q^{2j-1})}$$

*and*

$$\psi(-q) = \prod_{j=1}^{\infty} \frac{(1-q^j)(1-q^{4j})}{(1-q^{2j})}.$$

*Proof.* The first three results are restatements of (0.41), (0.42), and (0.45), respectively. The result for $\psi(-q)$ is obtained from the result for $\psi(q)$ by replacing $q$ with $-q$ and using (0.4).

□

**Lemma 3.6.** *The following identities hold*:

$$\varphi(q)\psi(q^2) = \psi^2(q) \tag{3.11}$$

*and*

$$\varphi(q)\varphi(-q) = \varphi^2(-q^2). \tag{3.12}$$

*Proof.* These are immediate from the results in Theorem 3.5.

□

**Theorem 3.7.**

$$\varphi^4(q) - \varphi^4(-q) = 16q\psi^4(q^2).$$

*Equivalently,*

$$\left(\sum_{j=-\infty}^{\infty} q^{j^2}\right)^4 - \left(\sum_{j=-\infty}^{\infty} (-1)^j q^{j^2}\right)^4 = \left(\sum_{j=-\infty}^{\infty} q^{(j+\frac{1}{2})^2}\right)^4.$$

*Proof.* By Lemmas 3.3 and 3.6, we have

$$\begin{aligned}
\varphi^4(q) - \varphi^4(-q) &= \left(\varphi^2(q) + \varphi^2(-q)\right)\left(\varphi^2(q) - \varphi^2(-q)\right) \\
&= 2\varphi^2(q^2) \times 8q\psi^2(q^4) \\
&= 16q\psi^4(q^2).
\end{aligned}$$

□

Theorem 3.7 has the following reformulation, which has been described by Jacobi as 'aequatio identica satis abstrusa' (nonobvious identity).

**Corollary 3.8 (aequatio identica satis abstrusa).**

$$\prod_{j=1}^{\infty}(1+q^{2j-1})^8 - \prod_{j=1}^{\infty}(1-q^{2j-1})^8 = 16q\prod_{j=1}^{\infty}(1+q^{2j})^8.$$

*Proof.* Use the results of Theorem 3.5 to express the result of Theorem 3.7 in terms of infinite products, viz.,

$$\prod_{j=1}^{\infty} \frac{(1-q^{2j})^{20}}{(1-q^j)^8(1-q^{4j})^8} - \prod_{j=1}^{\infty} \frac{(1-q^j)^8}{(1-q^{2j})^4} = 16q \prod_{j=1}^{\infty} \frac{(1-q^{4j})^8}{(1-q^{2j})^4}.$$

Divide both sides by $\prod_{j=1}^{\infty}(1-q^{2j})^4$ and observe that

$$\begin{aligned}\prod_{j=1}^{\infty} \frac{(1-q^{2j})^{16}}{(1-q^j)^8(1-q^{4j})^8} &= \prod_{j=1}^{\infty} \left(\frac{1-q^{2j}}{1-q^j}\right)^8 \left(\frac{1-q^{2j}}{1-q^{4j}}\right)^8 \\ &= \prod_{j=1}^{\infty} \left(\frac{1}{1-q^{2j-1}}\right)^8 (1-q^{4j-2})^8 \\ &= \prod_{j=1}^{\infty} (1+q^{2j-1})^8,\end{aligned}$$

$$\prod_{j=1}^{\infty} \frac{(1-q^j)^8}{(1-q^{2j})^8} = \prod_{j=1}^{\infty} (1-q^{2j-1})^8$$

and

$$16q \prod_{j=1}^{\infty} \frac{(1-q^{4j})^8}{(1-q^{2j})^8} = 16q \prod_{j=1}^{\infty} (1+q^{2j})^8$$

to complete the proof.

□

The next result will be useful in the study of elliptic functions to alternative bases in the next chapter, e.g., see Theorem 4.17 and its proof.

**Lemma 3.9.**

$$\left(\varphi^4(q) + 16q\psi^4(q^2)\right)^2 = \varphi^8(-q) + 64q\psi^8(q).$$

*Proof.* By Theorem 3.7, we have

$$\begin{aligned}\left(\varphi^4(q) + 16q\psi^4(q^2)\right)^2 &= \left(\varphi^4(-q) + 32q\psi^4(q^2)\right)^2 \\ &= \varphi^8(-q) + 64q\psi^4(q^2)\varphi^4(-q) + 1024q^2\psi^8(q^2) \\ &= \varphi^8(-q) + 64q\psi^4(q^2)\left(\varphi^4(-q) + 16q\psi^4(q^2)\right).\end{aligned}$$

Applying Theorem 3.7 again, and then using Lemma 3.6, we find that

$$\begin{aligned}\left(\varphi^4(q) + 16q\psi^4(q^2)\right)^2 &= \varphi^8(-q) + 64q\psi^4(q^2)\varphi^4(q)\\ &= \varphi^8(-q) + 64q\psi^8(q).\end{aligned}$$

□

The next result gives Lambert series expansions for the squares of Ramanujan's theta functions.

**Theorem 3.10.** *For any integer n, let $\chi_4(n)$ be defined by*

$$\chi_4(n) = \sin\frac{n\pi}{2} = \begin{cases} 1 & \text{if } n \equiv 1 \pmod 4,\\ -1 & \text{if } n \equiv 3 \pmod 4,\\ 0 & \text{otherwise.}\end{cases} \tag{3.13}$$

*Then*

$$\varphi^2(q) = 1 + 4\sum_{n=1}^{\infty} \chi_4(n)\frac{q^n}{1-q^n} \tag{3.14}$$

*and*

$$q\psi^2(q^4) = \sum_{n=1}^{\infty} \chi_4(n)\frac{q^n}{1-q^{2n}}. \tag{3.15}$$

*Proof.* Let $x = -1$ and $y = i$ in Theorem 0.27. The result simplifies to give

$$\prod_{n=1}^{\infty}\frac{(1-q^n)^2}{(1+q^n)^2} = 1 + 2\frac{(1-i)}{(1+i)}\sum_{m=1}^{\infty}\sum_{n=1}^{\infty} q^{mn}(-1)^m\left(i^n - i^{-n}\right).$$

The product on the left-hand side is just $\varphi^2(-q)$ by Theorem 3.5. Therefore, on simplifying the series on the right-hand side, we find that

$$\begin{aligned}\varphi^2(-q) &= 1 + 4\sum_{m=1}^{\infty}\sum_{n=1}^{\infty} q^{mn}(-1)^m \sin\frac{n\pi}{2}\\ &= 1 - 4\sum_{n=1}^{\infty}\frac{q^n}{1+q^n}\sin\frac{n\pi}{2}.\end{aligned} \tag{3.16}$$

On replacing $q$ with $-q$, and noting that terms in the series in (3.16) are nonzero if any only if $n$ is odd, we obtain (3.14).

Now we will prove (3.15). In Theorem 0.26, replace $q$ with $q^4$, then let $x = -q^2$ and $y = q^2$ and multiply both sides by $q/2$ to obtain

$$q\prod_{n=1}^{\infty}\frac{(1+q^{4n})^2(1-q^{4n})^2}{(1+q^{4n-2})^2(1-q^{4n-2})^2} = \frac{1}{2}\sum_{n=-\infty}^{\infty}\frac{(-1)^n q^{2n+1}}{1-q^{4n+2}}.$$

This is equivalent to

$$q\prod_{n=1}^{\infty}\frac{(1-q^{8n})^2}{(1-q^{8n-4})^2} = \sum_{n=0}^{\infty}\frac{(-1)^n q^{2n+1}}{1-q^{4n+2}} = \sum_{n=1}^{\infty}\chi_4(n)\frac{q^n}{1-q^{2n}}.$$

Now use the infinite product for $\psi(q)$ in Theorem 3.5 to complete the proof of (3.15).

□

There are many more proofs of (3.14) and (3.15) that come, for example, from manipulating Jacobi's triple product identity or specializing Ramanujan's summation formula or the identities in Chapter 1. In fact, most of the identities in this chapter have a plethora of proofs. We have attempted to select proofs that do not have too much of a complicated dependence on prior results, and we have tried to avoid proofs that use clever tricks. In the end, the choice of proof often comes down to a matter of personal preference.

## 3.3 The Borweins' cubic theta functions

We shall now establish some of the basic properties of the cubic theta functions $a(q)$, $b(q)$, and $c(q)$ given by Definition 3.2.

We begin with some analogues of Lemma 3.3. The results (3.9) and (3.10) can be rearranged to give

$$2\varphi^2(q^2) - \varphi^2(q) = \varphi^2(-q)$$

and

$$\varphi^2(q) - \varphi^2(q^2) = 4q\psi^2(q^4).$$

The results in the next lemma provide cubic analogues of these identities.

**Lemma 3.11.** *The following identities hold:*

$$3a(q^3) - a(q) = 2b(q) \tag{3.17}$$

*and*

$$a(q) - a(q^3) = 2c(q^3). \tag{3.18}$$

*Proof.* On using the series rearrangement

$$\sum_{j,k=-\infty}^{\infty} f(j,k) = \frac{1}{2} \sum_{j,k=-\infty}^{\infty} (f(j,k) + f(k,j))$$

we find that

$$b(q) = \sum_{j=-\infty}^{\infty} \sum_{k=-\infty}^{\infty} \omega^{j-k} q^{j^2+jk+k^2} = \sum_{j=-\infty}^{\infty} \sum_{k=-\infty}^{\infty} \cos \tfrac{2\pi(j-k)}{3}\, q^{j^2+jk+k^2}.$$

Since

$$\cos \frac{2\pi(j-k)}{3} = \begin{cases} 1 & \text{if } j \equiv k \pmod 3, \\ -1/2 & \text{otherwise,} \end{cases}$$

we separate the terms according to the residue class of $j-k$ modulo 3 to obtain

$$\begin{aligned} b(q) &= \sum_{j \equiv k \pmod 3} \sum q^{j^2+jk+k^2} - \frac{1}{2} \sum_{j \not\equiv k \pmod 3} \sum q^{j^2+jk+k^2} \\ &= \frac{3}{2} \sum_{j \equiv k \pmod 3} \sum q^{j^2+jk+k^2} - \frac{1}{2} \sum_{j=-\infty}^{\infty} \sum_{k=-\infty}^{\infty} q^{j^2+jk+k^2}. \end{aligned} \tag{3.19}$$

The first sum, on setting $j = 3n + k$, is

$$\begin{aligned} \frac{3}{2} \sum_{k=-\infty}^{\infty} \sum_{n=-\infty}^{\infty} q^{(3n+k)^2+(3n+k)k+k^2} &= \frac{3}{2} \sum_{k=-\infty}^{\infty} \sum_{n=-\infty}^{\infty} q^{9n^2+9nk+3k^2} \\ &= \frac{3}{2} \sum_{m=-\infty}^{\infty} \sum_{n=-\infty}^{\infty} q^{9n^2+9n(m-n)+3(m-n)^2} \\ &= \frac{3}{2} \sum_{m=-\infty}^{\infty} \sum_{n=-\infty}^{\infty} q^{3m^2+3mn+3n^2} \\ &= \frac{3}{2} a(q^3). \end{aligned}$$

On substituting this back into (3.19) we deduce

$$b(q) = \frac{3}{2} a(q^3) - \frac{1}{2} a(q)$$

and this proves (3.17).

By a similar procedure to the above, we find that

$$a(q) - b(q) = \frac{3}{2} \sum_{j \not\equiv k \pmod 3} \sum q^{j^2+jk+k^2} = 3 \sum_{j \equiv k+1 \pmod 3} \sum q^{j^2+jk+k^2}.$$

On setting $j = 3n + k + 1$ we obtain

$$\begin{aligned} a(q) - b(q) &= 3 \sum_{k=-\infty}^{\infty} \sum_{n=-\infty}^{\infty} q^{(3n+k+1)^2+(3n+k+1)k+k^2} \\ &= 3 \sum_{k=-\infty}^{\infty} \sum_{n=-\infty}^{\infty} q^{9n^2+9nk+3k^2+6n+3k+1}. \end{aligned}$$

Now put $k = m - n$ to deduce that

$$\begin{aligned} a(q) - b(q) &= 3 \sum_{m=-\infty}^{\infty} \sum_{n=-\infty}^{\infty} q^{9n^2+9n(m-n)+3(m-n)^2+6n+3(m-n)+1} \\ &= 3 \sum_{m=-\infty}^{\infty} \sum_{n=-\infty}^{\infty} q^{3m^2+3mn+3n^2+3m+3n+1} \\ &= 3c(q^3). \end{aligned} \tag{3.20}$$

On combining (3.17) and (3.20) to eliminate $b(q)$ we deduce (3.18).

□

The first identity in the next result is an analogue of (3.9).

**Lemma 3.12.** *The cubic theta functions satisfy the identities*

$$a(q) + a(-q) = 2a(q^4), \tag{3.21}$$

$$b(q) + b(-q) = 2b(q^4), \tag{3.22}$$

*and*

$$c(q) + c(-q) = 2c(q^4), \tag{3.23}$$

*where the principal value* $(-q)^{1/3} = -q^{1/3}$ *is used in the definition of* $c(-q)$.

*Proof.* By the definition of $a(q)$,

$$a(q) + a(-q) = \sum_{m=-\infty}^{\infty} \sum_{n=-\infty}^{\infty} \left( q^{m^2+mn+n^2} + (-q)^{m^2+mn+n^2} \right).$$

As $m^2 + mn + n^2$ is even if and only if $m$ and $n$ are both even, we deduce that

$$a(q) + a(-q) = 2 \sum_{m=-\infty}^{\infty} \sum_{n=-\infty}^{\infty} q^{(2m)^2+(2m)(2n)+(2n)^2} = 2a(q^4).$$

Next, by (3.17) and (3.21) we have

$$\begin{aligned} b(q) + b(-q) &= \frac{1}{2}\left(3a(q^3) - a(q)\right) + \frac{1}{2}\left(3a(-q^3) - a(-q)\right) \\ &= \frac{3}{2}\left(a(q^3) + a(-q^3)\right) - \frac{1}{2}\left(a(q) + a(-q)\right) \\ &= 3a(q^{12}) - a(q^4) \\ &= 2b(q^4). \end{aligned}$$

The identity (3.23) may be proved in a similar way by using (3.18) and (3.21).

□

Series manipulations can be used to prove the following useful result involving $c(q)$.

**Lemma 3.13.** *The cubic theta function $c(q)$ has the series representation given by*

$$c(q) = 3 \sum_{j=-\infty}^{\infty} \sum_{k=-\infty}^{\infty} q^{3j^2+3jk+3k^2+2j+k+\frac{1}{3}}.$$

*Proof.* Let

$$f(m,n) = \left(m + \frac{1}{3}\right)^2 + \left(m + \frac{1}{3}\right)\left(n + \frac{1}{3}\right) + \left(n + \frac{1}{3}\right)^2$$

and observe that

$$f(m,n) = f(m, -m-n-1).$$

We break the sum in the definition of $c(q)$ in (3.6) into three parts, depending on the residue class of $m - n$ modulo 3. First, note that

$$\sum_{m-n\equiv 1 \pmod 3}\sum q^{f(m,n)} = \sum_{m-n\equiv 1 \pmod 3}\sum q^{f(m,-m-n-1)} = \sum_{j-k\equiv 0 \pmod 3}\sum q^{f(j,k)}, \tag{3.24}$$

where in the last step we have made the change of variable $j = m$, $k = -m-n-1$, and so

$$j - k = 2m + n + 1 = 3m - (m-n) + 1 \equiv 0 \pmod 3.$$

Second, since $f(m,n)$ is symmetric in $m$ and $n$, we have

$$\sum_{m-n\equiv -1 \pmod 3}\sum q^{f(m,n)} = \sum_{m-n\equiv 1 \pmod 3}\sum q^{f(m,n)}. \tag{3.25}$$

Therefore, by (3.24) and (3.25),

$$c(q) = 3\sum_{m-n\equiv 0 \pmod 3}\sum q^{f(m,n)}.$$

It follows that

$$c(q) = 3\sum_{n=-\infty}^{\infty}\sum_{k=-\infty}^{\infty} q^{f(n+3k,n)}.$$

Making the change of variable $n = j - k$ we obtain

$$c(q) = 3\sum_{j=-\infty}^{\infty}\sum_{k=-\infty}^{\infty} q^{f(j+2k,j-k)} = 3\sum_{j=-\infty}^{\infty}\sum_{k=-\infty}^{\infty} q^{3j^2+3jk+3k^2+2j+k+\frac{1}{3}}.$$

□

The next goal is to establish infinite product representations for $b(q)$ and $c(q)$. There is no simple expression for $a(q)$ as an infinite product. The essence of the results is contained in the following lemma. We will only require the case $n = 3$, but as no more effort is required to prove the general case we state the result for any positive integer $n$.

**Lemma 3.14.** *Let $n$ be a positive integer and let $\omega = \exp(2\pi i/n)$. Then*

$$\sum_{m_1+\cdots+m_n=0} q^{\left(m_1^2+m_2^2+\cdots+m_n^2\right)/2}\,\omega^{m_1+2m_2+\cdots+nm_n} = \prod_{j=1}^{\infty}\frac{(1-q^j)^n}{(1-q^{nj})} \tag{3.26}$$

*and*

$$\sum_{m_1+\cdots+m_n=0} q^{n\left(m_1^2+m_2^2+\cdots+m_n^2\right)/2+m_1+2m_2+\cdots+nm_n} = \prod_{j=1}^{\infty}\frac{(1-q^{nj})^n}{(1-q^j)}. \tag{3.27}$$

*The sums are over all integers $m_1, m_2, \ldots, m_n$ that satisfy the condition $m_1+\cdots+m_n = 0$.*

*Proof.* Let

$$f(x) = \sum_{m_1,\ldots,m_n=-\infty}^{\infty} q^{(m_1^2+m_2^2+\cdots+m_n^2)/2}\,\omega^{m_1+2m_2+\cdots+nm_n}\,x^{m_1+m_2+\cdots+m_n}.$$

The term that is independent of $x$, i.e., the constant term of $f(x)$, is clearly given by the left-hand side of (3.26). On the other hand,

$$f(x) = \prod_{k=1}^{n} \left( \sum_{m=-\infty}^{\infty} q^{m^2/2} \omega^{km} x^m \right)$$

and by $n$ applications of the triple product identity (0.7) this is

$$f(x) = \prod_{k=1}^{n} \left( \prod_{j=1}^{\infty} (1 + q^{j-1/2} \omega^k x)(1 + q^{j-1/2} \omega^{-k} x^{-1})(1 - q^j) \right).$$

Since

$$\prod_{k=1}^{n} (1 - \omega^k X) = 1 - X^n,$$

this simplifies to

$$f(x) = \prod_{j=1}^{\infty} (1 + q^{n(j-1/2)} x^n)(1 + q^{n(j-1/2)} x^{-n})(1 - q^j)^n.$$

This may be expanded by the triple product identity (0.7) to give

$$f(x) = \prod_{j=1}^{\infty} \frac{(1 - q^j)^n}{(1 - q^{nj})} \times \sum_{k=-\infty}^{\infty} q^{nk^2/2} x^{nk}.$$

The constant term of $f(x)$ in this expression is just the right-hand side of (3.26). This completes the proof of (3.26).

Next, let

$$g(x) = \sum_{m_1,\ldots,m_n=-\infty}^{\infty} q^{n\left(\binom{m_1}{2}+\binom{m_2}{2}+\cdots+\binom{m_n}{2}\right)+m_1+2m_2+\cdots+nm_n} x^{m_1+m_2+\cdots+m_n},$$

where $\binom{m}{2} = m(m-1)/2$, for any integer $m$. The constant term of $g(x)$ is

$$\sum_{m_1+\cdots+m_n=0} q^{n\left(\binom{m_1}{2}+\binom{m_2}{2}+\cdots+\binom{m_n}{2}\right)+m_1+2m_2+\cdots+nm_n}$$

$$= \sum_{m_1+\cdots+m_n=0} q^{n\left(m_1^2+m_2^2+\cdots+m_n^2\right)/2+m_1+2m_2+\cdots+nm_n},$$

and this is the expression on the left-hand side of (3.27). On the other hand, application of the triple product identity in the form (0.15) gives

$$\begin{aligned}
g(x) &= \prod_{k=1}^{n}\left(\sum_{m=-\infty}^{\infty} q^{n\binom{m}{2}+km}x^m\right)\\
&= \prod_{k=1}^{n}\left(\prod_{j=1}^{\infty}(1+xq^{nj-n+k})(1+x^{-1}q^{nj-k})(1-q^{nj})\right)\\
&= \prod_{j=1}^{\infty}\frac{(1-q^{nj})^n}{(1-q^j)}\prod_{j=1}^{\infty}(1+xq^j)(1+x^{-1}q^{j-1})(1-q^j)\\
&= \prod_{j=1}^{\infty}\frac{(1-q^{nj})^n}{(1-q^j)}\sum_{m=-\infty}^{\infty} q^{\binom{m}{2}}x^{-m}.
\end{aligned}$$

The constant term of $g(x)$ in this expression is the right-hand side of (3.27), and this completes the proof of (3.27).

□

We are now ready for:

**Theorem 3.15.** *The following infinite product formulas hold*:

$$b(q) = \prod_{j=1}^{\infty}\frac{(1-q^j)^3}{(1-q^{3j})}, \tag{3.28}$$

$$c(q) = 3q^{1/3}\prod_{j=1}^{\infty}\frac{(1-q^{3j})^3}{(1-q^j)}. \tag{3.29}$$

*Proof.* Take $n = 3$ in (3.26) to obtain

$$\sum_{m_1+m_2+m_3=0} q^{(m_1^2+m_2^2+m_3^2)/2}\omega^{m_1+2m_2+3m_3} = \prod_{j=1}^{\infty}\frac{(1-q^j)^3}{(1-q^{3j})},$$

where $\omega = \exp(2\pi i/3)$. On putting $m_1 = j$, $m_2 = k$, and $m_3 = -j-k$, the left-hand side reduces to the series defining $b(q)$ given in (3.5). This proves (3.28).

Next, take $n = 3$ in (3.27) to obtain

$$\sum_{m_1+m_2+m_3=0} q^{3(m_1^2+m_2^2+m_3^2)/2+m_1+2m_2+3m_3} = \prod_{j=1}^{\infty}\frac{(1-q^{3j})^3}{(1-q^j)}.$$

Now set $m_1 = j, m_2 = k$ and $m_3 = -j - k$ to find that

$$\sum_{j=-\infty}^{\infty} \sum_{k=-\infty}^{\infty} q^{3j^2+3jk+3k^2-2j-k} = \prod_{j=1}^{\infty} \frac{(1-q^{3j})^3}{(1-q^j)}.$$

Replace $j$ and $k$ in the sum with $-j$ and $-k$, respectively, multiply by $3q^{1/3}$, and apply Lemma 3.13 to complete the proof of (3.29).

□

Formulas involving $a(-q)$, $b(-q)$, and $c(-q)$ were obtained in Lemma 3.12. The next result gives some different formulas for $b(-q)$ and $c(-q)$.

**Corollary 3.16.** *The following identities hold*:

$$b(-q) = \frac{b^3(q^2)}{b(q)b(q^4)}$$

*and*

$$c(-q) = -\frac{c^3(q^2)}{c(q)c(q^4)}.$$

*Proof.* By the infinite products in Theorem 3.15 and the elementary result (0.2),

$$\begin{aligned} b(-q) &= \prod_{j=1}^{\infty} \frac{(1-(-q)^j)^3}{(1-(-q)^{3j})} \\ &= \prod_{j=1}^{\infty} \frac{(1-q^{2j})^9(1-q^{3j})(1-q^{12j})}{(1-q^j)^3(1-q^{4j})^3(1-q^{6j})^3} \\ &= \frac{b^3(q^2)}{b(q)b(q^4)}. \end{aligned}$$

The result for $c(-q)$ may be proved in a similar way.

□

**Definition 3.17 ($n$-dissection).** For any integer $n \geq 2$, the $n$-dissection of a series $f(q) = \sum_{j=0}^{\infty} c_j q^j$ is the series rearrangement

$$\begin{aligned} f(q) = \sum_{j=0}^{\infty} c_{nj} q^{nj} + \sum_{j=0}^{\infty} c_{nj+1} q^{nj+1} + \sum_{j=0}^{\infty} c_{nj+2} q^{nj+2} \\ + \cdots + \sum_{j=0}^{\infty} c_{nj+n-1} q^{nj+n-1}. \end{aligned}$$

For example, the 2- and 3-dissections of Euler's product were given in Exercises 3 and 4, respectively, at the end of Chapter 0. As another example, the 4-dissection of $\varphi(q)$ and the 2-dissection of $\varphi^2(q)$ are given in Corollary 3.4. Yet another example is obtained by adding (3.17) and (3.18) to obtain

$$b(q) = a(q^3) - c(q^3). \tag{3.30}$$

Since $a(q^3)$ and $c(q^3)$ only contain powers of $q$ that are congruent to 0 and 1 modulo 3, respectively, this is the 3-dissection of $b(q)$. It is an analogue of the 2-dissection

$$\varphi^2(-q) = \varphi^2(q^2) - 4q\psi^2(q^4).$$

that follows from Corollary 3.4 by replacing $q$ with $-q$.

We will now use dissections to establish the following useful formula for $a(q)$.

**Theorem 3.18.**

$$a(q) = \frac{\sum_{n=-\infty}^{\infty} (-1)^n (6n+1) q^{(6n+1)^2/24}}{\sum_{n=-\infty}^{\infty} (-1)^n q^{(6n+1)^2/24}}.$$

Before starting the proof, it may be pointed out that, using the notation from (0.60), the result is equivalent to $a(q) = V_1$. See also Remark 0.17.

*Proof.* We compute the 3-dissection of $b(q)$ in another way. By (0.46),

$$\prod_{j=1}^{\infty}(1-q^j)^3 = \sum_{n=-\infty}^{\infty} (4n+1)q^{2n^2+n}. \tag{3.31}$$

Let us break the series in to three parts according to the residue class of $n$ modulo 3. We successively replace $n$ with $3n$, $-3n-2$, and $-3n-1$ to deduce

$$\prod_{j=1}^{\infty}(1-q^j)^3 = \sum_{n=-\infty}^{\infty} (12n+1)q^{18n^2+3n}$$
$$- \sum_{n=-\infty}^{\infty} (12n+7)q^{18n^2+21n+6} - \sum_{n=-\infty}^{\infty} (12n+3)q^{18n^2+9n+1}.$$

The first two series can be merged into a single series, and the last series can be summed by (3.31) to give

$$\prod_{j=1}^{\infty}(1-q^j)^3 = \sum_{n=-\infty}^{\infty} (-1)^n(6n+1)q^{(9n^2+3n)/2} - 3q\prod_{j=1}^{\infty}(1-q^{9j})^3.$$

We divide by $\prod_{j=1}^{\infty}(1-q^{3j})$ and make use of the infinite product for $b(q)$ given by (3.28), to deduce another form of the 3-dissection, viz.,

$$b(q)=\frac{\sum_{n=-\infty}^{\infty}(-1)^n(6n+1)q^{(9n^2+3n)/2}}{\prod_{j=1}^{\infty}(1-q^{3j})}-3q\prod_{j=1}^{\infty}\frac{(1-q^{9j})^3}{(1-q^{3j})}. \tag{3.32}$$

On comparing the 3-dissections (3.30) and (3.32), we deduce that

$$a(q)=\frac{\sum_{n=-\infty}^{\infty}(-1)^n(6n+1)q^{(3n^2+n)/2}}{\prod_{j=1}^{\infty}(1-q^{j})} \tag{3.33}$$

and we also recover the infinite product formula for $c(q)$, already proved in Lemma 3.15. Now apply (0.40) to the denominator in (3.33) to complete the proof. □

The next result is an analogue of Theorem 3.10.

**Theorem 3.19.** *The cubic theta functions have the following series representations*:

$$a(q)=1+6\sum_{j=1}^{\infty}\left(\frac{j}{3}\right)\frac{q^j}{1-q^j}, \tag{3.34}$$

$$b(q)=1+3\sum_{j=1}^{\infty}\left(\frac{j}{3}\right)\left(\frac{3q^{3j}}{1-q^{3j}}-\frac{q^j}{1-q^j}\right), \tag{3.35}$$

$$c(q^3)=3\sum_{j=1}^{\infty}\left(\frac{j}{3}\right)\left(\frac{q^j}{1-q^j}-\frac{q^{3j}}{1-q^{3j}}\right), \tag{3.36}$$

*where* $\left(\frac{j}{3}\right)$ *is the Legendre symbol defined by* (2.34).

*Proof.* We begin with (3.33) and use the triple product identity (0.15) to deduce

$$a(q)=\left(\prod_{j=1}^{\infty}\frac{1}{1-q^j}\right)\sum_{n=-\infty}^{\infty}(-1)^n(6n+1)q^{(3n^2+n)/2}$$

$$
\begin{aligned}
&= \left(\prod_{j=1}^{\infty} \frac{1}{1-q^j}\right) \left[ \frac{\mathrm{d}}{\mathrm{d}x} \sum_{n=-\infty}^{\infty} (-1)^n x^{6n+1} q^{(3n^2+n)/2} \right]_{x=1} \\
&= \left(\prod_{j=1}^{\infty} \frac{1}{1-q^j}\right) \left[ \frac{\mathrm{d}}{\mathrm{d}x} \left( x \prod_{j=1}^{\infty} (1-x^6 q^{3j-1})(1-x^{-6} q^{3j-2})(1-q^{3j}) \right) \right]_{x=1} \\
&= \left(\prod_{j=1}^{\infty} \frac{1}{1-q^j}\right) \times \left[ \prod_{j=1}^{\infty} (1-x^6 q^{3j-1})(1-x^{-6} q^{3j-2})(1-q^{3j}) \right. \\
&\qquad \left. \times \left\{ 1 + 6 \sum_{j=1}^{\infty} \left( \frac{x^{-6} q^{3j-2}}{1-x^{-6} q^{3j-2}} - \frac{x^6 q^{3j-1}}{1-x^6 q^{3j-1}} \right) \right\} \right]_{x=1} \\
&= 1 + 6 \sum_{j=1}^{\infty} \left(\frac{j}{3}\right) \frac{q^j}{1-q^j}.
\end{aligned}
$$

This proves (3.34). The identities (3.35) and (3.36) follow immediately from (3.34) on using Lemma 3.11.

□

The next result is the cubic analogue of Jacobi's identity in Theorem 3.7.

**Theorem 3.20.**

$$a^3(q) = b^3(q) + c^3(q).$$

*Proof.* From Lemma 3.11 we may deduce

$$b(q) = a(q^3) - c(q^3).$$

We have already observed that this is the 3-dissection of $b(q)$, and moreover, the terms $a(q^3)$ and $c(q^3)$ only contain powers of $q$ that are congruent to 0 and 1 modulo 3, respectively. Let $\omega = \exp(2\pi i/3)$. On successively replacing $q$ with $\omega q$ and $\omega^2 q$ and multiplying, we deduce

$$
\begin{aligned}
b(q)b(\omega q)b(\omega^2 q) &= \left(a(q^3) - c(q^3)\right)\left(a(q^3) - \omega c(q^3)\right)\left(a(q^3) - \omega^2 c(q^3)\right) \\
&= a^3(q^3) - c^3(q^3). \qquad (3.37)
\end{aligned}
$$

On the other hand, by the infinite product for $b(q)$ given by (3.28) and an application of (0.5), we have

$$b(q)b(\omega q)b(\omega^2 q) = \prod_{j=1}^{\infty} \frac{(1-q^j)^3(1-\omega^j q^j)^3(1-\omega^{2j} q^j)^3}{(1-q^{3j})^3}$$

$$= \prod_{j=1}^{\infty} \frac{(1-q^{3j})^9}{(1-q^{9j})^3}$$

$$= b^3(q^3). \tag{3.38}$$

Now combine (3.37) and (3.38) and replace $q^3$ with $q$ to complete the proof. □

## 3.4 Theta function identities: weight 1

In this section we continue to develop properties of Ramanujan's theta functions and the cubic theta functions. Many of the identities are consequences of the explicit forms of Halphen's identity in Theorems 1.12–1.15.

The next result links Ramanujan's theta functions $\varphi(q)$ and $\psi(q)$ with the cubic theta functions $a(q)$, $b(q)$, and $c(q)$. It is a fruitful source of applications and one of the most useful results in this chapter.

**Theorem 3.21.** *Let $\eta_k$ be the infinite product defined by* (0.39). *Then*

$$2a(q^2) - a(q) = \frac{\eta_1^6\eta_6}{\eta_2^3\eta_3^2} = \frac{b^2(q)}{b(q^2)} = \frac{\varphi^3(-q)}{\varphi(-q^3)}, \tag{3.39}$$

$$\frac{a(q)+a(q^2)}{2} = \frac{\eta_2^6\eta_3}{\eta_1^3\eta_6^2} = \frac{b^2(q^2)}{b(q)} = \frac{\psi^3(q)}{\psi(q^3)}, \tag{3.40}$$

$$\frac{a(q)+2a(q^2)}{3} = \frac{\eta_2\eta_3^6}{\eta_1^2\eta_6^3} = \frac{c^2(q)}{3c(q^2)} = \frac{\varphi^3(-q^3)}{\varphi(-q)} \tag{3.41}$$

*and*

$$\frac{a(q)-a(q^2)}{6} = \frac{\eta_1\eta_6^6}{\eta_2^2\eta_3^3} = \frac{c^2(q^2)}{3c(q)} = q\frac{\psi^3(q^3)}{\psi(q)}. \tag{3.42}$$

*Proof.* Let us begin by proving (3.39). The equalities

$$\frac{\eta_1^6\eta_6}{\eta_2^3\eta_3^2} = \frac{b^2(q)}{b(q^2)} = \frac{\varphi^3(-q)}{\varphi(-q^3)}$$

are immediate from the infinite product products in Theorems 3.5 and 3.15, so it remains to prove the first equality

$$2a(q^2) - a(q) = \frac{\eta_1^6\eta_6}{\eta_2^3\eta_3^2}. \tag{3.43}$$

Let $\omega = \exp(i\pi/3)$, take $(x, y, z) = (\omega^2, \omega^2, \omega^3)$ in Theorem 1.12 and follow the procedure in Section 1.7 to deduce

$$\frac{\eta_1^6\eta_6}{\eta_2^3\eta_3^2} = 1 + \sum_{j=0}^{\infty}\left(\frac{-6q^{6j+1}}{1-q^{6j+1}} + \frac{18q^{6j+2}}{1-q^{6j+2}} - \frac{18q^{6j+4}}{1-q^{6j+4}} + \frac{6q^{6j+5}}{1-q^{6j+5}}\right)$$
$$= 2\left(1 + 6\sum_{j=1}^{\infty}\left(\frac{j}{3}\right)\frac{q^{2j}}{1-q^{2j}}\right) - \left(1 + 6\sum_{j=1}^{\infty}\left(\frac{j}{3}\right)\frac{q^{j}}{1-q^{j}}\right).$$

Now apply the Lambert series for $a(q)$ given in Theorem 3.19 to obtain (3.43) and hence complete the proof of (3.39).

The identities in (3.40) and (3.41) can be proved by the same procedure on taking $(x, y, z) = (\omega, \omega, \omega^2)$ and $(x, y, z) = (\omega, \omega, \omega)$, respectively, in Theorem 1.12. For (3.42), replace $(q, a, b, c)$ with $(q^3, q^{1/2}, q^{1/2}, q)$ in Theorem 1.14.

□

**Theorem 3.22.** *The following quadratic transformation formulas hold*:

$$2q^2\frac{\psi^3(q^6)}{\psi(q^2)} = q\frac{\psi^3(-q^3)}{\psi(-q)} - q\frac{\psi^3(q^3)}{\psi(q)}, \tag{3.44}$$

$$2\frac{\psi^3(q^2)}{\psi(q^6)} = \frac{\psi^3(-q)}{\psi(-q^3)} + \frac{\psi^3(q)}{\psi(q^3)}, \tag{3.45}$$

$$\frac{\varphi^3(-q^6)}{\varphi(-q^2)} = \frac{\varphi^3(\pm q^3)}{\varphi(\pm q)} \pm 2q\frac{\psi^3(\pm q^3)}{\psi(\pm q)} \tag{3.46}$$

*and*

$$\frac{\varphi^3(-q^2)}{\varphi(-q^6)} = 2\frac{\psi^3(\pm q)}{\psi(\pm q^3)} - \frac{\varphi^3(\pm q)}{\varphi(\pm q^3)}. \tag{3.47}$$

*Each of* (3.46) *and* (3.47) *represents two identities depending on whether a plus sign or a minus sign is used throughout for the* $\pm$ *symbol.*

*Proof.* By (3.42),

$$q\frac{\psi^3(q^3)}{\psi(q)} = \frac{a(q) - a(q^2)}{6}.$$

Replacing $q$ with $-q$ and applying (3.21) gives

$$-q\frac{\psi^3(-q^3)}{\psi(-q)} = \frac{a(-q)-a(q^2)}{6} = \frac{-a(q)-a(q^2)+2a(q^4)}{6}.$$

Adding the last two identities gives

$$q\frac{\psi^3(q^3)}{\psi(q)} - q\frac{\psi^3(-q^3)}{\psi(-q)} = \frac{-a(q^2)+a(q^4)}{3} = -2q^2\frac{\psi^3(q^6)}{\psi(q^2)},$$

where (3.42) has been used to obtain the last equality. This proves the first result. The other identities may be proved in the same way, using the other identities in Theorem 3.21 in conjunction with (3.21).

□

The next results may be viewed as analogues of Theorem 3.21 in which $a(q^4)$ takes the place of $a(q^2)$. In light of Theorem 3.19, the following results also provide Lambert series expansions for $\varphi(q)\varphi(q^3)$ and $q\psi(q^2)\psi(q^6)$.

**Theorem 3.23.**

$$4a(q^4) - a(q) = 3\varphi(-q)\varphi(-q^3), \tag{3.48}$$

$$a(q) + 2a(q^4) = 3\varphi(q)\varphi(q^3) \tag{3.49}$$

*and*

$$a(q) - a(q^4) = 6q\psi(q^2)\psi(q^6). \tag{3.50}$$

*Hence,*

$$\varphi(q)\varphi(q^3) = 1 + 2\sum_{k=1}^{\infty}\left(\frac{k}{3}\right)\frac{q^k}{1-q^k} + 4\sum_{k=1}^{\infty}\left(\frac{k}{3}\right)\frac{q^{4k}}{1-q^{4k}} \tag{3.51}$$

*and*

$$q\psi(q^2)\psi(q^6) = \sum_{k=1}^{\infty}\left(\frac{q^{6k-5}}{1-q^{12k-10}} - \frac{q^{6k-1}}{1-q^{12k-2}}\right). \tag{3.52}$$

*Proof.* Let $\omega = \exp(i\pi/6)$, take $(x,y,z) = (\omega^2,\omega^5,\omega^6)$ in (1.26), and follow the procedure illustrated in Section 1.7 to deduce that

$$\prod_{n=1}^{\infty}\frac{(1-q^n)(1-q^{3n})}{(1+q^n)(1+q^{3n})} = 1 - 2\sum_{j=1}^{\infty}\left(\frac{j}{3}\right)\frac{q^j}{1-q^j} + 8\sum_{j=1}^{\infty}\left(\frac{j}{3}\right)\frac{q^{4j}}{1-q^{4j}}.$$

Theorem 3.5 implies that the left-hand side is $\varphi(-q)\varphi(-q^3)$, and from Theorem 3.19 we deduce that the right-hand side is $\frac{1}{3}(4a(q^4)-a(q))$. This proves (3.48).

The identity (3.49) follows from (3.48) by replacing $q$ with $-q$ and applying Lemma 3.12.

To prove (3.50), replace $(q, a, b, c)$ with $(q^6, q^{1/2}, q^2, q^{5/2})$ in Theorem 1.14 to deduce

$$q \prod_{n=1}^{\infty} \frac{(1-q^{4n})(1-q^{12n})}{(1-q^{4n-2})(1-q^{12n-6})} = \sum_{k=1}^{\infty} \left(\frac{k}{3}\right) \left( \frac{q^k}{1-q^k} - \frac{q^{4k}}{1-q^{4k}} \right). \tag{3.53}$$

The left-hand side is just $q\psi(q^2)\psi(q^6)$ by Theorem 3.5, while the series on the right-hand side is equal to $(a(q) - a(q^4))/6$ by Theorem 3.19. This completes the proof of (3.50).

The identity (3.51) is obtained by substituting the Lambert series for $a(q)$ from (3.34) in (3.49). It remains to prove (3.52). From (3.53), and noting that the product on the left-hand side is $q\psi(q^2)\psi(q^6)$, we deduce that

$$\begin{aligned} q\psi(q^2)\psi(q^6) &= \sum_{k=1}^{\infty} \left(\frac{k}{3}\right) \left( \left( \frac{q^k}{1-q^k} - \frac{q^{2k}}{1-q^{2k}} \right) + \left( \frac{q^{2k}}{1-q^{2k}} - \frac{q^{4k}}{1-q^{4k}} \right) \right) \\ &= \sum_{k=1}^{\infty} \left(\frac{k}{3}\right) \left( \frac{q^k}{1-q^{2k}} + \frac{q^{2k}}{1-q^{4k}} \right) \\ &= \sum_{k=1}^{\infty} \left( \frac{q^{6k-5}}{1-q^{12k-10}} - \frac{q^{6k-1}}{1-q^{12k-2}} \right). \end{aligned}$$

□

The following three consequences of Theorem 3.23 are worth noting. First, the identity (3.50) gives the 2-dissection of $a(q)$, viz.,

$$a(q) = a(q^4) + 6q\psi(q^2)\psi(q^6).$$

Second, the identities in Theorem 3.23 can be rearranged to give $a(q)$ and $a(q^4)$ in terms of Ramanujan's theta functions, namely

$$a(q) = \varphi(q)\varphi(q^3) + 4q\psi(q^2)\psi(q^6), \tag{3.54}$$

and

$$a(q^4) = \varphi(q)\varphi(q^3) - 2q\psi(q^2)\psi(q^6);$$

or alternatively,

$$a(q) = \varphi(-q)\varphi(-q^3) + 8q\psi(q^2)\psi(q^6) \tag{3.55}$$

and

$$a(q^4) = \varphi(-q)\varphi(-q^3) + 2q\psi(q^2)\psi(q^6). \tag{3.56}$$

Third, eliminating $a(q)$ and $a(q^4)$ from the identities in Theorem 3.23 gives

$$\varphi(q)\varphi(q^3) - \varphi(-q)\varphi(-q^3) = 4q\psi(q^2)\psi(q^6). \tag{3.57}$$

By Theorem 3.5, each term may be expressed as an infinite product. The resulting identity simplifies to

$$\prod_{j=1}^{\infty}(1+q^{2j-1})^2(1+q^{6j-3})^2 - \prod_{j=1}^{\infty}(1-q^{2j-1})^2(1-q^{6j-3})^2$$

$$= 2q\prod_{j=1}^{\infty}(1+q^{2j})^2(1+q^{6j})^2. \tag{3.58}$$

This may be regarded as a cubic analogue of Jacobi's 'aequatio identica satis abstrusa' in Corollary 3.8. A septic analogue will be given in Exercise 4 at the end of Chapter 7.

The results in the previous three theorems can be utilized to produce further identities.

**Theorem 3.24.** *The following identities hold*:

$$\varphi(q)\varphi(q^3) = \frac{\varphi^3(q^3)}{\varphi(q)} + 4q\frac{\psi^3(q^3)}{\psi(q)}$$

$$12q\psi(q^2)\psi(q^6) = \frac{\varphi^3(q)}{\varphi(q^3)} - \frac{\varphi^3(-q)}{\varphi(-q^3)}.$$

*Proof.* By Theorem 3.23,

$$\varphi(q)\varphi(q^3) = \frac{1}{3}(a(q) + 2a(q^4)) = \frac{1}{3}(a(q^2) + 2a(q^4)) + \frac{1}{3}(a(q) - a(q^2)).$$

Now apply Theorems 3.21 and 3.22 to deduce

$$\varphi(q)\varphi(q^3) = \frac{\varphi^3(-q^6)}{\varphi(-q^2)} + 2q\frac{\psi^3(q^3)}{\psi(q)} = \frac{\varphi^3(q^3)}{\varphi(q)} + 4q\frac{\psi^3(q^3)}{\psi(q)}.$$

This proves the first result. The other result may be proved in the same way. □

The next result is a useful companion to Theorems 3.21 and 3.23.

**Theorem 3.25.** *The following identities hold:*

$$\varphi^2(-q^3) + \varphi^2(-q) = 2\frac{\eta_1^2\eta_4\eta_6^7}{\eta_2^3\eta_3^2\eta_{12}^3}, \tag{3.59}$$

$$3\varphi^2(-q^3) - \varphi^2(-q) = 2\frac{\eta_2^7\eta_3^2\eta_{12}}{\eta_1^2\eta_4^3\eta_6^3}, \tag{3.60}$$

$$q\psi^2(q^4) + 3q^3\psi^2(q^{12}) = \frac{\eta_4^7\eta_6\eta_{24}^2}{\eta_2^3\eta_8^2\eta_{12}^3}, \tag{3.61}$$

$$q\psi^2(q^4) - q^3\psi^2(q^{12}) = \frac{\eta_2\eta_8^2\eta_{12}^7}{\eta_4^3\eta_6^3\eta_{24}^2}, \tag{3.62}$$

$$q\psi^2(q^4) - 3q^3\psi^2(q^{12}) = \frac{\eta_2^3\eta_8\eta_{24}}{\eta_4^2\eta_6}, \tag{3.63}$$

$$q\psi^2(q^4) + q^3\psi^2(q^{12}) = \frac{\eta_6^3\eta_8\eta_{24}}{\eta_2\eta_{12}^2}, \tag{3.64}$$

$$\varphi^2(-q^3) - \varphi^2(-q) = 4\frac{\eta_1\eta_3\eta_{12}^3}{\eta_4\eta_6^2} \tag{3.65}$$

*and*

$$3\varphi^2(-q^3) + \varphi^2(-q) = 4\frac{\eta_1\eta_3\eta_4^3}{\eta_2^2\eta_{12}}. \tag{3.66}$$

*Proof.* Throughout the proof, let $\omega = \exp(i\pi/6)$. If we replace $(x, y, z)$ with $(\omega, \omega^3, \omega^3)$ in (1.26) and simplify the infinite product, the result is

$$\frac{\eta_2^3\eta_3^2\eta_6}{\eta_1^2\eta_4\eta_{12}} = 1 + \sum_{j=1}^{\infty} h(j)\frac{q^j}{1-q^j}$$

where $h(j)$ is a function with period 12. If we compare coefficients of $q, q^2, \ldots, q^{12}$ on both sides, we find that

$$h(j) = \begin{cases} 2 & \text{if } j \equiv 1 \text{ or } 5 \pmod{12}, \\ -2 & \text{if } j \equiv 7 \text{ or } 11 \pmod{12}, \\ 0 & \text{otherwise.} \end{cases}$$

It follows that

$$\frac{\eta_2^3\eta_3^2\eta_6}{\eta_1^2\eta_4\eta_{12}} = 1 + 2\sum_{j=1}^{\infty}\chi_4(j)\frac{q^j}{1-q^j} + 2\sum_{j=1}^{\infty}\chi_4(j)\frac{q^{3j}}{1-q^{3j}}$$

where $\chi_4$ is the character defined by (3.13). By Theorem 3.10 this can be written as

$$\frac{\eta_2^3\eta_3^2\eta_6}{\eta_1^2\eta_4\eta_{12}} = \frac{1}{2}\left(\varphi^2(q) + \varphi^2(q^3)\right).$$

Now replace $q$ with $-q$ and use (0.2) to deduce (3.59).

The identity (3.60) can be proved in a similar way, by taking $(x, y, z) = (\omega^3, \omega^3, \omega^7)$ in (1.26). We omit the details as they are similar.

Let us now prove (3.61) and (3.62). Use Theorem 3.5 to write the identity (3.59) as

$$2\frac{\eta^2(\tau)\eta(4\tau)\eta^7(6\tau)}{\eta^3(2\tau)\eta^2(3\tau)\eta^3(12\tau)} = \frac{\eta^4(3\tau)}{\eta^2(6\tau)} + \frac{\eta^4(\tau)}{\eta^2(2\tau)}.$$

On replacing $\tau$ with $-1/24\tau$ and then applying the modular transformation given by Theorem 2.8, the result simplifies to

$$\frac{\eta^7(4\tau)\eta(6\tau)\eta^2(24\tau)}{\eta^3(2\tau)\eta^2(8\tau)\eta^3(12\tau)} = \frac{\eta^4(8\tau)}{\eta^2(4\tau)} + 3\frac{\eta^4(24\tau)}{\eta^2(12\tau)}.$$

By Theorem 3.5 this is equivalent to

$$\frac{\eta^7(4\tau)\eta(6\tau)\eta^2(24\tau)}{\eta^3(2\tau)\eta^2(8\tau)\eta^3(12\tau)} = q\psi^2(q^4) + 3q^3\psi^2(q^{12}).$$

This proves the identity (3.61). The identity (3.62) can be obtained from (3.60) by a similar procedure.

The identities (3.63) and (3.64) now follow from (3.61) and (3.62), respectively, on replacing $q$ with $iq$ and dividing both sides by $i$.

Finally, (3.65) and (3.66) follow from (3.63) and (3.64), respectively, by replacing $\tau$ with $-1/24\tau$ and applying the modular transformation. □

## 3.5 Theta function identities: weight 2

Many of the identities in this section are consequences of Weierstrass's identities from Theorems 1.17 and 1.18, or they follow by taking a product of two identities from Theorem 3.21. In general, the functions on one side of the identities are squares, or products of two functions, from the various identities in Section 3.4. Recall that $P(q)$ is Ramanujan's Eisenstein series defined by (0.61), that is,

$$P(q) = q\frac{\mathrm{d}}{\mathrm{d}q}\log\left(q\prod_{j=1}^{\infty}(1-q^j)^{24}\right) = 1 - 24\sum_{j=1}^{\infty}\frac{jq^j}{1-q^j}.$$

The results in the next theorem were known to Jacobi.

**Theorem 3.26.**

$$\varphi^4(q) = \frac{1}{3}\left(4P(q^4) - P(q)\right), \tag{3.67}$$

$$\varphi^4(-q) = \frac{1}{3}\left(P(q) - 6P(q^2) + 8P(q^4)\right), \tag{3.68}$$

*and*

$$q\psi^4(q^2) = \frac{1}{24}\left(-P(q) + 3P(q^2) - 2P(q^4)\right). \tag{3.69}$$

Before starting the proof, the reader is invited to compare the identity (3.67) with the result of Exercise 5b of Chapter 1.

*Proof.* Let $x = i$, $y = -1$ in (1.32) and simplify the resulting infinite product to obtain

$$\prod_{j=1}^{\infty} \frac{(1-q^j)^4}{(1+q^j)^4} = 1 + \sum_{j=1}^{\infty} h(j) \frac{jq^j}{1-q^j}$$

where $h(j)$ is a function with period 4. On equating coefficients of $q$, $q^2$, $q^3$, and $q^4$, we find that

$$h(j) = \begin{cases} -8 & \text{if } j \text{ is odd,} \\ 16 & \text{if } j \equiv 2 \pmod 4, \\ 0 & \text{if } j \equiv 0 \pmod 4. \end{cases}$$

It follows that

$$\prod_{j=1}^{\infty} \frac{(1-q^j)^4}{(1+q^j)^4} = 1 - 8\sum_{j=1}^{\infty} \frac{jq^j}{1-q^j} + 48\sum_{j=1}^{\infty} \frac{jq^{2j}}{1-q^{2j}} - 64\sum_{j=1}^{\infty} \frac{jq^{4j}}{1-q^{4j}}.$$

The left-hand side is just $\varphi^4(-q)$ by Theorem 3.5, and the right-hand side is $\frac{1}{3}\left(P(q) - 6P(q^2) + 8P(q^4)\right)$. Thus, we have proved (3.68).

The identity (3.67) is obtained from (3.68) by replacing $q$ with $-q$ and using (0.110). Finally, (3.69) is obtained by subtracting (3.68) from (3.67) and using Theorem 3.7, viz.,

$$q\psi^4(q^2) = \frac{1}{16}\left(\varphi^4(q) - \varphi^4(-q)\right).$$

This completes the proof of Theorem 3.26. □

**Theorem 3.27.**

$$\varphi^2(q)\varphi^2(q^3) = \frac{1}{6}\left(-P(q) + 2P(q^2) + 3P(q^3)\right.$$
$$\left.-4P(q^4) - 6P(q^6) + 12P(q^{12})\right), \tag{3.70}$$

$$\varphi^2(-q)\varphi^2(-q^3) = \frac{1}{6}\left(P(q) - 4P(q^2) - 3P(q^3) + 12P(q^6)\right), \tag{3.71}$$

$$q\psi^2(q)\psi^2(q^3) = \frac{1}{24}\left(-P(q) + P(q^2) + 3P(q^3) - 3P(q^6)\right). \tag{3.72}$$

*Proof.* See (1.42), (1.44), and (1.43), respectively.

□

The next two theorems will be useful in later applications. Their simple proofs illustrate some of the power of Theorem 3.21.

**Theorem 3.28.**

$$a^2(q) = \varphi^2(-q)\varphi^2(-q^3) + 16q\psi^2(q)\psi^2(q^3),$$

*and*

$$a^2(q^2) = \varphi^2(-q)\varphi^2(-q^3) + 4q\psi^2(q)\psi^2(q^3).$$

*Proof.* Multiply (3.39) by (3.41) and (3.40) by (3.42) to deduce, respectively, that

$$\varphi^2(-q)\varphi^2(-q^3) = \frac{1}{3}\left(4a^2(q^2) - a^2(q)\right)$$

and

$$q\psi^2(q)\psi^2(q^3) = \frac{1}{12}\left(a^2(q) - a^2(q^2)\right).$$

The required results now follow on taking appropriate linear combinations to eliminate either the $a^2(q^2)$ or the $a^2(q)$ term.

□

**Theorem 3.29.**

$$a(q)a(q^2) = b(q)b(q^2) + c(q)c(q^2).$$

*Proof.* Multiply (3.39) by (3.40) and (3.41) by (3.42) to deduce, respectively, that

$$b(q)b(q^2) = a^2(q^2) + \frac{a(q)a(q^2)}{2} - \frac{a^2(q)}{2}$$

and

$$c(q)c(q^2) = -a^2(q^2) + \frac{a(q)a(q^2)}{2} + \frac{a^2(q)}{2}.$$

The required result follows by addition.

□

The next result provides further Lambert series expansions and is an analogue of Theorem 3.27.

**Theorem 3.30.**

$$a(q)a(q^2) = \frac{1}{4}\left(6P(q^6) - 3P(q^3) + 2P(q^2) - P(q)\right), \tag{3.73}$$

$$b(q)b(q^2) = \frac{1}{8}\left(18P(q^6) - 9P(q^3) - 2P(q^2) + P(q)\right), \tag{3.74}$$

$$c(q)c(q^2) = \frac{3}{8}\left(-2P(q^6) + P(q^3) + 2P(q^2) - P(q)\right). \tag{3.75}$$

*Proof.* We begin by proving (3.74) and (3.75). Let $\omega = \exp(i\pi/3)$. The result for $b(q)b(q^2)$ may be obtained by taking $x = \omega, z = \omega^2$ in (1.26) and making use of the infinite product expansion for $b(q)$ in Theorem 3.15. The result for $c(q)c(q^2)$ may be obtained by replacing $(q, a, b)$ with $(q^3, q^{1/2}, q^{3/2})$ in (1.33), and making use of the infinite product for $c(q)$ in Theorem 3.15. In each case, the details are similar to the examples in Theorem 1.25 in Section 1.7.

The identity (3.73) follows by adding (3.74) and (3.75) and applying Theorem 3.29.

□

The last result in this section gives Lambert series expansions for the squares of the cubic theta functions.

**Theorem 3.31.** *Let* $\omega = \exp(2\pi i/3)$. *Then*

$$a^2(q) = \frac{1}{2}(3P(q^3) - P(q)), \tag{3.76}$$

$$b^2(q) = \frac{1}{8}(9P(q^9) + \omega^2 P(\omega q) + \omega P(\omega^2 q)), \tag{3.77}$$

$$c^2(q^3) = -\frac{1}{24}(P(q) + \omega P(\omega q) + \omega^2 P(\omega^2 q)). \tag{3.78}$$

Before starting the proof, the reader is invited to compare the identity (3.76) with the result of Exercise 5a of Chapter 1.

*Proof.* By Theorems 3.27 and 3.28, we have

$$a^2(q) = \varphi^2(-q)\varphi^2(-q^3) + 16q\psi^2(q)\psi^2(q^3) = \frac{1}{2}\left(3P(q^3) - P(q)\right)$$

and this proves (3.76).

From Lemma 3.11, we have

$$a^2(q) = \big(a(q^3) + 2c(q^3)\big)^2 = a^2(q^3) + 4a(q^3)c(q^3) + 4c^2(q^3).$$

On combining this with (3.76), we deduce

$$\frac{1}{2}\big(3P(q^3) - P(q)\big) = \frac{1}{2}\big(3P(q^9) - P(q^3)\big) + 4a(q^3)c(q^3) + 4c^2(q^3),$$

which may be rewritten as

$$P(q) = \big(4P(q^3) - 3P(q^9)\big) - 8a(q^3)c(q^3) - 8c^2(q^3). \tag{3.79}$$

Observe that (3.79) gives a 3-dissection of $P(q)$. For any function $f(q)$ that is analytic in a neighborhood of $q = 0$ with Taylor expansion

$$f(q) = \sum_{n=0}^{\infty} t_n q^n,$$

we have

$$\sum_{n=0}^{\infty} t_{3n} q^{3n} = \frac{1}{3}\big(f(q) + f(\omega q) + f(\omega^2 q)\big),$$

$$\sum_{n=0}^{\infty} t_{3n+1} q^{3n+1} = \frac{1}{3}\big(f(q) + \omega^2 f(\omega q) + \omega f(\omega^2 q)\big)$$

and

$$\sum_{n=0}^{\infty} t_{3n+2} q^{3n+2} = \frac{1}{3}\big(f(q) + \omega f(\omega q) + \omega^2 f(\omega^2 q)\big).$$

Thus, on extracting the terms in (3.79) for which the power of $q$ is congruent to 1 or 2 modulo 3, we deduce, respectively, that

$$-8a(q^3)c(q^3) = \frac{1}{3}\big(P(q) + \omega^2 P(\omega q) + \omega P(\omega^2 q)\big) \tag{3.80}$$

and

$$-8c^2(q^3) = \frac{1}{3}\big(P(q) + \omega P(\omega q) + \omega^2 P(\omega^2 q)\big). \tag{3.81}$$

In particular, from (3.81) we deduce (3.78).

It remains to prove (3.77). From Lemma 3.11,

$$b(q) = a(q^3) - c(q^3).$$

Squaring, and using (3.79) to eliminate the $c^2(q^3)$ term, we obtain

$$\begin{aligned} b^2(q) &= a^2(q^3) - 2a(q^3)c(q^3) + c^2(q^3) \\ &= a^2(q^3) - 3a(q^3)c(q^3) + \frac{1}{8}\left(-3P(q^9) + 4P(q^3) - P(q)\right). \end{aligned}$$

Now use (3.76) and (3.80) to deduce

$$\begin{aligned} b^2(q) &= \frac{1}{2}\left(3P(q^9) - P(q^3)\right) + \frac{1}{8}\left(P(q) + \omega^2 P(\omega q) + \omega P(\omega^2 q)\right) \\ &\quad + \frac{1}{8}\left(-3P(q^9) + 4P(q^3) - P(q)\right) \\ &= \frac{1}{8}\left(9P(q^9) + \omega^2 P(\omega q) + \omega P(\omega^2 q)\right). \end{aligned}$$

This proves (3.77) and completes the proof of the theorem. □

## 3.6 Theta function identities: weight 3

Many of the identities in this section are consequences of the identities in Theorems 1.21 and 1.22, or can be obtained by taking a product of three identities from Theorem 3.21. In general, the functions on one side of the identities are cubes, or products of three functions, from the various identities in Section 3.4. The most important identity in this category is

$$a^3(q) = b^3(q) + c^3(q),$$

proved in Theorem 3.20. Another topic that will be studied in this section is the vector space that is spanned by the set

$$\left\{b^3(q), c^3(q), b^3(q^2), c^3(q^2)\right\}.$$

Several different bases for this vector space will be given and the change of basis matrices will be determined. This will provide access to a large number of identities.

We begin a lemma that provides Lambert series expansions for various products of six theta functions.

**Lemma 3.32.** *Let* $\chi_4$ *be the character defined by* (3.13), *and let* $\eta_n$ *be defined by* (0.39). *The following identities hold*:

$$\varphi^2(-q)\varphi^4(-q^2) = \frac{\eta_1^4\eta_2^6}{\eta_4^4} = 1 - 4\sum_{j=1}^{\infty}\chi_4(j)\frac{j^2q^j}{1-q^j}, \tag{3.82}$$

$$q\psi^4(q)\psi^2(q^2) = \frac{\eta_2^6\eta_4^4}{\eta_1^4} = \sum_{j=1}^{\infty}\frac{j^2q^j}{1+q^{2j}}, \tag{3.83}$$

$$q\varphi^4(-q^2)\psi^2(q^4) = \frac{\eta_2^8\eta_8^4}{\eta_4^6} = \sum_{j=1}^{\infty}\chi_4(j)\frac{j^2q^j}{1-q^{2j}}, \tag{3.84}$$

$$q\varphi^2(-q)\psi^4(q^2) = \frac{\eta_1^4\eta_4^8}{\eta_2^6} = \sum_{j=1}^{\infty}(-1)^{j+1}\frac{j^2q^j}{1+q^{2j}}. \tag{3.85}$$

*Proof.* The first equality in each identity is a trivial consequence of the infinite product formulas for $\varphi$ and $\psi$ in Theorem 3.5. It remains to establish the second equality in each of (3.82)–(3.85).

The identity (3.82) may be obtained by taking $x = i$ in Theorem 1.21. The identities (3.83), (3.84), and (3.85) may be obtained by replacing $(q, a)$ with $(q^2, q)$, $(q, i)$, and $(-q^2, -q)$, respectively, in Theorem 1.22.

□

The functions in (3.82)–(3.85) are linearly independent, that is, the only linear relation of the form

$$c_1\varphi^2(-q)\varphi^4(-q^2) + c_2q\psi^4(q)\psi^2(q^2) + c_3q\varphi^4(-q^2)\psi^2(q^4) + c_4q\varphi^2(-q)\psi^4(q^2) = 0$$

is the trivial relation given by $c_1 = c_2 = c_3 = c_4 = 0$. The next two identities may be obtained by taking linear combinations of (3.82)–(3.85).

**Theorem 3.33.** *Let* $\chi_4$ *be the character defined by* (3.13), *and let* $\eta_n$ *be defined by* (0.39). *The following identities hold*:

$$\varphi^2(q)\varphi^4(-q^2) = \varphi^4(q)\varphi^2(-q) = \frac{\eta_2^{18}}{\eta_1^4\eta_4^8} = 1 + 4\sum_{j=1}^{\infty}\chi_4(j)\frac{j^2q^j}{1+q^j} \tag{3.86}$$

*and*

$$q\varphi^4(q^2)\psi^2(q^4) = q\varphi^2(q^2)\psi^4(q^2) = \frac{\eta_4^{18}}{\eta_2^8\eta_8^4} = \sum_{\substack{j=1 \\ j \text{ odd}}}^{\infty}\frac{j^2q^j}{1+q^{2j}}. \tag{3.87}$$

*Proof.* The identity (3.86) is obtained from (3.82) by replacing $q$ with $-q$. We will prove (3.87). The identities between the theta functions and the eta-quotients are easy consequences of the products in Theorem 3.5. It remains to establish the Lambert series. Clearly, the Lambert series in (3.87) is obtained by adding the Lambert series in (3.83) and (3.85) and dividing by 2. On adding the theta functions in (3.83) and (3.85) and dividing by 2, we obtain by using Lemmas 3.3 and 3.6,

$$\begin{aligned}\frac{1}{2}\left(q\psi^4(q)\psi^2(q^2)+q\varphi^2(-q)\psi^4(q^2)\right)&=\frac{1}{2}\left(q\varphi^2(q)\psi^4(q^2)+q\varphi^2(-q)\psi^4(q^2)\right)\\&=q\varphi^2(q^2)\psi^4(q^2).\end{aligned}$$

□

The next result considers further linear combinations of the functions in (3.82)–(3.85) that give rise to, among other things, Lambert series for the sixth powers of $\varphi$ and $\psi$.

**Theorem 3.34.** *Let* $\chi_4$ *be the character defined by* (3.13). *The following identities hold*:

$$\varphi^6(q)=1-4\sum_{j=1}^{\infty}\chi_4(j)\frac{j^2q^j}{1-q^j}+16\sum_{j=1}^{\infty}\frac{j^2q^j}{1+q^{2j}},\tag{3.88}$$

$$16q^3\psi^6(q^4)=\sum_{\substack{j=1\\ j\text{ odd}}}^{\infty}\frac{j^2q^j}{1+q^{2j}}-\sum_{j=1}^{\infty}\chi_4(j)\frac{j^2q^j}{1-q^{2j}},\tag{3.89}$$

$$\varphi^4(q)\varphi^2(q^2)=1-4\sum_{j=1}^{\infty}\chi_4(j)\frac{j^2q^{2j}}{1-q^{2j}}+8\sum_{j=1}^{\infty}\frac{j^2q^j}{1+q^{2j}},\tag{3.90}$$

$$\varphi^2(q)\varphi^4(q^2)=1-4\sum_{j=1}^{\infty}\chi_4(j)\frac{j^2q^{2j}}{1-q^{2j}}+4\sum_{j=1}^{\infty}\frac{j^2q^j}{1+q^{2j}}.\tag{3.91}$$

*Proof.* The idea is to express each of the functions $\varphi^6(q)$, $q^3\psi^6(q^4)$, $\varphi^4(q)\varphi^2(q^2)$, and $\varphi^2(q)\varphi^4(q^2)$ as linear combinations of the theta functions that occur in Lemma 3.32. To this end, repeated application of Lemmas 3.3 and 3.6 and Theorem 3.7 gives

$$\begin{aligned}\varphi^6(q)&=\varphi^2(q)\left(\varphi^4(-q)+16q\psi^4(q^2)\right)\\&=\varphi^2(-q)\varphi^4(-q^2)+16q\psi^4(q)\psi^2(q^2),\end{aligned}$$

$$16q^3\psi^6(q^4)=q\psi^2(q^4)\left(\varphi^4(q^2)-\varphi^4(-q^2)\right),$$

$$\begin{aligned}\varphi^4(q)\varphi^2(q^2) &= \varphi^2(q)\varphi^2(q^2)\left(\varphi^2(-q)+8q\psi^2(q^4)\right)\\ &= \varphi^4(-q^2)\varphi^2(q^2)+8q\varphi^2(q)\psi^4(q^2)\\ &= \varphi^4(-q^2)\left(\varphi^2(-q)+4q\psi^2(q^4)\right)+8q\psi^4(q)\psi^2(q^2)\\ &= \varphi^2(-q)\varphi^4(-q^2)+4q\varphi^4(-q^2)\psi^2(q^4)+8q\psi^4(q)\psi^2(q^2)\end{aligned}$$

and

$$\begin{aligned}\varphi^2(q)\varphi^4(q^2) &= \varphi^2(q)\varphi^2(q^2)\left(\varphi^2(-q)+4q\psi^2(q^4)\right)\\ &= \varphi^4(-q^2)\varphi^2(q^2)+4q\varphi^2(q)\psi^4(q^2)\\ &= \varphi^4(-q^2)\left(\varphi^2(-q)+4q\psi^2(q^4)\right)+4q\psi^4(q)\psi^2(q^2)\\ &= \varphi^2(-q)\varphi^4(-q^2)+4q\varphi^4(-q^2)\psi^2(q^4)+4q\psi^4(q)\psi^2(q^2).\end{aligned}$$

Now substitute the Lambert series from Lemma 3.32 to complete the proofs.

□

The next result gives Lambert series for the cubes of the cubic theta functions.

**Theorem 3.35.**

$$a^3(q) = 1 - 9\sum_{j=1}^{\infty}\left(\frac{j}{3}\right)\frac{j^2q^j}{1-q^j} + 27\sum_{j=1}^{\infty}\frac{j^2q^j(1-q^j)}{1-q^{3j}}, \tag{3.92}$$

$$b^3(q) = 1 - 9\sum_{j=1}^{\infty}\left(\frac{j}{3}\right)\frac{j^2q^j}{1-q^j}, \tag{3.93}$$

$$c^3(q) = 27\sum_{j=1}^{\infty}\frac{j^2q^j(1-q^j)}{1-q^{3j}}. \tag{3.94}$$

*Proof.* We prove (3.93) and (3.94) first. The identity (3.93) can be obtained by taking $x = \exp(2\pi i/3)$ in Theorem 1.21 and making use of the infinite product representation for $b(q)$ given by Theorem 3.15. The identity (3.94) may be obtained by replacing $(q, a)$ with $(q^{3/2}, q^{1/2})$ in Theorem 1.22 and applying Theorem 3.15. Finally, (3.92) follows by adding (3.93) and (3.94) and using Theorem 3.20.

□

Here is another set of functions that can be expanded in Lambert series.

**Theorem 3.36.**

$$\frac{\eta_1\eta_2^4\eta_3^5}{\eta_6^4} = 1 + \sum_{j=1}^{\infty}(-1)^j\left(\frac{j}{3}\right)\frac{j^2q^j}{1-q^j},$$

$$\frac{\eta_1^4\eta_2\eta_6^5}{\eta_3^4} = \sum_{j=1}^{\infty}\left(\frac{j}{3}\right)\frac{j^2q^j}{1-q^{2j}},$$

$$\frac{\eta_1^5\eta_3\eta_6^4}{\eta_2^4} = \sum_{j=1}^{\infty}(-1)^{j+1}\frac{j^2q^j(1-q^j)}{1-q^{3j}},$$

$$\frac{\eta_2^5\eta_3^4\eta_6}{\eta_1^4} = \sum_{j=1}^{\infty}\frac{j^2q^j(1-q^{4j})}{1-q^{6j}}.$$

*Proof.* The first result may be obtained by letting $x = \exp(i\pi/3)$ in Theorem 1.21. The second, third and fourth results are obtained by replacing $(q,a)$ with $(q, \exp(2\pi i/3))$, $(q^{3/2}, -q^{1/2})$ and $(q^3, q^2)$, respectively, in Theorem 1.22. □

Consider the four sets of functions $\{A_1, A_2, A_3, A_4\}$, $\{B_1, B_2, B_3, B_4\}$, $\{C_1, C_2, C_3, C_4\}$, and $\{D_1, D_2, D_3, D_4\}$, defined by

$$A_1 = a^3(q), \quad A_2 = a^2(q)a(q^2), \quad A_3 = a(q)a^2(q^2), \quad A_4 = a^3(q^2);$$

$$B_1 = b^3(q), \quad B_2 = b^3(q^2), \quad B_3 = c^3(q), \quad B_4 = c^3(q^2);$$

$$C_1 = \varphi^5(-q)\varphi(-q^3), \quad C_2 = \varphi(-q)\varphi^5(-q^3),$$

$$C_3 = q\psi^5(q)\psi(q^3), \quad C_4 = q^2\psi(q)\psi^5(q^3);$$

and

$$D_1 = \frac{\eta_1\eta_2^4\eta_3^5}{\eta_6^4}, \quad D_2 = \frac{\eta_1^4\eta_2\eta_6^5}{\eta_3^4}, \quad D_3 = \frac{\eta_1^5\eta_3\eta_6^4}{\eta_2^4}, \quad D_4 = \frac{\eta_2^5\eta_3^4\eta_6}{\eta_1^4}.$$

We will show that each of the four sets span the same vector space and determine the change of basis matrices. Lambert series for $A_1$, $B_1$, and $B_3$ (and hence also for $A_4$, $B_2$, and $B_4$) have been given in Theorem 3.35, and Lambert series for each of $D_1, D_2, D_3$, and $D_4$ have been given in Theorem 3.36. The change of basis formulas will enable Lambert series to be obtained for the remaining functions $A_2$, $A_3$, $C_1$, $C_2$, $C_3$, and $C_4$.

**Theorem 3.37.** *The following change of basis formulas hold:*

$$\begin{pmatrix} 2B_1 \\ 4B_2 \\ 2B_3 \\ 4B_4 \end{pmatrix} = \begin{pmatrix} 1 & -3 & 0 & 4 \\ -1 & 0 & 3 & 2 \\ 1 & 3 & 0 & -4 \\ 1 & 0 & -3 & 2 \end{pmatrix} \begin{pmatrix} A_1 \\ A_2 \\ A_3 \\ A_4 \end{pmatrix}, \tag{3.95}$$

$$\begin{pmatrix} 3C_1 \\ 9C_2 \\ 24C_3 \\ 72C_4 \end{pmatrix} = \begin{pmatrix} 1 & -2 & -4 & 8 \\ -1 & -2 & 4 & 8 \\ 1 & 1 & -1 & -1 \\ 1 & -1 & -1 & 1 \end{pmatrix} \begin{pmatrix} A_1 \\ A_2 \\ A_3 \\ A_4 \end{pmatrix}, \tag{3.96}$$

*and*

$$\begin{pmatrix} 6D_1 \\ 12D_2 \\ 18D_3 \\ 36D_4 \end{pmatrix} = \begin{pmatrix} -1 & -1 & 4 & 4 \\ -1 & 2 & 1 & -2 \\ -1 & 1 & 4 & -4 \\ 1 & 2 & -1 & -2 \end{pmatrix} \begin{pmatrix} A_1 \\ A_2 \\ A_3 \\ A_4 \end{pmatrix}. \tag{3.97}$$

*Proof.* All of the results (3.95)–(3.97) are fairly simple consequences of Theorem 3.21. We begin with (3.95). If we square (3.39) and multiply by (3.40), we deduce that

$$\left(\frac{b^2(q)}{b(q^2)}\right)^2 \left(\frac{b^2(q^2)}{b(q)}\right) = \frac{1}{2}\left(2a(q^2) - a(q)\right)^2 \left(a(q) + a(q^2)\right),$$

that is,

$$2b^3(q) = a^3(q) - 3a^2(q)a(q^2) + 4a^3(q^2).$$

This establishes the first row in the matrix in (3.95). The other rows are obtained similarly, by squaring one of (3.39) or (3.40) and multiplying by the other, or by squaring one of (3.41) or (3.42) and multiplying by the other.

The results in (3.96) may be obtained similarly, either by squaring one of (3.39) or (3.41) and multiplying by the other, or by squaring one of (3.40) or (3.42) and multiplying by the other. For example, squaring (3.39) and multiplying by (3.41) gives

$$\begin{aligned} \varphi^5(-q)\varphi(-q^3) &= \frac{1}{3}(2a(q^2) - a(q))^2(a(q) + 2a(q^2)) \\ &= \frac{1}{3}(a^3(q) - 2a^2(q)a(q^2) - 4a(q)a^2(q^2) + 8a^3(q^2)). \end{aligned}$$

This establishes the first row in the matrix in (3.96).

The results in (3.97) can be obtained by multiplying any three of the four identities (3.39)–(3.42). For example, on multiplying (3.39), (3.40), and (3.41), we find that

$$\left(\frac{\eta_1^6\eta_6}{\eta_2^3\eta_3^2}\right)\left(\frac{\eta_2^6\eta_3}{\eta_1^3\eta_6^2}\right)\left(\frac{\eta_2\eta_3^6}{\eta_1^2\eta_6^3}\right)=\frac{1}{6}\left(2a(q^2)-a(q)\right)\left(a(q)+a(q^2)\right)\left(a(q)+2a(q^2)\right),$$

that is,

$$6\frac{\eta_1\eta_2^4\eta_3^5}{\eta_6^4}=-a^3(q)-a^2(q)a(q^2)+4a(q)a^2(q^2)+4a^3(q^2).$$

This establishes the first row of the matrix in (3.97). The other rows in the matrix may be established similarly.

□

We end this section by listing the remaining change of basis formulas. As already mentioned, some of these will implicitly yield the remaining Lambert series expansions for the functions $A_2$, $A_3$, $C_1$, $C_2$, $C_3$ and $C_4$.

**Corollary 3.38.** *The following change of basis formulas hold:*

$$\begin{pmatrix}A_1\\A_2\\A_3\\A_4\end{pmatrix}=\begin{pmatrix}1&0&1&0\\-\frac{1}{3}&\frac{4}{3}&\frac{1}{3}&\frac{4}{3}\\\frac{1}{3}&\frac{2}{3}&\frac{1}{3}&-\frac{2}{3}\\0&1&0&1\end{pmatrix}\begin{pmatrix}B_1\\B_2\\B_3\\B_4\end{pmatrix}=\begin{pmatrix}-\frac{1}{2}&\frac{3}{2}&16&48\\\frac{1}{4}&\frac{3}{4}&16&-48\\-\frac{1}{2}&\frac{3}{2}&4&12\\\frac{1}{4}&\frac{3}{4}&4&-12\end{pmatrix}\begin{pmatrix}C_1\\C_2\\C_3\\C_4\end{pmatrix}$$

$$=\begin{pmatrix}1&-8&3&24\\1&4&-3&12\\1&-2&3&6\\1&1&-3&3\end{pmatrix}\begin{pmatrix}D_1\\D_2\\D_3\\D_4\end{pmatrix},$$

$$\begin{pmatrix}B_1\\B_2\\B_3\\B_4\end{pmatrix}=\begin{pmatrix}-\frac{1}{8}&\frac{9}{8}&-8&72\\-\frac{1}{8}&\frac{9}{8}&1&-9\\-\frac{3}{8}&\frac{3}{8}&24&-24\\\frac{3}{8}&-\frac{3}{8}&3&-3\end{pmatrix}\begin{pmatrix}C_1\\C_2\\C_3\\C_4\end{pmatrix}=\begin{pmatrix}1&-8&0&0\\1&1&0&0\\0&0&3&24\\0&0&-3&3\end{pmatrix}\begin{pmatrix}D_1\\D_2\\D_3\\D_4\end{pmatrix},$$

$$\begin{pmatrix} C_1 \\ C_2 \\ C_3 \\ C_4 \end{pmatrix} = \begin{pmatrix} \frac{1}{9} & \frac{8}{9} & -\frac{1}{3} & \frac{8}{3} \\ \frac{1}{9} & \frac{8}{9} & -\frac{1}{27} & \frac{8}{27} \\ \frac{1}{72} & -\frac{1}{72} & \frac{1}{24} & \frac{1}{24} \\ \frac{1}{72} & -\frac{1}{72} & \frac{1}{216} & \frac{1}{216} \end{pmatrix} \begin{pmatrix} B_1 \\ B_2 \\ B_3 \\ B_4 \end{pmatrix} = \begin{pmatrix} 1 & 0 & -9 & 0 \\ 1 & 0 & -1 & 0 \\ 0 & -\frac{1}{8} & 0 & \frac{9}{8} \\ 0 & -\frac{1}{8} & 0 & \frac{1}{8} \end{pmatrix} \begin{pmatrix} D_1 \\ D_2 \\ D_3 \\ D_4 \end{pmatrix}$$

*and*

$$\begin{pmatrix} D_1 \\ D_2 \\ D_3 \\ D_4 \end{pmatrix} = \begin{pmatrix} \frac{1}{9} & \frac{8}{9} & 0 & 0 \\ -\frac{1}{9} & \frac{1}{9} & 0 & 0 \\ 0 & 0 & \frac{1}{27} & -\frac{8}{27} \\ 0 & 0 & \frac{1}{27} & \frac{1}{27} \end{pmatrix} \begin{pmatrix} B_1 \\ B_2 \\ B_3 \\ B_4 \end{pmatrix} = \begin{pmatrix} -\frac{1}{8} & \frac{9}{8} & 0 & 0 \\ 0 & 0 & 1 & -9 \\ -\frac{1}{8} & \frac{1}{8} & 0 & 0 \\ 0 & 0 & 1 & -1 \end{pmatrix} \begin{pmatrix} C_1 \\ C_2 \\ C_3 \\ C_4 \end{pmatrix}.$$

## 3.7 Theta function identities: weight 4

The identities in this section are limiting cases of the identities in Theorems 1.21 and 1.22, or they can be obtained by taking a product of four identities from Theorem 3.21. The first result involves Ramanujan's Eisenstein series $Q(q)$ defined by (0.72), i.e.,

$$Q(q) = 1 + 240 \sum_{j=1}^{\infty} \frac{j^3 q^j}{1 - q^j}.$$

Recall the effect of changing $q$ to $-q$ is given by (0.111), viz.,

$$Q(-q) = -Q(q) + 18Q(q^2) - 16Q(q^4). \tag{3.98}$$

The next result is an analogue of Theorem 3.26.

**Theorem 3.39.** *The following identities hold*:

$$\varphi^8(q) = \frac{1}{15}(Q(q) - 2Q(q^2) + 16Q(q^4)), \tag{3.99}$$

$$\varphi^8(-q) = \frac{1}{15}(-Q(q) + 16Q(q^2)), \tag{3.100}$$

*and*

$$q\psi^8(q) = \frac{1}{240}(Q(q) - Q(q^2)). \tag{3.101}$$

*Proof.* Take the limit as $x \to -1$ in Theorem 1.21 to obtain

$$\prod_{j=1}^{\infty} \frac{(1-q^j)^8}{(1+q^j)^8} = 1 + 16 \sum_{j=1}^{\infty} (-1)^j \frac{j^3 q^j}{1-q^j}.$$

The left-hand side is $\varphi^8(-q)$ by Theorem 3.5, and on manipulating the series on the right-hand side we find that

$$\varphi^8(-q) = 1 - 16 \sum_{j=1}^{\infty} \frac{j^3 q^j}{1-q^j} + 32 \sum_{j=1}^{\infty} \frac{(2j)^3 q^{2j}}{1-q^{2j}} = \frac{1}{15}\left(-Q(q) + 16Q(q^2)\right).$$

This proves (3.100), and (3.99) is obtained by replacing $q$ with $-q$ and using (3.98).

In order to prove (3.101) we divide the result of Theorem 1.22 by $1-a^2$ and take the limit as $a \to 1$ to obtain

$$q \prod_{j=1}^{\infty} \frac{(1-q^{2j})^8}{(1-q^{2j-1})^8} = \sum_{j=1}^{\infty} \frac{j^3 q^j}{1-q^{2j}} = \sum_{j=1}^{\infty} \left( \frac{j^3 q^j}{1-q^j} - \frac{j^3 q^{2j}}{1-q^{2j}} \right).$$

On applying Theorem 3.5 to the left-hand side, we deduce (3.101). □

**Corollary 3.40.** *The following identities hold:*

$$Q(q) = \varphi^8(-q) + 256 q \psi^8(q) = \frac{\eta_1^{16}}{\eta_2^8} + 256 \frac{\eta_2^{16}}{\eta_1^8}$$

*and*

$$Q(q^2) = \varphi^8(-q) + 16 q \psi^8(q) = \frac{\eta_1^{16}}{\eta_2^8} + 16 \frac{\eta_2^{16}}{\eta_1^8}.$$

*Proof.* The first equality in each identity is immediate from Theorem 3.39. Then apply the infinite product formulas for theta functions in Theorem 3.5 to complete the proofs.

□

**Corollary 3.41.** *The following identity holds:*

$$2Q(q^2) = \left( \sum_{j=-\infty}^{\infty} q^{j^2} \right)^8 + \left( \sum_{j=-\infty}^{\infty} q^{(j+\frac{1}{2})^2} \right)^8 + \left( \sum_{j=-\infty}^{\infty} (-1)^j q^{j^2} \right)^8.$$

*Proof.* The right-hand side of the purported identity is

$$\varphi^8(q) + 256 q^2 \psi^8(q^2) + \varphi^8(-q),$$

and this in turn is equal, by Theorem 3.39, to

$$\frac{1}{15}(Q(q)-2Q(q^2)+16Q(q^4))+\frac{16}{15}(Q(q^2)-Q(q^4))+\frac{1}{15}(-Q(q)+16Q(q^2))$$
$$=2Q(q^2).$$

□

**Theorem 3.42.** *The following identities hold:*

$$a^4(q)=\frac{1}{10}(\varphi^8(-q)+9\varphi^8(-q^3)+256q\psi^8(q)+2304q^3\psi^8(q^3)), \tag{3.102}$$

$$a^4(q^2)=\frac{1}{10}(\varphi^8(-q)+9\varphi^8(-q^3)+16q\psi^8(q)+144q^3\psi^8(q^3)), \tag{3.103}$$

$$a^3(q)a(q^2)=\frac{1}{8}(-\varphi^8(-q)+9\varphi^8(-q^3))+16q\psi^8(q)-144q^3\psi^8(q^3), \tag{3.104}$$

*and*

$$a(q)a^3(q^2)=\frac{1}{8}(-\varphi^8(-q)+9\varphi^8(-q^3))+4q\psi^8(q)-36q^3\psi^8(q^3). \tag{3.105}$$

*Moreover,*

$$a^2(q)a^2(q^2)=\frac{1}{10}(\varphi^8(-q)+9\varphi^8(-q^3)+64q\psi^8(q)+576q^3\psi^8(q^3)) \tag{3.106}$$
$$+\frac{36}{5}q\prod_{j=1}^{\infty}(1-q^j)^2(1-q^{2j})^2(1-q^{3j})^2(1-q^{6j})^2.$$

*Proof.* We make use of Theorem 3.21 and write $a$ and $A$ for $a(q)$ and $a(q^2)$, respectively. Take the cube of (3.39) and multiply by (3.41) to obtain

$$\varphi(-q)^8=(2A-a)^3(a+2A)/3. \tag{3.107}$$

In a similar way, we may also deduce from Theorem 3.21 that

$$\varphi(-q^3)^8=(2A-a)(a+2A)^3/27, \tag{3.108}$$

$$q\psi(q)^8=(a+A)^3(a-A)/48 \tag{3.109}$$

and

$$q^3\psi(q^3)^8=(a+A)(a-A)^3/432. \tag{3.110}$$

The results (3.102)–(3.105) in Theorem 3.42 follow by taking appropriate linear combinations of (3.107)–(3.110).

Next, if we multiply each of (3.39)–(3.42), we obtain

$$q\prod_{j=1}^{\infty}(1-q^j)^2(1-q^{2j})^2(1-q^{3j})^2(1-q^{6j})^2$$

$$=\frac{1}{36}(2A-a)(a+2A)(a+A)(a-A)$$

$$=5a^2A^2-a^4-4A^4.$$

On using (3.102) and (3.103) to eliminate the $a^4$ and $A^4$ terms, we deduce (3.106). □

The next result gives Lambert series expansions for $a^4(q)$, $a^3(q)a(q^2)$, and $a(q)a^3(q^2)$. The corresponding result for $a^2(q)a^2(q^2)$ also involves an additional term.

**Theorem 3.43.** *The following identities hold*:

$$a^4(q)=\frac{1}{10}(Q(q)+9Q(q^3)),$$

$$a^3(q)a(q^2)=\frac{1}{40}(3Q(q)-8Q(q^2)-27Q(q^3)+72Q(q^6)),$$

$$a(q)a^3(q^2)=\frac{1}{40}(Q(q)-6Q(q^2)-9Q(q^3)+54Q(q^6))$$

*and*

$$a^4(q^2)=\frac{1}{10}(Q(q^2)+9Q(q^6)).$$

*Moreover,*

$$a(q)^2a(q^2)^2=\frac{1}{50}(Q(q)+4Q(q^2)+9Q(q^3)+36Q(q^6))$$

$$+\frac{36}{5}q\prod_{j=1}^{\infty}(1-q^j)^2(1-q^{2j})^2(1-q^{3j})^2(1-q^{6j})^2.$$

*Proof.* These all follow on substituting the results of Theorem 3.39 into Theorem 3.42. □

## 3.8 Representations of integers by quadratic forms and by quadratic polynomials

Many of the formulas derived in this chapter imply results about the number of representations of an integer by a quadratic form or by a quadratic polynomial. A sample of such results is given below. Some of the formulas in subsequent chapters will have similar representations, although we will not normally stop to point it out.

For a nonnegative integer $n$ and positive integer $k$, let $r_k(n)$ denote the number of solutions of

$$x_1^2 + x_2^2 + \cdots + x_k^2 = n$$

in integers, and let $t_k(n)$ denote the number of solutions of

$$\frac{x_1(x_1+1)}{2} + \frac{x_2(x_2+1)}{2} + \cdots + \frac{x_k(x_k+1)}{2} = n$$

in nonnegative integers. We say that $r_k(n)$ is the number of representations of $n$ as a sum of $k$ squares, and $t_k(n)$ is the number of representations of $n$ as a sum of $k$ triangular numbers. Moreover, $2^k\, t_k(n)$ counts the number of lattice points on the $k$-dimensional sphere

$$\left(x_1 + \frac{1}{2}\right)^2 + \cdots + \left(x_k + \frac{1}{2}\right)^2 = 2n + \frac{k}{4}.$$

The generating functions are given by

$$\sum_{n=0}^{\infty} r_k(n) q^n = \left(\sum_{j=-\infty}^{\infty} q^{j^2}\right)^k = \varphi^k(q)$$

and

$$\sum_{n=0}^{\infty} t_k(n) q^n = \left(\sum_{j=0}^{\infty} q^{j(j+1)/2}\right)^k = \psi^k(q).$$

For example, $r_2(5) = 8$, $r_2(9) = 4$, and $r_2(3) = 0$, because

$$5 = (\pm 1)^2 + (\pm 2)^2 = (\pm 2)^2 + (\pm 1)^2,$$
$$9 = (\pm 3)^2 + 0^2 = 0^2 + (\pm 3)^2$$

while $x_1^2 + x_2^2 = 3$ has no solutions in integers. Similarly, $t_2(6) = 3$ and $t_2(5) = 0$, because

$$6 = \frac{3\times 4}{2} + \frac{0\times 1}{2} = \frac{2\times 3}{2} + \frac{2\times 3}{2} = \frac{0\times 1}{2} + \frac{3\times 4}{2},$$

while

$$\frac{x_1(x_1+1)}{2} + \frac{x_2(x_2+1)}{2} = 5$$

has no solutions in integers.

For $k = 2$, 4, 6 and 8, formulas for $r_k(n)$ and $t_k(n)$ in terms of divisor sums can be obtained by equating coefficients of powers of $q$ in the identities in Theorems 3.10, 3.26, 3.34, and 3.39. Here are the results.

**Theorem 3.44.** *Let* $\chi_4$ *be defined by* (3.13). *The following identities hold*:

$$r_2(n) = 4\sum_{d|n} \chi_4(d), \qquad t_2(n) = \sum_{d|4n+1} \chi_4(d), \tag{3.111}$$

$$r_4(n) = 8\sum_{d|n,\ 4\nmid d} d, \qquad t_4(n) = \sum_{d|2n+1} d, \tag{3.112}$$

$$r_6(n) = 4\sum_{de=n} \chi_4(d)\,(4e^2 - d^2), \qquad t_6(n) = -\frac{1}{8}\sum_{d|4n+3} \chi_4(d)\,d^2, \tag{3.113}$$

$$r_8(n) = 16\sum_{d|n} (-1)^{n+d} d^3, \qquad t_8(n) = \sum_{\substack{de=n+1\\ e \text{ odd}}} d^3. \tag{3.114}$$

*Proof.* By the identity (3.14) in Theorem 3.10, we have

$$\varphi^2(q) = 1 + 4\sum_{j=1}^{\infty} \chi_4(j)\frac{q^j}{1-q^j} = 1 + 4\sum_{j=1}^{\infty}\sum_{k=1}^{\infty} \chi_4(j) q^{jk}.$$

Equating coefficients of $q^n$ on both sides gives

$$r_2(n) = 4\sum_{jk=n} \chi_4(j) = 4\sum_{d|n} \chi_4(d)$$

and this proves the first identity in (3.111).

Next, by the identity (3.15) in Theorem 3.10, we have

$$q\psi^2(q^4) = \sum_{j=1}^{\infty} \chi_4(j)\frac{q^j}{1-q^{2j}}.$$

This may be rewritten in the form

$$q\psi^2(q^4) = \sum_{j=0}^{\infty} (-1)^j \frac{q^{2j+1}}{1-q^{4j+2}} = \sum_{j=0}^{\infty}\sum_{k=0}^{\infty} (-1)^j q^{(2j+1)(2k+1)}.$$

Equating coefficients of $q^{4n+1}$ on both sides we obtain

$$t_2(n) = \sum_{(2j+1)(2k+1)=4n+1} (-1)^j.$$

As all of the divisors of $4n+1$ are odd, we may put $d = 2j+1$ and $e = 2k+1$, and write

$$t_2(n) = \sum_{de=4n+1} (-1)^{(d-1)/2} = \sum_{d|4n+1} \chi_4(d).$$

This proves the second identity in (3.111).

Now we will prove the identities in (3.112). By the identity (3.67) of Theorem 3.26, we have

$$\begin{aligned}
\varphi^4(q) &= \frac{1}{3}\left(4P(q^4) - P(q)\right) \\
&= 1 + 8\sum_{j=1}^{\infty}\left(\frac{jq^j}{1-q^j} - \frac{4jq^{4j}}{1-q^{4j}}\right) \\
&= 1 + 8\sum_{\substack{j=1 \\ 4\nmid j}}^{\infty} \frac{jq^j}{1-q^j} \\
&= 1 + 8\sum_{\substack{j=1 \\ 4\nmid j}}^{\infty}\sum_{k=1}^{\infty} jq^{jk}.
\end{aligned}$$

Equating coefficients of $q^n$ we deduce

$$r_4(n) = 8\sum_{\substack{jk=n \\ 4\nmid j}} j = 8\sum_{d|n,\ 4\nmid d} d.$$

To prove the other identity in (3.112), start with the identity (3.69) in Theorem 3.26 and manipulate to obtain

$$\begin{aligned}
q\psi^4(q^2) &= \frac{1}{24}\left(-P(q) + 3P(q^2) - 2P(q^4)\right) \\
&= \frac{1}{24}\left(-(P(q) - P(q^2)) + 2(P(q^2) - P(q^4))\right) \\
&= \sum_{j=1}^{\infty} j\left(\frac{q^j}{1-q^j} - \frac{q^{2j}}{1-q^{2j}}\right) - 2\sum_{j=1}^{\infty} j\left(\frac{q^{2j}}{1-q^{2j}} - \frac{q^{4j}}{1-q^{4j}}\right)
\end{aligned}$$

$$= \sum_{j=1}^{\infty} \frac{jq^j}{1-q^{2j}} - \sum_{j=1}^{\infty} \frac{2jq^{2j}}{1-q^{4j}}$$

$$= \sum_{j \text{ odd}} \frac{jq^j}{1-q^{2j}}$$

$$= \sum_{j=0}^{\infty} \sum_{k=0}^{\infty} (2j+1) q^{(2j+1)(2k+1)}.$$

Equating coefficients of $q^{2n+1}$ we obtain

$$t_4(n) = \sum_{(2j+1)(2k+1)=(2n+1)} (2j+1) = \sum_{d|2n+1} d.$$

The identities in (3.113) and (3.114) can be obtained from Theorems 3.34 and 3.39 by similar procedures. We omit the details as they are similar to the above. □

**Corollary 3.45 (Lagrange's four-square theorem).** *Every positive integer is a sum of four squares. That is,* $r_4(n) > 0$.

*Proof.* By (3.112) we have

$$r_4(n) = 8 \sum_{d|n,\, 4\nmid d} d \geq 8 > 0.$$

□

The corresponding formulas for $r_{2k}(n)$ and $t_{2k}(n)$ for $k \geq 5$ are more complicated in a way that we shall now describe. Ramanujan found that for any positive integer $m$,

$$\varphi^{4m}(q) = 1 + \frac{4m(-1)^{m+1}}{(2^{2m}-1)B_{2m}} \sum_{j=1}^{\infty} \frac{j^{2m-1}q^j}{1-(-q)^{m+j}} + \sum_{1 \leq j \leq (m-1)/2} c_{m,j}\, q^j E(q^2)^{24j-4m} E(-q)^{8m-24j}, \tag{3.115}$$

and

$$\varphi^{4m+2}(q) = 1 + \frac{4}{E_{2m}} \sum_{j=1}^{\infty} \chi_4(m) \frac{j^{2m}q^j}{1-q^j} - \frac{(-4)^{m+1}}{E_{2m}} \sum_{j=1}^{\infty} \frac{j^{2m}q^j}{1+q^{2j}} + \sum_{1 \leq j \leq m/2} d_{m,j} q^j E(-q)^{8m+4-24j} E(q^2)^{24j-4m-2}, \tag{3.116}$$

where $\chi_4$ is the character defined by (3.13), $c_{m,j}$ and $d_{m,j}$ are rational constants, $B_{2m}$ are the Bernoulli numbers defined by (0.64), and $E_{2m}$ are the Euler numbers defined by

$$\sec x = \sum_{m=0}^{\infty} \frac{(-1)^m E_{2m}}{(2m)!} x^{2m},$$

so that $E_0 = 1$, $E_2 = -1$, $E_4 = 5$, $E_6 = -61$ and $E_8 = 1385$.

The analogous results for sums of triangular numbers were also found by Ramanujan, and are given by the formulas

$$\frac{2^{2m}(2^{2m}-1)|B_{2m}|}{2m}\psi^{4m}(q^2) = \frac{1}{4^{m-1}}\sum_{j=1}^{\infty}\left(\frac{1+(-1)^{m+j}}{2}\right)\frac{j^{2m-1}q^{j-m}}{1-q^{2j}} + \sum_{1\le j\le (m-1)/2}\frac{(-1)^j c_{m,j}}{(256q^2)^j}E(q^2)^{24j-4m}E(q^4)^{8m-24j} \qquad (3.117)$$

and

$$|E_{2m}|\psi^{4m+2}(q^2) = \frac{1}{2^{4m}}\sum_{j=1}^{\infty}(2j-1)^{2m}q^{j-m-1}\left\{\frac{1}{1+q^{2j-1}} - \frac{(-1)^{m+j}}{1-q^{2j-1}}\right\} + \sum_{1\le j\le m/2}\frac{(-1)^j d_{m,j}}{(256q^2)^j}E(q^2)^{24j-4m-2}E(q^4)^{8m+4-24j}, \qquad (3.118)$$

where $m$ is any positive integer, and the coefficients $c_{m,j}$ and $d_{m,j}$ are the same numbers that occur in (3.115) and (3.116), respectively.

When $m = 1$ or $m = 2$, the condition $1 \le j \le (m-1)/2$ is empty, and it can be shown that the identity (3.115) reduces to the formulas for $\varphi^4(q)$ and $\varphi^8(q)$ in Theorems 3.26 and 3.39, respectively; see Exercise 14. Similarly, the identity (3.117) can be shown to reduce to the formulas for $q\psi^4(q^2)$ and $q\psi^8(q)$ in Theorems 3.26 and 3.39, respectively

In a similar way, the cases $m = 1$ of (3.116) and (3.118) reduce to the formulas for $\varphi^6(q)$ and $q^3\psi^6(q^4)$, respectively, in Theorem 3.34.

A proof of the identity (3.115) in the case $m = 3$ will be given in Exercise 2 at the end of Chapter 4.

Let us discuss the cases $m = 3$ and $m = 6$ of (3.115) in more detail. The values of the constants $c_{m,j}$ can be determined by expanding in powers of $q$ and equating coefficients, and this leads to the formulas

$$\varphi^{12}(q) = 1 + 8\sum_{j=1}^{\infty}\frac{j^5 q^j}{1+(-q)^j} + 16qE(q^2)^{12}$$

and

$$\varphi^{24}(q) = 1 + \frac{16}{691}\sum_{j=1}^{\infty}\frac{j^{11}q^j}{1-(-q)^j} + \frac{33152}{691}qE(-q)^{24} - \frac{65536}{691}q^2E(q^2)^{24}.$$

Equating coefficients of $q^n$ gives

$$r_{12}(n) = 8\sum_{d|n} d^5 - 512\sum_{d|\frac{n}{4}} d^5 + 16e_{12}(n) \tag{3.119}$$

and

$$r_{24}(n) = \frac{16}{691}\sum_{d|n}(-1)^{n+d}d^{11} + \frac{128}{691}\left((-1)^n 259\tau(n) - 512\tau(n/2)\right) \tag{3.120}$$

where $e_{12}(n)$ and $\tau(n)$ are the coefficients of $q^n$ in $qE(q^2)^{12}$ and $qE(q)^{24}$, respectively, and $\tau(n)$ is defined to be zero if $n$ is not a positive integer. The functions $e_{12}(n)$ and $\tau(n)$ have some remarkable properties. They are multiplicative, that is

$$e_{12}(mn) = e_{12}(m)e_{12}(n) \qquad \text{and} \qquad \tau(mn) = \tau(m)\tau(n) \tag{3.121}$$

if $m$ and $n$ are relatively prime. For prime powers, there are the three-term recurrence relations

$$e_{12}(p^{\lambda+1}) = e_{12}(p)e_{12}(p^{\lambda}) - p^5 e_{12}(p^{\lambda-1})$$

and

$$\tau(p^{\lambda+1}) = \tau(p)\tau(p^{\lambda}) - p^{11}\tau(p^{\lambda-1}). \tag{3.122}$$

Furthermore, Ramanujan proved that

$$|e_{12}(n)| = O(n^{3+\epsilon}) \qquad \text{and} \qquad |\tau(n)| = O(n^7) \quad \text{as } n \to \infty$$

for any positive number $\epsilon$ and conjectured that

$$|e_{12}(n)| \leq n^{5/2}d(n) \qquad \text{and} \qquad |\tau(n)| \leq n^{11/2}d(n) \tag{3.123}$$

where $d(n)$ is the number of divisors of $n$. The conjectures were proved by Deligne. Ramanujan's bounds on $|e_{12}(n)|$ and $|\tau(n)|$ in (3.123) imply that the dominant terms in the expressions for $r_{12}(n)$ and $r_{24}(n)$ in (3.119) and (3.120) are given by the divisor sums, while the terms involving $e_{12}(n)$ and $\tau(n)$ may be viewed as correction terms and have magnitudes approximately equal to the square roots of the dominant terms. A more detailed discussion of properties such as (3.121)–(3.123) is beyond the scope of this book and requires the theory of modular forms. The books by Hardy [174, Ch. 10] and Serre [267, Ch. 7] provide good introductions to these topics.

The next result is a cubic analogue of Theorem 3.44 in which $x^2+y^2$ is replaced with $x^2+xy+y^2$, that is, $\varphi^2(q)$ is replaced with the cubic theta function $a(q)$. Let $u_{2k}(n)$ denote the number of solutions of

$$(x_1^2+x_1y_1+y_1^2)+\cdots+(x_k^2+x_ky_k+y_k^2)=n$$

in integers. The generating function is given by

$$\sum_{n=0}^{\infty} u_{2k}(n)q^n=\left(\sum_{i=-\infty}^{\infty}\sum_{j=-\infty}^{\infty} q^{i^2+ij+j^2}\right)^k=a^k(q).$$

**Theorem 3.46.** *The following identities hold:*

$$u_2(n)=6\sum_{d\mid n}\left(\frac{d}{3}\right), \tag{3.124}$$

$$u_4(n)=12\sum_{d\mid n,\ 3\nmid d} d, \tag{3.125}$$

$$u_6(n)=9\sum_{de=n}\left(\frac{d}{3}\right)(3e^2-d^2), \tag{3.126}$$

$$u_8(n)=24\left(\sum_{d\mid n} d^3+9\sum_{d\mid \frac{n}{3}} d^3\right). \tag{3.127}$$

*Proof.* These follow from the formulas for $a(q)$, $a^2(q)$, $a^3(q)$, and $a^4(q)$ in Theorems 3.19, 3.31, 3.35, and 3.43, respectively. □

The analogous formulas for $u_{10}(n)$ and $u_{12}(n)$ are given by

$$u_{10}(n)=3\sum_{de=n}\left(\frac{d}{3}\right)(d^4+9e^4) \tag{3.128}$$

and

$$u_{12}(n)=\frac{252}{13}\left(\sum_{d\mid n} d^5-27\sum_{d\mid \frac{n}{3}} d^5\right)+\frac{216}{13}c(n) \tag{3.129}$$

where $c(n)$ is the coefficient of $q^n$ in $qE(q)^6E(q^3)^6$. A proof of (3.128) will be outlined in Exercise 17 at the end of the chapter, and (3.129) will be proved in Exercise 2c at the end of the next chapter. For more general results that give the value of $u_{2k}(n)$ for any positive integer $k$, see Theorem 11.2, the identity (11.20) and the exercises at the end of Chapter 11.

The results in the next theorem give formulas for the number of representations of an integer as a square plus three times a square, a triangular number plus three times a triangular number, a square plus seven times a square, and a triangular number plus seven times a triangular number. The notation on the left-hand side is from Fine's book [151] and is self-explanatory.

**Theorem 3.47.** *The following identities hold:*

$$\sum 1\ \left(n = x^2 + 3y^2\right) = \begin{cases} 2\sum_{d|n} \left(\frac{d}{3}\right) & \text{if } n \text{ is odd,} \\ 6\sum_{d|n} \left(\frac{d}{3}\right) & \text{if } n \text{ is even;} \end{cases}$$

$$\sum 1 \left(n = \frac{x(x+1)}{2} + \frac{3y(y+1)}{2};\ x,y \geq 0\right) = \sum_{d|(2n+1)} \left(\frac{d}{3}\right);$$

$$\sum 1\ \left(n = x^2 + 7y^2\right) = 2\sum_{d|n} (-1)^{n+d} \left(\frac{d}{7}\right)$$

*and*

$$\sum 1 \left(n = \frac{x(x+1)}{2} + \frac{7y(y+1)}{2};\ x,y \geq 0\right) = \sum_{\substack{d|(n+1) \\ d \text{ odd}}} \left(\frac{d}{7}\right).$$

*Proof.* By Theorem 3.23 and Lemma 3.12,

$$\varphi(q)\varphi(q^3) = \frac{1}{3}\left(a(q) + 2a(q^4)\right) = \frac{1}{3}\left(2a(q) + a(-q)\right).$$

Expanding both sides in powers of $q$ we obtain

$$\sum_{j=-\infty}^{\infty} \sum_{k=-\infty}^{\infty} q^{j^2+3k^2} = \frac{1}{3}\sum_{n=0}^{\infty} u_2(n)\left(2 + (-1)^n\right) q^n.$$

Now equate coefficients of $q^n$ and use the formula for $u_2(n)$ in (3.124) to deduce the first result.

Again by Theorem 3.23,

$$q\psi(q^2)\psi(q^6) = \frac{1}{6}\left(a(q) - a(q^4)\right).$$

Expanding both sides in powers of $q$ we deduce

$$\sum_{j=0}^{\infty}\sum_{k=0}^{\infty} q^{j(j+1)+3k(k+1)+1} = \frac{1}{6}\sum_{n=0}^{\infty} u_2(n)q^n - \frac{1}{6}\sum_{n=0}^{\infty} u_2(n)q^{4n}.$$

Now equate coefficients of $q^{2n+1}$. Since the second sum on the right contains only even powers of $q$, it follows that

$$\sum 1\ (2n+1 = j(j+1) + 3k(k+1) + 1;\ x, y \geq 0) = \frac{1}{6}\, u_2(2n+1),$$

that is,

$$\sum 1\left(n = \frac{x(x+1)}{2} + \frac{3y(y+1)}{2};\ x, y \geq 0\right) = \frac{1}{6}\, u_2(2n+1).$$

Now use the formula for $u_2(n)$ in (3.124) to deduce the second result.

The last two results can be obtained from Example 1.25 by equating coefficients of $q^n$ in

$$\sum_{j=-\infty}^{\infty}\sum_{k=-\infty}^{\infty} (-q)^{j^2+7k^2} = 1 + 2\sum_{j=1}^{\infty}\left(\frac{j}{7}\right)\frac{(-q)^j}{1-q^j}$$

and equating coefficients of $q^{n+1}$ in

$$\sum_{j=0}^{\infty}\sum_{k=0}^{\infty} q^{j(j+1)/2+7k(k+1)/2+1} = \sum_{\substack{j>0 \\ j \text{ odd}}}\left(\frac{j}{7}\right)\frac{q^j}{1-q^j},$$

respectively.

□

The next result provides analogues of the sum of four squares and sum of four triangular numbers results in (3.112).

**Theorem 3.48.** *Let n be a positive integer and write*

$$n = 2^\alpha \times 3^\beta \times m \quad \textit{where} \quad \gcd(m, 6) = 1.$$

*Then*

$$\sum 1\ \left(n = x^2 + y^2 + 3z^2 + 3t^2\right) = \begin{cases} 4\sum\limits_{d|m} d & \text{if } \alpha = 0, \\ 4(2^{\alpha+1} - 3)\sum\limits_{d|m} d & \text{if } \alpha \geq 1; \end{cases} \tag{3.130}$$

$$\sum 1\ \left(n = 1 + \frac{x(x+1)}{2} + \frac{y(y+1)}{2} + \frac{3z(z+1)}{2} + \frac{3t(t+1)}{2};\ x, y, z, t \geq 0\right)$$
$$= 2^{\alpha} \sum_{d|m} d; \tag{3.131}$$

*and*

$$\sum 1\ \left(n = x^2 + xy + y^2 + 2z^2 + 2zt + 2t^2\right) = 6(3^{\beta+1} - 2)\sum_{d|m} d. \tag{3.132}$$

*Proof.* Let $u(n)$ be defined by

$$\frac{1}{2}\left(3P(q^3) - P(q)\right) = \sum_{n=0}^{\infty} u(n)q^n.$$

Since

$$\sum_{n=0}^{\infty} u(n)q^n = 1 + 12\sum_{n=1}^{\infty}\left(\frac{jq^j}{1-q^j} - \frac{3jq^{3j}}{1-q^{3j}}\right) = 1 + 12\sum_{3\nmid j}\frac{jq^j}{1-q^j}$$

it follows that for a positive integer $n$,

$$u(n) = 12\sum_{d|n,\ 3\nmid d} d.$$

Next, let $f(n)$ be defined by

$$\varphi^2(q)\varphi^2(q^3) = \sum_{n=0}^{\infty} f(n)q^n.$$

We seek a formula for $f(n)$. By Theorem 3.27,

$$\varphi^2(-q)\varphi^2(-q^3) = -\frac{1}{6}\left(3P(q^3) - P(q)\right) + \frac{2}{3}\left(3P(q^6) - P(q^2)\right),$$

so expanding in powers of $q$ gives

$$\sum_{n=0}^{\infty}(-1)^n f(n)q^n = -\frac{1}{3}\sum_{n=0}^{\infty}u(n)q^n + \frac{4}{3}\sum_{n=0}^{\infty}u(n)q^{2n}. \tag{3.133}$$

If $n$ is odd, then equating coefficients gives

$$f(n) = \frac{1}{3}u(n) = 4\sum_{d|n,\,3\nmid d} d = 4\sum_{d|m} d. \tag{3.134}$$

If $n$ is even and nonzero, write $n = 2^\alpha k$ where $\alpha$ is a positive integer and $k$ is odd, and equate coefficients in (3.133) to obtain

$$\begin{aligned}
f(2^\alpha k) &= -\frac{1}{3}u(2^\alpha k) + \frac{4}{3}u(2^{\alpha-1}k) \\
&= -4\sum_{d|2^\alpha k,\,3\nmid d} d + 16\sum_{d|2^{\alpha-1}k,\,3\nmid d} d \\
&= -4(1+2+\cdots+2^\alpha)\sum_{d|m} d + 16(1+2+\cdots+2^{\alpha-1})\sum_{d|m} d \\
&= 4(2^{\alpha+1}-3)\sum_{d|m} d. \qquad (3.135)
\end{aligned}$$

On combining (3.134) and (3.135), we obtain (3.130).

To prove (3.131), use Theorem 3.27 and series manipulations to obtain

$$\begin{aligned}
q\psi^2(q)\psi^2(q^3) &= \sum_{j=1}^{\infty}\left(\frac{jq^j}{1-q^j} - \frac{jq^{2j}}{1-q^{2j}}\right) - 3\sum_{j=1}^{\infty}\left(\frac{jq^{3j}}{1-q^{3j}} - \frac{jq^{6j}}{1-q^{6j}}\right) \\
&= \sum_{j=1}^{\infty}\left(\frac{jq^j}{1-q^{2j}} - \frac{3jq^{3j}}{1-q^{6j}}\right) \\
&= \sum_{3\nmid j}\frac{jq^j}{1-q^{2j}} = \sum_{\substack{j=1 \\ 3\nmid j}}^{\infty}\sum_{k=1}^{\infty} jq^{j(2k-1)}.
\end{aligned}$$

The coefficient of $q^n$ on the right-hand side is

$$\sum_{\substack{de=n \\ 3\nmid d,\,2\nmid e}} d = 2^\alpha\sum_{d|m} d$$

and this implies (3.131).

The identity (3.132) can be proved using a similar procedure as for the proof of (3.130). Let $v(n)$ be defined by

$$2P(q^2) - P(q) = \sum_{n=0}^{\infty} v(n)q^n$$

and observe that

$$v(n) = 24 \sum_{d|n,\, d \text{ odd}} d.$$

Next, let $g(n)$ be defined by

$$a(q)a(q^2) = \sum_{n=0}^{\infty} g(n)q^n.$$

We seek a formula for $g(n)$. By Theorem 3.30, we have

$$a(q)q(q^2) = \frac{1}{4}\left(2P(q^2) - P(q)\right) + \frac{3}{4}\left(2P(q^6) - P(q^3)\right).$$

Expanding in powers of $q$, we find that

$$\sum_{n=0}^{\infty} g(n)q^n = \frac{1}{4}\sum_{n=0}^{\infty} v(n)q^n + \frac{3}{4}\sum_{n=0}^{\infty} v(n)q^{3n}.$$

Now equate coefficients of $q^n$. If $3 \nmid n$ then

$$g(n) = \frac{1}{4}v(n) = 6\sum_{\substack{d|n \\ d \text{ odd}}} d = 6\sum_{d|m} d.$$

If $3|n$, write $n = 3^\beta k$ where $\beta$ is a positive integer and $\gcd(3, k) = 1$. Then

$$\begin{aligned} g(3^\beta k) &= \frac{1}{4}v(3^\beta k) + \frac{3}{4}v(3^{\beta-1}k) \\ &= 6\sum_{\substack{d|3^\beta k \\ d \text{ odd}}} d + 18 \sum_{\substack{d|3^{\beta-1}k \\ d \text{ odd}}} d \end{aligned}$$

$$= 6(1 + 3 + \cdots + 3^{\beta}) \sum_{\substack{d|k \\ d \text{ odd}}} d + 18(1 + 3 + \cdots + 3^{\beta-1}) \sum_{\substack{d|k \\ d \text{ odd}}} d$$

$$= 6(3^{\beta+1} - 2) \sum_{d|m} d.$$

This completes the proof of (3.130).

□

## 3.9 Notes

The notation $\varphi(q)$ and $\psi(q)$ in Definition 3.1 is due to Ramanujan. It is introduced in his second notebook [253, Ch. 16] and used extensively by him.

Jacobi's description of the identity in Corollary 3.8 as 'aequatio identica satis abstrusa' (nonobvious identity) is from [189, p. 147]. It is a central result in the classical theory of Jacobian elliptic functions.

Many of the identities in this chapter that involve $\varphi(q)$ and $\psi(q)$ were given by Ramanujan in his second notebook [253]. For example, the results in Lemma 3.3–Theorem 3.7 occur in Chapter 16 as Entries 22 and 25; see Berndt's analysis [46, pp. 36–41]. The identity in Lemma 3.9 occurs in [50, (9.26)]; see also [48, p. 151].

The cubic theta functions $a(q)$, $b(q)$, and $c(q)$ and their main properties, including the basic cubic identity

$$a^3(q) = b^3(q) + c^3(q) \tag{3.136}$$

along with a cubic analogue of the arithmetic-geometric mean, were announced by J. M. Borwein and P. B. Borwein [63]. Proofs in [63] are sparse, so details of the proofs were given in work by the Borweins with F. Garvan [67].

The identities in Lemma 3.11 were used by the Borweins [63] to define $b(q)$ and $c(q)$ in terms of the function $a(q)$. Formulas for $b(q)$ and $c(q)$ in terms of the theta series (3.5) and (3.6) occur later in [63]. The identity (3.21) should be regarded as being known to Ramanujan; see [47, p. 171]. All three of (3.21)–(3.23) are given by the Borweins and Garvan in [67].

The infinite products for $b(q)$ and $c(q)$ in Theorem 3.15 were given by the Borweins and Garvan [67]. The proofs of Theorems 3.15 and 3.20 follow Garvan [158].

The Lambert series for $a(q)$ in (3.34) is often attributed to Lorenz, [226]. It is equivalent to a special case of a more general theorem of Dirichlet; see Landau [203, Theorem 204]. The Lambert series for $a(q)$ in (3.34) was known to Ramanujan, e.g., see [52, p. 196] or the lost notebook [25, p.402] and [254, pp. 355, 356]. The Lambert series for $b(q)$ and $c(q)$ in (3.35) and (3.36) were also known to Ramanujan; see [25, p. 406] and [254, p. 357]. The derivation of (3.34) and the proof of Theorem 3.18 have been reproduced from Hirschhorn [178].

Although Ramanujan did not give the basic cubic identity in Theorem 3.20 in precisely that form, the formula in his second notebook [253, Ch. 21, Entry 3] is equivalent to

$$\left(1+6\sum_{j=1}^{\infty}\left(\frac{j}{3}\right)\frac{q^j}{1-q^j}\right)^3=\prod_{j=1}^{\infty}\frac{(1-q^j)^9}{(1-q^{3j})^3}+27q\prod_{j=1}^{\infty}\frac{(1-q^{3j})^9}{(1-q^j)^3}.$$

The left-hand side is $a^3(q)$ by Theorem 3.19, and the infinite products on the right-hand side are $b^3(q)$ and $c^3(q)$, respectively, by Theorem 3.15.

Generalizations of the cubic theta functions have been introduced by M. Hirschhorn, F. Garvan, and J. Borwein [185]. Suppose $\omega = \exp(2\pi i/3)$ and let

$$a'(z;q)=\sum_{m=-\infty}^{\infty}\sum_{n=-\infty}^{\infty}q^{m^2+mn+n^2}z^n,$$

$$a(z;q)=\sum_{m=-\infty}^{\infty}\sum_{n=-\infty}^{\infty}q^{m^2+mn+n^2}z^{m-n},$$

$$b(z;q)=\sum_{m=-\infty}^{\infty}\sum_{n=-\infty}^{\infty}q^{m^2+mn+n^2}\omega^{m-n}z^n$$

and

$$c(z;q)=\sum_{m=-\infty}^{\infty}\sum_{n=-\infty}^{\infty}q^{m^2+mn+n^2+m+n+\frac{1}{3}}z^{m-n}.$$

Then [185, (1.25)]

$$a(z;q)^3=b(q)^2b(z^3;q)+c(z;q)^3$$

is a generalization that reduces to the cubic identity (3.136) when $z = 1$. Many of the other identities involving cubic theta functions have similar generalizations. For example, the identity [185, (1.26)]

$$a(z;q)a(z^2;q^2)=b(q^2)b(z^3;q)+c(z;q)c(z^2;q^2) \tag{3.137}$$

reduces to Theorem 3.29 when $z = 1$. Further results, including the companion identities

$$a'(z;q)^3=b(z;q)^3+c(q)^2c(z;q)$$

and

$$a'(z;q)a'(z;q^2) = b(z;q)b(z;q^2) + c(q)c(z;q^2) \tag{3.138}$$

have been given in [103]. Some of these results have been extended to include another variable by S. Bhargava [55] and R. Chapman [96].

The very interesting identity

$$a(z_1,z_2;q)^3 + d(z_1,z_2;q)^3 = b(z_1,z_2;q)^3 + c(z_1,z_2;q)^3$$

where

$$a(z_1,z_2;q) = \sum_{m=-\infty}^{\infty}\sum_{n=-\infty}^{\infty} q^{m^2+mn+n^2} z_1^m z_2^n,$$

$$b(z_1,z_2;q) = \sum_{m=-\infty}^{\infty}\sum_{n=-\infty}^{\infty} \omega^{m-n} q^{m^2+mn+n^2} z_1^m z_2^n,$$

$$c(z_1,z_2;q) = \sum_{m=-\infty}^{\infty}\sum_{n=-\infty}^{\infty} q^{(m+\frac{1}{3})^2+(m+\frac{1}{3})(n+\frac{1}{3})+(n+\frac{1}{3})^2} z_1^{(m+\frac{1}{3})} z_2^{(n+\frac{1}{3})},$$

$$d(z_1,z_2;q) = \sum_{m=-\infty}^{\infty}\sum_{n=-\infty}^{\infty} \omega^{m-n} q^{(m+\frac{1}{3})^2+(m+\frac{1}{3})(n+\frac{1}{3})+(n+\frac{1}{3})^2} z_1^{(m+\frac{1}{3})} z_2^{(n+\frac{1}{3})}$$

was discovered recently by D. Schultz [263, 264, p. 12]. It reduces to the basic cubic identity (3.136) when $z_1 = z_2 = 1$ because $d(1,1;q) = 0$.

The identities in Theorem 3.21 should be regarded as being known to Ramanujan. For example, see Entries 18.2.16–18.2.19 in [48, pp. 405–406]. These results also occur in the analysis by Berndt, Bhargava and Garvan, e.g., see (9.13), (9.12), (6.5), and (6.3) in [50] or in [48] for our (3.39)–(3.42), respectively. The Lambert series for $\varphi(q)\varphi(q^3)$ in Theorem 3.23 is due to Lorenz [226]. Another Lambert series for $\varphi(q)\varphi(q^3)$ was given even earlier in 1844 by Cauchy [72]; see Exercise 9. See also Dickson [135, pp. 20, 29]. The Lambert series for both $\varphi(q)\varphi(q^3)$ and $q\psi(q^2)\psi(q^6)$ in Theorem 3.23 are part of Entry 3 in Chapter 19 of Ramanujan's second notebook [253]. See Berndt's analyses [46, pp. 223–226] and [49, pp. 74–77] and also the works of Fine [151, (31.16), (31.17), (31.22)] and Hirschhorn [180].

The identities in Theorems 3.22, 3.24, and 3.25 should also be regarded as classical and known to Ramanujan. For example, (3.44) and (3.45) are Entries 36 and 37 in [48, p. 376]. The identity (3.54) occurs in Ramanujan's second notebook [253] at the bottom of p. 328; see also [46, p. 462] and [48, p. 93]. The identities (3.59), (3.60), and (3.65) are on p. 351 lines 3 and 4, p. 232 line $-10$, and p 350 line 14, respectively, in [46]. And (3.64) is [48, p. 110], while (3.66) can be obtained from Entry 44 in [47, p. 197].

The identities in Theorem 3.27 were known to Liouville [209]. The results were recorded as part of Entry 3 of Chapter 19 in Ramanujan's second notebook [253]; see Berndt's analysis [46, pp. 223–226].

Theorem 3.29 was proved by the Borweins and Garvan [67, Theorem 2.6]. They noted it is equivalent to the quadratic modular equation given by Ramanujan at the top of p. 259 of Ramanujan's second notebook [253]. See also Theorem 7.1 (i) in [48] or [50]. Another proof is by R. P. Lewis and Z.-G. Liu [207]. Generalizations are given by (3.137) and (3.138).

A result that is equivalent to the Lambert series for $a(q)a(q^2)$ in (3.73) was stated without proof by Liouville. For a proof, and reference to Liouville, see the work of A. Alaca, Ş. Alaca, and K. S. Williams [12, Theorem 13]. The Lambert series for $b(q)b(q^2)$ and $c(q)c(q^2)$ in (3.74) and (3.75) were given by N. Fine [151, equations 33.1 and 33.124 on p. 86]. See also Z.-G. Liu [220].

The Lambert series for $a^2(q)$, $a^3(q)$, and $a^4(q)$ in Theorems 3.31, 3.35, and 3.43 were given by Ramanujan in the lost notebook; see [25, pp. 402, 404] and [254, p. 356]. Further results and additional references are given in the book by K. S. Williams [302, Chapter 17].

The identities (3.82)–(3.84) were obtained by L. Carlitz [70] who used them to deduce the sum of six squares and sum of six triangular numbers identities in (3.88) and (3.89). The Lambert series for $c^3(q)$ was given by Ramanujan in his second notebook [253, p. 257] and also in the lost notebook [254, p.152]. Proofs of the Lambert series for $b^3(q)$ and $c^3(q)$ in Theorem 3.35 have been given by Carlitz [69]. The first identity in Theorem 3.36 was proved by Fine [151, p. 85, (32.71)].

The formulas for $\varphi^m(q)$ and $\psi^m(q)$ for $m \in \{2,4,6,8\}$ in Theorems 3.10, 3.26, 3.34, and 3.39 are classical. They were given by Jacobi [189, pp. 161, 164, 167, 170]. Jacobi's $\sqrt{2K/\pi}$ and $\sqrt{2kK/\pi}$ correspond to $\varphi(q)$ and $2q^{1/4}\psi(q^2)$, respectively. Jacobi [189, p. 239] noted that these formulas imply results about the number of representations of an integer as a sum of 2, 4, 6, or 8 squares or triangular numbers. Jacobi's results were made explicit by H. J. S. Smith [272, Section 127] and these are essentially the results we have given in Theorem 3.44. These days, proofs of Theorem 3.44 abound and it is impossible to cite all such references. Our favorite reference is to a proof that relies on Ramanujan's summation formula, in work with H. Y. Lam [123].

The sum of 10 squares result was known to Liouville [218], who expressed the result in term of divisor sums rather than as a theta function identity. A special case of the 10 squares identity had been given earlier by Eisenstein [144]. The number of representations of an even integer as a sum of 12 squares was known to Liouville [217] and the result for all integers was found by Glaisher [163]. The 14, 16, and 18 squares identities were given by Glaisher in a series of papers culminating in [162]. Ramanujan gave the formulas (3.115) and (3.116) for sums of an even number of squares in [249, Sec. 25]. He used the modular transformation to deduce the corresponding results for sums of an even number of triangular numbers given by (3.117) and (3.118); see [250, Sec. 12]. A proof of Ramanujan's results has been given by Mordell [233]. Another proof is in [102]. The analogous expansions for

$$\left(\sum_{m=-\infty}^{\infty}\sum_{n=-\infty}^{\infty} q^{m^2+mn+(p+1)n^2/4}\right)^k, \quad \text{for } p = 3, 7, 11 \text{ and } 23,$$

have been given in work with H. H. Chan [80]. That result is crucial to the development in Chapter 11; see Theorem 11.2. Further results, containing the expansions of

$$(\varphi(q)\varphi(q^p))^k, \quad \text{for } p = 3, 7, 11 \text{ and } 23,$$

as well as corresponding results for the function $\psi$, have been given in work with B. Kane and D. Ye [122].

There are other values of $(a, b)$ for which

$$\sum 1 \left(n = ax^2 + by^2\right)$$

and

$$\sum 1 \left(n = \frac{ax(x+1)}{2} + \frac{by(y+1)}{2};\ x, y \geq 0\right)$$

have similar formulas as divisor sums; the results in Theorem 3.47 correspond to the cases $(a, b) = (1, 3)$ and $(1, 7)$. See the Exercises at the end of this chapter for the cases $(a, b) = (1, 2)$, $(1, 4)$, $(1, 5)$, $(1, 6)$, $(2, 3)$, $(1, 15)$, and $(3, 5)$; and see Corollary 8.14 for the case $(a, b) = (1, 8)$. More general questions of this type are considered in the book by D. A. Cox [132] and the papers by J. G. Huard, P. Kaplan, and K. S. Williams [187], Z.-H. Sun [276], Sun and Williams [277, 278], and P. C. Toh [281].

The results in Theorem 3.48 were given by Liouville [209, 213, 214]. During the period 1859–1866, Liouville published more than 100 papers concerning representations of integers by more than 100 different quadratic forms. These papers were all published in Journal de Mathématiques Pures et Appliquées (Series 2, vols. 4–11), the journal he founded in 1836 and often called "Liouville's journal." Liouville never published proofs of these results. For an interesting discussion of Liouville's work on this topic, and requests from Hermite that Liouville publish proofs, see J. Lützen's book [228, pp. 227–230].

Liouville's identities have been proved and extended by many authors. We only mention P. S. Nazimoff [238] in 1884, K. Petr [242] in 1901, A. Walfisz [289]–[291] during 1938–1941 and more recently A. Alaca, Ş. Alaca, M. F. Lemire, and K. S. Williams in a series of papers including [6]– [10]. Liouville's original methods have been employed to prove many of Liouville's results in the book by Williams [302].

## 3.10 Exercises

1. Deduce the following identities from Jacobi's 'aequatio identica satis abstrusa' in Corollary 3.8:

   a. $q\psi^8(q) - q\psi^8(-q) = 16q^2\psi^8(q^2)$.
   b. $\varphi^8(-q^2) - \varphi^8(-q) = 16q\psi^8(-q)$.
   c. $E^8(q) - E^8(-q) = -16qE^8(q^4)$, where $E(q) = \prod_{j=1}^{\infty}(1-q^j)$.
   d. Conclude that the even part of $q\psi^8(q)$ is $8q^2\psi^8(q^2)$, and the odd part of $E^8(q)$ is $-8qE^8(q^4)$.

   Part c of this exercise and its interpretation in terms of the odd part of $E^8(q)$ were communicated to me by M. Hirschhorn.

2. Deduce the following identities from (3.57) and (3.58):

   a. $q\psi^2(q)\psi^2(q^3) - q\psi^2(-q)\psi^2(-q^3) = 4q^2\psi^2(q^2)\psi^2(q^6)$.
   b. $\varphi^2(-q^2)\varphi^2(-q^6) - \varphi^2(-q)\varphi^2(-q^3) = 4q\psi^2(-q)\psi^2(-q^3)$.
   c. $E^2(q)E^2(q^3) - E^2(-q)E^2(-q^3) = -4qE^2(q^4)E^2(q^{12})$.
   d. Interpret the results of Parts a and c in terms of the even and odd parts of various functions.

3. Prove the following identities:

   a. $\left(2P(q^2) - P(q)\right)^2 = \frac{1}{5}\left(Q(q) + 4Q(q^2)\right)$.
   b. $\left(\frac{3P(q^3) - P(q)}{2}\right)^2 = \frac{1}{10}\left(Q(q) + 9Q(q^3)\right)$.
   c. $\left(\frac{4P(q^4) - P(q)}{3}\right)^2 = \frac{1}{15}\left(Q(q) - 2Q(q^2) + 16Q(q^4)\right)$.

4. a. Prove that

$$b(q) = 1 + 3\sum_{j=1}^{\infty}\sum_{k=1}^{\infty}\left(\frac{j+k}{3}\right)q^{jk}.$$

   b. Prove that

$$\frac{c(q^3)}{3} = \sum_{j=0}^{\infty}\sum_{k=0}^{\infty}\left(q^{(3j+1)(3k+1)} - q^{(3j+2)(3k+2)}\right) = \sum_{j=-\infty}^{\infty}\left(\frac{j}{3}\right)\frac{q^{j^2}}{1-q^{3j}}.$$

5. Prove the following identities:

   a. $\frac{\eta_2^2\eta_9\eta_{18}}{\eta_1\eta_6} = \frac{1}{6}\left(a(q) + a(q^2) - a(q^3) - a(q^6)\right)$.

b. $\dfrac{\eta_1\eta_2\eta_9^2}{\eta_3\eta_{18}} = \dfrac{1}{6}\left(-a(q) - 2a(q^2) + 3a(q^3) + 6a(q^6)\right).$

c. $\dfrac{\eta_1^2\eta_9\eta_{18}}{\eta_2\eta_3} = \dfrac{1}{6}\left(a(q) - 2a(q^2) - a(q^3) + 2a(q^6)\right).$

d. $\dfrac{\eta_1\eta_2\eta_{18}^2}{\eta_6\eta_9} = \dfrac{1}{6}\left(a(q) - a(q^2) - 3a(q^3) + 3a(q^6)\right).$

Hints:

a. Replace $(q, a, b, c)$ with $(q^9, q, q^2, q^5)$ in (1.30).
b. Let $\omega = e^{i\pi/9}$ and take $(x, y, z) = (\omega^3, \omega^5, \omega^{11})$ in (1.26).
c. Use Lemma 3.11 and Theorem 3.15 to write the result of Part b in the form

$$\frac{\eta_1\eta_2\eta_9^2}{\eta_3\eta_{18}} = \frac{1}{3}\left(b(q) + 2b(q^2)\right) = \frac{1}{3}\left(\frac{\eta_1^3}{\eta_3} + \frac{2\eta_2^3}{\eta_6}\right).$$

Multiply by $3\eta_9\eta_{18}/\eta_1\eta_2$ and rearrange to obtain

$$\frac{\eta_1^2\eta_9\eta_{18}}{\eta_2\eta_3} = 3\frac{\eta_9^3}{\eta_3} - 2\frac{\eta_2^2\eta_9\eta_{18}}{\eta_1\eta_6}.$$

Now use the result of Part a along with Lemma 3.11 and Theorem 3.15 again to obtain the result of Part c.
d. Similar to the proof of Part c. Start with the identity from Part a and multiply by $\eta_1\eta_2/\eta_9\eta_{18}$.

6. Prove that $a^2(q) + 2b^2(q) = 3a^2(q^3) + 6c^2(q^3)$.
7. For $1 \le k \le 7$ and all positive integers $n$, prove that

$$r_k(8n+k) = 2^{k-1}\left\{2 + \binom{k}{4}\right\} t_k(n).$$

Thus for $1 \le k \le 7$, the study of $t_k(n)$ is reduced to the study of a subsequence of $r_k(n)$. See works with Barrucand and Hirschhorn [38] and with Hirschhorn [119]. Generalizations and extensions have been given in works with Adiga and Han [2] and with Baruah and Hirschhorn [41].
8. a. Prove that

$$\varphi(q)\varphi(q^2) = 1 + 2\sum_{j=1}^{\infty} \chi_{-2}(j)\frac{q^j}{1-q^j}$$

where $\chi_{-2}$ is the character defined by

$$\chi_{-2}(j) = \begin{cases} 1 & \text{if } j \equiv 1 \text{ or } 3 \pmod 8, \\ -1 & \text{if } j \equiv 5 \text{ or } 7 \pmod 8, \\ 0 & \text{otherwise.} \end{cases}$$

Hint: Replace $(q, x, y)$ with $(q^4, -q, -1)$ in Theorem 0.26. Then multiply by 2. It may be helpful to note that

$$1+2\sum_{k=1}^{\infty}\frac{q^k+q^{3k}}{1+q^{4k}}=1+2\sum_{k=1}^{\infty}\sum_{j=1}^{\infty}\chi_{-2}(j)q^{jk}=1+2\sum_{j=1}^{\infty}\chi_{-2}(j)\frac{q^j}{1-q^j}.$$

See Lorenz [226], Dickson [135, p. 29], Entry 8 in Chapter 17 of Ramanujan's second notebook [253], and Berndt's analysis [46, p. 114].

b. Use Theorem 0.26 to prove that

$$q^3\psi(q^8)\psi(q^{16})=\sum_{j=-\infty}^{\infty}\frac{q^{8j+3}}{1-q^{64j+24}}.$$

Hence, deduce that

$$q^3\psi(q^8)\psi(q^{16})=\sum_{j=0}^{\infty}\sum_{k=0}^{\infty}\left\{q^{(8j+1)(8k+3)}-q^{(8j+5)(8k+7)}\right\}.$$

See Fine [151, (31.19) and (31.23)]. Also see C. Adiga and K. Vasuki [4].

9. Prove Cauchy's result [71], [135, p. 20]:

$$\varphi(q)\varphi(q^3)=1+2\sum_{n=1}^{\infty}\frac{1+(-q)^n}{1+(-q)^{3n}}q^n.$$

Further results that imply Lambert series for $\varphi(q)\varphi(q^3)$ and $q\psi(q^2)\psi(q^6)$ are given in Theorem 3.23.

10. a. Prove that

$$\varphi(q)\varphi(q^4)=\frac{1}{2}\varphi(q)^2-\frac{1}{2}\varphi(q^2)^2+\varphi(q^4)^2.$$

Hence, use Theorem 3.10 to deduce that

$$\varphi(q)\varphi(q^4)=1+2\sum_{j=1}^{\infty}\chi_4(j)\left\{\frac{q^j}{1-q^j}-\frac{q^{2j}}{1-q^{2j}}+\frac{2q^{4j}}{1-q^{4j}}\right\}$$

where $\chi_4(j)=\sin(j\pi/2)$ is the character defined by (3.13). See p. 267 of Ramanujan's first notebook [253] and Berndt's analysis [48, p. 373].

b. Use Theorem 0.26 to prove that

$$q^5\psi(q^8)\psi(q^{32}) = \sum_{j=-\infty}^{\infty} \frac{q^{8j+5}}{1-q^{64j+40}}.$$

Hence, deduce that

$$q^5\psi(q^8)\psi(q^{32}) = \sum_{j=0}^{\infty}\sum_{k=0}^{\infty} \left\{q^{(8j+1)(8k+5)} - q^{(8j+3)(8k+7)}\right\}.$$

See p. 264 of Ramanujan's second notebook [253] and Berndt's analysis [47, p. 153]. See also Fine [151, (31.19) and (31.24)].

11. This exercise follows Berkovich and Yesilyurt [44]. See also Dickson [136, p. 84].

 a. Use Theorem 0.26 to prove the identities:

$$\sum_{n=-\infty}^{\infty} \frac{q^n}{1+q^{10n}} = \frac{f(q,q^9)}{f(-q,-q^9)f(1,q^{10})} \times \prod_{n=1}^{\infty}(1-q^{10n})^3,$$

$$\sum_{n=-\infty}^{\infty} \frac{q^{3n}}{1+q^{10n}} = \frac{f(q^3,q^7)}{f(-q^3,-q^7)f(1,q^{10})} \times \prod_{n=1}^{\infty}(1-q^{10n})^3,$$

$$\sum_{n=-\infty}^{\infty} \frac{q^{5n+1}}{1+q^{10n+2}} = q\frac{f(q^3,q^7)}{f(-q^5,-q^5)f(q^2,q^8)} \times \prod_{n=1}^{\infty}(1-q^{10n})^3$$

and

$$\sum_{n=-\infty}^{\infty} \frac{q^{5n+2}}{1+q^{10n+4}} = q^2\frac{f(q,q^9)}{f(-q^5,-q^5)f(q^4,q^6)} \times \prod_{n=1}^{\infty}(1-q^{10n})^3,$$

where $f(a,b)$ is Ramanujan's theta function studied in Section 0.4.

 b. Use Theorem 0.6 and the results of Part a to deduce that

$$\sum_{n=-\infty}^{\infty} \left(\frac{q^{5n+1}}{1+q^{10n+2}} - \frac{q^{5n+2}}{1+q^{10n+4}}\right) = \frac{\eta_1\eta_2\eta_{10}\eta_{20}}{\eta_4\eta_5},$$

$$\sum_{n=-\infty}^{\infty} \frac{q^n+q^{3n}}{1+q^{10n}} = \frac{\eta_2\eta_4\eta_5\eta_{10}}{\eta_1\eta_{20}},$$

$$\sum_{n=-\infty}^{\infty} \frac{q^n}{1+q^{10n}} - \sum_{n=-\infty}^{\infty} \frac{q^{5n+2}}{1+q^{10n+4}} = \frac{1}{2}\varphi(q)\varphi(q^5)$$

and

$$\sum_{n=-\infty}^{\infty} \frac{q^{3n}}{1+q^{10n}} + \sum_{n=-\infty}^{\infty} \frac{q^{5n+1}}{1+q^{10n+2}} = \frac{1}{2}\varphi(q)\varphi(q^5).$$

c. Use series rearrangements to show that

$$\sum_{n=-\infty}^{\infty} \frac{q^n+q^{3n}}{1+q^{10n}} = 1 + \sum_{n=1}^{\infty} \chi_{20}(n)\frac{q^n}{1-q^n}$$

and

$$\sum_{n=-\infty}^{\infty} \left(\frac{q^{5n+1}}{1+q^{10n+2}} - \frac{q^{5n+2}}{1+q^{10n+4}}\right) = \sum_{n=1}^{\infty} \chi_5(n)\frac{q^n}{1+q^{2n}}$$

where

$$\chi_5(n) = \left(\frac{n}{5}\right) = \begin{cases} 1 & \text{if } n \equiv 1 \text{ or } 4 \pmod 5, \\ -1 & \text{if } n \equiv 2 \text{ or } 3 \pmod 5, \\ 0 & \text{otherwise} \end{cases} \tag{3.139}$$

and

$$\chi_{20}(n) = \chi_4(n)\chi_5(n) = \begin{cases} 1 & \text{if } n \equiv 1,\ 3,\ 7 \text{ or } 9 \pmod{20}, \\ -1 & \text{if } n \equiv 11,\ 13,\ 17 \text{ or } 19 \pmod{20}, \\ 0 & \text{otherwise.} \end{cases}$$

d. Use the results of Parts b and c to deduce that

$$\varphi(q)\varphi(q^5) = 1 + \sum_{n=1}^{\infty} \chi_5(n)\frac{q^n}{1+q^{2n}} + \sum_{n=1}^{\infty} \chi_{20}(n)\frac{q^n}{1-q^n}.$$

e. Let $c(k)$ be defined by $\varphi(q)\varphi(q^5) = \sum_{k=0}^{\infty} c(k)q^k$. By separating the terms according to whether $n$ is even or odd, show that

$$\sum_{m=-\infty}^{\infty}\sum_{n=-\infty}^{\infty} q^{2m^2+2mn+3n^2} = \sum_{k=0}^{\infty} c(2k)q^k.$$

Deduce that

$$\sum_{m=-\infty}^{\infty}\sum_{n=-\infty}^{\infty} q^{2m^2+2mn+3n^2} = 1 - \sum_{n=1}^{\infty} \chi_5(n)\frac{q^n}{1+q^{2n}} + \sum_{n=1}^{\infty} \chi_{20}(n)\frac{q^n}{1-q^n}.$$

f. Use the method of Parts a and b above to prove that

$$\psi(q)\psi(q^5) = \sum_{n=-\infty}^{\infty} \frac{q^{3n}+q^{7n+1}}{1-q^{20n+5}} = \sum_{n=-\infty}^{\infty} \frac{q^n+q^{9n+6}}{1-q^{20n+15}}.$$

g. Deduce that

$$q^3\psi(q^4)\psi(q^{20})$$
$$= \sum_{j=0}^{\infty}\sum_{k=0}^{\infty}\{q^{(4j+1)(20k+3)} + q^{(4j+1)(20k+7)} - q^{(4j+3)(20k+13)} - q^{(4j+3)(20k+17)}\}$$
$$= \sum_{j=0}^{\infty}\sum_{k=0}^{\infty}\{q^{(4j+3)(20k+1)} + q^{(4j+3)(20k+9)} - q^{(4j+1)(20k+11)} - q^{(4j+1)(20k+19)}\}.$$

12. Parts of this exercise follow Berkovich and Yesilyurt [44]. See also Dickson [136, p. 84].

Let $A = \sum_{n=-\infty}^{\infty} \frac{q^n}{1+q^{12n}}$, $B = \sum_{n=-\infty}^{\infty} \frac{q^{5n}}{1+q^{12n}}$,

$C = \sum_{n=-\infty}^{\infty} \frac{q^{3n+1}}{1+q^{12n+4}}$ and $D = \sum_{n=-\infty}^{\infty} \frac{q^{9n+3}}{1+q^{12n+4}}$.

a. Use Theorem 0.26 to prove the identities:

$$A = \frac{f(q,q^{11})}{f(-q,-q^{11})f(1,q^{12})} \times \prod_{n=1}^{\infty}(1-q^{12n})^3,$$

$$B = \frac{f(q^5,q^7)}{f(\ q^5,\ q^7)f(1,q^{12})} \times \prod_{n=1}^{\infty}(1-q^{12n})^3,$$

$$C = q\frac{f(q^5,q^7)}{f(-q^3,-q^9)f(q^4,q^8)} \times \prod_{n=1}^{\infty}(1-q^{12n})^3$$

and

$$D = q^3 \frac{f(q^{-1}, q^{13})}{f(-q^3, -q^9)f(q^4, q^8)} \times \prod_{n=1}^{\infty} (1 - q^{12n})^3,$$

where $f(a, b)$ is Ramanujan's theta function studied in Section 0.4.

b. Use Theorem 0.6 and the results of Part a to deduce that

$$A - D = B + C = \frac{1}{2}\varphi(q)\varphi(q^6),$$

$$A - C = B + D = \frac{1}{2}\varphi(q^2)\varphi(q^3),$$

$$A - B = C + D = \frac{\eta_2\eta_3\eta_4\eta_{24}}{\eta_1\eta_8},$$

$$A + B = \frac{\eta_2\eta_3\eta_8\eta_{12}}{\eta_1\eta_{24}}. \tag{3.140}$$

and

$$C - D = \frac{\eta_1\eta_4\eta_6\eta_{24}}{\eta_3\eta_8}.$$

c. Prove that

$$\frac{1}{2}\left(3\varphi(q^3)\varphi(q^6) - \varphi(q)\varphi(q^2)\right) = \frac{\eta_1\eta_6\eta_8\eta_{12}}{\eta_3\eta_{24}}.$$

Hint: Let $\omega = \exp(i\pi/12)$, take $(x, y, z) = (\omega^5, \omega^7, \omega^{13})$ in Theorem 1.13, and make use of Exercise 8, above.

d. Use series rearrangements to show that

$$A + B = 1 + \sum_{n=1}^{\infty} \chi_{24}(n) \frac{q^n}{1 - q^n} \tag{3.141}$$

and

$$C - D = \sum_{n=1}^{\infty} \chi_3(n) \frac{q^n(1 - q^{2n})}{1 + q^{4n}}$$

where

$$\chi_{24}(n) = \begin{cases} 1 & \text{if } n \equiv 1,\ 5,\ 7 \text{ or } 11 \pmod{24}, \\ -1 & \text{if } n \equiv 13,\ 17,\ 19 \text{ or } 23 \pmod{24}, \\ 0 & \text{otherwise} \end{cases}$$

and $\chi_3(n) = \left(\frac{n}{3}\right)$ is the Legendre symbol.

e. Use the results of Parts b and d to deduce that

$$\varphi(q)\varphi(q^6) = 1 + \sum_{n=1}^{\infty} \chi_{24}(n)\frac{q^n}{1-q^n} + \sum_{n=1}^{\infty} \chi_3(n)\frac{q^n(1-q^{2n})}{1+q^{4n}}$$

and

$$\varphi(q^2)\varphi(q^3) = 1 + \sum_{n=1}^{\infty} \chi_{24}(n)\frac{q^n}{1-q^n} - \sum_{n=1}^{\infty} \chi_3(n)\frac{q^n(1-q^{2n})}{1+q^{4n}}.$$

f. Show that

$$q^5\psi(q^{16})\psi(q^{24}) = \frac{1}{2}\sum_{j=-\infty}^{\infty}\left\{\frac{q^{5(24j+1)}}{1-q^{8(24j+1)}} + \frac{q^{(24j+5)}}{1-q^{8(24j+5)}} + \frac{q^{3(24j+7)}}{1-q^{8(24j+7)}} + \frac{q^{7(24j+11)}}{1-q^{8(24j+11)}}\right\}$$

and

$$q^7\psi(q^{8})\psi(q^{48}) = \frac{1}{2}\sum_{j=-\infty}^{\infty}\left\{\frac{q^{7(24j+1)}}{1-q^{8(24j+1)}} + \frac{q^{3(24j+5)}}{1-q^{8(24j+5)}} + \frac{q^{(24j+7)}}{1-q^{8(24j+7)}} + \frac{q^{5(24j+11)}}{1-q^{8(24j+11)}}\right\}.$$

g. Combine the results of (3.140) and (3.141) to deduce the identity

$$\prod_{j=1}^{\infty}\frac{(1-q^{2j})(1-q^{3j})(1-q^{8j})(1-q^{12j})}{(1-q^j)(1-q^{24j})}$$
$$= 1 + \sum_{j=0}^{\infty}\left(\frac{q^{24j+1}}{1-q^{24j+1}} + \frac{q^{24j+5}}{1-q^{24j+5}} + \frac{q^{24j+7}}{1-q^{24j+7}} + \frac{q^{24j+11}}{1-q^{24j+11}}\right.$$
$$\left. - \frac{q^{24j+13}}{1-q^{24j+13}} - \frac{q^{24j+17}}{1-q^{24j+17}} - \frac{q^{24j+19}}{1-q^{24j+19}} - \frac{q^{24j+23}}{1-q^{24j+23}}\right).$$

This was noted as being "very beautiful" by Fine [151, xii].

13. Let $z_a$, $z_b$, $z_c$, $z_d$, $z_e$, and $z_f$ be defined by

$$z_a = z_a(q) = \frac{\eta_3^2\eta_5^2}{\eta_1\eta_{15}} \qquad z_b = z_b(q) = \frac{\eta_1^2\eta_{15}^2}{\eta_3\eta_5},$$

$$z_c = z_c(q) = \frac{\eta_1\eta_6\eta_{10}\eta_{15}}{\eta_2\eta_{30}}, \qquad z_d = z_d(q) = \frac{\eta_2\eta_3\eta_5\eta_{30}}{\eta_1\eta_{15}},$$

$$z_e = z_e(q) = \frac{\eta_2\eta_3\eta_5\eta_{30}}{\eta_6\eta_{10}} \quad \text{and} \quad z_f = z_f(q) = \frac{\eta_1\eta_6\eta_{10}\eta_{15}}{\eta_3\eta_5}.$$

a. Prove that

$$z_a = 1 + \sum_{j=1}^{\infty}\sum_{k=1}^{\infty} \chi_{15}(j) q^{jk}, \qquad z_b = \sum_{j=1}^{\infty}\sum_{k=1}^{\infty} \chi_3(j)\chi_5(k) q^{jk},$$

$$z_c = 1 + \sum_{j=1}^{\infty}\sum_{k=1}^{\infty} \chi_{15}(j)(-1)^j q^{jk}, \qquad z_d = \sum_{\substack{j=1 \\ j \text{ odd}}}^{\infty}\sum_{k=1}^{\infty} \chi_{15}(j) q^{jk},$$

$$z_e = \sum_{j=1}^{\infty}\sum_{k=1}^{\infty} \chi_3(j)\chi_5(k)(-1)^{j+1} q^{jk} \quad \text{and} \quad z_f = \sum_{\substack{j=1 \\ j \text{ odd}}}^{\infty}\sum_{k=1}^{\infty} \chi_3(j)\chi_5(k) q^{jk},$$

where $\chi_3(j) = \left(\frac{j}{3}\right)$ and $\chi_5(j) = \left(\frac{j}{5}\right)$ are the Legendre symbols given by (2.34) and (3.139), and

$$\chi_{15}(j) = \chi_3(j)\chi_5(j) = \begin{cases} 1 & \text{if } j \equiv 1, 2, 4 \text{ or } 8 \pmod{15}, \\ -1 & \text{if } j \equiv 7, 11, 13 \text{ or } 14 \pmod{15}, \\ 0 & \text{otherwise.} \end{cases}$$

Hints: Let $\omega = \exp(i\pi/15)$.
For $z_a$, take $(x, y, z) = (\omega^2, \omega^4, \omega^8)$ in Theorem 1.12.
For $z_b$, replace $q$ with $q^{5/2}$ in Theorem 1.15, then let $(a, b, c)=(q^{1/2}, q, \omega^{10})$.
For $z_c$, take $(x, y, z) = (\omega^7, \omega^{11}, \omega^{13})$ in Theorem 1.12.
For $z_d$, replace $q$ with $q^{30}$ in Theorem 1.13, then let $(x, y, z) = (q^{-1}, q^{11}, q^{13})$.
For $z_e$, replace $q$ with $q^{5/2}$ in Theorem 1.15, then let $(a, b, c) = (q^{1/2}, q, \omega^5)$.
For $z_f$, use the product definitions and the series expansions proved so far to show that

$$\begin{aligned} z_f(q) = \frac{\eta_1\eta_2\eta_{15}\eta_{30}}{\eta_3\eta_5\eta_6\eta_{10}} \times z_a(q^2) &= \frac{\eta_1\eta_2\eta_{15}\eta_{30}}{\eta_3\eta_5\eta_6\eta_{10}} \times \frac{1}{2}\left(z_c(q) + z_a(q)\right) \\ &= \frac{1}{2} \times \left(\frac{\eta_1^2\eta_{15}^2}{\eta_3\eta_5} + \frac{\eta_2\eta_3\eta_5\eta_{30}}{\eta_6\eta_{10}}\right) \\ &= \frac{1}{2}\left(z_b(q) + z_e(q)\right). \end{aligned}$$

b. Show that

$$\begin{aligned}
\varphi(-q^3)\varphi(-q^5) &= z_c + z_e,\\
\varphi(-q)\varphi(-q^{15}) &= z_c - z_e,\\
2q\psi(q^3)\psi(q^5) &= z_d + z_f,\\
q^2\psi(q)\psi(q^{15}) &= z_d - z_f.
\end{aligned}$$

Hints: Use the results of Part a to show that

$$z_a(q) = z_a(q^2) + z_d(q) \quad \text{and} \quad z_c(q) = z_a(q^2) - z_d(q).$$

Then write each side in terms of eta-quotients and multiply by $\eta_1\eta_{15}/\eta_6\eta_{10}$ to deduce the first two results. The third and fourth results can be proved in a similar way by first establishing the identities

$$2z_a(q^2) = z_a(q) + z_c(q) \quad \text{and} \quad 2z_d(q) = z_a(q) - z_c(q)$$

and multiplying by $\eta_2\eta_{30}/\eta_3\eta_5$.

c. Show that

$$\sum_{j=-\infty}^{\infty}\sum_{k=-\infty}^{\infty} q^{j^2+jk+4k^2} = \varphi(q)\varphi(q^{15}) + 4q^4\psi(q^2)\psi(q^{30})$$

and

$$\sum_{j=-\infty}^{\infty}\sum_{k=-\infty}^{\infty} q^{2j^2+jk+2k^2} = \varphi(q^3)\varphi(q^5) + 4q^2\psi(q^6)\psi(q^{10}).$$

Hint: For the first result, break the double sum into two according to whether $k$ is even or odd. For the second result, use the series rearrangement

$$\sum_{j=-\infty}^{\infty}\sum_{k=-\infty}^{\infty} c(j,k) = \sum_{j=-\infty}^{\infty}\sum_{k=-\infty}^{\infty} c(j+k, j-k) + \sum_{j=-\infty}^{\infty}\sum_{k=-\infty}^{\infty} c(j+k+1, j-k).$$

d. Use the series in Part a to show that

$$z_c(-q) + 2z_d(q^2) = z_a(q) \quad \text{and} \quad -z_e(-q) - 2z_f(q^2) = z_b(q).$$

Hence, use the results of Parts b and c to show that

$$\sum_{j=-\infty}^{\infty}\sum_{k=-\infty}^{\infty} q^{j^2+jk+4k^2} = z_a + z_b$$

and

$$\sum_{j=-\infty}^{\infty}\sum_{k=-\infty}^{\infty} q^{2j^2+jk+2k^2} = z_a - z_b.$$

The results of Parts a, b, and d can be used to write the number of representations of a positive integer by each of the quadratic forms

$$x^2+15y^2, \quad 3x^2+5x^2, \quad x^2+xy+4y^2 \quad \text{and} \quad 2x^2+xy+2y^2,$$

and by each of the quadratic functions

$$\frac{x(x+1)}{2}+\frac{15y(y+1)}{2} \quad \text{and} \quad \frac{3x(x+1)}{2}+\frac{5y(y+1)}{2},$$

in terms of divisor sums. See [112].

14. a. Show that the identity (3.67) may be expressed in the form

$$\left(\sum_{j=-\infty}^{\infty} q^{j^2}\right)^4 = 1+8\sum_{j=1}^{\infty}\frac{jq^j}{1+(-q)^j}.$$

b. Show that the identity (3.100) may be expressed in the form

$$\left(\sum_{j=-\infty}^{\infty} q^{j^2}\right)^8 = 1+16\sum_{j=1}^{\infty}\frac{j^3q^j}{1-(-q)^j}.$$

c. Observe that these identities give the special cases $m = 1$ and $m = 2$ of Ramanujan's identity (3.115).

15. Prove that

$$\varphi(q)\varphi^4(q^2)\varphi(q^4) = \varphi^2(q^4)\varphi^4(-q^4)+2q\varphi^2(q)\psi^4(q^2).$$

Hence, use Lemma 3.32 to deduce that

$$\varphi(q)\varphi^4(q^2)\varphi(q^4) = 1-4\sum_{j=1}^{\infty}\chi_4(j)\frac{j^2q^{4j}}{1-q^{4j}}+2\sum_{j=1}^{\infty}\frac{j^2q^j}{1+q^{2j}}.$$

See the work of Alaca, Alaca, and Williams [15] for this result and others, and for references to Liouville.

16. a. Let $\omega = \exp(2\pi i/3)$. Show that

$$a(q)a(q^3) = \frac{1}{12}\left(9P(q^9)+6P(q^3)-2P(q)+\omega P(\omega q)+\omega^2 P(\omega^2 q)\right).$$

Hint: Square both sides of the identity $a(q) - a(q^3) = 2c(q^3)$ from Lemma 3.11. Then apply Theorem 3.31.

b. Prove that

$$a(q)a(q^4) = \frac{1}{4}\left(-P(q)+3P(q^2)+3P(q^3)-4P(q^4)-9P(q^6)+12P(q^{12})\right).$$

Hint: By Lemma 3.12, $a(-q) = 2a(q^4) - a(q)$. Square both sides, then apply Theorem 3.31 and (0.110).

The analogous result for $a(q)a(q^2)$ has been given in Theorem 3.30. Further results for $a(q)a(q^6)$ and $a(q^2)a(q^3)$ can be found in [12].

17. The goal of this exercise is to obtain a Lambert series expansion for $a^5(q)$ that implies the identity for $u_{10}(n)$ in (3.128). Let $h$ be defined by

$$h(\theta) = h(\theta, \tau) = \frac{1}{2i} f(\theta, 2\pi, 2\pi\tau, 3)$$

where $f$ is the elliptic function defined by (2.37).

a. Use Theorem 2.20 to show that

$$h(\theta) = \frac{1}{6} + \sum_{j=1}^{\infty} \frac{q^j - q^{2j}}{1 - q^{3j}} \cos j\theta.$$

b. Deduce that

$$h(\theta) = \sum_{n=0}^{\infty} E^0_{2n+1}(\tau, 3)\,(-1)^n \frac{\theta^{2n}}{(2n)!}$$

where $E^0_{2n+1}(\tau, 3)$ is defined by (2.43), viz.,

$$E^0_{2n+1}(\tau, 3) = \begin{cases} \dfrac{1}{6} + \displaystyle\sum_{j=1}^{\infty} \frac{q^j - q^{2j}}{1 - q^{3j}} & \text{if } n = 0, \\ \displaystyle\sum_{j=1}^{\infty} \frac{j^{2n}(q^j - q^{2j})}{1 - q^{3j}} & \text{if } n \geq 1. \end{cases}$$

c. Show that

$$h(\theta) = \frac{1}{6} + \frac{1}{2}\sum_{n=1}^{\infty} \left(\frac{n}{3}\right)\left(\frac{q^n e^{i\theta}}{1 - q^n e^{i\theta}} + \frac{q^n e^{-i\theta}}{1 - q^n e^{-i\theta}}\right).$$

d. Use the result of Part c to deduce that $h$ is an elliptic function with periods $2\pi$ and $2\pi\tau$. Show that $h$ has simple poles at the points $\theta = 2\pi m + 2\pi n\tau$ where $n \not\equiv 0 \pmod 3$ and that these are the only singularities of $h$.

e. Show that

$$(h(\theta) - h(0))\,(\wp(\theta, 3\tau) - \wp(2\pi\tau, 3\tau)) = -\frac{1}{2}E_3^0(\tau, \chi_3).$$

Hint: Show that the left-hand side is an elliptic function with no singularities, and hence is a constant independent of $\theta$. The constant may be evaluated by expanding the left-hand side in powers of $\theta$ using the result of Part b and the expansion

$$\wp(\theta, 3\tau) = \frac{1}{\theta^2} + O(\theta^2).$$

f. Show that

$$E_5^0(\tau, 3) = -12\wp(2\pi\tau, 3\tau) \times E_3^0(\tau, 3).$$

Hint: Expand the result of Part e in powers of $\theta$ and equate coefficients of $\theta^2$.

g. Use (1.16) to show that

$$\wp(2\pi\tau, 3\tau) = -\frac{1}{24}\left(3P(q^3) - P(q)\right).$$

h. From Parts f and g it follows that

$$E_5^0(\tau, 3) = \frac{1}{2}\left(3P(q^3) - P(q)\right) \times E_3^0(\tau, 3).$$

Apply the transformation $\tau \to -1/3\tau$ and make use of (2.19) and Theorem 2.21 to deduce

$$E_5^\infty(\tau, 3) = -\frac{3}{2}\left(3P(q^3) - P(q)\right) \times E_3^\infty(\tau, 3).$$

i. Use (2.43), (2.44) and Theorem 3.35 to show that

$$E_3^0(\tau, 3) = \sum_{j=1}^{\infty} \frac{j^2(q^j - q^{2j})}{1 - q^{3j}} = \frac{1}{27}c^3(q)$$

and

$$E_3^\infty(\tau, 3) = -\frac{1}{9} + \sum_{j=1}^{\infty} \left(\frac{j}{3}\right) \frac{j^2 q^j}{1 - q^j} = -\frac{1}{9}b^3(q).$$

j. By Theorem 3.31 we have $\frac{1}{2}(3P(q^3) - P(q)) = a^2(q)$. Use this, and the results of Parts h and i, to deduce that

$$E_5^0(\tau, 3) = \frac{1}{27}a^2(q)c^3(q)$$

and

$$E_5^{\infty}(\tau,3) = \frac{1}{3}a^2(q)b^3(q).$$

Conclude that

$$\begin{aligned} a^5(q) &= 3E_5^{\infty}(\tau,3) + 27E_5^{0}(\tau,3) \\ &= 1 + 3\sum_{j=1}^{\infty}\left(\frac{j}{3}\right)\frac{j^4q^j}{1-q^j} + 27\sum_{j=1}^{\infty}\frac{j^4(q^j-q^{2j})}{1-q^{3j}} \end{aligned}$$

and show that this implies (3.128). See H. Petersson [241, p. 90].

18. Let $h$ be as for the previous exercise.

 a. Use the result of Part e of the previous exercise, along with the differential equation for the Weierstrass elliptic function from (1.77), to prove that

$$\left(\frac{dh}{d\theta}\right)^2 = -2\left(h - \frac{a}{6}\right)\left(2h^3 + ah^2 - \frac{b^3}{27}\right)$$

 where $a = a(q)$ and $b = b(q)$. See Berndt [48, p. 140], Berndt, Bhargava, and Garvan [50, Eq. (8.32)], Chan and Liu [88, Eq. (2.28)] and [105, Th. 12.2].

 b. Show that

$$\frac{d^2h}{d\theta^2} = -8h^3 - 2ah^2 + \frac{a^2h}{3} + \frac{b^3}{27}.$$

 c. Use the result from Part b, the series expansion from Part b of the previous exercise, and induction on $n$ to show that for any positive integer $n$, there are unique rational constants $k_{n,j}$ such that

$$\sum_{j=1}^{\infty}\frac{j^{2n}(q^j-q^{2j})}{1-q^{3j}} = \sum_{\substack{j\geq 1\\ 3j\leq 2n+1}} k_{n,j}\, a(q)^{2n+1-3j}\, c(q)^{3j}.$$

 See the previous exercise for the cases $n = 1$ and $n = 2$.

 d. Hence, deduce Ramanujan's identities

$$\sum_{j=1}^{\infty}\frac{j^2q^j}{1+q^j+q^{2j}} = \frac{z^3x}{27},$$

$$\sum_{j=1}^{\infty} \frac{j^4 q^j}{1+q^j+q^{2j}} = \frac{z^5 x}{27},$$

$$\sum_{j=1}^{\infty} \frac{j^6 q^j}{1+q^j+q^{2j}} = \frac{z^7}{27} x \left(1 + \frac{4x}{3}\right)$$

and

$$\sum_{j=1}^{\infty} \frac{j^8 q^j}{1+q^j+q^{2j}} = \frac{z^9}{27} x \left(1 + 8x + \frac{80x^2}{81}\right),$$

where $z = a(q)$ and $x = c^3(q)/a^3(q)$. These results were given by Ramanujan in his second notebook [253, p. 257] (where the last formula contains a slight misprint that is corrected here). For other proofs and extensions, see Berndt [48, p. 143], Berndt, Bhargava, and Garvan [50, Th. 8.7], Chan and Liu [88, Cor. 2.4] and [105, Th. 13.11].

19. Let $h$ be as for Exercise 17.

a. Use Part c of Exercise 17 to show that

$$\frac{dh}{d\theta} = \frac{i}{2} \sum_{n=-\infty}^{\infty} \left(\frac{n}{3}\right) \frac{q^n e^{i\theta}}{(1-q^n e^{i\theta})^2}.$$

Deduce that

$$\begin{aligned}
\frac{dh}{d\theta} &= \frac{i}{2} \left(\wp(\theta - 2\pi\tau, 3\tau) - \wp(\theta + 2\pi\tau, 3\tau)\right) \\
&= -q(q;q)_\infty (q^3;q^3)_\infty^3 \\
&\quad \times \sin\theta \frac{(q^3 e^{2i\theta}, q^3 e^{-2i\theta}; q^3)_\infty (q^3 e^{i\theta}, q^3 e^{-i\theta}; q^3)_\infty^2}{(qe^{i\theta}, qe^{-i\theta}; q)_\infty^2} \\
&= -\frac{1}{2} \times \frac{\eta_3^9}{\eta_1^3} \times \frac{s(2\theta, 3\tau) s(\theta, 3\tau)^2}{s(\theta, \tau)^2}.
\end{aligned}$$

b. Show that

$$\left(h(\theta) + \frac{a(q)}{12}\right)^2 = \frac{1}{4} \left(\wp(\theta, 3\tau) - \wp(\theta, \tau)\right) + \frac{a^2(q)}{16}.$$

See [48, 50, 88, 105] and [221] for further results.

# Chapter 4
# Levels 1, 2, 3, and 4: Jacobi's Inversion Theorem and Ramanujan's Alternative Theories

## 4.1 Introduction

One of the fundamental results in the theory of elliptic functions is the inversion theorem of Jacobi. We start by writing Jacobi's aequatio identica satis abstrusa in Corollary 3.8 as

$$16q\prod_{j=1}^{\infty}\frac{(1+q^{2j})^8}{(1+q^{2j-1})^8}+\prod_{j=1}^{\infty}\frac{(1-q^{2j-1})^8}{(1+q^{2j-1})^8}=1$$

and let $x_4$ be defined by

$$x_4=16q\prod_{j=1}^{\infty}\frac{(1+q^{2j})^8}{(1+q^{2j-1})^8}$$

so that

$$1-x_4=\prod_{j=1}^{\infty}\frac{(1-q^{2j-1})^8}{(1+q^{2j-1})^8}.$$

Jacobi's inversion theorem provides an explicit formula for the inverse function that gives $q$ as a function of $x_4$. To write down the inverse function, we will need the ${}_2F_1$ hypergeometric function that is defined for $|x|<1$ by

$${}_2F_1\left(\begin{matrix}a,\ b\\ c\end{matrix};x\right)=\sum_{n=0}^{\infty}\frac{(a)_n(b)_n}{(c)_n n!}x^n,$$



S. Cooper, *Ramanujan's Theta Functions*, DOI 10.1007/978-3-319-56172-1_5

where $(a)_n$ is the shifted factorial defined by

$$(a)_n = \begin{cases} 1 & \text{if } n = 0, \\ a(a+1)(a+2)\cdots(a+n-1) & \text{if } n \in \mathbb{Z}^+. \end{cases}$$

For typesetting, we will sometimes use the single-line notation ${}_2F_1(a, b;\, c;\, x)$.

Here is Jacobi's inversion theorem.

**Theorem 4.1 (Inversion theorem, level 4).** *Suppose $0 \le q < 1$ and let $x_4$ be defined by*

$$x_4 = 16q \prod_{j=1}^{\infty} \frac{(1+q^{2j})^8}{(1+q^{2j-1})^8}. \tag{4.1}$$

*As $q$ increases from 0 to 1, $x_4$ increases from 0 to 1. The inverse function $q = q(x_4)$ exists and is given by the explicit formula*

$$q = \exp\left(-\pi \frac{{}_2F_1\left(\frac{1}{2}, \frac{1}{2}; 1; 1-x_4\right)}{{}_2F_1\left(\frac{1}{2}, \frac{1}{2}; 1; x_4\right)}\right). \tag{4.2}$$

*Moreover, the hypergeometric function of $x_4$ can be expressed as a function of $q$ by means of the formula*

$${}_2F_1\left(\frac{1}{2}, \frac{1}{2}; 1; x_4\right) = \left(\sum_{j=-\infty}^{\infty} q^{j^2}\right)^2. \tag{4.3}$$

The parameters $k$ and $q$, where $k = x_4^{1/2}$, are called the modulus and nome, respectively. In the statement of Theorem 4.1, the nome $q$ is the primary variable, and $x_4$ is defined in terms of the nome by (4.1). The result to be proved is the formula (4.2) that expresses the nome in terms of $x_4$. It is possible to do it the other way around, and take $x_4$ as the primary variable and use (4.2) to define the nome $q$ in terms of $x_4$. Then, the result to be proved would be (4.1) that expresses $x_4$ in terms of the nome.

Both approaches are taken in the literature. Jacobi deduces (4.1) from (4.2), whereas Whittaker and Watson deduce (4.2) from (4.1). We have elected to follow the approach given in Theorem 4.1 in order to make use of the extensive theory of theta functions developed in the previous chapter. Moreover, this approach easily extends to three other cases, introduced by Ramanujan, that we shall now briefly discuss.

More than 80 years after Jacobi's work, Ramanujan gave three new analogues of Theorem 4.1. The three theories are now collectively known as 'Ramanujan's theories of elliptic functions to alternative bases', whereas Jacobi's Theorem 4.1 is called 'the classical theory'. Each of the classical and alternative theories involves two functions $x_\ell$ and $X_\ell$ that are related by

$$\alpha_\ell X_\ell = x_\ell(1 - x_\ell)$$

where $\alpha_\ell$ is a constant that depends only on $\ell$, and $\ell = 1, 2, 3$ or $4$. It turns out—see Part 1 of Theorem 4.20—that the function $X_\ell$ satisfies an involution of the form

$$X_\ell(e^{-2\pi\sqrt{t/\ell}}) = X_\ell(e^{-2\pi/\sqrt{t\ell}}), \qquad t > 0,$$

and the number $\ell$ coincides with the level of the modular function $X_\ell$. Each theory shall be referred to by its level. Jacobi's classical theory in Theorem 4.1 corresponds to the level $\ell = 4$, and Ramanujan's three alternative theories correspond to levels $\ell = 1, 2$ and 3.

Before stating the results, it will be useful to recall Ramanujan's Eisenstein series defined by (0.71)–(0.73):

$$P(q) = 1 - 24\sum_{j=1}^{\infty}\frac{jq^j}{1-q^j}, \qquad Q(q) = 1 + 240\sum_{j=1}^{\infty}\frac{j^3q^j}{1-q^j},$$

and

$$R(q) = 1 - 504\sum_{j=1}^{\infty}\frac{j^5q^j}{1-q^j};$$

and the theta functions given by Definitions 3.1 and 3.2:

$$\varphi(q) = \sum_{j=-\infty}^{\infty} q^{j^2}, \qquad \psi(q) = \sum_{j=0}^{\infty} q^{j(j+1)/2},$$

$$a(q) = \sum_{j=-\infty}^{\infty}\sum_{k=-\infty}^{\infty} q^{j^2+jk+k^2}, \qquad b(q) = \sum_{j=-\infty}^{\infty}\sum_{k=-\infty}^{\infty} \omega^{j-k} q^{j^2+jk+k^2},$$

and

$$c(q) = \sum_{j=-\infty}^{\infty}\sum_{k=-\infty}^{\infty} q^{(j+\frac{1}{3})^2+(j+\frac{1}{3})(k+\frac{1}{3})+(k+\frac{1}{3})^2}.$$

Here are the main results for Ramanujan's theories of elliptic functions to alternative bases.

**Theorem 4.2 (Inversion theorem, level 1).** *Suppose* $0 \le q < 1$ *and let* $x_1$ *be defined by*

$$x_1 = \frac{1}{2}\left(1 - \frac{R(q)}{Q^{3/2}(q)}\right).$$

*As* $q$ *increases from* 0 *to* 1, $x_1$ *increases from* 0 *to* 1. *The inverse function* $q = q(x_1)$ *exists, and is given by the explicit formula*

$$q = \exp\left(-2\pi \frac{{}_2F_1\left(\frac{1}{6}, \frac{5}{6}; 1; 1 - x_1\right)}{{}_2F_1\left(\frac{1}{6}, \frac{5}{6}; 1; x_1\right)}\right).$$

*Moreover, the hypergeometric function of* $x_1$ *can be expressed as a function of* $q$ *by means of the formula*

$${}_2F_1\left(\frac{1}{6}, \frac{5}{6}; 1; x_1\right) = Q^{1/4}(q).$$

**Theorem 4.3 (Inversion theorem, level 2).** *Suppose* $0 \le q < 1$ *and let* $x_2$ *be defined by*

$$x_2 = \frac{64q\psi^8(q)}{\varphi^8(-q) + 64q\psi^8(q)}.$$

*As* $q$ *increases from* 0 *to* 1, $x_2$ *increases from* 0 *to* 1. *The inverse function* $q = q(x_2)$ *exists, and is given by the explicit formula*

$$q = \exp\left(-\frac{2\pi}{\sqrt{2}} \frac{{}_2F_1\left(\frac{1}{4}, \frac{3}{4}; 1; 1 - x_2\right)}{{}_2F_1\left(\frac{1}{4}, \frac{3}{4}; 1; x_2\right)}\right).$$

*Moreover, the hypergeometric function of* $x_2$ *can be expressed as a function of* $q$ *by means of the formula*

$${}_2F_1\left(\frac{1}{4}, \frac{3}{4}; 1; x_2\right) = \left(\varphi^8(-q) + 64q\psi^8(q)\right)^{1/4}.$$

**Theorem 4.4 (Inversion theorem, level 3).** *Suppose* $0 \le q < 1$ *and let* $x_3$ *be defined by*

$$x_3 = \frac{c^3(q)}{a^3(q)}.$$

*As $q$ increases from 0 to 1, $x_3$ increases from 0 to 1. The inverse function $q = q(x_3)$ exists and is given by the explicit formula*

$$q = \exp\left(-\frac{2\pi}{\sqrt{3}}\frac{{}_2F_1\left(\frac{1}{3},\frac{2}{3};1;1-x_3\right)}{{}_2F_1\left(\frac{1}{3},\frac{2}{3};1;x_3\right)}\right). \tag{4.4}$$

*Moreover, the hypergeometric function of $x_3$ can be expressed as a function of $q$ by means of the formula*

$${}_2F_1\left(\frac{1}{3},\frac{2}{3};1;x_3\right) = a(q). \tag{4.5}$$

The goal of this chapter is to prove Theorems 4.1–4.4. The definitions of $x_1$, $x_2$, $x_3$, and $x_4$ in Theorems 4.1–4.4 may appear to be pulled out of the air without any motivation. This has been done for efficiency. In Section 4.4 it will be seen in retrospect how to put all four theories into a single formulation.

We shall work through the classical theory in the next section. Then, the alternative theories will be analyzed by the same procedure in Section 4.3.

The chapter ends with an application to the arithmetic-geometric mean and some of its analogues.

## 4.2 Jacobi's classical theory: level 4

In this section we will prove Theorem 4.1. For simplicity, it will be assumed that $0 \le q < 1$. The limiting case $q \to 1^-$ will also be considered. The general case $|q| < 1$, $q \in \mathbb{C}$, is also of significant interest, and further details can be found in the book by Whittaker and Watson [299, Sec. 21.7].

Our starting point will be the definitions of $x_4$ and $z_4$ given by

$$x_4 = 16q\prod_{j=1}^{\infty}\frac{(1+q^{2j})^8}{(1+q^{2j-1})^8} \tag{4.6}$$

and

$$z_4 = \left(\sum_{j=-\infty}^{\infty} q^{j^2}\right)^2. \tag{4.7}$$

In terms of Ramanujan's theta functions, by Theorems 3.5 and 3.7 we have

$$x_4 = \frac{16q\psi^4(q^2)}{\varphi^4(q)}, \qquad 1 - x_4 = \frac{\varphi^4(-q)}{\varphi^4(q)} \qquad \text{and} \qquad z_4 = \varphi^2(q). \tag{4.8}$$

The results in the next lemma will be handy.

**Lemma 4.5.** *The following identities hold*:

$$1 - x_4 = \prod_{j=1}^{\infty} \frac{(1 - q^{2j-1})^8}{(1 + q^{2j-1})^8} \tag{4.9}$$

*and*

$$z_4^2 = \frac{4P(q^4) - P(q)}{3}. \tag{4.10}$$

*Proof.* The identity (4.9) is just a rearrangement of Jacobi's 'aequatio identica satis abstrusa' identity in Corollary 3.8. Equation (4.10) is the sum of four squares identity that was proved in Theorem 3.26. □

The next lemma expresses each of $x_4$, $1 - x_4$, and $z_4$ in terms of eta-quotients.

**Lemma 4.6.** *The following infinite product representations hold*:

$$x_4 = 16q \prod_{j=1}^{\infty} \frac{(1 - q^j)^8 (1 - q^{4j})^{16}}{(1 - q^{2j})^{24}}, \tag{4.11}$$

$$1 - x_4 = \prod_{j=1}^{\infty} \frac{(1 - q^j)^{16} (1 - q^{4j})^8}{(1 - q^{2j})^{24}} \tag{4.12}$$

*and*

$$z_4 = \prod_{j=1}^{\infty} \frac{(1 - q^{2j})^{10}}{(1 - q^j)^4 (1 - q^{4j})^4}. \tag{4.13}$$

*Hence,*

$$16q \prod_{j=1}^{\infty} (1 - q^j)^{24} = z_4^{12} x_4 (1 - x_4)^4, \tag{4.14}$$

$$16^2 q^2 \prod_{j=1}^{\infty} (1 - q^{2j})^{24} = z_4^{12} x_4^2 (1 - x_4)^2 \tag{4.15}$$

*and*

$$16^4 q^4 \prod_{j=1}^{\infty} (1-q^{4j})^{24} = z_4^{12} x_4^4 (1-x_4). \tag{4.16}$$

*Proof.* The identities (4.11) and (4.12) are just rearrangements of the infinite products in (4.6) and (4.9), respectively. The identity (4.13) was proved earlier as part of Theorem 3.5. Finally, the infinite products (4.14)–(4.16) are direct consequences of (4.11)–(4.13). □

**Theorem 4.7.** *The following differentiation formula holds:*

$$q\frac{\mathrm{d}x_4}{\mathrm{d}q} = z_4^2 x_4 (1-x_4).$$

*Proof.* By (4.11) and (4.12),

$$\frac{x_4}{1-x_4} = 16q \prod_{j=1}^{\infty} \frac{(1-q^{4j})^8}{(1-q^j)^8}. \tag{4.17}$$

Take logarithms and apply the differential operator $q\frac{\mathrm{d}}{\mathrm{d}q}$ to obtain

$$\left(\frac{1}{x_4} + \frac{1}{1-x_4}\right) q\frac{\mathrm{d}x_4}{\mathrm{d}q} = 1 + 8\sum_{j=1}^{\infty} \frac{jq^j}{1-q^j} - 32\sum_{j=1}^{\infty} \frac{jq^{4j}}{1-q^{4j}},$$

that is,

$$\frac{1}{x_4(1-x_4)} q\frac{\mathrm{d}x_4}{\mathrm{d}q} = \frac{4P(q^4) - P(q)}{3}.$$

The result now follows on applying (4.10). □

The next result describes some basic properties of $x_4$ as a function of $q$.

**Theorem 4.8.** *Let* $x_4 = x_4(q)$ *be defined by* (4.6). *The following properties hold.*

1. *For any positive real number* $t$, *we have* $x_4(e^{-\pi t}) + x_4(e^{-\pi/t}) - 1$.
2. $x_4(e^{-\pi}) = \frac{1}{2}$.
3. *As* $q$ *increases from* 0 *to* 1, $x_4(q)$ *strictly increases from* 0 *to* 1.

*Proof.* Let $\tau = it/2$ so that $q = \exp(2\pi i\tau) = e^{-\pi t}$.

Let $p = \exp(-2\pi i/\tau) = e^{-\pi/t}$. By (4.11) and (4.12),

$$\frac{x_4(q)}{1-x_4(q)} = 16q \prod_{j=1}^{\infty} \frac{(1-q^{4j})^8}{(1-q^j)^8} = 16\left(\frac{\eta(4\tau)}{\eta(\tau)}\right)^8 = 16\left(\frac{\eta(2it)}{\eta(it/2)}\right)^8.$$

On applying the modular transformation (2.8) we deduce that

$$\frac{x_4(q)}{1-x_4(q)} = \frac{1}{16}\left(\frac{\eta(i/2t)}{\eta(2i/t)}\right)^8 = \frac{1-x_4(p)}{x_4(p)},$$

which simplifies to $x_4(q) + x_4(p) = 1$. This proves Property 1, and Property 2 is obtained by setting $t = 1$.

It remains to prove Property 3. The definition (4.6) implies $x_4 = 0$ when $q = 0$. Next, it follows from (4.9) that $1 - x_4$ is a decreasing function for $0 < q < 1$ and $\lim_{q\to 1^-}(1 - x_4(q)) = 0$. This establishes Property 3.

□

The next two lemmas express values of the Eisenstein series $P$ and $Q$ in terms of $x_4$ and $z_4$.

**Lemma 4.9.** *Let $x_4$ and $z_4$ be defined by* (4.6) *and* (4.7). *Then*

$$P(q) = \frac{12}{z_4} q\frac{\mathrm{d}z_4}{\mathrm{d}q} + (1-5x_4)\, z_4^2, \tag{4.18}$$

$$P(q^2) = \frac{6}{z_4} q\frac{\mathrm{d}z_4}{\mathrm{d}q} + (1-2x_4)\, z_4^2 \tag{4.19}$$

*and*

$$P(q^4) = \frac{3}{z_4} q\frac{\mathrm{d}z_4}{\mathrm{d}q} + \left(1-\frac{5x_4}{4}\right) z_4^2. \tag{4.20}$$

*Proof.* By (4.14), we have

$$16q\prod_{j=1}^{\infty}(1-q^j)^{24} = z_4^{12}x_4(1-x_4)^4.$$

Take logarithms, apply the operator $q\frac{\mathrm{d}}{\mathrm{d}q}$, and use Theorem 4.7 to obtain

$$P(q) = \frac{12}{z_4} q\frac{\mathrm{d}z_4}{\mathrm{d}q} + \left(\frac{1}{x_4} - \frac{4}{1-x_4}\right) q\frac{\mathrm{d}x_4}{\mathrm{d}q} = \frac{12}{z_4} q\frac{\mathrm{d}z_4}{\mathrm{d}q} + (1-5x_4)z_4^2.$$

This proves (4.18). The identities (4.19) and (4.20) can be proved in the same way, starting with (4.15) and (4.16), respectively. □

**Lemma 4.10.** *Let $x_4$ and $z_4$ be defined by* (4.6) *and* (4.7). *Then*

$$Q(q) = z_4^4(1 + 14x_4 + x_4^2),$$

$$Q(q^2) = z_4^4(1 - x_4 + x_4^2)$$

*and*

$$Q(q^4) = z_4^4 \left(1 - x_4 + \frac{x_4^2}{16}\right).$$

*Proof.* By Theorem 3.39,

$$Q(q) = \frac{\eta_1^{16}}{\eta_2^8} + 256\frac{\eta_2^{16}}{\eta_1^8} \qquad \text{and} \qquad Q(q^2) = \frac{\eta_1^{16}}{\eta_2^8} + 16\frac{\eta_2^{16}}{\eta_1^8},$$

and replacing $q$ with $q^2$ in the last formula gives

$$Q(q^4) = \frac{\eta_2^{16}}{\eta_4^8} + 16\frac{\eta_4^{16}}{\eta_2^8}.$$

Now apply (4.14)–(4.16) to express the eta-quotients in terms of $z_4$ and $x_4$. The result is

$$Q(q) = z_4^4(1 - x_4)^2 + 16z_4^4x_4 = z_4^4(1 + 14x_4 + x_4^2),$$
$$Q(q^2) = z_4^4(1 - x_4)^2 + z_4^4x_4 = z_4^4(1 - x_4 + x_4^2)$$

and

$$Q(q^4) = z_4^4(1 - x_4) + \frac{z_4^4x_4^2}{16} = z_4^4\left(1 - x_4 + \frac{x_4^2}{16}\right).$$

□

The next result shows that $z_4$ satisfies a second order linear differential equation with respect to $x_4$.

**Theorem 4.11.** *Let $x_4$ and $z_4$ be defined by* (4.6) *and* (4.7). *Then*

$$\frac{\mathrm{d}}{\mathrm{d}x_4}\left(x_4(1 - x_4)\frac{\mathrm{d}z_4}{\mathrm{d}x_4}\right) = \frac{z_4}{4}. \tag{4.21}$$

*Proof.* By the chain rule and Theorem 4.7,

$$\frac{\mathrm{d}}{\mathrm{d}x_4} = q\frac{\mathrm{d}}{\mathrm{d}q}\Big/ q\frac{\mathrm{d}x_4}{\mathrm{d}q} = \frac{1}{z_4^2x_4(1 - x_4)}q\frac{\mathrm{d}}{\mathrm{d}q}.$$

Two applications of this gives

$$\frac{\mathrm{d}}{\mathrm{d}x_4}\left(x_4(1 - x_4)\frac{\mathrm{d}z_4}{\mathrm{d}x_4}\right) = \frac{1}{z_4^2x_4(1 - x_4)}q\frac{\mathrm{d}}{\mathrm{d}q}\left(\frac{1}{z_4^2}q\frac{\mathrm{d}z_4}{\mathrm{d}q}\right). \tag{4.22}$$

The identity (4.18) may be used to rewrite (4.22) as

$$\frac{\mathrm{d}}{\mathrm{d}x_4}\left(x_4(1-x_4)\frac{\mathrm{d}z_4}{\mathrm{d}x_4}\right) = \frac{1}{12z_4^2x_4(1-x_4)}q\frac{\mathrm{d}}{\mathrm{d}q}\left(\frac{P(q)}{z_4} - (1-5x_4)z_4\right).$$

The derivative on the right-hand side may be calculated by the standard rules of calculus, using Theorems 0.21, 4.7, and Lemma 4.9 to compute the derivatives of $P(q)$, $x_\ell$, and $z_\ell$, respectively. The result simplifies to

$$\frac{\mathrm{d}}{\mathrm{d}x_4}\left(x_4(1-x_4)\frac{\mathrm{d}z_4}{\mathrm{d}x_4}\right) = \frac{5z_4}{12} + \frac{z_4^4(1-5x_4)^2 - Q(q)}{144z_4^3x_4(1-x_4)}.$$

On applying Lemma 4.10 to eliminate the $Q(q)$ term we complete the proof. □

We will need the following simple description of the behavior of the hypergeometric function as $x \to 1$.

**Lemma 4.12.** *Suppose* $0 < a < 1$. *Then*

$$\lim_{x\to 1^-} {}_2F_1(a, 1-a; 1; x) = +\infty.$$

*Proof.* We use a term by term comparison with the logarithmic series

$$-\log(1-x) = \sum_{j=1}^{\infty} \frac{x^j}{j}.$$

For $j \geq 2$, we have

$$\begin{aligned}
\frac{(a)_j(1-a)_j}{j!^2} &= \prod_{i=1}^{j} \frac{(i+a-1)(i-a)}{i^2} \\
&= a(1-a)\prod_{i=2}^{j} \frac{i^2 - i + a(1-a)}{i^2} \\
&> a(1-a)\prod_{i=2}^{j} \frac{i^2 - i}{i^2} \\
&= a(1-a)\prod_{i=2}^{j}\left(1 - \frac{1}{i}\right) \\
&= \frac{a(1-a)}{j}.
\end{aligned}$$

Therefore,

$$\begin{aligned}
\lim_{x\to 1^-} {}_2F_1(a,1-a;1;x) &= \lim_{x\to 1^-}\left(1+a(1-a)x+\sum_{j=2}^{\infty}\frac{(a)_j(1-a)_j}{j!^2}x^j\right)\\
&\geq \lim_{x\to 1^-}\left(1+a(1-a)x+a(1-a)\sum_{j=2}^{\infty}\frac{x^j}{j}\right)\\
&= \lim_{x\to 1^-}\left(1-a(1-a)\log(1-x)\right)\\
&= +\infty.
\end{aligned}$$

□

The next result gives a formula for $z_4$ as a function of $x_4$.

**Theorem 4.13.** *Let $x_4$ and $z_4$ be defined by* (4.6) *and* (4.7). *Then*

$$z_4 = {}_2F_1\left(\frac{1}{2},\frac{1}{2};1;x_4\right).$$

*Proof.* The general solution of (4.21) is given by

$$z_4 = A\,{}_2F_1\left(\frac{1}{2},\frac{1}{2};1;x_4\right) + B\,{}_2F_1\left(\frac{1}{2},\frac{1}{2};1;1-x_4\right)$$

where $A$ and $B$ are constants. To determine the values of $A$ and $B$, use the fact that as $q\to 0^+$, $x_4\to 0^+$ and $z_4\to 1$. It follows that

$$1 = A + B\lim_{x_4\to 0+} {}_2F_1\left(\frac{1}{2},\frac{1}{2};1;1-x_4\right).$$

Appealing to Lemma 4.12 we deduce that $B = 0$ and therefore $A = 1$. This completes the proof. □

The next goal is to find a formula for the inverse function $q = q_4(x_4)$, which will be given in Theorem 4.15. We first need a technical lemma.

**Lemma 4.14.** *Let $x_4$ and $z_4$ be given by* (4.6) *and* (4.7). *Let*

$$z_4^* = z_4(q)\log q.$$

*Then $z_4^*$ is also a solution of the differential equation* (4.21). *That is,*

$$\frac{\mathrm{d}}{\mathrm{d}x_4}\left(x_4(1-x_4)\frac{\mathrm{d}z_4^*}{\mathrm{d}x_4}\right) = \frac{z_4^*}{4}.$$

*Proof.* Let $u = \log q$. Then

$$\begin{aligned} x_4(1-x_4)\frac{\mathrm{d}z_4^*}{\mathrm{d}x_4} &= x_4(1-x_4)\frac{\mathrm{d}}{\mathrm{d}x_4}(z_4\log q) \\ &= x_4(1-x_4)\log q\frac{\mathrm{d}z_4}{\mathrm{d}x_4} + x_4(1-x_4)\frac{z_4}{q}\frac{\mathrm{d}q}{\mathrm{d}x_4} \\ &= ux_4(1-x_4)\frac{\mathrm{d}z_4}{\mathrm{d}x_4} + \frac{1}{z_4}, \end{aligned} \tag{4.23}$$

where Theorem 4.7 has been used to obtain the last line. Therefore, on applying $\mathrm{d}/\mathrm{d}x_4$ to (4.23) and making use of Theorems 4.7 and 4.11, we deduce that

$$\begin{aligned} &\frac{\mathrm{d}}{\mathrm{d}x_4}\left(x_4(1-x_4)\frac{\mathrm{d}z_4^*}{\mathrm{d}x_4}\right) \\ &= \frac{\mathrm{d}}{\mathrm{d}x_4}\left(ux_4(1-x_4)\frac{\mathrm{d}z_4}{\mathrm{d}x_4} + \frac{1}{z_4}\right) \\ &= u\frac{\mathrm{d}}{\mathrm{d}x_4}\left(x_4(1-x_4)\frac{\mathrm{d}z_4}{\mathrm{d}x_4}\right) + x_4(1-x_4)\frac{\mathrm{d}z_4}{\mathrm{d}x_4}\frac{\mathrm{d}u}{\mathrm{d}x_4} - \frac{1}{z_4^2}\frac{\mathrm{d}z_4}{\mathrm{d}x_4} \\ &= \frac{1}{4}uz_4 + x_4(1-x_4)\frac{\mathrm{d}z_4}{\mathrm{d}x_4}\frac{1}{q}\frac{\mathrm{d}q}{\mathrm{d}x_4} - \frac{1}{z_4^2}\frac{\mathrm{d}z_4}{\mathrm{d}x_4} \\ &= \frac{1}{4}z_4^*. \end{aligned}$$

□

We are now ready to prove the inversion formula.

**Theorem 4.15.** *Suppose* $0 \le q < 1$ *and let* $x_4$ *be defined by* (4.6), *i.e.,*

$$x_4 = 16q\prod_{j=1}^{\infty}\frac{(1+q^{2j})^8}{(1+q^{2j-1})^8}.$$

*Then*

$$q = \exp\left(-\pi\,\frac{{}_2F_1\left(\frac{1}{2},\frac{1}{2};1;1-x_4\right)}{{}_2F_1\left(\frac{1}{2},\frac{1}{2};1;x_4\right)}\right).$$

*Proof.* Since $z_4^*$ is a solution of the differential equation (4.21), it follows that

$$z_4^* = C\,{}_2F_1\left(\frac{1}{2},\frac{1}{2};1;x_4\right) + D\,{}_2F_1\left(\frac{1}{2},\frac{1}{2};1;1-x_4\right)$$

for some constants $C$ and $D$. Since $z_4^* = z_4 \log q$, we may use Theorem 4.13 to deduce that

$$\log q = C + D\frac{{}_2F_1\left(\frac{1}{2}, \frac{1}{2}; 1; 1 - x_4\right)}{{}_2F_1\left(\frac{1}{2}, \frac{1}{2}; 1; x_4\right)}. \tag{4.24}$$

Now let $q \to 1^-$ so that $x_4 \to 1^-$ by Theorem 4.8. It follows from Lemma 4.12 that

$$0 = C + 0,$$

and hence $C = 0$. Now let $q = e^{-\pi}$. By Theorem 4.8 we have $x_4 = 1/2$ and therefore (4.24) simplifies to give $D = -\pi$. Using these values of $C$ and $D$ in (4.24) and exponentiating, we complete the proof. □

It may be noted that Theorem 4.1 has now been proved: the identity (4.2) is true by Theorem 4.15, while (4.3) holds by Theorem 4.13.

We end this section with an evaluation of Ramanujan's theta functions. By analogy with

$$\pi = 2\int_0^1 \frac{\mathrm{d}t}{\sqrt{1-t^2}},$$

let $\tilde{\omega}$ be defined by

$$\tilde{\omega} = 2\int_0^1 \frac{\mathrm{d}t}{\sqrt{1-t^4}}. \tag{4.25}$$

**Corollary 4.16.** *Ramanujan's theta functions have the special values*

$$\varphi^2(e^{-\pi}) = \frac{\sqrt{2}\,\tilde{\omega}}{\pi}, \qquad \varphi^2(-e^{-\pi}) = \frac{\tilde{\omega}}{\pi} \qquad \textit{and} \qquad e^{-\pi/2}\psi^2(e^{-2\pi}) = \frac{\tilde{\omega}}{4\pi}.$$

*Proof.* Let $x_4 = \frac{1}{2}$. Then, by Theorem 4.1 it follows that $q = e^{-\pi}$ and

$$\varphi^2(e^{-\pi}) = {}_2F_1\left(\begin{matrix}\frac{1}{2}, \frac{1}{2} \\ 1\end{matrix}; \frac{1}{2}\right).$$

The hypergeometric function can be expressed as an integral. By Exercise 4, followed by the changes of variables given by $x = \sin\theta$ and $t^2 = x^2/(2-x^2)$, we have

$${}_2F_1\left(\begin{matrix}\frac{1}{2}, \frac{1}{2} \\ 1\end{matrix}; \frac{1}{2}\right) = \frac{2}{\pi}\int_0^{\pi/2} \frac{\mathrm{d}\theta}{\sqrt{1 - \frac{1}{2}\sin^2\theta}}$$

$$= \frac{2}{\pi} \int_0^1 \frac{dx}{\sqrt{(1-x^2)(1-\frac{1}{2}x^2)}}$$

$$= \frac{2\sqrt{2}}{\pi} \int_0^1 \frac{dt}{\sqrt{1-t^4}} = \frac{\sqrt{2}}{\pi} \times \tilde{\omega}. \tag{4.26}$$

It follows that $\varphi^2(e^{-\pi}) = \sqrt{2}\,\tilde{\omega}/\pi$. Next, substituting the values $x_4 = \frac{1}{2}$ and $q = e^{-\pi}$ in (4.8) gives

$$\frac{1}{2} = \frac{16e^{-\pi}\psi^4(e^{-2\pi})}{\varphi^4(e^{-\pi})} \quad \text{and} \quad \frac{1}{2} = \frac{\varphi^4(-e^{-\pi})}{\varphi^4(e^{-\pi})}.$$

The values of $\varphi^2(-e^{-\pi})$ and $e^{-\pi/2}\psi^2(e^{-2\pi})$ can now be deduced from the value of $\varphi^2(e^{-\pi})$. □

## 4.3 Ramanujan's alternative theories: levels 1, 2, and 3

The objective of this section is to carry out an analysis for levels 1, 2, and 3 that is similar to that in the previous section for level 4. Although the main focus is on levels 1, 2,and 3, the results for level 4 will be inserted, where appropriate, for comparison. In the next section we will extract the common essence from all four theories and state a single theorem that encapsulates them all.

We start with some definitions. For $\ell = 1, 2, 3$ or $4$, let $x_\ell$ and $z_\ell$ be defined by

$$x_1 = \frac{1}{2}\left(1 - \frac{R(q)}{Q^{3/2}(q)}\right), \qquad z_1 = Q^{1/4}(q), \tag{4.27}$$

$$x_2 = \frac{64q\psi^8(q)}{\varphi^8(-q) + 64q\psi^8(q)}, \qquad z_2 = \left(\varphi^8(-q) + 64q\psi^8(q)\right)^{1/4}, \tag{4.28}$$

$$x_3 = \frac{c^3(q)}{a^3(q)}, \qquad z_3 = a(q), \tag{4.29}$$

$$x_4 = \frac{16q\psi^4(q^2)}{\varphi^4(q)}, \qquad z_4 = \varphi^2(q). \tag{4.30}$$

The definition of $x_4$ given by (4.30) is equivalent to the previous definition (4.6) by (4.8). From Theorems 3.7 and 3.20, it follows that

$$1 - x_3 = \frac{b^3(q)}{a^3(q)} \tag{4.31}$$

and

$$1 - x_4 = \frac{\varphi^4(-q)}{\varphi^4(q)}. \tag{4.32}$$

The next result gives the connection of each $z_\ell$ with the level $\ell$.

**Theorem 4.17.** *The following identities hold*:

$$z_\ell^2 = \frac{\ell P(q^\ell) - P(q)}{\ell - 1}, \qquad \textit{for } \ell = 2, 3, 4 \tag{4.33}$$

*and*

$$z_\ell^4 = \frac{\ell^2 Q(q^\ell) + Q(q)}{\ell^2 + 1}, \qquad \textit{for } \ell = 1, 2, 3. \tag{4.34}$$

*Proof.* We proceed on a case-by-case basis. By Lemma 3.9,

$$z_2^2 = \left(\varphi^8(-q) + 64q\psi^8(q)\right)^{1/2} = \varphi^4(q) + 16q\psi^4(q^2).$$

Now apply Theorem 3.26 to obtain

$$\begin{aligned} z_2^2 &= \frac{1}{3}\left(4P(q^4) - P(q)\right) + \frac{2}{3}\left(-2P(q^4) + 3P(q^2) - P(q^4)\right) \\ &= 2P(q^2) - P(q). \end{aligned}$$

This proves (4.33) for the case $\ell = 2$. Next, by Theorem 3.31,

$$z_3^2 = a^2(q) = \frac{1}{2}\left(3P(q^3) - P(q)\right),$$

which proves (4.33) in the case $\ell = 3$. The identity (4.33) is true for $\ell = 4$ by (4.10).

The identity (4.34) is true when $\ell = 1$ by the definition of $z_1$. Next, by Theorem 3.39,

$$\begin{aligned} z_2^4 &= \varphi^8(-q) + 64q\psi^8(q) \\ &= \frac{1}{15}\left(-Q(q) + 16Q(q^2)\right) + \frac{4}{15}\left(Q(q) - Q(q^2)\right) \\ &= \frac{4Q(q^2) + Q(q)}{5}. \end{aligned}$$

This proves (4.34) in the case $\ell = 2$. Finally when $\ell = 3$, Theorem 3.43 gives

$$z_3^4 = a^4(q) = \frac{9Q(q^3) + Q(q)}{10}.$$

□

The identity (4.34) does not hold when $\ell = 4$; the corresponding result in this case, as can be seen from Theorem 3.39, is

$$z_4^4 = \frac{1}{15}\left(Q(q) - 2Q(q^2) + 16Q(q^4)\right).$$

We will not require this.

The next result gives a connection between each function $x_\ell$ and its level $\ell$.

**Theorem 4.18.** *The following identities hold*:

$$x_1(1-x_1) = \frac{432q\prod_{j=1}^{\infty}(1-q^j)^{24}}{Q^3(q)} \tag{4.35}$$

*and*

$$\frac{x_\ell}{1-x_\ell} = \ell^{6/(\ell-1)}\, q \prod_{j=1}^{\infty}\left(\frac{1-q^{j\ell}}{1-q^j}\right)^{24/(\ell-1)} \quad \textit{for } \ell = 2, 3 \textit{ or } 4. \tag{4.36}$$

*Proof.* We again proceed on a case-by-case basis. By the definition of $x_1$ and Corollary 0.23,

$$\begin{aligned} x_1(1-x_1) &= \frac{1}{4}\left(1 - \frac{R(q)}{Q^{3/2}(q)}\right)\left(1 + \frac{R(q)}{Q^{3/2}(q)}\right) \\ &= \frac{Q^3(q) - R^2(q)}{4Q^3(q)} \\ &= \frac{432q\prod_{j=1}^{\infty}(1-q^j)^{24}}{Q^3(q)}, \end{aligned}$$

and this proves (4.35). Next, by the definition of $x_2$,

$$\frac{x_2}{1-x_2} = 64q\frac{\psi^8(q)}{\varphi^8(-q)}.$$

Now apply Theorem 3.5 to deduce

$$\frac{x_2}{1-x_2} = 64q\prod_{j=1}^{\infty}\left(\frac{1-q^{2j}}{1-q^j}\right)^{24},$$

which gives the result for $\ell = 2$. Next, by (4.29) and (4.31),

$$\frac{x_3}{1-x_3} = \frac{c^3(q)}{b^3(q)}.$$

Now apply Theorem 3.15 to deduce

$$\frac{x_3}{1-x_3} = 27q\prod_{j=1}^{\infty}\left(\frac{1-q^{3j}}{1-q^j}\right)^{12}.$$

Finally, the identity (4.36) holds when $\ell = 4$ by (4.17). □

All four theories satisfy a common differentiation formula which has been given in the case $\ell = 4$ by Theorem 4.7.

**Theorem 4.19.** *Let* $\ell = 1, 2, 3,$ *or* 4. *The following differentiation formula holds*:

$$q\frac{\mathrm{d}x_\ell}{\mathrm{d}q} = z_\ell^2 x_\ell(1-x_\ell).$$

*Proof.* Consider first the case $\ell = 1$. By (4.35),

$$x_1(1-x_1) = \frac{432q\prod_{j=1}^{\infty}(1-q^j)^{24}}{Q^3(q)}.$$

Take the logarithm and apply the operator $q\frac{\mathrm{d}}{\mathrm{d}q}$ to obtain

$$\left(\frac{1}{x_1} - \frac{1}{1-x_1}\right) q\frac{\mathrm{d}x_1}{\mathrm{d}q} = q\frac{\mathrm{d}}{\mathrm{d}q}\log\left(q\prod_{j=1}^{\infty}(1-q^j)^{24}\right) - \frac{3}{Q(q)}q\frac{\mathrm{d}Q(q)}{\mathrm{d}q}.$$

On using (0.61) and (0.79) this simplifies to

$$\frac{(1-2x_1)}{x_1(1-x_1)}q\frac{\mathrm{d}x_1}{\mathrm{d}q} = P(q) - \frac{P(q)Q(q)-R(q)}{Q(q)} = \frac{R(q)}{Q(q)}.$$

Therefore,

$$q\frac{\mathrm{d}x_1}{\mathrm{d}q} = \frac{R(q)}{Q(q)}\frac{x_1(1-x_1)}{(1-2x_1)} = z_1^2 x_1(1-x_1),$$

where the definitions of $x_1$ and $z_1$ in terms of $Q$ and $R$ have been used to obtain the last step.

Next, if $\ell = 2$, 3, or 4, then by (4.36) we deduce that

$$\frac{x_\ell}{1-x_\ell} = \ell^{6/(\ell-1)}q\prod_{j=1}^{\infty}\left(\frac{1-q^{j\ell}}{1-q^j}\right)^{24/(\ell-1)}.$$

Take the logarithm and apply the differential operator $q\frac{\mathrm{d}}{\mathrm{d}q}$ to obtain

$$\left(\frac{1}{x_\ell}+\frac{1}{1-x_\ell}\right)q\frac{\mathrm{d}x_\ell}{\mathrm{d}q}=\frac{\ell P(q^\ell)-P(q)}{\ell-1}.$$

This simplifies to the required result on applying (4.33). □

The next result lists some basic properties of the functions $x=x_\ell(q)$ on the interval $0\le q\le 1$. Results for the case $\ell=4$ have already been given in Theorem 4.8.

**Theorem 4.20.** *Let $\ell=1, 2, 3$ or $4$. Then*

1. *For any positive real number $t$, we have $x_\ell(e^{-2\pi t/\sqrt{\ell}})+x_\ell(e^{-2\pi/t\sqrt{\ell}})=1$.*
2. *$x_\ell(e^{-2\pi/\sqrt{\ell}})=\frac{1}{2}$.*
3. *As $q$ increases from $0$ to $1$, $x_\ell(q)$ strictly increases from $0$ to $1$.*

*Proof.* Consider first the case $\ell=1$. By Theorems 4.18 and 4.19,

$$\frac{\mathrm{d}x_1}{\mathrm{d}q}=\frac{z_1^2x_1(1-x_1)}{q}=\frac{432z_1^2}{Q^3(q)}\prod_{j=1}^{\infty}(1-q^j)^{24}. \tag{4.37}$$

Since

$$Q(q)=1+240\sum_{j=1}^{\infty}\sum_{k=1}^{\infty}j^3q^{jk},$$

the expression on the far right-hand side of (4.37) is positive for $0<q<1$. It follows that the function $x_1=x_1(q)$ is increasing on this interval. Next, let $q=\exp(-2\pi t)$ and $p=\exp(-2\pi/t)$. By (2.15) and (4.35),

$$\begin{aligned}x_1(q)(1-x_1(q))&=\frac{Q^3(q)-R^2(q)}{4Q^3(q)}\\&=\frac{Q^3(p)-R^2(p)}{4Q^3(p)}\\&=x_1(p)(1-x_1(p)).\end{aligned}$$

This may be rearranged to give

$$(x_1(q)-x_1(p))(x_1(q)+x_1(p)-1)=0,$$

so either $x_1(q)=x_1(p)$ or $x_1(q)+x_1(p)=1$. Since $x_1(q)$ is an increasing function of $q$, it follows that $x_1(e^{-2\pi t})$ and $x_1(e^{-2\pi/t})$ are decreasing and increasing functions of $t$, respectively. Therefore $x_1(q)$ and $x_1(p)$ are not the same function and we must have

$$x_1(q)+x_1(p)=1.$$

This proves Property 1, and Property 2 follows by taking $t = 1$. It remains to finish the proof of Property 3. We have already shown that $x_1(q)$ is an increasing function on the interval $0 < q < 1$. From the definition (4.27) we have $x_1(0) = 0$. Furthermore,

$$\begin{aligned}\lim_{q\to 1^-} x_1(q) &= \lim_{t\to 0+} x_1(e^{-2\pi t})\\ &= \lim_{t\to 0+} 1 - x_1(e^{-2\pi/t})\\ &= \lim_{q\to 0+} 1 - x_1(q)\\ &= 1.\end{aligned}$$

It follows that $x_1$ increases from 0 to 1 as $q$ increases from 0 to 1.

The remaining cases $\ell = 2$, 3, and 4 can be handled simultaneously. Let $q = \exp(-2\pi t/\sqrt{\ell})$ and $p = \exp(-2\pi/t\sqrt{\ell})$. By (4.36) and the modular transformation (2.8) we have

$$\begin{aligned}\frac{x_\ell(q)}{1-x_\ell(q)} &= \ell^{6/(\ell-1)}\left(\frac{\eta(it\sqrt{\ell})}{\eta(it/\sqrt{\ell})}\right)^{24/(\ell-1)}\\ &= \frac{1}{\ell^{6/(\ell-1)}}\left(\frac{\eta(i/t\sqrt{\ell})}{\eta(i\sqrt{\ell}/t)}\right)^{24/(\ell-1)}\\ &= \frac{1-x_\ell(p)}{x_\ell(p)},\end{aligned}$$

which simplifies to $x_\ell(q)+x_\ell(p) = 1$. This proves Property 1, and Property 2 follows on setting $t = 1$. It remains to prove Property 3. On rearranging the identity (4.36), we find that

$$1 - x_\ell(q) = \frac{1}{1+\ell^{6/(\ell-1)}q\prod_{j=1}^{\infty}\left(\frac{1-q^{j\ell}}{1-q^j}\right)^{24/(\ell-1)}}. \tag{4.38}$$

On the interval $0 < q < 1$, $(1-q^{j\ell})/(1-q^j)$ is an increasing function, and it follows from (4.38) that $1 - x_\ell(q)$ is a decreasing function of $q$. Thus $x_\ell(q)$ is an increasing function on the interval. By definition, $x_\ell(0) = 0$, while (4.38) implies $\lim_{q\to 1^-}(1 - x_\ell(q)) = 0$. It follows that $x_\ell$ increases from 0 to 1 as $q$ increases from 0 to 1. This proves Property 3 and completes the proof of Theorem 4.20. □

The results in the next theorem involve eta-products.

**Theorem 4.21.** *Let $\ell = 1, 2, 3$ or $4$ and let $x_\ell$ and $z_\ell$ be given by* (4.27)–(4.30). *In the case $\ell = 1$ we have*

$$z_1^{12}x_1(1-x_1) = 432q\prod_{j=1}^{\infty}(1-q^j)^{24}. \tag{4.39}$$

*For $\ell = 2$, 3, or 4, we have*

$$z_\ell^{12}x_\ell(1-x_\ell)^\ell = \ell^{6/(\ell-1)}q\prod_{j=1}^{\infty}(1-q^j)^{24} \tag{4.40}$$

*and*

$$z_\ell^{12}x_\ell^\ell(1-x_\ell) = \ell^{6\ell/(\ell-1)}q^\ell\prod_{j=1}^{\infty}(1-q^{\ell j})^{24}. \tag{4.41}$$

*In the case $\ell = 4$ there is an additional result*

$$z_4^{12}x_4^2(1-x_4)^2 = 256q^2\prod_{j=1}^{\infty}(1-q^{2j})^{24}.$$

*Proof.* We proceed on a case-by-case basis. The result (4.39) for the level $\ell = 1$ follows immediately from (4.27) and (4.35). For the case $\ell = 2$, apply the definition (4.28) and the infinite products in Theorem 3.5 to deduce that

$$z_2^{12}x_2(1-x_2)^2 = 64q\psi^8(q)\varphi^{16}(-q) = 64q\prod_{j=1}^{\infty}(1-q^j)^{24}$$

and

$$z_2^{12}x_2^2(1-x_2) = 64^2q^2\psi^{16}(q)\varphi^8(-q) = 64^2q^2\prod_{j=1}^{\infty}(1-q^{2j})^{24}.$$

This proves the result for the case $\ell = 2$.

When $\ell = 3$, (4.29), (4.31), and Theorem 3.15 give

$$z_3^{12}x_3(1-x_3)^3 = b^9(q)c^3(q) = 27q\prod_{j=1}^{\infty}(1-q^j)^{24}$$

and

$$z_3^{12}x_3^3(1-x_3) = b^3(q)c^9(q) = 27^3q^3\prod_{j=1}^{\infty}(1-q^{3j})^{24}.$$

This proves the result for the level $\ell = 3$. The results for level $\ell = 4$ were proved in Theorem 4.6. □

The results in the next lemma give the results of differentiation of $z_\ell$.

**Lemma 4.22.** *Let $\ell = 1, 2, 3$ or $4$ and let $x_\ell$ and $z_\ell$ be given by* (4.27)–(4.30). *Then*

$$\frac{12}{z_\ell} q \frac{\mathrm{d}z_\ell}{\mathrm{d}q} + (1 - (\ell+1)x_\ell)\, z_\ell^2 = P(q) \tag{4.42}$$

*and*

$$\frac{12}{z_\ell} q \frac{\mathrm{d}z_\ell}{\mathrm{d}q} + (\ell - (\ell+1)x_\ell)\, z_\ell^2 = \ell P(q^\ell). \tag{4.43}$$

*Proof.* The results in the case $\ell = 4$ were proved in detail in Lemma 4.9. The results for the other values of $\ell$ can be proved by the same procedure: take the logarithm of each identity (4.39)–(4.41), apply the differential operator $q\frac{\mathrm{d}}{\mathrm{d}q}$ and make use of Theorem 4.19. □

We are now ready to prove that $z_\ell$ satisfies a second order linear differential equation with respect to $x_\ell$. It will be convenient to use the following definition.

**Definition 4.23.** For $\ell = 1, 2, 3$ or 4, let $s_\ell$ be defined by

$$s_\ell = \begin{cases} 1/6 & \text{if } \ell = 1, \\ 1/4 & \text{if } \ell = 2, \\ 1/3 & \text{if } \ell = 3, \\ 1/2 & \text{if } \ell = 4. \end{cases}$$

The number $1/s_\ell$ is sometimes called the signature.

**Theorem 4.24.** *Let $\ell = 1, 2, 3,$ or $4$ and let $x_\ell$ and $z_\ell$ be given by* (4.27)–(4.30). *Let $s_\ell$ be as in Definition 4.23. Then*

$$\frac{\mathrm{d}}{\mathrm{d}x_\ell}\left(x_\ell(1-x_\ell)\frac{\mathrm{d}z_\ell}{\mathrm{d}x_\ell}\right) = s_\ell(1-s_\ell)z_\ell. \tag{4.44}$$

*Proof.* The result has already been proved for the level $\ell = 4$ in Theorem 4.11. Therefore we may assume $\ell = 1, 2,$ or 3. By the chain rule and Theorem 4.19,

$$\frac{\mathrm{d}}{\mathrm{d}x_\ell} = q\frac{\mathrm{d}}{\mathrm{d}q} \Big/ q\frac{\mathrm{d}x_\ell}{\mathrm{d}q} = \frac{1}{z_\ell^2 x_\ell(1-x_\ell)} q\frac{\mathrm{d}}{\mathrm{d}q}.$$

Two applications of this gives

$$\frac{\mathrm{d}}{\mathrm{d}x_\ell}\left(x_\ell(1-x_\ell)\frac{\mathrm{d}z_\ell}{\mathrm{d}x_\ell}\right)=\frac{1}{z_\ell^2 x_\ell(1-x_\ell)}q\frac{\mathrm{d}}{\mathrm{d}q}\left(\frac{1}{z_\ell^2}q\frac{\mathrm{d}z_\ell}{\mathrm{d}q}\right). \tag{4.45}$$

Using (4.42) in (4.45), we find that

$$\frac{\mathrm{d}}{\mathrm{d}x_\ell}\left(x_\ell(1-x_\ell)\frac{\mathrm{d}z_\ell}{\mathrm{d}x_\ell}\right)=\frac{1}{12z_\ell^2 x_\ell(1-x_\ell)}q\frac{\mathrm{d}}{\mathrm{d}q}\left(\frac{P(q)}{z_\ell}-(1-(\ell+1)x_\ell)z_\ell\right).$$

The derivative on the right-hand side may be calculated by the standard rules of calculus, using Theorems 0.21, 4.19, and Lemma 4.22 to compute the derivatives of $P(q)$, $x_\ell$, and $z_\ell$, respectively. The result simplifies to

$$\frac{\mathrm{d}}{\mathrm{d}x_\ell}\left(x_\ell(1-x_\ell)\frac{\mathrm{d}z_\ell}{\mathrm{d}x_\ell}\right)=\frac{(\ell+1)z_\ell}{12}+\frac{z_\ell^4(1-(\ell+1)x_\ell)^2-Q(q)}{144z_\ell^3 x_\ell(1-x_\ell)}. \tag{4.46}$$

On the other hand, using (4.43) in (4.45) we obtain

$$\frac{\mathrm{d}}{\mathrm{d}x_\ell}\left(x_\ell(1-x_\ell)\frac{\mathrm{d}z_\ell}{\mathrm{d}x_\ell}\right)=\frac{1}{12z_\ell^2 x_\ell(1-x_\ell)}q\frac{\mathrm{d}}{\mathrm{d}q}\left(\frac{\ell P(q^\ell)}{z_\ell}-(\ell-(\ell+1)x_\ell)z_\ell\right).$$

The derivative on the right-hand side may be calculated using Theorems 0.21, 4.19, and Lemma 4.22, and the result simplifies to

$$\frac{\mathrm{d}}{\mathrm{d}x_\ell}\left(x_\ell(1-x_\ell)\frac{\mathrm{d}z_\ell}{\mathrm{d}x_\ell}\right)=\frac{(\ell+1)z_\ell}{12}+\frac{z_\ell^4(\ell-(\ell+1)x_\ell)^2-\ell^2 Q(q^\ell)}{144z_\ell^3 x_\ell(1-x_\ell)}. \tag{4.47}$$

Adding (4.46) and (4.47) and dividing by 2, we obtain

$$\frac{\mathrm{d}}{\mathrm{d}x_\ell}\left(x_\ell(1-x_\ell)\frac{\mathrm{d}z_\ell}{\mathrm{d}x_\ell}\right)=\frac{(\ell+1)(11-\ell)z_\ell}{144}+\frac{z_\ell^4(1+\ell^2)-Q(q)-\ell^2 Q(q^\ell)}{288z_\ell^3 x_\ell(1-x_\ell)}.$$

With the help of (4.34), this simplifies to

$$\frac{\mathrm{d}}{\mathrm{d}x_\ell}\left(x_\ell(1-x_\ell)\frac{\mathrm{d}z_\ell}{\mathrm{d}x_\ell}\right)=\frac{(\ell+1)(11-\ell)}{144}z_\ell.$$

Comparing this with the statement of the theorem, we complete the proof. □

The next result gives a formula for $z_\ell$ as a function of $x_\ell$.

**Theorem 4.25.** *Let* $\ell = 1, 2, 3$ *or* 4 *and let* $x_\ell$ *and* $z_\ell$ *be given by* (4.27)–(4.30). *Let* $s_\ell$ *be defined by Definition 4.23. Then*

$$z_\ell = {}_2F_1\left(s_\ell, 1-s_\ell; 1; x_\ell\right).$$

*Proof.* The general solution of (4.44) is given by

$$z_\ell = A_\ell\, {}_2F_1\left(s_\ell, 1-s_\ell; 1; x_\ell\right) + B_\ell\, {}_2F_1\left(s_\ell, 1-s_\ell; 1; 1-x_\ell\right)$$

for some constants $A_\ell$ and $B_\ell$. To determine the constants, use the fact that as $q \to 0^+$, $x_\ell \to 0^+$ and $z_\ell \to 1$. Hence

$$1 = A_\ell + B_\ell \lim_{x\to 0+} {}_2F_1\left(s_\ell, 1-s_\ell; 1; 1-x_\ell\right).$$

Appealing to Lemma 4.12 we find that $A_\ell = 1$ and $B_\ell = 0$. This completes the proof. □

The next goal is to find an explicit formula for the inverse function $q = q_\ell(x_\ell)$, which will be given in Theorem 4.27. We first need a technical lemma.

**Lemma 4.26.** *Let $\ell = 1, 2, 3$ or 4 and let $x_\ell$ and $z_\ell$ be given by* (4.27)–(4.30). *Let $s_\ell$ be defined by Definition 4.23. Let*

$$z_\ell^* = z_\ell(q)\log q.$$

*Then $z_\ell^*$ is also a solution of the differential equation* (4.44). *That is,*

$$\frac{\mathrm{d}}{\mathrm{d}x_\ell}\left(x_\ell(1-x_\ell)\frac{\mathrm{d}z_\ell^*}{\mathrm{d}x}\right) = s_\ell(1-s_\ell)z_\ell^*. \tag{4.48}$$

*Proof.* This result has been proved in the case of level $\ell = 4$ in Lemma 4.14. The details for the other levels $\ell = 1, 2$, and 3 are the same. □

We are now ready for the inversion formula.

**Theorem 4.27.** *Let $q$ be a real number in the interval $0 \le q < 1$. For $\ell = 1, 2, 3$ or 4, let $x_\ell$ and $z_\ell$ be given by* (4.27)–(4.30). *Let $s_\ell$ be defined by Definition 4.23.Then*

$$q = \exp\left(\frac{-2\pi}{\sqrt{\ell}}\frac{{}_2F_1\left(s_\ell, 1-s_\ell; 1; 1-x_\ell\right)}{{}_2F_1\left(s_\ell, 1-s_\ell; 1; x_\ell\right)}\right).$$

*Proof.* The proof is nearly identical to the proof of Theorem 4.15. The general solution of the differential equation (4.48) is given by

$$z_\ell^* = C_\ell\, {}_2F_1\left(s_\ell, 1-s_\ell; 1; x_\ell\right) + D_\ell\, {}_2F_1\left(s_\ell, 1-s_\ell; 1; 1-x_\ell\right)$$

for some constants $C_\ell$ and $D_\ell$. Since $z_\ell^* = z_\ell \log q$, we may use Theorem 4.25 to deduce that

$$\log q = C_\ell + D_\ell \frac{{}_2F_1\left(s_\ell, 1-s_\ell; 1; 1-x_\ell\right)}{{}_2F_1\left(s_\ell, 1-s_\ell; 1; x_\ell\right)}. \tag{4.49}$$

As $q \to 1^-$, $x_\ell \to 1^-$ by Theorem 4.20, and so Lemma 4.12 implies

$$0 = C_\ell + 0,$$

therefore $C_\ell = 0$. Now let $q = e^{-2\pi/\sqrt{\ell}}$. By Theorem 4.20, $x_\ell = 1/2$, and therefore $D_\ell = -2\pi/\sqrt{\ell}$. Using these values of $C_\ell$ and $D_\ell$ in (4.49) and exponentiating, we complete the proof. □

Observe that Theorems 4.2, 4.3, and 4.4 are true by the results in Theorems 4.25 and 4.27.

## 4.4 All four theories

We are now in a position to give a general statement that encapsulates all of Theorems 4.1–4.4.

**Theorem 4.28 (All four theories).** *Suppose* $0 \le q < 1$. *Let* $\ell = 1$, 2, 3 *or* 4 *and let* $z_\ell$ *be defined by*

$$z_\ell^2 = \frac{\ell P(q^\ell) - P(q)}{\ell - 1} \quad \text{if } \ell = 2, 3 \text{ or } 4, \tag{4.50}$$

*or alternatively by*

$$z_\ell^4 = \frac{\ell^2 Q(q^\ell) + Q(q)}{\ell^2 + 1} \quad \text{if } \ell = 1, 2 \text{ or } 3 \tag{4.51}$$

*with the condition that* $z_\ell = 1$ *when* $q = 0$. *Observe that the functions* $z_1, \ldots, z_4$ *defined by* (4.50) *or* (4.51) *are the same as the functions in* (4.27)–(4.30), *by Theorem 4.17. Let* $x_\ell$ *be the solution of the first order differential equation*

$$q\frac{\mathrm{d}x_\ell}{\mathrm{d}q} = z_\ell^2 x_\ell (1 - x_\ell)$$

*with initial condition*

$$x_\ell(e^{-2\pi/\sqrt{\ell}}) = \frac{1}{2}.$$

*Let* $s_\ell$ *be defined by*

$$s_\ell = \begin{cases} 1/6 & \text{if } \ell = 1, \\ 1/4 & \text{if } \ell = 2, \\ 1/3 & \text{if } \ell = 3, \\ 1/2 & \text{if } \ell = 4. \end{cases}$$

*Then $x_\ell$ is given by*

$$x_1(1-x_1) = \frac{432q\prod_{j=1}^{\infty}(1-q^j)^{24}}{Q^3(q)}$$

*and*

$$\frac{x_\ell}{1-x_\ell} = \ell^{6/(\ell-1)}\, q \prod_{j=1}^{\infty}\left(\frac{1-q^{j\ell}}{1-q^j}\right)^{24/(\ell-1)}$$

*for $\ell = 2, 3$ or $4$. Moreover,*

$$z_\ell = {}_2F_1\left(s_\ell, 1-s_\ell; 1; x_\ell\right)$$

*and*

$$q = \exp\left(\frac{-2\pi}{\sqrt{\ell}} \frac{{}_2F_1\left(s_\ell, 1-s_\ell; 1; 1-x_\ell\right)}{{}_2F_1\left(s_\ell, 1-s_\ell; 1; x_\ell\right)}\right).$$

*Proof.* This is an assemblage of results from Theorems 4.17, 4.18, 4.19, 4.25, and 4.27. □

The next theorem provides another way of writing all four theories in a common form. It involves squaring the ${}_2F_1$ hypergeometric series in the previous theorem and is motivated by Ramanujan's identity [248, (25)] and by Ramanujan's series for $1/\pi$ in [248, Sections 13 and 14]. It also serves as a prototype for extensions to other levels that will be studied in subsequent chapters. The ${}_3F_2$ hypergeometric function is defined for $|x| < 1$ by

$${}_3F_2\left(\begin{matrix} a,\ b,\ c \\ d,\ e \end{matrix}; x\right) = \sum_{n=0}^{\infty} \frac{(a)_n(b)_n(c)_n}{(d)_n(e)_n n!} x^n.$$

We will need the following result, which expresses the square of a certain ${}_2F_1$ function in terms of ${}_3F_2$ functions.

**Lemma 4.29.** *The following identities hold:*

$$\left\{{}_2F_1\left(\begin{matrix} a,\ b \\ a+b+\frac{1}{2} \end{matrix}; x\right)\right\}^2 = {}_3F_2\left(\begin{matrix} 2a,\ 2b,\ a+b \\ 2a+2b,\ a+b+\frac{1}{2} \end{matrix}; x\right), \tag{4.52}$$

*and*

$${}_2F_1\left(\begin{matrix} 2a,\ 2b \\ a+b+\frac{1}{2} \end{matrix}; x\right) = {}_2F_1\left(\begin{matrix} a,\ b \\ a+b+\frac{1}{2} \end{matrix}; 4x(1-x)\right). \tag{4.53}$$

*Hence,*

$$\left\{{}_2F_1\left(\begin{matrix}2a,\ 2b\\ a+b+\frac{1}{2}\end{matrix};x\right)\right\}^2 = {}_3F_2\left(\begin{matrix}2a,\ 2b,\ a+b\\ 2a+2b,\ a+b+\frac{1}{2}\end{matrix};4x(1-x)\right). \tag{4.54}$$

*Proof.* See Exercise 8 at the end of the chapter. □

The identity in (4.52) is called Clausen's identity. The formula that we shall utilize is (4.54), which is the result of combining (4.52) and (4.53).

**Theorem 4.30.** *For $\ell = 1, 2, 3$ or $4$, let $Z_\ell$ and $X_\ell$ be defined by*

$$Z_\ell = \begin{cases} Q^{1/2}(q), & \text{if } \ell = 1, \\ \dfrac{\ell P(q^\ell) - P(q)}{\ell - 1} & \text{if } \ell = 2, 3 \text{ or } 4 \end{cases} \tag{4.55}$$

*and*

$$X_\ell = \begin{cases} \left(\dfrac{\eta_1^2\eta_\ell^2}{Z_\ell}\right)^{12/(\ell+1)} & \text{if } \ell = 1, 2 \text{ or } 3, \\ \left(\dfrac{\eta_1^4\eta_4^4}{\eta_2^4 Z_4}\right)^2 & \text{if } \ell = 4. \end{cases} \tag{4.56}$$

*Then each $Z_\ell$ can be expressed in terms of theta functions by the formulas:*

$$Z_1^2 = \frac{1}{2}\left(\left(\sum_{j=-\infty}^{\infty} q^{j^2/2}\right)^8 + \left(\sum_{j=-\infty}^{\infty} (-1)^j q^{j^2/2}\right)^8 + \left(\sum_{j=-\infty}^{\infty} q^{(j+1/2)^2/2}\right)^8\right),$$

$$Z_2 = \frac{1}{2}\left(\left(\sum_{j=-\infty}^{\infty} q^{j^2/2}\right)^4 + \left(\sum_{j=-\infty}^{\infty} (-1)^j q^{j^2/2}\right)^4\right),$$

$$Z_3 = \left(\sum_{j=-\infty}^{\infty}\sum_{k=-\infty}^{\infty} q^{j^2+jk+k^2}\right)^2,$$

$$Z_4 = \left(\sum_{j=-\infty}^{\infty} q^{j^2}\right)^4.$$

*And each $Z_\ell$ may be expressed as functions of $X_\ell$ by the power series:*

$$Z_1 = {}_3F_2\left({\frac{1}{6}, \frac{1}{2}, \frac{5}{6} \atop 1,\ 1}; 1728X_1\right) = \sum_{j=0}^{\infty}\binom{6j}{3j}\binom{3j}{j}\binom{2j}{j}X_1^j,$$

$$Z_2 = {}_3F_2\left({\frac{1}{4}, \frac{1}{2}, \frac{3}{4} \atop 1,\ 1}; 256X_2\right) = \sum_{j=0}^{\infty}\binom{4j}{2j}\binom{2j}{j}^2 X_2^j,$$

$$Z_3 = {}_3F_2\left({\frac{1}{3}, \frac{1}{2}, \frac{2}{3} \atop 1,\ 1}; 108X_3\right) = \sum_{j=0}^{\infty}\binom{3j}{j}\binom{2j}{j}^2 X_3^j,$$

$$Z_4 = {}_3F_2\left({\frac{1}{2}, \frac{1}{2}, \frac{1}{2} \atop 1,\ 1}; 64X_4\right) = \sum_{j=0}^{\infty}\binom{2j}{j}^3 X_4^j.$$

*Moreover, the functions $Z_\ell$ and $X_\ell$ defined by* (4.55) *and* (4.56) *are related to the functions $z_\ell$ and $x_\ell$ in Theorem 4.28 by the formulas*

$$Z_\ell = z_\ell^2 \quad \textit{for } \ell = 1, 2, 3, 4 \tag{4.57}$$

*and*

$$X_1 = \frac{x_1(1-x_1)}{432}, \quad X_2 = \frac{x_2(1-x_2)}{64}, \quad X_3 = \frac{x_3(1-x_3)}{27} \tag{4.58}$$

*and*

$$X_4 = \frac{x_4(1-x_4)}{16}. \tag{4.59}$$

*In particular,*

$$X_1 = \frac{1}{j(\tau)}$$

*and*

$$X_4 = q\prod_{j=1}^{\infty}\frac{(1-q^j)^{24}(1-q^{4j})^{24}}{(1-q^{2j})^{48}} \tag{4.60}$$

*where $j(\tau)$ is the modular $j$-invariant given by* (2.22).

*Proof.* First we will prove the formulas for $Z_1$, $Z_2$, $Z_3$, and $Z_4$ in terms of theta functions. The result for $Z_1^2$ in terms of theta functions is immediate from Corollary 3.41. Next, by the definition of $Z_2$ and Theorem 3.26,

$$Z_2 = 2P(q^2) - P(q) = \frac{1}{2}\left(\varphi^4(q^{1/2}) + \varphi^4(-q^{1/2})\right).$$

Similarly, by the definition of $Z_3$ and Theorem 3.31,

$$Z_3 = \frac{1}{2}\left(3P(q^3) - P(q)\right) = a^2(q).$$

And, by the definition of $Z_4$ and Theorem 3.26 again,

$$Z_4 = \frac{1}{3}\left(4P(q^4) - P(q)\right) = \varphi^4(q).$$

The required results for $Z_2$, $Z_3$, and $Z_4$ follow from the above by using the definitions of $\varphi(q)$ and $a(q)$ in Definitions 3.1 and 3.2.

Before proving the formulas that express $Z_\ell$ as power series in $X_\ell$, we first prove the identities (4.57)–(4.59) that relate $Z_\ell$ and $X_\ell$ to the functions $z_\ell$ and $x_\ell$. Clearly, from the definitions (4.50), (4.51), and (4.55) we have

$$Z_\ell = z_\ell^2 \quad \text{for } \ell = 1, 2, 3, 4$$

and this is (4.57). Next, we prove (4.58) and (4.59). The proof is on a case-by-case basis. By (4.27), (4.35), and (4.57),

$$x_1(1-x_1) = 432\frac{\eta_1^{24}}{Q^3(q)} = 432\frac{\eta_1^{24}}{z_1^{12}} = 432\,\frac{\eta_1^{24}}{Z_1^6} = 432X_1.$$

By (4.28), the infinite product formulas for theta functions in Theorem 3.5, and (4.57),

$$x_2(1-x_2) = \frac{64q\psi^8(q)\varphi^8(-q)}{(\varphi^8(-q) + 64q\psi^8(q))^2} = 64\frac{\eta_1^8\eta_2^8}{z_2^8} = 64\left(\frac{\eta_1^2\eta_2^2}{Z_2}\right)^4 = 64X_2.$$

By (4.29), (4.31), the infinite product formulas for cubic theta functions in Theorem 3.15 and (4.57),

$$x_3(1-x_3) = \frac{b^3(q)c^3(q)}{a^6(q)} = 27\frac{\eta_1^6\eta_3^6}{z_3^6} = 27\left(\frac{\eta_1^2\eta_3^2}{Z_3}\right)^3 = 27X_3.$$

By (4.30), (4.32), the infinite product formulas for theta functions in Theorem 3.5 (or, simply by Lemma 4.6),

$$x_4(1-x_4) = \frac{16q\psi^4(q^2)\varphi^4(-q)}{\varphi^8(q)} = 16\frac{\eta_1^8\eta_4^8}{\eta_2^8 z_4^4} = 16\left(\frac{\eta_1^4\eta_4^4}{\eta_2^4 Z_4}\right)^2 = 16X_4.$$

The expansions of $Z_\ell$ in powers of $X_\ell$ can now be obtained from the corresponding expansions of $z_\ell$ in powers of $x_\ell$ in Theorem 4.28 by squaring and applying Clausen's identity. Here are the details for $Z_1$. By Theorem 4.28, (4.52), and (4.57),

$$Z_1 = z_1^2 = \left\{ {}_2F_1\left( \begin{matrix} \frac{1}{6}, \frac{5}{6} \\ 1 \end{matrix}; x_1 \right) \right\}^2 = {}_3F_2\left( \begin{matrix} \frac{1}{6}, \frac{1}{2}, \frac{5}{6} \\ 1, 1 \end{matrix}; 4x_1(1-x_1) \right).$$

By (4.58) this is

$$\begin{aligned} Z_1 &= {}_3F_2\left( \begin{matrix} \frac{1}{6}, \frac{1}{2}, \frac{5}{6} \\ 1, 1 \end{matrix}; 1728X_1 \right) \\ &= \sum_{n=0}^{\infty} \frac{(\frac{1}{6})_n(\frac{5}{6})_n(\frac{1}{2})_n}{n!^3}(1728X_1)^n \\ &= \sum_{n=0}^{\infty} \frac{(\frac{1}{6})_n(\frac{2}{6})_n(\frac{3}{6})_n(\frac{4}{6})_n(\frac{5}{6})_n(1)_n}{(\frac{1}{3})_n(\frac{2}{3})_n(1)_n} \times \frac{1}{n!^3} \times (1728X_1)^n \\ &= \sum_{n=0}^{\infty} \frac{(6n)!/6^{6n}}{(3n)!/3^{3n}} \times \frac{1}{n!^3} \times (1728X_1)^n \\ &= \sum_{n=0}^{\infty} \binom{6n}{3n}\binom{3n}{n}\binom{2n}{n} X_1^n. \end{aligned}$$

The analogous results for $Z_2$, $Z_3$, and $Z_4$ are obtained in a similar way.

Lastly, by the definitions (4.55) and (4.56),

$$X_1 = \frac{\eta^{24}(\tau)}{Q^3(q)} = \frac{1}{j(\tau)}.$$

And, by (4.11), (4.12) and (4.59),

$$X_4 = \frac{x_4(1-x_4)}{16} = q\prod_{j=1}^{\infty} \frac{(1-q^j)^{24}(1-q^{4j})^{24}}{(1-q^{2j})^{48}}.$$

□

A further property common to all four theories is outlined in Exercise 9 at the end of the chapter.

## 4.5 Parameterizations

We will now show that the Eisenstein series $Q$ and $R$ can be expressed as polynomials in $x_\ell$ and $z_\ell$, for $\ell = 1, 2, 3$ or 4. In addition, the modular $j$-invariant can be expressed as a rational function of any of the $x_\ell$.

**Theorem 4.31.** *Let $Q$ and $R$ be Ramanujan's Eisenstein series defined by* (0.72) *and* (0.73). *For $\ell = 1, 2, 3$, or 4, let $x_\ell$ and $z_\ell$ be given by* (4.27)–(4.30). *Then, the following parameterizations hold:*

*Level 1:*

$$Q(q) = z_1^4, \qquad R(q) = z_1^6(1 - 2x_1).$$

*Level 2:*

$$Q(q) = z_2^4(1 + 3x_2), \qquad R(q) = z_2^6(1 - 9x_2),$$

$$Q(q^2) = z_2^4(1 - \frac{3}{4}x_2), \qquad R(q^2) = z_2^6(1 - \frac{9}{8}x_2).$$

*Level 3:*

$$Q(q) = z_3^4(1 + 8x_3), \qquad R(q) = z_3^6(1 - 20x_3 - 8x_3^2),$$

$$Q(q^3) = z_3^4(1 - \frac{8}{9}x_3), \qquad R(q^3) = z_3^6(1 - \frac{4}{3}x_3 + \frac{8}{27}x_3^2).$$

*Level 4:*

$$Q(q) = z_4^4(1 + 14x_4 + x_4^2), \qquad R(q) = z_4^6(1 + x_4)(1 - 34x_4 + x_4^2),$$

$$Q(q^2) = z_4^4(1 - x_4 + x_4^2), \qquad R(q^2) = z_4^6(1 + x_4)(1 - 2x_4)(1 - \frac{1}{2}x_4),$$

$$Q(q^4) = z_4^4(1 - x_4 + \frac{1}{16}x_4^2), \qquad R(q^4) = z_4^6(1 - \frac{1}{2}x_4)(1 - x_4 - \frac{1}{32}x_4^2).$$

*Proof.* Let us first prove the results that involve the Eisenstein series $Q$. The results for levels 1 and 4 follow directly from (4.27) and Lemma 4.10, respectively. It remains to prove the results for levels 2 and 3. Combining (4.44) and (4.46), we obtain

$$s_\ell(1-s_\ell)z_\ell = \frac{(\ell+1)z_\ell}{12} + \frac{(1-(\ell+1)x_\ell)^2 z_\ell}{144x_\ell(1-x_\ell)} - \frac{Q(q)}{144z_\ell^3 x_\ell(1-x_\ell)},$$

where the value of $s_\ell$ is given by (4.23). Rearranging, we obtain

$$Q(q) = z_\ell^4\left\{12(\ell+1-12s_\ell(1-s_\ell))x_\ell(1-x_\ell) + (1-(\ell+1)x_\ell)^2\right\}. \tag{4.61}$$

Similarly, from (4.44) and (4.47) we find that

$$Q(q^\ell) = \frac{z_\ell^4}{\ell^2}\left\{12(\ell+1-12s_\ell(1-s_\ell))x_\ell(1-x_\ell) + (\ell-(\ell+1)x_\ell)^2\right\}. \tag{4.62}$$

The remaining results that involve $Q(q)$ and $Q(q^\ell)$ now follow by taking $\ell = 2$ and $\ell = 3$ in (4.61) and (4.62).

Let us now prove the results for the Eisenstein series $R$. The result for level $\ell = 1$ follows directly from (4.27). Now suppose $\ell = 2$. By Corollary 0.23, Theorem 4.21, and the results for $Q(q)$ and $Q(q^2)$ just proved,

$$\begin{aligned} R^2(q) &= Q^3(q) - 1728q\prod_{j=1}^{\infty}(1-q^j)^{24} \\ &= z_2^{12}(1+3x_2)^3 - 27z_2^{12}x_2(1-x_2)^2 \\ &= z_2^{12}(1-9x_2)^2 \end{aligned}$$

and

$$\begin{aligned} R^2(q^2) &= Q^3(q^2) - 1728q^2\prod_{j=1}^{\infty}(1-q^{2j})^{24} \\ &= z_2^{12}(1-\frac{3}{4}x_2)^3 - \frac{27}{64}z_2^{12}x_2^2(1-x_2) \\ &= z_2^{12}(1-\frac{9}{8}x_2)^2. \end{aligned}$$

Taking square roots, and comparing the values of both sides when $q = 0$ to determine the sign, we find that

$$R(q) = z_2^6(1-9x_2) \quad \text{and} \quad R(q^2) = z_2^6(1-\frac{9}{8}x_2).$$

The results for $\ell = 3$ and $\ell = 4$ may be obtained in a similar way. □

The analogous results for the Eisenstein series $P$ were given in Lemmas 4.9 and 4.22; those results involve the derivative $q\frac{dz_\ell}{dq}$ as well as $z_\ell$ and $x_\ell$.

Recall the modular $j$-invariant was defined by (2.21), that is,

$$j(\tau) = \frac{1728\,Q^3(q)}{Q^3(q) - R^2(q)},$$

where $q = \exp(2\pi i\tau)$. The next result records parameterizations of the modular $j$-invariant in terms of $x_1$, $x_2$, $x_3$ and $x_4$.

**Theorem 4.32.** *Let* $\ell = 1, 2, 3$ *or* $4$ *and let* $x_\ell$ *be given by* (4.27)–(4.30). *The following parameterizations hold*:

*Level 1*:

$$j(\tau) = \frac{432}{x_1(1-x_1)}.$$

*Level 2*:

$$j(\tau) = \frac{2^6(1+3x_2)^3}{x_2(1-x_2)^2}, \qquad j(2\tau) = \frac{2^{12}(1-\frac{3}{4}x_2)^3}{x_2^2(1-x_2)}.$$

*Level 3*:

$$j(\tau) = \frac{3^3(1+8x_3)^3}{x_3(1-x_3)^3}, \qquad j(3\tau) = \frac{3^9(1-\frac{8}{9}x_3)^3}{x_3^3(1-x_3)}.$$

*Level 4*:

$$j(\tau) = \frac{2^4(1+14x_4+x_4^2)^3}{x_4(1-x_4)^4}, \qquad j(2\tau) = \frac{2^8(1-x_4+x_4^2)^3}{x_4^2(1-x_4)^2}$$

*and*

$$j(4\tau) = \frac{2^{16}(1-x_4+\frac{1}{16}x_4^2)^3}{x_4^4(1-x_4)}.$$

*Proof.* By the definition of the modular $j$-invariant and Corollary 0.23,

$$j(\tau) = \frac{Q^3(q)}{q \prod_{j=1}^{\infty} (1 - q^j)^{24}}.$$

The claimed results are now immediate from Theorems 4.21 and 4.31. □

## 4.6 The arithmetic-geometric mean

Let $\alpha_0$ and $\beta_0$ be nonnegative real numbers and define sequences $\{\alpha_n\}$ and $\{\beta_n\}$ by

$$\alpha_{n+1} = \frac{\alpha_n + \beta_n}{2}, \qquad \beta_{n+1} = \sqrt{\alpha_n \beta_n}. \tag{4.63}$$

We will show that the sequences $\{\alpha_n\}$ and $\{\beta_n\}$ have a common limit as $n \to \infty$, called the arithmetic-geometric mean of $\alpha_0$ and $\beta_0$, and identify the value of the limit.

The arithmetic-geometric mean inequality

$$\frac{\alpha + \beta}{2} \geq \sqrt{\alpha\beta}, \qquad \text{where } \alpha,\ \beta \geq 0,$$

implies

$$\alpha_n \geq \beta_n \quad \text{for all } n \geq 1.$$

We deliberately do not assume anything about the relative values of $\alpha_0$ and $\beta_0$. It follows that for $n \geq 1$,

$$\alpha_{n+1} = \frac{\alpha_n + \beta_n}{2} \leq \alpha_n$$

and

$$\beta_{n+1} = \sqrt{\alpha_n \beta_n} \geq \beta_n.$$

Therefore,

$$\beta_n \leq \beta_{n+1} \leq \alpha_{n+1} \leq \alpha_n \qquad \text{for } n \geq 1.$$

It follows that $\{\alpha_n\}_{n \geq 1}$ is a nonincreasing sequence that is bounded below by any element $\beta_n$, $n \geq 1$. Hence, the sequence $\{\alpha_n\}$ converges. The sequence $\{\beta_n\}$ converges for a similar reason.

Next, for $n \geq 1$ we have

$$0 \leq \alpha_{n+1} - \beta_{n+1} \leq \alpha_{n+1} - \beta_n = \frac{\alpha_n - \beta_n}{2}. \tag{4.64}$$

It follows that $\{\alpha_n\}$ and $\{\beta_n\}$ converge to a common limit, called the arithmetic-geometric mean (AGM) of $\alpha_0$ and $\beta_0$ and denoted by $M(\alpha_0, \beta_0)$. Thus,

$$M(\alpha_0, \beta_0) = \lim_{n\to\infty} \alpha_n = \lim_{n\to\infty} \beta_n. \tag{4.65}$$

The convergence is much more rapid than is suggested by (4.64). For,

$$\alpha_{n+1}^2 - \beta_{n+1}^2 = \left(\frac{\alpha_n + \beta_n}{2}\right)^2 - \alpha_n\beta_n = \left(\frac{\alpha_n - \beta_n}{2}\right)^2.$$

If we assume $\alpha_0\beta_0 \neq 0$ then

$$\begin{aligned}\alpha_{n+1} - \beta_{n+1} &= \frac{1}{4(\alpha_{n+1} + \beta_{n+1})}(\alpha_n - \beta_n)^2 \\ &= \frac{1}{8\alpha_{n+2}}(\alpha_n - \beta_n)^2 \\ &\leq \frac{1}{8M(\alpha_0, \beta_0)}(\alpha_n - \beta_n)^2.\end{aligned}$$

It follows that $\{\alpha_n\}$ and $\{\beta_n\}$ converge quadratically to $M(\alpha_0, \beta_0)$.

Clearly,

$$M(\alpha_0, \beta_0) = M(\beta_0, \alpha_0), \tag{4.66}$$

$$M(\alpha_0, \beta_0) = 0 \quad \text{if } \alpha_0 = 0 \text{ or } \beta_0 = 0 \tag{4.67}$$

and

$$M(\lambda\alpha_0, \lambda\beta_0) = \lambda M(\alpha_0, \beta_0) \quad \text{for any } \lambda \geq 0. \tag{4.68}$$

By (4.66)–(4.68), it suffices to find the value of $M(1, x)$ for $0 < x < 1$.

The next lemma provides a step to identifying $M(\alpha_0, \beta_0)$.

**Lemma 4.33.** *Suppose $0 \leq q < 1$ and consider the arithmetic-geometric mean iteration that starts with*

$$\alpha_0 = \varphi^2(q) \qquad \text{and} \qquad \beta_0 = \varphi^2(-q).$$

*Then*

$$\alpha_n = \varphi^2(q^{2^n}) \qquad \text{and} \qquad \beta_n = \varphi^2(-q^{2^n}).$$

*Hence,*

$$M(\alpha_0, \beta_0) = M(\varphi^2(q), \varphi^2(-q)) = 1.$$

*Proof.* By Lemmas 3.3 and 3.6,

$$\alpha_1 = \frac{\alpha_0 + \beta_0}{2} = \frac{\varphi^2(q) + \varphi^2(-q)}{2} = \varphi^2(q^2)$$

and

$$\beta_1 = \sqrt{\alpha_0\beta_0} = \sqrt{\varphi^2(q)\varphi^2(-q)} = \varphi^2(-q^2).$$

By induction it follows that

$$\alpha_n = \varphi^2(q^{2^n}) \qquad \text{and} \qquad \beta_n = \varphi^2(-q^{2^n}).$$

Taking the limit as $n \to \infty$ gives

$$M(\varphi^2(q), \varphi^2(-q)) = \lim_{n\to\infty} \alpha_n = \lim_{n\to\infty} \varphi^2(q^{2^n}) = \varphi^2(0) = 1.$$

□

The next result expresses the value of $M(1,x)$ in terms of the hypergeometric function.

**Theorem 4.34.** *Suppose* $0 < x < 1$. *Then*

$$M(1,x) = \frac{1}{{}_2F_1\left(\frac{1}{2}, \frac{1}{2}; 1; 1-x^2\right)}.$$

*Proof.* By (4.68) and Lemma 4.33,

$$M\left(1, \frac{\varphi^2(-q)}{\varphi^2(q)}\right) = \frac{1}{\varphi^2(q)} M(\varphi^2(q), \varphi^2(-q)) = \frac{1}{\varphi^2(q)}. \tag{4.69}$$

Also, by (4.32),

$$\frac{\varphi^2(-q)}{\varphi^2(q)} = \sqrt{1-x_4},$$

and by (4.30) and Theorem 4.25 we have

$$\varphi^2(q) = z_4 = {}_2F_1(\tfrac{1}{2}, \tfrac{1}{2}; 1; x_4).$$

Using these in (4.69), we find that

$$M(1, \sqrt{1-x_4}) = \frac{1}{{}_2F_1(\frac{1}{2}, \frac{1}{2}; 1; x_4)}.$$

Now replace $x_4$ with $1-x^2$ to complete the proof. □

**Corollary 4.35 (Landen's transformation).** *The following transformation formulas hold*:

$$ {}_2F_1\left(\begin{matrix}\frac{1}{2}, \frac{1}{2}\\ 1\end{matrix}; 1-\left(\frac{1-x}{1+x}\right)^2\right) = (1+x)\,{}_2F_1\left(\begin{matrix}\frac{1}{2}, \frac{1}{2}\\ 1\end{matrix}; x^2\right) $$

*and*

$$ {}_2F_1\left(\begin{matrix}\frac{1}{2}, \frac{1}{2}\\ 1\end{matrix}; \left(\frac{1-x}{1+x}\right)^2\right) = \frac{1+x}{2}\,{}_2F_1\left(\begin{matrix}\frac{1}{2}, \frac{1}{2}\\ 1\end{matrix}; 1-x^2\right). $$

*Proof.* Consider the arithmetic-geometric mean iteration starting with $\alpha_0 = 1+x$ and $\beta_0 = 1-x$. Then

$$ \alpha_1 = \frac{\alpha_0+\beta_0}{2} = 1 \quad \text{and} \quad \beta_1 = \sqrt{\alpha_0\beta_0} = \sqrt{1-x^2}. $$

Since $M(\alpha_0, \beta_0) = M(\alpha_1, \beta_1)$, this gives

$$ M(1+x, 1-x) = M(1, \sqrt{1-x^2}). $$

On the other hand, by (4.68)

$$ M(1+x, 1-x) = (1+x)M\left(1, \frac{1-x}{1+x}\right). $$

It follows that

$$ M(1, \sqrt{1-x^2}) = (1+x)M\left(1, \frac{1-x}{1+x}\right). $$

Now apply Theorem 4.34 to obtain

$$ \frac{1}{{}_2F_1(\frac{1}{2}, \frac{1}{2}; 1; x^2)} = \frac{1+x}{{}_2F_1(\frac{1}{2}, \frac{1}{2}; 1; 1-(\frac{1-x}{1+x})^2)} $$

and this implies the first result. The second result may be obtained from the first by replacing $x$ with $(1-x)/(1+x)$. □

*Example 4.36 (Gauss).*

$$ \frac{1}{M(1, \sqrt{2})} = \frac{2}{\pi}\int_0^1 \frac{\mathrm{d}t}{\sqrt{1-t^4}}. $$

*Proof.* By the homogeneous property (4.68) and Theorem 4.34,

$$\frac{1}{M(1,\sqrt{2})} = \frac{1}{\sqrt{2}M(1,1/\sqrt{2})} = \frac{1}{\sqrt{2}}\,{}_2F_1\left(\begin{matrix}\frac{1}{2},\frac{1}{2}\\1\end{matrix};\frac{1}{2}\right).$$

On the other hand, from (4.26),

$${}_2F_1\left(\begin{matrix}\frac{1}{2},\frac{1}{2}\\1\end{matrix};\frac{1}{2}\right) = \frac{2\sqrt{2}}{\pi}\int_0^1 \frac{\mathrm{d}t}{\sqrt{1-t^4}}$$

and the result follows. □

## 4.7 Other iterative means

In view of the role played by theta functions in Lemma 4.33, it is natural to look through the theta function identities in Chapter 3 to see if there are other iterative means that can be identified. There are three, one for each of the levels $\ell = 2$, 3, and 4.

Let us begin with the analogous cubic mean. By Lemma 3.11,

$$\frac{a(q)+2b(q)}{3} = a(q^3) \tag{4.70}$$

and

$$\frac{a(q)-b(q)}{3} = c(q^3). \tag{4.71}$$

An expression for $b(q^3)$ in terms of $a(q)$ and $b(q)$ can be obtained using the cubic identity

$$a^3(q) = b^3(q) + c^3(q) \tag{4.72}$$

from Theorem 3.20. This gives, using (4.70)–(4.72),

$$\begin{aligned} b^3(q^3) &= a^3(q^3) - c^3(q^3) \\ &= \left(\frac{a(q)+2b(q)}{3}\right)^3 - \left(\frac{a(q)-b(q)}{3}\right)^3 \\ &= \frac{a^2(q)+a(q)b(q)+b^2(q)}{3} \times b(q). \end{aligned} \tag{4.73}$$

The identities (4.70) and (4.73) suggest considering the iteration

$$\alpha_{n+1} = \frac{\alpha_n + 2\beta_n}{3}, \qquad \beta_{n+1} = \sqrt[3]{\frac{\alpha_n^2 + \alpha_n\beta_n + \beta_n^2}{3} \times \beta_n}.$$

This can be rephrased in another way by letting $\gamma_n$ be defined by

$$\alpha_n^3 = \beta_n^3 + \gamma_n^3$$

and considering the iteration

$$\alpha_{n+1} = \frac{\alpha_n + 2\beta_n}{3}, \qquad \gamma_{n+1} = \frac{\alpha_n - \beta_n}{3}, \qquad \beta_{n+1} = \left(\alpha_{n+1}^3 - \gamma_{n+1}^3\right)^{1/3}. \tag{4.74}$$

It turns out that we will be able to analyze this iteration, as well as two others that are defined as follows. In all cases it is assumed that the initial values $\alpha_0$ and $\beta_0$ are nonnegative real numbers. Here are the three iterations; the iteration in (4.74) is restated below as the mean $M_3$.

**Mean $M_2$:** Let

$$\alpha_{n+1} = \frac{\alpha_n + 3\beta_n}{4}, \qquad \gamma_{n+1} = \frac{\alpha_n - \beta_n}{4}, \qquad \beta_{n+1} = \left(\alpha_{n+1}^2 - \gamma_{n+1}^2\right)^{1/2}$$

and define

$$M_2(\alpha_0, \beta_0) = \lim_{n\to\infty} \alpha_n.$$

**Mean $M_3$:** Let

$$\alpha_{n+1} = \frac{\alpha_n + 2\beta_n}{3}, \qquad \gamma_{n+1} = \frac{\alpha_n - \beta_n}{3}, \qquad \beta_{n+1} = \left(\alpha_{n+1}^3 - \gamma_{n+1}^3\right)^{1/3}$$

and define

$$M_3(\alpha_0, \beta_0) = \lim_{n\to\infty} \alpha_n.$$

**Mean $M_4$:** Let

$$\alpha_{n+1} = \frac{\alpha_n + \beta_n}{2}, \qquad \gamma_{n+1} = \frac{\alpha_n - \beta_n}{2}, \qquad \beta_{n+1} = \left(\alpha_{n+1}^4 - \gamma_{n+1}^4\right)^{1/4}$$

and define

$$M_4(\alpha_0, \beta_0) = \lim_{n\to\infty} \alpha_n.$$

The arithmetic-geometric mean defined by (4.63) and (4.65) can be formulated in a similar way.

**Arithmetic geometric mean $M$:** Let

$$\alpha_{n+1} = \frac{\alpha_n + \beta_n}{2}, \qquad \gamma_{n+1} = \frac{\alpha_n - \beta_n}{2}, \qquad \beta_{n+1} = \left(\alpha_{n+1}^2 - \gamma_{n+1}^2\right)^{1/2}$$

and define

$$M(a_0, b_0) = \lim_{n\to\infty} \alpha_n.$$

The iterative mean $M$ was analyzed in the previous section; we shall now analyze $M_2, M_3$, and $M_4$.

The means $M$ and $M_4$ are symmetric, that is,

$$M(\alpha_0, \beta_0) = M(\beta_0, \alpha_0) \quad \text{and} \quad M_4(\alpha_0, \beta_0) = M_4(\beta_0, \alpha_0)$$

for all $\alpha_0, \beta_0 \geq 0$. The means $M_2$ and $M_3$ are not symmetric. We shall initially restrict the study to $0 \leq \beta_0 \leq \alpha_0$.

Convergence of each of the iterations may be established as follows. It is easy to check for each mean $M_2, M_3$, and $M_4$ that

$$\beta_n \leq \beta_{n+1} \leq \alpha_{n+1} \leq \alpha_n \qquad \text{for } n \geq 1. \tag{4.75}$$

It follows that $\{\alpha_n\}_{n\geq 1}$ is a nonincreasing sequence that is bounded below by any element $\beta_n$. Hence, the sequence $\{\alpha_n\}$ converges. The sequence $\{\beta_n\}$ converges for a similar reason. Furthermore, by (4.75) together with the definitions of $\alpha_{n+1}$,

$$0 \leq \alpha_{n+1} - \beta_{n+1} \leq \alpha_{n+1} - \beta_n = \begin{cases} \frac{\alpha_n - \beta_n}{4} & \text{if } \ell = 2, \\ \frac{\alpha_n - \beta_n}{3} & \text{if } \ell = 3, \\ \frac{\alpha_n - \beta_n}{2} & \text{if } \ell = 4. \end{cases}$$

Therefore $\lim_{n\to\infty}(\alpha_n - \beta_n) = 0$, so the sequences $\{\alpha_n\}$ and $\{\beta_n\}$ converge to the same limit. Moreover, we have

$$\alpha_{n+1}^\ell - \beta_{n+1}^\ell = \gamma_{n+1}^\ell = \begin{cases} \left(\frac{\alpha_n - \beta_n}{4}\right)^2 & \text{if } \ell = 2, \\ \left(\frac{\alpha_n - \beta_n}{3}\right)^3 & \text{if } \ell = 3, \\ \left(\frac{\alpha_n - \beta_n}{2}\right)^4 & \text{if } \ell = 4. \end{cases}$$

It follows that the rate of convergence for the means $M_2, M_3$, and $M_4$ is quadratic, cubic, and quartic, respectively.

In order to identify the value of $M_\ell(\alpha_0, \beta_0)$, suppose $0 \leq q < 1$ and define $a_\ell$, $b_\ell$, and $c_\ell$ by

$$b_2(q) = \frac{\eta_1^8}{\eta_2^4}, \qquad c_2(q) = 8\frac{\eta_2^8}{\eta_1^4}, \qquad a_2^2(q) = b_2^2(q) + c_2^2(q);$$

$$b_3(q) = \frac{\eta_1^3}{\eta_3}, \qquad c_3(q) = 3\frac{\eta_3^3}{\eta_1}, \qquad a_3^3(q) = b_3^3(q) + c_3^3(q);$$

$$b_4(q) = \frac{\eta_1^2}{\eta_2}, \qquad c_4(q) = 2\frac{\eta_4^2}{\eta_2}, \qquad a_4^4(q) = b_4^4(q) + c_4^4(q).$$

All of the above functions may be expressed in terms of theta functions as follows.

**Lemma 4.37.** *The following formulas hold:*

$$\begin{aligned} a_2(q) &= \varphi^4(q) + 16q\psi^4(q^2), & b_2(q) &= \varphi^4(-q), & c_2(q) &= 8q^{1/2}\psi^4(q), \\ a_3(q) &= a(q), & b_3(q) &= b(q), & c_3(q) &= c(q), \\ a_4(q) &= \varphi(q), & b_4(q) &= \varphi(-q), & c_4(q) &= 2q^{1/4}\psi(q^2). \end{aligned}$$

*In terms of $x_\ell$ and $z_\ell$ defined by* (4.28)–(4.30), *we also have*

$$\left(\frac{c_\ell(q)}{a_\ell(q)}\right)^\ell = x_\ell(q), \qquad \left(\frac{b_\ell(q)}{a_\ell(q)}\right)^\ell = 1 - x_\ell(q), \quad \textit{for} \quad \ell = 2, 3, 4,$$

*and*

$$a_2(q) = z_2^2(q), \quad a_3(q) = z_3(q), \quad a_4(q) = z_4^{1/2}(q).$$

*Proof.* The formulas for $b_\ell(q)$ and $c_\ell(q)$ are immediate consequences of the definitions and the infinite product formulas in Lemma 3.5 and Theorem 3.15. Next, by Lemma 3.9,

$$a_2^2(q) = b_2^2(q) + c_2^2(q) = \varphi^8(-q) + 64q\psi^8(q) = \left(\varphi^4(q) + 16q\psi^4(q^2)\right)^2,$$

and this proves the result for $a_2(q)$. The results for $a_3(q)$ and $a_4(q)$ follow from Theorems 3.20 and 3.7, respectively. The expressions in terms of $x_\ell$ and $z_\ell$ are immediate from (4.28) to (4.30). □

The next result provides quadratic, cubic, and quartic functional equations for theta functions.

**Lemma 4.38.** *The following identities hold:*

$$\frac{a_2(q) + 3b_2(q)}{4} = a_2(q^2), \qquad \frac{a_2(q) - b_2(q)}{4} = c_2(q^2), \tag{4.76}$$

$$\frac{a_3(q) + 2b_3(q)}{3} = a_3(q^3), \qquad \frac{a_3(q) - b_3(q)}{3} = c_3(q^3), \tag{4.77}$$

$$\frac{a_4(q) + b_4(q)}{2} = a_4(q^4), \qquad \frac{a_4(q) - b_4(q)}{2} = c_4(q^4). \tag{4.78}$$

*Proof.* Let us first consider the functions $a_2$, $b_2$, and $c_2$. Applying Lemma 4.37 and then Theorem 3.7, we find that

$$\begin{aligned} a_2(q) + 3b_2(q) &= \varphi^4(q) + 16q\psi^4(q^2) + 3\varphi^4(-q) \\ &= 2\varphi^4(q) + 2\varphi^4(-q) \\ &= 2\left(\varphi^2(q) - \varphi^2(-q)\right)^2 + 4\varphi^2(q)\varphi^2(-q). \end{aligned}$$

To this we successively apply Theorem 3.3, Lemma 3.6, Theorem 3.7, and then Lemma 4.37, to obtain

$$\begin{aligned} a_2(q) + 3b_2(q) &= 128q^2\psi^4(q^4) + 4\varphi^4(-q^2) \\ &= 4\varphi^4(q^2) + 64q^2\psi^4(q^4) \\ &= 4a_2(q^2). \end{aligned}$$

This proves the first part of (4.76). Next, by Lemma 4.37 and Theorem 3.7,

$$a_2(q) - b_2(q) = \varphi^4(q) + 16q\psi^4(q^2) - \varphi^4(-q) = 32q\psi^4(q^2) = 4c_2(q^2).$$

This completes the proof of (4.76).

The results in (4.77) follow readily from Lemmas 3.11 and 4.37. The results in (4.78) follow immediately from Lemmas 3.3 and 4.37. □

We are now ready to identify the values of the means $M_\ell$ for $\ell = 2$, 3 or 4. Since

$$M_\ell(\alpha_0, \beta_0) = 0 \qquad \text{if } \alpha_0 = 0 \text{ or } \beta_0 = 0$$

and

$$M_\ell(\lambda\alpha_0, \lambda\beta_0) = \lambda M_\ell(\alpha_0, \beta_0) \quad \text{for any } \lambda \geq 0,$$

it is enough to study $M_\ell(1, x)$ for $0 < x < 1$.

**Theorem 4.39.** *Let $x$ be a positive real number. Then*

$$\begin{aligned} M_2(1,x) &= \left(\frac{1}{{}_2F_1(\frac{1}{4}, \frac{3}{4}; 1; 1 - x^2)}\right)^2, \\ M_3(1,x) &= \frac{1}{{}_2F_1(\frac{1}{3}, \frac{2}{3}; 1; 1 - x^3)} \end{aligned}$$

*and*

$$M_4(1,x) = \left(\frac{1}{{}_2F_1(\frac{1}{2},\frac{1}{2};1;1-x^4)}\right)^{1/2}.$$

*Proof.* Let $\ell = 2$, 3, or 4. Since

$$M_\ell(\alpha_n,\beta_n) = M_\ell(\alpha_{n+1},\beta_{n+1}),$$

it follows from Lemma 4.38 that

$$M_\ell(a_\ell(q),b_\ell(q)) = M_\ell(a_\ell(q^\ell),b_\ell(q^\ell)).$$

This may be iterated to give

$$M_\ell(a_\ell(q),b_\ell(q)) = M_\ell(a_\ell(q^{(\ell^n)}),b_\ell(q^{(\ell^n)})), \text{ for } n = 1,2,3,\ldots.$$

Letting $n \to \infty$ gives

$$M_\ell(a_\ell(q),b_\ell(q)) = M_\ell(a_\ell(0),b_\ell(0)) = M_\ell(1,1) = 1.$$

By the homogeneity property of $M_\ell$, this may be rewritten as

$$M_\ell\left(1,\frac{b_\ell(q)}{a_\ell(q)}\right) = \frac{1}{a_\ell(q)}.$$

Lemma 4.37 may be used to express $a_\ell$ and $b_\ell$ in terms of $x_\ell$ and $z_\ell$. The result is

$$M_2(1,(1-x_2)^{1/2}) = \frac{1}{z_2^2},$$

$$M_3(1,(1-x_3)^{1/3}) = \frac{1}{z_3}$$

and

$$M_4(1,(1-x_4)^{1/4}) = \frac{1}{z_4^{1/2}}.$$

By Theorem 4.25, each of $z_2$, $z_3$, and $z_4$ may be expressed as hypergeometric functions, viz.,

$$M_2(1,(1-x_2)^{1/2}) = \left(\frac{1}{{}_2F_1(\frac{1}{4},\frac{3}{4};1;x_2)}\right)^2,$$

$$M_3(1,(1-x_3)^{1/3}) = \frac{1}{{}_2F_1(\frac{1}{3},\frac{2}{3};1;x_3)}$$

and

$$M_4(1,(1-x_4)^{1/4}) = \left(\frac{1}{{}_2F_1(\frac{1}{2},\frac{1}{2};1;x_4)}\right)^{1/2}.$$

In each of the three identities above, replace $x_2, x_3, x_4$ with $1-x^2$, $1-x^3$, and $1-x^4$, respectively, to complete the proof. □

## 4.8 Notes

Jacobi published his results in his book *"Fundamenta Nova Theoriae Functionum Ellipticarum"*, called Fundamenta Nova for short, in 1829. The book has been reprinted as part of the Volume 1 of Jacobi's collected works [189, pp. 49–239].

Jacobi's formulation of the inversion theorem goes as follows. To start with, suppose $0 < k < 1$ and let $K$ and $K'$ be defined as in [189, §17, p. 82] by

$$K = \int_0^{\pi/2} \frac{d\varphi}{\sqrt{1-k^2\sin^2\varphi}} \quad \text{and} \quad K' = \int_0^{\pi/2} \frac{d\varphi}{\sqrt{1-(1-k^2)\sin^2\varphi}}. \tag{4.79}$$

Here, $k$ is called the modulus, and $K$ and $K'$ are called the complete elliptic integral of the first kind and the complementary complete elliptic integral, respectively. Next, the nome $q$ is defined as a function of the modulus $k$ by [189, §35, p. 142], as

$$q = \exp\left(\frac{-\pi K'}{K}\right).$$

The formula for the inverse function that gives $k$ in terms of $q$ is given in [189, §36, p. 146], by

$$k = 4\sqrt{q}\prod_{j=1}^{\infty}\frac{(1+q^{2j})^4}{(1+q^{2j-1})^4}.$$

Furthermore, the complete elliptic integral $K$ is expressed as a function of $q$ by [189, §36, p. 146], as

$$\frac{2K}{\pi} = \prod_{j=1}^{\infty}\frac{(1-q^{2j})^2(1+q^{2j-1})^2}{(1-q^{2j-1})^2(1+q^{2j})^2}. \tag{4.80}$$

Jacobi's results (4.79)–(4.80) are equivalent to our statement of Theorem 4.1, with $x = k^2$ and $z = 2K/\pi$, since $K = \frac{\pi}{2}\,{}_2F_1(\frac{1}{2}, \frac{1}{2}; 1; k^2)$; see Exercise 4.

A different and concise approach to Jacobi's inversion theorem is outlined at the beginning of Chapter 17 of Ramanujan's second notebook [253]. Like Jacobi, Ramanujan takes (4.2) as a definition of the nome $q$ in terms of $x_4$, and then deduces the formulas (4.1) and (4.3). The details have been worked out in full by B. C. Berndt; see [46, pp. 87–102] or [49, pp. 109–132].

The inversion formula for the level 1 theory in Theorem 4.2 was known to R. Fricke in 1916 [147, p. 21 (26)], [155, pp. 279–339, esp. p. 336].

Ramanujan's results for the theories to alternative bases were alluded to in his 1914 paper *"Modular equations and approximations to* $\pi$," [248], and some more specific formulas, but no proofs, were recorded in the unorganized pages of Ramanujan's second notebook [253, pp. 257–262]. In the first edition of Ramanujan's Collected Papers (see p. 336 of the 1927 edition of the Collected Papers, or [255, p. 360]) Mordell wrote:

*"It is unfortunate that Ramanujan has not developed in detail the corresponding theories referred to in Section 14 (of [248])."*

In his presidential address to the London Mathematical Society, Watson [297] wrote:

*"There are developments of functions analogous to elliptic functions which I have not seen elsewhere."*

Despite the interest of Mordell and Watson, Ramanujan's alternative theories do not seem to have been investigated until the work of Venkatachaliengar in 1988, [288, Chapter 5]. Significant advances were made by J. M. Borwein and P. B. Borwein [63], and all of Ramanujan's results in [253, pp. 257–262] were systematically analyzed in the seminal work of B. C. Berndt, S. Bhargava, and F. Garvan [50]; see also [48, Chapter 33]. It may be mentioned that the connection between cubic theta functions and the hypergeometric function in (4.5) was given by the Borweins in [63, Theorem 2.3], while the cubic inversion formula (4.4) is in [48, p. 99]. The approach taken in this chapter is based on [110].

Example 4.36 was originally obtained numerically by Gauss, who recorded in his diary on May 30, 1799, that $M(1, \sqrt{2}) = \pi/\tilde{\omega}$ to 11 decimal places, where $\tilde{\omega} = 2\displaystyle\int_0^1 \frac{\mathrm{d}t}{\sqrt{1-t^4}}$. Gauss wrote:

*"We have established that the arithmetic-geometric mean between* $1$ *and* $\sqrt{2}$ *is* $\pi/\tilde{\omega}$ *to the eleventh decimal place; the demonstration of this fact will surely open an entirely new field of analysis."*

The work by Cox [132] provides a clear and detailed analysis of Gauss' work on this topic. The arithmetic-geometric mean for complex numbers is also considered in [132]. The wide-ranging connections of the arithmetic-geometric mean with topics in analytic number theory, including the high precision computation of $\pi$, are presented in the book by J. M. Borwein and P. B. Borwein [60]. The means $M_2$, $M_3$, $M_4$, and other means have been studied by J. Borwein, D. Bailey, and R. Girgensohn in [59, pp. 263–268], the Borweins [62], the Borweins and F. Garvan [66], and by J. Borwein and F. Garvan [68].

## 4.9 Exercises

1. a. Use (0.2) and (4.14)–(4.16) to prove that

$$q\prod_{j=1}^{\infty}(1-(-q)^j)^{24}$$

$$=q\prod_{j=1}^{\infty}(1-q^j)^{24}+48q^2\prod_{j=1}^{\infty}(1-q^{2j})^{24}+4096q^4\prod_{j=1}^{\infty}(1-q^{4j})^{24}.$$

   b. Deduce that

$$\tau(2n)+24\tau(n)+2^{11}\tau\left(\frac{n}{2}\right)=0,$$

   where $\tau(x)$ is defined to be 0 if $x$ is not an integer. Observe that this gives (3.122) in the case $p=2$.
   See Andrews and Berndt [27, p. 115].

2. a. Use Theorem 4.31 to show that

$$\left(\sum_{j=-\infty}^{\infty}q^{j^2}\right)^{12}=\frac{1}{63}\left(64R(q^4)-R(q)\right)+16q\prod_{j=1}^{\infty}(1-q^{2j})^{12}.$$

   See K. S. Williams [300].

   b. Show that the result of Part a may be expressed in the form

$$\left(\sum_{j=-\infty}^{\infty}q^{j^2}\right)^{12}=1+8\sum_{j=1}^{\infty}\frac{j^5q^j}{1+(-q)^j}+16q\prod_{j=1}^{\infty}(1-q^{2j})^{12}.$$

   This is the case $m=3$ of Ramanujan's identity (3.115).

   c. Use Theorem 4.31 to show that

$$\left(\sum_{j=-\infty}^{\infty}\sum_{k=-\infty}^{\infty}q^{j^2+jk+k^2}\right)^6$$

$$=\frac{1}{26}\left(27R(q^3)-R(q)\right)+\frac{216}{13}q\prod_{j=1}^{\infty}(1-q^j)^6(1-q^{3j})^6,$$

   and show that this implies (3.129). See H. Petersson [241, p. 90].

3. For any positive integer $n$, let $P_n = P(q^n)$ and $Q_n = Q(q^n)$. Prove that

$$q\frac{\mathrm{d}}{\mathrm{d}q}(P_1 - P_2) = \frac{1}{6}(P_1 - P_2)^2 - \frac{1}{10}(Q_1 - Q_2)$$

and

$$q\frac{\mathrm{d}}{\mathrm{d}q}(P_1 - P_3) = \frac{1}{8}(P_1 - P_3)^2 - \frac{1}{10}(Q_1 - Q_3).$$

4. Recall that the gamma function may be defined for $\mathrm{Re}(a) > 0$ by

$$\Gamma(a) = \int_0^\infty e^{-t} t^{a-1} dt.$$

It has the special values $\Gamma(1) = 1$ and $\Gamma(1/2) = \sqrt{\pi}$. The gamma function satisfies the recurrence relation $\Gamma(a+1) = a\Gamma(a)$. Hence, $\Gamma(n+1) = n!$ for any nonnegative integer $n$. The following integral, denoted by $B(a,b)$ and called the beta function, can be evaluated in terms of the gamma function by means of the formula

$$B(a,b) = \int_0^1 t^{a-1}(1-t)^{b-1} dt = \frac{\Gamma(a)\Gamma(b)}{\Gamma(a+b)}, \quad \text{for } \mathrm{Re}(a), \mathrm{Re}(b) > 0.$$

A full and detailed account of properties of the gamma and beta functions can be found in the book by G. Andrews, R. Askey, and R. Roy [24, Chapter 1].

a. Prove the binomial theorem: $(1-x)^{-a} = \sum_{n=0}^{\infty} \frac{(a)_n}{n!} x^n$, for $|x| < 1$.
Hint: Either use Taylor's theorem, or use term by term differentiation to show that the series on the right-hand side satisfies the differential equation $(1-x)f'(x) = af(x)$ and initial condition $f(0) = 1$.
b. By expanding the integrand by the binomial theorem and integrating term by term, show that for $|x| < 1$,

$$\int_0^{\pi/2} \frac{\mathrm{d}\theta}{\sqrt{1 - x\sin^2\theta}} = \frac{\pi}{2}\, {}_2F_1\left(\begin{matrix} \frac{1}{2}, \frac{1}{2} \\ 1 \end{matrix}; x\right).$$

c. Assuming $\mathrm{Re}(c) > \mathrm{Re}(b) > 0$ and $|x| < 1$, prove Euler's integral representation

$$\frac{\Gamma(c)}{\Gamma(b)\Gamma(c-b)} \int_0^1 t^{b-1}(1-t)^{c-b-1}(1-xt)^{-a} dt = {}_2F_1\left(\begin{matrix} a, b \\ c \end{matrix}; x\right).$$

Hint: Expand $(1-xt)^{-a}$ by the binomial theorem and integrate term by term.
d. Observe that the result in Part b is the special case $a = b = \frac{1}{2}$, $c = 1$, of the result of Part c, via the change of variable $t = \sin^2\theta$.

5. By analogy with (4.25), let $\tilde{\omega}_3$ be defined by

$$\tilde{\omega}_3 = 2\int_0^1 \frac{dt}{\sqrt{1-t^3}}.$$

a. Show that

$$\tilde{\omega}_3 = \int_0^1 \left(\frac{2}{1-u^3}\right)^{2/3} du.$$

b. Show that

$$\tilde{\omega}_3 = \frac{2}{3}B\left(\frac{1}{3},\frac{1}{2}\right) = \frac{2^{2/3}}{3}B\left(\frac{1}{3},\frac{1}{3}\right)$$

where $B(a, b)$ is the beta function defined in the previous exercise.

c. Show that the cubic theta functions have special values given by

$$a(e^{-2\pi/\sqrt{3}}) = \frac{\sqrt{27}}{4\pi}\,\tilde{\omega}_3$$

and

$$b(e^{-2\pi/\sqrt{3}}) = c(e^{-2\pi/\sqrt{3}}) = \frac{\sqrt{27}}{4\sqrt[3]{2}\,\pi}\,\tilde{\omega}_3.$$

Hint: Take $x_3 = \frac{1}{2}$ in Theorem 4.4 and apply Euler's integral representation from Exercise 4 to obtain

$$\begin{aligned} a(e^{-2\pi/\sqrt{3}}) &= {}_2F_1\left(\begin{matrix}\frac{1}{3},\frac{2}{3}\\ 1\end{matrix};\frac{1}{2}\right) \\ &= \frac{1}{\Gamma(\frac{1}{3})\Gamma(\frac{2}{3})}\int_0^1 t^{-1/3}(1-t)^{-2/3}(1-\tfrac{t}{2})^{-1/3}dt. \end{aligned}$$

The integral can be expressed in terms of the integral in Part a by the change of variable $t = 2u^{3/2}/(1+u^{3/2})$.

For a more general result that gives the evaluation of certain ${}_2F_1$ hypergeometric functions at $x = \frac{1}{2}$, see Andrews, Askey, and Roy [24, p. 148].

d. Show that

$$\frac{1}{M_3(1, 1/\sqrt[3]{2})} = \frac{\sqrt{27}}{4\pi}\,\tilde{\omega}_3.$$

6. a. Show that

$$a\left(e^{-\pi/\sqrt{3}}\right) = 2a\left(e^{-4\pi/\sqrt{3}}\right).$$

Hint: Take $p = 3$, $n = 1$, $\tau = i/2\sqrt{3}$ in the last part of Theorem 2.21. Then use (2.47) and (3.34) to write $a(q) = 6E_1^0(\tau, 3) = 6E_1^\infty(\tau, 3)$.

b. Use the result of Part a to deduce that $a\left(-e^{-\pi/\sqrt{3}}\right) = 0$.

Hint: By Lemma 3.12, we have

$$a(q) + a(-q) = 2a(q^4).$$

c. Use Theorem 3.21 to show that

$$\frac{b^3(q)}{b^3(q^2)} = \frac{4a(q^2) - 2a(q)}{a(q^2) + a(q)},$$

and hence deduce that

$$\frac{b\left(-e^{-\pi/\sqrt{3}}\right)}{b\left(e^{-2\pi/\sqrt{3}}\right)} = \sqrt[3]{4}.$$

Hint: Use Theorem 3.21.

d. Use the result of Part c, together with the values in Exercise 5, to deduce that

$$b\left(-e^{-\pi/\sqrt{3}}\right) = \frac{\sqrt{3}}{2\pi} B\left(\frac{1}{3}, \frac{1}{3}\right) = \frac{\sqrt{3}}{2\pi} \times \frac{\Gamma\left(\frac{1}{3}\right)^2}{\Gamma\left(\frac{2}{3}\right)} = \frac{3}{4\pi^2} \times \Gamma\left(\frac{1}{3}\right)^3.$$

The values of $a\left(-e^{-\pi/\sqrt{3}}\right)$ and $b\left(-e^{-\pi/\sqrt{3}}\right)$ are used in the study of Dixon's elliptic functions; see Lemma 1.52.

7. Use the case $\ell = 1$ of Theorem 4.28 and the transformation formula (4.53) to show that

$$(Q(q))^{1/4} = {}_2F_1\left(\begin{matrix}\frac{1}{12}, \frac{5}{12} \\ 1\end{matrix}; \frac{1728}{j(\tau)}\right),$$

where $j(\tau)$ is given by (2.21).

8. The hypergeometric function ${}_pF_q$ is defined by

$${}_pF_q\left(\begin{matrix}a_1, \ldots, a_p \\ b_1, \ldots, b_q\end{matrix}; x\right) = \sum_{n=0}^{\infty} \frac{(a_1)_n(a_2)_n \cdots (a_p)_n}{(b_1)_n(b_2)_n \cdots (b_q)_n} \frac{x^n}{n!}. \tag{4.81}$$

We will be mainly interested in the case $p = q + 1$.

a. Suppose $p = q + 1$. Use the ratio test to show that the series in (4.81) converges if $|x| < 1$ and diverges if $|x| > 1$. More sophisticated tests can be applied in the case $|x| = 1$, e.g., see [24, p. 62].

b. For any complex number $m$ let $D_m$ be the differential operator defined by

$$D_m(y) = x\frac{\mathrm{d}y}{\mathrm{d}x} + my,$$

where $y$ is any differentiable function of $x$. Show that

$$D_{a_j}\left({}_pF_q\left(\begin{matrix} a_1, \ \ldots, \ a_p \\ b_1, \ \ldots, \ b_q \end{matrix}; x\right)\right) = a_j \, {}_pF_q\left(\begin{matrix} a_1, \ \ldots, \ a_j+1, \ \ldots, \ a_p \\ b_1, \ \ldots, \ b_q \end{matrix}; x\right),$$

$$D_{b_k-1}\left({}_pF_q\left(\begin{matrix} a_1, \ \ldots, \ a_p \\ b_1, \ \ldots, \ b_q \end{matrix}; x\right)\right) = (b_k-1) \, {}_pF_q\left(\begin{matrix} a_1, \ \ldots, \ a_p \\ b_1, \ \ldots, \ b_k-1, \ \ldots, \ b_q \end{matrix}; x\right)$$

and

$$\frac{\mathrm{d}}{\mathrm{d}x}\left({}_pF_q\left(\begin{matrix} a_1, \ \ldots, \ a_p \\ b_1, \ \ldots, \ b_q \end{matrix}; x\right)\right) = \frac{a_1 \cdots a_p}{b_1 \cdots b_q} \, {}_pF_q\left(\begin{matrix} a_1+1, \ \ldots, \ a_p+1 \\ b_1+1, \ \ldots, \ b_q+1 \end{matrix}; x\right).$$

c. Use the results of Part b to show that the hypergeometric function $y = {}_pF_q(a_1, \ldots, a_p; \, b_1, \ldots, b_q; \, x)$ satisfies the differential equation

$$D_{a_1} D_{a_2} \cdots D_{a_p}(y) = D_{b_1} D_{b_2} \cdots D_{b_q}\left(\frac{\mathrm{d}y}{\mathrm{d}x}\right).$$

d. Take $p = 2$ and $q = 1$ in the result of Part c to show that the function $y = {}_2F_1(a, b; \, c; \, x)$ satisfies the differential equation

$$x(1-x)\frac{\mathrm{d}^2y}{\mathrm{d}x^2} + (c - (a+b+1)x)\frac{\mathrm{d}y}{\mathrm{d}x} - aby = 0.$$

e. Prove that the functions

$$y_1 = \left\{{}_2F_1\left(\begin{matrix} a, \ b \\ a+b+\frac{1}{2} \end{matrix}; x\right)\right\}^2$$

and

$$y_2 = {}_3F_2\left(\begin{matrix} 2a, \ 2b, \ a+b \\ 2a+2b, \ a+b+\frac{1}{2} \end{matrix}; x\right)$$

both satisfy the differential equation

$$x^2(1-x)\frac{d^3y}{dx^3} + 3x\left(a+b+\frac{1}{2}-(a+b+1)x\right)\frac{d^2y}{dx^2}$$
$$+\left((a+b)(2a+2b+1)-(2a^2+2b^2+8ab+3a+3b+1)x\right)\frac{dy}{dx}$$
$$-4ab(a+b)y=0$$

and that the first three terms in the Taylor expansion of each function are given by

$$1+\frac{4ab}{2a+2b+1}x+\frac{4ab\,(2b+1)\,(2a+1)\,(a+b+1)}{(2a+2b+1)^2\,(2a+2b+3)}x^2.$$

Deduce that $y_1 = y_2$ and hence prove Clausen's identity (4.52). See [24, p. 116].
Hint: Look ahead to Exercise 4 at the end of Chapter 6 for a method to find the differential equation satisfied by $y_1$.

f. Prove the transformation formula (4.53) by finding a second order linear differential equation that is satisfied by both sides, and by showing that the first two terms in the Taylor expansions about $x = 0$ agree.

9. For $\ell = 1, 2, 3,$ or $4$, let $Z_\ell$ and $X_\ell$ be as for Theorem 4.30 and let $B_\ell$ be defined by

$$B_\ell = \frac{1}{Z_\ell}q\frac{d}{dq}\log X_\ell.$$

a. Show that

$$B_\ell = (1-4\alpha_\ell X_\ell)^{1/2}$$

where $\alpha_\ell = 432, 64, 27,$ or $16$ according to whether $\ell = 1, 2, 3,$ or $4$, respectively.

b. Show that $Z_\ell$ satisfies the following second order nonlinear differential equation with respect to $X_\ell$:

$$B_\ell X_\ell\frac{d}{dX_\ell}\left(B_\ell X_\ell\frac{dZ_\ell}{dX_\ell}\right)-\frac{1}{2Z_\ell}\left(B_\ell X_\ell\frac{dZ_\ell}{dX_\ell}\right)^2 = 2\beta_\ell Z_\ell X_\ell$$

where $\beta_\ell = 60, 12, 6,$ or $4$ according to whether $\ell = 1, 2, 3,$ or $4$, respectively.

The numbers $\alpha_\ell$ and $\beta_\ell$ will be studied again later, in a wider context; see Table 6.4.

10. Use properties of the means $M_2$, $M_3$, and $M_4$ to prove the following analogues of Landen's transformation in Corollary 4.35:

a. $${}_2F_1\left(\begin{matrix}\frac{1}{4}, \frac{3}{4}\\ 1\end{matrix}; \left(\frac{1-x}{1+3x}\right)^2\right) = \frac{\sqrt{1+3x}}{2}\,{}_2F_1\left(\begin{matrix}\frac{1}{4}, \frac{3}{4}\\ 1\end{matrix}; 1-x^2\right),$$
$${}_2F_1\left(\begin{matrix}\frac{1}{4}, \frac{3}{4}\\ 1\end{matrix}; 1-\left(\frac{1-x}{1+3x}\right)^2\right) = \sqrt{1+3x}\,{}_2F_1\left(\begin{matrix}\frac{1}{4}, \frac{3}{4}\\ 1\end{matrix}; x^2\right).$$

b. $${}_2F_1\left(\begin{matrix}\frac{1}{3}, \frac{2}{3}\\ 1\end{matrix}; \left(\frac{1-x}{1+2x}\right)^3\right) = \frac{(1+2x)}{3}\,{}_2F_1\left(\begin{matrix}\frac{1}{3}, \frac{2}{3}\\ 1\end{matrix}; 1-x^3\right),$$
$${}_2F_1\left(\begin{matrix}\frac{1}{3}, \frac{2}{3}\\ 1\end{matrix}; 1-\left(\frac{1-x}{1+2x}\right)^3\right) = (1+2x)\,{}_2F_1\left(\begin{matrix}\frac{1}{3}, \frac{2}{3}\\ 1\end{matrix}; x^3\right).$$

c. $${}_2F_1\left(\begin{matrix}\frac{1}{2}, \frac{1}{2}\\ 1\end{matrix}; \left(\frac{1-x}{1+x}\right)^4\right) = \frac{(1+x)^2}{4}\,{}_2F_1\left(\begin{matrix}\frac{1}{2}, \frac{1}{2}\\ 1\end{matrix}; 1-x^4\right),$$
$${}_2F_1\left(\begin{matrix}\frac{1}{2}, \frac{1}{2}\\ 1\end{matrix}; 1-\left(\frac{1-x}{1+x}\right)^4\right) = (1+x)^2\,{}_2F_1\left(\begin{matrix}\frac{1}{2}, \frac{1}{2}\\ 1\end{matrix}; x^4\right).$$

11. a. Show that

$$\frac{1}{M(a,b)} = \frac{2}{\pi}\int_0^{\pi/2} \frac{d\theta}{(a^2\cos^2\theta + b^2\sin^2\theta)^{1/2}}.$$

Hint: By symmetry of $M$, we may assume that $a \geq b > 0$. Then, by the homogeneous property and Theorem 4.34, we find that

$$\frac{1}{M(a,b)} = \frac{1}{aM(1,b/a)} = \frac{1}{a}\,{}_2F_1\left(\begin{matrix}\frac{1}{2}, \frac{1}{2}\\ 1\end{matrix}; 1-\frac{b^2}{a^2}\right).$$

Now use the result of Exercise 4 Part b.

b. Show that

$$\frac{1}{M(a,b)} = \frac{1}{\pi}\int_{-\infty}^{\infty} \frac{dt}{\sqrt{(a^2+t^2)(b^2+t^2)}}.$$

Hint: Make the change of variable $t = b\tan\theta$ in the integral in Part a.

c. Prove that

$$\frac{1}{M_3(a,b)} = \frac{\sqrt{27}}{2\pi}\int_0^{\infty} \frac{t\,dt}{\sqrt[3]{(a^3+t^3)(b^3+t^3)^2}}.$$

Note that neither $M_3(a,b)$ nor the integral is symmetric in $a$ and $b$.

See Borwein, Bailey and Girgensohn [59, Section 6.1].

# Chapter 5
# Level 5: The Rogers–Ramanujan Continued Fraction

## 5.1 The Rogers–Ramanujan continued fraction

This chapter is about the function $r = r(q)$ defined by

$$r(q) = q^{1/5} \prod_{j=1}^{\infty} \frac{(1 - q^{5j-4})(1 - q^{5j-1})}{(1 - q^{5j-3})(1 - q^{5j-2})}. \tag{5.1}$$

We will show in Theorem 5.8 that $r(q)$ arises naturally in the 5-dissection of Euler's product

$$\prod_{j=1}^{\infty}(1 - q^j) = \sum_{j=-\infty}^{\infty} (-1)^j q^{j(3j-1)/2}.$$

More than one hundred years ago, L. J. Rogers and S. Ramanujan independently discovered that $r$ has an expansion as a continued fraction given by

$$r(q) = \cfrac{q^{1/5}}{1 + \cfrac{q}{1 + \cfrac{q^2}{1 + \cfrac{q^3}{1 + \cdots}}}} \tag{5.2}$$

and for this reason, $r$ is called the Rogers–Ramanujan continued fraction.

The factor $q^{1/5}$ that occurs in the definition (5.1) is similar to the factor $q^{1/24}$ in Dedekind's eta function defined by (2.20). A simple explanation for why the $q^{1/5}$ is there is given after Theorem 2.11: it completes the square in the quadratic exponents when the numerator and denominator are expanded as series by Jacobi's triple product identity. More importantly, the factor occurs in a natural and inevitable way in the modular transformation formula in Theorems 2.11 and 5.10. Consequently,



S. Cooper, *Ramanujan's Theta Functions*, DOI 10.1007/978-3-319-56172-1_6

most of the results for $r(q)$ are best stated if the factor $q^{1/5}$ is included; for example, see Theorem 5.8 and the material in this chapter from Section 5.5 and onwards. Occasionally, for example in Theorem 5.7, the factor of $q^{1/5}$ is not needed and can be omitted.

An infinite continued fraction simply means the limit of a finite continued fraction, that is

$$\cfrac{a_1}{b_1 + \cfrac{a_2}{b_2 + \cfrac{a_3}{b_3 + \cdots}}} = \lim_{n\to\infty} \cfrac{a_1}{b_1 + \cfrac{a_2}{b_2 + \cfrac{a_3}{\ddots + \cfrac{\ddots}{b_{n-1} + \cfrac{a_n}{b_n}}}}}. \tag{5.3}$$

The infinite continued fraction on the left-hand side of (5.3) is said to converge if the limit on the right-hand side of (5.3) exists. Various criteria for convergence in terms of the properties of $a_n$ and $b_n$ are known, e.g., see the book by Lorentzen and Waadeland [225].

The remainder of this section is devoted to showing that the function $r(q)$ defined by the product (5.1) has the continued fraction expansion given by (5.2). The first step will be to establish a finite continued fraction for a quotient of two polynomials. We begin by establishing a three-term recurrence relation.

**Lemma 5.1.** *Let $n$ be a nonnegative integer. For each integer $r$ with $0 \le r \le n$, let $F_r$ be defined by*

$$F_r = \sum_{k=0}^{\infty} q^{k(r+k)} \begin{bmatrix} n+1-k-r \\ k \end{bmatrix}_q x^k \tag{5.4}$$

*where $\begin{bmatrix} n \\ k \end{bmatrix}_q$ is the $q$-binomial coefficient given by* (0.105)*. Then,*

$$F_n = 1, \qquad F_{n-1} = 1 + xq^n$$

*and*

$$F_r - F_{r+1} = xq^{r+1} F_{r+2}, \qquad 0 \le r \le n-2.$$

Before starting the proof, observe that by (0.106), the $q$-binomial coefficient is zero if $k > n+1-k-r$. This implies that the sum in (5.4) contains only finitely many nonzero terms. Precisely,

$$F_r = \sum_{k=0}^{\lfloor (n+1-r)/2 \rfloor} q^{k(r+k)} \begin{bmatrix} n+1-k-r \\ k \end{bmatrix}_q x^k,$$

and $F_r$ is a polynomial in $x$ of degree $\lfloor (n+1-r)/2 \rfloor$ with coefficients that are polynomials in $q$.

*Proof.* The values of $F_n$ and $F_{n-1}$ are obvious from the definition of $F_r$. Next, for $0 \le r \le n-2$ we have

$$\begin{aligned}
F_r - F_{r+1} &= \left(1 + \sum_{k=1}^{\infty} q^{k(r+k)} \begin{bmatrix} n+1-k-r \\ k \end{bmatrix}_q x^k\right) \\
&\quad - \left(1 + \sum_{k=1}^{\infty} q^{k(r+1+k)} \begin{bmatrix} n-k-r \\ k \end{bmatrix}_q x^k\right) \\
&= \sum_{k=1}^{\infty} q^{k(r+k)} \left(\begin{bmatrix} n+1-k-r \\ k \end{bmatrix}_q - q^k \begin{bmatrix} n-k-r \\ k \end{bmatrix}_q\right) x^k \\
&= \sum_{k=1}^{\infty} q^{k(r+k)} \begin{bmatrix} n-k-r \\ k-1 \end{bmatrix}_q x^k
\end{aligned}$$

where the last step follows by the recurrence relation for binomial coefficients in Exercise 16 of Chapter 0. Shifting the summation index gives

$$\begin{aligned}
F_r - F_{r+1} &= \sum_{k=0}^{\infty} q^{(k+1)(r+1+k)} \begin{bmatrix} n-k-r-1 \\ k \end{bmatrix}_q x^{k+1} \\
&= xq^{r+1} \sum_{k=0}^{\infty} q^{k(r+2+k)} \begin{bmatrix} n-k-r-1 \\ k \end{bmatrix}_q x^k \\
&= xq^{r+1} F_{r+2}.
\end{aligned}$$

□

The recurrence relation in the previous lemma can be iterated to produce a continued fraction.

**Theorem 5.2.** *Let $F_0$ and $F_1$ be as for Lemma 5.1. Then*

$$\frac{F_0}{F_1} = 1 + \cfrac{xq}{1 + \cfrac{xq^2}{\ddots + \cfrac{\ddots}{1 + \cfrac{xq^{n-1}}{1 + \cfrac{xq^n}{1}}}}}.$$

*Proof.* Taking $r = 0$ and then $r = 1$ in the recurrence relation in Lemma 5.1 we obtain

$$\frac{F_0}{F_1} = 1 + \frac{xq}{F_1/F_2} = 1 + \cfrac{xq}{1 + \cfrac{xq^2}{F_2/F_3}}.$$

The procedure can be continued to give

$$\frac{F_0}{F_1} = 1 + \cfrac{xq}{1 + \cfrac{xq^2}{\ddots + \cfrac{\ddots}{1 + \cfrac{xq^{n-1}}{F_{n-1}/F_n}}}} = 1 + \cfrac{xq}{1 + \cfrac{xq^2}{\ddots + \cfrac{\ddots}{1 + \cfrac{xq^{n-1}}{1 + \cfrac{xq^n}{1}}}}}.$$

□

For the remainder of the section, it will be useful to employ the compact notation for products given by (0.113).

**Theorem 5.3.** *The following identity, that expresses an infinite continued fraction as a quotient of infinite series, holds*:

$$1 + \cfrac{xq}{1 + \cfrac{xq^2}{1 + \cfrac{xq^3}{1 + \cdots}}} = \frac{\displaystyle\sum_{k=0}^{\infty} \frac{q^{k^2} x^k}{(q;q)_k}}{\displaystyle\sum_{k=0}^{\infty} \frac{q^{k^2+k} x^k}{(q;q)_k}}.$$

*In particular, the case $x = 1$ gives*

$$1 + \cfrac{q}{1 + \cfrac{q^2}{1 + \cfrac{q^3}{1 + \cdots}}} = \frac{\displaystyle\sum_{k=0}^{\infty} \frac{q^{k^2}}{(q;q)_k}}{\displaystyle\sum_{k=0}^{\infty} \frac{q^{k^2+k}}{(q;q)_k}}. \tag{5.5}$$

*Proof.* Let $n \to \infty$ in Theorem 5.2 and observe that

$$\lim_{n\to\infty} \begin{bmatrix} n+1-k-r \\ k \end{bmatrix}_q = \frac{1}{(q;q)_k}.$$

□

It will now be shown that each of the infinite series in the quotient on the right-hand side of (5.5) can be expressed as infinite products. We will need:

**Lemma 5.4.** *Let*

$$G(x) = 1 + \sum_{j=1}^{\infty} (-1)^j \, q^{(5j^2-j)/2} \, x^{2j} \, (1 - q^{2j} x) \, \frac{(qx; q)_{j-1}}{(q; q)_j}. \tag{5.6}$$

*Then*

$$G(1) = \prod_{j=1}^{\infty} (1 - q^{5j-3})(1 - q^{5j-2})(1 - q^{5j})$$

*and*

$$(1 - q)\, G(q) = \prod_{j=1}^{\infty} (1 - q^{5j-4})(1 - q^{5j-1})(1 - q^{5j}).$$

*Moreover, the following three-term functional equation holds:*

$$\frac{G(x)}{1 - qx} = G(qx) + qx(1 - q^2 x) G(q^2 x).$$

*Proof.* Substituting the values $x = 1$ and $x = q$ in (5.6) we obtain

$$G(1) = 1 + \sum_{j=1}^{\infty} (-1)^j q^{(5j^2-j)/2} (1 + q^j) = \sum_{j=-\infty}^{\infty} (-1)^j q^{(5j^2-j)/2}$$

and

$$(1 - q) G(q) = 1 - q + \sum_{j=1}^{\infty} (-1)^j q^{(5j^2+3j)/2} (1 - q^{2j+1}) = \sum_{j=-\infty}^{\infty} (-1)^j q^{(5j^2+3j)/2}.$$

Jacobi's triple product identity (0.15) can now be applied to give the required infinite products.

Let us prove the recurrence relation. The proof involves some clever manipulations. Split the series by writing

$$1 - q^{2j} x = q^j (1 - x q^j) + (1 - q^j)$$

to obtain

$$G(x) = 1 + \sum_{j=1}^{\infty} (-1)^j \, q^{(5j^2-j)/2} \, x^{2j} \, q^j \, \frac{(qx; q)_j}{(q; q)_j}$$

$$+ \sum_{j=1}^{\infty} (-1)^j \, q^{(5j^2-j)/2} \, x^{2j} \, \frac{(qx; q)_{j-1}}{(q; q)_{j-1}}.$$

On replacing $j$ with $j+1$ in the second sum, we obtain

$$\begin{aligned}G(x) &= 1+\sum_{j=1}^{\infty}(-1)^j\, q^{(5j^2-j)/2}\, x^{2j}\, q^j\, \frac{(qx;q)_j}{(q;q)_j}\\ &\quad -\sum_{j=0}^{\infty}(-1)^j\, q^{(5j^2+9j+4)/2}\, x^{2j+2}\, \frac{(qx;q)_j}{(q;q)_j}\\ &= 1-q^2x^2+\sum_{j=1}^{\infty}(-1)^j\, q^{(5j^2+j)/2}\, x^{2j}\,(1-q^{4j+2}x^2)\,\frac{(qx;q)_j}{(q;q)_j}. \qquad (5.7)\end{aligned}$$

Now consider $G(x)/(1-qx)-G(qx)$. Using (5.7) for $G(x)/(1-qx)$ and (5.6) for $G(qx)$, we obtain

$$\begin{aligned}&\frac{G(x)}{1-qx}-G(qx)\\ &= 1+qx+\sum_{j=1}^{\infty}(-1)^j\, q^{(5j^2+j)/2}\, x^{2j}\,(1-q^{4j+2}x^2)\,\frac{(q^2x;q)_{j-1}}{(q;q)_j}\\ &\quad -1-\sum_{j=1}^{\infty}(-1)^j\, q^{(5j^2-j)/2}\, q^{2j}\, x^{2j}\,(1-q^{2j+1}x)\,\frac{(q^2x;q)_{j-1}}{(q;q)_j}\\ &= qx+\sum_{j=1}^{\infty}(-1)^j\, q^{(5j^2+j)/2}\, x^{2j}\,\left\{q^{3j+1}x(1-q^{j+1}x)+(1-q^j)\right\}\,\frac{(q^2x;q)_{j-1}}{(q;q)_j}.\end{aligned}$$

Split the series again, according to the two terms within the braces, to obtain

$$\begin{aligned}\frac{G(x)}{1-qx}-G(qx) &= qx+\sum_{j=1}^{\infty}(-1)^j\, q^{(5j^2+j)/2+3j+1}\, x^{2j+1}\,\frac{(q^2x;q)_j}{(q;q)_j}\\ &\quad +\sum_{j=1}^{\infty}(-1)^j\, q^{(5j^2+j)/2}\, x^{2j}\,\frac{(q^2x;q)_{j-1}}{(q;q)_{j-1}}.\end{aligned}$$

In the last sum, separate out the term corresponding to $j=1$ and then replace $j$ with $j+1$ for the remaining terms to obtain

$$\begin{aligned}&\frac{G(x)}{1-qx}-G(qx)\\ &= qx+\sum_{j=1}^{\infty}(-1)^j\, q^{(5j^2+7j+2)/2}\, x^{2j+1}\,\frac{(q^2x;q)_j}{(q;q)_j}\\ &\quad -q^3x^2-\sum_{j=1}^{\infty}(-1)^j\, q^{(5j^2+11j+6)/2}\, x^{2j+2}\,\frac{(q^2x;q)_j}{(q;q)_j}\end{aligned}$$

$$= qx(1-q^2x) + qx\sum_{j=1}^{\infty}(-1)^j\, q^{(5j^2+7j)/2}\, x^{2j}\,(1-q^{2j+2}x)\,\frac{(q^2x;q)_j}{(q;q)_j}$$

$$= qx(1-q^2x)\left(1+\sum_{j=1}^{\infty}(-1)^j\, q^{(5j^2-j)/2}\, q^{4j}\, x^{2j}\,(1-q^{2j+2}x)\,\frac{(q^3x;q)_{j-1}}{(q;q)_j}\right)$$

$$= qx(1-q^2x)\,G(q^2x).$$

□

The next result gives another expression for $G(x)$.

**Lemma 5.5.** *Let $G(x)$ be defined by* (5.6). *Then*

$$G(x) = \left(1+\sum_{k=1}^{\infty}\frac{q^{k^2}x^k}{(1-q)(1-q^2)\cdots(1-q^k)}\right)\left(\prod_{j=1}^{\infty}(1-q^jx)\right).$$

*Proof.* Let $H(x) = \dfrac{G(x)}{\prod_{j=1}^{\infty}(1-q^jx)}$. Then the three-term functional equation in Lemma 5.4 implies

$$H(x) = H(qx) + qxH(q^2x).$$

It follows that the coefficients in the expansion

$$H(x) = \sum_{k=0}^{\infty} h_k x^k$$

satisfy the recurrence relation

$$h_k = \frac{q^{2k-1}}{1-q^k}\,h_{k-1}, \qquad k \geq 1.$$

Solving the recurrence relation and noting that $H(0) = 1$ we obtain

$$H(x) = 1+\sum_{k=1}^{\infty}\frac{q^{k^2}x^k}{(1-q)(1-q^2)\cdots(1-q^k)}.$$

□

The preceding results can be assembled to give:

**Theorem 5.6 (Rogers–Ramanujan identities).** *The following results hold*:

$$1+\sum_{k=1}^{\infty}\frac{q^{k^2}}{(1-q)(1-q^2)\cdots(1-q^k)}=\prod_{j=1}^{\infty}\frac{1}{(1-q^{5j-4})(1-q^{5j-1})}$$

*and*

$$1+\sum_{k=1}^{\infty}\frac{q^{k^2+k}}{(1-q)(1-q^2)\cdots(1-q^k)}=\prod_{j=1}^{\infty}\frac{1}{(1-q^{5j-3})(1-q^{5j-2})}.$$

*Proof.* Take $x=1$ in Lemma 5.5 and use the value of $G(1)$ given by Lemma 5.4 to obtain

$$\begin{aligned}1+\sum_{k=1}^{\infty}&\frac{q^{k^2}}{(1-q)(1-q^2)\cdots(1-q^k)}\\ &=\prod_{j=1}^{\infty}\frac{(1-q^{5j-3})(1-q^{5j-2})(1-q^{5j})}{(1-q^j)}=\prod_{j=1}^{\infty}\frac{1}{(1-q^{5j-4})(1-q^{5j-1})}.\end{aligned}$$

This proves the first result. The second result follows in the same way by taking $x=q$ in Lemma 5.5 and using the value of $(1-q)\,G(q)$ given by Lemma 5.4. □

Combinatorial interpretations of the Rogers–Ramanujan identities will be given in Exercise 5 at the end of the chapter.

We are now ready to show that the function $r(q)$ defined by the infinite product (5.1) has an expansion as a continued fraction given by (5.2).

**Theorem 5.7.** *The following identity holds*:

$$\prod_{j=1}^{\infty}\frac{(1-q^{5j-4})(1-q^{5j-1})}{(1-q^{5j-3})(1-q^{5j-2})}=\cfrac{1}{1+\cfrac{q}{1+\cfrac{q^2}{1+\cfrac{q^3}{1+\cdots}}}}.$$

*Proof.* Substitute the results of Theorem 5.6 into the identity (5.5) and take reciprocals. □

## 5.2 Properties of the Rogers–Ramanujan continued fraction

In this section we develop some of the main properties of the function $r(q)$. The first result relates the Rogers–Ramanujan continued fraction to Dedekind's eta function.

**Theorem 5.8.** *The function* $r = r(q)$ *defined by* (5.1) *satisfies the identities*

$$\frac{1}{r(q^5)} - 1 - r(q^5) = \frac{1}{q} \prod_{n=1}^{\infty} \frac{1 - q^j}{1 - q^{25j}} \tag{5.8}$$

*and*

$$\frac{1}{r^5(q)} - 11 - r^5(q) = \frac{1}{q} \prod_{n=1}^{\infty} \frac{(1 - q^j)^6}{(1 - q^{5j})^6}. \tag{5.9}$$

Before starting the proof, observe that (5.8) can be written in the form

$$\prod_{j=1}^{\infty} (1 - q^j) = A(q^5) - qB(q^5) - q^2 C(q^5)$$

where

$$A(q) = \prod_{j=1}^{\infty} \frac{(1 - q^{5j-3})(1 - q^{5j-2})(1 - q^{5j})}{(1 - q^{5j-4})(1 - q^{5j-1})},$$

$$B(q) = \prod_{j=1}^{\infty} (1 - q^{5j})$$

and

$$C(q) = \prod_{j=1}^{\infty} \frac{(1 - q^{5j-4})(1 - q^{5j-1})(1 - q^{5j})}{(1 - q^{5j-3})(1 - q^{5j-2})}.$$

Hence, (5.8) gives a 5-dissection of Euler's product.

*Proof.* We start by replacing $q$ with $q^{24}$ in Euler's product in Theorem 0.9:

$$q \prod_{j=1}^{\infty} (1 - q^{24j}) = \sum_{n=-\infty}^{\infty} (-1)^n q^{(6n+1)^2}. \tag{5.10}$$

Break the sum on the right into five cases according to the residue class of $n$ modulo 5 and write

$$\sum_{n=-\infty}^{\infty}(-1)^n q^{(6n+1)^2} = S_0 + S_1 + S_2 + S_3 + S_4$$

where

$$S_0 = \sum_{n=5k}(-1)^n q^{(6n+1)^2} = q^{81}\sum_{k=-\infty}^{\infty}(-1)^k q^{100(3k+1)^2} q^{-180(3k+1)}, \tag{5.11}$$

$$S_1 = \sum_{n=5k+1}(-1)^n q^{(6n+1)^2} = -q^9\sum_{k=-\infty}^{\infty}(-1)^k q^{100(3k+1)^2} q^{-60(3k+1)}, \tag{5.12}$$

$$S_2 = \sum_{n=5k+2}(-1)^n q^{(6n+1)^2} = q^9\sum_{k=-\infty}^{\infty}(-1)^k q^{100(3k+1)^2} q^{60(3k+1)}, \tag{5.13}$$

$$S_3 = \sum_{n=5k+3}(-1)^n q^{(6n+1)^2} = -q^{81}\sum_{k=-\infty}^{\infty}(-1)^k q^{100(3k+1)^2} q^{180(3k+1)} \tag{5.14}$$

and

$$S_4 = \sum_{n=5k+4}(-1)^n q^{(6n+1)^2} = \sum_{k=-\infty}^{\infty}(-1)^k q^{25(6k+5)^2} = -\sum_{k=-\infty}^{\infty}(-1)^k q^{25(6k+1)^2}. \tag{5.15}$$

The summation index $k$ has been replaced with $-k-1$ to obtain the very last sum in (5.15). The identities (5.11) and (5.14) may be combined by the quintuple product identity (0.34) to give

$$\begin{aligned}
&\sum_{n\equiv 0 \text{ or } 3 \pmod 5}(-1)^n q^{(6n+1)^2} \\
&= q^{81}\sum_{k=-\infty}^{\infty} q^{100(3k+1)^2}\left((-q^{180})^{3k+1} - (-q^{180})^{-3k-1}\right) \\
&= q\prod_{j=1}^{\infty}(1+q^{600j-120})(1+q^{600j-480}) \\
&\qquad\qquad \times (1-q^{1200j-840})(1-q^{1200j-360})(1-q^{600j})
\end{aligned}$$

$$= q \prod_{j=1}^{\infty} \frac{(1-q^{1200j-240})(1-q^{1200j-960})}{(1-q^{600j-120})(1-q^{600j-480})} \times (1-q^{1200j-840})(1-q^{1200j-360})(1-q^{600j})$$

$$= q \prod_{j=1}^{\infty} \frac{(1-q^{600j-240})(1-q^{600j-360})(1-q^{600j})}{(1-q^{600j-120})(1-q^{600j-480})}$$

$$= \frac{q^{25}}{r(q^{120})} \prod_{j=1}^{\infty} (1-q^{600j}). \tag{5.16}$$

A similar calculation applied to (5.12) and (5.13) gives

$$\sum_{n\equiv 1 \text{ or } 2 \pmod 5} (-1)^n q^{(6n+1)^2} = -q^{25}\, r(q^{120}) \prod_{j=1}^{\infty} (1-q^{600j}). \tag{5.17}$$

Lastly, by (5.15) and Euler's product in Theorem 0.9,

$$\sum_{n\equiv 4 \pmod 5} (-1)^n q^{(6n+1)^2} = -\sum_{k=-\infty}^{\infty} (-1)^k q^{25(6k+1)^2} = -q^{25} \prod_{j=1}^{\infty} (1-q^{600j}). \tag{5.18}$$

On adding (5.16), (5.17), and (5.18) and substituting the result back into (5.10) we deduce that

$$q \prod_{j=1}^{\infty} (1-q^{24j}) = \left( \frac{1}{r(q^{120})} - r(q^{120}) - 1 \right) q^{25} \prod_{j=1}^{\infty} (1-q^{600j}).$$

On dividing both sides by $q^{25} \prod_{j=1}^{\infty}(1-q^{600j})$ and then replacing $q^{24}$ with $q$, we obtain (5.8).

The identity (5.9) follows from (5.8) by successively replacing $q$ with $\omega^m q$, where $\omega = \exp(2\pi i/5)$ and $m \in \{0, 1, 2, 3, 4\}$, and multiplying the resulting five equalities together. Here are the details. Let

$$\alpha = \frac{1+\sqrt{5}}{2} \quad \text{and} \quad \beta = \frac{1-\sqrt{5}}{2}$$

so that

$$1 - x - x^2 = (1-\alpha x)(1-\beta x),$$

and note that $\alpha\beta = -1$ and $\alpha^5 + \beta^5 = 11$. The left-hand side of (5.8) is

$$q^{-1}\prod_{j=1}^{\infty}\frac{(1-q^{25j-20})(1-q^{25j-5})}{(1-q^{25j-15})(1-q^{25j-10})} - 1 - q\prod_{j=1}^{\infty}\frac{(1-q^{25j-20})(1-q^{25j-5})}{(1-q^{25j-15})(1-q^{25j-10})},$$

so replacing $q$ with $\omega^j q$, where $j \in \{0, 1, 2, 3, 4\}$, and multiplying, we find that

$$\begin{aligned}
&\prod_{j=0}^{4}\left(\frac{1}{\omega^j r(q^5)} - 1 - \omega^j r(q^5)\right) \\
&= \prod_{j=0}^{4}\frac{1}{\omega^j r(q^5)}\left(1-\alpha\omega^j r(q^5)\right)\left(1-\beta\omega^j r(q^5)\right) \\
&= \frac{1}{r^5(q^5)}\left(1-\alpha^5 r^5(q^5)\right)\left(1-\beta^5 r^5(q^5)\right) \\
&= \frac{1}{r^5(q^5)} - 11 - r^5(q^5). \qquad (5.19)
\end{aligned}$$

On the other hand, replacing $q$ with $\omega^j q$ in the right-hand side of (5.8) and multiplying with the help of (0.6) in the case $m = 5$, we deduce that

$$\prod_{j=0}^{4}\left(\frac{1}{\omega^j q}\frac{E(\omega^j q)}{E(q^{25})}\right) = \frac{1}{q^5}\frac{E(q^5)^6}{E(q^{25})^6}, \qquad (5.20)$$

where $E(q)$ is Euler's product. On equating (5.19) and (5.20) and then replacing $q^5$ with $q$, we obtain (5.9). □

**Corollary 5.9.** *As $q$ increases from 0 to 1, the function $r(q)$ defined by the infinite product in* (5.1) *strictly increases from* 0 *to* $(\sqrt{5}-1)/2$.

*Proof.* For $0 < q < 1$ let $f(q)$ be the function defined by

$$f(q) = \frac{1}{q}\prod_{j=1}^{\infty}\frac{1-q^j}{1-q^{25j}} = \frac{1}{q}\prod_{j=1}^{\infty}\frac{1}{1+q^j+q^{2j}+\cdots+q^{24j}}.$$

Clearly, $f$ is strictly decreasing on the interval $0 < q < 1$ and

$$\lim_{q\to 0+} f(q) = +\infty \qquad \text{and} \qquad \lim_{q\to 1-} f(q) = 0.$$

By the first part of Theorem 5.8,

$$\frac{1}{r(q^5)} - 1 - r(q^5) = f(q)$$

and hence, on solving for $r(q)$, we find that

$$r(q) = \frac{1}{2}\left(-1 - f(q^{1/5}) + \sqrt{(1 + f(q^{1/5}))^2 + 4}\right)$$

where the plus sign is used in front of the square root term, and not the minus sign, because $f(q)$ and $r(q)$ are both positive when $0 < q < 1$. It is easy to check that as $x$ increases from 0 to $+\infty$, the function

$$\frac{1}{2}\left(-1 - x + \sqrt{(1 + x)^2 + 4}\right)$$

strictly decreases from $(\sqrt{5} - 1)/2$ to 0. The result now follows. □

The next result gives the effect of two transformation formulas for $r(q)$. The formulas would be more complicated to write down without the factor $q^{1/5}$ in the definition (5.1).

**Theorem 5.10.** *Suppose* $\operatorname{Re} t > 0$ *and let* $\alpha = (1 + \sqrt{5})/2$. *The following modular transformation formulas hold*:

$$r(e^{-2\pi/t}) = \frac{1 - \alpha r(e^{-2\pi t})}{\alpha + r(e^{-2\pi t})} \tag{5.21}$$

*and*

$$r^5(e^{-2\pi/5t}) = \frac{1 - \alpha^5 r^5(e^{-2\pi t})}{\alpha^5 + r^5(e^{-2\pi t})}. \tag{5.22}$$

*Furthermore, the following evaluations hold*:

$$r(e^{-2\pi}) = \sqrt{\alpha^2 + 1} - \alpha = \sqrt{\frac{5 + \sqrt{5}}{2}} - \frac{1 + \sqrt{5}}{2} \tag{5.23}$$

*and*

$$r(e^{-2\pi/\sqrt{5}}) = \left(\sqrt{\alpha^{10} + 1} - \alpha^5\right)^{1/5}. \tag{5.24}$$

*Proof.* Let $q = e^{-2\pi t}$ and $p = e^{-2\pi/t}$, and write $r = r(q)$, $R = r(p)$. By (5.8) and the definition of Dedekind's eta function (2.20) we have

$$\frac{1}{r} - 1 - r = \frac{1}{q^{1/5}} \prod_{j=1}^{\infty} \frac{(1 - q^{j/5})}{(1 - q^{5j})} = \frac{\eta(it/5)}{\eta(5it)}$$

and

$$\frac{1}{R} - 1 - R = \frac{1}{p^{1/5}} \prod_{j=1}^{\infty} \frac{(1-p^{j/5})}{(1-p^{5j})} = \frac{\eta(i/5t)}{\eta(5i/t)}.$$

By the modular transformation (2.8) we have

$$\frac{\eta(it/5)}{\eta(5it)} = 5 \times \frac{\eta(5i/t)}{\eta(i/5t)}$$

and it follows that

$$\left(\frac{1}{r} - 1 - r\right)\left(\frac{1}{R} - 1 - R\right) = 5.$$

This may be rearranged to give

$$\left(1 - \frac{1}{2}(r+R) - rR\right)^2 = \frac{5}{4}(r+R)^2$$

and it follows that

$$1 - \frac{1}{2}(r+R) - rR = \pm\frac{\sqrt{5}}{2}(r+R). \tag{5.25}$$

The sign may be determined by substituting the value $t = 1$. Then $p = q = e^{-2\pi}$, so $r = R$ and (5.25) reduces to

$$1 - r - r^2 = \pm\sqrt{5}\,r. \tag{5.26}$$

By Corollary 5.9 we know that if $0 < q < 1$ then $0 < r < \frac{\sqrt{5}-1}{2}$ and therefore $0 < 1 - r - r^2 < 1$. It follows that the plus sign must be selected in (5.26) and hence also in (5.25). The identity (5.21) then follows by rearrangement of (5.25).

The identity (5.22) may be proved by the same method starting from (5.9). We omit the details as they are similar.

The identity (5.21) may be written in the symmetric form

$$\left(\alpha + r\big(e^{-2\pi t}\big)\right)\left(\alpha + r\big(e^{-2\pi/t}\big)\right) = 1 + \alpha^2. \tag{5.27}$$

The evaluation (5.23) is obtained by setting $t = 1$, solving for $r(e^{-2\pi})$ and choosing the positive solution.

In a similar way, replacing $t$ with $\sqrt{t/5}$ in (5.22) and rearranging, we obtain the symmetric form

$$\left(\alpha^5 + r^5\left(e^{-2\pi\sqrt{t/5}}\right)\right)\left(\alpha^5 + r^5\left(e^{-2\pi/\sqrt{5t}}\right)\right) = 1 + \alpha^{10}. \tag{5.28}$$

The evaluation (5.24) is obtained from this by setting $t = 1$, solving for $r(e^{-2\pi/\sqrt{5}})$, and selecting the positive solution. □

The next result gives a quintic transformation formula for $r(q)$.

**Theorem 5.11.** *Let $u = r(q)$ and $v = r(q^5)$. Then*

$$u^5 = v\frac{1 - 2v + 4v^2 - 3v^3 + v^4}{1 + 3v + 4v^2 + 2v^3 + v^4} \tag{5.29}$$

*and*

$$v = \frac{1 - \alpha\left(\dfrac{1 - \alpha^5 u^5}{\alpha^5 + u^5}\right)^{1/5}}{\alpha + \left(\dfrac{1 - \alpha^5 u^5}{\alpha^5 + u^5}\right)^{1/5}}. \tag{5.30}$$

*Proof.* Let $q = e^{-2\pi t}$. Replace $t$ with $5t$ in (5.21) and combine the result with (5.22) to deduce

$$\left(\frac{1 - \alpha v}{\alpha + v}\right)^5 = \frac{1 - \alpha^5 u^5}{\alpha^5 + u^5}.$$

If we solve for $u^5$ we get (5.29), and if we solve for $v$ we get (5.30). □

## 5.3 Product to series identities

In this section we establish some product-to-sum identities for weight one and weight two modular forms. We begin with identities for weight one modular forms that may be compared with the results in Theorems 3.10 and 3.19.

**Theorem 5.12.** *The following identities hold:*

$$\prod_{j=1}^{\infty} \frac{(1 - q^j)^2}{(1 - q^{5j-4})^5(1 - q^{5j-1})^5}$$
$$= 1 + 3\sum_{j=1}^{\infty}\left(\frac{q^{5j-4}}{1 - q^{5j-4}} - \frac{q^{5j-1}}{1 - q^{5j-1}}\right) + \sum_{j=1}^{\infty}\left(\frac{q^{5j-3}}{1 - q^{5j-3}} - \frac{q^{5j-2}}{1 - q^{5j-2}}\right) \tag{5.31}$$

*and*

$$q\prod_{j=1}^{\infty}\frac{(1-q^j)^2}{(1-q^{5j-3})^5(1-q^{5j-2})^5}$$
$$=\sum_{j=1}^{\infty}\left(\frac{q^{5j-4}}{1-q^{5j-4}}-\frac{q^{5j-1}}{1-q^{5j-1}}\right)-3\sum_{j=1}^{\infty}\left(\frac{q^{5j-3}}{1-q^{5j-3}}-\frac{q^{5j-2}}{1-q^{5j-2}}\right). \tag{5.32}$$

*Proof.* Replace $q$ with $q^5$ in the second explicit form of Halphen's identity in Theorem 1.13 and then put $x = y = z = q$ to deduce (5.31). The identity (5.32) may be obtained by replacing $q$ with $q^{5/2}$ in the third explicit form of Halphen's identity in Theorem 1.14 and then letting $a = b = c = q^{1/2}$. □

The next result gives Lambert series expansions for weight two modular forms.

**Theorem 5.13.** *The following identities hold*:

$$\prod_{j=1}^{\infty}\frac{(1-q^j)^5}{(1-q^{5j})}=1-5\sum_{j=1}^{\infty}\left(\frac{j}{5}\right)\frac{jq^j}{1-q^j} \tag{5.33}$$

*and*

$$q\prod_{j=1}^{\infty}\frac{(1-q^{5j})^5}{(1-q^j)}=\sum_{j=1}^{\infty}\left(\frac{j}{5}\right)\frac{q^j}{(1-q^j)^2}, \tag{5.34}$$

*where the values of the Legendre symbol are given by*

$$\left(\frac{j}{5}\right)=\begin{cases}1 & \text{if } j\equiv 1 \text{ or } 4 \pmod 5,\\ -1 & \text{if } j\equiv 2 \text{ or } 3 \pmod 5,\\ 0 & \text{if } j\equiv 0 \pmod 5.\end{cases}$$

*Proof.* Let $\omega = \exp(2\pi i/5)$, take $x = \omega$, $y = \omega^2$ in the first explicit form of Weierstrass' identity in Theorem 1.17, and follow the procedure in Example 1.26 to obtain (5.33).

Next, replace $(q, a, b)$ with $(q^{5/2}, q^{1/2}, q)$ in the second explicit form of Weierstrass' identity in Theorem 1.18 to obtain

$$q\prod_{j=1}^{\infty}\frac{(1-q^{5j})^5}{(1-q^j)}=\sum_{k=1}^{\infty}\frac{kq^k(1-q^k)(1-q^{2k})}{1-q^{5k}}=\sum_{k=1}^{\infty}\sum_{j=1}^{\infty}\left(\frac{j}{5}\right)k\,q^{jk}.$$

Now interchange the order of summation and sum over $k$ to deduce (5.34). □

## 5.4 Ramanujan's modulo 5 partition identity

The identity (5.34) leads to a short proof of one of Ramanujan's partition congruences.

**Corollary 5.14.** *Let $p(n)$ denote the number of partitions of $n$. Then,*

$$\sum_{n=0}^{\infty} p(5n+4)q^n = 5\prod_{j=1}^{\infty}\frac{(1-q^{5j})^5}{(1-q^j)^6}.$$

*In particular, $p(5n+4)$ is divisible by 5.*

*Proof.* Start with the identity (5.34) in the form

$$q\prod_{j=1}^{\infty}\frac{(1-q^{5j})^5}{(1-q^j)} = \sum_{j=1}^{\infty}\sum_{k=1}^{\infty}\left(\frac{j}{5}\right) k\,q^{jk}. \tag{5.35}$$

The left-hand side may be expanded using Theorem 0.11 to give

$$\left(\sum_{n=0}^{\infty} p(n)q^{n+1}\right)\left(\prod_{j=1}^{\infty}(1-q^{5j})^5\right) = \sum_{j=1}^{\infty}\sum_{k=1}^{\infty}\left(\frac{j}{5}\right) k\,q^{jk}.$$

Equate terms on both sides for which the power of $q$ is a multiple of 5 to obtain

$$\left(\sum_{n=0}^{\infty} p(5n+4)q^{5n+5}\right)\left(\prod_{j=1}^{\infty}(1-q^{5j})^5\right) = \sum_{j=1}^{\infty}\sum_{k=1}^{\infty}\left(\frac{j}{5}\right)(5k)\,q^{5jk}.$$

The right-hand side may be summed by (5.35) to give

$$\left(\sum_{n=0}^{\infty} p(5n+4)q^{5n+5}\right)\left(\prod_{j=1}^{\infty}(1-q^{5j})^5\right) = 5q^5\prod_{j=1}^{\infty}\frac{(1-q^{25j})^5}{(1-q^{5j})}.$$

Now replace $q^5$ with $q$ and rearrange to complete the proof. □

## 5.5 The six level 5 functions

We will now introduce six modular functions of level 5 that come in three pairs.

1. Let $x_5$ and $z_5$ be defined by

$$x_5(q) = q\prod_{j=1}^{\infty}\frac{(1-q^{5j-4})^5(1-q^{5j-1})^5}{(1-q^{5j-3})^5(1-q^{5j-2})^5} \tag{5.36}$$

and

$$z_5(q) = \prod_{j=1}^{\infty} \frac{(1-q^j)^2}{(1-q^{5j-4})^5(1-q^{5j-1})^5}. \tag{5.37}$$

Observe that $x_5$ is the ratio of the infinite products that occur in Theorem 5.12, and $z_5$ is one of those infinite products. The function $x_5$ is also the fifth power of the Rogers–Ramanujan continued fraction $r(q)$.

2. Let $w_5$ and $y_5$ be the eta-quotients defined by

$$w_5 = q \prod_{j=1}^{\infty} \frac{(1-q^{5j})^6}{(1-q^j)^6} \qquad \text{and} \qquad y_5 = \prod_{j=1}^{\infty} \frac{(1-q^j)^5}{(1-q^{5j})}. \tag{5.38}$$

Here, $w_5$ is the ratio of the infinite products that occur in Theorem 5.13 and $y_5$ is one of those infinite products.

3. Finally, let $X_5$ and $Z_5$ be defined by

$$Z_5 = \frac{1}{4}\left(5P(q^5) - P(q)\right) \qquad \text{and} \qquad X_5 = \frac{1}{Z_5^2} \times q \prod_{j=1}^{\infty} (1-q^j)^4(1-q^{5j})^4 \tag{5.39}$$

where $P$ is Ramanujan's weight two Eisenstein series defined by (0.71).

The functions $x_5$, $w_5$, and $X_5$ are modular functions. The function $z_5$ is a modular form of weight one, while $y_5$ and $Z_5$ are modular forms of weight two. The following examples, that will be proved later in the chapter, indicate why the functions have been defined in pairs. In Theorem 5.29, it will be shown that

$$\frac{\mathrm{d}}{\mathrm{d}x_5}\left(x_5(1-11x_5-x_5^2)\frac{\mathrm{d}z_5}{\mathrm{d}x_5}\right) = (3+x_5)z_5.$$

This is an analogue of the differential equation in Theorem 4.24

$$\frac{\mathrm{d}}{\mathrm{d}x_\ell}\left(x_\ell(1-x_\ell)\frac{\mathrm{d}z_\ell}{\mathrm{d}x_\ell}\right) = s_\ell(1-s_\ell)z_\ell$$

that holds for levels $\ell = 1, 2, 3$, and 4. It will be shown further that

$$z_5 = \sum_{n=0}^{\infty} \left\{ \sum_{j=0}^{n} \binom{n}{j}^2 \binom{n+j}{j} \right\} x_5^n, \tag{5.40}$$

and the series on the right is an analogue of the ${}_2F_1$ hypergeometric function. Moreover, an analogue of Theorem 4.30 will be shown to be

$$Z_5 = \sum_{n=0}^{\infty} \left\{ \binom{2n}{n} \sum_{j=0}^{n} \binom{n}{j}^2 \binom{n+j}{j} \right\} X_5^n. \tag{5.41}$$

Finally, the expansion of $y_5$ in powers of $w_5$ will be shown to be

$$\frac{\eta_1^5}{\eta_5} = \sum_{n=0}^{\infty} \left\{ \sum_{j=0}^{n} (-1)^{n+j} \binom{n}{j}^3 \binom{4n-5j}{3n} \right\} \left( \frac{\eta_5}{\eta_1} \right)^{6j}. \tag{5.42}$$

We now record the basic properties of the functions $w_5$, $x_5$, $y_5$, $z_5$, $X_5$, and $Z_5$.

**Theorem 5.15.** *The following identities hold:*

$$\frac{1}{w_5} = \frac{1}{x_5} - 11 - x_5 \tag{5.43}$$

*and*

$$x_5 z_5^2 = w_5 y_5. \tag{5.44}$$

*Hence,*

$$\frac{1}{x_5} + x_5 = \frac{1}{w_5}\sqrt{1 + 22w_5 + 125w_5^2} \tag{5.45}$$

*and*

$$z_5^2(1 + x_5^2) = y_5\sqrt{1 + 22w_5 + 125w_5^2}. \tag{5.46}$$

*Proof.* The first identity is just a restatement of (5.9). The second identity can be obtained by routine manipulations of infinite products. For,

$$\begin{aligned} x_5 z_5^2 &= q \prod_{j=1}^{\infty} \frac{(1-q^{5j-4})^5(1-q^{5j-1})^5}{(1-q^{5j-3})^5(1-q^{5j-2})^5} \times \prod_{j=1}^{\infty} \frac{(1-q^j)^4}{(1-q^{5j-4})^{10}(1-q^{5j-1})^{10}} \\ &= q \prod_{j=1}^{\infty} \frac{(1-q^j)^4}{(1-q^{5j-4})^5(1-q^{5j-3})^5(1-q^{5j-2})^5(1-q^{5j-1})^5}. \end{aligned}$$

On using

$$\prod_{j=1}^{\infty}(1-q^{5j-4})(1-q^{5j-3})(1-q^{5j-2})(1-q^{5j-1}) = \prod_{j=1}^{\infty} \frac{1-q^j}{1-q^{5j}},$$

the result simplifies to

$$\begin{aligned} x_5 z_5^2 &= q \prod_{j=1}^{\infty} \frac{(1-q^{5j})^5}{(1-q^j)} \\ &= \left( q \prod_{j=1}^{\infty} \frac{(1-q^{5j})^6}{(1-q^j)^6} \right) \times \left( \prod_{j=1}^{\infty} \frac{(1-q^j)^5}{(1-q^{5j})} \right) = w_5 y_5. \end{aligned}$$

The third identity follows from the first identity by simple manipulation:

$$\begin{aligned}\frac{1}{x_5}+x_5&=\sqrt{\left(\frac{1}{x_5}-x_5\right)^2+4}\\&=\sqrt{\left(\frac{1}{w_5}+11\right)^2+4}\\&=\frac{1}{w_5}\sqrt{1+22w_5+125w_5^2}.\end{aligned}$$

This proves (5.45). The last identity (5.46) follows by taking the product of (5.44) and (5.45). □

The next result is an analogue of Theorem 4.21.

**Theorem 5.16.** *The following identities hold*:

$$y_5^6 w_5 = q\prod_{j=1}^{\infty}(1-q^j)^{24}$$

*and*

$$y_5^6 w_5^5 = q^5\prod_{j=1}^{\infty}(1-q^{5j})^{24}.$$

*Proof.* These are immediate from the definitions in (5.38). □

**Theorem 5.17.** *The following differentiation formulas hold*:

$$q\frac{\mathrm{d}}{\mathrm{d}q}\log x_5 = y_5$$

*and*

$$q\frac{\mathrm{d}}{\mathrm{d}q}\log w_5 = Z_5.$$

*Proof.* Using the definition of $x_5$ we find that

$$\begin{aligned}q\frac{\mathrm{d}}{\mathrm{d}q}\log x_5 &= q\frac{\mathrm{d}}{\mathrm{d}q}\log\left(q\prod_{j=1}^{\infty}\frac{(1-q^{5j-4})^5(1-q^{5j-1})^5}{(1-q^{5j-3})^5(1-q^{5j-2})^5}\right)\\&=1-5\sum_{j=1}^{\infty}\left(\frac{j}{5}\right)\frac{jq^j}{1-q^j}.\end{aligned}$$

The series may be converted to a product by Theorem 5.13 to give

$$q\frac{\mathrm{d}}{\mathrm{d}q}\log x_5 = \prod_{j=1}^{\infty}\frac{(1-q^j)^5}{(1-q^{5j})} = y_5.$$

The proof of the second result is even simpler. By the definitions of $w_5$ and $Z_5$ together with (0.61),

$$q\frac{\mathrm{d}}{\mathrm{d}q}\log w_5 = q\frac{\mathrm{d}}{\mathrm{d}q}\log\left(\frac{\eta_5^6}{\eta_1^6}\right) = \frac{1}{4}\left(5P(q^5) - P(q)\right) = Z_5.$$

□

The results in the next theorem are analogues of the differentiation formula (4.19)

$$q\frac{\mathrm{d}x_\ell}{\mathrm{d}q} = z_\ell^2 x_\ell(1-x_\ell)$$

that holds for $\ell \in \{1,2,3,4\}$.

**Theorem 5.18.** *The following differentiation formulas hold:*

$$q\frac{\mathrm{d}x_5}{\mathrm{d}q} = z_5^2 x_5(1-11x_5-x_5^2) \tag{5.47}$$

*and*

$$q\frac{\mathrm{d}w_5}{\mathrm{d}q} = y_5\, w_5\sqrt{1+22w_5+125w_5^2}. \tag{5.48}$$

The corresponding formula for $q\dfrac{\mathrm{d}X_5}{\mathrm{d}q}$ will be given below, in Theorem 5.21.

*Proof.* Using Theorems 5.15 and 5.17 we find that

$$q\frac{\mathrm{d}x_5}{\mathrm{d}q} = x_5 y_5 = \frac{x_5^2 z_5^2}{w_5} = z_5^2 x_5(1-11x_5-x_5^2).$$

This proves the first result. Next, by (5.43) and the chain rule,

$$q\frac{\mathrm{d}w_5}{\mathrm{d}q} = -w_5^2\, q\frac{\mathrm{d}}{\mathrm{d}q}\frac{1}{w_5} = -w_5^2\, q\frac{\mathrm{d}}{\mathrm{d}q}\left(\frac{1}{x_5} - 11 - x_5\right) = w_5^2\left(\frac{1}{x_5^2}+1\right) q\frac{\mathrm{d}x_5}{\mathrm{d}q}.$$

On using (5.47) and Theorem 5.15, the result simplifies to

$$q\frac{\mathrm{d}w_5}{\mathrm{d}q} = w_5^2\left(\frac{1}{x_5^2}+1\right) z_5^2 x_5(1-11x_5-x_5^2) = y_5\, w_5\sqrt{1+22w_5+125w_5^2}.$$

□

The next result relates the weight two modular forms $y_5$ and $Z_5$ to the weight one modular form $z_5$.

**Theorem 5.19.** *The following results hold*:

$$y_5 = z_5^2(1 - 11x_5 - x_5^2) \tag{5.49}$$

*and*

$$Z_5 = y_5\sqrt{1 + 22w_5 + 125w_5^2} = z_5^2(1 + x_5^2). \tag{5.50}$$

*Proof.* The first result is an immediate consequence of the results in Theorem 5.15. We will prove the second result. By Theorem 5.17 and Theorem 5.18 followed by an application of (5.46), we find that

$$Z_5 = q\frac{\mathrm{d}}{\mathrm{d}q}\log w_5 = y_5\sqrt{1 + 22w_5 + 125w_5^2} = z_5^2(1 + x_5^2).$$

□

The next result gives an algebraic relation for the modular functions $w_5$, $x_5$, and $X_5$.

**Theorem 5.20.** *The following identities hold*:

$$X_5 = \frac{w_5}{1 + 22w_5 + 125w_5^2} = \frac{x_5(1 - 11x_5 - x_5^2)}{(1 + x_5^2)^2}.$$

*Proof.* By the definition of $X_5$ and the results of Theorems 5.16 and 5.19,

$$X_5 = \frac{\eta_1^4\eta_5^4}{Z_5^2} = \frac{(y_5^6w_5)^{1/6}(y_5^6w_5^5)^{1/6}}{y_5^2(1 + 22w_5 + 125w_5^2)} = \frac{w_5}{1 + 22w_5 + 125w_5^2}.$$

This proves the first equality. The second equality is obtained from the first by using (5.43). □

**Theorem 5.21.** *The following differentiation formula holds*:

$$q\frac{\mathrm{d}X_5}{\mathrm{d}q} = Z_5X_5\sqrt{1 - 44X_5 - 16X_5^2}.$$

*Proof.* By Theorem 5.20,

$$\frac{1}{X_5} = \frac{1}{w_5} + 22 + 125w_5. \tag{5.51}$$

Taking the derivative and multiplying by $-1$ gives

$$\frac{1}{X_5^2}\,q\frac{\mathrm{d}X_5}{\mathrm{d}q} = \left(\frac{1}{w_5^2} - 125\right)q\frac{\mathrm{d}w_5}{\mathrm{d}q}. \tag{5.52}$$

By Theorem 5.17 we have

$$q\frac{dw_5}{dq} = w_5 Z_5 \tag{5.53}$$

and by (5.51) we have

$$\begin{aligned}\frac{1}{w_5} - 125w_5 &= \sqrt{\left(\frac{1}{w_5} + 125w_5\right)^2 - 500} \\ &= \sqrt{\left(\frac{1}{X_5} - 22\right)^2 - 500} \\ &= \sqrt{\frac{1}{X_5^2} - \frac{44}{X_5} - 16}. \end{aligned} \tag{5.54}$$

The proof may be completed by substituting (5.53) and (5.54) into (5.52). □

Because of their frequent occurrence above, and in the coming sections, let us define $A_5$ and $B_5$ by

$$A_5 = \sqrt{1 + 22w_5 + 125w_5^2} \tag{5.55}$$

and

$$B_5 = \sqrt{1 - 44X_5 - 16X_5^2}. \tag{5.56}$$

## 5.6 Eisenstein series

This section is devoted to obtaining expressions for the Eisenstein series $P(q)$, $P(q^5)$, $Q(q)$, $Q(q^5)$, $R(q)$, and $R(q^5)$ in terms of the eta-quotients $w_5$ and $y_5$. We begin with the Eisenstein series $P$.

**Theorem 5.22.** *Let $w_5$, $y_5$, and $A_5$ be given by* (5.38) *and* (5.55) *and let $P$ be Ramanujan's Eisenstein series defined by* (0.71). *Then*

$$A_5 y_5 = \frac{1}{4}\left(5P(q^5) - P(q)\right) \tag{5.57}$$

*and*

$$A_5 w_5 \frac{dy_5}{dw_5} = \frac{5}{24}\left(P(q) - P(q^5)\right). \tag{5.58}$$

*Proof.* The first result is simply a restatement of the identity (5.50), using the definition of $Z_5$. We will prove the second result. By Theorem 5.18 and the definitions of $w_5$, $y_5$, and $A_5$ we deduce that $w_5 = 0$ when $q = 0$, and

$$\frac{\mathrm{d}w_5}{\mathrm{d}q} = \frac{A_5 w_5 y_5}{q} = 1 + O(q)$$

in a neighborhood of $q = 0$. Hence, the inverse function $q = q(w_5)$ exists in a neighborhood of $w_5 = 0$ and so the function $y_5$ may be regarded as a function of $w_5$. By direct calculation we find that

$$\begin{aligned} q\frac{\mathrm{d}y_5}{\mathrm{d}q} = y_5\, q\frac{\mathrm{d}}{\mathrm{d}q}\log y_5 &= y_5\, q\frac{\mathrm{d}}{\mathrm{d}q}\log\prod_{j=1}^{\infty}\frac{(1-q^j)^5}{(1-q^{5j})} \\ &= 5y_5\sum_{j=1}^{\infty}\left(\frac{jq^{5j}}{1-q^{5j}} - \frac{jq^j}{1-q^j}\right) \\ &= \frac{5y_5}{24}\left(P(q) - P(q^5)\right). \end{aligned}$$

Hence, by the chain rule and Theorem 5.18, we obtain

$$\frac{\mathrm{d}y_5}{\mathrm{d}w_5} = q\frac{\mathrm{d}y_5}{\mathrm{d}q}\bigg/ q\frac{\mathrm{d}w_5}{\mathrm{d}q} = \frac{5(P(q) - P(q^5))}{24A_5 w_5}.$$

This gives (5.58). □

The next goal is to establish parameterizations for each of $Q(q)$, $Q(q^5)$, $R(q)$, and $R(q^5)$ in terms of $w_5$ and $y_5$. We first prove a preliminary result.

**Theorem 5.23.** *Let $w_5$, $y_5$, and $A_5$ be given by* (5.38) *and* (5.55) *and let $Q$ and $R$ be Ramanujan's Eisenstein series defined by* (0.72) *and* (0.73). *Then*

$$25Q(q^5) - Q(q) = 24y_5^2(1 - 125w_5^2) \tag{5.59}$$

*and*

$$125R(q^5) - R(q) = 18000A_5 w_5^2 y_5^3 + A_5 y_5\left(125Q(q^5) - Q(q)\right).$$

*Proof.* From (5.57),

$$12\left(5P(q^5) - P(q)\right) = 48A_5 y_5.$$

Applying $q\frac{\mathrm{d}}{\mathrm{d}q}$ we obtain

$$12q\frac{\mathrm{d}}{\mathrm{d}q}(5P(q^5) - P(q)) = 48\left(A_5\frac{\mathrm{d}y_5}{\mathrm{d}w_5} + (11 + 125w_5)\frac{y_5}{A_5}\right) q\frac{\mathrm{d}w_5}{\mathrm{d}q}.$$

The derivatives may be computed using Theorems 0.21, 5.18, and 5.22. The result is

$$\begin{aligned}&25P(q^5)^2 - P(q)^2 - 25Q(q^5) + Q(q)\\&= 48\left(\frac{5}{24w_5}\left(P(q) - P(q^5)\right) + (11 + 125w_5)\frac{y_5}{A_5}\right)A_5w_5y_5.\end{aligned}$$

On completing the squares for $P(q)$ and $P(q^5)$, we obtain

$$25Q(q^5) - Q(q) = \left(5P(q^5) + A_5y_5\right)^2 - (P(q) + 5A_5y_5)^2 + 24y_5^2(1 - 125w_5^2).$$

By (5.57) this simplifies to give (5.59).

Next, applying $q\frac{\mathrm{d}}{\mathrm{d}q}$ to (5.59) and multiplying by 3 we find that

$$3q\frac{\mathrm{d}}{\mathrm{d}q}\left(25Q(q^5) - Q(q)\right) = \left(144y_5\frac{\mathrm{d}y_5}{\mathrm{d}w_5}(1 - 125w_5^2) - 18000y_5^2w_5\right)q\frac{\mathrm{d}w_5}{\mathrm{d}q}.$$

The derivatives may be computed using Theorems 0.21, 5.18, and 5.22. The result simplifies to

$$\begin{aligned}&125P(q^5)Q(q^5) - P(q)Q(q) - 125R(q^5) + R(q)\\&= 30y_5^2(1 - 125w_5^2)\left(P(q) - P(q^5)\right) - 18000A_5w_5^2y_5^3.\end{aligned}$$

By (5.59) this is equivalent to

$$\begin{aligned}&125P(q^5)Q(q^5) - P(q)Q(q) - 125R(q^5) + R(q)\\&= \frac{5}{4}\left(25Q(q^5) - Q(q)\right)\left(P(q) - P(q^5)\right) - 18000A_5w_5^2y_5^3.\end{aligned}$$

Rearranging, we find

$$125R(q^5) - R(q) = 18000A_5w_5^2y_5^3 + \frac{1}{4}\left(5P(q^5) - P(q)\right)\left(125Q(q^5) - Q(q)\right).$$

Finally, using (5.57) we complete the proof. □

The results in Theorem 5.16 may be rephrased, using Corollary 0.23, as

$$Q^3(q) - R^2(q) = 1728y_5^6w_5 \tag{5.60}$$

and

$$Q^3(q^5) - R^2(q^5) = 1728y_5^6w_5^5, \tag{5.61}$$

while from Theorem 5.23 we have

$$25Q(q^5) - Q(q) = 24y_5^2(1 - 125w_5^2) \tag{5.62}$$

and

$$125R(q^5) - R(q) = 18000A_5w_5^2y_5^3 + A_5y_5\left(125Q(q^5) - Q(q)\right). \tag{5.63}$$

The identities (5.60)–(5.63) give a system of four nonlinear equations that may be solved to give $Q(q)$, $Q(q^5)$, $R(q)$, and $R(q^5)$ in terms of $y_5$ and $w_5$. The solution is given in the next result.

**Theorem 5.24.** *Let $w_5$, $y_5$, and $A_5$ be given by* (5.38) *and* (5.55) *and let $Q$ and $R$ be Ramanujan's Eisenstein series defined by* (0.72) *and* (0.73). *Then*

$$Q(q) = y_5^2(1 + 250w_5 + 3125w_5^2),$$
$$Q(q^5) = y_5^2(1 + 10w_5 + 5w_5^2),$$
$$R(q) = y_5^3A_5(1 - 500w_5 - 15625w_5^2)$$

*and*

$$R(q^5) = y_5^3A_5(1 + 4w_5 - w_5^2).$$

*Proof.* Let

$$u_1 = \frac{Q(q)}{y_5^2}, \quad u_2 = \frac{Q(q^5)}{y_5^2}, \quad u_3 = \frac{R(q)}{A_5y_5^3} \quad \text{and} \quad u_4 = \frac{R(q^5)}{A_5y_5^3}, \tag{5.64}$$

and observe that when $q = 0$ we have $w_5 = 0$, $y_5 = 1$, and $u_1 = u_2 = u_3 = u_4 = 1$. The identities (5.60)–(5.63) become

$$u_1^3 - (1 + 22w_5 + 125w_5^2)u_3^2 = 1728w_5, \tag{5.65}$$
$$u_2^3 - (1 + 22w_5 + 125w_5^2)u_4^2 = 1728w_5^5, \tag{5.66}$$
$$-u_1 + 25u_2 = 24(1 - 125w_5^2) \tag{5.67}$$

and

$$u_1 - 125u_2 - u_3 + 125u_4 = 18000w_5^2, \tag{5.68}$$

respectively. Let us seek a solution of (5.65)–(5.68) that expresses each of $u_1$, $u_2$, $u_3$, and $u_4$ as power series in $w_5$. We begin by determining the linear approximations, and seek solutions of the form

$$u_j = 1 + a_jw_5 + O(w_5^2), \quad 1 \le j \le 4. \tag{5.69}$$

If we substitute (5.69) into (5.65)–(5.68) and compare the first order terms, i.e., the coefficients of $w_5$, we obtain the linear system

$$\begin{pmatrix} 3 & 0 & -2 & 0 \\ 0 & 3 & 0 & -2 \\ -1 & 25 & 0 & 0 \\ 1 & -125 & -1 & 125 \end{pmatrix} \begin{pmatrix} a_1 \\ a_2 \\ a_3 \\ a_4 \end{pmatrix} = \begin{pmatrix} 1750 \\ 22 \\ 0 \\ 0 \end{pmatrix}.$$

The solution is given by $a_1 = 250$, $a_2 = 10$, $a_3 = -500$, and $a_4 = 4$. Next, we determine the quadratic approximations, and set

$$u_j = 1 + a_j w_5 + b_j w_5^2 + O(w_5^3), \quad 1 \le j \le 4. \tag{5.70}$$

If we substitute (5.70) into (5.65)–(5.68) and compare coefficients of $w_5^2$, we obtain the linear system

$$\begin{pmatrix} 3 & 0 & -2 & 0 \\ 0 & 3 & 0 & -2 \\ -1 & 25 & 0 & 0 \\ 1 & -125 & -1 & 125 \end{pmatrix} \begin{pmatrix} b_1 \\ b_2 \\ b_3 \\ b_4 \end{pmatrix} = \begin{pmatrix} 40625 \\ 17 \\ -3000 \\ 18000 \end{pmatrix}.$$

The solution is given by $b_1 = 3125$, $b_2 = 5$, $b_3 = -15625$, $b_4 = -1$.

It turns out that the cubic, quartic, and higher approximations are identical to the quadratic approximations. That is because the quadratic functions

$$u_1 = 1 + 250w_5 + 3125w_5^2, \tag{5.71}$$

$$u_2 = 1 + 10w_5 + 5w_5^2, \tag{5.72}$$

$$u_3 = 1 - 500w_5 - 15625w_5^2, \tag{5.73}$$

$$u_4 = 1 + 4w_5 - w_5^2 \tag{5.74}$$

are in fact solutions of (5.65)–(5.68), as can be checked by direct substitution. The implicit function theorem implies this is the unique solution of (5.65)–(5.68) that satisfies $u_1 = u_2 = u_3 = u_4 = 1$ when $w_5 = 0$. To see this, write (5.65)–(5.68) as a vector equation $\mathbf{F}(w_5, \mathbf{u}) = \mathbf{g}(w_5)$, where $\mathbf{u} = (u_1, u_2, u_3, u_4)$ and

$$\mathbf{F}(w_5, \mathbf{u}) = \mathbf{F}(w_5, u_1, u_2, u_3, u_4) = \begin{pmatrix} u_1^3 - (1 + 22w_5 + 125w_5^2)u_3^2 \\ u_2^3 - (1 + 22w_5 + 125w_5^2)u_4^2 \\ -u_1 + 25u_2 \\ u_1 - 125u_2 - u_3 + 125u_4 \end{pmatrix}.$$

The Jacobian determinant is

$$\left|\frac{\partial \mathbf{F}}{\partial \mathbf{u}}\right|_{w_5=0} = \begin{vmatrix} 3 & 0 & -2 & 0 \\ 0 & 3 & 0 & -2 \\ -1 & 25 & 0 & 0 \\ 1 & -125 & -1 & 125 \end{vmatrix} = -200 \neq 0.$$

On substituting (5.71)–(5.74) into (5.64) we complete the proof of Theorem 5.24. □

The results of Theorem 5.24 can be expressed in terms of $x_5$ and $z_5$ to give an analogue of the parameterizations in Theorem 4.31.

**Theorem 5.25.** *Let $x_5$ and $z_5$ be defined by* (5.36) *and* (5.37), *and let $Q$ and $R$ be Ramanujan's Eisenstein series defined by* (0.72) *and* (0.73). *Then*

$$\begin{aligned} Q(q) &= z_5^4(1+228x_5+494x_5^2-228x_5^3+x_5^4), \\ Q(q^5) &= z_5^4(1-12x_5+14x_5^2+12x_5^3+x_5^4), \\ R(q) &= z_5^6(1+x_5^2)(1-522x_5-10006x_5^2+522x_5^3+x_5^4) \end{aligned}$$

*and*

$$R(q^5) = z_5^6(1+x_5^2)(1-18x_5+74x_5^2+18x_5^3+x_5^4).$$

*Proof.* Use the results of Theorem 5.15 to express the results of Theorem 5.24 in terms of $x_5$ and $z_5$. □

Recall the modular $j$-invariant was defined by (2.21), viz.,

$$j(\tau) = \frac{1728\, Q^3(q)}{Q^3(q)-R^2(q)},$$

where $q = \exp(2\pi i\tau)$. The next result gives parameterizations of the $j$-invariant in terms of the modular functions $w_5$ and $x_5$.

**Theorem 5.26.** *Let $x_5$ and $w_5$ be defined by* (5.36) *and* (5.38). *Then*

$$j(\tau) = \frac{(1+250w_5+3125w_5^2)^3}{w_5} = \frac{(1+228x_5+494x_5^2-228x_5^3+x_5^4)^3}{x_5(1-11x_5-x_5^2)^5}$$

*and*

$$j(5\tau) = \frac{(1+10w_5+5w_5^2)^3}{w_5^5} = \frac{(1-12x_5+14x_5^2+12x_5^3+x_5^4)^3}{x_5^5(1-11x_5-x_5^2)}.$$

*Proof.* Simply substitute the results of Theorems 5.24 and 5.25 into the definition of $j(\tau)$. □

Since $x_5 = r^5$, Theorem 5.26 expresses each of $j(\tau)$ and $j(5\tau)$ as rational functions of the Rogers–Ramanujan continued fraction.

## 5.7 Differential equations

The goal of this section is to use the parameterizations of Ramanujan's Eisenstein series in terms of $w_5$ and $y_5$ from the previous section together with Ramanujan's differential equations to derive:

1. a second order linear differential equation for $z_5$ in terms of $x_5$;
2. a third order linear differential equation for $y_5$ in terms of $w_5$;
3. a third order linear differential equation for $Z_5$ in terms of $X_5$.

For once and for all in this section, let us recall that $w_5, x_5, y_5, z_5, X_5, Z_5, A_5$, and $B_5$ are defined by (5.36)–(5.39), (5.55), and (5.56).

We begin by noting some relations for differential operators.

**Lemma 5.27.** *The following identities for differential operators hold:*

$$q\frac{\mathrm{d}}{\mathrm{d}q} = A_5\, w_5\, y_5\, \frac{\mathrm{d}}{\mathrm{d}w_5} = B_5\, X_5\, Z_5\, \frac{\mathrm{d}}{\mathrm{d}X_5}$$

*and*

$$x_5\, \frac{\mathrm{d}}{\mathrm{d}x_5} = A_5\, w_5\, \frac{\mathrm{d}}{\mathrm{d}w_5}.$$

*Proof.* By the chain rule and (5.48),

$$q\frac{\mathrm{d}}{\mathrm{d}q} = q\,\frac{\mathrm{d}w_5}{\mathrm{d}q}\,\frac{\mathrm{d}}{\mathrm{d}w_5} = A_5\, w_5\, y_5\, \frac{\mathrm{d}}{\mathrm{d}w_5}. \tag{5.75}$$

Similarly, the chain rule and Theorem 5.21 give

$$q\frac{\mathrm{d}}{\mathrm{d}q} = q\,\frac{\mathrm{d}X_5}{\mathrm{d}q}\,\frac{\mathrm{d}}{\mathrm{d}X_5} = B_5\, X_5\, Z_5\, \frac{\mathrm{d}}{\mathrm{d}X_5}.$$

Finally, by the chain rule and (5.47) and (5.49),

$$q\frac{\mathrm{d}}{\mathrm{d}q} = q\,\frac{\mathrm{d}x_5}{\mathrm{d}q}\,\frac{\mathrm{d}}{\mathrm{d}x_5} = z_5^2\, x_5\,(1 - x_5 - x_5^2)\,\frac{\mathrm{d}}{\mathrm{d}x_5} = y_5\, x_5\, \frac{\mathrm{d}}{\mathrm{d}x_5}. \tag{5.76}$$

By combining (5.75) and (5.76), we deduce

$$y_5\, x_5\, \frac{\mathrm{d}}{\mathrm{d}x_5} = A_5\, w_5\, y_5\, \frac{\mathrm{d}}{\mathrm{d}w_5}$$

and this gives the last result. □

The next result gives second order nonlinear differential equations for the weight two modular forms $y_5$ and $Z_5$ in terms of the modular functions $w_5$ and $X_5$, respectively.

**Theorem 5.28.** *The following differential equations hold:*

$$A_5 w_5 \frac{\mathrm{d}}{\mathrm{d}w_5}\left(A_5 w_5 \frac{\mathrm{d}y_5}{\mathrm{d}w_5}\right) = \frac{1}{2y_5}\left(A_5 w_5 \frac{\mathrm{d}y_5}{\mathrm{d}w_5}\right)^2 - \frac{5}{2} w_5 (2+25w_5) y_5 \tag{5.77}$$

*and*

$$B_5 X_5 \frac{\mathrm{d}}{\mathrm{d}X_5}\left(B_5 X_5 \frac{\mathrm{d}Z_5}{\mathrm{d}X_5}\right) = \frac{1}{2Z_5}\left(B_5 X_5 \frac{\mathrm{d}Z_5}{\mathrm{d}X_5}\right)^2 + 6X_5(1+X_5)Z_5. \tag{5.78}$$

*Proof.* From (5.58), we have

$$A_5 w_5 \frac{\mathrm{d}y_5}{\mathrm{d}y_5} = \frac{5}{24}\left(P(q) - P(q^5)\right).$$

Applying the differential operator $q\frac{\mathrm{d}}{\mathrm{d}q}$ and using Lemma 5.27 we obtain

$$A_5 w_5 y_5\, \frac{\mathrm{d}}{\mathrm{d}w_5}\left(A_5 w_5 \frac{\mathrm{d}y_5}{\mathrm{d}w_5}\right) = \frac{5}{24} q \frac{\mathrm{d}}{\mathrm{d}q}\left(P(q) - P(q^5)\right).$$

The derivatives on the right-hand side may be computed using (0.78), and the result is

$$A_5 w_5 y_5 \frac{\mathrm{d}}{\mathrm{d}w_5}\left(A_5 w_5 \frac{\mathrm{d}y_5}{\mathrm{d}w_5}\right) = \frac{5}{288}\left(P(q)^2 - 5P(q^5)^2 - Q(q) + 5Q(q^5)\right). \tag{5.79}$$

The right-hand side of (5.79) may be expressed in terms of $y_5$ and $w_5$ as follows. From Theorem 5.22,

$$P(q) = A_5 y_5 + 6A_5 w_5 \frac{\mathrm{d}y_5}{\mathrm{d}w_5} \quad \text{and} \quad P(q^5) = A_5 y_5 + \frac{6}{5} A_5 w_5 \frac{\mathrm{d}y_5}{\mathrm{d}w_5}.$$

Using these in (5.79) and then making use of Theorem 5.24 to express $Q(q)$ and $Q(q^5)$ in terms of $w_5$ and $y_5$, we obtain

$$A_5 w_5 y_5 \frac{\mathrm{d}}{\mathrm{d}w_5}\left(A_5 w_5 \frac{\mathrm{d}y_5}{\mathrm{d}w_5}\right)$$
$$= \frac{5}{288}\left(-4A_5^2 y_5^2 + \frac{144}{5}\left(A_5 w_5 \frac{\mathrm{d}y_5}{\mathrm{d}w_5}\right)^2 + y_5^2(4 - 200w_5 - 3100w_5^2)\right).$$

Now divide both sides by $y_5$ and use the fact that $A_5^2 = 1 + 22w_5 + 125w_5^2$ to obtain the first result.

The second result can be obtained from the first result by a change of variables. We begin by applying the differential operator $q\,\mathrm{d}/\mathrm{d}q$ to the result $Z_5 = A_5\,y_5$ in Theorem 5.19. On making use of the identities in Lemma 5.27, the result simplifies to

$$B_5\,X_5\,\frac{\mathrm{d}Z_5}{\mathrm{d}X_5} = A_5\,w_5\,\frac{\mathrm{d}y_5}{\mathrm{d}w_5} + y_5\,\frac{w_5(11 + 125w_5)}{A_5}. \tag{5.80}$$

On applying the operator $q\,\mathrm{d}/\mathrm{d}q$ to (5.80) and again making use of the identities in Lemma 5.27, the result simplifies to

$$B_5 X_5 Z_5 \frac{\mathrm{d}}{\mathrm{d}X_5}\left(B_5 X_5 \frac{\mathrm{d}Z_5}{\mathrm{d}X_5}\right)$$
$$= A_5 w_5 y_5 \frac{\mathrm{d}}{\mathrm{d}w_5}\left(A_5 w_5 \frac{\mathrm{d}y_5}{\mathrm{d}w_5}\right) + w_5^2(11 + 125w_5)\,y_5\,\frac{\mathrm{d}y_5}{\mathrm{d}w_5}$$
$$+ \frac{w_5(11 + 371w_5 + 4125w_5^2 + 15625w_5^3)}{1 + 22w_5 + 125w_5^2}\,y_5^2. \tag{5.81}$$

From (5.80) and (5.81) we deduce that

$$B_5 X_5 Z_5 \frac{\mathrm{d}}{\mathrm{d}X_5}\left(B_5 X_5 \frac{\mathrm{d}Z_5}{\mathrm{d}X_5}\right) - \frac{1}{2}\left(B_5\,X_5\,\frac{\mathrm{d}Z_5}{\mathrm{d}X_5}\right)^2$$
$$= A_5 w_5 y_5 \frac{\mathrm{d}}{\mathrm{d}w_5}\left(A_5 w_5 \frac{\mathrm{d}y_5}{\mathrm{d}w_5}\right) - \frac{1}{2}\left(A_5\,w_5\,\frac{\mathrm{d}y_5}{\mathrm{d}w_5}\right)^2$$
$$+ y_5^2\,\frac{w_5(22 + 621w_5 + 5500w_5^2 + 15625w_5^3)}{2(1 + 22w_5 + 125w_5^2)}.$$

The derivatives on the right-hand side may be eliminated using (5.77). Then, the remaining terms may be expressed in terms of $Z_5$ and $X_5$ by using Theorems 5.19 and 5.20. The result is (5.78). □

The next result gives a second order linear differential equation that is satisfied by $z_5$ with respect to $x_5$.

**Theorem 5.29.** *The following differential equation holds:*

$$\frac{\mathrm{d}}{\mathrm{d}x_5}\left(x_5(1 - 11x_5 - x_5^2)\frac{\mathrm{d}z_5}{\mathrm{d}x_5}\right) = (3 + x_5)z_5.$$

*Proof.* The idea is to make a change of variables in the differential equation (5.77) from $w_5$ and $y_5$ to $x_5$ and $z_5$, respectively. By (5.49) we have

$$(1 - 11x_5 - x_5^2)z_5^2 = y_5 \tag{5.82}$$

and from Lemma 5.27 we have

$$A_5 w_5 \frac{\mathrm{d}}{\mathrm{d}w_5} = x_5 \frac{\mathrm{d}}{\mathrm{d}x_5}. \tag{5.83}$$

On using (5.82) and (5.83) in (5.77) we obtain

$$\begin{aligned}
&x_5 \frac{\mathrm{d}}{\mathrm{d}x_5}\left(x_5 \frac{\mathrm{d}}{\mathrm{d}x_5}\left((1 - 11x_5 - x_5^2)z_5^2\right)\right) \\
&= \frac{1}{2z_5^2(1 - 11x_5 - x_5^2)}\left(x_5 \frac{\mathrm{d}}{\mathrm{d}x_5}\left((1 - 11x_5 - x_5^2)z_5^2\right)\right)^2 \\
&\quad - \frac{5x_5(2 + 3x_5 - 2x_5^2)}{2(1 - 11x_5 - x_5^2)} z_5^2.
\end{aligned}$$

Now divide both sides by $2x_5z_5$. After some straightforward but lengthy simplifications, we complete the proof. □

Here are third order linear differential equations satisfied by $y_5$ and $Z_5$.

**Theorem 5.30.** *The following differential equations hold*:

$$\frac{\mathrm{d}}{\mathrm{d}w_5}\left(A_5 w_5 \frac{\mathrm{d}}{\mathrm{d}w_5}\left(A_5 w_5 \frac{\mathrm{d}y_5}{\mathrm{d}w_5}\right)\right) + 5w_5(2 + 25w_5)\frac{\mathrm{d}y_5}{\mathrm{d}w_5} + 5(1 + 25w_5)y_5 = 0$$

*and*

$$\frac{\mathrm{d}}{\mathrm{d}X_5}\left(B_5 X_5 \frac{\mathrm{d}}{\mathrm{d}X_5}\left(B_5 X_5 \frac{\mathrm{d}Z_5}{\mathrm{d}X_5}\right)\right) - 12X_5(1 + X_5)\frac{\mathrm{d}Z_5}{\mathrm{d}X_5} - 6(1 + 2X_5)Z_5 = 0.$$

*Proof.* Apply $\mathrm{d}/\mathrm{d}w_5$ to the result of Theorem 5.28 to obtain

$$\begin{aligned}
&\frac{\mathrm{d}}{\mathrm{d}w_5}\left(A_5 w_5 \frac{\mathrm{d}}{\mathrm{d}w_5}\left(A_5 w_5 \frac{\mathrm{d}y_5}{\mathrm{d}w_5}\right)\right) \\
&= -\frac{1}{2y_5^2}\frac{\mathrm{d}y_5}{\mathrm{d}w_5}\left(A_5 w_5 \frac{\mathrm{d}y_5}{\mathrm{d}w_5}\right)^2 + \frac{1}{y_5}\left(A_5 w_5 \frac{\mathrm{d}y_5}{\mathrm{d}w_5}\right)\frac{\mathrm{d}}{\mathrm{d}w_5}\left(A_5 w_5 \frac{\mathrm{d}y_5}{\mathrm{d}w_5}\right) \\
&\quad - \frac{5}{2}w_5(2 + 25w_5)\frac{\mathrm{d}y_5}{\mathrm{d}w_5} - 5(1 + 25w_5)y_5
\end{aligned}$$

$$= \frac{1}{y_5}\frac{dy_5}{dw_5}\left\{A_5 w_5 \frac{d}{dw_5}\left(A_5 w_5 \frac{dy_5}{dw_5}\right) - \frac{1}{2y_5}\left(A_5 w_5 \frac{dy_5}{dw_5}\right)^2\right\}$$
$$- \frac{5}{2}w_5(2+25w_5)\frac{dy_5}{dw_5} - 5(1+25w_5)y_5.$$

The quantity in braces may be simplified by Theorem 5.28 to give

$$\frac{d}{dw_5}\left(A_5 w_5 \frac{d}{dw_5}\left(A_5 w_5 \frac{dy_5}{dw_5}\right)\right)$$
$$= \frac{1}{y_5}\frac{dy_5}{dw_5}\left\{-\frac{5}{2}w_5(2+25w_5)y_5\right\} - \frac{5}{2}w_5(2+25w_5)\frac{dy_5}{dw_5} - 5(1+25w_5)y_5,$$

and on simplifying we obtain the first differential equation. The second differential equation may be proved in the same way. □

The differential equations in Theorem 5.30 may be expanded and written as

$$w_5^2(1+22w_5+125w_5^2)\frac{d^3y_5}{dw_5^3} + 3w_5(1+33w_5+250w_5^2)\frac{d^2y_5}{dw_5^2}$$
$$+(1+76w_5+875w_5^2)\frac{dy_5}{dw_5} + 5(1+25w_5)y_5 = 0$$

and

$$X_5^2(1-44X_5-16X_5^2)\frac{d^3Z_5}{dX_5^3} + 3X_5(1-66X_5-32X_5^2)\frac{d^2Z_5}{dX_5^2}$$
$$+(1-144X_5-108X_5^2)\frac{dZ_5}{dX_5} - 6(1+2X_5)Z_5 = 0.$$

## 5.8 Series expansions

The differential equations in Theorems 5.29 and 5.30 can be solved to give $z_5$ (or $y_5$ or $Z_5$) as a power series in $x_5$ (or $w_5$ or $X_5$, respectively). We begin with the solution of the differential equation in Theorem 5.29.

**Theorem 5.31.** *Let* $x_5$ *and* $z_5$ *be given by* (5.36) *and* (5.37). *Let* $\{t_5(n)\}$ *be the sequence defined by the initial condition*

$$t_5(0) = 1 \tag{5.84}$$

*and recurrence relation*

$$(n+1)^2 t_5(n+1) = (11n^2+11n+3)t_5(n) + n^2 t_5(n-1), \quad n \geq 0. \tag{5.85}$$

*Then in a neighborhood of $q = 0$, the identity*

$$z_5 = \sum_{n=0}^{\infty} t_5(n) x_5^n \tag{5.86}$$

*holds.*

Before starting the proof, note that the single initial condition $t_5(0) = 1$ in (5.84) is enough to start the recurrence relation. That is, the value of $t_5(1)$ can be determined by taking $n = 0$ in the recurrence relation (5.85), and this gives $t_5(1) = 3$.

*Proof.* By the definition (5.36) we have

$$x_5 = q \prod_{j=1}^{\infty} \frac{(1-q^{5j-4})^5(1-q^{5j-1})^5}{(1-q^{5j-3})^5(1-q^{5j-2})^5} = q + O(q^2)$$

while the definition (5.37) and Theorem 5.12 imply

$$\begin{aligned} z_5 &= \prod_{j=1}^{\infty} \frac{(1-q^j)^2}{(1-q^{5j-4})^5(1-q^{5j-1})^5} \\ &= 1 + 3\sum_{j=1}^{\infty}\left(\frac{q^{5j-4}}{1-q^{5j-4}} - \frac{q^{5j-1}}{1-q^{5j-1}}\right) + \sum_{j=1}^{\infty}\left(\frac{q^{5j-3}}{1-q^{5j-3}} - \frac{q^{5j-2}}{1-q^{5j-2}}\right) \\ &= 1 + 3q + O(q^2). \end{aligned}$$

It follows that the inverse function $q = q(x_5)$ exists and is analytic in a neighborhood of $x_5 = 0$. Consequently, the function $z_5 = z_5(x_5)$ exists and is analytic in a neighborhood of $x_5 = 0$ and

$$z_5 = 1 + 3x_5 + O(x_5^2) \tag{5.87}$$

near $x_5 = 0$. On substituting the series expansion $z_5 = \sum_{n=0}^{\infty} t_5(n) x_5^n$ into the differential equation in Theorem 5.29 and equating coefficients of $x_5^n$ we deduce that $t_5(n)$ satisfies the three-term recurrence relation (5.85). From the expansion (5.87) it follows that $t_5(0) = 1$ and $t_5(1) = 3$. The condition $t_5(1) = 3$ can be dropped as it is implied by taking $n = 0$ in the recurrence relation (5.85). □

The analogue of Theorem 5.31 for the weight two modular forms $y_5$ and $Z_5$ is as follows.

**Theorem 5.32.** *Let $w_5$, $y_5$, $X_5$ and $Z_5$ be defined by* (5.38) *and* (5.39). *Let $\{s_5(n)\}$ and $\{T_5(n)\}$ be the sequences defined by the initial conditions*

$$s_5(0) = 1, \qquad T_5(0) = 1 \tag{5.88}$$

*and recurrence relations*

$$(n+1)^3 s_5(n+1) = -(2n+1)(11n^2+11n+5)s_5(n) - 125n^3 s_5(n-1) \tag{5.89}$$

*and*

$$(n+1)^3 T_5(n+1) = 2(2n+1)(11n^2+11n+3)T_5(n) + 4n(4n^2-1)T_5(n-1), \tag{5.90}$$

*for $n \geq 0$. Then in a neighborhood of $q = 0$, the identities*

$$y_5 = \sum_{n=0}^{\infty} s_5(n) w_5^n \qquad \textit{and} \qquad Z_5 = \sum_{n=0}^{\infty} T_5(n) X_5^n \tag{5.91}$$

*hold.*

For the same reason as given for Theorem 5.31, the single initial condition given by (5.88) is sufficient to start each recurrence relation.

*Proof.* Apply the same procedure as in the proof of Theorem 5.31 to the differential equations in Theorem 5.30. □

## 5.9 The method of creative telescoping

In this section we will exhibit explicit formulas for $t_5(n)$, $s_5(n)$, and $T_5(n)$ as sums of products of binomial coefficients. A consequence is that $t_5(n)$, $s_5(n)$, and $T_5(n)$ are all integers. Although integrality can be deduced by considering the expansions of (5.86) and (5.91) in powers of $q$, the integrality of $t_5(n)$, $s_5(n)$, and $T_5(n)$ is not obvious from the recurrence relations (5.85), (5.89), and (5.90).

We begin with an explicit formula for $t_5(n)$.

**Theorem 5.33.** *Let $\{t_5(n)\}$ be the sequence defined by the recurrence relation* (5.85) *and initial condition* (5.84)*, that is, $t_5(0) = 1$ and*

$$(n+1)^2 t_5(n+1) = (11n^2+11n+3)t_5(n) + n^2 t_5(n-1), \quad \textit{for } n \geq 0.$$

*Let*

$$\sigma(n) = \sum_{k=0}^{n} \binom{n}{k}^2 \binom{n+k}{k}.$$

*Then*

$$t_5(n) = \sigma(n). \tag{5.92}$$

*Proof.* It is easy to check that $t_5(0) = \sigma(0) = 1$ and $t_5(1) = \sigma(1) = 3$. We will show that $\sigma(n)$ satisfies the same recurrence relation as $t_5(n)$ given by (5.85) and this will establish that $t_5(n) = \sigma(n)$ for all positive integers $n$. Let

$$\lambda_{n,k} = \binom{n}{k}^2 \binom{n+k}{k}$$

so that

$$\sigma(n) = \sum_{k=0}^{n} \lambda_{n,k}. \tag{5.93}$$

Simple calculations give

$$\frac{\lambda_{n-1,k}}{\lambda_{n,k}} = \frac{(n-k)^2}{n(n+k)}, \qquad \frac{\lambda_{n+1,k}}{\lambda_{n,k}} = \frac{(n+1)(n+1+k)}{(n+1-k)^2}$$

and

$$\frac{\lambda_{n,k-1}}{\lambda_{n,k}} = \frac{k^3}{(n+1-k)^2(n+k)}.$$

We seek a rational function of $n$ and $k$, that we denote by $L_{n,k}$, with the property that

$$\begin{aligned}(n+1)^2\lambda_{n+1,k} &- (11n^2+11n+3)\lambda_{n,k} - n^2\lambda_{n-1,k} \\ &= L_{n,k}\lambda_{n,k} - L_{n,k-1}\lambda_{n,k-1}.\end{aligned} \tag{5.94}$$

Multiplying both sides by $(n+k)(n+1-k)^2/\lambda_{n,k}$, we find that $L_{n,k}$ must satisfy

$$\begin{aligned}&(n+1)^3(n+k+1)(n+k) \\ &-(11n^2+11n+3)(n+1-k)^2(n+k) - n(n-k)^2(n+1-k)^2 \\ &= (n+1-k)^2(n+k)L_{n,k} - k^3L_{n,k-1}.\end{aligned} \tag{5.95}$$

The polynomial $L_{n,k}$ defined by

$$L_{n,k} = k^2 + 6nk + 3k - 11n^2 - 9n - 2$$

satisfies (5.95), and hence satisfies (5.94). Summing (5.94) over the values of $k$ for which $1 \le k \le n-1$ and using (5.93) we deduce

$$\begin{aligned}&(n+1)^2\sigma(n+1) - (11n^2+11n+3)\sigma(n) - n^2\sigma(n-1) \\ &= (n+1)^2\left(\lambda_{n+1,n+1} + \lambda_{n+1,n} + \lambda_{n+1,0}\right) \\ &\quad - \left(11n^2+11n+3\right)\left(\lambda_{n,n} + \lambda_{n,0}\right) - n^2\lambda_{n-1,0} \\ &\qquad + L_{n,n-1}\lambda_{n,n-1} - L_{n,0}\lambda_{n,0}\end{aligned}$$

$$= \left\{(n+1)^2\lambda_{n+1,0} - (11n^2+11n+3)\lambda_{n,0} - n^2\lambda_{n-1,0} - L_{n,0}\lambda_{n,0}\right\}$$
$$+ \left\{(n+1)^2(\lambda_{n+1,n+1} + \lambda_{n+1,n}) - (11n^2+11n+3)\lambda_{n,n} + L_{n,n-1}\lambda_{n,n-1}\right\}$$
$$= 0,$$

since each term in braces simplifies to 0. Thus, we have shown that $\sigma(n)$ satisfies the same recurrence relation and initial conditions as $t_5(n)$. This completes the proof. □

There is an algorithm for finding the function $L_{n,k}$, called Zeilberger's algorithm, that can be automated to a computer program. For example, this has been implemented in the computer algebra system Maple, and typing the commands:

```
> with(sumtools):
> sumrecursion(binomial(n, k)^2*binomial(n+k, n),
    k, t(n));
```

produces the recurrence relation

$$-(n-1)^2\, t(n-2) - \left(11\,n^2 - 11\,n + 3\right) t(n-1) + t(n)\, n^2$$

that is equal to zero. Given an explicit formula for $t(n)$ as a sum that involves binomial coefficients, it is possible to determine the recurrence relation, but not the other way around. The only way to find an explicit formula such as the binomial sum in (5.92) that satisfies a given recurrence relation is to search a database such as the one maintained by Sloane [271].

Here is an explicit formula for the terms in the sequence $\{s_5(n)\}$.

**Theorem 5.34.** *Let* $\{s_5(n)\}$ *be the sequence defined by the recurrence relation* (5.89) *and initial condition* (5.88), *that is,* $s_5(0) = 1$ *and*

$$(n+1)^3 s_5(n+1) = -(2n+1)(11n^2+11n+5)s_5(n) - 125n^3 s_5(n-1) \quad \textit{for } n \geq 0.$$

*Then*

$$s_5(n) = \sum_{k=0}^{n} (-1)^{n+k} \binom{n}{k}^3 \binom{4n-5k}{3n}. \tag{5.96}$$

*Proof.* Enter the given binomial sum into a computer algebra system such as Maple and have the computer determine its recurrence relation. On checking that the recurrence relation and initial conditions are the same as those given by (5.89) and (5.88) we complete the proof. □

It is convenient to phrase the identity in Theorem 5.34 in terms of binomial coefficients of the form $\binom{n}{k}$ with $0 \leq k \leq n$. To accomplish this, break the sum in (5.96) into three, to give

$$\begin{aligned}
s_5(n) &= \sum_{k=0}^{n} (-1)^{n+k} \binom{n}{k}^3 \binom{4n-5k}{3n} \\
&= \left( \sum_{0 \le 5k \le n} + \sum_{n < 5k \le 4n} + \sum_{4n < 5k \le 5n} \right) (-1)^{n+k} \binom{n}{k}^3 \binom{4n-5k}{3n}.
\end{aligned}$$

The middle sum is 0 because $0 \le 4n - 5k < 3n$ and so $\binom{4n-5k}{3n} = 0$ by the definition of the binomial coefficient in (0.103). We apply (0.104) to the last sum and then replace $k$ with $n - k$ to obtain

$$\begin{aligned}
s_5(n) &= \sum_{k=0}^{n} (-1)^{n+k} \binom{n}{k}^3 \binom{4n-5k}{3n} \\
&= \sum_{0 \le 5k \le n} (-1)^{n+k} \binom{n}{k}^3 \binom{4n-5k}{3n} \\
&\quad + \sum_{4n < 5k \le 5n} (-1)^{n+k} \binom{n}{k}^3 (-1)^{3n} \binom{5k-n-1}{3n} \\
&= \sum_{0 \le k \le n/5} (-1)^{n+k} \binom{n}{k}^3 \binom{4n-5k}{3n} \\
&\quad + \sum_{0 \le k < n/5} (-1)^{n+k} \binom{n}{k}^3 \binom{4n-5k-1}{3n}.
\end{aligned}$$

This is the form given by Almkvist, van Straten, and Zudilin in [18].

Here is an explicit formula for $T_5(n)$.

**Theorem 5.35.** *Let* $\{T_5(n)\}$ *be the sequence defined by the recurrence relation* (5.90) *and initial condition* (5.88), *that is,* $T_5(0) = 1$ *and*

$$(n+1)^3 T_5(n+1) = 2(2n+1)(11n^2 + 11n + 3) T_5(n) + 4n(4n^2 - 1) T_5(n-1)$$

*for* $n \ge 0$. *Let* $\{t_5(n)\}$ *be the sequence studied in Theorem 5.33. Then*

$$T_5(n) = \binom{2n}{n} t_5(n) = \binom{2n}{n} \sum_{k=0}^{n} \binom{n}{k}^2 \binom{n+k}{k}.$$

*Proof.* Let $u(n) = \binom{2n}{n} t_5(n)$. We have already proved that the sequence $\{t_5(n)\}$ satisfies the recurrence relation

$$(n+1)^2 t_5(n+1) = (11n^2 + 11n + 3)t_5(n) + n^2 t_5(n-1).$$

Multiply both sides of the recurrence relation by $2(2n+1)\binom{2n}{n}$ to deduce that

$$(n+1)^3 u(n+1) = 2(2n+1)(11n^2 + 11n + 3)u(n) + 4n(4n^2 - 1)u(n-1)$$

for $n \geq 0$. Since $u(0) = 1 = T_5(0)$ and $u(1) = 6 = T_5(1)$, it follows that

$$T_5(n) = u(n) = \binom{2n}{n} t_5(n)$$

for all $n \geq 0$. □

The result that $T_5(n) = \binom{2n}{n} t_5(n)$ is not a coincidence. This will be studied in detail in Section 5.11.

## 5.10 Zagier's sporadic sequences

In the previous section it was shown that the sequence $\{t_5(n)\}$ defined by

$$t_5(n) = \sum_{k=0}^{n} \binom{n}{k}^2 \binom{n+k}{k}$$

satisfies the initial condition $t_5(0) = 0$ and three-term recurrence relation

$$(n+1)^2 t_5(n+1) = (11n^2 + 11n + 3)t_5(n) + n^2 t_5(n-1), \quad n \geq 0.$$

As another example, it was known to Franel [153] in the 19th century that the sequence defined by

$$t_c(n) = \sum_{k=0}^{n} \binom{n}{k}^3$$

satisfies the three-term recurrence relation

$$(n+1)^2 t_c(n+1) = (7n^2 + 7n + 2)t_c(n) + 8n^2 t_c(n-1), \quad n \geq 0.$$

Although it is clear from the formulas involving sums of binomial coefficients that the sequences $\{t_5(n)\}$ and $\{t_c(n)\}$ are integer valued, the integrality is not at all obvious from the recurrence relations. This prompts:

*Question 5.36.* For which integers $\alpha$, $\beta$, $\gamma$ does the sequence $\{t(n)\}$, defined by $t(0) = 1$ and

$$(n+1)^2 t(n+1) = (\alpha n^2 + \alpha n + \beta) t(n) + \gamma\, n^2 t(n-1) \quad \text{for } n \geq 0,$$

take only integer values?

In an attempt to answer the question, Zagier [307] performed a computer search for other integral sequences that satisfy the three-term recurrence relation. The search yielded six sequences that are not either terminating (two successive terms that are zero), polynomial, or the product of a polynomial with a multinomial coefficient. They are called Zagier's sporadic sequences. The values of the parameters $\alpha$, $\beta$, $\gamma$, along with explicit formulas for the terms $t(n)$ as sums involving binomial or multinomial coefficients, are listed in Table 5.1. The multinomial coefficient is defined by

$$\binom{k_1 + k_2 + \cdots + k_m}{k_1,\ k_2, \ldots, k_m} = \frac{(k_1 + k_2 + \cdots + k_m)!}{k_1! k_2! \cdots k_m!} \tag{5.97}$$

for nonnegative integers $k_1,\ k_2, \ldots, k_m$. Each of the explicit formulas in the table can be verified using Zeilberger's algorithm, as described in the previous section. The companion sequences $s(n)$ in Table 5.1 will be discussed in the next section.

## 5.11 Two analogues of Clausen's identity

By Theorem 5.19, the weight two modular forms $y_5$ and $Z_5$ are related to the weight one modular form $z_5$ by the formulas

$$y_5 = z_5^2(1 - 11x_5 - x_5^2) \quad \text{and} \quad Z_5 = z_5^2(1 + x_5^2). \tag{5.98}$$

By Theorems 5.15 and 5.20,

$$w_5 = \frac{x_5}{1 - 11x_5 - x_5^2} \quad \text{and} \quad X_5 = \frac{x_5(1 - 11x_5 - x_5^2)}{(1 + x_5^2)^2}. \tag{5.99}$$

Furthermore, by Theorems 5.31, 5.32, and 5.35, the expansions

$$z_5 = \sum_{n=0}^{\infty} t_5(n) x_5^n, \quad y_5 = \sum_{n=0}^{\infty} s_5(n) w_5^n, \quad \text{and} \quad Z_5 = \sum_{n=0}^{\infty} \binom{2n}{n} t_5(n) X_5^n \tag{5.100}$$

**Table 5.1** Solutions to recurrence relations

| $(\alpha,\beta,\gamma)$ | $t(n)$ | $s(n)$ |
|---|---|---|
| $(11,3,1)$ | $\sum_j \binom{n}{j}^2\binom{n+j}{j}$ | $\sum_j (-1)^{j+n}\binom{n}{j}^3\binom{4n-5j}{3n}$ |
| $(-17,-6,-72)$ | $\sum_{j,\ell} (-8)^{n-j}\binom{n}{j}\binom{j}{\ell}^3$ | $\sum_j \binom{n}{j}^2\binom{n+j}{j}^2$ |
| $(10,3,-9)$ | $\sum_j \binom{n}{j}^2\binom{2j}{j}$ | $(-1)^n\sum_j \binom{n}{j}^2\binom{2j}{j}\binom{2n-2j}{n-j}$ |
| $(7,2,8)$ | $\sum_j \binom{n}{j}^3$ | $\sum_j (-3)^{n-3j}\binom{n+j}{j,j,j,j,n-3j}$ |
| $(-12,-4,-32)$ | $(-1)^n\sum_j \binom{n}{j}\binom{2j}{j}\binom{2n-2j}{n-j}$ | $\sum_j \binom{n}{j}^2\binom{2j}{n}^2$ |
| $(-9,-3,-27)$ | $\sum_j (-3)^{n-3j}\binom{n}{j,j,j,n-3j}$ | $\sum_{j,\ell} \binom{n}{j}^2\binom{n}{\ell}\binom{j}{\ell}\binom{j+\ell}{n}$ |

$$(n+1)^2 t(n+1) = (\alpha n^2+\alpha n+\beta)t(n)+\gamma n^2 t(n-1),$$

$$(n+1)^3 s(n+1) = -(2n+1)(\alpha n^2+\alpha n+\alpha-2\beta)s(n)-(\alpha^2+4\gamma)n^3 s(n-1),$$

Initial conditions: $t(0)=s(0)=1$.

hold, where $t_5(n)$ and $s_5(n)$ are given respectively by the initial conditions $t_5(0) = s_5(0) = 1$ and recurrence relations

$$(n+1)^2 t_5(n+1) = (11n^2+11n+3)t_5(n)+n^2 t_5(n-1)$$

and

$$(n+1)^3 s_5(n+1) = -(2n+1)(11n^2+11n+5)s_5(n)-125n^3 s_5(n-1)$$

for $n \geq 0$. The identities in (5.98) and (5.99) can be used in (5.100) to give

$$\left(\sum_{n=0}^{\infty} t_5(n)x_5^n\right)^2 = \frac{1}{1-11x_5-x_5^2}\sum_{n=0}^{\infty} s_5(n)\left(\frac{x_5}{1-11x_5-x_5^2}\right)^n \tag{5.101}$$

and

$$\left(\sum_{n=0}^{\infty} t_5(n)x_5^n\right)^2 = \frac{1}{1+x_5^2}\sum_{n=0}^{\infty}\binom{2n}{n} t_5(n)\left(\frac{x_5(1-11x_5-x_5^2)}{(1+x_5^2)^2}\right)^n. \tag{5.102}$$

We will now show that (5.101) and (5.102) can be extended to the general sequence $\{t(n)\}$ and its companion sequence $\{s(n)\}$ defined by

$$(n+1)^2 t(n+1) = (\alpha n^2 + \alpha n + \beta) t(n) + \gamma\, n^2 t(n-1),$$

and

$$(n+1)^3 s(n+1) = -(2n+1)(\alpha n^2 + \alpha n + \alpha - 2\beta) s(n) - (\alpha^2 + 4\gamma) n^3 s(n-1)$$

for $n \geq 0$, together with initial conditions $t(0) = s(0) = 1$. This leads to two analogues of Clausen's identity for the square of a ${}_2F_1$ hypergeometric function.

So that all of the definitions are in one place, let us record the following.

**Definition 5.37.** Let $\alpha$, $\beta$ and $\gamma$ be fixed constants.

1. Let $\delta = \alpha^2 + 4\gamma$.
2. Suppose $z$ is a function of $x$. Let $w$, $A$, and $y$ be the functions of $x$ defined by
$$w = \frac{x}{1 - \alpha x - \gamma x^2}, \qquad A = \sqrt{1 + 2\alpha w + \delta w^2},$$
and
$$y = (1 - \alpha x - \gamma x^2) z^2.$$
3. Let $X$, $B$, and $Z$ be the functions of $w$, and hence functions of $x$, defined by
$$X = \frac{w}{1 + 2\alpha w + \delta w^2}, \qquad B = \sqrt{1 - 4\alpha X - 16\gamma X^2} \qquad \text{and} \qquad Z = yA.$$

The results in the next two lemmas follow from the definitions by basic calculus and require no further comment.

**Lemma 5.38.** *The following identities hold*:

$$\frac{\mathrm{d}w}{\mathrm{d}x} = \frac{1 + \gamma x^2}{(1 - \alpha x - \gamma x^2)^2},$$

$$A = \frac{1 + \gamma x^2}{1 - \alpha x - \gamma x^2}$$

*and*

$$x \frac{\mathrm{d}}{\mathrm{d}x} = A\, w \frac{\mathrm{d}}{\mathrm{d}w}.$$

**Lemma 5.39.** *The following identities hold*:

$$\frac{\mathrm{d}X}{\mathrm{d}w} = \frac{1 - \delta w^2}{(1 + 2\alpha w + \delta w^2)^2},$$

$$B = \frac{1 - \delta w^2}{1 + 2\alpha w + \delta w^2}$$

*and*

$$Awy\frac{\mathrm{d}}{\mathrm{d}w} = BXZ\frac{\mathrm{d}}{\mathrm{d}X}.$$

The sequence $\{t(n)\}$ can be studied by considering the differential equation satisfied by its generating function; the next result gives the differential equation.

**Lemma 5.40.** *Suppose $t(n)$ satisfies the recurrence relation*

$$(n+1)^2 t(n+1) = (\alpha n^2 + \alpha n + \beta)t(n) + \gamma n^2 t(n-1) \quad \textit{for } n \geq 0.$$

*Then the generating function*

$$z = \sum_{n=0}^{\infty} t(n)x^n$$

*satisfies the second order linear differential equation*

$$\frac{\mathrm{d}}{\mathrm{d}x}\left(x(1-\alpha x-\gamma x^2)\frac{\mathrm{d}z}{\mathrm{d}x}\right) = (\beta+\gamma x)z. \tag{5.103}$$

*Proof.* Substitute the series expansion $z = \sum_{n=0}^{\infty} t(n)x^n$ into the left- and right-hand sides of the differential equation. Compare coefficients of $x^n$ to see that $z$ is a solution of the differential equation if and only if the recurrence relation holds. □

The differential equation (5.103) may be compared with the differential equation in Theorem 4.24, namely

$$\frac{\mathrm{d}}{\mathrm{d}x_\ell}(x_\ell(1-x_\ell)) = s_\ell(1-s_\ell)z_\ell, \quad \text{where } \ell \in \{1,2,3,4\}. \tag{5.104}$$

The differential equation (5.104) has three regular singular points. Riemann found that second order differential equations with three regular singular points can be solved in terms of the hypergeometric function, e.g., see [24, Sec. 2.3], [299, Sec. 14.3]. In contrast, the differential equation (5.103) has four regular singular points and is harder to analyze. Any second order linear differential equation with four regular singular points and no other singular points can be reduced to a differential equation called the Heun equation. A detailed analysis of the Heun equation and its solutions has been given by Maier [230].

The next theorem gives the result of a change of variable in a differential equation.

**Theorem 5.41.** *Let $\alpha$, $\beta$, and $\gamma$ be constants and let $\delta = \alpha^2 + 4\gamma$. Suppose $z$ satisfies the following second order linear differential equation with respect to $x$:*

$$\frac{\mathrm{d}}{\mathrm{d}x}\left(x(1-\alpha x-\gamma x^2)\frac{\mathrm{d}z}{\mathrm{d}x}\right) = (\beta+\gamma x)z. \tag{5.105}$$

*If $w$, $A$, and $y$ are related to $x$ and $z$ by the formulas in Definition 5.37, then $y$ satisfies the following second order nonlinear differential equation with respect to $w$:*

$$Aw\frac{\mathrm{d}}{\mathrm{d}w}\left(Aw\frac{\mathrm{d}y}{\mathrm{d}w}\right) = \frac{1}{2y}\left(Aw\frac{\mathrm{d}y}{\mathrm{d}w}\right)^2 + \left((2\beta-\alpha)w - \frac{\delta w^2}{2}\right)y. \tag{5.106}$$

*Moreover, $y$ satisfies the following third order linear differential equation with respect to $w$:*

$$\begin{aligned}&\frac{\mathrm{d}}{\mathrm{d}w}\left(Aw\frac{\mathrm{d}}{\mathrm{d}w}\left(Aw\frac{\mathrm{d}y}{\mathrm{d}w}\right)\right)\\ &\quad + \left((2\alpha-4\beta)w+\delta w^2\right)\frac{\mathrm{d}y}{\mathrm{d}w} + (\alpha-2\beta+\delta w)\,y = 0.\end{aligned} \tag{5.107}$$

*Equivalently,*

$$\begin{aligned}&w^2(1+2\alpha w+\delta w^2)\frac{\mathrm{d}^3y}{\mathrm{d}w^3} + 3w(1+3\alpha w+2\delta w^2)\frac{\mathrm{d}^2y}{\mathrm{d}w^2}\\ &\quad + (1+4(2\alpha-\beta)w+7\delta w^2)\frac{\mathrm{d}y}{\mathrm{d}w} + (\alpha-2\beta+\delta w)y = 0.\end{aligned} \tag{5.108}$$

*Another equivalent form is*

$$\left\{D^3 + w(2D+1)(\alpha D^2+\alpha D+\alpha-2\beta) + \delta(D+1)^3\right\}y = 0 \tag{5.109}$$

*where $D = w\dfrac{\mathrm{d}}{\mathrm{d}w}$.*

*Proof.* By Definition 5.37 and Lemma 5.38 we find that

$$\begin{aligned}&Aw\frac{\mathrm{d}}{\mathrm{d}w}\left(Aw\frac{\mathrm{d}y}{\mathrm{d}w}\right)\\ &= x\frac{\mathrm{d}}{\mathrm{d}x}\left(x\frac{\mathrm{d}}{\mathrm{d}x}\left\{(1-\alpha x-\gamma x^2)z^2\right\}\right)\\ &= x\frac{\mathrm{d}}{\mathrm{d}x}\left(2x(1-\alpha x-\gamma x^2)z\frac{\mathrm{d}z}{\mathrm{d}x} - (\alpha x+2\gamma x^2)z^2\right)\end{aligned}$$

$$= 2zx\frac{\mathrm{d}}{\mathrm{d}x}\left(x(1-\alpha x-\gamma x^2)\frac{\mathrm{d}z}{\mathrm{d}x}\right)+2x^2(1-\alpha x-\gamma x^2)\left(\frac{\mathrm{d}z}{\mathrm{d}x}\right)^2$$
$$-(\alpha x+4\gamma x^2)z^2-2x(\alpha x+2\gamma x^2)z\frac{\mathrm{d}z}{\mathrm{d}x}.$$

On making use of the differential equation (5.105), this simplifies to

$$Aw\frac{\mathrm{d}}{\mathrm{d}w}\left(Aw\frac{\mathrm{d}y}{\mathrm{d}w}\right)$$
$$=\left((2\beta-\alpha)x-2\gamma x^2\right)z^2-2x^2(\alpha+2\gamma x)z\frac{\mathrm{d}z}{\mathrm{d}x}+2x^2(1-\alpha x-\gamma x^2)\left(\frac{\mathrm{d}z}{\mathrm{d}x}\right)^2. \tag{5.110}$$

On the other hand, by Definition 5.37 and Lemma 5.38 we also have

$$\frac{1}{2y}\left(Aw\frac{\mathrm{d}y}{\mathrm{d}w}\right)^2$$
$$=\frac{1}{2(1-\alpha x-\gamma x^2)z^2}\left(x\frac{\mathrm{d}}{\mathrm{d}x}\{(1-\alpha x-\gamma x^2)z^2\}\right)^2$$
$$=\frac{1}{2(1-\alpha x-\gamma x^2)z^2}\left\{2x(1-\alpha x-\gamma x^2)z\frac{\mathrm{d}z}{\mathrm{d}x}-(\alpha x+2\gamma x^2)z^2\right\}^2$$
$$=\frac{(\alpha x+2\gamma x^2)^2}{2(1-\alpha x-\gamma x^2)}z^2-2x^2(\alpha+2\gamma x)z\frac{\mathrm{d}z}{\mathrm{d}x}+2x^2(1-\alpha x-\gamma x^2)\left(\frac{\mathrm{d}z}{\mathrm{d}x}\right)^2. \tag{5.111}$$

Subtracting the result of (5.111) from (5.110) we obtain

$$Aw\frac{\mathrm{d}}{\mathrm{d}w}\left(Aw\frac{\mathrm{d}y}{\mathrm{d}w}\right)-\frac{1}{2y}\left(Aw\frac{\mathrm{d}y}{\mathrm{d}w}\right)^2$$
$$=\left((2\beta-\alpha)x-2\gamma x^2-\frac{(\alpha x+2\gamma x^2)^2}{2(1\quad\alpha x\quad\gamma x^2)}\right)z^2$$
$$=\left((2\beta-\alpha)x-\frac{\delta x^2}{2(1-\alpha x-\gamma x^2)}\right)z^2$$
$$=\left((2\beta-\alpha)w-\frac{\delta w^2}{2}\right)y,$$

where the last step is obtained by applying Definition 5.37. This proves (5.106).

Now we will prove (5.107). Apply $\mathrm{d}/\mathrm{d}w$ to (5.106) to obtain

$$\begin{aligned}
&\frac{\mathrm{d}}{\mathrm{d}w}\left(Aw\frac{\mathrm{d}}{\mathrm{d}w}\left(Aw\frac{\mathrm{d}y}{\mathrm{d}w}\right)\right)\\
&= -\frac{1}{2y^2}\frac{\mathrm{d}y}{\mathrm{d}w}\left(Aw\frac{\mathrm{d}y}{\mathrm{d}w}\right)^2+\frac{1}{y}\left(Aw\frac{\mathrm{d}y}{\mathrm{d}w}\right)\frac{\mathrm{d}}{\mathrm{d}w}\left(Aw\frac{\mathrm{d}y}{\mathrm{d}w}\right)\\
&\qquad+\left((2\beta-\alpha)w-\frac{\delta w^2}{2}\right)\frac{\mathrm{d}y}{\mathrm{d}w}+(2\beta-\alpha-\delta w)y\\
&= \frac{1}{y}\frac{\mathrm{d}y}{\mathrm{d}w}\left\{Aw\frac{\mathrm{d}}{\mathrm{d}w}\left(Aw\frac{\mathrm{d}y}{\mathrm{d}w}\right)-\frac{1}{2y}\left(Aw\frac{\mathrm{d}y}{\mathrm{d}w}\right)^2\right\}\\
&\qquad+\left((2\beta-\alpha)w-\frac{\delta w^2}{2}\right)\frac{\mathrm{d}y}{\mathrm{d}w}+(2\beta-\alpha-\delta w)y.
\end{aligned}$$

The quantity in braces may be simplified using (5.106) to give

$$\begin{aligned}
&\frac{\mathrm{d}}{\mathrm{d}w}\left(Aw\frac{\mathrm{d}}{\mathrm{d}w}\left(Aw\frac{\mathrm{d}y}{\mathrm{d}w}\right)\right)\\
&= \frac{1}{y}\frac{\mathrm{d}y}{\mathrm{d}w}\left((2\beta-\alpha)w-\frac{\delta w^2}{2}\right)y\\
&\qquad+\left((2\beta-\alpha)w-\frac{\delta w^2}{2}\right)\frac{\mathrm{d}y}{\mathrm{d}w}+(2\beta-\alpha-\delta w)y\\
&= \left((4\beta-2\alpha)w-\delta w^2\right)\frac{\mathrm{d}y}{\mathrm{d}w}+(2\beta-\alpha-\delta w)y.
\end{aligned}$$

This proves (5.107). The last two equations (5.108) and (5.109) can be obtained from (5.107) by simple calculations. □

The next theorem gives the result of a another change of variable in a differential equation.

**Theorem 5.42.** *Let $\alpha$, $\beta$, and $\gamma$ be constants and let $\delta = \alpha^2 + 4\gamma$. Suppose $z$ satisfies the following second order linear differential equation with respect to $x$:*

$$\frac{\mathrm{d}}{\mathrm{d}x}\left(x(1-\alpha x-\gamma x^2)\frac{\mathrm{d}z}{\mathrm{d}x}\right)=(\beta+\gamma x)z. \tag{5.112}$$

*If $X$, $B$, and $Z$ are given by Definition 5.37, then $Z$ satisfies the following second order nonlinear differential equation with respect to $X$:*

$$BX\frac{\mathrm{d}}{\mathrm{d}X}\left(BX\frac{\mathrm{d}Z}{\mathrm{d}X}\right)=\frac{1}{2Z}\left(BX\frac{\mathrm{d}Z}{\mathrm{d}X}\right)^2+2X\left(\beta+3\gamma X\right)Z. \tag{5.113}$$

*Moreover, Z satisfies the following third order linear differential equation with respect to X:*

$$\frac{\mathrm{d}}{\mathrm{d}X}\left(BX\frac{\mathrm{d}}{\mathrm{d}X}\left(BX\frac{\mathrm{d}Z}{\mathrm{d}X}\right)\right)$$
$$-4X(\beta+3\gamma X)\frac{\mathrm{d}Z}{\mathrm{d}X}-2(\beta+6\gamma X)Z=0. \tag{5.114}$$

*Equivalently,*

$$X^2(1-4\alpha X-16\gamma X^2)\frac{\mathrm{d}^3Z}{\mathrm{d}X^3}+3X(1-6\alpha X-32\gamma X^2)\frac{\mathrm{d}^2Z}{\mathrm{d}X^2}$$
$$+(1-4(3\alpha+\beta)X-108\gamma X^2)\frac{\mathrm{d}Z}{\mathrm{d}X}-2(\beta+6\gamma X)Z=0. \tag{5.115}$$

*Proof.* By the chain rule using Lemma 5.39, we have

$$BX\frac{\mathrm{d}Z}{\mathrm{d}X}=\frac{1}{Z}\left(BXZ\frac{\mathrm{d}Z}{\mathrm{d}X}\right)=\frac{1}{yA}\left(Awy\frac{\mathrm{d}}{\mathrm{d}w}(yA)\right)=w\frac{\mathrm{d}}{\mathrm{d}w}(yA).$$

Hence,

$$BX\frac{\mathrm{d}Z}{\mathrm{d}X}=wA\frac{\mathrm{d}y}{\mathrm{d}w}+\frac{w(\alpha+\delta w)}{A}y.$$

Apply the operator $BXZ\dfrac{\mathrm{d}}{\mathrm{d}X}$. By Lemma 5.39,

$$BXZ\frac{\mathrm{d}}{\mathrm{d}X}\left(BX\frac{\mathrm{d}Z}{\mathrm{d}X}\right)$$
$$=Awy\frac{\mathrm{d}}{\mathrm{d}w}\left(wA\frac{\mathrm{d}y}{\mathrm{d}w}\right)+Awy\frac{\mathrm{d}}{\mathrm{d}w}\left(\frac{w(\alpha+\delta w)}{A}y\right)$$
$$=Awy\frac{\mathrm{d}}{\mathrm{d}w}\left(wA\frac{\mathrm{d}y}{\mathrm{d}w}\right)+y\frac{\mathrm{d}y}{\mathrm{d}w}w^2(\alpha+\delta w)+Awy^2\frac{\mathrm{d}}{\mathrm{d}w}\left(\frac{aw+\delta w^2}{A}\right).$$

And,

$$\frac{1}{2}\left(BX\frac{\mathrm{d}Z}{\mathrm{d}X}\right)^2=\frac{1}{2}\left(wA\frac{\mathrm{d}y}{\mathrm{d}w}\right)^2+y\frac{\mathrm{d}y}{\mathrm{d}w}w^2(\alpha+\delta w)+\frac{w^2(\alpha+\delta w)^2}{2A^2}y^2.$$

Subtracting and simplifying, we obtain

$$BXZ\frac{\mathrm{d}}{\mathrm{d}X}\left(BX\frac{\mathrm{d}Z}{\mathrm{d}X}\right)-\frac{1}{2}\left(BX\frac{\mathrm{d}Z}{\mathrm{d}X}\right)^2=\frac{2w(\beta+(3\gamma+2\alpha\beta)w+\beta\delta w^2)}{A^2}y^2.$$

The right-hand side may be expressed in terms of $X$ and $Z$ by the formulas in Definition 5.37. This gives (5.113).

The other differential equations (5.114) and (5.115) can be deduced from (5.113) by following the same procedure as in the proof of Theorem 5.41. We omit the details as they are similar. □

Here are the analogues of Clausen's identity.

**Theorem 5.43.** *Let $\{t(n)\}$ and $\{s(n)\}$ be sequences that satisfy the initial conditions*

$$t(0) = s(0) = 1$$

*and the three-term recurrence relations*

$$(n+1)^2 t(n+1) = (\alpha n^2 + \alpha n + \beta) t(n) + \gamma n^2 t(n-1)$$

*and*

$$(n+1)^3 s(n+1) = -(2n+1)(\alpha n^2 + \alpha n + \alpha - 2\beta) s(n) - (\alpha^2 + 4\gamma) n^3 s(n-1)$$

*for $n \geq 0$. Then in a neighborhood of $x = 0$, the following identities hold:*

$$\left(\sum_{n=0}^{\infty} t(n) x^n\right)^2 = \frac{1}{1 - \alpha x - \gamma x^2} \sum_{n=0}^{\infty} s(n) \left(\frac{x}{1 - \alpha x - \gamma x^2}\right)^n \tag{5.116}$$

*and*

$$\left(\sum_{n=0}^{\infty} t(n) x^n\right)^2 = \frac{1}{1 + \gamma x^2} \sum_{n=0}^{\infty} \binom{2n}{n} t(n) \left(\frac{x(1 - \alpha x - \gamma x^2)}{(1 + \gamma x^2)^2}\right)^n. \tag{5.117}$$

*Proof.* If $t(n)$ satisfies the given recurrence relation and initial conditions, then the generating function $z$ defined by

$$z = \sum_{n=0}^{\infty} t(n) x^n = 1 + \beta x + O(x^2)$$

satisfies the differential equation (5.112). Let $w$ and $y$ be defined by Definition 5.37, that is,

$$y = (1 - \alpha x - \gamma x^2) z^2 \qquad \text{and} \qquad w = \frac{x}{1 - \alpha x - \gamma x^2}.$$

When $x = 0$ we have $w = 0$, $z = 1$, and $y = 1$. Since $w$ is an analytic function of $x$ in a neighborhood of $x = 0$ and $\left.\frac{dw}{dx}\right|_{x=0} \neq 0$, it follows that

$$y = \sum_{n=0}^{\infty} s(n) w^n$$

in a neighborhood of $w = 0$, for some coefficients $s(n)$, and $s(0) = 1$. By Theorem 5.41, $y$ satisfies a third order differential equation with respect to $w$ given by 5.108. It follows that the coefficients $s(n)$ satisfy the three-term recurrence relation in the statement of the theorem. Moreover,

$$\begin{aligned} y &= (1 - \alpha x - \gamma x^2)z^2 \\ &= (1 - \alpha x - \gamma x^2)(1 + \beta x + O(x^2))^2 \\ &= 1 + (2\beta - \alpha)x + O(x^2) \\ &= 1 + (2\beta - \alpha)w + O(w^2) \quad \text{as } w \to 0, \end{aligned}$$

so $s(0) = 1$ and $s(1) = 2\beta - \alpha$. The value $s(1) = 2\beta - \alpha$ is consistent with the value computed by taking $n = 0$ in the recurrence relation, so it can be dropped and replaced with the stipulation that the recurrence relation must also hold for $n = 0$. This completes the proof of (5.116).

The identity (5.117) can be proved by the same method using the differential equation (5.115). □

## 5.12 Connection with hypergeometric functions

Let $t_5(n)$ and $s_5(n)$ be the sequences defined in Section 5.8 and let $f_5(x)$ and $G_5(x)$ be the respective generating functions, so that

$$f_5(x) = \sum_{n=0}^{\infty} t_5(n)x^n = 1 + 3x + 19x^2 + 147x^3 + 1251x^4 + 11253x^5 + \cdots \tag{5.118}$$

and

$$G_5(x) = \sum_{n=0}^{\infty} s_5(n)x^n = 1 - 5x + 35x^2 - 275x^3 + 2275x^4 - 19255x^5 + \cdots . \tag{5.119}$$

We will show that $f_5$ and $G_5$ satisfy quintic transformation formulas. We shall also obtain expressions for $f_5$ and $G_5$ in terms of hypergeometric functions.

By Theorems 5.31 and 5.32 we have

$$z_5 = f_5(x_5) \qquad \text{and} \qquad y_5 = G_5(w_5), \tag{5.120}$$

where $x_5$, $w_5$, $z_5$, and $y_5$ are given by (5.36)–(5.38).

The next result shows that $f_5$ and $G_5$ are interrelated.

**Theorem 5.44.** *The functions $f_5(x)$ and $G_5(x)$ satisfy the following identity in a neighborhood of $x = 0$:*

$$(f_5(x))^2 = \frac{1}{1-11x-x^2} G_5\left(\frac{x}{1-11x-x^2}\right).$$

*Proof.* Start with (5.49) in the form

$$z_5^2 = \frac{1}{1-11x_5-x_5^2} y_5.$$

and use (5.120) to obtain

$$(f_5(x_5))^2 = \frac{1}{1-11x_5-x_5^2} G_5(w_5).$$

Finally, use (5.43) to express $w_5$ in terms of $x_5$ and then replace all occurrences of $x_5$ with $x$ to complete the proof. □

To help with further calculations, we will need:

**Lemma 5.45.** *Let $u = r(q)$ and $v = r(q^5)$. Let $x_5$ and $z_5$ be defined by* (5.36) *and* (5.37). *Then*

$$\frac{1}{u^{25}(1-11u^5-u^{10})} = \frac{(1+3v+4v^2+2v^3+v^4)^{12}}{v^5(1-11v^5-v^{10})^5} \tag{5.121}$$

*and*

$$\frac{z_5(q)}{z_5(q^5)} = 1+3x_5(q^5)^{1/5}+4x_5(q^5)^{2/5}+2x_5(q^5)^{3/5}+x_5(q^5)^{4/5}. \tag{5.122}$$

*Proof.* From Theorem 5.11 we have

$$u^5 = v\frac{1-2v+4v^2-3v^3+v^4}{1+3v+4v^2+2v^3+v^4}. \tag{5.123}$$

It follows that

$$1-11u^5-u^{10} = \frac{(1-v-v^2)^5}{(1+3v+4v^2+2v^3+v^4)^2}.$$

Therefore,

$$\begin{aligned}\frac{1}{u^{25}(1-11u^5-u^{10})} &= \frac{(1+3v+4v^2+2v^3+v^4)^7}{v^5(1-2v+4v^2-3v^3+v^4)^5(1-v-v^2)^5}\\ &= \frac{(1+3v+4v^2+2v^3+v^4)^{12}}{v^5(1-11v^5-v^{10})^5},\end{aligned}$$

where the last step follows by the factorization

$$1 - 11v^5 - v^{10} = (1 - v - v^2)(1 - 2v + 4v^2 - 3v^3 + v^4) \times (1 + 3v + 4v^2 + 2v^3 + v^4).$$

This proves (5.121). Now we will prove (5.122). By (5.44) we find that

$$\begin{aligned}\left(\frac{z_5(q)}{z_5(q^5)}\right)^{12} &= \left(\frac{w_5(q)\,y_5(q)}{x_5(q)}\right)^6 \Big/ \left(\frac{w_5(q^5)\,y_5(q^5)}{x_5(q^5)}\right)^6 \\ &= \left(\frac{y_5(q)^6\,w_5(q)^5}{y_5(q^5)^6\,w_5(q^5)}\right) \times \left(\frac{x_5(q^5)^6\,w_5(q)}{x_5(q)^6\,w_5(q^5)^5}\right).\end{aligned} \tag{5.124}$$

From Theorem 5.16 we have

$$\frac{y_5(q)^6\,w_5(q)^5}{y_5(q^5)^6\,w_5(q^5)} = 1,$$

while Theorem 5.15 implies

$$\frac{x_5(q^5)^6\,w_5(q)}{x_5(q)^6\,w_5(q^5)^5} = \frac{x_5(q^5)^6\left(\dfrac{1}{x_5(q^5)} - 11 - x_5(q^5)\right)^5}{x_5(q)^6\left(\dfrac{1}{x_5(q)} - 11 - x_5(q)\right)}.$$

Hence, (5.124) simplifies to

$$\left(\frac{z_5(q)}{z_5(q^5)}\right)^{12} = \frac{x_5(q^5)\left(1 - 11x_5(q^5) - x_5(q^5)^2\right)^5}{x_5(q)^5\left(1 - 11x_5(q) - x_5(q)^2\right)} = \frac{v^5(1 - 11v^5 - v^{10})^5}{u^{25}(1 - 11u^5 - u^{10})},$$

where we have used the definition of $x_5$ in (5.36) to write $x_5(q) = u^5$ and $x_5(q^5) = v^5$ in the last step. The proof may be completed by appealing to (5.121) and then expressing the result in terms of $x_5$. □

Here is the quintic transformation formula for $f_5$.

**Theorem 5.46.** *Let $f_5$ be the function defined by* (5.118). *The following functional equation holds in a neighborhood of $x = 0$:*

$$f_5(x^5) = \frac{1}{1 + 3x + 4x^2 + 2x^3 + x^4}\, f_5\left(x\,\frac{1 - 2x + 4x^2 - 3x^3 + x^4}{1 + 3x + 4x^2 + 2x^3 + x^4}\right).$$

*Proof.* By Theorem 5.31,

$$z_5(q) = \sum_{n=0}^{\infty} t_5(n)\,x_5(q)^n.$$

Hence, with $u$ and $v$ as for Lemma 5.45, we have

$$z_5(q) = \sum_{n=0}^{\infty} t_5(n) u^{5n} \qquad \text{and} \qquad z_5(q^5) = \sum_{n=0}^{\infty} t_5(n) v^{5n}.$$

Substituting these expansions in (5.122) we obtain

$$\sum_{n=0}^{\infty} t_5(n)\, u^{5n} = \left(1 + 3v + 4v^2 + 2v^3 + v^4\right) \sum_{n=0}^{\infty} t_5(n)\, v^{5n},$$

that is,

$$f_5(u^5) = \left(1 + 3v + 4v^2 + 2v^3 + v^4\right) f_5(v^5).$$

The proof may be completed by using (5.123) to express $u^5$ in terms of $v$, and then replacing all occurrences of the variable $v$ with $x$. □

The corresponding quadratic transformation formula for $f_5(x)$ is given in (13.63).

**Theorem 5.47.** *Let $G_5$ be defined by* (5.119). *The following functional equation holds in a neighborhood of $x = 0$:*

$$G_5\left(\frac{x^5}{1 - 11x^5 - x^{10}}\right) = \frac{1 - 11x^5 - x^{10}}{(1 - x - x^2)^5}\, G_5\left(\frac{x(1 - 11x^5 - x^{10})}{(1 - x - x^2)^6}\right).$$

*Proof.* Applying Theorems 5.44 and 5.46, we obtain

$$\frac{1}{1 - 11x^5 - x^{10}}\, G_5\left(\frac{x^5}{1 - 11x^5 - x^{10}}\right) = \left(f_5(x^5)\right)^2$$

$$= \frac{1}{(1 + 3x + 4x^2 + 2x^3 + x^4)^2} \left\{ f_5\left( x\, \frac{1 - 2x + 4x^2 - 3x^3 + x^4}{1 + 3x + 4x^2 + 2x^3 + x^4} \right) \right\}^2.$$

The proof is completed by applying Theorem 5.44 again, noting that if

$$t = x\, \frac{1 - 2x + 4x^2 - 3x^3 + x^4}{1 + 3x + 4x^2 + 2x^3 + x^4}$$

then

$$\frac{1}{1 - 11t - t^2} = \frac{(1 + 3x + 4x^2 + 2x^3 + x^4)^2}{(1 - x - x^2)^5}$$

and

$$\frac{t}{1 - 11t - t^2} = x\, \frac{(1 - 11x^5 - x^{10})}{(1 - x - x^2)^6}.$$

□

The next result gives expressions for $f_5$ and $G_5$ in terms of hypergeometric functions.

**Theorem 5.48.** *The functions* $f_5(x)$ *and* $G_5(x)$ *defined by* (5.118) *and* (5.119) *have representations in terms of hypergeometric functions given by*:

$$f_5(x) = (1+228x+494x^2-228x^3+x^4)^{-1/4} \times {}_2F_1\left(\begin{matrix}\frac{1}{12}, \frac{5}{12}\\ 1\end{matrix}; \frac{1728x(1-11x-x^2)^5}{(1+228x+494x^2-228x^3+x^4)^3}\right) \tag{5.125}$$

$$= (1-12x+14x^2+12x^3+x^4)^{-1/4} \times {}_2F_1\left(\begin{matrix}\frac{1}{12}, \frac{5}{12}\\ 1\end{matrix}; \frac{1728x^5(1-11x-x^2)}{(1-12x+14x^2+12x^3+x^4)^3}\right) \tag{5.126}$$

*and*

$$G_5(x) = \frac{1}{\sqrt{1+250x+3125x^2}}\, {}_3F_2\left(\begin{matrix}\frac{1}{6}, \frac{1}{2}, \frac{5}{6}\\ 1, 1\end{matrix}; \frac{1728x}{(1+250x+3125x^2)^3}\right) \tag{5.127}$$

$$= \frac{1}{\sqrt{1+10x+5x^2}}\, {}_3F_2\left(\begin{matrix}\frac{1}{6}, \frac{1}{2}, \frac{5}{6}\\ 1, 1\end{matrix}; \frac{1728x^5}{(1+10x+5x^2)^3}\right). \tag{5.128}$$

*Proof.* By the parameterization for $Q(q)$ in Theorem 5.24,

$$Q(q) = y_5^2(1+250w_5+3125w_5^2). \tag{5.129}$$

By Theorem 5.15, $w_5$ and $y_5$ may be expressed in terms of $x_5$ and $z_5$ to give

$$Q(q) = z_5^4(1+228x_5+494x_5^2-228x_5^3+x_5^4),$$

and by (5.120) this is equivalent to

$$Q(q) = (f_5(x_5))^4\,(1+228x_5+494x_5^2-228x_5^3+x_5^4). \tag{5.130}$$

Next, by Exercise 7 of Chapter 4 and the parameterization of $j$ given by Theorem 5.26, we have

$$Q^{1/4}(q) = {}_2F_1\left(\begin{matrix}\frac{1}{12}, \frac{5}{12}\\ 1\end{matrix}; \frac{1728}{j(\tau)}\right) = {}_2F_1\left(\begin{matrix}\frac{1}{12}, \frac{5}{12}\\ 1\end{matrix}; \frac{1728x_5(1-11x_5-x_5^2)^5}{(1+228x_5+494x_5^2-228x_5^3+x_5^4)^3}\right). \tag{5.131}$$

On combining (5.130) and (5.131) we obtain (5.125). The identity (5.126) may be proved in the same way, by using $Q(q^5)$ and $j(5\tau)$ in place of $Q(q)$ and $j(\tau)$, respectively.

Now we will prove (5.127). By Theorem 4.30,

$$Q^{1/2}(q) = {}_3F_2\left( \begin{matrix} \frac{1}{6}, \frac{1}{2}, \frac{5}{6} \\ 1, 1 \end{matrix} ; \frac{1728}{j(\tau)} \right).$$

Use (5.129) on the left and apply Theorem 5.26 to the right-hand side to obtain

$$y_5 \sqrt{1 + 250w_5 + 3125w_5^2} = {}_3F_2\left( \begin{matrix} \frac{1}{6}, \frac{1}{2}, \frac{5}{6} \\ 1, 1 \end{matrix} ; \frac{1728w_5}{(1 + 250w_5 + 3125w_5^2)^3} \right).$$

Since $y_5 = G_5(x_5)$ by (5.120), this gives (5.127). The identity (5.128) may be proved similarly by using $Q(q^5)$ and $j(5\tau)$ in place of $Q(q)$ and $j(\tau)$, respectively. □

The last result in this section is a quintic transformation formula for the ${}_2F_1$ hypergeometric function.

**Theorem 5.49.** *The following hypergeometric transformation formula holds in a neighborhood of* $y = 0$:

$$\frac{1}{\sqrt[4]{1 + 250y + 3125y^2}} \, {}_2F_1\left( \begin{matrix} \frac{1}{12}, \frac{5}{12} \\ 1 \end{matrix} ; \frac{1728y}{(1 + 250y + 3125y^2)^3} \right)$$
$$= \frac{1}{\sqrt[4]{1 + 10y + 5y^2}} \, {}_2F_1\left( \begin{matrix} \frac{1}{12}, \frac{5}{12} \\ 1 \end{matrix} ; \frac{1728y^5}{(1 + 10y + 5y^2)^3} \right).$$

*Proof.* Let

$$y = \frac{x}{1 - 11x - x^2}$$

and note that

$$\frac{y}{(1 + 250y + 3125y^2)^3} = \frac{x(1 - 11x - x^2)^5}{(1 + 228x + 494x^2 - 228x^3 + x^4)^3},$$
$$\frac{y^5}{(1 + 10y + 5y^2)^3} = \frac{x^5(1 - 11x - x^2)}{(1 - 12x + 14x^2 + 12x^3 + x^4)^3}$$

and

$$\frac{1 + 10y + 5y^2}{1 + 250y + 3125y^2} = \frac{1 - 12x + 14x^2 + 12x^3 + x^4}{1 + 228x + 494x^2 - 228x^3 + x^4}.$$

The result now follows from (5.125) and (5.126) by making this change of variable. □

Further results, connecting the functions studied in this chapter with hypergeometric functions, will be given in Theorem 13.1.

## 5.13 Notes

The Rogers–Ramanujan continued fraction appears in work of L. J. Rogers published in 1894, [256].

The evaluation of $r(e^{-2\pi})$ in (5.23) along with the quintic transformation formula (5.29) are two of the formulas in letters written by Ramanujan that captured Hardy's attention. Hardy [174, pp. 7, 9] described them as *"extremely difficult theorems"* and wrote:
*"(they) defeated me completely; I had never seen anything in the least like them before. A single look at them is enough to show that they could only be written down by a mathematician of the highest class. They must be true because, if they were not true, no one would have had the imagination to invent them."*

The Rogers–Ramanujan continued fraction occurs in Ramanujan's notebooks, and there is a wealth of material in the lost notebook; e.g., see the books by Andrews and Berndt [25, Chapters 1–5] and by Berndt [49, pp. 156–157]. An expository account of the Rogers–Ramanujan continued fraction that describes the connection with modular forms and the icosahedron has been given by W. Duke [142].

The finite form of the Rogers–Ramanujan continued fraction in Theorem 5.2 was given by Ramanujan in his second notebook [253, Chapter 16, Entry 16]. The proof that we have given via the three-term recurrence relation in Lemma 5.1 is the same as the proof of Berndt [46, p. 33].

The Rogers–Ramanujan identities are due independently to Rogers [256] and Ramanujan [253, Chapter 16, Entry 38]. See the discussion by Berndt [46, pp. 77–79] for more information and references to many proofs. The proof we have given of Theorem 5.6, that uses Lemmas 5.4 and 5.5, closely follows Ramanujan's proof in [252]. Ramanujan's argument has been generalized by Selberg [266]; see the exercises at the end of Chapter 8 for some examples. Finite forms of the Rogers–Ramanujan identities are given by the formulas

$$\sum_{j=0}^{n} \begin{bmatrix} n \\ j \end{bmatrix}_q q^{j^2} = \frac{(q;q)_n}{(q;q)_{2n}} \sum_{j=-n}^{n} (-1)^j \begin{bmatrix} 2n \\ n+j \end{bmatrix}_q q^{\binom{j}{2}+2j^2}$$

and

$$\sum_{j=0}^{n} \begin{bmatrix} n \\ j \end{bmatrix}_q q^{j^2+j} = \frac{(q;q)_n}{(q;q)_{2n+1}} \sum_{j=-n}^{n+1} (-1)^j \begin{bmatrix} 2n+1 \\ n+j \end{bmatrix}_q q^{\binom{j}{2}+2j^2-j}.$$

With the help of Jacobi's triple product identity, they reduce to the Rogers–Ramanujan identities in the limiting case $n \to \infty$. See Chu and Di Claudio [97, pp. 52–55].

The identities in Theorem 5.8 that relate the Rogers–Ramanujan continued fraction to Dedekind's eta function appear in Ramanujan's second notebook [253, Chapter 19, Entry 11 (iii)] and in the lost notebook [254, pp. 238, 239]. Another proof will be given in Exercise 10, below.

The beautiful symmetric forms of the transformation formulas for the Rogers–Ramanujan continued fraction in (5.27) and (5.28) were known to Ramanujan. See Andrews and Berndt [25, pp. 90, 91] for more information and precise references to Ramanujan's work.

The identities in Theorem 5.13 are in Ramanujan's second notebook [253, Chapter 19, Entry 9 (v)], and also in Ramanujan's lost notebook [254, pp. 139, 354, 357]. The proof we have given is essentially the same as the proof by J. M. Dobbie [140]. For other proofs and references, see Bailey [35, 36], Andrews and Berndt [25, pp. 406, 407], Berndt [46, pp. 257–261], Hirschhorn [177] and the commentary by Berndt in [255, pp. 372, 373].

The identity in Corollary 5.14 was singled out by Hardy [255, p. xxxv] for special praise:
*"if I had to select one formula from all Ramanujan's work, I would agree with Major MacMahon in selecting a formula from [251], viz.*

$$p(4) + p(9)q + p(14)q^2 + \cdots = 5\frac{\{(1-q^5)(1-q^{10})(1-q^{15})\cdots\}^5}{\{(1-q)(1-q^2)(1-q^3)\cdots\}^6}\text{''}. \tag{5.132}$$

The proof of Corollary 5.14 that we have given is from Ramanujan's lost notebook [254, p. 139]. Other proofs are outlined in the exercises below.

The identity (5.40) in the form

$$1 + 3\sum_{j=1}^{\infty}\left(\frac{q^{5j-4}}{1-q^{5j-4}} - \frac{q^{5j-1}}{1-q^{5j-1}}\right) + \sum_{j=1}^{\infty}\left(\frac{q^{5j-3}}{1-q^{5j-3}} - \frac{q^{5j-2}}{1-q^{5j-2}}\right)$$
$$= \sum_{n=0}^{\infty}\left\{\sum_{j=0}^{n}\binom{n}{j}^2\binom{n+j}{j}\right\}\left(q\prod_{j=1}^{\infty}\frac{(1-q^{5j-4})^5(1-q^{5j-1})^5}{(1-q^{5j-3})^5(1-q^{5j-2})^5}\right)^n$$

was first proved, using the theory of modular forms, by F. Beukers [54].

Theorem 5.24, that expresses various Eisenstein series in terms of Dedekind's eta function, was given by Ramanujan in the lost notebook [254, pp. 50, 51]. The method of proof we have given is based on work with P. C. Toh [124]; references to other proofs are given in [124].

Zagier's discovery of the sporadic sequences is described in [307]. The binomial sums for $t(n)$ in Table 5.1 were given by Zagier, op. cit. In the case $(a, b, c) = (-12, -4, -32)$ we have

$$t(n) = (-1)^n\sum_j 4^{n-2j}\binom{n}{2j}\binom{2j}{j}^2 = (-1)^n\sum_j\binom{n}{j}\binom{2j}{j}\binom{2n-2j}{n-j};$$

the first binomial sum was given by Zagier, and the second binomial sum was given by Almkvist, van Straten, and Zudilin [18]. The numbers $(-2)^n t(n)$ in the case

$(a, b, c) = (-12, -4, -32)$ have been studied by Larcombe and French [204] and are sometimes called the Catalan–Larcombe–French numbers. The binomial sums for $s(n)$ in Table 5.1 were given by Almkvist et al, op. cit. The recurrence relation in Theorem 5.34 appears in a two-page abstract by T. Sato [260]. The analogues of Clausen's identity in (5.116) and (5.117) were first given in work by Almkvist, van Straten, and Zudilin [18] and by Chan, Tanigawa, Yang, and Zudilin [90], respectively. The sequences $s(n)$ and $t(n)$ satisfy various congruence and supercongruence properties; we only mention the works of I. Gessel [161], and A. Malik and A. Straub [231].

## 5.14 Exercises

1. Use (5.21) and (5.24) to prove that

$$r(e^{-2\pi\sqrt{5}}) = \frac{\sqrt{5}}{1+\sqrt[5]{5^{3/4}\left(\frac{\sqrt{5}-1}{2}\right)^{5/2}-1}} - \frac{\sqrt{5}+1}{2}.$$

This is one of the results in Ramanujan's second letter to Hardy; see, e.g., [52, p. 57] or [255, p. xxviii].

2. This exercise outlines a proof of Theorem 5.7, i.e.,

$$\prod_{j=1}^{\infty} \frac{(1-q^{5j-4})(1-q^{5j-1})}{(1-q^{5j-3})(1-q^{5j-2})} = \cfrac{1}{1+\cfrac{q}{1+\cfrac{q^2}{1+\cfrac{q^3}{1+\cdots}}}},$$

that does not use the Rogers–Ramanujan identities. It follows Ramanujan [252] and requires properties of the function $G(x)$ studied in Lemma 5.4 and goes as follows. Let

$$K(x) = \frac{G(x)}{(1-qx)G(qx)}.$$

a. Show that the three-term functional equation for $G$ in Lemma 5.4 implies

$$K(x) = 1 + \frac{qx}{K(qx)}.$$

b. Deduce that

$$K(x) = 1 + \cfrac{qx}{1 + \cfrac{q^2x}{1 + \cfrac{q^3x}{1 + \cdots}}}.$$

c. Use Parts a and b and the values of $G(1)$ and $(1-q)\,G(q)$ in Lemma 5.4 to deduce that

$$\cfrac{1}{1 + \cfrac{q}{1 + \cfrac{q^2}{1 + \cfrac{q^3}{1 + \cdots}}}} = \frac{1}{K(1)} = \prod_{j=1}^{\infty} \frac{(1-q^{5j-4})(1-q^{5j-1})}{(1-q^{5j-3})(1-q^{5j-2})}.$$

3. This exercise shows how it is possible to give a rapid proof of Theorem 5.3. It is based on lectures of R. Askey, [31].

a. Let $\ell$ be the value of the Rogers–Ramanujan continued fraction in the case $q = 1$, that is,

$$\ell = \cfrac{1}{1 + \cfrac{1}{1 + \cfrac{1}{1 + \cdots}}}.$$

Observe that $\ell = 1/(1+\ell)$. Solve, and choose the positive solution, to deduce that $\ell = (\sqrt{5}-1)/2$.

b. Generalize the result of Part a to show that

$$\cfrac{1}{1 + \cfrac{x}{1 + \cfrac{x}{1 + \cfrac{x}{1 + \cdots}}}} = \frac{\sqrt{1+4x}-1}{2x}.$$

c. A further generalization is the continued fraction $\ell(x,q)$ defined by

$$\ell(x,q) = \cfrac{1}{1 + \cfrac{xq}{1 + \cfrac{xq^2}{1 + \cfrac{xq^3}{1 + \cdots}}}}.$$

Show that

$$\ell(x,q) = \frac{1}{1 + xq\ell(xq,q)}.$$

d. Assume that

$$\ell(x,q) = \frac{h(xq,q)}{h(x,q)} \quad \text{for some function} \quad h(x,q) = \sum_{k=0}^{\infty} c_k(q)x^k.$$

Show that

$$h(x,q) = h(xq,q) + xqh(xq^2,q)$$

and deduce that

$$c_k(q) = \frac{q^{2k-1}}{1-q^k} c_{k-1}(q) = \frac{q^{k^2}}{(q;q)_k} c_0(q).$$

e. Conclude that

$$\cfrac{1}{1 + \cfrac{xq}{1 + \cfrac{xq^2}{1 + \cfrac{xq^3}{1 + \cdots}}}} = \frac{\displaystyle\sum_{k=0}^{\infty} \frac{q^{k^2+k}x^k}{(q;q)_k}}{\displaystyle\sum_{k=0}^{\infty} \frac{q^{k^2}x^k}{(q;q)_k}}.$$

4. This exercise shows how the Rogers–Ramanujan identities can be obtained as special cases of more general results. It also gives some insight into the origin of the function $G(x)$; in (5.6), the definition was plucked from the air without motivation. Let $a_0, a_1, \ldots, a_r$, and $b_1, \ldots, b_r$ be complex numbers with the property that $b_j \notin \{1, q^{-1}, q^{-2}, \ldots\}$ for all $j = 1, 2, \ldots, r$. The basic hypergeometric series is defined by

$$_{r+1}\phi_r\left[\begin{matrix} a_0, a_1, \ldots, a_r \\ b_1, \ldots, b_r \end{matrix} \,\middle|\, q; z\right] = \sum_{j=0}^{\infty} \frac{(a_0;q)_j (a_1;q)_j \cdots (a_r;q)_j}{(q;q)_j (b_1;q)_j \cdots (b_r;q)_j} z^j.$$

The series converges for $|z| < 1$.

a. Use the $q$-binomial theorem (Theorem 0.29) to show that

$$_1\phi_0\left[\begin{matrix} a \\ - \end{matrix} \,\middle|\, q; z\right] = \frac{(ax;q)_\infty}{(x;q)_\infty}.$$

b. Watson's transformation formula is the result

$$
{}_8\phi_7\left[\begin{matrix} a, q\sqrt{a}, -q\sqrt{a}, & b, & c, & d, & e, & q^{-n} \\ \sqrt{a}, & -\sqrt{a}, & qa/b, & qa/c, & qa/d, & qa/e, & aq^{n+1} \end{matrix}\,\middle|\, q; \frac{q^{2+n}a^2}{bcde}\right]
$$

$$
= \frac{(qa;q)_\infty (qa/bc;q)_\infty}{(qa/b;q)_\infty (qa/c;q)_\infty}\, {}_4\phi_3\left[\begin{matrix} q^{-n}, & b, & c, & qa/de \\ & qa/d, & qa/e, & q^{-n}bc/a \end{matrix}\,\middle|\, q;q\right].
$$

For a proof and further information, see the expository and historical article by G. E. Andrews [23]. Take the limit as $b, c, d, e, n \to \infty$ to show that

$$
\sum_{j=0}^{\infty} \frac{q^{j^2}a^j}{(q;q)_j} = \frac{1}{(aq;q)_\infty} \sum_{j=0}^{\infty} (-1)^j \frac{1-q^{2j}a}{1-a} \frac{(a;q)_j}{(q;q)_j} q^{5\binom{j}{2}+2j} a^{2j}.
$$

Observe that the series on the right-hand side is the function $G$ defined by (5.6).

c. Let $a = 1$, rearrange the terms on the right-hand side to form a bilateral infinite series, and apply Jacobi's triple product identity, to prove the first of the Rogers–Ramanujan identities.

d. Let $a = q$ in the result of Part b, rearrange the terms on the right-hand side to form a bilateral infinite series, and apply Jacobi's triple product identity, to prove the second of the Rogers–Ramanujan identities.

5. The goal of this exercise is to give a combinatorial interpretation, in terms of partitions, for the Rogers–Ramanujan identities in Theorem 5.6.

a. Show that the coefficient of $q^N$ in

$$
\frac{q^{n^2}}{(1-q)(1-q^2)\cdots(1-q^n)}
$$

is equal to the number of partitions of $N$ into exactly $n$ parts where the parts differ by at least two.
Hint: Justify the following sequence of equalities.

$$
\begin{aligned}
&\text{coeff. of } q^N \text{ in } \frac{q^{n^2}}{(1-q)(1-q^2)\cdots(1-q^n)} \\
&= \text{coeff. of } q^{N-n^2} \text{ in } \frac{1}{(1-q)(1-q^2)\cdots(1-q^n)} \\
&= \text{no. of partitions of } N-n^2 \text{ into parts not exceeding } n \\
&= \text{no. of partitions of } N-n^2 \text{ into at most } n \text{ parts} \\
&= \#\left\{\lambda_1 + \cdots + \lambda_j = N - n^2 : \lambda_1 \ge \cdots \ge \lambda_j > 0,\ j \le n\right\}
\end{aligned}
$$

$$\begin{aligned}
&= \#\left\{\lambda_1 + \cdots + \lambda_n = N - n^2 : \ \lambda_1 \geq \cdots \geq \lambda_n \geq 0\right\} \\
&= \#\{(\lambda_1 + 2n - 1) + (\lambda_2 + 2n - 3) + \cdots + (\lambda_n + 1) = N : \\
&\qquad\qquad\qquad\qquad \lambda_1 \geq \cdots \geq \lambda_n \geq 0\} \\
&= \#\left\{\mu_1 + \cdots + \mu_n = N : \ \mu_j - \mu_{j+1} \geq 2, \ \mu_n \geq 1\right\}.
\end{aligned}$$

b. Deduce from the identity

$$1 + \sum_{k=1}^{\infty} \frac{q^{k^2}}{(1-q)(1-q^2)\cdots(1-q^k)} = \prod_{j=1}^{\infty} \frac{1}{(1-q^{5j-4})(1-q^{5j-1})}$$

that the number of partitions of $N$ into parts that differ by at least two equals the number of partitions of $N$ into parts that are congruent to either 1 or 4 modulo 5.

c. Deduce from the identity

$$1 + \sum_{k=1}^{\infty} \frac{q^{k^2+k}}{(1-q)(1-q^2)\cdots(1-q^k)} = \prod_{j=1}^{\infty} \frac{1}{(1-q^{5j-3})(1-q^{5j-2})}$$

that the number of partitions of $N$ into parts that differ by at least two with smallest part at least two equals the number of partitions of $N$ into parts that are congruent to either 2 or 3 modulo 5.

It is instructive to compare the results of Parts b and c in the exercise above with Euler's identity in Theorem 0.12, which can be phrased as: the number of partitions of $N$ into parts that differ by at least one equals the number of partitions of $N$ into parts that are congruent to 1 modulo 2.

6. This exercise is based on work by M. Hirschhorn [179].

a. Prove the following 5-dissection with respect to the variable $a$:

$$a\frac{(a, qa^{-1}, a^2, qa^{-2}, q; q)_\infty}{(q^5; q^5)_\infty} = \sum_{j=1}^{4} \left(\frac{j}{5}\right) a^j \frac{(q^j a^5, q^{5-j} a^{-5}; q^5)_\infty}{(q^j, q^{5-j}; q^5)_\infty}.$$

This identity is sometimes called the septuple product identity.
Hint: Use two applications of Jacobi's triple product identity to expand the product $(a, qa^{-1}, q; q)_\infty (a^2, qa^{-2}, q; q)_\infty$ in powers of $a$. Then 5-dissect the series and use Jacobi's triple product to convert each of the resulting series to a product.

b. Show that the result of Part a reduces to (5.8) on replacing $q$ with $q^5$ and setting $a = q$.

7. a. Prove that

$$\frac{1}{2}\left(q\prod_{j=1}^{\infty}\frac{(1-q^{5j})^5}{(1-q^j)}\right)\left(\frac{s(\theta,5\tau)s(2\theta,5\tau)}{s(\theta,\tau)}\right)=\sum_{j=1}^{\infty}\frac{q^j-q^{2j}-q^{3j}+q^{4j}}{1-q^{5j}}\sin j\theta$$

and

$$\frac{1}{2}\left(\prod_{j=1}^{\infty}\frac{(1-q^{j})^5}{(1-q^{5j})}\right)\left(\frac{s(\theta,\tau)s(2\theta,\tau)}{s(5\theta,5\tau)}\right)=\frac{\sin\frac{\theta}{2}\sin\theta}{\sin\frac{5\theta}{2}}-\sum_{j=1}^{\infty}\left(\frac{j}{5}\right)\frac{q^j}{1-q^j}\sin j\theta,$$

where $s(\theta,\tau)$ is the function defined by (0.51) and studied in Chapter 1. Hint: The right-hand side of the first identity is

$$\frac{i}{2}\sum_{j=1}^{\infty}\frac{(q^{j/2}-q^{-j/2})(q^j-q^{-j})(e^{i\theta}-e^{-i\theta})}{q^{5j/2}-q^{-5j/2}},$$

and this can be summed by Theorem 1.15. For the second identity, let $\omega=e^{2\pi i/5}$ and take $x=\omega e^{i\theta}$, $y=\omega^2 e^{-i\theta}$, $z=\omega^3 e^{-i\theta}$ in Theorem 1.12.

b. Show that the functions in Part a are constant multiples of $f(\theta,2\pi,2\pi\tau,5)$ and $g(\theta,2\pi,2\pi\tau,5)$, where $f$ and $g$ are the functions in Theorem 2.20. Show that they are elliptic functions of order 4, and determine their periods, zeros, and poles.

c. Show that the identities in Part a reduce to the results in Theorem 5.13 in the limiting case $\theta\to 0$.

d. Show that taking $\theta=2\pi/3$ in the identities in Part a gives the results

$$q\prod_{j=1}^{\infty}\frac{(1-q^j)^2(1-q^{15j})^2}{(1-q^{3j})(1-q^{5j})}=\sum_{j=1}^{\infty}\sum_{k=1}^{\infty}\left(\frac{j}{3}\right)\left(\frac{k}{5}\right)q^{jk}$$

and

$$\prod_{j=1}^{\infty}\frac{(1-q^{3j})^2(1-q^{5j})^2}{(1-q^j)(1-q^{15j})}=1+\sum_{j=1}^{\infty}\sum_{k=1}^{\infty}\left(\frac{j}{3}\right)\left(\frac{j}{5}\right)q^{jk}.$$

Further results of this type are in Exercise 13 of Chapter 3.
See [112] and Liu [223].

8. This exercise follows Ramanujan [251].

a. Show that the identity (5.8) may be written in the form

$$E(q)=E(q^{25})\left(a-q-\frac{q^2}{a}\right)$$

where $E(q)$ is Euler's product and

$$a = \prod_{j=1}^{\infty} \frac{(1-q^{25j-15})(1-q^{25j-10})}{(1-q^{25j-20})(1-q^{25j-5})}.$$

As $a$ is a function of $q^5$, this gives the 5-dissection of $E(q)$.

b. Let $\omega = \exp(2\pi i/5)$. Use the result of Part a to deduce that

$$E(\omega q)E(\omega^2 q)E(\omega^3 q)E(\omega^4 q) = E(q^{25})^4 \left( \frac{a^5 - 11q^5 - \dfrac{q^{10}}{a^5}}{a - q - \dfrac{q^2}{a}} \right).$$

Hint: See (5.19).

c. Deduce that the 5-dissection of $1/E(q)$ is given by

$$\frac{1}{E(q)} = \frac{E(q^{25})^5}{E(q^5)^6} \left\{ \left(a^4 - \frac{3q^5}{a}\right) + q\left(a^3 + \frac{2q^5}{a^2}\right) + q^2\left(2a^2 - \frac{q^5}{a^3}\right) + q^3\left(3a + \frac{q^5}{a^4}\right) + 5q^4 \right\}.$$

Hint: Apply (0.6) to the left-hand side of Part b, and use long division on the right-hand side.

d. Deduce that

$$\sum_{n=0}^{\infty} p(5n+4)q^n = 5 \prod_{j=1}^{\infty} \frac{(1-q^{5j})^5}{(1-q^j)^6}.$$

Hint: Expand the result of Part c in powers of $q$, and extract the terms in which the powers are congruent to 4 modulo 5.

9. The shifted Fibonacci numbers are defined by $f_0 = f_1 = 1$ and

$$f_{n+1} = f_n + f_{n-1}, \quad \text{for } n \geq 1.$$

a. Show that $\displaystyle\sum_{n=0}^{\infty} f_n x^n = \frac{1}{1-x-x^2}$.

b. Deduce that

$$f_n = \frac{1}{\sqrt{5}} \left(\alpha^{n+1} - \beta^{n+1}\right),$$

where $\alpha = \frac{1}{2}\left(1+\sqrt{5}\right)$ and $\beta = \frac{1}{2}\left(1-\sqrt{5}\right)$.

Hint: By partial fractions,

$$\frac{1}{1-x-x^2} = \frac{1}{\sqrt{5}}\left(\frac{\alpha}{1-\alpha x} - \frac{\beta}{1-\beta x}\right).$$

Now expand by geometric series.

c. Use the result of Part b to show that

$$\sum_{n=0}^{\infty} f_{5n+4}x^n = \frac{5}{1-11x-x^2}.$$

Hence, deduce that $f_{5n+4}$ is divisible by 5.

d. Use the result of Exercise 8 Part a with

$$b = \frac{1}{a} = \prod_{j=1}^{\infty} \frac{(1-q^{25j-20})(1-q^{25j-5})}{(1-q^{25j-15})(1-q^{25j-10})}$$

to show that

$$\sum_{n=0}^{\infty} p(n)q^n = \frac{1}{E(q)} = \frac{1}{E(q^{25})}\left(\frac{b}{1-bq-b^2q^2}\right)$$

$$= \frac{1}{E(q^{25})}\sum_{n=0}^{\infty} f_n q^n b^{n+1}.$$

Since $f_{5n+4}$ is divisible by 5, it follows that $p(5n+4)$ is also divisible by 5. The connection of the divisibility of $p(5n+4)$ to the divisibility of $f_{5n+4}$ is due to M. Hirschhorn [181].

10. Let $\omega = \exp(2\pi i/5)$, $\alpha = \frac{1}{2}(1+\sqrt{5})$, and $\beta = \frac{1}{2}(1-\sqrt{5})$.

a. Observe that the functions $\dfrac{\sin(\pi(4n+1)/5)}{\sin(\pi/5)}$ and $\dfrac{\sin(2\pi(4n+1)/5)}{\sin(2\pi/5)}$ are periodic functions of the integer variable $n$ with period 5 and have the values in Table 5.2.

**Table 5.2** Table for Exercise 10 of Chapter 5

| $n$ | 0 | 1 | 2 | 3 | 4 |
|---|---|---|---|---|---|
| $\dfrac{\sin(\pi(4n+1)/5)}{\sin(\pi/5)}$ | 1 | 0 | $-1$ | $\alpha$ | $-\alpha$ |
| $\dfrac{\sin(2\pi(4n+1)/5)}{\sin(2\pi/5)}$ | 1 | 0 | $-1$ | $\beta$ | $-\beta$ |

b. Show that

$$\prod_{j=1}^{\infty}(1-\omega q^j)(1-\omega^4 q^j)(1-q^j)$$

$$=\prod_{j=1}^{\infty}(1-q^{25j-15})(1-q^{25j-10})(1-q^{25j})$$

$$-\alpha q\prod_{j=1}^{\infty}(1-q^{25j-20})(1-q^{25j-15})(1-q^{25j})$$

and

$$\prod_{j=1}^{\infty}(1-\omega^2 q^j)(1-\omega^3 q^j)(1-q^j)$$

$$=\prod_{j=1}^{\infty}(1-q^{25j-15})(1-q^{25j-10})(1-q^{25j})$$

$$-\beta q\prod_{j=1}^{\infty}(1-q^{25j-20})(1-q^{25j-15})(1-q^{25j}).$$

Hint: To prove the first result, take $x=\omega$ in Theorem 0.3 and collect terms in the series according to the residue class of the summation index $n$, modulo 5. The terms corresponding to $n\equiv 0$ or $2 \pmod 5$ can be combined into a single series that can be summed by the Jacobi triple product identity (0.15). The terms corresponding to $n\equiv 3$ or $4 \pmod 5$ can be handled similarly.

The second identity can be proved by the same procedure, starting by substituting the value $x=\omega^2$ in Theorem 0.3.

c. Let $r=r(q)$ be the Rogers–Ramanujan continued fraction. Use the results of Part b to deduce that

$$\frac{1}{\sqrt{r}}-\alpha\sqrt{r}=\frac{1}{q^{1/10}}\prod_{j=1}^{\infty}\frac{(1-q^j)^{1/2}}{(1-q^{5j})^{1/2}}\times\frac{1}{(1+\alpha q^{j/5}+q^{2j/5})}$$

and

$$\frac{1}{\sqrt{r}}-\beta\sqrt{r}=\frac{1}{q^{1/10}}\prod_{j=1}^{\infty}\frac{(1-q^j)^{1/2}}{(1-q^{5j})^{1/2}}\times\frac{1}{(1+\beta q^{j/5}+q^{2j/5})}.$$

d. Take the product of the identities in Part c to deduce that

$$\frac{1}{r}-1-r=\frac{1}{q^{1/5}}\prod_{j=1}^{\infty}\frac{(1-q^{j/5})}{(1-q^j)}.$$

e. Prove that

$$\left(\frac{1}{\sqrt{r}}\right)^5 - \left(\alpha\sqrt{r}\right)^5 = \frac{1}{q^{1/2}}\prod_{j=1}^{\infty}\frac{(1-q^j)^{1/2}}{(1-q^{5j})^{1/2}} \times \frac{1}{(1+\alpha q^j+q^{2j})^5}$$

and

$$\left(\frac{1}{\sqrt{r}}\right)^5 - \left(\beta\sqrt{r}\right)^5 = \frac{1}{q^{1/2}}\prod_{j=1}^{\infty}\frac{(1-q^j)^{1/2}}{(1-q^{5j})^{1/2}} \times \frac{1}{(1+\beta q^j+q^{2j})^5}.$$

f. Deduce that

$$\frac{1}{r^5} - 11 - r^5 = \frac{1}{q}\prod_{j=1}^{\infty}\frac{(1-q^j)^5}{(1-q^{5j})^6}.$$

The identities in Parts c and e were given by Ramanujan in the lost notebook [254, pp. 206]. They provide factorizations, as well as alternative proofs, of the identities in Theorem (5.8). Further factorizations and references to other proofs have been given in work with Hirschhorn [121].

11. a. Prove the fifth order modular equation:

$$\frac{1}{q}\prod_{j=1}^{\infty}\frac{(1-q^j)^6}{(1-q^{5j})^6} = v^5 + 5v^4 + 15v^3 + 25v^2 + 25v,$$

where

$$v = \frac{1}{q^{1/5}}\prod_{j=1}^{\infty}\frac{(1-q^{j/5})}{(1-q^{5j})}.$$

Hint: By Theorem 5.8,

$$\frac{1}{r} - r = v + 1 \qquad \text{and} \qquad \frac{1}{q}\prod_{j=1}^{\infty}\frac{(1-q^j)^6}{(1-q^{5j})^6} = \frac{1}{r^5} - r^5 - 11,$$

where $r = r(q)$. Use the elementary identity

$$\frac{1}{r^5} - r^5 = \left(\frac{1}{r} - r\right)^5 + 5\left(\frac{1}{r} - r\right)^3 + 5\left(\frac{1}{r} - r\right)$$

to eliminate $r$.

b. Replace $q$ with $q^5$ in Part a to deduce that if $X = \eta_1/\eta_5$ and $Y = \eta_5/\eta_{25}$, then

$$Y^6 = X^5Y^5 + 5X^4Y^4 + 15X^3Y^3 + 25X^2Y^2 + 25XY.$$

# Chapter 6
# Level 6: Ramanujan's Cubic Continued Fraction

## 6.1 Introduction

This chapter centers around the function $r_c = r_c(q)$ defined by

$$r_c(q) = q^{1/3} \prod_{j=1}^{\infty} \frac{(1-q^{6j-5})(1-q^{6j-1})}{(1-q^{6j-3})^2} = \frac{\eta_1 \eta_6^3}{\eta_2 \eta_3^3}. \tag{6.1}$$

Ramanujan found an expression for $r_c$ as a continued fraction:

$$r_c(q) = \cfrac{q^{1/3}}{1 + \cfrac{q+q^2}{1 + \cfrac{q^2+q^4}{1 + \cfrac{q^3+q^6}{1+\cdots}}}}. \tag{6.2}$$

The function $r_c(q)$ has come to be known as Ramanujan's cubic continued fraction, or cubic continued fraction for short. It has many properties that are analogous to those of the Rogers–Ramanujan continued fraction studied in Chapter 5. The function $r_c(q)$ satisfies a natural cubic transformation formula—see (6.22)—and this is the reason for the name "cubic" continued fraction. It will be useful to study the cubic continued fraction in conjunction with two other level 6 modular functions $r_a$ and $r_b$ defined by

$$r_a(q) = \frac{\eta_2 \eta_6^5}{\eta_1^5 \eta_3} \qquad \text{and} \qquad r_b(q) = \frac{\eta_1^2 \eta_6^4}{\eta_2^4 \eta_3^2}.$$

The subscripts $a$, $b$, and $c$ are used simply to distinguish the functions $r_a$, $r_b$, and $r_c$ from each other and have no other special meaning. Other functions with subscripts



S. Cooper, *Ramanujan's Theta Functions*, DOI 10.1007/978-3-319-56172-1_7

$a$, $b$, or $c$ will be encountered in this chapter; they will have some connection with $r_a$, $r_b$ or $r_c$, respectively.

Many of the results in this chapter are analogues of results in the previous chapter. The significant difference is that there will be three analogues for each of the functions $w_5$, $x_5$, $y_5$, $z_5$, $X_5$, and $Z_5$ from the previous chapter.

## 6.2 Proof of the continued fraction expansion

Let us prove that the function $r_c$ defined by the infinite product in (6.1) has the continued fraction expansion given by (6.2). We begin by proving a general expansion.

**Theorem 6.1.** *The following identity holds for* $|abq| < |\lambda|$:

$$\frac{1+\sum_{k=1}^{\infty}\left(\prod_{j=1}^{k}\frac{(a+\lambda q^{j-1})(b+\lambda q^{j-1})}{(1-q^j)}\right)\left(\frac{q}{\lambda}\right)^k}{1+\sum_{k=1}^{\infty}\left(\prod_{j=1}^{k}\frac{(a+\lambda q^{j-1})(b+\lambda q^{j})}{(1-q^j)}\right)\left(\frac{q}{\lambda}\right)^k}$$

$$= 1+\cfrac{aq+\lambda q}{1+\cfrac{bq+\lambda q^2}{1+\cfrac{aq^2+\lambda q^3}{1+\cfrac{bq^2+\lambda q^4}{1+\cfrac{\cdots}{\cfrac{\cdots}{1+\cfrac{aq^n+\lambda q^{2n-1}}{1+\cfrac{bq^n+\lambda q^{2n}}{1+\cdots}}}}}}}}.$$

*Proof.* Let $f(s)$ and $g(s)$ be defined by

$$f(s) = 1+\sum_{k=1}^{\infty}\left(\prod_{j=1}^{k}\frac{(a+\lambda q^{j+s-1})(b+\lambda q^{j+s-1})}{(1-q^j)}\right)\left(\frac{q}{\lambda}\right)^k$$

and

$$g(s) = 1 + \sum_{k=1}^{\infty} \left( \prod_{j=1}^{k} \frac{(a + \lambda q^{j+s-1})(b + \lambda q^{j+s})}{(1 - q^j)} \right) \left(\frac{q}{\lambda}\right)^k.$$

It is straightforward to check that

$$f(s) - g(s) = (aq^{s+1} + \lambda q^{2s+1}) f(s+1)$$

and

$$g(s) - f(s+1) = (bq^{s+1} + \lambda q^{2s+2}) g(s+1).$$

It follows that

$$\begin{aligned} \frac{f(s)}{g(s)} &= 1 + \cfrac{aq^{s+1} + \lambda q^{2s+1}}{\cfrac{g(s)}{f(s+1)}} \\ &= 1 + \cfrac{aq^{s+1} + \lambda q^{2s+1}}{1 + \cfrac{bq^{s+1} + \lambda q^{2s+2}}{\cfrac{f(s+1)}{g(s+1)}}}. \end{aligned}$$

By iteration, it follows that

$$\frac{f(0)}{g(0)} = 1 + \cfrac{aq + \lambda q}{1 + \cfrac{bq + \lambda q^2}{1 + \cfrac{aq^2 + \lambda q^3}{1 + \cfrac{bq^2 + \lambda q^4}{1 + \cfrac{\cdots}{\cfrac{\cdots}{1 + \cfrac{aq^s + \lambda q^{2s-1}}{1 + \cfrac{bq^s + \lambda q^{2s}}{\cfrac{f(s)}{g(s)}}}}}}}}}.$$

Now let $s \to \infty$ to complete the proof. □

Observe that when $a = b = 0$, the result in Theorem 6.1 reduces to Theorem 5.3. Here is another specialization.

**Theorem 6.2.**

$$\frac{1+2\sum_{j=1}^{\infty}\frac{(1+q)(1+q^2)(1+q^3)\cdots(1+q^{2j-1})}{(1-q^2)(1-q^4)\cdots(1-q^{2j})}q^j}{1+\sum_{j=1}^{\infty}\frac{(1+q)(1+q^2)(1+q^3)\cdots(1+q^{2j})}{(1-q^2)(1-q^4)\cdots(1-q^{2j})}q^j}$$

$$=1+\cfrac{q+q^2}{1+\cfrac{q^2+q^4}{1+\cfrac{q^3+q^6}{1+\cfrac{q^4+q^8}{1+\cdots}}}}.$$

*Proof.* Replace $q$ with $q^2$ in Theorem 6.1, then put $a=1/q$ and $b=\lambda=1$. □

It remains to convert the left-hand side of the identity in Theorem 6.2 to a product. This will be accomplished in three lemmas. For $|t|<1$, let $h(a,t,q)$ be defined by

$$h(a,t,q)=1+\sum_{j=1}^{\infty}\frac{(1-aq^{j+1})(1-aq^{j+2})\cdots(1-aq^{2j})}{(1-q)(1-q^2)\cdots(1-q^j)}t^j. \tag{6.3}$$

**Lemma 6.3.** *The following identities hold:*

$$h(-1,q,q)=1+\sum_{j=1}^{\infty}\frac{(1+q)(1+q^2)(1+q^3)\cdots(1+q^{2j})}{(1-q^2)(1-q^4)\cdots(1-q^{2j})}q^j$$

*and*

$$h(-q^{-1},q,q)+q\,h(-q,q,q)$$
$$=1+2\sum_{j=1}^{\infty}\frac{(1+q)(1+q^2)(1+q^3)\cdots(1+q^{2j-1})}{(1-q^2)(1-q^4)\cdots(1-q^{2j})}q^j.$$

*Proof.* In the definition (6.3), take $a=-1$, $t=q$ and multiply numerator and denominator by $(1+q)(1+q^2)\cdots(1+q^j)$. This gives the first result.

Now we will prove the second result. Using the compact notation for products given by (0.113), we may write

$$h(-q^{-1},q,q)+q\,h(-q,q,q)=1+\sum_{j=1}^{\infty}\frac{(-q^j;q)_j}{(q;q)_j}q^j+q\sum_{j=0}^{\infty}\frac{(-q^{j+2};q)_j}{(q;q)_j}q^j.$$

Shifting the summation index in the second sum and then combining terms we find that

$$\begin{aligned}
&h(-q^{-1}, q, q) + q\,h(-q, q, q) \\
&= 1 + \sum_{j=1}^{\infty} \frac{(-q^j; q)_j}{(q; q)_j} q^j + \sum_{j=1}^{\infty} \frac{(-q^{j+1}; q)_{j-1}}{(q; q)_{j-1}} q^j \\
&= 1 + \sum_{j=1}^{\infty} \frac{(-q^{j+1}; q)_{j-1}}{(q; q)_j} \{(1 + q^j) + (1 - q^j)\} q^j \\
&= 1 + 2\sum_{j=1}^{\infty} \frac{(-q^{j+1}; q)_{j-1}}{(q; q)_j} q^j .
\end{aligned}$$

The proof is completed by multiplying the numerator and denominator by $(-q; q)_j$. □

**Lemma 6.4.** *The following identity holds*:

$$h(a, t, q) = \frac{1}{(t; q)_\infty} \sum_{k=0}^{\infty} \frac{(t; q)_k}{(q; q)_k} q^{k(3k+1)/2} (-at)^k.$$

*Proof.* The numerator of the series in (6.3) may be expanded by the finite form of the $q$-binomial theorem in Theorem 0.31 to give

$$\begin{aligned}
h(a, t, q) &= \sum_{j=0}^{\infty} \frac{(aq^{j+1}; q)_j}{(q; q)_j} t^j \\
&= \sum_{j=0}^{\infty} \frac{t^j}{(q; q)_j} \sum_{k=0}^{j} \begin{bmatrix} j \\ k \end{bmatrix}_q q^{k(k-1)/2} (-aq^{j+1})^k.
\end{aligned}$$

Interchange the order of summation and then put $j = m + k$ to obtain

$$\begin{aligned}
h(a, t, q) &= \sum_{k=0}^{\infty} q^{k(k+1)/2} (-a)^k \sum_{j=k}^{\infty} \begin{bmatrix} j \\ k \end{bmatrix}_q \frac{(q^k t)^j}{(q; q)_j} \\
&= \sum_{k=0}^{\infty} q^{k(k+1)/2} (-a)^k \sum_{m=0}^{\infty} \begin{bmatrix} m+k \\ k \end{bmatrix}_q \frac{(q^k t)^{m+k}}{(q; q)_{m+k}}.
\end{aligned}$$

Substituting the value of the $q$-binomial coefficient and simplifying we deduce that

$$h(a, t, q) = \sum_{k=0}^{\infty} q^{k(3k+1)/2} \frac{(-at)^k}{(q; q)_k} \sum_{m=0}^{\infty} \frac{(q^k t)^m}{(q; q)_m}.$$

The series involving $m$ can be summed by Theorem 0.30 to give

$$h(a,t,q) = \sum_{k=0}^{\infty} q^{k(3k+1)/2} \frac{(-at)^k}{(q;q)_k} \times \frac{1}{(q^k t;q)_\infty}$$

$$= \frac{1}{(t;q)_\infty} \sum_{k=0}^{\infty} \frac{(t;q)_k}{(q;q)_k} q^{k(3k+1)/2} (-at)^k,$$

where the last step was obtained by using

$$\frac{1}{(q^k t;q)_\infty} = \frac{(t;q)_k}{(t;q)_\infty}.$$

□

**Lemma 6.5.** *The following identities hold*:

$$h(-1,q,q) = \prod_{j=1}^{\infty} \frac{(1-q^{6j})^2}{(1-q^j)(1-q^{3j})}$$

*and*

$$h(-q^{-1},q,q) + q\,h(-q,q,q) = \prod_{j=1}^{\infty} \frac{(1-q^{2j})(1-q^{3j})^2}{(1-q^j)^2(1-q^{6j})}.$$

*Proof.* By Lemma 6.4,

$$h(-1,q,q) = \frac{1}{(q;q)_\infty} \sum_{k=0}^{\infty} q^{3k(k+1)/2}.$$

The series may be summed by (0.45) with $q^3$ in place of $q$, and this gives the first result.

Next, by Lemma 6.4 again,

$$h(-q^{-1},q,q) + q\,h(-q,q,q)$$

$$= \frac{1}{(q;q)_\infty} \sum_{k=0}^{\infty} q^{k(3k+1)/2} + \frac{1}{(q;q)_\infty} \sum_{k=0}^{\infty} q^{(3k^2+5k+2)/2}.$$

On replacing $k$ with $-1-k$ in the second sum we find that the two series may be combined into a single bilateral series to give

$$h(-q^{-1},q,q) + q\,h(-q,q,q) = \frac{1}{(q;q)_\infty} \sum_{k=-\infty}^{\infty} q^{k(3k+1)/2}.$$

Now apply Jacobi's triple product identity (0.7) to deduce

$$h(-q^{-1},q,q)+q\,h(-q,q,q)=\prod_{j=1}^{\infty}\frac{(1+q^{3j-2})(1+q^{3j-1})(1-q^{3j})}{(1-q^j)}.$$

The proof may be completed by routine manipulations of infinite products. □

We are now ready for the main result of this section.

**Theorem 6.6.** *The expressions for $r_c(q)$ in (6.1) and (6.2) represent the same function. That is,*

$$\frac{\eta_1\eta_6^3}{\eta_2\eta_3^3}=\cfrac{q^{1/3}}{1+\cfrac{q+q^2}{1+\cfrac{q^2+q^4}{1+\cfrac{q^3+q^6}{1+\cdots}}}}.$$

*Proof.* By Lemmas 6.3 and 6.4, the left-hand side of the identity in Theorem 6.2 is

$$\frac{1+2\displaystyle\sum_{j=1}^{\infty}\frac{(1+q)(1+q^2)(1+q^3)\cdots(1+q^{2j-1})}{(1-q^2)(1-q^4)\cdots(1-q^{2j})}q^j}{1+\displaystyle\sum_{j=1}^{\infty}\frac{(1+q)(1+q^2)(1+q^3)\cdots(1+q^{2j})}{(1-q^2)(1-q^4)\cdots(1-q^{2j})}q^j}$$

$$=\frac{h(-q^{-1},q,q)+qh(-q,q,q)}{h(-1,q,q)}$$

$$=\prod_{j=1}^{\infty}\frac{(1-q^{2j})(1-q^{3j})^2}{(1-q^j)^2(1-q^{6j})}\Big/\frac{(1-q^{6j})^3}{(1-q^j)(1-q^{3j})}$$

$$=\prod_{j=1}^{\infty}\frac{(1-q^{2j})(1-q^{3j})^3}{(1-q^j)(1-q^{6j})^3}.$$

Therefore, the identity in Theorem 6.2 takes the form

$$\prod_{j=1}^{\infty}\frac{(1-q^{2j})(1-q^{3j})^3}{(1-q^j)(1-q^{6j})^3}=1+\cfrac{q+q^2}{1+\cfrac{q^2+q^4}{1+\cfrac{q^3+q^6}{1+\cfrac{q^4+q^8}{1+\cdots}}}}.$$

Now take reciprocals and then multiply by $q^{1/3}$. □

## 6.3 Basic properties of Ramanujan's cubic continued fraction

We now begin a systematic study of the functions $r_a$, $r_b$, and $r_c$ defined by

$$r_a(q) = \frac{\eta_2\eta_6^5}{\eta_1^5\eta_3}, \qquad r_b(q) = \frac{\eta_1^2\eta_6^4}{\eta_2^4\eta_3^2} \qquad \text{and} \qquad r_c(q) = \frac{\eta_1\eta_6^3}{\eta_2\eta_3^3}. \tag{6.4}$$

By analogy with (5.36), let us define $x_a$, $x_b$, and $x_c$ by

$$x_a(q) = r_a(q), \quad x_b(q) = r_b^2(q) \quad \text{and} \quad x_c(q) = r_c^3(q), \tag{6.5}$$

so that

$$x_a(q) = \frac{\eta_2\eta_6^5}{\eta_1^5\eta_3}, \qquad x_b(q) = \frac{\eta_1^4\eta_6^8}{\eta_2^8\eta_3^4} \qquad \text{and} \qquad x_c(q) = \frac{\eta_1^3\eta_6^9}{\eta_2^3\eta_3^9}. \tag{6.6}$$

The starting point of the theory is the following lemma.

**Lemma 6.7.** *The functions $r_a$, $r_b$, and $r_c$ defined by* (6.4) *satisfy the identities*:

$$1 + 8r_a(q) = \frac{\eta_2^4\eta_3^8}{\eta_1^8\eta_6^4}, \qquad 1 + 9r_a(q) = \frac{\eta_2^9\eta_3^3}{\eta_1^9\eta_6^3}, \tag{6.7}$$

$$1 - 3r_b(q^2) = \frac{\eta_1^3\eta_{12}}{\eta_3\eta_4^3}, \qquad 1 + r_b(q^2) = \frac{\eta_2^2\eta_3^3\eta_{12}}{\eta_1\eta_4^3\eta_6^2}, \tag{6.8}$$

$$1 - 2r_c(q^3) = \frac{\eta_1^2\eta_{18}}{\eta_2\eta_9^2}, \qquad 1 + r_c(q^3) = \frac{\eta_2^2\eta_3\eta_{18}}{\eta_1\eta_6\eta_9^2}. \tag{6.9}$$

*Proof.* We begin by proving the identities in (6.7). Multiplying (3.59) by (3.65) we find that

$$\varphi^4(-q^3) - \varphi^4(-q) = 8\frac{\eta_1^3\eta_6^5}{\eta_2^3\eta_3}$$

while multiplying (3.61) by (3.63) gives

$$q^2\psi^4(q^4) - 9q^6\psi^4(q^{12}) = \frac{\eta_4^5\eta_{24}^3}{\eta_8\eta_{12}^3}.$$

The left-hand sides may be expressed in terms of eta functions by means of Theorem 3.5 to give

$$\frac{\eta_3^8}{\eta_6^4} - \frac{\eta_1^8}{\eta_2^4} = 8\frac{\eta_1^3\eta_6^5}{\eta_2^3\eta_3} \qquad \text{and} \qquad \frac{\eta_8^8}{\eta_4^4} - 9\frac{\eta_{24}^8}{\eta_{12}^4} = \frac{\eta_4^5\eta_{24}^3}{\eta_8\eta_{12}^3}.$$

Now multiply by $\eta_2^4/\eta_1^8$ and $\eta_8\eta_{12}^3/\eta_4^5\eta_{24}^3$, respectively to deduce

$$\frac{\eta_2^4\eta_3^8}{\eta_1^8\eta_6^4} - 1 = 8\frac{\eta_2\eta_6^5}{\eta_1^5\eta_3} \quad \text{and} \quad \frac{\eta_8^9\eta_{12}^3}{\eta_4^9\eta_{24}^3} - 9\frac{\eta_8\eta_{24}^5}{\eta_4^5\eta_{12}} = 1,$$

that is,

$$\frac{\eta_2^4\eta_3^8}{\eta_1^8\eta_6^4} - 1 = 8r_a(q) \quad \text{and} \quad \frac{\eta_8^9\eta_{12}^3}{\eta_4^9\eta_{24}^3} - 9r_a(q^4) = 1.$$

On replacing $q^4$ with $q$ in the very last identity, we obtain the results in (6.7).

To prove the results in (6.8), begin with (3.63) and (3.64):

$$q\psi^2(q^4) - 3q^3\psi^2(q^{12}) = \frac{\eta_2^3\eta_8\eta_{24}}{\eta_4^2\eta_6} \quad \text{and} \quad q\psi^2(q^4) + q^3\psi^2(q^{12}) = \frac{\eta_6^3\eta_8\eta_{24}}{\eta_2\eta_{12}^2}.$$

Divide each identity by $q\psi^2(q^4)$ and use Theorem 3.5 to convert the theta functions to eta-quotients to obtain

$$1 - 3\frac{\eta_4^2\eta_{24}^4}{\eta_8^4\eta_{12}^2} = \frac{\eta_2^3\eta_{24}}{\eta_6\eta_8^3} \quad \text{and} \quad 1 + \frac{\eta_4^2\eta_{24}^4}{\eta_8^4\eta_{12}^2} = \frac{\eta_4^2\eta_6^3\eta_{24}}{\eta_2\eta_8^3\eta_{12}^2},$$

that is,

$$1 - 3r_b(q^4) = \frac{\eta_2^3\eta_{24}}{\eta_6\eta_8^3} \quad \text{and} \quad 1 + r_b(q^4) = \frac{\eta_4^2\eta_6^3\eta_{24}}{\eta_2\eta_8^3\eta_{12}^2}.$$

On replacing $q^4$ with $q^2$ we deduce the identities in (6.8).

To prove the results in (6.9) we apply the trisection

$$\sum_{j=-\infty}^{\infty} a_j = \sum_{j=-\infty}^{\infty} a_{3j} + \sum_{j=-\infty}^{\infty} a_{3j+1} + \sum_{j=-\infty}^{\infty} a_{3j-1}$$

to obtain

$$\sum_{j=-\infty}^{\infty} (-1)^j q^{j^2} = \sum_{j=-\infty}^{\infty} (-1)^j q^{9j^2} - 2\sum_{j=-\infty}^{\infty} (-1)^j q^{(3j+1)^2} \tag{6.10}$$

and

$$\sum_{j=-\infty}^{\infty} q^{(j+1/2)^2} = \sum_{j=-\infty}^{\infty} q^{(3j+3/2)^2} + 2\sum_{j=-\infty}^{\infty} q^{(3j+1/2)^2}. \tag{6.11}$$

Applying Jacobi's triple product identity (0.7) to each sum in (6.10) and (6.11) and expressing the resulting infinite products in terms of eta-quotients, we find that

$$\frac{\eta_1^2}{\eta_2} = \frac{\eta_9^2}{\eta_{18}} - 2\frac{\eta_3\eta_{18}^2}{\eta_6\eta_9} \quad \text{and} \quad 2\frac{\eta_4^2}{\eta_2} = 2\frac{\eta_{36}^2}{\eta_{18}} + 2\frac{\eta_{12}\eta_{18}^2}{\eta_6\eta_{36}}, \tag{6.12}$$

respectively. Multiplying the first identity in (6.12) by $\eta_{18}/\eta_9^2$, we obtain the first identity in (6.9). Similarly, multiplying the second identity in (6.12) by $\eta_6\eta_{36}/2\eta_{12}\eta_{18}^2$ and replacing $q^2$ with $q$, we obtain the second identity in (6.9).

□

The next results are direct consequences of Lemma 6.7.

**Lemma 6.8.** *The functions $x_a$, $x_b$, and $x_c$ satisfy the identities*

$$1 + 8x_a(q) = \frac{\eta_2^4\eta_3^8}{\eta_1^8\eta_6^4}, \qquad 1 + 9x_a(q) = \frac{\eta_2^9\eta_3^3}{\eta_1^9\eta_6^3}, \tag{6.13}$$

$$1 - 9x_b(q) = \frac{\eta_1^9\eta_6^3}{\eta_2^9\eta_3^3}, \qquad 1 - x_b(q) = \frac{\eta_1\eta_3^5}{\eta_2^5\eta_6}, \tag{6.14}$$

$$1 - 8x_c(q) = \frac{\eta_1^8\eta_6^4}{\eta_2^4\eta_3^8}, \qquad 1 + x_c(q) = \frac{\eta_2^5\eta_6}{\eta_1\eta_3^5}. \tag{6.15}$$

*Proof.* Since $x_a(q) = r_a(q)$, the results in (6.13) have already been proved in (6.7). Let us prove (6.14). The first identity in (6.8), when written out in full as an infinite product, is

$$1 - 3q\prod_{j=1}^{\infty}\frac{(1-q^{2j})^2(1-q^{12j})^4}{(1-q^{4j})^4(1-q^{6j})^2} = \prod_{j=1}^{\infty}\frac{(1-q^j)^3(1-q^{12j})}{(1-q^{3j})(1-q^{4j})^3}. \tag{6.16}$$

Replacing $q$ with $-q$ and using (0.2) we find that

$$1+3q\prod_{j=1}^{\infty}\frac{(1-q^{2j})^2(1-q^{12j})^4}{(1-q^{4j})^4(1-q^{6j})^2} = \prod_{j=1}^{\infty}\frac{(1-q^{2j})^9(1-q^{3j})(1-q^{12j})^2}{(1-q^j)^3(1-q^{4j})^6(1-q^{6j})^3}. \tag{6.17}$$

If we multiply (6.16) and (6.17) we find that

$$1 - 9r_b^2(q^2) = \prod_{j=1}^{\infty}\frac{(1-q^{2j})^9(1-q^{12j})^3}{(1-q^{4j})^9(1-q^{6j})^3},$$

and on replacing $q^2$ with $q$ we obtain the first identity in (6.14). The second identity in (6.14) may be obtained by applying a similar procedure starting with the other identity in (6.8).

Now we will prove the identities in (6.15). The first identity in (6.9), when written in full as an infinite product, is

$$1 - 2q \prod_{j=1}^{\infty} \frac{(1-q^{3j})(1-q^{18j})^3}{(1-q^{6j})(1-q^{9j})^3} = \prod_{j=1}^{\infty} \frac{(1-q^j)^2(1-q^{18j})}{(1-q^{2j})(1-q^{9j})^2}.$$

Replace $q$ with $e^{2\pi im/3} q$ where $m \in \{0,1,2\}$, and multiply the three resulting identities together. The result simplifies, on making use of (0.5) on the right-hand side, to give

$$1 - 8r_c^3(q^3) = \prod_{j=1}^{\infty} \frac{(1-q^{3j})^8(1-q^{18j})^4}{(1-q^{6j})^4(1-q^{9j})^8}.$$

On replacing $q^3$ with $q$ we obtain the first identity in (6.15). The second identity in (6.15) can be obtained by applying a similar procedure to the other identity in (6.9). □

**Theorem 6.9.** *The functions $x_a$, $x_b$, and $x_c$ given by* (6.6) *are related by*

$$\frac{1}{x_b} - \frac{1}{x_c} = 1, \qquad \frac{1}{x_c} - \frac{1}{x_a} = 8 \qquad \textit{and} \qquad \frac{1}{x_b} - \frac{1}{x_a} = 9.$$

*Proof.* By (6.13)–(6.15) we have

$$(1 - x_b)(1 + x_c) = 1, \quad (1 + 8x_a)(1 - 8x_c) = 1$$

and

$$(1 + 9x_a)(1 - 9x_b) = 1.$$

The results follow by rearrangement. □

**Corollary 6.10.** *The functions $x_a = x_a(q)$, $x_b = x_b(q)$, and $x_c = x_c(q)$ are all increasing functions of $q$ on the interval $0 \le q < 1$. Moreover,*

$$x_a(0) = x_b(0) = x_c(0) = 0$$

*and*

$$\lim_{q \to 1^-} x_a(q) = +\infty, \quad \lim_{q \to 1^-} x_b(q) = \frac{1}{9}, \quad \textit{and} \quad \lim_{q \to 1^-} x_c(q) = \frac{1}{8}.$$

*Proof.* From (6.6),

$$\begin{aligned} x_a(q) &= q\prod_{n=1}^{\infty}\frac{(1-q^{2n})}{(1-q^n)}\times\frac{(1-q^{6n})}{(1-q^{3n})}\times\frac{(1-q^{6n})^4}{(1-q^n)^4} \\ &= q\prod_{n=1}^{\infty}(1+q^n)(1+q^{3n})(1+q^n+q^{2n}+q^{3n}+q^{4n}+q^{5n})^4. \end{aligned}$$

This reveals that $x_a$ is an increasing function of $q$ on the interval $0 \le q < 1$, and moreover $x_a(0) = 0$ and $\lim_{q\to 1^-} x_a(q) = +\infty$.

Next, from Theorem 6.9 we have

$$x_b = \frac{x_a}{1+9x_a} \qquad \text{and} \qquad x_c = \frac{x_a}{1+8x_a}.$$

The required properties of $x_b$ and $x_c$ now follow from the corresponding properties of $x_a$. □

The next results are analogues of Theorem 5.8.

**Theorem 6.11.** *Let $r_a$, $r_b$, and $r_c$ be defined by* (6.4), *let $x_a$, $x_b$, and $x_c$ be given by* (6.6), *and note that $r_a(q) = x_a(q)$. The following identities hold*:

$$\frac{1}{r_a(q)} + 17 + 72r_a(q) = \frac{\eta_2^{12}\eta_3^{12}}{\eta_1^{12}\eta_6^{12}},$$

$$\frac{1}{r_b(q^2)} - 2 - 3r_b(q^2) = \frac{\eta_1^2\eta_3^2}{\eta_4^2\eta_{12}^2}, \qquad \frac{1}{x_b(q)} - 10 + 9x_b(q) = \frac{\eta_1^6\eta_3^6}{\eta_2^6\eta_6^6},$$

*and*

$$\frac{1}{r_c(q^3)} - 1 - 2r_c(q^3) = \frac{\eta_1\eta_2}{\eta_9\eta_{18}}, \qquad \frac{1}{x_c(q)} - 7 - 8x_c(q) = \frac{\eta_1^4\eta_2^4}{\eta_3^4\eta_6^4}.$$

*Proof.* By (6.4) and (6.7),

$$\begin{aligned} \frac{1}{r_a(q)} + 17 + 72r_a(q) &= \frac{(1+8r_q(q))(1+9r_a(q))}{r_a(q)} \\ &= \left(\frac{\eta_2^4\eta_3^8}{\eta_1^8\eta_6^4}\right)\times\left(\frac{\eta_2^9\eta_3^3}{\eta_1^9\eta_6^3}\right)\Big/\left(\frac{\eta_2\eta_6^5}{\eta_1^5\eta_3}\right) \\ &= \frac{\eta_2^{12}\eta_3^{12}}{\eta_1^{12}\eta_6^{12}}. \end{aligned}$$

The other results follow similarly from the definitions (6.4), (6.5), and (6.6) together with Lemmas 6.7 and 6.8. □

## 6.4 Transformation formulas

The modular transformations in the next result are analogues of the results in Theorem 5.10.

**Theorem 6.12.** *Let $r_a$, $r_b$, and $r_c$ be defined by* (6.4), *let $x_a$, $x_b$, and $x_c$ be given by* (6.6), *and note that $r_a(q) = x_a(q)$. The following transformation formulas hold*:

$$r_a(e^{-\pi/3t}) = \frac{1}{72r_a(e^{-2\pi t})},$$

$$r_b(e^{-2\pi/3t}) = \frac{\frac{1}{3} - r_b(e^{-2\pi t})}{1 + r_b(e^{-2\pi t})}, \qquad x_b(e^{-\pi/3t}) = \frac{\frac{1}{9} - x_b(e^{-2\pi t})}{1 - x_b(e^{-2\pi t})},$$

$$r_c(e^{-\pi/t}) = \frac{\frac{1}{2} - r_c(e^{-2\pi t})}{1 + r_c(e^{-2\pi t})}, \qquad x_c(e^{-\pi/3t}) = \frac{\frac{1}{8} - x_c(e^{-2\pi t})}{1 + x_c(e^{-2\pi t})}.$$

*Proof.* By Theorem 2.8 we have

$$\eta\left(\frac{i}{t}\right) = \sqrt{t}\,\eta(it), \qquad \text{for } \mathrm{Re}(t) > 0.$$

Therefore, by the definition of $r_a$ in (6.4), we obtain

$$r_a(e^{-\pi/3t}) = \frac{\eta(i/3t)\eta^5(i/t)}{\eta^5(i/6t)\eta(i/2t)} = \frac{1}{72}\frac{\eta(3it)\eta^5(it)}{\eta^5(6it)\eta(2it)} = \frac{1}{72r_a(e^{-2\pi t})}.$$

This proves the first result. By a similar procedure, using the definition of $r_b$, we obtain

$$r_b(e^{-2\pi/3t}) = \frac{\eta^2(i/3t)\eta^4(2i/t)}{\eta^4(2i/3t)\eta^2(i/t)} = \frac{1}{3} \times \frac{\eta^2(3it)\eta^4(it/2)}{\eta^4(3it/2)\eta^2(it)}.$$

The eta functions on the right-hand side can be rearranged so that the identities (6.8) in Lemma 6.7 can be applied. This gives

$$\begin{aligned} r_b(e^{-2\pi/3t}) &= \frac{1}{3} \times \frac{\eta^3(it/2)\eta(6it)}{\eta(3it/2)\eta^3(2it)} \Big/ \frac{\eta^2(it)\eta^3(3it/2)\eta(6it)}{\eta(it/2)\eta^3(2it)\eta^2(3it)} \\ &= \frac{1}{3} \times \frac{1 - 3r_b(e^{-2\pi t})}{1 + r_b(e^{-2\pi t})}. \end{aligned}$$

The result for $r_c$ can be proved in the same way.

Finally, the results for $x_b$ and $x_c$ can be proved by the same procedure, using the identities in Lemma 6.8. We omit the details as they are similar. □

By considering the fixed points of the transformation formulas in Theorem 6.12, the exact values of $r_a(q)$, $r_b(q)$, and $r_c(q)$ at the points $q = e^{-2\pi/\sqrt{6}}$, $q = e^{-2\pi/\sqrt{3}}$, and $q = e^{-2\pi/\sqrt{2}}$ can be determined. See Exercise 1 at the end of the chapter.

The results in Theorem 6.12 can be used to produce quadratic and cubic transformation formulas that are analogues of the quintic transformation formulas for the Rogers–Ramanujan continued fraction in Theorem 5.11. It will be convenient to introduce some notation especially for this. Let us write

$$u_a = r_a(q), \qquad u_b = r_b(q), \qquad u_c = r_c(q), \tag{6.18}$$

$$v_a = r_a(q^2), \qquad v_b = r_b(q^2), \qquad v_c = r_c(q^2), \tag{6.19}$$

$$w_a = r_a(q^3), \qquad w_b = r_b(q^3), \qquad w_c = r_c(q^3). \tag{6.20}$$

The quadratic transformation is best stated in terms of the function $r_b(q)$ and the cubic transformation is most simply stated in terms of $r_c(q)$.

**Theorem 6.13.** *The following quadratic transformation formulas hold:*

$$u_b^2 = v_b \frac{(1-v_b)}{(1+3v_b)} \qquad \textit{and} \qquad v_b = \frac{1-\left(\dfrac{1-9u_b^2}{1-u_b^2}\right)^{1/2}}{3+\left(\dfrac{1-9u_b^2}{1-u_b^2}\right)^{1/2}}. \tag{6.21}$$

*And the following cubic transformation formulas hold:*

$$u_c^3 = w_c \frac{(1-w_c+w_c^2)}{(1+2w_c+4w_c^2)} \qquad \textit{and} \qquad w_c = \frac{1-\left(\dfrac{1-8u_c^3}{1+u_c^3}\right)^{1/3}}{2+\left(\dfrac{1-8u_c^3}{1+u_c^3}\right)^{1/3}}. \tag{6.22}$$

*Proof.* By Theorem 6.12,

$$r_b(e^{-2\pi/3t}) = \frac{\frac{1}{3}-r_b(e^{-2\pi t})}{1+r_b(e^{-2\pi t})} \quad \text{and} \quad r_b^2(e^{-\pi/3t}) = \frac{\frac{1}{9}-r_b^2(e^{-2\pi t})}{1-r_b^2(e^{-2\pi t})}.$$

Replace $t$ with $2t$ in the first identity and substitute the result into the second identity to obtain

$$\left(\frac{\frac{1}{3}-v_b}{1+v_b}\right)^2 = \frac{\frac{1}{9}-u_b^2}{1-u_b^2}.$$

The formulas in (6.21) now follow on solving either for $u_b^2$ or for $v_b$.

Now we will prove (6.22). By Theorem 6.12,

$$r_c(e^{-\pi/t}) = \frac{\frac{1}{2} - r_c(e^{-2\pi t})}{1 + r_c(e^{-2\pi t})} \quad \text{and} \quad r_c^3(e^{-\pi/3t}) = \frac{\frac{1}{8} - r_c^3(e^{-2\pi t})}{1 + r_c^3(e^{-2\pi t})}.$$

If we replace $t$ with $3t$ in the first identity and substitute the result into the second identity, we obtain

$$\left(\frac{\frac{1}{2} - w_c}{1 + w_c}\right)^3 = \frac{\frac{1}{8} - u_c^3}{1 + u_c^3}.$$

The formulas in (6.22) now follow on solving either for $u_c^3$ or for $w_c$. □

Quadratic transformation formulas for $u_a$ and $u_c$, as well as cubic transformation formulas for $u_a$ and $u_b$, can be obtained from Theorem 6.12 by making a change of variables using Theorem 6.9. See Exercise 2.

## 6.5 Weight one modular forms

In this section we consider the weight one modular forms $z_a$, $z_b$, $z_c$, and $z_d$ defined by

$$z_a = \frac{\eta_1^6\eta_6}{\eta_2^3\eta_3^2}, \qquad z_b = \frac{\eta_2^6\eta_3}{\eta_1^3\eta_6^2}, \qquad z_c = \frac{\eta_2\eta_3^6}{\eta_1^2\eta_6^3} \quad \text{and} \quad z_d = \frac{\eta_1\eta_6^6}{\eta_2^2\eta_3^3}. \tag{6.23}$$

They will be studied in conjunction with the modular functions $x_a$, $x_b$, and $x_c$ defined in (6.6), that is,

$$x_a = \frac{\eta_2\eta_6^5}{\eta_1^5\eta_3}, \qquad x_b = \frac{\eta_1^4\eta_6^8}{\eta_2^8\eta_3^4} \quad \text{and} \quad x_c = \frac{\eta_1^3\eta_6^9}{\eta_2^3\eta_3^9}. \tag{6.24}$$

The eta-quotients in (6.23) have been studied in Theorem 3.21. Hence we may immediately deduce the results in the next theorem.

**Theorem 6.14.** *The following identities hold:*

$$z_a = 2a(q^2) - a(q),$$
$$2z_b = a(q) + a(q^2),$$
$$3z_c = a(q) + 2a(q^2)$$

*and*

$$6z_d = a(q) - a(q^2).$$

*Proof.* These are just restatements of results in Theorem 3.21. □

The results in Theorem 6.14 together with the Lambert series for $a(q)$ in Theorem 3.19 may be used to produce the following Lambert series expansions:

$$z_a = 1 - 6\sum_{j=1}^{\infty}\left(\frac{q^{6j-5}}{1-q^{6j-5}} - \frac{q^{6j-1}}{1-q^{6j-1}}\right) + 18\sum_{j=1}^{\infty}\left(\frac{q^{6j-4}}{1-q^{6j-4}} - \frac{q^{6j-2}}{1-q^{6j-2}}\right),$$

$$z_b = 1 + 3\sum_{j=1}^{\infty}\left(\frac{q^{6j-5}}{1-q^{6j-5}} - \frac{q^{6j-1}}{1-q^{6j-1}}\right),$$

$$z_c = 1 + 2\sum_{j=1}^{\infty}\left(\frac{q^{6j-5}}{1-q^{6j-5}} + \frac{q^{6j-4}}{1-q^{6j-4}} - \frac{q^{6j-2}}{1-q^{6j-2}} - \frac{q^{6j-1}}{1-q^{6j-1}}\right)$$

and

$$z_d = \sum_{j=1}^{\infty}\left(\frac{q^{6j-5}}{1-q^{6j-5}} - \frac{q^{6j-1}}{1-q^{6j-1}}\right) - 2\sum_{j=1}^{\infty}\left(\frac{q^{6j-4}}{1-q^{6j-4}} - \frac{q^{6j-2}}{1-q^{6j-2}}\right).$$

As the next result suggests, the function $z_d$ plays a different role from the others. The identities in (6.25) indicate that $z_a$, $z_b$, and $z_c$ are all analogues of the function $z_5$ defined in (5.37), while $z_d$ is the analogue of $x_5 z_5$.

**Theorem 6.15.** *The following identities hold*:

$$x_a z_a = x_b z_b = x_c z_c = z_d, \tag{6.25}$$

$$\frac{1}{9}(z_b - z_a) = \frac{1}{8}(z_c - z_a) = z_b - z_c = z_d$$

*and*

$$z_a = 9z_c - 8z_b.$$

*Proof.* The identities in (6.25) follow directly from the eta-quotients given by (6.23) and (6.24). The other results are simple consequences of Theorem 6.14. □

The next result gives representations for the cubic theta function $a(q)$ in terms of the level 6 functions.

**Theorem 6.16.** *The following identities hold*:

$$a(q) = (1 + 12x_a)z_a = (1 + 3x_b)z_b = (1 + 4x_c)z_c$$

*and*

$$a(q^2) = (1 + 6x_a)z_a = (1 - 3x_b)z_b = (1 - 2x_c)z_c.$$

*Proof.* Eliminating the $a(q^2)$ term from the identities in Theorem 6.14 we deduce that

$$a(q) = z_a + 12z_d = z_b + 3z_d = z_c + 4z_d,$$

while eliminating the $a(q)$ term gives

$$a(q^2) = z_a + 6z_d = z_b - 3z_d = z_c - 2z_d.$$

The required results now follow by using (6.25) to replace each occurrence of $z_d$ with either $x_a z_a$, $x_b z_b$, or $x_c z_c$, as appropriate. □

In order to prove the results in the next three theorems, recall from Theorem 3.30 that

$$a(q)a(q^2) = \frac{1}{4}\left(6P(q^6) - 3P(q^3) + 2P(q^2) - P(q)\right) \tag{6.26}$$

while from Theorem 3.31 we have

$$a^2(q) = \frac{1}{2}\left(3P(q^3) - P(q)\right) \tag{6.27}$$

and hence

$$a^2(q^2) = \frac{1}{2}\left(3P(q^6) - P(q^2)\right). \tag{6.28}$$

**Theorem 6.17.** *The following identities hold:*

$$\begin{aligned}
z_a^2 &= \frac{1}{2}\left(P(q) - 8P(q^2) + 9P(q^3)\right), \\
z_b^2 &= \frac{1}{8}\left(-2P(q) + P(q^2) + 9P(q^6)\right), \\
z_c^2 &= \frac{1}{6}\left(-P(q) - P(q^3) + 8P(q^3)\right), \\
z_d^2 &= \frac{1}{24}\left(-P(q^2) + 2P(q^3) - P(q^6)\right).
\end{aligned}$$

*Proof.* By Theorem 6.14,

$$z_a^2 = \left(2a(q^2) - a(q)\right)^2 = 4a^2(q^2) - 4a(q)a(q^2) + a^2(q).$$

On applying (6.26)–(6.28) we obtain

$$\begin{aligned} z_a^2 &= 2(3P(q^6) - P(q^2)) - (6P(q^6) - 3P(q^3) + 2P(q^2) - P(q)) \\ &\qquad + \frac{1}{2}(3P(q^3) - P(q)) \\ &= \frac{1}{2}(P(q) - 8P(q^2) + 9P(q^3)). \end{aligned}$$

This proves the first result. The proofs of the other results are similar. □

**Theorem 6.18.** *The following identities hold*:

$$\begin{aligned} (1 - 72x_a^2)z_a^2 &= \frac{1}{2}\left(P(q) - 2P(q^2) - 3P(q^3) + 6P(q^6)\right), \\ (1 - 9x_b^2)z_b^2 &= \frac{1}{4}\left(-P(q) + 2P(q^2) - 3P(q^3) + 6P(q^6)\right), \\ (1 + 8x_c^2)z_c^2 &= \frac{1}{6}\left(-P(q) - 2P(q^2) + 3P(q^3) + 6P(q^6)\right). \end{aligned}$$

*Proof.* By (6.25), the left-hand sides are equal to $z_a^2 - 72z_d^2$, $z_b^2 - 9z_d^2$ and $z_c^2 + 8z_d^2$, respectively. The required results now follow from Theorem 6.17. □

Lambert series for the functions $z_j z_k$ with $j \neq k$ can also be given. Here is the complete list.

**Theorem 6.19.** *The following series expansions hold*:

$$\begin{aligned} z_a z_b &= \frac{1}{8}\left(P(q) - 2P(q^2) - 9P(q^3) + 18P(q^6)\right), \\ z_a z_c &= \frac{1}{6}\left(P(q) - 4P(q^2) - 3P(q^3) + 12P(q^6)\right), \\ z_b z_c &= \frac{1}{24}\left(-5P(q) + 2P(q^2) - 3P(q^3) + 30P(q^6)\right), \\ z_a z_d &= \frac{1}{24}\left(-P(q) + 10P(q^2) - 15P(q^3) + 6P(q^6)\right), \\ z_b z_d &= \frac{1}{24}\left(-P(q) + P(q^2) + 3P(q^3) - 3P(q^6)\right), \\ z_c z_d &= \frac{1}{24}\left(-P(q) + 2P(q^2) + P(q^3) - 2P(q^6)\right). \end{aligned}$$

*Proof.* The proof is similar to the proof of Theorem 6.17. By Theorem 6.14 and (6.26)–(6.28) we have

$$\begin{aligned}
z_a z_b &= \frac{1}{2}\left(2a(q^2) - a(q)\right)\left(a(q) + a(q^2)\right) \\
&= a^2(q^2) + \frac{1}{2}a(q)a(q^2) - \frac{1}{2}a^2(q) \\
&= \frac{1}{2}\left(3P(q^6) - P(q^2)\right) + \frac{1}{8}\left(6P(q^6) - 3P(q^3) + 2P(q^2) - P(q)\right) \\
&\quad - \frac{1}{4}\left(3P(q^3) - P(q)\right) \\
&= \frac{1}{8}\left(P(q) - 2P(q^2) - 9P(q^3) + 18P(q^6)\right).
\end{aligned}$$

This proves the first result. The other results may be proved in a similar way. □

Four of the six results in Theorem 6.19 have already been encountered in Theorems 3.27 and 3.30. For, by comparing the infinite products in Theorem 3.21 and (6.23), we have

$$z_a z_c = \varphi^2(-q)\varphi^2(-q^3), \qquad z_b z_d = q\psi^2(q)\psi^2(q^3),$$

$$z_a z_b = b(q)b(q^2) \quad \text{and} \quad z_c z_d = \frac{1}{9}c(q)c(q^2),$$

and these may be compared with (3.71), (3.72), (3.74), and (3.75). All six functions in Theorem 6.19 will occur again in Theorem 6.20.

The right-hand sides of the identities in Theorems 6.17–6.19 involve expressions of the form

$$c_1 P(q) + c_2 P(q^2) + c_3 P(q^3) + c_4 P(q^6)$$

where in each case $6c_1 + 3c_2 + 2c_3 + c_6 = 0$. Additional results will be given below in Theorem 6.28 and the discussion that comes after, where the four-dimensional space

$$\left\{c_1 P(q) + c_2 P(q^2) + c_3 P(q^3) + c_4 P(q^6) \,\middle|\, c_1, c_2, c_3, c_4 \in \mathbb{C}\right\}$$

will be studied.

Numerous identities can be deduced by exploiting the linear relations among the right-hand sides of the results in Theorems 6.17 and 6.19. We only mention the results

$$z_d^2 = z_a^2 + 8z_b^2 - 9z_c^2$$

and

$$\begin{aligned}
z_a^2 &= z_b z_c + 64 z_b z_d - 81 z_c z_d,\\
z_b^2 &= z_a z_c - \tfrac{1}{8} z_a z_d + \tfrac{81}{8} z_c z_d,\\
z_c^2 &= z_a z_b - \tfrac{1}{9} z_a z_d + \tfrac{64}{9} z_b z_d,\\
z_d^2 &= -\tfrac{1}{8} z_a z_b + \tfrac{1}{9} z_a z_c + \tfrac{1}{72} z_b z_c.
\end{aligned}$$

## 6.6 Weight two eta-quotients

This section is about the weight two modular forms $y_a$, $y_b$, and $y_c$ defined by

$$y_a = q\frac{\mathrm{d}}{\mathrm{d}q}\log x_a, \quad y_b = q\frac{\mathrm{d}}{\mathrm{d}q}\log x_b, \quad \text{and} \quad y_c = q\frac{\mathrm{d}}{\mathrm{d}q}\log x_c, \tag{6.29}$$

where $x_a$, $x_b$, and $x_c$ are given by (6.6). They will be studied in conjunction with the modular functions $w_a$, $w_b$, and $w_c$ defined by

$$w_a = \frac{\eta_1^{12}\eta_6^{12}}{\eta_2^{12}\eta_3^{12}}, \qquad w_b = \frac{\eta_2^{6}\eta_6^{6}}{\eta_1^{6}\eta_3^{6}} \quad \text{and} \quad w_c = \frac{\eta_3^{4}\eta_6^{4}}{\eta_1^{4}\eta_2^{4}}. \tag{6.30}$$

We begin by showing that each of $y_a$, $y_b$, and $y_c$ can be expressed as eta-quotients.

**Theorem 6.20.** *The following infinite product expansions hold:*

$$\begin{aligned}
y_a &= z_b z_c = \frac{\eta_2^7\eta_3^7}{\eta_1^5\eta_6^5}, & w_a y_a &= z_a z_d = \frac{\eta_1^7\eta_6^7}{\eta_2^5\eta_3^5},\\
y_b &= z_a z_c = \frac{\eta_1^4\eta_3^4}{\eta_2^2\eta_6^2}, & w_b y_b &= z_b z_d = \frac{\eta_2^4\eta_6^4}{\eta_1^2\eta_3^2},\\
y_c &= z_a z_b = \frac{\eta_1^3\eta_2^3}{\eta_3\eta_6}, & w_c y_c &= z_c z_d = \frac{\eta_3^3\eta_6^3}{\eta_1\eta_2}.
\end{aligned}$$

*Proof.* By a direct calculation using the definitions (6.29), (6.6), and (0.61), we have

$$\begin{aligned}
y_a &= q\frac{\mathrm{d}}{\mathrm{d}q}\log\frac{\eta_2\eta_6^5}{\eta_1^5\eta_3} = \frac{1}{24}\left(-5P(q) + 2P(q^2) - 3P(q^3) + 30P(q^6)\right),\\
y_b &= q\frac{\mathrm{d}}{\mathrm{d}q}\log\frac{\eta_1^4\eta_6^8}{\eta_2^8\eta_3^4} = \frac{1}{6}\left(P(q) - 4P(q^2) - 3P(q^3) + 12P(q^6)\right),
\end{aligned}$$

$$y_c = q\frac{\mathrm{d}}{\mathrm{d}q}\log\frac{\eta_1^3\eta_6^9}{\eta_2^3\eta_3^9} = \frac{1}{8}\left(P(q) - 2P(q^2) - 9P(q^3) + 18P(q^6)\right).$$

Appealing to Theorem 6.19 it follows that

$$y_a = z_b z_c, \qquad y_b = z_a z_c \qquad \text{and} \qquad y_c = z_a z_b.$$

The expressions for $y_a$, $y_b$, and $y_c$ in terms of eta-quotients can now be obtained from the definitions of $z_a$, $z_b$, and $z_c$ as eta-quotients given in (6.23).

The results for $w_a y_a$, $w_b y_b$, and $w_c y_c$ can be obtained by comparing eta function representations of the various functions: eta-quotients for $z_a$, $z_b$, $z_c$, and $z_d$ are given by (6.23); eta-quotients for $w_a$, $w_b$, and $w_c$ are given in (6.30); and eta-quotients for $y_a$, $y_b$, and $y_c$ have been proved above as part of this theorem. □

For reference, a summary of the eta-quotients encountered so far is given in Table 6.1.

The next result relates the modular functions $w_a$, $w_b$, and $w_c$ defined in (6.30) to the modular functions $x_a$, $x_b$, and $x_c$ given by (6.6).

**Table 6.1** Eta-quotients, $f = \eta_1^{e1}\eta_2^{e2}\eta_3^{e3}\eta_6^{e6}$

| $f$ | $e_1$ | $e_2$ | $e_3$ | $e_6$ | weight $= \frac{1}{2}(e_1+e_2+e_3+e_6)$ | reference |
|---|---|---|---|---|---|---|
| $x_a$ | −5 | 1 | −1 | 5 | 0 | (6.6) |
| $x_b$ | 4 | −8 | −4 | 8 | 0 | |
| $x_c$ | 3 | −3 | −9 | 9 | 0 | |
| $1+8x_a$ | −8 | 4 | 8 | −4 | 0 | Lemma 6.8 |
| $1+9x_a$ | −9 | 9 | 3 | −3 | 0 | |
| $1-x_b$ | 1 | −5 | 5 | −1 | 0 | |
| $1-9x_b$ | 9 | −9 | −3 | 3 | 0 | |
| $1+x_c$ | −1 | 5 | −5 | 1 | 0 | |
| $1-8x_c$ | 8 | −4 | −8 | 4 | 0 | |
| $w_a$ | 12 | −12 | −12 | 12 | 0 | (6.30) |
| $w_b$ | −6 | 6 | −6 | 6 | 0 | |
| $w_c$ | −4 | −4 | 4 | 4 | 0 | |
| $z_a$ | 6 | −3 | −2 | 1 | 1 | (6.23) |
| $z_b$ | −3 | 6 | 1 | −2 | 1 | |
| $z_c$ | −2 | 1 | 6 | −3 | 1 | |
| $z_d$ | 1 | −2 | −3 | 6 | 1 | |
| $y_a$ | −5 | 7 | 7 | −5 | 2 | Theorem 6.20 |
| $y_b$ | 4 | −2 | 4 | −2 | 2 | |
| $y_c$ | 3 | 3 | −1 | −1 | 2 | |
| $w_a y_a$ | 7 | −5 | −5 | 7 | 2 | |
| $w_b y_b$ | −2 | 4 | −2 | 4 | 2 | |
| $w_c y_c$ | −1 | −1 | 3 | 3 | 2 | |

**Theorem 6.21.** *The following identities hold.*

$$w_a = \frac{x_a}{(1+8x_a)(1+9x_a)},$$

$$w_b = \frac{x_b}{(1-x_b)(1-9x_b)},$$

$$w_c = \frac{x_c}{(1+x_c)(1-8x_c)},$$

*and*

$$w_a = \frac{x_b(1-9x_b)}{1-x_b} = \frac{x_c(1-8x_c)}{1+x_c},$$

$$w_b = \frac{x_a(1+9x_a)}{1+8x_a} = \frac{x_c(1+x_c)}{1-8x_c},$$

$$w_c = \frac{x_a(1+8x_a)}{1+9x_a} = \frac{x_b(1-x_b)}{1-9x_b}.$$

*Proof.* The first three results are just restatements of results in Theorem 6.11. The remaining three results follow by applying Theorem 6.9. Alternatively, all of the results can be proved by referring to the eta-quotients in Table 6.1. □

The next result relates the weight two modular forms $y_a$, $y_b$, and $y_c$ to the squares of the weight one modular forms $z_a$, $z_b$, $z_c$, and $z_d$.

**Theorem 6.22.** *The following identities hold*:

$$x_a w_a y_a = x_b w_b y_b = x_c w_c y_c = z_d^2 = \left(\frac{\eta_1 \eta_6^6}{\eta_2^2 \eta_3^3}\right)^2,$$

*and*

$$x_a z_a^2 = w_a y_a, \quad x_b z_b^2 = w_b y_b \quad \textit{and} \quad x_c z_c^2 = w_c y_c.$$

*Hence,*

$$y_a = (1+8x_a)(1+9x_a)z_a^2,$$

$$y_b = (1-x_b)(1-9x_b)z_b^2,$$

$$y_c = (1+x_c)(1-8x_c)z_c^2.$$

*Proof.* The simplest way is to compare infinite products in Table 6.1. □

For future reference, we record the following expansions in terms of weight two Eisenstein series.

**Theorem 6.23.** *The following series expansions hold*:

$$\begin{aligned}
y_a &= \frac{1}{24}\left(-5P(q)+2P(q^2)-3P(q^3)+30P(q^6)\right),\\
y_b &= \frac{1}{6}\left(P(q)-4P(q^2)-3P(q^3)+12P(q^6)\right),\\
y_c &= \frac{1}{8}\left(P(q)-2P(q^2)-9P(q^3)+18P(q^6)\right),\\
w_a y_a &= \frac{1}{24}\left(-P(q)+10P(q^2)-15P(q^3)+6P(q^6)\right),\\
w_b y_b &= \frac{1}{24}\left(-P(q)+P(q^2)+3P(q^3)-3P(q^6)\right),\\
w_c y_c &= \frac{1}{24}\left(-P(q)+2P(q^2)+P(q^3)-2P(q^6)\right).
\end{aligned}$$

*Proof.* The results are immediate from Theorems 6.19 and 6.20. □

## 6.7 Weight two Eisenstein series

This section is about the weight two modular forms $Z_a$, $Z_b$, and $Z_c$ defined by

$$Z_a = q\frac{\mathrm{d}}{\mathrm{d}q}\log w_a, \quad Z_b = q\frac{\mathrm{d}}{\mathrm{d}q}\log w_b, \quad \text{and} \quad Z_c = q\frac{\mathrm{d}}{\mathrm{d}q}\log w_c, \tag{6.31}$$

where $w_a$, $w_b$, and $w_c$ are given by (6.30). They will be studied in conjunction with the modular functions $X_a$, $X_b$, and $X_c$ defined by

$$X_a = \frac{\eta_1^2\eta_2^2\eta_3^2\eta_6^2}{Z_a^2}, \qquad X_b = \frac{\eta_1^2\eta_2^2\eta_3^2\eta_6^2}{Z_b^2} \quad \text{and} \quad X_c = \frac{\eta_1^2\eta_2^2\eta_3^2\eta_6^2}{Z_c^2}. \tag{6.32}$$

For future reference, let $A_a$, $A_b$, and $A_c$ be defined by

$$A_a = \sqrt{1-34w_a+w_a^2}, \qquad A_b = \sqrt{1+20w_b+64w_b^2},$$

and

$$A_c = \sqrt{1+14w_c+81w_c^2}.$$

We begin by noting the expansions in terms of Ramanujan's Eisenstein series $P(q)$.

**Theorem 6.24.** *The following series expansions hold*:

$$Z_a = \frac{1}{2}\left(P(q) - 2P(q^2) - 3P(q^3) + 6P(q^6)\right),$$

$$Z_b = \frac{1}{4}\left(-P(q) + 2P(q^2) - 3P(q^3) + 6P(q^6)\right),$$

$$Z_c = \frac{1}{6}\left(-P(q) - 2P(q^2) + 3P(q^3) + 6P(q^6)\right).$$

*Proof.* By the definition of $w_a$ in (6.30) and a direct calculation, we have

$$Z_a = q\frac{\mathrm{d}}{\mathrm{d}q}\log w_a = q\frac{\mathrm{d}}{\mathrm{d}q}\log\frac{\eta_1^{12}\eta_6^{12}}{\eta_2^{12}\eta_3^{12}} = \frac{1}{2}\left(P(q) - 2P(q^2) - 3P(q^3) + 6P(q^6)\right).$$

The other two results may be proved by the same procedure. □

**Theorem 6.25.** *The following identities hold*:

$$Z_a = (1 - 72x_a^2)z_a^2, \quad Z_b = (1 - 9x_b^2)z_b^2, \quad \textit{and} \quad Z_c = (1 + 8x_c^2)z_c^2.$$

*Proof.* This is immediate from Theorems 6.18 and 6.24. □

The next result gives expressions for the weight two Eisenstein series $Z_a$, $Z_b$, and $Z_c$ in terms of eta functions. The expressions are not eta-quotients, but are algebraic functions of eta functions.

**Theorem 6.26.** *The following results hold*:

$$Z_a = A_a\, y_a, \qquad Z_b = A_b\, y_b \qquad \textit{and} \qquad Z_c = A_c\, y_c.$$

*Proof.* By the definition of $Z_a$ together with the chain rule, we find that

$$Z_a = q\frac{\mathrm{d}}{\mathrm{d}q}\log w_a = \left(\frac{x_a}{w_a}\right)\left(\frac{\mathrm{d}w_a}{\mathrm{d}x_a}\right)\left(q\frac{\mathrm{d}}{\mathrm{d}q}\log x_a\right). \tag{6.33}$$

By Theorem 6.21,

$$\frac{1}{w_a} = \frac{1}{x_a} + 17 + 72x_a,$$

and differentiation gives

$$\left(\frac{x_a}{w_a}\right)\left(\frac{\mathrm{d}w_a}{\mathrm{d}x_a}\right) = w_a\left(\frac{1}{x_a} - 72x_a\right). \tag{6.34}$$

Also, by the definition of $y_a$ in (6.29),

$$q\frac{\mathrm{d}}{\mathrm{d}q}\log x_a = y_a. \tag{6.35}$$

Using (6.34) and (6.35) in (6.33) we obtain

$$Z_a = y_a w_a\left(\frac{1}{x_a} - 72x_a\right).$$

By Theorem 6.21 again,

$$\begin{aligned}
Z_a = y_a w_a\left(\frac{1}{x_a} - 72x_a\right) &= y_a w_a\sqrt{\left(\frac{1}{x_a} + 72x_a\right)^2 - 288} \\
&= y_a w_a\sqrt{\left(\frac{1}{w_a} - 17\right)^2 - 288} \\
&= y_a\sqrt{1 - 34w_a + w_a^2} \\
&= A_a\, y_a.
\end{aligned}$$

The proofs of the other two results are similar. □

The modular functions $X_a$, $X_b$ and $X_c$ can be expressed in terms of $w_a$, $w_b$ and $w_c$, or in terms of $x_a$, $x_b$ and $x_c$, as follows.

**Theorem 6.27.** *The following identities hold*:

$$\begin{aligned}
X_a &= \frac{w_a}{1 - 34w_a + w_a^2} = \frac{x_a(1+8x_a)(1+9x_a)}{(1-72x_a^2)^2}, \\
X_b &= \frac{w_b}{1 + 20w_b + 64w_b^2} = \frac{x_b(1-x_b)(1-9x_b)}{(1-9x_b^2)^2}, \\
X_c &= \frac{w_c}{1 + 14w_c + 81w_c^2} = \frac{x_c(1+x_c)(1-8x_c)}{(1+8x_c^2)^2}.
\end{aligned}$$

*Proof.* Referring to the results in Table 6.1 we find that

$$w_a y_a^2 = \eta_1^2\eta_2^2\eta_3^2\eta_6^2.$$

Hence, using the definition of $X_a$ along with the result for $Z_a$ in Theorem 6.26, we deduce

$$X_a = \frac{\eta_1^2\eta_2^2\eta_3^2\eta_6^2}{Z_a^2} = \frac{w_a^2 y_a^2}{A_a^2 y_a^2} = \frac{w_a}{1 - 34w_a + w_a^2}. \tag{6.36}$$

Next, by Theorem 6.21,

$$w_a = \frac{x_a}{(1+8x_a)(1+9x_a)}. \tag{6.37}$$

Substituting (6.37) into (6.36) we obtain the required expression for $X_a$ in terms of $x_a$. The results for $X_b$ and $X_c$ can be proved by the same procedure. □

The next result is an analogue of (5.58).

**Theorem 6.28.** *The following expansions in terms of the weight two Eisenstein series $P(q)$ hold:*

$$\begin{aligned} A_a w_a \frac{\mathrm{d}y_a}{\mathrm{d}w_a} &= \frac{1}{24}\left(-5P(q)+14P(q^2)+21P(q^3)-30P(q^6)\right), \\ A_b w_b \frac{\mathrm{d}y_b}{\mathrm{d}w_b} &= \frac{1}{6}\left(P(q)-P(q^2)+3P(q^3)-3P(q^6)\right), \\ A_c w_c \frac{\mathrm{d}y_c}{\mathrm{d}w_c} &= \frac{1}{8}\left(P(q)+2P(q^2)-P(q^3)-2P(q^6)\right). \end{aligned}$$

*Proof.* By the chain rule,

$$\begin{aligned} A_a w_a \frac{\mathrm{d}y_a}{\mathrm{d}w_a} &= A_a w_a \left(q\frac{\mathrm{d}y_a}{\mathrm{d}q}\right) \Big/ \left(q\frac{\mathrm{d}w_a}{\mathrm{d}q}\right) \\ &= A_a y_a \left(q\frac{\mathrm{d}}{\mathrm{d}q}\log y_a\right) \Big/ \left(q\frac{\mathrm{d}}{\mathrm{d}q}\log w_a\right). \end{aligned} \tag{6.38}$$

Let us calculate the derivatives on the right-hand side of (6.38). By direct calculation using the eta-quotient for $y_a$ in Theorem 6.20 we find that

$$q\frac{\mathrm{d}}{\mathrm{d}q}\log y_a = q\frac{\mathrm{d}}{\mathrm{d}q}\log\frac{\eta_2^7\eta_3^7}{\eta_1^5\eta_6^5} = \frac{1}{24}\left(-5P(q)+14P(q^2)+21P(q^3)-30P(q^6)\right). \tag{6.39}$$

Also, by the definition (6.31) and Theorem 6.26,

$$q\frac{\mathrm{d}}{\mathrm{d}q}\log w_a = Z_a = y_a A_a. \tag{6.40}$$

On using (6.39) and (6.40) in (6.38) we deduce

$$A_a w_a \frac{\mathrm{d}y_a}{\mathrm{d}w_a} = \frac{1}{24}\left(-5P(q)+14P(q^2)+21P(q^3)-30P(q^6)\right).$$

The other two results may be proved in a similar way. □

By now, we have amassed a large number of expansions in terms of the weight two Eisenstein series $P(q)$. They are summarized in Table 6.2. It is also useful to link the Eisenstein series to the eta-quotients $w_a$, $w_b$, $w_c$, $y_a$, $y_b$, and $y_c$ as follows.

**Theorem 6.29.** *The following identities, expressed as matrix products, hold*:

$$\begin{pmatrix} y_a \\ w_a y_a \\ A_a y_a \\ A_a w_a \frac{\mathrm{d}y_a}{\mathrm{d}w_a} \end{pmatrix} = \frac{1}{24} \begin{pmatrix} -5 & 2 & -3 & 30 \\ -1 & 10 & -15 & 6 \\ 12 & -24 & -36 & 72 \\ -5 & 14 & 21 & -30 \end{pmatrix} \begin{pmatrix} P(q) \\ P(q^2) \\ P(q^3) \\ P(q^6) \end{pmatrix},$$

$$\begin{pmatrix} y_b \\ w_b y_b \\ A_b y_b \\ A_b w_b \frac{\mathrm{d}y_b}{\mathrm{d}w_b} \end{pmatrix} = \frac{1}{24} \begin{pmatrix} 4 & -16 & -12 & 48 \\ -1 & 1 & 3 & -3 \\ -6 & 12 & -18 & 36 \\ 4 & -4 & 12 & -12 \end{pmatrix} \begin{pmatrix} P(q) \\ P(q^2) \\ P(q^3) \\ P(q^6) \end{pmatrix}$$

*and*

$$\begin{pmatrix} y_c \\ w_c y_c \\ A_c y_c \\ A_c w_c \frac{\mathrm{d}y_c}{\mathrm{d}w_c} \end{pmatrix} = \frac{1}{24} \begin{pmatrix} 3 & -6 & -27 & 54 \\ -1 & 2 & 1 & -2 \\ -4 & -8 & 12 & 24 \\ 3 & 6 & -3 & -6 \end{pmatrix} \begin{pmatrix} P(q) \\ P(q^2) \\ P(q^3) \\ P(q^6) \end{pmatrix}.$$

*Each matrix has nonzero determinant, so each of* $P(q)$, $P(q^2)$, $P(q^3)$, *and* $P(q^6)$ *may be expressed as a linear combination of* $y_j$, $w_j y_j$, $A_j y_j$, *and* $A_j w_j \frac{\mathrm{d}y_j}{\mathrm{d}w_j}$, *for any* $j \in \{a, b, c\}$.

*Proof.* Let $j \in \{a, b, c\}$. The Eisenstein series for $y_j$ and $w_j y_j$ have been given in Theorem 6.23. The Eisenstein series for $A_j y_j$ follow from Theorems 6.24 and 6.26. Finally, the results for $A_j w_j \frac{\mathrm{d}y_j}{\mathrm{d}w_j}$ are given by Theorem 6.28. □

Numerous identities can be deduced by exploiting the linear relations that hold among the Eisenstein series that are listed in Table 6.2. Here are some examples.

**Theorem 6.30.** *The following identities hold*:

$$y_c \quad y_b = w_a y_a, \quad y_a - y_c - 8 w_b y_b, \quad y_a - y_b - 9 w_c y_c,$$

*and*

$$(1-w_a) y_a = (1+8w_b) y_b, \quad (1+w_a) y_a = (1+9w_c) y_c, \quad (1-8w_b) y_b = (1-9w_c) y_c.$$

*Hence,*

$$w_a y_a + 8 w_b y_b = 9 w_c y_c.$$

**Table 6.2** Eisenstein series, $f = c_1P(q) + c_2P(q^2) + c_3P(q^3) + c_6P(q^6)$

| $f$ | $c_1$ | $c_2$ | $c_3$ | $c_6$ | reference |
|---|---|---|---|---|---|
| $2z_a^2$ | 1 | −8 | 9 | 0 | Theorem 6.17 |
| $8z_b^2$ | −2 | 1 | 0 | 9 | |
| $6z_c^2$ | −1 | 0 | −1 | 8 | |
| $24z_d^2$ | 0 | −1 | 2 | −1 | |
| $24y_a = 24z_bz_c$ | −5 | 2 | −3 | 30 | Theorems 6.19 and 6.23 |
| $6y_b = 6z_az_c$ | 1 | −4 | −3 | 12 | |
| $8y_c = 8z_az_b$ | 1 | −2 | −9 | 18 | |
| $24w_ay_a = 24z_az_d$ | −1 | 10 | −15 | 6 | |
| $24w_by_b = 24z_bz_d$ | −1 | 1 | 3 | −3 | |
| $24w_cy_c = 24z_cz_d$ | −1 | 2 | 1 | −2 | |
| $2Z_a = 2A_ay_a = 2(1-72x_a^2)z_a^2$ | 1 | −2 | −3 | 6 | Theorems 6.24–6.26 |
| $4Z_b = 4A_by_b = 4(1-9x_b^2)z_b^2$ | −1 | 2 | −3 | 6 | |
| $6Z_c = 6A_cy_c = 6(1+8x_c^2)z_c^2$ | −1 | −2 | 3 | 6 | |
| $24A_aw_a\,\mathrm{d}y_a/\mathrm{d}w_a$ | −5 | 14 | 21 | −30 | Theorem 6.28 |
| $6A_bw_b\,\mathrm{d}y_b/\mathrm{d}w_b$ | 1 | −1 | 3 | −3 | |
| $8A_cw_c\,\mathrm{d}y_c/\mathrm{d}w_c$ | 1 | 2 | −1 | −2 | |

*Furthermore,*

$$6P(q^6) - P(q) = (5 - w_a)y_a, \quad 3P(q^3) - 2P(q^2) = (1 - 5w_a)y_a,$$

$$\frac{3P(q^3) - P(q)}{2} = (1 + 16w_b)y_b, \quad \frac{3P(q^6) - P(q^2)}{2} = (1 + 4w_b)y_b,$$

$$2P(q^2) - P(q) = (1 + 27w_c)y_c, \quad 2P(q^6) - P(q^3) = (1 + 3w_c)y_c.$$

*Proof.* These are all immediate from Table 6.2. □

The last result in this section involves Ramanujan's Eisenstein series $Q(q)$ defined by (0.72), viz.,

$$Q(q) = 1 + 240\sum_{j=1}^{\infty}\frac{j^3q^j}{1-q^j}.$$

**Theorem 6.31.**

$$(1 - w_a^2)y_a^2 = \frac{1}{24}\left(Q(q) - 4Q(q^2) - 9Q(q^3) + 36Q(q^6)\right),$$

$$(1 - 64w_b^2)y_b^2 = \frac{1}{30}\left(-Q(q) + 4Q(q^2) - 9Q(q^3) + 36Q(q^6)\right),$$

$$(1 - 81w_c^2)y_c^2 = \frac{1}{40}\left(-Q(q) - 4Q(q^2) + 9Q(q^3) + 36Q(q^6)\right).$$

*Proof.* Write $a$ and $A$ for $a(q)$ and $a(q^2)$, respectively. By Theorems 6.14 and 6.20 we have

$$\begin{aligned}(1-w_a^2)y_a^2 &= z_b z_c - z_a z_d \\ &= \frac{1}{36}\left\{(a+A)^2(a+2A)^2 - (2A-a)^2(a-A)^2\right\} \\ &= \frac{1}{36}\left(a^3A + 2aA^3\right).\end{aligned}$$

Now apply Theorem 3.43 to express $a^3A$ and $aA^3$ as linear combinations of $Q(q)$, $Q(q^2)$, $Q(q^3)$, and $Q(q^6)$. This proves the first result, and the other two results may be proved in a similar way. □

## 6.8 Differential equations for weight one modular forms

The goal of this section is to show that each of $z_a$, $z_b$, and $z_c$ satisfies a second order linear differential equation with respect to $x_a$, $x_b$, and $x_c$, respectively. This can be accomplished by changing variables in the differential equation in Theorem 4.24 when the level is 2 or 3. We shall work through the details using the level 3 theory below. An alternative method, using the level 2 theory, is outlined in Exercise 13.

Let us recall the main details of the level 3 theory. The functions $x_3$ and $z_3$ are defined by (4.29), that is,

$$x_3 = \frac{c^3(q)}{a^3(q)} \quad \text{and} \quad z_3 = a(q).$$

By Theorem 4.24, $z_3$ satisfies a second order linear differential equation with respect to $x_3$, viz.,

$$\frac{\mathrm{d}}{\mathrm{d}x_3}\left(x_3(1-x_3)\frac{\mathrm{d}z_3}{\mathrm{d}x_3}\right) = \frac{2z_3}{9}. \tag{6.41}$$

The next lemma gives the changes of variables that we will require.

**Lemma 6.32.** *The level 3 and level 6 modular functions and modular forms are related by*

$$x_3 = \frac{27x_a(1+8x_a)^2}{(1+12x_a)^3} = \frac{27x_b(1-x_b)^2}{(1+3x_b)^3} = \frac{27x_c}{(1+4x_c)^3}, \tag{6.42}$$

$$1-x_3 = \frac{1+9x_a}{(1+12x_a)^3} = \frac{(1-9x_b)^2}{(1+3x_b)^3} = \frac{(1+x_c)(1-8x_c)^2}{(1+4x_c)^3} \tag{6.43}$$

*and*

$$z_3 = (1+12x_a)z_a = (1+3x_b)z_b = (1+4x_c)z_c. \tag{6.44}$$

*Proof.* By the definition of $x_3$ and Theorem 3.20,

$$x_3 = \frac{c^3(q)}{a^3(q)} = \frac{c^3(q)}{b^3(q)+c^3(q)}.$$

Now apply the infinite products in Theorem 3.15 to obtain

$$x_3 = \frac{27\eta_3^9/\eta_1^3}{\eta_1^9/\eta_3^3 + 27\eta_3^9/\eta_1^3} = \frac{27}{\eta_1^{12}/\eta_3^{12}+27}. \tag{6.45}$$

It is easy to check, using Table 6.1, that

$$\frac{\eta_1^{12}}{\eta_3^{12}} = \frac{1+9x_a}{x_a(1+8x_a)^2} = \frac{(1-9x_b)^2}{x_b(1-x_b)^2} = \frac{(1+x_c)(1-8x_c)^2}{x_c}. \tag{6.46}$$

On substituting the results from (6.46) into (6.45) we deduce the identities in (6.42). The results in (6.43) follow from the corresponding results in (6.42) by simple calculation. The results in (6.44) were proved as part of Theorem 6.16. □

**Lemma 6.33.** *The following relations for differential operators hold*:

$$\begin{aligned} x_3(1-x_3)\frac{\mathrm{d}}{\mathrm{d}x_3} &= \frac{x_a(1+8x_a)(1+9x_a)}{(1+12x_a)^2}\frac{\mathrm{d}}{\mathrm{d}x_a} \\ &= \frac{x_b(1-x_b)(1-9x_b)}{(1+3x_b)^2}\frac{\mathrm{d}}{\mathrm{d}x_b} \\ &= \frac{x_c(1+x_c)(1-8x_c)}{(1+4x_c)^2}\frac{\mathrm{d}}{\mathrm{d}x_c}. \end{aligned}$$

*Proof.* By (6.42) and the quotient rule we find that

$$\frac{\mathrm{d}x_3}{\mathrm{d}x_a} = \frac{27(1+8x_a)}{(1+12x_a)^4}.$$

Using this, along with (6.42), (6.43), and the chain rule, we obtain

$$\begin{aligned} x_3(1-x_3)\frac{\mathrm{d}}{\mathrm{d}x_3} &= x_3(1-x_3)\frac{\mathrm{d}x_a}{\mathrm{d}x_3}\frac{\mathrm{d}}{\mathrm{d}x_a} \\ &= \frac{27x_a(1+8x_a)^2}{(1+12x_a)^3} \times \frac{1+9x_a}{(1+12x_a)^3} \times \frac{(1+12x_a)^4}{27(1+8x_a)} \times \frac{\mathrm{d}}{\mathrm{d}x_a} \end{aligned}$$

$$= \frac{x_a(1+8x_a)(1+9x_a)}{(1+12x_a)^2}\frac{\mathrm{d}}{\mathrm{d}x_a}.$$

This proves the first identity. The other identities can be proved by the same method. □

Here are the differential equations for $z_a$, $z_b$, and $z_c$ in terms of $x_a$, $x_b$, and $x_c$.

**Theorem 6.34.** *The following differential equations hold*:

$$\frac{\mathrm{d}}{\mathrm{d}x_a}\left(x_a(1+8x_a)(1+9x_a)\frac{\mathrm{d}z_a}{\mathrm{d}x_a}\right) + 6(1+12x_a)z_a = 0,$$

$$\frac{\mathrm{d}}{\mathrm{d}x_b}\left(x_b(1-x_b)(1-9x_b)\frac{\mathrm{d}z_b}{\mathrm{d}x_b}\right) - 3(1-3x_b)z_b = 0,$$

$$\frac{\mathrm{d}}{\mathrm{d}x_c}\left(x_c(1+x_c)(1-8x_c)\frac{\mathrm{d}z_c}{\mathrm{d}x_c}\right) - 2(1+4x_c)z_c = 0.$$

*Proof.* Multiply both sides of the differential equation (6.41) by $x_3(1-x_3)$ to obtain

$$x_3(1-x_3)\frac{\mathrm{d}}{\mathrm{d}x_3}\left(x_3(1-x_3)\frac{\mathrm{d}z_3}{\mathrm{d}x_3}\right) = \frac{2}{9}x_3(1-x_3)z_3.$$

By Lemmas 6.32 and 6.33 this is equivalent to

$$\frac{x_a(1+8x_a)(1+9x_a)}{(1+12x_a)^2}\frac{\mathrm{d}}{\mathrm{d}x_a}\left(\frac{x_a(1+8x_a)(1+9x_a)}{(1+12x_a)^2}\frac{\mathrm{d}}{\mathrm{d}x_a}((1+12x_a)z_a)\right)$$
$$= \frac{2}{9}\times\frac{27x_a(1+8x_a)^2}{(1+12x_a)^3}\times\frac{(1+9x_a)}{(1+12x_a)^3}\times(1+12x_a)z_a.$$

This simplifies to

$$\frac{\mathrm{d}}{\mathrm{d}x_a}\left(\frac{x_a(1+8x_a)(1+9x_a)}{(1+12x_a)}\frac{\mathrm{d}z_a}{\mathrm{d}x_a} + \frac{12x_a(1+8x_a)(1+9x_a)}{(1+12x_a)^2}z_a\right)$$
$$= \frac{6(1+8x_a)}{(1+12x_a)^3}z_a,$$

that is,

$$\frac{1}{1+12x_a}\frac{\mathrm{d}}{\mathrm{d}x_a}\left(x_a(1+8x_a)(1+9x_a)\frac{\mathrm{d}z_a}{\mathrm{d}x_a}\right)$$
$$+ 12\frac{\mathrm{d}}{\mathrm{d}x_a}\left(\frac{x_a(1+8x_a)(1+9x_a)}{(1+12x_a)^2}\right)z_a = \frac{6(1+8x_a)}{(1+12x_a)^3}z_a.$$

This can be simplified further to give the first result. The other two results may be obtained in a similar way. □

The differential equations in Theorem 6.34 can be used to expand $z_a$, $z_b$, and $z_c$ in powers of $x_a$, $x_b$, and $x_c$, respectively. We begin by finding recurrence relations for the coefficients.

**Theorem 6.35.** *Let $x_a$, $x_b$, $x_c$, $z_a$, $z_b$, and $z_c$ be as in Table 6.1. Let $t_a(n)$, $t_b(n)$, and $t_c(n)$ be the sequences defined by the recurrence relations*

$$(n+1)^2 t_a(n+1) = -(17n^2+17n+6)t_a(n) - 72n^2 t_a(n-1), \tag{6.47}$$

$$(n+1)^2 t_b(n+1) = (10n^2+10n+3)t_b(n) - 9n^2 t_b(n-1), \tag{6.48}$$

$$(n+1)^2 t_c(n+1) = (7n^2+7n+2)t_c(n) + 8n^2 t_c(n-1) \tag{6.49}$$

*for $n \geq 0$, with initial condition*

$$t_a(0) = t_b(0) = t_c(0) = 1. \tag{6.50}$$

*Then*

$$z_a = \sum_{n=0}^{\infty} t_a(n) x_a^n, \quad z_b = \sum_{n=0}^{\infty} t_b(n) x_b^n \quad \textit{and} \quad z_c = \sum_{n=0}^{\infty} t_c(n) x_c^n.$$

As was the case for Theorem 5.31, the single initial condition in (6.50) is enough to start each recurrence relation in (6.47)–(6.49).

*Proof.* Let us begin with the sequence $t_a(n)$. The functions $x_a$ and $z_a$ are analytic functions of $q$ in a neighborhood of $q = 0$. From the eta-quotients in Table 6.1,

$$x_a = q + O(q^2) \qquad \text{and} \qquad z_a = 1 - 6q + O(q^2).$$

It follows that the inverse function $q = q(x_a)$ exists and is analytic in a neighborhood of $x_a = 0$, so $z_a$ is an analytic function of $x_a$ in a neighborhood of $x_a = 0$, and

$$z_a = 1 - 6x_a + O(x_a^2).$$

The differential equation for $z_a$ in Theorem 6.34 has a unique solution in powers of $x_a$ because the indicial equation has a repeated root 0. On substituting the power series expansion

$$z_a = \sum_{n=0}^{\infty} t_a(n) x_a^n$$

into the differential equation and equating coefficients we deduce the recurrence relation (6.47) for $n \geq 1$. On noting that $t_a(0) = 1$ and $t_a(1) = -6$, it follows that the recurrence relation (6.47) also holds for $n = 0$, and the initial condition also holds.

The results for $t_b(n)$ and $t_c(n)$ may be proved in a similar way. □

The next result gives explicit formulas for $t_b(n)$ and $t_c(n)$.

**Theorem 6.36.** *The sequences $t_b(n)$ and $t_c(n)$ defined by recurrence relations and initial conditions* (6.48)–(6.50) *are given by*

$$t_b(n) = \sum_{\substack{j+k+\ell=n \\ j,k,\ell\geq 0}} \left(\frac{n!}{j!k!\ell!}\right)^2$$

*and*

$$t_c(n) = \sum_{\substack{j+k=n \\ j,k\geq 0}} \left(\frac{n!}{j!k!}\right)^3.$$

*That is, $t_b(n)$ is the sum of the squares of the trinomial coefficients, and $t_c(n)$ is the sum of the cubes of the binomial coefficients.*

*Proof.* Let $\sigma_b(n)$ and $\sigma_c(n)$ be the sequences defined by

$$\sigma_b(n) = \sum_{\substack{j+k+\ell=n \\ j,k,\ell\geq 0}} \left(\frac{n!}{j!k!\ell!}\right)^2 \quad \text{and} \quad \sigma_c(n) = \sum_{\substack{j+k=n \\ j,k\geq 0}} \left(\frac{n!}{j!k!}\right)^3.$$

Their generating functions are given by

$$\left(\sum_{j=0}^{\infty} \frac{x^j}{j!^2}\right)^3 = \sum_{n=0}^{\infty} \frac{\sigma_b(n)}{n!^2} x^n \tag{6.51}$$

and

$$\left(\sum_{j=0}^{\infty} \frac{x^j}{j!^3}\right)^2 = \sum_{n=0}^{\infty} \frac{\sigma_c(n)}{n!^3} x^n. \tag{6.52}$$

Let $g_b$, $g_c$, $h_b$, and $h_c$ be the functions defined by

$$g_b = \sum_{j=0}^{\infty} \frac{x^j}{j!^2}, \quad g_c = \sum_{j=0}^{\infty} \frac{x^j}{j!^3}, \quad h_b = g_b^3 \quad \text{and} \quad h_c = g_c^2.$$

It is easy to check that $g_b$ and $g_c$ satisfy the differential equations

$$xg_b'' + g_b' - g_b = 0$$

and

$$x^2 g_c''' + 3x g_c'' + g_c' - g_c = 0.$$

It follows—see Exercise 4—that $h_b$ and $h_c$ are solutions of

$$x^3 h_b'''' + 6x^2 h_b''' + (7x - 10x^2)h_b'' + (1 - 20x)h_b' + (-3 + 9x)h_b = 0 \tag{6.53}$$

and

$$\begin{aligned} x^4 h_c^{(6)} + 14x^3 h_c^{(5)} + 55x^2 h_c'''' + (65x - 7x^2)h_c''' \\ +(16 - 28x)h_c'' - 16h_c' - 8h_c = 0. \end{aligned} \tag{6.54}$$

If we substitute the generating functions from (6.51) and (6.52) into the differential equations (6.53) and (6.54), we find that $\sigma_b(n)$ and $\sigma_c(n)$ satisfy the same recurrence relations as $t_b(n)$ and $t_c(n)$ in (6.48) and (6.49), respectively. Furthermore, the sequences have the same initial conditions, namely

$$t_b(0) = \sigma_b(0) = 1, \quad t_b(1) = \sigma_b(1) = 3,$$
$$t_c(0) = \sigma_c(0) = 1, \quad t_c(1) = \sigma_c(1) = 2.$$

It follows that $t_b(n) = \sigma_b(n)$ and $t_c(n) = \sigma_c(n)$ for all nonnegative integers $n$. This completes the proof. □

It is possible to give an expression for $t_b(n)$ as a single sum. We have

$$t_b(n) = \sum_{j=0}^{n}\sum_{k=0}^{n-j} \frac{n!^2}{j!^2 k!^2 (n-j-k)!^2} = \sum_{j=0}^{n} \binom{n}{j}^2 \sum_{k=0}^{n-j} \binom{n-j}{k}^2.$$

Now replace $j$ with $n - j$ to deduce

$$t_b(n) = \sum_{j=0}^{n} \binom{n}{j}^2 \sum_{k=0}^{j} \binom{j}{k}^2 = \sum_{j=0}^{n} \binom{n}{j}^2 \binom{2j}{j},$$

where the last step follows by a well-known identity—see Exercise 7.
There is no analogous known formula for $t_a(n)$ as a single sum. However there are explicit formulas for $t_a(n)$ as double sums, viz.,

$$t_a(n) = \sum_{k=0}^{n} (-8)^{n-k} \binom{n}{k} \sum_{j=0}^{k} \binom{k}{j}^3 = \sum_{k=0}^{n} (-9)^{n-k} \binom{n}{k} \sum_{j=0}^{k} \binom{k}{j}^2 \binom{2j}{j}.$$

These are particular instances of the next theorem.

**Theorem 6.37.** *The coefficients $t_a(n)$, $t_b(n)$, and $t_c(n)$ are interrelated by the formulas*:

$$t_b(n) = \sum_{k=0}^{n} \binom{n}{k} t_c(k), \qquad t_c(n) = \sum_{k=0}^{n} (-1)^{n-k} \binom{n}{k} t_b(k),$$

$$t_c(n) = \sum_{k=0}^{n} 8^{n-k} \binom{n}{k} t_a(k), \quad t_a(n) = \sum_{k=0}^{n} (-8)^{n-k} \binom{n}{k} t_c(k),$$

$$t_b(n) = \sum_{k=0}^{n} 9^{n-k} \binom{n}{k} t_a(k), \quad t_a(n) = \sum_{k=0}^{n} (-9)^{n-k} \binom{n}{k} t_b(k).$$

*Proof.* We will only prove the very last result, as the others can be proved by the same technique. By Theorem 6.15,

$$x_a z_a = x_b z_b.$$

Therefore, by Theorem 6.35 we have

$$\sum_{n=0}^{\infty} t_a(n) x_a^{n+1} = \sum_{n=0}^{\infty} t_b(n) x_b^{n+1}. \tag{6.55}$$

By Theorem 6.9 we also have

$$x_b = \frac{x_a}{1 + 9x_a},$$

so (6.55) becomes

$$\begin{aligned} \sum_{n=0}^{\infty} t_a(n) x_a^{n+1} &= \sum_{n=0}^{\infty} t_b(n) \left( \frac{x_a}{1 + 9x_a} \right)^{n+1} \\ &= \sum_{n=0}^{\infty} t_b(n) x_a^{n+1} \sum_{k=0}^{\infty} \binom{n+k}{k} (-9x_a)^k, \end{aligned}$$

where the binomial theorem has been used to obtain the last step. Equating coefficients of $x_a^{n+1}$ we obtain

$$t_a(n) = \sum_{k=0}^{n} (-9)^{n-k} \binom{n}{k} t_b(k).$$

□

The results of this section may be summarized by substituting the results of Theorems 6.36 and 6.37 into Theorem 6.35 to produce the identities

$$z_a = \sum_{n=0}^{\infty} t_a(n) x_a^n, \qquad z_b = \sum_{n=0}^{\infty} t_b(n) x_b^n \qquad \text{and} \qquad z_c = \sum_{n=0}^{\infty} t_c(n) x_c^n,$$

that is,

$$\frac{\eta_1^6 \eta_6}{\eta_2^3 \eta_3^2} = \sum_{n=0}^{\infty} t_a(n) \left( \frac{\eta_2 \eta_6^5}{\eta_1^5 \eta_3} \right)^n, \tag{6.56}$$

$$\frac{\eta_2^6 \eta_3}{\eta_1^3 \eta_6^2} = \sum_{n=0}^{\infty} t_b(n) \left( \frac{\eta_1^4 \eta_6^8}{\eta_2^8 \eta_3^4} \right)^n \tag{6.57}$$

and

$$\frac{\eta_2 \eta_3^6}{\eta_1^2 \eta_3^3} = \sum_{n=0}^{\infty} t_c(n) \left( \frac{\eta_1^3 \eta_6^9}{\eta_2^3 \eta_3^9} \right)^n, \tag{6.58}$$

where

$$t_b(n) = \sum_{j=0}^{n} \binom{n}{j}^2 \binom{2j}{j}, \qquad t_c(n) = \sum_{j=0}^{n} \binom{n}{j}^3 \tag{6.59}$$

and

$$t_a(n) = \sum_{k=0}^{n} (-9)^{n-k} \binom{n}{k} t_b(k) = \sum_{k=0}^{n} (-8)^{n-k} \binom{n}{k} t_c(k). \tag{6.60}$$

Recurrence relations for the sequences $\{t_a(n)\}$, $\{t_b(n)\}$, and $\{t_c(n)\}$ were given in Theorem 6.35.

## 6.9 Apéry, Domb, and Almkvist–Zudilin numbers

In the previous section, expansions for the weight one modular forms $z_a$, $z_b$, and $z_c$ in powers of $x_a$, $x_b$, and $x_c$, respectively, were found. In this section we will use the analogues of Clausen's identity to produce expansions of the weight two modular forms $y_a$, $y_b$, $y_c$, $Z_a$, $Z_b$, and $Z_c$ in powers of $w_a$, $w_b$, $w_c$, $X_a$, $X_b$, and $X_c$, respectively.

We begin with expansions of the eta-quotients $y_a$, $y_b$, and $y_c$ in powers of the eta-quotients $w_a$, $w_b$, and $w_c$.

**Theorem 6.38.** *The following expansions hold*:

$$y_a = \sum_{n=0}^{\infty} s_a(n) w_a^n, \qquad y_b = \sum_{n=0}^{\infty} s_b(n) w_b^n \qquad \text{and} \qquad y_c = \sum_{n=0}^{\infty} s_c(n) w_c^n;$$

*that is,*

$$\frac{\eta_2^7 \eta_3^7}{\eta_1^5 \eta_6^5} = \sum_{n=0}^{\infty} s_a(n) \left( \frac{\eta_1 \eta_6}{\eta_2 \eta_3} \right)^{12n}, \tag{6.61}$$

$$\frac{\eta_1^4 \eta_3^4}{\eta_2^2 \eta_6^2} = \sum_{n=0}^{\infty} s_b(n) \left( \frac{\eta_2 \eta_6}{\eta_1 \eta_3} \right)^{6n} \tag{6.62}$$

*and*

$$\frac{\eta_1^3 \eta_2^3}{\eta_3 \eta_6} = \sum_{n=0}^{\infty} s_c(n) \left( \frac{\eta_3 \eta_6}{\eta_1 \eta_2} \right)^{4n}, \tag{6.63}$$

*where*

$$s_a(n) = \sum_{j=0}^{n} \binom{n}{j}^2 \binom{n+j}{j}^2, \qquad s_b(n) = (-1)^n \sum_{j=0}^{n} \binom{n}{j}^2 \binom{2j}{j} \binom{2n-2j}{n-j} \tag{6.64}$$

*and*

$$s_c(n) = \sum_{j=0}^{\lfloor n/3 \rfloor} (-3)^{n-3j} \binom{n+j}{j,j,j,j,n-3j}, \tag{6.65}$$

*and where the multinomial coefficient is defined by* (5.97).

The numbers $s_a(n)$, $(-1)^n s_b(n)$ and $s_c(n)$ are called the Apéry numbers, Domb numbers, and Almkvist–Zudilin numbers, respectively.

*Proof.* The method of creative telescoping can be used to show that the sequences in (6.64) and (6.65) satisfy the following recurrence relations:

$$(n+1)^3 s_a(n+1) = (2n+1)(17n^2+17n+5) s_a(n) - n^3 s_a(n-1), \tag{6.66}$$

$$(n+1)^3 s_b(n+1) = -(2n+1)(10n^2+10n+4) s_b(n) - 64n^3 s_b(n-1), \tag{6.67}$$

and

$$(n+1)^3 s_c(n+1) = -(2n+1)(7n^2+7n+3) s_c(n) - 81n^3 s_c(n-1), \tag{6.68}$$

that is,

$$(n+1)^3 s(n+1) = -(2n+1)(\alpha n^2 + \alpha n + \alpha - 2\beta)s(n) - \delta n^3 s(n-1)$$

where $(\alpha, \beta, \gamma) = (-17, -6, -72)$, $(10, 3, -9)$, or $(7, 2, 8)$, respectively, and $\delta = \alpha^2 + 4\gamma$. The recurrence relations hold for $n \geq 0$ and the single initial condition

$$s_a(0) = s_b(0) = s_c(0) = 1$$

is enough to start each sequence.

Let us first consider the case $(\alpha, \beta, \gamma) = (-17, -6, -72)$. By Theorem 6.35, the expansion

$$z_a = \sum_{n=0}^{\infty} t_a(n) x_a^n$$

holds, where the sequence $\{t_a(n)\}$ satisfies the recurrence relation

$$(n+1)^2 t(n+1) = (\alpha n^2 + \alpha n + \beta)t(n) + \gamma n^2 t(n-1)$$

and initial condition $t_a(0) = 1$. On squaring and applying the generalized Clausen identity (5.116) it follows that

$$(1 - \alpha x_a - \gamma x_a^2) z_a^2 = \sum_{n=0}^{\infty} s_a(n) \left( \frac{x_a}{1 - \alpha x_a - \gamma x_a^2} \right)^n,$$

that is,

$$(1 + 17x_a + 72x_a^2) z_a^2 = \sum_{n=0}^{\infty} s_a(n) \left( \frac{x_a}{1 + 17x_a + 72x_a^2} \right)^n.$$

By Theorems 6.21 and 6.22, this is equivalent to

$$y_a = \sum_{n=0}^{\infty} s_a(n) w_a^n.$$

Now apply (6.30) and Theorem 6.20 to express $w_a$ and $y_a$ in terms of eta-quotients. This gives (6.61).

The other results (6.62) and (6.63) may be proved by the same procedure, by taking $(\alpha, \beta, \gamma) = (10, 3, -9)$ or $(7, 2, 8)$, respectively. □

The next result gives expansions of $Z_a$, $Z_b$, and $Z_c$ in powers of $X_a$, $X_b$, and $X_c$, respectively.

**Theorem 6.39.** *Let* $\{t_a(n)\}$, $\{t_b(n)\}$, *and* $\{t_c(n)\}$ *be the sequences defined by the recurrence relations* (6.47)–(6.49), *or equivalently by the formulas* (6.59) *and* (6.60). *The following expansions hold*:

$$Z_a = \sum_{n=0}^{\infty} \binom{2n}{n} t_a(n) X_a^n, \quad Z_b = \sum_{n=0}^{\infty} \binom{2n}{n} t_b(n) X_b^n, \quad \textit{and} \quad Z_c = \sum_{n=0}^{\infty} \binom{2n}{n} t_c(n) X_c^n.$$

*That is,*

$$\frac{6P_6 - 3P_3 - 2P_2 + P_1}{2} = \sum_{n=0}^{\infty} \binom{2n}{n} t_a(n) \left( \frac{2\eta_1\eta_2\eta_3\eta_6}{6P_6 - 3P_3 - 2P_2 + P_1} \right)^{2n}, \tag{6.69}$$

$$\frac{6P_6 - 3P_3 + 2P_2 - P_1}{4} = \sum_{n=0}^{\infty} \binom{2n}{n} t_b(n) \left( \frac{4\eta_1\eta_2\eta_3\eta_6}{6P_6 - 3P_3 + 2P_2 - P_1} \right)^{2n} \tag{6.70}$$

*and*

$$\frac{6P_6 + 3P_3 - 2P_2 - P_1}{6} = \sum_{n=0}^{\infty} \binom{2n}{n} t_c(n) \left( \frac{6\eta_1\eta_2\eta_3\eta_6}{6P_6 + 3P_3 - 2P_2 - P_1} \right)^{2n}. \tag{6.71}$$

*Proof.* Let us prove the results that involve $t_a(n)$. By Theorem 6.35,

$$z_a = \sum_{n=0}^{\infty} t_a(n) x_a^n.$$

Square both sides and apply the generalized Clausen identity (5.117) to deduce

$$(1 + \gamma x_a^2) z_a^2 = \sum_{n=0}^{\infty} \binom{2n}{n} t_a(n) \left( \frac{x_a(1 - \alpha x_a - \gamma x_a^2)}{(1 + \gamma x_a^2)} \right)^n,$$

where $(\alpha, \beta, \gamma) = (-17, -6, -72)$. By Theorems 6.25 and 6.27, this is equivalent to

$$Z_a = \sum_{n=0}^{\infty} \binom{2n}{n} t_a(n) X_a^n.$$

Now use the formulas for $X_a$ and $Z_a$ in (6.32) and Theorem 6.24 to deduce (6.69).
The other results may be proved in the same way. □

## 6.10 Summary

It is useful to gather the main results obtained so far in the form of the summary given in Tables 6.3 and 6.4. The results in Table 6.3 for levels 8 and 9 will be obtained in Chapters 8 and 9. Here is an explanation to go with the tables.

Let $(\alpha, \beta, \gamma)$ be any of the six triples in Table 6.3. Consider the sequences $\{t(n)\}$ and $\{s(n)\}$ defined by the recurrence relations

$$(n+1)^2 t(n+1) = (\alpha n^2 + \alpha n + \beta) t(n) + \gamma n^2 t(n-1)$$

and

$$(n+1)^3 s(n+1) = -(2n+1)(\alpha n^2 + \alpha n + \alpha - 2\beta) s(n) - (\alpha^2 + 4\gamma) n^3 s(n-1)$$

and initial conditions $t(0) = s(0) = 1$. The single initial condition involving the term corresponding to $n = 0$ is enough to start each sequence. Expressions for the sequences $\{t(n)\}$ and $\{s(n)\}$ in terms of sums of binomial coefficients are given in Table 5.1.

The generating functions are given by

$$z = \sum_{n=0}^{\infty} t(n) x^n, \quad y = \sum_{n=0}^{\infty} s(n) w^n \quad \text{and} \quad Z = \sum_{n=0}^{\infty} \binom{2n}{n} t(n) X^n.$$

The power series variables $x$, $w$, and $X$ are related by

$$w = \frac{x}{1 - \alpha x - \gamma x^2} \quad \text{and} \quad X = \frac{w}{1 + 2\alpha w + (\alpha^2 + 4\gamma) w^2}. \tag{6.72}$$

The functions $z$, $y$, and $Z$ are related by

$$x z^2 = w y \quad \text{and} \quad X^{1/2} Z = w^{1/2} y. \tag{6.73}$$

In addition, for the uniformizing functions in Tables 6.3 and 6.4, the following relations also hold:

$$y = q \frac{\mathrm{d}}{\mathrm{d}q} \log x, \qquad Z = q \frac{\mathrm{d}}{\mathrm{d}q} \log w.$$

The cases of levels 1, 2, 3, and 4 are also covered by the theory above by the parameter sets $(\alpha, \beta, \gamma) = (432, 60, 0)$, $(64, 12, 0)$, $(27, 6, 0)$, and $(16, 4, 0)$, respectively. The theory behind these results was developed in Chapter 4. In particular, see Theorems 4.25, 4.28 and 4.30. Since $\gamma = 0$, the recurrence relation for $t(n)$ simplifies to a two-term relation that can be solved as a product of factorials. Moreover, the identities (6.72) and (6.73) together with the condition $\gamma = 0$ imply

$$X = x(1 - \alpha x) \qquad \text{and} \qquad Z = z^2.$$

The results for levels 1, 2, 3, and 4 are summarized in Table 6.4.

## 6.11 Connection with hypergeometric functions

Let $t_a(n)$, $t_b(n)$, and $t_c(n)$ be defined by the recurrence relations (6.47)–(6.49) and initial condition (6.50), or equivalently by the binomial sums (6.59) and (6.60). Consider the generating functions $f_a, f_b$, and $f_c$ defined by

$$f_a(x) = \sum_{n=0}^{\infty} t_a(n)x^n = 1 - 6x + 42x^2 - 312x^3 + 2394x^4 - 18756x^5 + \cdots,$$

$$f_b(x) = \sum_{n=0}^{\infty} t_b(n)x^n = 1 + 3x + 15x^2 + 93x^3 + 639x^4 + 4653x^5 + \cdots$$

and

$$f_c(x) = \sum_{n=0}^{\infty} t_c(n)x^n = 1 + 2x + 10x^2 + 56x^3 + 346x^4 + 2252x^5 + \cdots.$$

The series converge for $|x| < 1/9$, $|x| < 1/9$, and $|x| < 1/8$, respectively. Numerous interrelations and transformation formulas hold, and each function can be represented in terms of hypergeometric functions in several ways. We begin with:

**Theorem 6.40.** *The functions $f_a$, $f_b$, and $f_c$ are inter-related in a neighborhood of $x = 0$ by*

$$\begin{aligned} f_a(x) &= \frac{1}{1+9x} f_b\left(\frac{x}{1+9x}\right) = \frac{1}{1+8x} f_c\left(\frac{x}{1+8x}\right), \\ f_b(x) &= \frac{1}{1-9x} f_a\left(\frac{x}{1-9x}\right) = \frac{1}{1-x} f_c\left(\frac{x}{1-x}\right), \end{aligned}$$

$$f_c(x) = \frac{1}{1-8x} f_a\left(\frac{x}{1-8x}\right) = \frac{1}{1+x} f_b\left(\frac{x}{1+x}\right).$$

*Proof.* In the course of the proof of Theorem 6.37, it was shown that the identity $x_a z_a = x_b z_b$ from Theorem 6.15 implies

$$\sum_{n=0}^{\infty} t_a(n)x^{n+1} = \sum_{n=0}^{\infty} t_b(n)\left(\frac{x}{1+9x}\right)^{n+1}.$$

This is equivalent to

$$f_a(x) = \frac{1}{1+9x} f_b\left(\frac{x}{1+9x}\right).$$

This proves the first result. The other identities can be proved in the same way. □

**Table 6.3** Uniformizing functions, levels 5, 6, 8, and 9

| $(\alpha, \beta, \gamma)$<br>Level | $z$<br>$x$ | $y$<br>$w$ | $Z$<br>$X$ | Reference |
|---|---|---|---|---|
| $(11, 3, 1)$ | $\dfrac{(q;q)_\infty^2}{(q,q^4;q^5)_\infty^5}$ | $\dfrac{\eta_1^5}{\eta_5}$ | $\dfrac{1}{4}(5P_5 - P_1)$ | Sec. 5.5 |
| 5 | $q\dfrac{(q,q^4;q^5)_\infty^5}{(q^2,q^3;q^5)_\infty^5}$ | $\dfrac{\eta_5^6}{\eta_1^6}$ | $\dfrac{\eta_1^4\eta_5^4}{Z^2}$ | |
| $(-17, -6, -72)$ | $\dfrac{\eta_1^6\eta_6}{\eta_2^3\eta_3^2}$ | $\dfrac{\eta_2^7\eta_3^7}{\eta_1^5\eta_6^5}$ | $\dfrac{1}{2}(6P_6 - 3P_3 - 2P_2 + P_1)$ | Secs. 6.8 and 6.9 |
| 6 | $\dfrac{\eta_2\eta_6^5}{\eta_1^5\eta_3}$ | $\dfrac{\eta_1^{12}\eta_6^{12}}{\eta_2^{12}\eta_3^{12}}$ | $\dfrac{\eta_1^2\eta_2^2\eta_3^2\eta_6^2}{Z^2}$ | |
| $(10, 3, -9)$ | $\dfrac{\eta_2^6\eta_3}{\eta_1^3\eta_6^2}$ | $\dfrac{\eta_1^4\eta_3^4}{\eta_2^2\eta_6^2}$ | $\dfrac{1}{4}(6P_6 - 3P_3 + 2P_2 - P_1)$ | Secs. 6.8 and 6.9 |
| 6 | $\dfrac{\eta_1^4\eta_6^8}{\eta_2^8\eta_3^4}$ | $\dfrac{\eta_2^6\eta_6^6}{\eta_1^6\eta_3^6}$ | $\dfrac{\eta_1^2\eta_2^2\eta_3^2\eta_6^2}{Z^2}$ | |
| $(7, 2, 8)$ | $\dfrac{\eta_2\eta_3^6}{\eta_1^2\eta_6^3}$ | $\dfrac{\eta_1^3\eta_2^3}{\eta_3\eta_6}$ | $\dfrac{1}{6}(6P_6 + 3P_3 - 2P_2 - P_1)$ | Secs. 6.8 and 6.9 |
| 6 | $\dfrac{\eta_1^3\eta_6^9}{\eta_2^3\eta_3^9}$ | $\dfrac{\eta_3^4\eta_6^4}{\eta_1^4\eta_2^4}$ | $\dfrac{\eta_1^2\eta_2^2\eta_3^2\eta_6^2}{Z^2}$ | |
| $(-12, -4, -32)$ | $\dfrac{\eta_1^4}{\eta_2^2}$ | $\dfrac{\eta_2^6\eta_4^6}{\eta_1^4\eta_8^4}$ | $\dfrac{1}{3}(8P_8 - 4P_4 - 2P_2 + P_1)$ | Sec. 8.8 |
| 8 | $\dfrac{\eta_2^2\eta_8^4}{\eta_1^4\eta_4^2}$ | $\dfrac{\eta_1^8\eta_8^8}{\eta_2^8\eta_4^8}$ | $\dfrac{\eta_2^4\eta_4^4}{Z_8^2}$ | |
| $(-9, -3, -27)$ | $\dfrac{\eta_1^3}{\eta_3}$ | $\dfrac{\eta_3^{10}}{\eta_1^3\eta_9^3}$ | $\dfrac{1}{4}(9P_9 - 6P_3 + P_1)$ | Sec. 9.2 |
| 9 | $\dfrac{\eta_9^3}{\eta_1^3}$ | $\dfrac{\eta_1^6\eta_9^6}{\eta_3^{12}}$ | $\dfrac{\eta_3^8}{Z_9^2}$ | |

The formulas in Theorem 6.40 provide analytic continuations of each function from a disc to a half-plane. For example, the left-hand side of the identity

$$f_a(x) = \frac{1}{1+8x} f_c\left(\frac{x}{1+8x}\right) \tag{6.74}$$

converges if $|x| < 1/9$, while the right-hand side is valid for $|x/(1+8x)| < 1/8$, that is, $\text{Re}(x) > -1/16$. By analytic continuation, the identity (6.74) may be used to define $f_a(x)$ in the half-plane $\text{Re}(x) > -1/16$. Similar observations can be made for the other two functions.

The next result is an analogue of Theorem 5.46.

**Table 6.4** Uniformizing functions, levels 1, 2, 3, and 4

| $(\alpha,\beta,\gamma)$<br>Level | $z$<br>$x$ | $y$<br>$w$ | $Z$<br>$X$ | $t(n)$ |
|---|---|---|---|---|
| $(432,60,0)$ | $Z^{1/2}$ | $\left(\frac{Q^{3/2}+R}{2}\right)\times\frac{1}{Z^2}$ | $Q^{1/2}$ | $\binom{6n}{3n}\binom{3n}{n}$ |
| 1 | $\left(\frac{Q^{3/2}-R}{864}\right)\times\frac{1}{Z^3}$ | $\frac{1}{432}\left(\frac{Q^{3/2}-R}{Q^{3/2}+R}\right)$ | $\frac{\eta_1^{24}}{Z^6}$ | |
| $(64,12,0)$ | $Z^{1/2}$ | $\frac{\eta_1^{16}}{\eta_2^8}\times\frac{1}{Z}$ | $2P_2-P_1$ | $\binom{4n}{2n}\binom{2n}{n}$ |
| 2 | $\frac{\eta_2^{16}}{\eta_1^8}\times\frac{1}{Z^2}$ | $\frac{\eta_2^{24}}{\eta_1^{24}}$ | $\frac{\eta_1^8\eta_2^8}{Z^4}$ | |
| $(27,6,0)$ | $Z^{1/2}$ | $\frac{\eta_1^9}{\eta_3^3}\times\frac{1}{Z^{1/2}}$ | $\frac{1}{2}(3P_3-P_1)$ | $\binom{3n}{n}\binom{2n}{n}$ |
| 3 | $\frac{\eta_3^9}{\eta_1^3}\times\frac{1}{Z^{3/2}}$ | $\frac{\eta_3^{12}}{\eta_1^{12}}$ | $\frac{\eta_1^6\eta_3^6}{Z^3}$ | |
| $(16,4,0)$ | $Z^{1/2}$ | $\frac{\eta_1^8}{\eta_2^4}$ | $\frac{1}{3}(4P_4-P_1)$ | $\binom{2n}{n}^2$ |
| 4 | $\frac{\eta_4^8}{\eta_2^4}\times\frac{1}{Z}$ | $\frac{\eta_4^8}{\eta_1^8}$ | $\frac{\eta_1^8\eta_4^8}{\eta_2^8}\times\frac{1}{Z^2}$ | |

**Theorem 6.41.** *The following quadratic and cubic transformation formulas hold in a neighborhood of* $x=0$:

$$f_b(x^2)=\frac{1}{1+3x}f_b\left(\frac{x(1-x)}{1+3x}\right)$$

*and*

$$f_c(x^3)=\frac{1}{1+2x+4x^2}f_c\left(\frac{x(1-x+x^2)}{1+2x+4x^2}\right).$$

*Proof.* From Theorem 6.16 we have

$$a(q)=(1+3x_b(q))z_b(q)\qquad\text{and}\qquad a(q^2)=(1-3x_b(q))z_b(q).$$

Since $z_b(q)=f_b(x_b(q))$, it follows that

$$(1+3x_b(q^2))f_b(x_b(q^2))=(1-3x_b(q))f_b(x_b(q)).$$

By the quadratic transformation in Theorem 6.13, and using the notation

$$v_b=r_b(q^2)=(x_b(q^2))^{1/2},$$

we deduce further that

$$(1+3v_b^2)f_b(v_b^2) = \left(1 - \frac{3v_b(1-v_b)}{1+3v_b}\right) f_b\left(\frac{v_b(1-v_b)}{1+3v_b}\right).$$

The first result follows by replacing $v_b$ with $x$ and simplifying.

To prove the second result, we refer to Table 6.2 to deduce that

$$\frac{\eta_1^4}{\eta_2^2} = (1-8x_c(q))^{3/4} z_c(q) \qquad \text{and} \qquad \frac{\eta_3^4}{\eta_6^2} = (1-8x_c(q))^{1/4} z_c(q).$$

Since $z_c(q) = f_c(x_c(q))$, it follows that

$$(1-8x_c(q^3))^{3/4} f_c(x_c(q^3)) = (1-8x_c(q))^{1/4} f_c(x_c(q)).$$

By the cubic transformation in Theorem 6.13, and using the notation

$$w_3 = r_c(q^3) = (r_c(q^3))^{1/3},$$

we deduce that

$$f_c(w_c^3)(1-8w_c^3)^{3/4} = \left(1 - \frac{8w_c(1-w_c+w_c^2)}{1+2w_c+4w_c^2}\right)^{1/4} f_c\left(\frac{w_c(1-w_c+w_c^2)}{1+2w_c+4w_c^2}\right).$$

The proof may be completed by simplifying the first factor on the right-hand side and replacing $w_c$ with $x$. □

The next two theorems give representations for $f_a$, $f_b$, and $f_c$ in terms of ${}_2F_1$ hypergeometric functions. The simplest formulas involve hypergeometric functions from the level 2 and level 3 theories.

**Theorem 6.42.** *For $|x|$ sufficiently small, the functions $f_a$, $f_b$, and $f_c$ have the following representations in terms of the ${}_2F_1(\frac{1}{4}, \frac{3}{4}; 1; \cdot)$ function:*

$$\begin{aligned} f_a(x) &= \frac{1}{\sqrt{1+36x+216x^2}}\, {}_2F_1\left(\begin{matrix}\frac{1}{4}, \frac{3}{4} \\ 1\end{matrix}; \frac{64x(1+9x)^3}{(1+36x+216x^2)^2}\right) \\ &= \frac{1}{\sqrt{1+12x+24x^2}}\, {}_2F_1\left(\begin{matrix}\frac{1}{4}, \frac{3}{4} \\ 1\end{matrix}; \frac{64x^3(1+9x)}{(1+12x+24x^2)^2}\right), \end{aligned}$$

$$\begin{aligned} f_b(x) &= \frac{1}{\sqrt{1+18x-27x^2}}\, {}_2F_1\left(\begin{matrix}\frac{1}{4}, \frac{3}{4} \\ 1\end{matrix}; \frac{64x}{(1+18x-27x^2)^2}\right) \\ &= \frac{1}{\sqrt{1-6x-3x^2}}\, {}_2F_1\left(\begin{matrix}\frac{1}{4}, \frac{3}{4} \\ 1\end{matrix}; \frac{64x^3}{(1-6x-3x^2)^2}\right) \end{aligned}$$

*and*

$$f_c(x) = \frac{1}{\sqrt{1+20x-8x^2}}\,{}_2F_1\left(\begin{matrix}\frac{1}{4},\ \frac{3}{4}\\ 1\end{matrix};\frac{64x(1+x)^3}{(1+20x-8x^2)^2}\right)$$
$$= \frac{1}{\sqrt{1-4x-8x^2}}\,{}_2F_1\left(\begin{matrix}\frac{1}{4},\ \frac{3}{4}\\ 1\end{matrix};\frac{64x^3(1+x)}{(1-4x-8x^2)^2}\right).$$

*Proof.* We shall first prove the results for the function $f_b$. Recall the level 2 case of Theorem 4.28, viz.,

$$z_2(q) = {}_2F_1\left(\begin{matrix}\frac{1}{4},\ \frac{3}{4}\\ 1\end{matrix};x_2(q)\right) \tag{6.75}$$

where

$$z_2(q)^2 = 2P(q^2) - P(q) \qquad \text{and} \qquad \frac{x_2(q)}{1-x_2(q)} = 64\,\frac{\eta_2^{24}}{\eta_1^{24}}.$$

The idea is to change variables in (6.75) from $x_2$ and $z_2$ to $x_b$ and $z_b$.

From Table 6.2 and Theorem 6.15 we deduce

$$2P(q^2) - P(q) = z_b^2 + 18z_bz_d - 27z_d^2 = (1+18x_b-27x_b^2)\,z_b^2 \tag{6.76}$$

and

$$2P(q^6) - P(q^3) = z_b^2 - 6z_bz_d - 3z_d^2 = (1-6x_b-3x_b^2)\,z_b^2. \tag{6.77}$$

From Table 6.1 we find that

$$\frac{\eta_2^{24}}{\eta_1^{24}} = \frac{x_b}{(1-x_b)(1-9x_b)^3} \qquad \text{and} \qquad \frac{\eta_6^{24}}{\eta_3^{24}} = \frac{x_b^3}{(1-x_b)^3(1-9x_b)}.$$

Hence,

$$x_2(q) = \frac{64\eta_2^{24}/\eta_1^{24}}{1+64\eta_2^{24}/\eta_1^{24}} = \frac{64x_b(1-x_b)^{-1}(1-9x_b)^{-3}}{1+64x_b(1-x_b)^{-1}(1-9x_b)^{-3}}$$
$$= \frac{64x_b}{(1+18x_b-27x_b^2)^2}, \tag{6.78}$$

and similarly,

$$x_2(q^3) = \frac{64x_b^3}{(1-6x_b-3x_b^2)^2}. \tag{6.79}$$

Using (6.76) and (6.78) in (6.75) we obtain

$$z_b\sqrt{1+18x_b-27x_b^2} = {}_2F_1\left(\begin{matrix}\frac{1}{4},\frac{3}{4}\\1\end{matrix};\frac{64x_b}{(1+18x_b-27x_b^2)^2}\right).$$

Since $z_b = f_b(x_b)$, this proves the first formula involving $f_b$. The other formula for $f_b$ can be obtained by replacing $q$ with $q^3$ in (6.75) and using (6.77) and (6.79).

The formulas that involve $f_a$ and $f_c$ can be deduced from the formulas for $f_b$ by further changes of variable using Theorem 6.40. □

**Theorem 6.43.** *For $|x|$ sufficiently small, the following representations in terms of the ${}_2F_1(\frac{1}{3},\frac{2}{3};1;\cdot)$ function hold:*

$$\begin{aligned} f_a(x) &= \frac{1}{1+12x}\,{}_2F_1\left(\begin{matrix}\frac{1}{3},\frac{2}{3}\\1\end{matrix};\frac{27x(1+8x)^2}{(1+12x)^3}\right)\\ &= \frac{1}{1+6x}\,{}_2F_1\left(\begin{matrix}\frac{1}{3},\frac{2}{3}\\1\end{matrix};\frac{27x^2(1+8x)}{(1+6x)^3}\right), \end{aligned}$$

$$\begin{aligned} f_b(x) &= \frac{1}{1+3x}\,{}_2F_1\left(\begin{matrix}\frac{1}{3},\frac{2}{3}\\1\end{matrix};\frac{27x(1-x)^2}{(1+3x)^3}\right)\\ &= \frac{1}{1-3x}\,{}_2F_1\left(\begin{matrix}\frac{1}{3},\frac{2}{3}\\1\end{matrix};\frac{27x^2(1-x)}{(1-3x)^3}\right), \end{aligned}$$

*and*

$$\begin{aligned} f_c(x) &= \frac{1}{1+4x}\,{}_2F_1\left(\begin{matrix}\frac{1}{3},\frac{2}{3}\\1\end{matrix};\frac{27x}{(1+4x)^3}\right)\\ &= \frac{1}{1-2x}\,{}_2F_1\left(\begin{matrix}\frac{1}{3},\frac{2}{3}\\1\end{matrix};\frac{27x^2}{(1-2x)^3}\right). \end{aligned}$$

*Proof.* We prove the results for $f_c$ first. By the level 3 case of Theorem 4.28 and (4.29),

$$z_3(q) = {}_2F_1\left(\begin{matrix}\frac{1}{3},\frac{2}{3}\\1\end{matrix};x_3(q)\right) \tag{6.80}$$

where

$$z_3(q) = a(q) \qquad \text{and} \qquad \frac{x_3(q)}{1-x_3(q)} = 27\,\frac{\eta_3^{12}}{\eta_1^{12}}.$$

By Theorem 6.16,

$$a(q) = (1 + 4x_c)z_c \qquad \text{and} \qquad a(q^2) = (1 - 2x_c)z_c. \tag{6.81}$$

From Table 6.1 we find that

$$\frac{\eta_3^{12}}{\eta_1^{12}} = \frac{x_c}{(1 + x_c)(1 - 8x_c)^2} \qquad \text{and} \qquad \frac{\eta_6^{12}}{\eta_2^{12}} = \frac{x_c^2}{(1 + x_c)^2(1 - 8x_c)}.$$

It follows that

$$x_3(q) = \frac{27x_c}{(1 + 4x_c)^3} \qquad \text{and} \qquad x_3(q^2) = \frac{27x_c^2}{(1 - 2x_c)^3}. \tag{6.82}$$

Using (6.81) and (6.82) in (6.80) we obtain

$$(1 + 4_c)z_c = {}_2F_1\left(\begin{matrix} \frac{1}{3}, \frac{2}{3} \\ 1 \end{matrix}; \frac{27x_c}{(1 + 4x_c)^3}\right).$$

Since $z_c = f_c(x_c)$, this proves the first identity for $f_c$. The other identity for $f_c$ can be proved by replacing $q$ with $q^2$ in (6.80) and using the identities for $a(q^2)$ and $x_3(q^2)$ in (6.81) and (6.82). Finally, the formulas for $f_a$ and $f_b$ can be deduced from the formulas for $f_c$ by changing variables using Theorem 6.40. □

We now briefly discuss the generating functions for the Apéry, Domb, and Almkvist–Zudilin numbers. Let $G_a$, $G_b$, and $G_c$ be defined by

$$G_a(x) = \sum_{n=0}^{\infty} s_a(n)x^n = 1 + 5x + 73x^2 + 1445x^3 + 33001x^4 + \cdots, \tag{6.83}$$

$$G_b(x) = \sum_{n=0}^{\infty} s_b(n)x^n = 1 - 4x + 28x^2 - 256x^3 + 2716x^4 + \cdots \tag{6.84}$$

and

$$G_c(x) = \sum_{n=0}^{\infty} s_c(n)x^n = 1 - 3x + 9x^2 - 3x^3 - 279x^4 + \cdots, \tag{6.85}$$

where $s_a(n)$, $s_b(n)$, and $s_c(n)$ are given by the binomial sums (6.64) and (6.65), or by the recurrence relations (6.66)–(6.68). The series converge for $|x| < 17 - 12\sqrt{2}$, $|x| < 1/16$, and $|x| < 1/9$, respectively.

Properties of $G_a$, $G_b$, and $G_c$ are inherited from the corresponding properties of $f_a, f_b$, and $f_c$ by means of the following result.

**Theorem 6.44.** *The following identities hold*:

$$(f_a(x))^2 = \frac{1}{(1+8x)(1+9x)}\, G_a\left(\frac{x}{(1+8x)(1+9x)}\right),$$

$$(f_b(x))^2 = \frac{1}{(1-x)(1-9x)}\, G_b\left(\frac{x}{(1-x)(1-9x)}\right)$$

*and*

$$(f_c(x))^2 = \frac{1}{(1+x)(1-8x)}\, G_c\left(\frac{x}{(1+x)(1-8x)}\right).$$

*Proof.* We start with the identity

$$x_a z_a^2 = w_a y_a$$

from Theorem 6.22. By Theorems 6.35 and 6.38, this may be written in the form

$$x_a\,(f_a(x_a))^2 = w_a\, G_a(w_a).$$

Now use Theorem 6.21 to express $w_a$ in terms if $x_a$, and then replace all occurrences of $x_a$ with $x$. This proves the first identity, and the other two results can be proved by the same procedure. □

Numerous identities for $G_a$, $G_b$, and $G_c$ can be constructed by combining the identities of Theorem 6.44 with other results; some are outlined in the exercises at the end of the chapter. The next two theorems offer nice representations of the functions $G_b$ and $G_c$ in terms of ${}_3F_2$ hypergeometric functions.

**Theorem 6.45.** *The following identities hold in a neighborhood of* $x = 0$:

$$\begin{aligned} G_b(x) &= \frac{1}{1+16x}\, {}_3F_2\left(\begin{matrix}\frac{1}{3}, \frac{1}{2}, \frac{2}{3} \\ 1, 1\end{matrix}; \frac{108x}{(1+16x)^3}\right) \\ &= \frac{1}{1+4x}\, {}_3F_2\left(\begin{matrix}\frac{1}{3}, \frac{1}{2}, \frac{2}{3} \\ 1, 1\end{matrix}; \frac{108x^2}{(1+4x)^3}\right). \end{aligned}$$

*Proof.* The case $\ell = 3$ of Theorem 4.30 gives

$$Z_3 = {}_3F_2\left(\begin{matrix}\frac{1}{3}, \frac{1}{2}, \frac{2}{3} \\ 1, 1\end{matrix}; \frac{108\eta_1^6\eta_3^6}{Z_3^3}\right) \tag{6.86}$$

where

$$Z_3 = \frac{3P(q^3) - P(q)}{2}.$$

By Theorem 6.30 we have

$$\frac{3P(q^3)-P(q)}{2}=(1+16w_b)\,y_b \quad\text{and}\quad \frac{3P(q^6)-P(q^2)}{2}=(1+4w_b)\,y_b, \tag{6.87}$$

and from Table 6.1 we have

$$\eta_1^6\eta_3^6=y_b^3\,w_b \qquad\text{and}\qquad \eta_2^6\eta_6^6=y_b^3\,w_b^2. \tag{6.88}$$

Using the first parts of (6.87) and (6.88) in (6.86) we obtain

$$(1+16w_b)y_b={}_3F_2\left(\begin{matrix}\frac{1}{3},\ \frac{1}{2},\ \frac{2}{3}\\ 1,\ 1\end{matrix};\frac{108w_b}{(1+16w_b)^3}\right).$$

Since $y_b = G_b(w_b)$, this gives the first identity. The second identity in the theorem can be obtained by first replacing $q$ with $q^3$ in (6.86) and then making use of the second parts of (6.87) and (6.88). □

**Theorem 6.46.** *The following identities hold in a neighborhood of* $x = 0$:

$$\begin{aligned}G_c(x)&=\frac{1}{1+27x}\,{}_3F_2\left(\begin{matrix}\frac{1}{4},\ \frac{1}{2},\ \frac{3}{4}\\ 1,\ 1\end{matrix};\frac{256x}{(1+27x)^4}\right)\\ &=\frac{1}{1+3x}\,{}_3F_2\left(\begin{matrix}\frac{1}{4},\ \frac{1}{2},\ \frac{3}{4}\\ 1,\ 1\end{matrix};\frac{256x^3}{(1+3x)^4}\right).\end{aligned}$$

*Proof.* The proof is similar to the proof of the previous theorem. The case $\ell = 2$ of Theorem 4.30 gives

$$Z_2={}_3F_2\left(\begin{matrix}\frac{1}{4},\ \frac{1}{2},\ \frac{3}{4}\\ 1,\ 1\end{matrix};\frac{256\eta_1^8\eta_2^8}{Z_2^4}\right) \tag{6.89}$$

where

$$Z_2=2P(q^2)-P(q).$$

By Theorem 6.30 we have

$$2P(q^2)-P(q)=(1+27w_c)\,y_c \quad\text{and}\quad 2P(q^6)-P(q^3)=(1+3w_c)\,y_c, \tag{6.90}$$

and from Table 6.1 we have

$$\eta_1^8\eta_2^8=y_c^4\,w_c \qquad\text{and}\qquad \eta_3^8\eta_6^8=y_c^4\,w_c^3. \tag{6.91}$$

Using the first parts of (6.90) and (6.91) in (6.89) we obtain

$$(1+27w_c)y_c = {}_3F_2\left(\begin{matrix}\frac{1}{4}, \frac{1}{2}, \frac{3}{4} \\ 1, 1\end{matrix}; \frac{256w_c}{(1+27w_c)^4}\right).$$

Since $y_c = G_c(w_c)$, this gives the first identity. The second identity can be obtained by the same procedure, by replacing $q$ with $q^2$ in (6.89) and making use of the second parts of (6.90) and (6.91). □

By now it is clear that numerous other expressions can be obtained for $G_a$, $G_b$, or $G_c$ in terms of the hypergeometric functions from theories of levels 1, 2, 3, or 4. The results in the two theorems above have been selected because they have particularly simple forms.

Further results, connecting the functions studied in this chapter with hypergeometric functions, will be given in Theorem 13.3.

## 6.12 Notes

The continued fraction in Theorem 6.1 appears in Ramanujan's lost notebook [254, p. 41]. Ramanujan actually evaluated the continued fraction as a quotient of two other infinite series; the equivalence of the two formulas has been established by G. Bhatnagar [57, Sec. 6], by using a $q$-extension of Pfaff's transformation formula, e.g., see [253, Chapter 16, Entry 8]. The proof of Theorem 6.1 that we have given follows Bhatnagar [57]. Bhatnagar's paper also contains an interesting discussion about convergence of continued fractions. Ramanujan's continued fraction has been studied in depth by G. E. Andrews [20, 21]. The convergents, that is, the values of

$$1+\cfrac{aq+\lambda q}{1+\cfrac{bq+\lambda q^2}{1+\cfrac{\cdots}{\cfrac{\cdots}{1+\cfrac{aq^n+\lambda q^{2n-1}}{1}}}}} \quad \text{and} \quad 1+\cfrac{aq+\lambda q}{1+\cfrac{bq+\lambda q^2}{1+\cfrac{\cdots}{\cfrac{\cdots}{1+\cfrac{aq^n+\lambda q^{2n-1}}{1+\cfrac{bq^n+\lambda q^{2n}}{1}}}}}}$$

have been found by M. D. Hirschhorn [176]. Further references for the continued fraction in Theorem 6.1 have been given by Andrews and Berndt in [25, Sec. 6.2, especially p. 150].

The proof of Lemma 6.4 follows Fine [151, p. 53, (25.94)]. The proof of the first identity in Lemma 6.5 is also due to Fine [151, p. 53, (25.95)].

Ramanujan's cubic continued fraction in Theorem 6.6 appears in Ramanujan's second letter to Hardy [174, p. xxviii], in Ramanujan's second notebook [253,

Chapter 20, Entry 1], and in Ramanujan's lost notebook [254, p. 366]. The first published proof is due to Watson [296]. References to other proofs are given by Berndt in [46, p. 347]. The proof we have given requires only the finite form of the $q$-binomial theorem and Jacobi's triple product identity.

The cubic transformation formula for $r_c(q)$ in Theorem 6.13 was given by H. H. Chan, [77]. Another proof has been given by C. Adiga, T. Kim, M. S. M. Naika, and H. S. Madhusudhan [3].

The expansions of the weight one modular forms $z_a$, $z_b$, and $z_c$ in powers of $x_a$, $x_b$, and $x_c$, respectively, in Theorem 6.35 and (6.56)–(6.58), are due to Zagier [307].

The simple proof of Theorem 6.36, that yields recurrence relations for the sums of the squares of the trinomial coefficients and the sums of the cubes of the binomial coefficients, is due to Barrucand [37].

The Apéry numbers were introduced and used by R. Apéry to prove that the number $\zeta(3) := \sum_{n=1}^{\infty} \frac{1}{n^3}$ is irrational. See the account by van der Poorten [285]. The connection of the Apéry numbers with modular forms was indicated in Sato's abstract [260]. The Domb numbers seem to have been first studied by C. Domb [141]. The Almkvist–Zudilin numbers appear in work by G. Almkvist and W. Zudilin [19, Sec. 7, Eq. ($\delta$)]. The identity (6.62) involving Domb numbers is due to H. H. Chan, S. H. Chan, and Z.-G. Liu [79], and the identity (6.63) involving the Almkvist–Zudilin numbers is due to Chan and H. Verrill [92]. The series $\sum_{n=0}^{\infty} \binom{2n}{n} t_b(n) X^n$ that occurs in (6.70) was first studied by M. Rogers [257]. The identities (6.69)–(6.71) are given by Chan, Tanigawa, Yang and Zudilin [90]. The last identity for $f_c(x)$ in Theorem 6.43 is originally due to M. N. Lalin and M. D. Rogers [202, Corollary 2.7]. The generating functions for the Apéry, Domb, and Almkvist–Zudilin numbers in (6.83)–(6.85) have been studied by Chan and Zudilin [94]. The generating functions will be studied again, in more detail, in Chapter 13.

## 6.13 Exercises

1. Prove that the evaluations in Table 6.5 hold.
   Hint: The values of $r_a(e^{-2\pi/\sqrt{6}})$, $r_b(e^{-2\pi/\sqrt{3}})$, and $r_c(e^{-2\pi/\sqrt{2}})$ can be obtained from Theorem 6.12 by considering the fixed points in the respective transformation formulas and finding the positive solution. The other evaluations can then be deduced using Theorem 6.9.
2. Refer to the definitions in (6.18)–(6.20) and to Theorem 6.13.

   a. Prove the following quadratic transformation formulas:

**Table 6.5** Table for Exercise 1 of Chapter 6

| $q$ | $r_a(q)$ | $r_b(q)$ | $r_c(q)$ |
|---|---|---|---|
| $e^{-2\pi/\sqrt{6}}$ | $\frac{\sqrt{2}}{12}$ | $\frac{\sqrt{6}-\sqrt{3}}{3}$ | $\frac{1}{2}\left(6\sqrt{2}-8\right)^{1/3}$ |
| $e^{-2\pi/\sqrt{3}}$ | $\frac{\sqrt{3}-1}{24}$ | $\frac{2\sqrt{3}-3}{3}$ | $\frac{1}{2}\left(3\sqrt{3}-5\right)^{1/3}$ |
| $e^{-2\pi/\sqrt{2}}$ | $\frac{\sqrt{6}-2}{36}$ | $\frac{\sqrt{3}-\sqrt{2}}{6}$ | $\frac{\sqrt{6}-2}{2}$ |

$$9+\frac{1}{v_a}=\left\{\frac{3\sqrt{1+8u_a}+1}{\sqrt{1+8u_a}-1}\right\}^2$$

and

$$1+\frac{1}{v_c^3}=\left\{\frac{\sqrt{1-8u_c^3}+3}{\sqrt{1-8u_c^3}-1}\right\}^2.$$

b. Prove the following cubic transformation formulas:

$$8+\frac{1}{w_a}=\left\{\frac{2\sqrt[3]{1+9u_a}+1}{\sqrt[3]{1+9u_a}-1}\right\}^3$$

and

$$1-\frac{1}{w_b^2}=\left\{\frac{\sqrt[3]{1-9u_b^2}+2}{\sqrt[3]{1-9u_b^2}-1}\right\}^3.$$

3. Let $p_{1,2}(n)$ and $p_{1,1,3,3}(n)$ be defined by

$$\sum_{n=0}^{\infty}p_{1,2}(n)q^n=\prod_{j=1}^{\infty}\frac{1}{(1-q^j)(1-q^{2j})}$$

and

$$\sum_{n=0}^{\infty}p_{1,1,3,3}(n)q^n=\prod_{j=1}^{\infty}\frac{1}{(1-q^j)^2(1-q^{3j})^2}.$$

a. Show that $p_{1,2}(n)$ is the number of finite sequences of positive integers

$$(\lambda_1,\lambda_2,\ldots,\lambda_r\mid\mu_1,\mu_2,\ldots,\mu_s)$$

such that

$$(\lambda_1 + \cdots + \lambda_r) + (\mu_1 + \cdots + \mu_s) = n,$$
$$\lambda_1 \geq \lambda_2 \geq \cdots \geq \lambda_r, \qquad \mu_1 \geq \mu_2 \geq \cdots \geq \mu_s,$$

and $\mu_1, \ldots, \mu_s$ are all even. Equivalently, $p_{1,2}(n)$ is the number of partitions of $n$ into parts with two colors, say red and blue, such that the blue parts are all even. For example, $p_{1,2}(3) = 4$, the sequences being

$$(3\,|\,),\ (2,1\,|\,),\ (1,1,1\,|\,),\ (1\,|\,2).$$

Give a similar interpretation for $p_{1,1,3,3}(n)$.

b. Prove the following analogue of Corollary 5.14,

$$\sum_{n=0}^{\infty} p_{1,2}(3n+2)q^n = 3\prod_{j=1}^{\infty} \frac{(1-q^{3j})^3(1-q^{6j})^3}{(1-q^j)^4(1-q^{2j})^4}$$

and deduce that $p_{1,2}(3n+2)$ is divisible by 3.
Hint: Combine the identities for $w_c y_c$ in Theorems 6.20 and 6.23 to show that

$$q\prod_{j=1}^{\infty} \frac{(1-q^{3j})^3(1-q^{6j})^3}{(1-q^j)(1-q^{2j})} = \frac{1}{24}\left(-P(q) + 2P(q^2) + P(q^3) - 2P(q^6)\right)$$
$$= \sum_{k \text{ odd},}\ \sum_{j \not\equiv 0 \pmod 3} k\, q^{jk}.$$

Now use the technique in the proof of Corollary 5.14 and equate powers of $q$ that are multiples of 3.

c. Prove that

$$\sum_{n=0}^{\infty} p_{1,1,3,3}(2n+1)q^n = 2\prod_{j=1}^{\infty} \frac{(1-q^{2j})^4(1-q^{6j})^4}{(1-q^j)^6(1-q^{3j})^6}$$

and deduce that $p_{1,1,3,3}(2n+1)$ is even.
Hint: Combine the identities for $w_b y_b$ in Theorems 6.20 and 6.23 to show that

$$q\prod_{j=1}^{\infty} \frac{(1-q^{2j})^4(1-q^{6j})^4}{(1-q^j)^2(1-q^{3j})^2} = \frac{1}{24}\left(-P(q) + P(q^2) + 3P(q^3) - 3P(q^6)\right)$$
$$= \sum_{k \text{ odd},}\ \sum_{j \not\equiv 0 \pmod 3} j\, q^{jk}.$$

Now equate even powers of $q$.

The identities in Parts b and c are analogues of Hardy's favorite identity of Ramanujan (5.132). For generalizations and extensions of the results in Parts b and c, see works of H.-C. Chan [74, 75, 76] and the work of H. H. Chan and P. C. Toh [91].

4. a. Let $m$ and $n$ be positive integers. Suppose $y = y(x)$ satisfies a linear differential equation of order $m$. Let $Y(x) = (y(x))^n$ be the $n$th power of $y$. Show that $Y$ satisfies a linear differential equation of order at most $\binom{m+n-1}{n}$.

   Hint: Use induction to show that the derivative $\dfrac{d^k Y}{dx^k}$ may be expressed as a linear combination of terms of the form

$$(y)^{a_1} \times \left(\frac{dy}{dx}\right)^{a_2} \times \left(\frac{d^2y}{dx^2}\right)^{a_3} \times \cdots \times \left(\frac{d^{m-1}y}{dx^{m-1}}\right)^{a_m},$$

   where $a_1, \ldots, a_m$ are nonnegative integers that satisfy

$$a_1 + a_2 + \cdots + a_m = n. \tag{6.92}$$

   The number of sequences $\langle a_1, a_2, \ldots, a_m \rangle$ of nonnegative integers that satisfy the property (6.92) is $\binom{m+n-1}{n}$. Therefore, the set of functions

$$\left\{ \frac{d^k Y}{dx^k} : \; 0 \le k \le \binom{m+n-1}{n} \right\}$$

   is linearly dependent.

   b. Suppose $y$ is a solution of the second order linear differential equation

$$y''(x) + f_1(x)y'(x) + f_2(x)y(x) = 0.$$

   Show that the function $Y(x) = (y(x))^2$ is a solution of the third order linear differential equation

$$Y''' + 3f_1 Y'' + (f_1' + 4f_2 + 2f_1^2)Y' + (2f_2' + 4f_1 f_2)Y = 0.$$

   c. Suppose that $y$ is a solution of the second order linear differential equation

$$xy'' + y' - y = 0.$$

   Show that the function $Y = (y(x))^3$ is a solution of

$$x^3 Y'''' + 6x^2 Y''' + (7x - 10x^2)Y'' + (1 - 20x)Y' + (-3 + 9x)Y = 0.$$

d. Suppose that $y$ is a solution of the second order linear differential equation

$$x^2 y''' + 3xy'' + y' - y = 0.$$

Show that the function $Y = (y(x))^2$ is a solution of

$$x^4 Y^{(6)} + 14x^3 Y^{(5)} + 55x^2 Y'''' + (65x - 7x^2) Y'''$$
$$+ (16 - 28x) Y'' - 16Y' - 8Y = 0.$$

Parts b, c, and d are particular instances of the special cases $(m, n) = (2, 2)$, $(2, 3)$, and $(3, 2)$ respectively, of the result of Part a.

5. a. Use Corollary 3.40 and Table 6.1 to prove that:

$$Q(q) = z_a^3 z_c + 256 z_b^3 z_d, \qquad Q(q^2) = z_a^3 z_c + 16 z_b^3 z_d,$$
$$Q(q^3) = z_a z_c^3 + 256 z_b z_d^3, \qquad Q(q^6) = z_a z_c^3 + 16 z_b z_d^3.$$

b. Use Part a and Theorem 6.15 to deduce that for $j \in \{1, 2, 3, 6\}$ and $k \in \{a, b, c\}$,

$$Q(q^j) = z_k^4 \times \text{ a polynomial of degree 4 in } x_k.$$

6. a. Use Corollary 0.23 and the result of Exercise 5 to prove that:

$$R(q) = \sqrt{z_a^3 z_c + 64 z_b^3 z_d} \times (z_a^3 z_c - 512 z_b^3 z_d),$$
$$R(q^2) = \sqrt{z_a^3 z_c + 64 z_b^3 z_d} \times (z_a^3 z_c - 8 z_b^3 z_d),$$
$$R(q^3) = \sqrt{z_a z_c^3 + 64 z_b z_d^3} \times (z_a z_c^3 - 512 z_b z_d^3),$$
$$R(q^6) = \sqrt{z_a z_c^3 + 64 z_b z_d^3} \times (z_a z_c^3 - 8 z_b z_d^3).$$

b. Show that

$$z_a^3 z_c + 64 z_b^3 z_d = z_a^4 (1 + 36 x_a + 216 x_a^2)^2$$

and

$$z_a z_c^3 + 64 z_b z_d^3 = z_a^4 (1 + 12 x_a + 24 x_a^2)^2.$$

c. Deduce that for $j \in \{1, 2, 3, 6\}$ and $k \in \{a, b, c\}$,

$$R(q^j) = z_k^6 \times \text{ a polynomial of degree 6 in } x_k.$$

7. a. Prove the identity

$$\sum_{k=0}^{n} \binom{n}{k}^2 = \binom{2n}{n}$$

by comparing the coefficients of $x^n$ on both sides in the expansions of

$$(1+x)^n(1+x)^n = (1+x)^{2n}.$$

b. Equate coefficients of $x^n$ in the identities for $f_c(x)$ in Theorem 6.43, to show that

$$\begin{aligned}\sum_{k=0}^{n} \binom{n}{k}^3 &= \sum_{k=0}^{n} \binom{n+2k}{k, k, k, n-k} (-4)^{n-k} \\ &= \sum_{k=0}^{\lfloor n/2 \rfloor} \binom{n+k}{k, k, k, n-2k} 2^{n-2k}\end{aligned}$$

where the multinomial coefficient is given by (5.97).

c. Show that

$$\sum_{k=0}^{n} \binom{n}{k}^3 = \sum_{k=0}^{\lfloor n/2 \rfloor} \binom{n}{2k}\binom{2k}{k}\binom{2n-2k}{n-k}.$$

See Moreau [235].

d. Show that the Domb numbers, defined by (6.64), are given by

$$s_b(n) = \sum_k (-1)^k \binom{n}{k}\binom{2k}{n}\binom{2k}{k}\binom{2n-2k}{n-k}.$$

See Z.-W. Sun [279].

8. Let $u(n)$ be defined by

$$\frac{1}{\sqrt{1-16x}} \sum_{n=0}^{\infty} \binom{2n}{n}^2 x^n = \sum_{n=0}^{\infty} u(n)\, x^n.$$

a. Show that $(n+1)^2 u(n+1) = 4(8n^2+8n+3)u(n) - 256\, n^2\, u(n-1)$.

b. Show that

$$u(n) = \binom{2n}{n}^{-1} \sum_{k=0}^{n} \binom{2k}{k}\binom{4k}{2k}\binom{2n-2k}{n-k}\binom{4n-4k}{2n-2k}$$
$$= \sum_{k=0}^{n} 4^{n-k} \binom{2k}{k}^2 \binom{2n-2k}{n-k}.$$

See A143583 in Sloane's encyclopedia [271].

9. Show that

$$\frac{\eta_1^8}{\eta_2^4} = \sum_{n=0}^{\infty} s(n) \left(\frac{\eta_4}{\eta_1}\right)^{8n},$$

where $s(n)$ is given by

$$s(n) = (-1)^n \sum_{j=0}^{n} \binom{2j}{j}^2 \binom{2n-2j}{n-j}^2.$$

Hint: This is the expansion of $y$ in terms of $w$ in the case of level 4; see Table 6.4. See A036917 in Sloane's encyclopedia [271].

10. a. Prove that $t_a(n)$ equals the coefficient of $x^n y^n z^n$ in

$$\frac{1}{1 - xy^2 - yz^2 - zx^2 - x^2y - y^2z - z^2x + 6xyz}.$$

b. Prove that $t_c(n)$ equals the coefficient of $x^n y^n z^n$ in

$$\frac{1}{1 - x - y - x + 4xyz}.$$

See Straub and Zudilin [274] for further examples.

11. a. Show that

$$G_a(w_a) = (1 - x_b)\,(f_b(x_b))^2 = (1 + x_c)\,(f_c(x_c))^2,$$
$$G_b(w_b) = (1 + 8x_a)\,(f_a(x_a))^2 = (1 - 8x_c)\,(f_c(x_c))^2,$$

and

$$G_c(w_c) = (1 + 9x_a)\,(f_a(x_a))^2 = (1 - 9x_b)\,(f_b(x_b))^2.$$

b. Hence, use Theorem 6.21 to deduce that

$$\frac{1}{1+x}\,G_a\left(\frac{x(1-8x)}{1+x}\right)=\frac{1}{1-8x}\,G_b\left(\frac{x(1+x)}{1-8x}\right),$$
$$\frac{1}{1-x}\,G_a\left(\frac{x(1-9x)}{1-x}\right)=\frac{1}{1-9x}\,G_c\left(\frac{x(1-x)}{1-9x}\right)$$

and

$$\frac{1}{1+8x}\,G_b\left(\frac{x(1+9x)}{1+8x}\right)=\frac{1}{1+9x}\,G_c\left(\frac{x(1+8x)}{1+9x}\right).$$

See Chan and Zudilin [94].

12. a. Show that

$$G_a(x)=\sum_{n=0}^{\infty}\binom{2n}{n}t_b(n)\frac{x^n}{(1+x)^{2n+1}}=\sum_{n=0}^{\infty}\binom{2n}{n}t_c(n)\frac{x^n}{(1-x)^{2n+1}},$$

$$G_b(x)=\sum_{n=0}^{\infty}\binom{2n}{n}t_a(n)\frac{x^n}{(1-8x)^{2n+1}}=\sum_{n=0}^{\infty}\binom{2n}{n}t_c(n)\frac{x^n}{(1+8x)^{2n+1}},$$

$$G_c(x)=\sum_{n=0}^{\infty}\binom{2n}{n}t_a(n)\frac{x^n}{(1-9x)^{2n+1}}=\sum_{n=0}^{\infty}\binom{2n}{n}t_b(n)\frac{x^n}{(1+9x)^{2n+1}}.$$

b. By comparing coefficients of $x^n$ in the results of Part a, deduce that

$$s_a(n)=\sum_{k=0}^{n}\binom{n}{k}\binom{n+k}{k}(-1)^{n-k}t_b(k)=\sum_{k=0}^{n}\binom{n}{k}\binom{n+k}{k}t_c(k),$$

$$s_b(n)=\sum_{k=0}^{n}\binom{n}{k}\binom{n+k}{k}(-8)^{n-k}t_c(k)=\sum_{k=0}^{n}\binom{n}{k}\binom{n+k}{k}8^{n-k}t_a(k),$$

$$s_c(n)=\sum_{k=0}^{n}\binom{n}{k}\binom{n+k}{k}(-9)^{n-k}t_b(k)=\sum_{k=0}^{n}\binom{n}{k}\binom{n+k}{k}9^{n-k}t_a(k).$$

See Strehl [275].

13. Recall the level 2 modular forms $x_2$ and $z_2$ defined by (4.28):

$$x_2=\frac{64q\psi^8(q)}{\varphi^8(-q)+64q\psi^8(q)}\qquad\text{and}\qquad z_2=\left(\varphi^8(-q)+64q\psi^8(q)\right)^{1/4}.$$

By Theorem 4.24 we have

$$\frac{\mathrm{d}}{\mathrm{d}x_2}\left(x_2(1-x_2)\frac{\mathrm{d}z_2}{\mathrm{d}x_2}\right) = \frac{3z_2}{16}. \tag{6.93}$$

a. Prove the following analogue of Lemma 6.32:

$$\begin{aligned} x_2 &= \frac{64x_a(1+9x_a)^3}{(1+36x_a+216x_a^2)^2} \\ &= \frac{64x_b}{(1+18x_b-27x_b^2)^2} = \frac{64x_c(1+x_c)^3}{(1+20x_c-8x_c^2)^2}, \end{aligned}$$

$$\begin{aligned} 1-x_2 &= \frac{1+8x_a}{(1+36x_a+216x_a^2)^2} \\ &= \frac{(1-x_b)(1-9x_b)}{(1+18x_b-27x_b^2)^2} = \frac{(1-8x_c)^3}{(1+20x_c-8x_c^2)^2} \end{aligned}$$

and

$$\begin{aligned} z_2 &= \sqrt{1+36x_a+216x_a^2}\, z_a \\ &= \sqrt{1+18x_b-27x_b^2}\, z_b = \sqrt{1+20x_c-8x_c^2}\, z_c. \end{aligned}$$

b. Prove the following analogue of Lemma 6.33:

$$\begin{aligned} x_2(1-x_2)\frac{\mathrm{d}}{\mathrm{d}x_2} &= \frac{x_a(1+8x_a)(1+9x_a)}{(1+36x_a+216x_a^2)}\frac{\mathrm{d}}{\mathrm{d}x_a} \\ &= \frac{x_b(1-x_b)(1-9x_b)}{(1+18x_b-27x_b^2)}\frac{\mathrm{d}}{\mathrm{d}x_b} \\ &= \frac{x_c(1+x_c)(1-8x_c)}{(1+20x_c-8x_c^2)}\frac{\mathrm{d}}{\mathrm{d}x_c}. \end{aligned}$$

c. Use the results of Parts a and b to change variables in (6.93) and give another proof of Theorem 6.34.

14. Use Theorem 6.40 to prove that for $\theta$ sufficiently large,

$$\sum_{n=0}^{\infty} t_a(n) \left(\frac{\operatorname{csch}\theta}{3}\right)^{2n+2} = \sum_{n=0}^{\infty} t_b(n) \left(\frac{\operatorname{sech}\theta}{3}\right)^{2n+2},$$

$$\sum_{n=0}^{\infty} t_a(n) \left(\frac{\operatorname{csch}\theta}{2\sqrt{2}}\right)^{2n+2} = \sum_{n=0}^{\infty} t_c(n) \left(\frac{\operatorname{sech}\theta}{2\sqrt{2}}\right)^{2n+2}$$

and

$$\sum_{n=0}^{\infty} t_c(n) \,(\operatorname{csch}\theta)^{2n+2} = \sum_{n=0}^{\infty} t_b(n) \,(\operatorname{sech}\theta)^{2n+2}.$$

15. This exercise derives a level 18 theory from the level 6 theory for the functions $z_c$ and $x_c$.

Let $u$, $v$, and $w$ be the eta-quotients defined by

$$u = \frac{\eta_3\eta_{18}^3}{\eta_6\eta_9^3}, \quad v = \frac{\eta_2^6\eta_3^{12}\eta_{18}^6}{\eta_1^6\eta_6^{12}\eta_9^6}, \quad \text{and} \quad w = \frac{\eta_1^2\eta_2^2\eta_9^2\eta_{18}^2}{\eta_3^4\eta_6^4}.$$

Let $x$ and $X$ be the modular functions defined by

$$x = u^3 \quad \text{and} \quad X = \frac{w}{(1+3w)^2}.$$

Let $z$ and $Z$ be the weight one and weight two modular forms, respectively, defined by

$$z = \frac{\eta_6\eta_9^6}{\eta_3^2\eta_{18}^3}$$

and

$$Z = q\frac{\mathrm{d}}{\mathrm{d}q}\log v = \frac{1}{4}\left(18P_{18} - 9P_9 - 12P_6 + 6P_3 + 2P_2 - P_1\right)$$

where $P_n$ is given by (0.75). Observe that $x = x_c(q^3)$ and $z = z_c(q^3)$; see (6.23) and (6.24).

a. Use (6.9) and (6.15) to show that

$$1 - 2u = \frac{\eta_1^2\eta_{18}}{\eta_2\eta_9^2}, \qquad 1 + u = \frac{\eta_2^2\eta_3\eta_{18}}{\eta_1\eta_6\eta_9^2}$$

and

$$1 - 8u^3 = \frac{\eta_3^8 \eta_{18}^4}{\eta_6^4 \eta_9^8}, \qquad 1 + u^3 = \frac{\eta_6^5 \eta_{18}}{\eta_3 \eta_9^5}.$$

b. Deduce that

$$v = \frac{u(1+u)^2(1-8u^3)}{(1-2u)^2(1+u^3)},$$
$$w = \frac{u(1+u)^2(1-2u)^2}{(1+u^3)(1-8u^3)}$$

and

$$X = \frac{u(1+u^3)(1-8u^3)}{(1+2u-2u^2)^4}.$$

c. Show that

$$1 - 28X + 192X^2$$
$$= \frac{(1+2u^2)^2(1-4u-2u^2)^2(1-2u+6u^2+4u^3+4u^2)^2}{(1+2u-2u^2)^8}$$

and

$$\frac{\mathrm{d}X}{\mathrm{d}u} = \frac{(1+2u^2)(1-4u-2u^2)(1-2u+6u^2+4u^3+4u^4)}{(1+2u-2u^2)^5}.$$

d. Use the results of Part c to deduce that

$$(1+2u-2u^2)\frac{\mathrm{d}}{\mathrm{d}u} = \sqrt{1-28X+192X^2}\,\frac{\mathrm{d}}{\mathrm{d}X}.$$

e. Show that

$$q\frac{\mathrm{d}}{\mathrm{d}q}\log X = Z\sqrt{1-28X+192X^2}$$

and

$$Z = (1+2u-2u^2)^3 z^2.$$

f. From Theorem 6.34,

$$\frac{d}{dw}\left(w(1+w)(1-8w)\frac{dz}{dw}\right) = 2(1+4w)z.$$

Use this to show that

$$D^2Z - \frac{1}{2Z}(DZ)^2 = 6X(1-15X)\,Z,$$

where

$$D = X\sqrt{1-28X+192X^2}\,\frac{d}{dX}.$$

g. Deduce that

$$X^2(1-28X+192X^2)\frac{d^3Z}{dX^3} + 3X(1-42X+384X^2)\frac{d^2Z}{dX^2}$$

$$+(1-96t+1332X^2)\frac{dZ}{dX} - 6(1-30X)Z = 0.$$

h. Show that

$$Z = \sum_{k=0}^{\infty} T_{18}(k)X^k$$

where the coefficients satisfy the recurrence relation

$$(k+1)^3T_{18}(k+1) = 2(2k+1)(7k^2+7k+3)T_{18}(k) - 12k(16k^2-1)T_{18}(k-1)$$

and initial condition $T_{18}(0) = 1$. Show that

$$T_{18}(k) = \sum_{j=0}^{k}(-1)^j\binom{k}{j}\binom{2j}{j}\binom{2k-2j}{k-j}\binom{2k-3j}{k}.$$

See [114] for further details.

16. Let $L_1 = P(q)$, $L_2 = 2P(q^2)$, $L_3 = 3P(q^3)$, and $L_6 = 6P(q^6)$. Prove that

$$(L_1+L_3)^2 + (L_2+L_6)^2 = 4(L_1L_2+L_3L_6).$$

# Chapter 7
# Level 7

## 7.1 Introduction

The theory for level 7 parallels the level 5 theory in many ways and it is recommended that this chapter be read in conjunction with Chapter 5. There are some differences and complications. The first significant difference is that the 7-dissection of Euler's product involves three functions whose product is $-1$, whereas the corresponding 5-dissection involves just two functions, namely the Rogers–Ramanujan continued fraction and its reciprocal. Another significant difference is that the natural eta-quotient for level 7 has weight 3, whereas the eta-quotient for level 5 has weight 2. That is one of the reasons why the level 7 theory does not arise in Zagier's examples in Table 6.3.

## 7.2 The 7-dissection of Euler's product: $\alpha$, $\beta$ and $\gamma$

Let $\alpha$, $\beta$, $\gamma$, and $v$ be defined by

$$\alpha = \alpha(q) = \quad q^{-2/7} \prod_{j=1}^{\infty} \frac{(1-q^{7j-5})(1-q^{7j-2})}{(1-q^{7j-6})(1-q^{7j-1})}, \tag{7.1}$$

$$\beta = \beta(q) = -q^{-1/7} \prod_{j=1}^{\infty} \frac{(1-q^{7j-4})(1-q^{7j-3})}{(1-q^{7j-5})(1-q^{7j-2})}, \tag{7.2}$$

$$\gamma = \gamma(q) = \quad q^{3/7} \prod_{j=1}^{\infty} \frac{(1-q^{7j-6})(1-q^{7j-1})}{(1-q^{7j-4})(1-q^{7j-3})} \tag{7.3}$$

 
S. Cooper, *Ramanujan's Theta Functions*, DOI 10.1007/978-3-319-56172-1_8

and

$$v = v(q) = q^{-2/7} \prod_{j=1}^{\infty} \frac{(1-q^{j/7})}{(1-q^{7j})}. \tag{7.4}$$

The functions $\alpha$, $\beta$, and $\gamma$ play the role that the Rogers–Ramanujan continued fraction and its reciprocal did for the level 5 theory. They arise naturally in the 7-dissection of Euler's product, as we will now show.

**Theorem 7.1.** *For $\alpha$, $\beta$, $\gamma$ and $v$ defined by* (7.1)–(7.4),

$$\alpha\beta\gamma = -1 \tag{7.5}$$

*and*

$$\alpha + \beta + \gamma = v + 1. \tag{7.6}$$

Analogous results for $\alpha\beta + \beta\gamma + \gamma\alpha$, $\alpha^7 + \beta^7 + \gamma^7$ and $\alpha^7\beta^7 + \beta^7\gamma^7 + \gamma^7\alpha^7$ are proved in Theorems 7.3 and 7.8.

Before starting the proof, observe that if we multiply (7.6) by $q^{2/7}\prod_{j=1}^{\infty}(1-q^{7j})$ and replace $q$ with $q^7$, the result may be expressed in the form

$$\prod_{j=1}^{\infty}(1-q^j) = G(q^7) - qH(q^7) - q^2I(q^7) + q^5J(q^7)$$

where

$$G(q) = \prod_{j=1}^{\infty} \frac{(1-q^{7j-5})(1-q^{7j-2})(1-q^{7j})}{(1-q^{7j-6})(1-q^{7j-1})},$$

$$H(q) = \prod_{j=1}^{\infty} \frac{(1-q^{7j-4})(1-q^{7j-3})(1-q^{7j})}{(1-q^{7j-5})(1-q^{7j-2})},$$

$$I(q) = \prod_{j=1}^{\infty}(1-q^{7j})$$

and

$$J(q) = \prod_{j=1}^{\infty} \frac{(1-q^{7j-6})(1-q^{7j-1})(1-q^{7j})}{(1-q^{7j-4})(1-q^{7j-3})}.$$

Thus, (7.6) gives a 7-dissection for Euler's product.

*Proof.* The identity (7.5) is a trivial consequence of the definitions (7.1)–(7.3). We will prove (7.6). Replace $q$ with $q^{24}$ in the formula for $\eta_1$ in Theorem 0.9 to obtain

$$q\prod_{j=1}^{\infty}(1-q^{24j}) = \sum_{n=-\infty}^{\infty}(-1)^n q^{(6n+1)^2}. \tag{7.7}$$

Break the sum on the right into seven cases according to the residue class of $n$ modulo 7, to obtain

$$\sum_{n=-\infty}^{\infty}(-1)^n q^{(6n+1)^2} = S_0 + S_1 + S_2 + S_3 + S_4 + S_5 + S_6$$

where

$$S_2 = \sum_{n=7k+2}(-1)^n q^{(6n+1)^2} = q^{36}\sum_{k=-\infty}^{\infty}(-1)^k q^{49(6k+1)^2+84(6k+1)}, \tag{7.8}$$

$$S_0 = \sum_{n=7k}(-1)^n q^{(6n+1)^2} = q^{36}\sum_{k=-\infty}^{\infty}(-1)^k q^{49(6k+1)^2-84(6k+1)}, \tag{7.9}$$

$$S_3 = \sum_{n=7k+3}(-1)^n q^{(6n+1)^2} = -q^{144}\sum_{k=-\infty}^{\infty}(-1)^k q^{49(6k+1)^2+168(6k+1)}, \tag{7.10}$$

$$S_6 = \sum_{n=7k-1}(-1)^n q^{(6n+1)^2} = -q^{144}\sum_{k=-\infty}^{\infty}(-1)^k q^{49(6k+1)^2-168(6k+1)}, \tag{7.11}$$

$$S_4 = \sum_{n=7k+4}(-1)^n q^{(6n+1)^2} = q^{324}\sum_{k=-\infty}^{\infty}(-1)^k q^{49(6k+1)^2+252(6k+1)}, \tag{7.12}$$

$$S_5 = \sum_{n=7k-2}(-1)^n q^{(6n+1)^2} = q^{324}\sum_{k=-\infty}^{\infty}(-1)^k q^{49(6k+1)^2-252(6k+1)} \tag{7.13}$$

and

$$S_1 = \sum_{n=7k+1}(-1)^n q^{(6n+1)^2} = -\sum_{k=-\infty}^{\infty}(-1)^k q^{49(6k+1)^2}. \tag{7.14}$$

The identities (7.8)–(7.13) may be combined in pairs and converted to an infinite product by the quintuple product identity (0.36), which we write in the equivalent form

$$\sum_{n=-\infty}^{\infty} (-1)^n q^{(6n+1)^2} \left(x^{6n+1} + x^{-6n-1}\right) = \frac{q}{x} \frac{(x^4, q^{24}x^{-4}, q^{24}; q^{24})_\infty}{(x^2, q^{24}x^{-2}; q^{24})_\infty}.$$

Thus, (7.8) and (7.9) may be combined to give

$$\begin{aligned} S_2 + S_0 &= \sum_{n \equiv 0 \text{ or } 2 \pmod 7} (-1)^n q^{(6n+1)^2} \\ &= q^{36} \sum_{k=-\infty}^{\infty} (-1)^n q^{49(6k+1)^2} \left((q^{84})^{6k+1} + (q^{84})^{-6k-1}\right) \\ &= q \frac{(q^{336}, q^{840}; q^{1176})_\infty}{(q^{168}, q^{1008}; q^{1176})_\infty} \times (q^{1176}; q^{1176})_\infty. \end{aligned} \tag{7.15}$$

Similar calculations applied to (7.10) and (7.11) yield

$$S_3 + S_6 = -q^{25} \frac{(q^{504}, q^{672}; q^{1176})_\infty}{(q^{336}, q^{840}; q^{1176})_\infty} \times (q^{1176}; q^{1176})_\infty \tag{7.16}$$

while (7.12) and (7.13) lead to

$$S_4 + S_5 = q^{121} \frac{(q^{168}, q^{1008}; q^{1176})_\infty}{(q^{504}, q^{672}; q^{1176})_\infty} \times (q^{1176}; q^{1176})_\infty. \tag{7.17}$$

Lastly, by (7.7) and (7.14), we have

$$S_1 = -\sum_{k=-\infty}^{\infty} (-1)^k q^{49(6k+1)^2} = -q^{49} (q^{1176}; q^{1176})_\infty. \tag{7.18}$$

On adding (7.15)–(7.18) and substituting the result back into (7.7) we deduce that

$$\begin{aligned} q(q^{24}; q^{24})_\infty &= (q^{1176}; q^{1176})_\infty \\ &\times \left\{ q \frac{(q^{336}, q^{840}; q^{1176})_\infty}{(q^{168}, q^{1008}; q^{1176})_\infty} - q^{25} \frac{(q^{504}, q^{672}; q^{1176})_\infty}{(q^{336}, q^{840}; q^{1176})_\infty} - q^{49} + q^{121} \frac{(q^{168}, q^{1008}; q^{1176})_\infty}{(q^{504}, q^{672}; q^{1176})_\infty} \right\}. \end{aligned}$$

On dividing both sides by $q^{49}(q^{1176}; q^{1176})_\infty$ and then replacing $q^{168}$ with $q$, we obtain (7.6). □

The cube of Euler's product may also be 7-dissected and this leads to the next theorem.

**Theorem 7.2.** *Let $\alpha$, $\beta$, and $\gamma$ be defined by* (7.1)–(7.3). *Then*

$$\alpha\beta^2 - \alpha^2 + \gamma = 0, \tag{7.19}$$

$$\beta\gamma^2 - \beta^2 + \alpha = 0, \tag{7.20}$$

*and*

$$\gamma\alpha^2 - \gamma^2 + \beta = 0. \tag{7.21}$$

*Proof.* Let us write

$$\prod_{j=1}^{\infty}(1-q^j) = J_0 + J_1 + J_2 + J_3 + J_4 + J_5 + J_6 \tag{7.22}$$

and

$$\prod_{j=1}^{\infty}(1-q^j)^3 = K_0 + K_1 + K_2 + K_3 + K_4 + K_5 + K_6 \tag{7.23}$$

where $J_r$ and $K_r$ are power series that contain only terms in which the power of $q$ is congruent to $r$ modulo 7. By (7.6) we have

$$J_0 = q^2\,\alpha(q^7)\prod_{j=1}^{\infty}(1-q^{49j}), \tag{7.24}$$

$$J_1 = q^2\,\beta(q^7)\prod_{j=1}^{\infty}(1-q^{49j}), \tag{7.25}$$

$$J_5 = q^2\,\gamma(q^7)\prod_{j=1}^{\infty}(1-q^{49j}), \tag{7.26}$$

$$J_2 = -q^2\prod_{j=1}^{\infty}(1-q^{49j}) \tag{7.27}$$

and $J_3 = J_4 = J_6 = 0$. By (0.46) we have

$$\prod_{j=1}^{\infty}(1-q^j)^3 = \sum_{j=-\infty}^{\infty}(4j+1)q^{2j^2+j}$$

and since $2j^2 + j \equiv 0, 1, 3$ or $6 \pmod 7$, it follows that $K_2 = K_4 = K_5 = 0$. Thus, (7.22) and (7.23) simplify to

$$\prod_{j=1}^{\infty}(1-q^j) = J_0 + J_1 + J_2 + J_5 \tag{7.28}$$

and

$$\prod_{j=1}^{\infty}(1-q^j)^3 = K_0 + K_1 + K_3 + K_6, \tag{7.29}$$

respectively. Now take the cube of (7.28) and collect the terms according to the powers of $q$ modulo 7 to obtain

$$\begin{aligned}\prod_{j=1}^{\infty}(1-q^j)^3 &= (J_0^3 + 3J_1^2J_5 + 6J_0J_2J_5) + (3J_0^2J_1 + 6J_1J_2J_5 + J_5^3)\\ &\quad + 3(J_0J_1^2 + J_0^2J_2 + J_2^2J_5) + (6J_0J_1J_2 + J_1^3 + 3J_0J_5^2)\\ &\quad + 3(J_0J_2^2 + J_1^2J_2 + J_1J_5^2) + 3(J_0^2J_5 + J_1J_2^2 + J_2J_5^2)\\ &\quad + (J_2^3 + 6J_0J_1J_5).\end{aligned} \tag{7.30}$$

On comparing terms in the expansions (7.29) and (7.30) it follows that

$$J_0J_1^2 + J_0^2J_2 + J_2^2J_5 = J_0J_2^2 + J_1^2J_2 + J_1J_5^2 = J_0^2J_5 + J_1J_2^2 + J_2J_5^2 = 0.$$

On using (7.24)–(7.27) to express $J_0$, $J_1$, $J_2$, and $J_5$ in terms of $\alpha$, $\beta$, and $\gamma$, we complete the proof. □

A different proof of Theorem 7.2 is outlined in Exercise 1 at the end of the chapter.

Theorem 7.1 provides expressions for two symmetric functions of $\alpha$, $\beta$, and $\gamma$, namely

$$\alpha\beta\gamma = -1 \qquad \text{and} \qquad \alpha + \beta + \gamma = v + 1, \tag{7.31}$$

where $v$ is the eta-quotient defined by (7.4). The next result provides the corresponding expression for the symmetric function $\alpha\beta + \beta\gamma + \gamma\alpha$.

**Theorem 7.3.** *Let $\alpha$, $\beta$, $\gamma$, and $v$ defined by* (7.1)–(7.4). *Then*

$$\alpha\beta + \beta\gamma + \gamma\alpha = -\frac{1}{2}\left(3v + 4 + \sqrt{4v^3 + 21v^2 + 28v}\right).$$

*Proof.* Let $M = \alpha\beta + \beta\gamma + \gamma\alpha$. Multiply (7.19), (7.20), and (7.21) by $\alpha$, $\beta$, and $\gamma$, respectively, and add to deduce that

$$(\alpha^2\beta^2 + \beta^2\gamma^2 + \gamma^2\alpha^2) - (\alpha^3 + \beta^3 + \gamma^3) + (\alpha\beta + \beta\gamma + \gamma\alpha) = 0. \tag{7.32}$$

Now,

$$\alpha^2\beta^2 + \beta^2\gamma^2 + \gamma^2\alpha^2 = (\alpha\beta + \beta\gamma + \gamma\alpha)^2 - 2(\alpha + \beta + \gamma)\alpha\beta\gamma$$

and

$$\alpha^3 + \beta^3 + \gamma^3 = (\alpha + \beta + \gamma)^3 - 3(\alpha + \beta + \gamma)(\alpha\beta + \beta\gamma + \gamma\alpha) + 3\alpha\beta\gamma.$$

By (7.31) and the definition of $M$, these simplify to

$$\alpha^2\beta^2 + \beta^2\gamma^2 + \gamma^2\alpha^2 = M^2 + 2(v + 1)$$

and

$$\alpha^3 + \beta^3 + \gamma^3 = (v + 1)^3 - 3(v + 1)M - 3,$$

respectively. Hence, (7.32) simplifies to

$$M^2 + 2(v + 1) - (v + 1)^3 + 3(v + 1)M + 3 + M = 0.$$

Solving the quadratic equation for $M$ we obtain

$$M = -\frac{1}{2}\left(3v + 4 \pm \sqrt{4v^3 + 21v^2 + 28v}\right). \tag{7.33}$$

To determine the sign of the square root we expand both sides in powers of $q$ and compare the lowest powers. From (7.1)–(7.4) we have

$$\alpha = q^{-2/7} + \cdots, \quad \beta = -q^{-1/7} + \cdots, \quad \gamma = q^{3/7} + \cdots, \quad v = q^{-2/7} + \cdots$$

and so

$$M = \alpha\beta + \beta\gamma + \gamma\alpha = -q^{-3/7} + \cdots$$

and

$$\sqrt{4v^3 + 21v^2 + 28v} = 2q^{-3/7} + \cdots.$$

It follows that the lowest power of $q$ that occurs on either side of (7.33) is $-q^{-3/7}$. On comparing coefficients of $q^{-3/7}$ in (7.33), we deduce that the plus sign in front of the square root should be selected. □

## 7.3 The allied functions $r_1$, $r_2$, and $r_3$

To help develop further properties of $\alpha$, $\beta$, and $\gamma$, we introduce the functions $r_1$, $r_2$, and $r_3$ defined by

$$r_1 = \alpha^3\beta, \quad r_2 = \beta^3\gamma \quad \text{and} \quad r_3 = \gamma^3\alpha. \tag{7.34}$$

We will require two simple lemmas.

**Lemma 7.4.** *The functions $r_1$, $r_2$ and $r_3$ satisfy the identities*

$$r_1 r_2 r_3 = 1, \tag{7.35}$$

$$r_1 r_2 = -r_1 - 1, \quad r_2 r_3 = -r_2 - 1 \quad \textit{and} \quad r_3 r_1 = -r_3 - 1. \tag{7.36}$$

*Proof.* From the definition (7.34) and the identity (7.5) we have

$$r_1 r_2 r_3 = (\alpha^3\beta)(\beta^3\gamma)(\gamma^3\alpha) = (\alpha\beta\gamma)^4 = 1.$$

Next, by the definition (7.34) and the identity (7.20) we have

$$r_1 r_2 = \alpha^3\beta^4\gamma = \alpha^4\beta^2\gamma + \alpha^3\beta^3\gamma^3.$$

This may be simplified using the property $\alpha\beta\gamma = -1$ from (7.5) to give

$$r_1 r_2 = -\alpha^3\beta - 1 = r_1 - 1,$$

where the definition of $r_1$ has been used in the very last step. The other identities for $r_2 r_3$ and $r_3 r_1$ may be proved by the same method. □

**Lemma 7.5.** *Let $\alpha$, $\beta$, $\gamma$, $r_1$, $r_2$, and $r_3$ be defined by* (7.1)–(7.3) *and* (7.34). *Then*

$$\alpha^2\beta^3 = r_1 + 1, \qquad \beta^2\gamma^3 = r_2 + 1, \qquad \gamma^2\alpha^3 = r_3 + 1, \tag{7.37}$$

$$\alpha\beta^5 = r_1 - r_2 + 1, \qquad \beta\gamma^5 = r_2 - r_3 + 1, \qquad \gamma\alpha^5 = r_3 - r_1 + 1, \tag{7.38}$$

$$\alpha^7 = r_1^2 - r_1 + r_3 + 1, \quad \beta^7 = r_2^2 - r_2 + r_1 + 1, \quad \gamma^7 = r_3^2 - r_3 + r_2 + 1 \tag{7.39}$$

*and*

$$\alpha^7\beta^7 = r_1(r_1+1)^2, \qquad \beta^7\gamma^7 = r_2(r_2+1)^2, \qquad \gamma^7\alpha^7 = r_3(r_3+1)^2. \tag{7.40}$$

*Proof.* We prove the first identity in each of (7.37)–(7.40); the others can be proved by the same method using symmetry.
First, by the definitions of $r_1$ and $r_2$ together with Lemma 7.4, we have

$$\alpha^2\beta^3 = (\alpha^3\beta)(\beta^3\gamma)/(\alpha\beta\gamma) = -r_1 r_2 = r_1 + 1.$$

Second, using both parts of Lemma 7.4 and the definitions of $r_1$ and $r_2$, we find that

$$\alpha\beta^5 = \alpha^3\beta^7\gamma^2 = (\alpha^3\beta)(\beta^3\gamma)^2 = r_1r_2^2 = (-r_1 - 1)r_2 = r_1 - r_2 + 1.$$

Third, by (7.19) we have

$$\alpha^7 = \alpha^5(\alpha\beta^2 + \gamma) = \alpha^6\beta^2 + \alpha^5\gamma.$$

Now apply the definition of $r_1$ and the identity (7.38) to obtain

$$\alpha^7 = r_1^2 + (r_3 - r_1 + 1).$$

Fourth, by the definition of $r_1$ and (7.37) we have

$$\alpha^7\beta^7 = (\alpha^3\beta)(\alpha^2\beta^3)^2 = r_1(r_1 + 1)^2.$$

□

The next result relates $r_1$, $r_2$, and $r_3$ to Dedekind's eta function.

**Theorem 7.6.** *The functions $r_1$, $r_2$, and $r_3$ defined by* (7.34) *satisfy the identities*

$$r_1 + r_2 + r_3 + 8 = -\frac{1}{q}\prod_{j=1}^{\infty}\frac{(1-q^j)^4}{(1-q^{7j})^4} \tag{7.41}$$

*and*

$$r_1r_2 + r_2r_3 + r_3r_1 - 5 = \frac{1}{q}\prod_{j=1}^{\infty}\frac{(1-q^j)^4}{(1-q^{7j})^4}. \tag{7.42}$$

*Proof.* Write (7.6) in the form

$$v = \alpha + \beta + \gamma - 1.$$

Replace $q^{1/7}$ with $\omega^m q^{1/7}$ where $\omega = \exp(2\pi i/7)$ and $m \in \{0, 1, 2, 3, 4, 5, 6\}$, and multiply the resulting seven equalities together. The left-hand side is, by (0.6),

$$\prod_{m=0}^{6}\left(\frac{1}{\omega^{2m}q^{2/7}}\prod_{j=1}^{\infty}\frac{(1-\omega^m q^{j/7})}{(1-q^{7j})}\right) = \frac{1}{q^2}\prod_{j=1}^{\infty}\frac{(1-q^j)^8}{(1-q^{7j})^8}. \tag{7.43}$$

The right-hand side is

$$\prod_{m=0}^{6}(\alpha\omega^{5m}+\beta\omega^{6m}+\gamma\omega^{3m}-1)=\begin{vmatrix} -1 & 0 & 0 & \gamma & 0 & \alpha & \beta \\ \beta & -1 & 0 & 0 & \gamma & 0 & \alpha \\ \alpha & \beta & -1 & 0 & 0 & \gamma & 0 \\ 0 & \alpha & \beta & -1 & 0 & 0 & \gamma \\ \gamma & 0 & \alpha & \beta & -1 & 0 & 0 \\ 0 & \gamma & 0 & \alpha & \beta & -1 & 0 \\ 0 & 0 & \gamma & 0 & \alpha & \beta & -1 \end{vmatrix}$$

$$\begin{aligned} &= \alpha^7+\beta^7+\gamma^7+7\alpha\beta\gamma(\alpha^3\beta+\beta^3\gamma+\gamma^3\alpha)+7(\alpha^5\gamma+\beta^5\alpha+\gamma^5\beta) \\ &\quad +7\alpha^2\beta^2\gamma^2+14(\alpha^3\gamma^2+\beta^3\alpha^2+\gamma^3\beta^2)+7(\alpha^3\beta+\beta^3\gamma+\gamma^3\alpha) \\ &\quad +14\alpha\beta\gamma-1, \end{aligned} \tag{7.44}$$

on evaluating the determinant. On combining (7.43) and (7.44) and simplifying using (7.5) we obtain

$$\begin{aligned} &\frac{1}{q^2}\prod_{j=1}^{\infty}\frac{(1-q^j)^8}{(1-q^{7j})^8} \\ &= \alpha^7+\beta^7+\gamma^7+7(\alpha^5\gamma+\beta^5\alpha+\gamma^5\beta)+14(\alpha^3\gamma^2+\beta^3\alpha^2+\gamma^3\beta^2)-8. \end{aligned}$$

By (7.37)–(7.39) this becomes

$$\begin{aligned} &\frac{1}{q^2}\prod_{j=1}^{\infty}\frac{(1-q^j)^8}{(1-q^{7j})^8} \\ &= r_1^2+r_2^2+r_3^2+14(r_1+r_2+r_3)+58 \\ &= (r_1+r_2+r_3)^2-2(r_1r_2+r_2r_3+r_3r_1)+14(r_1+r_2+r_3)+58. \end{aligned}$$

By (7.36) this simplifies further to

$$\begin{aligned} \frac{1}{q^2}\prod_{j=1}^{\infty}\frac{(1-q^j)^8}{(1-q^{7j})^8} &= (r_1+r_2+r_3)^2+16(r_1+r_2+r_3)+64 \\ &= (r_1+r_2+r_3+8)^2, \end{aligned}$$

hence

$$\frac{1}{q}\prod_{j=1}^{\infty}\frac{(1-q^j)^4}{(1-q^{7j})^4}=\pm(r_1+r_2+r_3+8). \tag{7.45}$$

In order to determine the sign, we use (7.1) and (7.3) to deduce that the first terms in the $q$-expansions are given by

$$r_1 = -\frac{1}{q} + \cdots, \qquad r_2 = -1 + \cdots \quad \text{and} \quad r_3 = q + \cdots.$$

It follows that the minus sign should be used in (7.45), and this proves (7.41). Then (7.42) follows from (7.41) by using (7.36). □

The results in Theorem 7.6 may be expressed in terms of $\alpha$, $\beta$, and $\gamma$, as follows.

**Corollary 7.7.** *Let $\eta_k$, $\alpha$, $\beta$, and $\gamma$ be defined by* (0.39) *and* (7.1)–(7.3). *Then,*

$$\alpha^3\beta + \beta^3\gamma + \gamma^3\alpha = -8 - \frac{\eta_1^4}{\eta_7^4} \tag{7.46}$$

*and*

$$\alpha^3\gamma^2 + \beta^3\alpha^2 + \gamma^3\beta^2 = -5 - \frac{\eta_1^4}{\eta_7^4}. \tag{7.47}$$

Here are two further results that relate symmetric functions of $\alpha$, $\beta$, and $\gamma$ to Dedekind's eta function.

**Theorem 7.8.** *If $\alpha$, $\beta$, and $\gamma$ are given by* (7.1)–(7.3), *then*

$$\alpha^7 + \beta^7 + \gamma^7 = 57 + 14\frac{\eta_1^4}{\eta_7^4} + \frac{\eta_1^8}{\eta_7^8} \tag{7.48}$$

*and*

$$\alpha^7\beta^7 + \beta^7\gamma^7 + \gamma^7\alpha^7 = -\left(289 + 126\frac{\eta_1^4}{\eta_7^4} + 19\frac{\eta_1^8}{\eta_7^8} + \frac{\eta_1^{12}}{\eta_7^{12}}\right).$$

*Proof.* By (7.39) we have

$$\begin{aligned}\alpha^7 + \beta^7 + \gamma^7 &= r_1^2 + r_2^2 + r_3^2 + 3\\ &= (r_1 + r_2 + r_3)^2 - 2(r_1r_2 + r_2r_3 + r_3r_1) + 3.\end{aligned}$$

On applying Theorem 7.6 this becomes

$$\alpha^7 + \beta^7 + \gamma^7 = \left(-8 - \frac{\eta_1^4}{\eta_7^4}\right)^2 - 2\left(5 + \frac{\eta_1^4}{\eta_7^4}\right) + 3 = 57 + 14\frac{\eta_1^4}{\eta_7^4} + \frac{\eta_1^8}{\eta_7^4}.$$

This proves the first result. Next, by (7.40) and some rearrangements, we have

$$\begin{aligned}&\alpha^7\beta^7 + \beta^7\gamma^7 + \gamma^7\alpha^7 = r_1(r_1+1)^2 + r_2(r_2+1)^2 + r_3(r_3+1)^2\\ &= (r_1^3 + r_2^3 + r_3^3) + 2(r_1^2 + r_2^2 + r_3^2) + (r_1 + r_2 + r_3)\end{aligned}$$

$$= (r_1 + r_2 + r_3)^3 - 3(r_1 + r_2 + r_3)(r_1 r_2 + r_2 r_3 + r_3 r_1) + 3 r_1 r_2 r_3$$
$$+ 2(r_1 + r_2 + r_3)^2 - 4(r_1 r_2 + r_2 r_3 + r_3 r_1) + (r_1 + r_2 + r_3).$$

Now apply (7.35), (7.41), and (7.42) and simplify to complete the proof. □

The next goal is to derive the seventh order modular equation. We will need two lemmas.

**Lemma 7.9.** *Let* $\alpha$, $\beta$, *and* $\gamma$ *be defined by* (7.1)–(7.3). *Let*

$$L = \alpha + \beta + \gamma \qquad \text{and} \qquad M = \alpha\beta + \beta\gamma + \gamma\alpha. \tag{7.49}$$

*Then*

$$\alpha^2\beta + \beta^2\gamma + \gamma^2\alpha = LM + 3 + L - L^2 + 2M \tag{7.50}$$

*and*

$$\alpha^3\beta + \beta^3\gamma + \gamma^3\alpha = L^2M - 2M^2 - M - LM + L^2 - 3. \tag{7.51}$$

*Proof.* By elementary algebra we have

$$\begin{aligned} &\alpha^2\beta + \beta^2\gamma + \gamma^2\alpha \\ &= (\alpha + \beta + \gamma)(\alpha\beta + \beta\gamma + \gamma\alpha) - 3\alpha\beta\gamma - (\alpha\beta^2 + \beta\gamma^2 + \gamma\alpha^2). \end{aligned}$$

By the definitions of $L$ and $M$ together with (7.5) and Theorem 7.2 this simplifies to

$$\begin{aligned} &\alpha^2\beta + \beta^2\gamma + \gamma^2\alpha \\ &= LM + 3 - (\alpha^2 - \gamma + \beta^2 - \alpha + \gamma^2 - \beta) \\ &= LM + 3 + L - (\alpha^2 + \beta^2 + \gamma^2). \end{aligned}$$

By elementary algebra again, together with the definitions of $L$ and $M$, this simplifies further to

$$\begin{aligned} &\alpha^2\beta + \beta^2\gamma + \gamma^2\alpha \\ &= LM + 3 + L - (\alpha + \beta + \gamma)^2 + 2(\alpha\beta + \beta\gamma + \gamma\alpha) \\ &= LM + 3 + L - L^2 + 2M. \end{aligned}$$

This proves (7.50).

Again by elementary algebra, (7.5) and the definitions of $L$ and $M$ we have

$$\begin{aligned}
&\alpha^3\beta + \beta^3\gamma + \gamma^3\alpha \\
&= (\alpha^2+\beta^2+\gamma^2)(\alpha\beta+\beta\gamma+\gamma\alpha) - \alpha\beta\gamma(\alpha+\beta+\gamma) - (\alpha\beta^3+\beta\gamma^3+\gamma\alpha^3) \\
&= (L^2-2M)M + L - (\alpha\beta^3+\beta\gamma^3+\gamma\alpha^3).
\end{aligned}$$

Now apply Theorem 7.2 and the definition of $M$ to obtain

$$\begin{aligned}
&\alpha^3\beta + \beta^3\gamma + \gamma^3\alpha \\
&= (L^2-2M)M + L - (\alpha^2\beta - \gamma\beta + \beta^2\gamma - \alpha\gamma + \gamma^2\alpha - \beta\alpha) \\
&= (L^2-2M)M + L + M - (\alpha^2\beta + \beta^2\gamma + \gamma^2\alpha).
\end{aligned}$$

The proof of (7.51) may be completed by substituting the result of (7.50). □

**Lemma 7.10.** *Let $v$ be defined by* (7.4), *that is,* $v = \eta_{1/7}/\eta_7$. *Then*

$$\frac{\eta_1^4}{\eta_7^4} = \frac{1}{2}\left(7v(v^2+5v+7) + (v^2+7v+7)\sqrt{4v^3+21v^2+28v}\right) \tag{7.52}$$

*and*

$$\begin{aligned}
\frac{\eta_1^8}{\eta_7^8} = \frac{1}{2}\Big(&2v^7 + 63v^6 + 532v^5 + 2009v^4 + 3724v^3 + 3087v^2 + 686v \\
&+ 7v(v^2+5v+7)(v^2+7v+7)\sqrt{4v^3+21v^2+28v}\Big).
\end{aligned} \tag{7.53}$$

*Proof.* By Corollary 7.7 and Lemma 7.9 we have

$$\begin{aligned}
\frac{\eta_1^4}{\eta_7^4} &= -(\alpha^3\beta + \beta^3\gamma + \gamma^3\alpha + 8) \\
&= -(L^2M - 2M^2 - M - LM + L^2 + 5),
\end{aligned}$$

where $L$ and $M$ are defined by (7.49). Expressions for $L$ and $M$ in terms of $v$ are given by Theorems 7.1 and 7.3, respectively. On using these in the above and simplifying we obtain (7.52). The identity (7.53) may be obtained by squaring (7.52). □

The next result may be compared with the fifth order modular equation in Exercise 11 of Chapter 5.

**Theorem 7.11 (Seventh order modular equation).** *Let* $v = \eta_{1/7}/\eta_7$. *Then*

$$\frac{\eta_1^8}{\eta_7^8} - 7v(v^2+5v+7)\frac{\eta_1^4}{\eta_7^4}$$
$$= v^7 + 7v^6 + 21v^5 + 49v^4 + 147v^3 + 343v^2 + 343v.$$

*Hence, if* $X = \eta_1/\eta_7$ *and* $Y = \eta_7/\eta_{49}$, *then*

$$Y^8 - 7(X^3Y^3 + 5X^2Y^2 + 7XY)\,Y^4$$
$$= X^7Y^7 + 7X^6Y^6 + 21X^5Y^5 + 49X^4Y^4 + 147X^3Y^3 + 343X^2Y^2 + 343XY.$$

*Proof.* The first identity follows from Lemma 7.10 by eliminating the terms that involve square roots. The second identity is obtained from the first by replacing $q$ with $q^7$. □

## 7.4 Ramanujan's modulo 7 partition identity

In this section we prove two identities relating series to eta-quotients, and use them to deduce Ramanujan's partition congruence modulo 7.

**Theorem 7.12.** *The following identities hold*:

$$\sum_{j=1}^{\infty}\left(\frac{j}{7}\right)\frac{q^j(1+q^j)}{(1-q^j)^3} = \eta_1^3\eta_7^3 + 8\frac{\eta_7^7}{\eta_1}, \tag{7.54}$$

$$8 - 7\sum_{j=1}^{\infty}\left(\frac{j}{7}\right)\frac{j^2q^j}{(1-q^j)} = 8\frac{\eta_1^7}{\eta_7} + 49\eta_1^3\eta_7^3. \tag{7.55}$$

*Proof.* In Theorem 1.22 replace $q$ with $q^{7/2}$ and then successively let $a = q^{5/2}, q^{3/2}$, and $q^{1/2}$ to obtain

$$q\frac{(q^2,q^5;q^7)_\infty(q^7;q^7)_\infty^6}{(q,q^6;q^7)_\infty^4} = \sum_{j=1}^{\infty}\frac{j^2(q^j - q^{6j})}{1-q^{7j}}$$
$$= \sum_{j=1}^{\infty}\sum_{k=1}^{\infty} j^2\left(q^{j(7k-6)} - q^{j(7k-1)}\right), \tag{7.56}$$

$$q^2\frac{(q^3,q^4;q^7)_\infty(q^7;q^7)_\infty^6}{(q^2,q^5;q^7)_\infty^4} = \sum_{j=1}^{\infty}\frac{j^2(q^{2j}-q^{5j})}{1-q^{7j}}$$
$$= \sum_{j=1}^{\infty}\sum_{k=1}^{\infty} j^2\left(q^{j(7k-5)}-q^{j(7k-2)}\right) \qquad (7.57)$$

and

$$q^3\frac{(q,q^6;q^7)_\infty(q^7;q^7)_\infty^6}{(q^3,q^4;q^7)_\infty^4} = \sum_{j=1}^{\infty}\frac{j^2(q^{3j}-q^{4j})}{1-q^{7j}}$$
$$= \sum_{j=1}^{\infty}\sum_{k=1}^{\infty} j^2\left(q^{j(7k-4)}-q^{j(7k-3)}\right). \qquad (7.58)$$

By the definitions (7.1)–(7.3) we see that the products on the left-hand sides of (7.56)–(7.58) are

$$\frac{\eta_7^7}{\eta_1}\frac{\alpha^2}{\gamma}, \qquad \frac{\eta_7^7}{\eta_1}\frac{\beta^2}{\alpha} \qquad \text{and} \qquad -\frac{\eta_7^7}{\eta_1}\frac{\gamma^2}{\beta},$$

respectively. Therefore, on adding (7.56) and (7.57) and subtracting (7.58), we deduce

$$\frac{\eta_7^7}{\eta_1}\left(\frac{\alpha^2}{\gamma}+\frac{\beta^2}{\alpha}+\frac{\gamma^2}{\beta}\right) = \sum_{j=1}^{\infty}\sum_{k=1}^{\infty} j^2\left(\frac{k}{7}\right)q^{jk} = \sum_{k=1}^{\infty}\left(\frac{k}{7}\right)\frac{q^k(1+q^k)}{(1-q^k)^3}, \qquad (7.59)$$

where the last equality is obtained by interchanging the order of summation and using

$$\sum_{j=1}^{\infty} j^2x^j = \frac{x(1+x)}{(1-x)^3}.$$

By Theorem 7.1 and Corollary 7.7, we have

$$\frac{\eta_7^7}{\eta_1}\left(\frac{\alpha^2}{\gamma}+\frac{\beta^2}{\alpha}+\frac{\gamma^2}{\beta}\right) = \frac{\eta_7^7}{\eta_1}\left(\frac{\alpha^3\beta+\beta^3\gamma+\gamma^3\alpha}{\alpha\beta\gamma}\right) = \eta_1^3\eta_7^3 + 8\frac{\eta_7^7}{\eta_1}. \qquad (7.60)$$

On combining (7.59) and (7.60), we obtain (7.54).

The identity (7.55) can be deduced from (7.54) by a modular transformation as follows. The left-hand side of (7.54) is

$$\sum_{j=1}^{\infty}\left(\frac{j}{7}\right)\frac{q^j(1+q^j)}{(1-q^j)^3}=\sum_{j=1}^{\infty}\sum_{k=1}^{\infty}\left(\frac{j}{7}\right)k^2q^{jk}=\sum_{k=1}^{\infty}k^2\sum_{j=1}^{\infty}\left(\frac{j}{7}\right)q^{jk}$$

$$=\sum_{k=1}^{\infty}\frac{k^2}{1-q^{7k}}\sum_{j=1}^{6}\left(\frac{j}{7}\right)q^{jk}=E_3^0(\tau,7),$$

where $E_3^0(\tau,7)$ is the Eisenstein series defined by (2.43). Hence the identity (7.54) can be expressed in the form

$$E_3^0(\tau,7)=\eta^3(\tau)\eta^3(7\tau)+8\,\frac{\eta^7(7\tau)}{\eta(\tau)}.$$

Now replace $\tau$ with $-1/7\tau$ to obtain

$$E_3^0\left(\frac{-1}{7\tau},7\right)=\eta^3\left(\frac{-1}{7\tau}\right)\eta^3\left(\frac{-1}{\tau}\right)+8\,\frac{\eta^7\left(\frac{-1}{\tau}\right)}{\eta\left(\frac{-1}{7\tau}\right)}.$$

The left-hand side may be transformed using Theorem 2.21, and Theorem 2.8 can be used to transform the eta-quotients on the right. This gives, after multiplying by $-i\sqrt{7}/\tau^3$ and simplifying,

$$-7\,E_3^{\infty}(\tau,7)=49\eta^3(\tau)\eta^3(7\tau)+8\,\frac{\eta^7(\tau)}{\eta(7\tau)}.$$

Writing each side in terms of $q$ using the definitions (2.20) and (2.44) we obtain

$$\frac{7B_{3,7}}{6}-7\sum_{j=1}^{\infty}\left(\frac{j}{7}\right)\frac{j^2q^j}{1-q^j}=49q\prod_{j=1}^{\infty}(1-q^j)^3(1-q^{7j})^3+8\prod_{j=1}^{\infty}\frac{(1-q^j)^7}{(1-q^{7j})},$$

where $B_{3,7}$ is a generalized Bernoulli number given by (2.42). Letting $q=0$ we deduce that $\frac{7}{6}B_{3/7}=8$, and this completes the proof of (7.55). □

The next result is the level 7 analogue of Hardy's favorite result of Ramanujan in Corollary 5.14 and (5.132).

**Corollary 7.13.** *Let $p(n)$ denote the number of partitions of $n$. Then,*

$$\sum_{n=0}^{\infty}p(7n+5)q^n=7\prod_{j=1}^{\infty}\frac{(1-q^{7j})^3}{(1-q^j)^4}+49q\prod_{j=1}^{\infty}\frac{(1-q^{7j})^7}{(1-q^j)^8}.$$

*In particular, $p(7n+5)$ is divisible by 7.*

*Proof.* Start with the identity (7.54) in the form

$$\sum_{j=1}^{\infty}\sum_{k=1}^{\infty}\left(\frac{j}{7}\right) k^2 q^{jk} = q\prod_{j=1}^{\infty}(1-q^j)^3(1-q^{7j})^3 + 8q^2\prod_{j=1}^{\infty}\frac{(1-q^{7j})^7}{(1-q^j)}. \tag{7.61}$$

Two of the products on the right-hand side may be expanded using (0.46) and Theorem 0.11 to give

$$\begin{aligned}\sum_{j=1}^{\infty}\sum_{k=1}^{\infty}\left(\frac{j}{7}\right) k^2 q^{jk} &= q\left(\sum_{n=-\infty}^{\infty}(4n+1)\, q^{2n^2+n}\right)\left(\prod_{j=1}^{\infty}(1-q^{7j})^3\right)\\ &\quad + 8\left(\sum_{n=0}^{\infty}p(n)q^{n+2}\right)\left(\prod_{j=1}^{\infty}(1-q^{7j})^7\right).\end{aligned}$$

Equate terms on both sides for which the power of $q$ is a multiple of 7, and note that $2n^2+n+1 \equiv 0 \pmod 7$ if and only if $n=-7k-2$, to deduce that

$$\begin{aligned}49\sum_{j=1}^{\infty}\sum_{k=1}^{\infty}\left(\frac{j}{7}\right) k^2 q^{7jk} &= -7q^7\left(\sum_{k=-\infty}^{\infty}(4k+1)\, q^{49(2k^2+k)}\right)\left(\prod_{j=1}^{\infty}(1-q^{7j})^3\right)\\ &\quad + 8\left(\sum_{n=0}^{\infty}p(7n+5)q^{7n+7}\right)\left(\prod_{j=1}^{\infty}(1-q^{7j})^7\right).\end{aligned}$$

The left-hand side may be summed using (7.61), and the series in the first term on the right-hand side can be summed using (0.46). This gives

$$\begin{aligned}&49\left(q^7\prod_{j=1}^{\infty}(1-q^{7j})^3(1-q^{49j})^3 + 8q^{14}\prod_{j=1}^{\infty}\frac{(1-q^{49j})^7}{(1-q^{7j})}\right)\\ &= -7q^7\prod_{j=1}^{\infty}(1-q^{7j})^3(1-q^{49j})^3 + 8\left(\sum_{n=0}^{\infty}p(7n+5)q^{7n+7}\right)\left(\prod_{j=1}^{\infty}(1-q^{7j})^7\right).\end{aligned}$$

The proof may be completed by replacing $q^7$ with $q$, simplifying, and rearranging. □

## 7.5 Weight one-half theta functions

In this section we briefly introduce three weight one-half theta functions and establish their connection with Klein's quartic curve. Let $h_1$, $h_2$, and $h_3$ be defined either as products or by series by

$$h_1 = q^{1/56}(q^3, q^4, q^7; q^7)_\infty = \sum_{j=-\infty}^{\infty} (-1)^j q^{(14j+1)^2/56},$$

$$h_2 = -q^{25/56}(q, q^6, q^7; q^7)_\infty = \sum_{j=-\infty}^{\infty} (-1)^{j+1} q^{(14j+5)^2/56}$$

and

$$h_3 = q^{9/56}(q^2, q^5, q^7; q^7)_\infty = \sum_{j=-\infty}^{\infty} (-1)^j q^{(14j+3)^2/56}.$$

Each product is equal to the corresponding series by Jacobi's triple product identity. The functions $\alpha$, $\beta$, and $\gamma$ may be expressed in terms of $h_1$, $h_2$, and $h_3$ by

$$\alpha = -\frac{h_3}{h_2}, \qquad \beta = -\frac{h_1}{h_3} \quad \text{and} \quad \gamma = -\frac{h_2}{h_1}.$$

Conversely, the $h_j$ may be expressed in terms of $\alpha$, $\beta$, and $\gamma$ and Dedekind's eta function by the formulas

$$h_1^3 = (\alpha\beta^2)(\eta_1\eta_7^2), \quad h_2^3 = (\beta\gamma^2)(\eta_1\eta_7^2) \quad \text{and} \quad h_3^3 = (\gamma\alpha^2)(\eta_1\eta_7^2).$$

**Theorem 7.14.** *The following identities hold*:

$$\frac{h_1^2}{h_3} + \frac{h_2^2}{h_1} + \frac{h_3^2}{h_2} = 0 \tag{7.62}$$

*and*

$$\frac{h_3}{h_1^2} + \frac{h_1}{h_2^2} + \frac{h_2}{h_3^2} = \frac{\eta_1}{\eta_7^2}\left(\frac{\eta_1^4}{\eta_7^4} + 13 + 49\frac{\eta_7^4}{\eta_1^4}\right)^{1/3}. \tag{7.63}$$

*Proof.* Replace $q$ with $q^7$ and then take $(b, c, d, e) = (q, q^3, q^3, q^3)$ in the explicit form of the Weierstrass three-term identity in Theorem 1.20 to deduce that

$$\begin{aligned}(q^2, q^2, q^2, q^3, q^4, q^5, q^5, q^5; q^7)_\infty - (q, q^3, q^3, q^3, q^4, q^4, q^4, q^6; q^7)_\infty \\ = q(q, q, q, q^2, q^5, q^6, q^6, q^6; q^7)_\infty.\end{aligned} \tag{7.64}$$

This is equivalent to

$$\frac{(q^3,q^4;q^7)_\infty^2}{(q^2,q^5;q^7)_\infty}+q\frac{(q,q^6;q^7)_\infty^2}{(q^3,q^4;q^7)_\infty}-\frac{(q^2,q^5;q^7)_\infty^2}{(q,q^6;q^7)_\infty}=0$$

which in turn is equivalent to (7.62).

The identity (7.63) is simply an aggregation of two results that will be proved in the next two sections and is left to Exercise 3 at the end of the chapter. The identity (7.63) is not used in the proofs of any subsequent identities, so there is no circular reasoning. □

The identity (7.62) may be written in the form

$$h_1^3h_2+h_2^3h_3+h_3^3h_1=0.$$

This is the equation of Klein's quartic curve in homogeneous form; see [206] for more information.

## 7.6 Weight one theta functions

In this section we establish some of the fundamental properties of the septic theta function defined by

$$z_7(q)=\sum_{m=-\infty}^{\infty}\sum_{n=-\infty}^{\infty}q^{m^2+mn+2n^2}. \tag{7.65}$$

We begin by recalling the following Lambert series expansions.

**Lemma 7.15.** *Let $\varphi$ and $\psi$ be Ramanujan's theta functions given by Definition 3.1. Then*

$$\varphi(q)\varphi(q^7)=1+2\sum_{j=1}^{\infty}\left(\frac{j}{7}\right)\frac{q^j}{1-(-q)^j}$$

*and*

$$q\psi(q)\psi(q^7)=\sum_{\substack{j>0\\ j\text{ odd}}}\left(\frac{j}{7}\right)\frac{q^j}{1-q^j}.$$

*Proof.* See Example 1.25. □

The next result gives a Lambert series expansion for $z_7(q)$. It is the analogue of results for $\varphi^2(q)$ and $a(q)$ in (3.14) and (3.34), respectively.

**Theorem 7.16.** *The following identity holds:*

$$z_7(q) = 1 + 2\sum_{j=1}^{\infty}\left(\frac{j}{7}\right)\frac{q^j}{1-q^j}.$$

*Proof.* On breaking the sum into the separate cases according to whether $n$ is even or odd, we find that

$$\begin{aligned} z_7(q) &= \sum_{m=-\infty}^{\infty}\sum_{n=-\infty}^{\infty} q^{m^2+mn+2n^2} \\ &= \sum_{m=-\infty}^{\infty}\sum_{n=-\infty}^{\infty} q^{m^2+2mn+8n^2} + \sum_{m=-\infty}^{\infty}\sum_{n=-\infty}^{\infty} q^{m^2+m(2n+1)+2(2n+1)^2} \\ &= \sum_{m=-\infty}^{\infty}\sum_{n=-\infty}^{\infty} q^{(m+n)^2+7n^2} + \sum_{m=-\infty}^{\infty}\sum_{n=-\infty}^{\infty} q^{(m+n+\frac{1}{2})^2+7(n+\frac{1}{2})^2}. \end{aligned}$$

On interchanging the order of summation and setting $j = m + n$ this becomes

$$\begin{aligned} z_7(q) &= \sum_{n=-\infty}^{\infty}\sum_{j=-\infty}^{\infty} q^{j^2+7n^2} + \sum_{n=-\infty}^{\infty}\sum_{j=-\infty}^{\infty} q^{(j+\frac{1}{2})^2+7(n+\frac{1}{2})^2} \\ &= \varphi(q)\varphi(q^7) + 4q^2\psi(q^2)\psi(q^{14}). \end{aligned} \tag{7.66}$$

By Lemma 7.15 we have

$$\varphi(q)\varphi(q^7) = 1 + 2\sum_{\substack{j>0 \\ j \text{ odd}}}\left(\frac{j}{7}\right)\frac{q^j}{1+q^j} + 2\sum_{\substack{j>0 \\ j \text{ even}}}\left(\frac{j}{7}\right)\frac{q^j}{1-q^j}$$

and

$$4q^2\psi(q^2)\psi(q^{14}) = 4\sum_{\substack{j>0 \\ j \text{ odd}}}\left(\frac{j}{7}\right)\frac{q^{2j}}{1-q^{2j}} = 2\sum_{\substack{j>0 \\ j \text{ odd}}}\left(\frac{j}{7}\right)\left(\frac{q^j}{1-q^j} - \frac{q^j}{1+q^j}\right),$$

and on using these results in (7.66) we complete the proof. □

The next result may be compared with the identities in (3.55) and (3.56), viz.,

$$a(q) = \varphi(-q)\varphi(-q^3) + 8q\psi(q^2)\psi(q^6)$$

and

$$a(q^4) = \varphi(-q)\varphi(-q^3) + 2q\psi(q^2)\psi(q^6).$$

**Theorem 7.17.** *The following identities hold*:

$$z_7(q) = \varphi(-q)\varphi(-q^7) + 4q\psi(q)\psi(q^7)$$

*and*

$$z_7(q^2) = \varphi(-q)\varphi(-q^7) + 2q\psi(q)\psi(q^7).$$

*Proof.* By Lemma 7.15 and Theorem 7.16 we have

$$\begin{aligned}
&\varphi(-q)\varphi(-q^7) + 4q\psi(q)\psi(q^7) \\
&= 1 + 2\sum_{j=1}^{\infty}\left(\frac{j}{7}\right)\frac{(-q)^j}{1-q^j} + 4\sum_{\substack{j>0 \\ j \text{ odd}}}\left(\frac{j}{7}\right)\frac{q^j}{1-q^j} \\
&= 1 + 2\sum_{j=1}^{\infty}\left(\frac{j}{7}\right)\frac{q^j}{1-q^j} \\
&= z_7(q)
\end{aligned}$$

and

$$\begin{aligned}
&\varphi(-q)\varphi(-q^7) + 2q\psi(q)\psi(q^7) \\
&= 1 + 2\sum_{j=1}^{\infty}\left(\frac{j}{7}\right)\frac{(-q)^j}{1-q^j} + 2\sum_{\substack{j>0 \\ j \text{ odd}}}\left(\frac{j}{7}\right)\frac{q^j}{1-q^j} \\
&= 1 + 2\sum_{\substack{j>0 \\ j \text{ even}}}\left(\frac{j}{7}\right)\frac{q^j}{1-q^j} \\
&= 1 + 2\sum_{j-1}^{\infty}\left(\frac{j}{7}\right)\frac{q^{2j}}{1-q^{2j}}, \quad \text{since } \left(\frac{2j}{7}\right) = \left(\frac{j}{7}\right), \\
&= z_7(q^2).
\end{aligned}$$

□

The next result gives a representation of $z_7(q)$ as a sum of infinite products.

**Theorem 7.18.**

$$z_7(q) = \left( \frac{(q^3,q^4;q^7)_\infty}{(q,q^6;q^7)_\infty^2} - q\frac{(q,q^6;q^7)_\infty}{(q^2,q^5;q^7)_\infty^2} + q\frac{(q^2,q^5;q^7)_\infty}{(q^3,q^4;q^7)_\infty^2} \right) \times (q^7;q^7)_\infty^2.$$

*Proof.* In Theorem 1.13 replace $q$ with $q^7$ and put $(x,y,z) = (q,q^2,q^3)$ to obtain

$$\frac{(q^3,q^4,q^7,q^7;q^7)_\infty}{(q,q^6;q^7)_\infty^2} = 1 + \sum_{k=1}^{\infty} \left( \frac{2q^{(7k-6)k}}{1-q^{7k-6}} + \frac{q^{(7k-5)k}}{1-q^{7k-5}} + \frac{q^{(7k-4)k}}{1-q^{7k-4}} \right.$$
$$\left. - \frac{q^{(7k-3)k}}{1-q^{7k-3}} - \frac{q^{(7k-2)k}}{1-q^{7k-2}} - \frac{2q^{(7k-1)k}}{1-q^{7k-1}} \right). \tag{7.67}$$

Next, replace $q$ with $q^{7/2}$ in Theorem 1.14 and take $(a,b,c) = (q^{1/2},q^{1/2},q^{3/2})$ to obtain

$$q\frac{(q,q^6,q^7,q^7;q^7)_\infty}{(q^2,q^5;q^7)_\infty^2} = \sum_{k=1}^{\infty} \left( \frac{q^{(7k-6)k}}{1-q^{7k-6}} - \frac{2q^{(7k-5)k}}{1-q^{7k-5}} + \frac{q^{(7k-4)k}}{1-q^{7k-4}} \right.$$
$$\left. - \frac{q^{(7k-3)k}}{1-q^{7k-3}} + \frac{2q^{(7k-2)k}}{1-q^{7k-2}} - \frac{q^{(7k-1)k}}{1-q^{7k-1}} \right). \tag{7.68}$$

Further, replace $q$ with $q^{7/2}$ in Theorem 1.14 and take $(a,b,c) = (q^{1/2},q,q)$ to obtain

$$q\frac{(q^2,q^5,q^7,q^7;q^7)_\infty}{(q^3,q^4;q^7)_\infty^2} = \sum_{k=1}^{\infty} \left( \frac{q^{(7k-6)k}}{1-q^{7k-6}} - \frac{q^{(7k-5)k}}{1-q^{7k-5}} - \frac{2q^{(7k-4)k}}{1-q^{7k-4}} \right.$$
$$\left. + \frac{2q^{(7k-3)k}}{1-q^{7k-3}} + \frac{q^{(7k-2)k}}{1-q^{7k-2}} - \frac{q^{(7k-1)k}}{1-q^{7k-1}} \right). \tag{7.69}$$

On adding (7.67) and (7.69) and then subtracting (7.68) we deduce that

$$\left( \frac{(q^3,q^4;q^7)_\infty}{(q,q^6;q^7)_\infty^2} - q\frac{(q,q^6;q^7)_\infty}{(q^2,q^5;q^7)_\infty^2} + q\frac{(q^2,q^5;q^7)_\infty}{(q^3,q^4;q^7)_\infty^2} \right) \times (q^7;q^7)_\infty^2$$
$$= 1 + 2\sum_{j=1}^{\infty} \left( \frac{j}{7} \right) \frac{q^j}{1-q^j}.$$

Now apply Theorem 7.16 to the right-hand side to complete the proof. □

## 7.7 The weight two Eisenstein series and weight three eta-quotient

In the previous section we studied the weight one modular form defined by the theta series

$$z_7 = \sum_{m=-\infty}^{\infty} \sum_{n=-\infty}^{\infty} q^{m^2+mn+2n^2},$$

or equivalently by the Eisenstein series

$$z_7 = 1 + 2\sum_{j=1}^{\infty} \left(\frac{j}{7}\right) \frac{q^j}{1-q^j}.$$

The two formulas for $z_7$ are equivalent, by Theorem 7.16.

In this section we will study four functions that come in two pairs:

1. Let $Z_7$ and $X_7$ be defined by

$$Z_7 = \frac{1}{6}\left(7P(q^7) - P(q)\right) \quad \text{and} \quad X_7 = \frac{1}{Z_7^{3/2}} \times q \prod_{j=1}^{\infty} (1-q^j)^3(1-q^{7j})^3 \qquad (7.70)$$

   where $P$ is Ramanujan's weight two Eisenstein series defined by (0.61).
2. Let $y_7$ and $w_7$ be the eta-quotients defined by

$$y_7 = \prod_{j=1}^{\infty} \frac{(1-q^j)^7}{(1-q^{7j})} \quad \text{and} \quad w_7 = q \prod_{j=1}^{\infty} \frac{(1-q^{7j})^4}{(1-q^j)^4}. \qquad (7.71)$$

The function $Z_7$ is a modular form of weight 2 while $X_7$ is a modular function. They are analogues of the functions $Z$ and $X$ for levels 5, 6, 8, and 9 in Table 6.3. The functions $y_7$ and $w_7$ are also analogues of the eta-quotients $y$ and $w$ for levels 5, 6, 8, and 9 that appear in Table 6.3. The significant difference is that $y_7$ is a modular form of weight 3, while the functions $y$ in Table 6.3 are modular forms weight 2. This is one of the reasons why the level 7 theory is different from the theory for Zagier's six sporadic examples in Table 6.3.

One may also ask about the modular function to accompany the weight one modular form $z_7$. It turns out that the appropriate modular function is $x_7 = X_7$, as we will see in Theorem 7.30. Therefore, in the remainder of this chapter, the three pairs of functions to be studied are

$$(x_7, z_7), \quad (X_7, Z_7) \quad \text{and} \quad (w_7, y_7), \quad \text{where } x_7 = X_7.$$

Whereas there were six distinct functions in the level 5 and 6 theories, e.g., see the functions denoted by $x$, $z$, $X$, $Z$, $w$, and $y$ in Table 6.3, there are only five distinct functions in the level 7 theory.

We will now develop the properties of $w_7$, $X_7$, $y_7$, and $Z_7$ and give their connections with the weight one modular form $z_7$. We begin with a trivial property.

**Theorem 7.19.** *The following identities hold:*

$$y_7^4 w_7 = q \prod_{j=1}^{\infty} (1-q^j)^{24}$$

*and*

$$y_7^4 w_7^7 = q^7 \prod_{j=1}^{\infty} (1-q^{7j})^{24}.$$

*Proof.* This is immediate from the definitions of $w_7$ and $y_7$. □

The next result relates $z_7$ and $Z_7$.

**Theorem 7.20.** *The weight two Eisenstein series $Z_7$ is related to the weight one modular form $z_7$ by the formula*

$$z_7^2 = Z_7.$$

*That is,*

$$\left(1+2\sum_{j=1}^{\infty}\left(\frac{j}{7}\right)\frac{q^j}{1-q^j}\right)^2 = \frac{1}{6}\left(7P(q^7)-P(q)\right).$$

*Proof.* Take $(\alpha,\beta,\gamma) = (\frac{2\pi}{7}, \frac{4\pi}{7}, -\frac{6\pi}{7})$ in Theorem 1.24 to obtain

$$\left(a_0+\sum_{j=1}^{\infty}\frac{b_j\, q^j}{1-q^j}\right)^2 = c_0+\sum_{j=1}^{\infty}\frac{d_j\, j\, q^j}{1-q^j} \tag{7.72}$$

where

$$a_0 = \cot\frac{\pi}{7}+\cot\frac{2\pi}{7}-\cot\frac{3\pi}{7},$$

$$b_j = 4\left(\sin\frac{2\pi j}{7}+\sin\frac{4\pi j}{7}-\sin\frac{6\pi j}{7}\right), \quad j\geq 1,$$

$$c_0 = \csc^2 \frac{\pi}{7} + \csc^2 \frac{2\pi}{7} + \csc^2 \frac{3\pi}{7} - 1$$

and

$$d_j = 8 \left( 3 - \cos \frac{2\pi j}{7} - \cos \frac{4\pi j}{7} - \cos \frac{6\pi j}{7} \right), \quad j \geq 1.$$

The values of $a_0$ and $b_j$ may be determined using Lemma 2.19. For, a simple calculation gives

$$a_0 = \frac{1}{2} \sum_{r=1}^{6} \left( \frac{r}{6} \right) \cot \frac{r\pi}{7} = -\frac{1}{2} \lim_{u \to 0} \sum_{r=1}^{6} \left( \frac{r}{6} \right) \cot \left( u - \frac{r\pi}{7} \right).$$

Now apply (2.40) and then use L'Hôpital's rule to deduce that

$$\begin{aligned} a_0 &= -\frac{1}{2} \lim_{u \to 0} 2\sqrt{7} \frac{(e^{2iu} + e^{4iu} - e^{6iu} + e^{8iu} - e^{10iu} - e^{12iu})}{(e^{14iu} - 1)} \\ &= -\sqrt{7} \times \left( \frac{2 + 4 - 6 + 8 - 10 - 12}{14} \right) = \sqrt{7}. \end{aligned}$$

Next, by the periodicity of the exponential function, we have

$$\begin{aligned} b_j &= 4 \left( \sin \frac{2\pi j}{7} + \sin \frac{4\pi j}{7} - \sin \frac{6\pi j}{7} \right) \\ &= -2i \left( e^{2\pi i j/7} + e^{4\pi i j/7} - e^{6\pi i j/7} + e^{-6\pi i j/7} - e^{-4\pi i j/7} - e^{-2\pi i j/7} \right) \\ &= -2i \sum_{r=1}^{6} \left( \frac{r}{7} \right) e^{2\pi i r j/7}. \end{aligned}$$

Now apply the Gauss sum (2.39) to deduce

$$b_j = 2\sqrt{7} \left( \frac{j}{7} \right).$$

Substituting the values of $a_0$ and $b_j$ in (7.72) we obtain

$$7 \left( 1 + 2 \sum_{j=1}^{\infty} \left( \frac{j}{7} \right) \frac{q^j}{1 - q^j} \right)^2 = c_0 + \sum_{j=1}^{\infty} \frac{d_j \, j \, q^j}{1 - q^j}. \tag{7.73}$$

It remains to determine $c_0$ and $d_j$. Letting $q = 0$ we deduce that $c_0 = 7$. The values of $d_j$ may be found by expanding both sides as far as $q^7$, comparing coefficients, and noting that $d_j$ is periodic with period 7. We find that

$$d_j = \begin{cases} 0 & \text{if } j \equiv 0 \pmod 7, \\ 28 & \text{otherwise.} \end{cases}$$

Therefore, the identity (7.73) simplifies to

$$\left(1 + 2\sum_{j=1}^{\infty}\left(\frac{j}{7}\right)\frac{q^j}{1-q^j}\right)^2 = 1 + 4\sum_{j \not\equiv 0 \pmod 7}\frac{jq^j}{1-q^j}$$
$$= \frac{1}{6}\left(7P(q^7) - P(q)\right).$$

This completes the proof. □

The next result expresses the cube of the weight one modular form $z_7$ as a sum of eta-quotients.

**Theorem 7.21.** *The following identity holds*:

$$z_7^3 = (1 + 13w_7 + 49w_7^2)y_7.$$

*Proof.* By the definitions of $\alpha$, $\beta$, and $\gamma$ in (7.1)–(7.3), the result of Theorem 7.18 may be written as

$$z_7(q) = q^{2/3}\frac{(q^7;q^7)_\infty^{1/3}}{(q;q)_\infty^{1/3}} \times \left(\frac{\alpha^{1/3}}{\gamma^{4/3}} + \frac{\beta^{1/3}}{\alpha^{4/3}} + \frac{\gamma^{1/3}}{\beta^{4/3}}\right) \times (q^7;q^7)_\infty^2.$$

Taking the cube we obtain

$$z_7^3(q) = \left\{\frac{\alpha}{\gamma^4} + \frac{\beta}{\alpha^4} + \frac{\gamma}{\beta^4} + 3\left(\frac{\alpha^{1/3}}{\beta^{8/3}\gamma^{2/3}} + \frac{\beta^{1/3}}{\gamma^{8/3}\alpha^{2/3}} + \frac{\gamma^{1/3}}{\alpha^{8/3}\beta^{2/3}}\right)\right.$$
$$\left. +3\left(\frac{\alpha^{2/3}}{\beta^{4/3}\gamma^{7/3}} + \frac{\beta^{2/3}}{\gamma^{4/3}\alpha^{7/3}} + \frac{\gamma^{2/3}}{\alpha^{4/3}\beta^{7/3}}\right) + \frac{6}{\alpha\beta\gamma}\right\} \times \frac{\eta_7^7}{\eta_1}.$$

By (7.5) we have $\alpha\beta\gamma = -1$, so the above simplifies to

$$z_7^3(q) = \left\{\alpha^5\beta^4 + \beta^5\gamma^4 + \gamma^5\alpha^4 + 3\left(\alpha^3\gamma^2 + \beta^3\alpha^2 + \gamma^3\beta^2\right)\right.$$
$$\left. -3\left(\alpha^3\beta + \beta^3\gamma + \gamma^3\alpha\right) - 6\right\} \times \frac{\eta_7^7}{\eta_1}.$$

Repeated application of (7.5) and (7.19)–(7.21) — see Exercise 2 — gives

$$\begin{aligned}&\alpha^5\beta^4+\beta^5\gamma^4+\gamma^5\alpha^4\\&=(\alpha^7+\beta^7+\gamma^7)-(\alpha^3\gamma^2+\beta^3\alpha^2+\gamma^3\beta^2)+2(\alpha^3\beta+\beta^3\gamma+\gamma^3\alpha).\end{aligned}\tag{7.74}$$

It follows that

$$\begin{aligned}z_7^3(q)=\big\{&(\alpha^7+\beta^7+\gamma^7)+2(\alpha^3\gamma^2+\beta^3\alpha^2+\gamma^3\beta^2)\\&-(\alpha^3\beta+\beta^3\gamma+\gamma^3\alpha)-6\big\}\times\frac{\eta_7^7}{\eta_1}.\end{aligned}$$

Now apply (7.46), (7.47), and (7.48) to complete the proof. □

The next result relates the modular functions $w_7$ and $X_7$.

**Theorem 7.22.** *The following identity holds*:

$$X_7=\frac{w_7}{1+13w_7+49w_7^2}.$$

*Proof.* By the definitions of $y_7$ and $w_7$ we have

$$w_7y_7=q\prod_{j=1}^{\infty}(1-q^j)^3(1-q^{7j})^3=\eta_1^3\eta_7^3.$$

Moreover, by Theorems 7.20 and 7.21 we have

$$Z_7^{3/2}=z_7^3=(1+13w_7+49w_7^2)y_7.$$

Using these in the definition of $X_7$ we obtain

$$X_7=\frac{\eta_1^3\eta_7^3}{Z_7^{3/2}}=\frac{w_7}{1+13w_7+49w_7^2}.$$

□

We end this section by calculating some derivatives.

**Theorem 7.23.** *The following differentiation formulas hold*:

$$q\frac{\mathrm{d}}{\mathrm{d}q}\log w_7=Z_7=z_7^2\tag{7.75}$$

*and*

$$q\frac{d}{dq}\log X_7 = Z_7\sqrt{1-26X_7-27X_7^2}. \tag{7.76}$$

*Proof.* By a direct calculation using the definitions of $w_7$ and $Z_7$, together with (0.61), we have

$$q\frac{d}{dq}\log w_7 = q\frac{d}{dq}\log\frac{\eta_7^4}{\eta_1^4} = \frac{1}{6}\left(7P(q^7)-P(q)\right) = Z_7.$$

Since $Z_7 = z_7^2$ by Theorem 7.20, this proves the first result.

Next, by Theorem 7.22 and the chain rule, we have

$$\begin{aligned}q\frac{d}{dq}\log X_7 &= q\frac{d}{dq}\log\frac{w_7}{1+13w_7+49w_7^2}\\ &= q\frac{d}{dq}\log w_7 - q\frac{d}{dq}\log(1+13w_7+49w_7^2)\\ &= q\frac{d}{dq}\log w_7 - \frac{w_7(13+98w_7)}{1+13w_7+49w_7^2}\,q\frac{d}{dq}\log w_7.\end{aligned}$$

Using (7.75), this becomes

$$q\frac{d}{dq}\log X_7 = Z_7\left(1-\frac{w_7(13+98w_7)}{1+13w_7+49w_7^2}\right) = Z_7\left(\frac{1-49w_7^2}{1+13w_7+49w_7^2}\right). \tag{7.77}$$

On the other hand, by Theorem 7.22 we have

$$1+X_7 = 1+\frac{w_7}{1+13w_7+49w_7^2} = \frac{(1+7w_7)^2}{1+13w_7+49w_7^2}$$

and

$$1-27X_7 = 1-\frac{27w_7}{1+13w_7+49w_7^2} = \frac{(1-7w_7)^2}{1+13w_7+49w_7^2}.$$

It follows that

$$\frac{1-49w_7^2}{1+13w_7+49w_7^2} = \sqrt{(1+X_7)(1-27X_7)}. \tag{7.78}$$

Using this in (7.77) we complete the proof of the second result. □

In view of (7.76) and by analogy with (5.56) we define $B_7 = B_7(q)$ by

$$B_7 = \sqrt{1-26X_7-27X_7^2}. \tag{7.79}$$

By Theorem 7.22 and (7.78) we have

$$B_7 = \frac{1 - 49w_7^2}{1 + 13w_7 + 49w_7^2} = \frac{w_7}{X_7}\frac{\mathrm{d}X_7}{\mathrm{d}w_7}. \tag{7.80}$$

**Theorem 7.24.** *The following differentiation formulas hold:*

$$q\frac{\mathrm{d}}{\mathrm{d}q}\log y_7 = \frac{7}{24}\left(P(q) - P(q^7)\right) \tag{7.81}$$

*and*

$$q\frac{\mathrm{d}}{\mathrm{d}q}\log z_7 = \frac{7}{72}\left(P(q) - P(q^7)\right) + \frac{w_7}{3}\left(\frac{13 + 98w_7}{1 + 13w_7 + 49w_7^2}\right)z_7^2. \tag{7.82}$$

*Hence,*

$$w_7 z_7 \frac{\mathrm{d}z_7}{\mathrm{d}w_7} = \frac{7}{72}\left(P(q) - P(q^7)\right) + \frac{w_7}{3}\left(\frac{13 + 98w_7}{1 + 13w_7 + 49w_7^2}\right)z_7^2. \tag{7.83}$$

*Proof.* From the definition of $y_7$ in (7.70) and using (0.61) we have

$$q\frac{\mathrm{d}}{\mathrm{d}q}\log y_7 = q\frac{\mathrm{d}}{\mathrm{d}q}\log\left(\prod_{j=1}^{\infty}\frac{(1-q^j)^7}{(1-q^{7j})}\right) = \frac{7}{24}\left(P(q) - P(q^7)\right),$$

and this is (7.81). Next, from Theorem 7.21 we have

$$z_7^3 = (1 + 13w_7 + 49w_7^2)y_7.$$

Applying logarithmic differentiation we obtain

$$3q\frac{\mathrm{d}}{\mathrm{d}q}\log z_7 = \frac{w_7(13 + 98w_7)}{1 + 13w_7 + 49w_7^2}\, q\frac{\mathrm{d}}{\mathrm{d}q}\log w_7 + q\frac{\mathrm{d}}{\mathrm{d}q}\log y_7.$$

Now apply (7.75) and (7.81) to evaluate the derivatives on the right-hand side and divide by 3 to obtain (7.82). Finally, by the chain rule and (7.75) we have

$$w_7 z_7 \frac{\mathrm{d}z_7}{\mathrm{d}w_7} = z_7^2\; q\frac{\mathrm{d}}{\mathrm{d}q}\log z_7 \Big/ q\frac{\mathrm{d}}{\mathrm{d}q}\log w_7 = q\frac{\mathrm{d}}{\mathrm{d}q}\log z_7.$$

Hence, (7.83) follows from (7.82). □

## 7.8 Eisenstein series

This section is devoted to obtaining expressions for the Eisenstein series $P$, $Q$, and $R$ in terms of the eta-quotients $w_7$ and $y_7$. It is slightly more convenient to record the results in terms of the modular function $w_7$ and the weight one modular form $z_7$. The results can be converted to eta-quotients using Theorem 7.21.

**Theorem 7.25.** *Let $z_7$ and $w_7$ be given by* (7.65) *and* (7.71) *and let $P(q)$ be Ramanujan's Eisenstein series defined by* (0.71)*. Then*

$$P(q) = \left(\frac{1-39w_7-343w_7^2}{1+13w_7+49w_7^2}\right) z_7^2 + 12 w_7 z_7 \frac{dz_7}{dw_7}$$

*and*

$$P(q^7) = \left(\frac{1+\frac{39}{7}w_7-7w_7^2}{1+13w_7+49w_7^2}\right) z_7^2 + \frac{12}{7} w_7 z_7 \frac{dz_7}{dw_7}.$$

*Proof.* By Theorems 7.20 and 7.24 we have

$$z_7^2 = \frac{1}{6}\left(7P(q^7)-P(q)\right)$$

and

$$w_7 z_7 \frac{dz_7}{dw_7} = \frac{7}{72}\left(P(q)-P(q^7)\right) + \frac{w_7}{3}\left(\frac{13+98w_7}{1+13w_7+49w_7^2}\right) z_7^2.$$

The required results are obtained by solving for $P(q)$ or $P(q^7)$, respectively. □

**Theorem 7.26.** *Let $z_7$ and $w_7$ be given by* (7.65) *and* (7.71) *and let $Q(q)$ and $R(q)$ be Ramanujan's Eisenstein series defined by* (0.72) *and* (0.73)*. Then*

$$49Q(q^7) - Q(q) = 48\left(\frac{1-49w_7^2}{1+13w_7+49w_7^2}\right) z_7^4 \tag{7.84}$$

*and*

$$343R(q^7) - R(q) = \left(343Q(q^7)-Q(q)\right) z_7^2 + \frac{48w_7(-13+196w_7+4459w_7^2+19208w_7^3)}{(1+13w_7+49w_7^2)^2} z_7^6. \tag{7.85}$$

*Proof.* By Theorem 7.20 we have

$$z_7^2 = \frac{1}{6}\left(7P(q^7) - P(q)\right).$$

Applying $q\frac{\mathrm{d}}{\mathrm{d}q}$ and multiplying by 72 we obtain

$$144 z_7\, q\frac{\mathrm{d}z_7}{\mathrm{d}q} = 12q\frac{\mathrm{d}}{\mathrm{d}q}\left(7P(q^7) - P(q)\right).$$

The derivatives may be evaluated using (0.78) and (7.82). The result is

$$\begin{aligned}14(P(q) - P(q^7))z_7^2 + 48\left(\frac{13w_7 + 98w_7^2}{1 + 13w_7 + 49w_7^2}\right) z_7^4\\ = 49(P^2(q^7) - Q(q^7)) - (P^2(q) - Q(q)).\end{aligned}$$

Rearranging and completing the squares for $P(q)$ and $P(q^7)$ we obtain

$$\begin{aligned}&\left(7P(q^7) + z_7^2\right)^2 - \left(P(q) + 7z_7^2\right)^2 + 48z_7^4\\ &\qquad = 49Q(q^7) - Q(q) + 48\left(\frac{13w_7 + 98w_7^2}{1 + 13w_7 + 49w_7^2}\right) z_7^4. \qquad (7.86)\end{aligned}$$

Observe that

$$\begin{aligned}&\left(7P(q^7) + z_7^2\right)^2 - \left(P(q) + 7z_7^2\right)^2\\ &= \left(7P(q^7) - P(q) - 6z_7^2\right)\left(7P(q^7) + P(q) + 8z_7^2\right) = 0,\end{aligned}$$

by Theorem 7.20. Therefore, the identity (7.86) simplifies to

$$49Q(q^7) - Q(q) = 48\left(1 - \frac{13w_7 + 98w_7^2}{1 + 13w_7 + 49w_7^2}\right) z_7^4$$

and this proves (7.84).

Next, use Theorem 7.21 to write (7.84) in the form

$$49Q(q^7) - Q(q) = 48 z_7 y_7 (1 - 49w_7^2). \qquad (7.87)$$

Multiply by 3 and apply $q\frac{\mathrm{d}}{\mathrm{d}q}$ to obtain

$$3q\frac{\mathrm{d}}{\mathrm{d}q}\left(49Q(q^7) - Q(q)\right) = 144q\frac{\mathrm{d}(z_7 y_7)}{\mathrm{d}q}(1-49w_7^2) - 14122 z_7 y_7 w_7 q\frac{\mathrm{d}w_7}{\mathrm{d}q}. \qquad (7.88)$$

By Theorem 7.21, (7.81), (7.82), and logarithmic differentiation we find that

$$
\begin{aligned}
\frac{1}{z_7 y_7} q \frac{\mathrm{d}(z_7 y_7)}{dq} &= q\frac{\mathrm{d}}{dq}\log z_7 + q\frac{\mathrm{d}}{dq}\log y_7 \\
&= \frac{7}{18}\left(P(q) - P(q^7)\right) + \frac{w_7}{3}\left(\frac{13+98w_7}{1+13w_7+49w_7^2}\right) z_7^2 \\
&= \frac{7}{18}(P(q) - P(q^7)) + \frac{y_7 w_7}{3z_7}(13+98w_7).
\end{aligned}
$$

On substituting this back into (7.88) and using (0.79) and (7.75) to compute the derivatives of $Q(q)$, $Q(q^7)$, and $w_7$, we obtain

$$
\begin{aligned}
&343P(q^7)Q(q^7) - P(q)Q(q) - 343R(q^7) + R(q) \\
&= 56 z_7 y_7 (P(q) - P(q^7))(1 - 49w_7^2) \\
&\quad + 48 y_7^2 w_7 (13 + 98 w_7)(1 - 49w_7^2) - 14112 z_7^3 y_7 w_7^2.
\end{aligned}
$$

By Theorem 7.21 and (7.87), this is equivalent to

$$
\begin{aligned}
&343P(q^7)Q(q^7) - P(q)Q(q) - 343R(q^7) + R(q) \\
&= \frac{7}{6}(P(q) - P(q^7))(49Q(q^7) - Q(q)) \\
&\quad + 48 y_7^2 w_7 (13 + 98 w_7)(1 - 49w_7^2) - 14112 y_7^2 w_7^2 (1 + 13w_7 + 49w_7^2).
\end{aligned}
$$

Rearranging, we obtain

$$
\begin{aligned}
343R(q^7) - R(q) &= \frac{1}{6}(7P(q^7) - P(q))(343Q(q^7) - Q(q)) \\
&\quad + 48 y_7^2 w_7(-13 + 196 w_7 + 4459 w_7^2 + 19208 w_7^3).
\end{aligned}
$$

The proof of (7.85) may be completed by applying Theorem 7.21. □

The next result expresses the Eisenstein series $Q(q)$, $Q(q^7)$, $R(q)$, and $R(q^7)$ in terms of the eta-quotient $w_7$ and the weight one theta series $z_7$.

**Theorem 7.27.**

$$
Q(q) = \left(\frac{1 + 245w_7 + 2401w_7^2}{1 + 13w_7 + 49w_7^2}\right) z_7^4,
$$

$$
Q(q^7) = \left(\frac{1 + 5w_7 + w_7^2}{1 + 13w_7 + 49w_7^2}\right) z_7^4,
$$

$$
R(q) = \frac{(1 - 490w_7 - 21609w_7^2 - 235298w_7^3 - 823543w_7^4)}{(1 + 13w_7 + 49w_7^2)^2} z_7^6
$$

*and*

$$R(q^7) = \frac{(1 + 14w_7 + 63w_7^2 + 70w_7^3 - 7w_7^4)}{(1 + 13w_7 + 49w_7^2)^2} z_7^6.$$

*Proof.* From Corollary 0.23 and Theorem 7.19 we have

$$Q^3(q) - R^2(q) = 1728y_7^4 w_7, \tag{7.89}$$
$$Q^3(q^7) - R^2(q^7) = 1728y_7^4 w_7^7. \tag{7.90}$$

Let

$$u_1 = \frac{Q(q)}{y_7 z_7}, \quad u_2 = \frac{Q(q^7)}{y_7 z_7}, \quad u_3 = \frac{R(q)}{y_7^2}, \quad u_4 = \frac{R(q^7)}{y_7^2} \tag{7.91}$$

and note that Theorem 7.21 may be used to convert all occurrences of $y_7$ into expressions that involve $w_7$ and $z_7$. When $q = 0$ we have $w_7 = 0$, $y_7 = 1$ and $u_1 = u_2 = u_3 = u_4 = 1$. The results of Theorem 7.26 and (7.89) and (7.90) may be written in the equivalent forms

$$\begin{aligned} -u_1 + 49u_2 &= 48(1 - 49w_7^2), \\ bu_1 - 343bu_2 - u_3 + 343u_4 &= 48w_7(-13 + 196w_7 + 4459w_7^2 + 19208w_7^3), \\ bu_1^3 - u_3^2 &= 1728w_7, \\ bu_2^3 - u_4^2 &= 1728w_7^7, \end{aligned}$$

respectively, where $b = 1 + 13w_7 + 49w_7^2$. This system may be solved by the same method that was used to solve (5.65)–(5.68). This time, approximations as far as the quartic need to be computed. The result is

$$u_1 = 1 + 245w_7 + 2401w_7^2, \tag{7.92}$$
$$u_2 = 1 + 5w_7 + w_7^2, \tag{7.93}$$
$$u_3 = 1 - 490w_7 - 21609w_7^2 - 235298w_7^3 - 823543w_7^4, \tag{7.94}$$
$$u_4 = 1 + 14w_7 + 63w_7^2 + 70w_7^3 - 7w_7^4. \tag{7.95}$$

It is easily checked that this is in fact a solution, and the implicit function theorem guarantees that the solution is unique. Substituting (7.92)–(7.95) into (7.91) we complete the proof of Theorem 7.27. □

Recall from (2.21) that the modular $j$-invariant is defined for $\mathrm{Im}(\tau) > 0$ by

$$j(\tau) = \frac{1728Q(q)^3}{Q(q)^3 - R(q)^2}, \quad \text{where} \quad q = \exp(2\pi i\tau).$$

The next result gives parameterizations of the $j$-invariant as rational functions of the eta-quotient $w_7$.

**Theorem 7.28.** *Let $w_7$ be defined by* (7.71). *Then*

$$j(\tau) = \frac{(1 + 13w_7 + 49w_7^2)(1 + 245w_7 + 2401w_7^2)^3}{w_7}$$

*and*

$$j(7\tau) = \frac{(1 + 13w_7 + 49w_7^2)(1 + 5w_7 + w_7^2)^3}{w_7^7}.$$

*Proof.* Use the results of Theorem 7.27 in the definition of the $j$-invariant. □

## 7.9 Differential equations

In this section we will show that $z_7$ and $Z_7$ satisfy linear differential equations with respect to $X_7$ of orders 2 and 3, respectively. We begin by finding a differential equation satisfied by $z_7$ with respect to $w_7$.

**Theorem 7.29.** *The following differential equation for $z_7$ with respect to $w_7$ holds*:

$$\frac{\mathrm{d}}{\mathrm{d}w_7}\left(w_7\frac{\mathrm{d}z_7}{\mathrm{d}w_7}\right) = \frac{(1 + 16w_7 + 49w_7^2)}{(1 + 13w_7 + 49w_7^2)^2} \times 2z_7.$$

*Proof.* By (7.83) we have

$$w_7\frac{\mathrm{d}z_7}{\mathrm{d}w_7} = \frac{7}{72z_7}\left(P(q) - P(q^7)\right) + \frac{w_7}{3}\left(\frac{13 + 98w_7}{1 + 13w_7 + 49w_7^2}\right)z_7. \tag{7.96}$$

Consider the operator $\mathrm{d}/\mathrm{d}w_7$. By (7.75),

$$\frac{\mathrm{d}}{\mathrm{d}w_7} = \frac{1}{w_7 z_7^2}\, q\frac{\mathrm{d}}{\mathrm{d}q}.$$

Applying the operator to (7.96) we deduce that

$$\begin{aligned}
&\frac{\mathrm{d}}{\mathrm{d}w_7}\left(w_7\frac{\mathrm{d}z_7}{\mathrm{d}w_7}\right)\\
&=\frac{7}{72w_7z_7^3}q\frac{\mathrm{d}}{\mathrm{d}q}\left(P(q)-P(q^7)\right)-\frac{7}{72z_7^2}\left(\frac{\mathrm{d}z_7}{\mathrm{d}w_7}\right)\left(P(q)-P(q^7)\right)\\
&\quad+\frac{w_7}{3}\left(\frac{13+98w_7}{1+13w_7+49w_7^2}\right)\frac{\mathrm{d}z_7}{\mathrm{d}w_7}+\frac{(13+196w_7+637w_7^2)}{(1+13w_7+49w_7^2)^2}\frac{z_7}{3}. \qquad (7.97)
\end{aligned}$$

The derivative of $P(q)-P(q^7)$ can be computed by (0.78). Then, all terms in (7.97) that involve $P(q)$, $P(q^7)$, $Q(q)$, and $Q(q^7)$ can be expressed in terms of $w_7$, $z_7$, and $\mathrm{d}z_7/\mathrm{d}w_7$ by Theorems 7.25 and 7.27. The required result follows after a lengthy, but straightforward, simplification. □

**Theorem 7.30.** *The following differential equation for $z_7$ with respect to $X_7$ holds:*

$$X_7(1-26X_7-27X_7^2)\frac{\mathrm{d}^2z_7}{\mathrm{d}X_7^2}+(1-39X_7-54X_7^2)\frac{\mathrm{d}z_7}{\mathrm{d}X_7}-2(1+3X_7)z_7=0.$$

*Proof.* By (7.22) we have

$$X_7=\frac{w_7}{1+13w_7+49w_7^2}$$

and it follows that

$$\frac{w_7(1+16w_7+49w_7^2)}{(1+13w_7+49w_7^2)^2}=X_7(1+3X_7). \qquad (7.98)$$

Moreover, by (7.79) and (7.80) we have

$$w_7\frac{\mathrm{d}}{\mathrm{d}w_7}=X_7\sqrt{1-26X_7-27X_7^2}\frac{\mathrm{d}}{\mathrm{d}X_7}. \qquad (7.99)$$

Hence, on multiplying the differential equation in Theorem 7.29 by $w_7$ and making use of (7.98) and (7.99), we deduce that

$$\begin{aligned}
&X_7\sqrt{1-26X_7-27X_7^2}\,\frac{\mathrm{d}}{\mathrm{d}X_7}\left(X_7\sqrt{1-26X_7-27X_7^2}\,\frac{\mathrm{d}z_7}{\mathrm{d}X_7}\right)\\
&\qquad=2X_7(1+3X_7)z_7.
\end{aligned}$$

This simplifies to the required differential equation. □

**Theorem 7.31.** *The following differential equation for $Z_7$ with respect to $X_7$ holds*:

$$X_7^2(1-26X_7-27X_7^2)\frac{\mathrm{d}^3Z_7}{\mathrm{d}X_7^3}+3X_7(1-39X_7-54X_7^2)\frac{\mathrm{d}^2Z_7}{\mathrm{d}X_7^2}$$
$$+(1-86X_7-186X_7^2)\frac{\mathrm{d}Z_7}{\mathrm{d}X_7}-4(1+6X_7)Z_7=0.$$

*Proof.* Since $Z_7 = z_7^2$, this result can be obtained from Theorem 7.30 using Exercise 4 b of Chapter 6. □

The differential equations in Theorems 7.30 and 7.31 can be used to expand each of $z_7$ and $Z_7$ in powers of $X_7$.

**Theorem 7.32.** *Let $t_7(n)$ and $T_7(n)$ be the sequences defined by the recurrence relations*

$$(n+1)^2t_7(n+1)=(26n^2+13n+2)t_7(n)+(27n^2-27n+6)t_7(n-1)$$

*and*

$$(n+1)^3T_7(n+1)=(2n+1)(13n^2+13n+4)T_7(n)+3n(9n^2-1)T_7(n-1)$$

*and initial conditions $t_7(0)=T_7(0)=1$, $t_7(-1)=0$. Then*

$$z_7=\sum_{n=0}^{\infty}t_7(n)X_7^n \qquad \text{and} \qquad Z_7=\sum_{n=0}^{\infty}T_7(n)X_7^n.$$

*Hence,*

$$\left(\sum_{n=0}^{\infty}t_7(n)X_7^n\right)^2=\sum_{n=0}^{\infty}T_7(n)X_7^n.$$

*Furthermore,*

$$T_7(n)=\sum_{k=0}^{n}\binom{n}{k}^2\binom{2k}{n}\binom{n+k}{k}=\sum_{k=0}^{n}(-1)^{n+k}\binom{3n+1}{n-k}\binom{n+k}{k}^3.$$

*Proof.* The recurrence relations are obtained by substituting the series expansions

$$z_7=\sum_{n=0}^{\infty}t_7(n)X_7^n \qquad \text{and} \qquad Z_7=\sum_{n=0}^{\infty}T_7(n)X_7^n.$$

into the differential equations in Theorems 7.30 and 7.31 and equating coefficients. The formula for $T_7(n)$ may be verified by using the method of creative telescoping. □

At the time of writing, no similar formula for $t_7(n)$ as a sum of binomial coefficients is known. It is possible that no simple formula exists; see O'Brien's thesis [p. 75, 239]. The numbers $t_7(n)$ are integers; this is not obvious from the recurrence relation, but it is clear from the expansion of $z_7$ in powers of $X_7$.

The next result expresses the generating function for $T_7(n)$ in terms of hypergeometric functions.

**Theorem 7.33.** *The following identities hold in a neighborhood of* $x = 0$:

$$\sum_{n=0}^{\infty}\left\{\sum_{j=0}^{n}\binom{n}{j}^2\binom{2j}{n}\binom{j+n}{j}\right\}\left(\frac{x}{1+13x+49x^2}\right)^n$$

$$=\sqrt{\frac{1+13x+49x^2}{1+245x+2401x^2}}\;{}_3F_2\left(\begin{matrix}\frac{1}{6},\ \frac{1}{2},\ \frac{5}{6}\\ 1,\ 1\end{matrix};\frac{1728x}{(1+13x+49x^2)(1+245x+2401x^2)^3}\right)$$

$$=\sqrt{\frac{1+13x+49x^2}{1+5x+x^2}}\;{}_3F_2\left(\begin{matrix}\frac{1}{6},\ \frac{1}{2},\ \frac{5}{6}\\ 1,\ 1\end{matrix};\frac{1728x^7}{(1+13x+49x^2)(1+5x+x^2)^3}\right).$$

*Proof.* On the one hand, by Theorems 7.20 and 7.32, we have

$$z_7^2 = Z_7 = \sum_{n=0}^{\infty}\left\{\sum_{j=0}^{n}\binom{n}{j}^2\binom{2j}{n}\binom{j+n}{j}\right\}\left(\frac{w_7}{1+13w_7+49w_7^2}\right)^n. \tag{7.100}$$

On the other hand, by Theorem 4.30 we have

$$Q^{1/2}(q) = {}_3F_2\left(\begin{matrix}\frac{1}{6},\ \frac{1}{2},\ \frac{5}{6}\\ 1,\ 1\end{matrix};1728X_1(q)\right)$$

where $X_1(q) = 1/j(\tau)$. Hence, by Theorems 7.27 and 7.28 we obtain

$$z_7^2\left(\frac{1+245w_7+2401w_7^2}{1+13w_7+49w_7^2}\right)^{1/2}$$

$$= {}_3F_2\left(\begin{matrix}\frac{1}{6},\ \frac{1}{2},\ \frac{5}{6}\\ 1,\ 1\end{matrix};\frac{1728w_7}{(1+13w_7+49w_7^2)(1+245w_7+2401w_7^2)^3}\right),$$

while replacing $q$ with $q^7$ and applying a similar procedure leads to

$$z_7^2\left(\frac{1+5w_7+w_7^2}{1+13w_7+49w_7^2}\right)^{1/2}$$
$$={}_3F_2\left(\begin{matrix}\frac{1}{6},\ \frac{1}{2},\ \frac{5}{6}\\ 1,\ 1\end{matrix};\frac{1728w_7^7}{(1+13w_7+49w_7^2)(1+5w_7+w_7^2)^3}\right).$$

The proof may be completed by solving the previous two equations for $z_7^2$ and combining with (7.100), and then replacing all occurrences of $w_7$ with $x$. □

## 7.10 Notes

The formulas involving $\alpha$, $\beta$, and $\gamma$ in Sections 7.2 and 7.3 appear in Ramanujan's second notebook [253, Ch. 19, Entry 18, also p. 300] and in Ramanujan's lost notebook [254, pp. 242–243]. Several of the results have been proved by Watson [298], Kolberg [197], Garvan [157], and Berndt [46, pp. 305–313]. Further results have been given by Liu [219] and [222].

A generalization of the 5- and 7-dissections in (5.8) and (7.6) to an $n$-dissection, where $n$ is relatively prime to 6, has been given by Ramanathan [247], Evans [148], and Berndt [46, p. 274].

The deduction of the partition congruence identity in Corollary 7.13 from Theorem 7.12 is given in Ramanujan's lost notebook [254, p. 145]. For references to other proofs of Theorem 7.12 and Corollary 7.13, see Berndt's commentary [255, pp. 372–373]. A generalization of Corollary 7.13 has been given by Garvan [157, Theorem (1.6)].

The identities (7.62) and (7.63) were given by Ramanujan [253, p. 300]. The proof of (7.62) that we have given is from W. Chu and L. Di Claudio [97, pp. 136–137]. Other proofs of (7.62) and (7.63) have been given by Berndt and Zhang [53] (see also Berndt [47, pp. 176–184]) and Liu [222, (1.21) and (1.27)]. The connection of (7.62) with the Klein quartic has been described by Lachaud [201]. More information about the Klein quartic is given in the series of articles published in [206].

Lemma 7.15 was known to Ramanujan [253, Ch. 19, Entry 17] and proved by Berndt [46, pp. 303–304]. Our proof is slightly different from Berndt's. The identity in Theorem 7.18 was given by Liu [219, Theorem 5]. Theorems 7.20 and 7.21 are Entry 5 (i) in Chapter 21 of Ramanujan's second notebook [253]. See also [46, p. 467]. Generalizations of Theorem 7.21 have been given in work with D. Ye [126].

Results equivalent to Theorem 7.27 were given by Ramanujan in the lost notebook [254, p. 53]. The proof we have given here is from work with P. C. Toh [124] where references to other proofs are also given. Theorems 7.31 and 7.32 were proved in [114]. The sequence $T_7(n)$ was discovered by an experimental search for integer valued sequences satisfying a particular type of recurrence relation. The formulas for $T_7(n)$ as sums of binomial coefficients are due to Zudilin; see [114] and [308] for more details. The asymptotic behavior of $T_7(n)$ has been analyzed by Hirschhorn [183]. Some congruence properties of the sequence $t_7(n)$ have been conjectured in work with H. H. Chan and F. Sica [83]. Further properties of the sequence $t_7(n)$ have been investigated by L. O'Brien [239]. Theorem 7.33 was given by Zudilin [308].

## 7.11 Exercises

1. Use (7.64) to show that

$$\alpha^2\beta^3 - \alpha^3\beta = \beta^2\gamma^3 - \beta^3\gamma = \gamma^2\alpha^3 - \gamma^3\alpha = 1$$

where $\alpha$, $\beta$, and $\gamma$ are defined by (7.1)–(7.3). Hence give another proof of the identities in Theorem 7.2. See W. Chu and L. Di Claudio [97, pp. 136–137].

2. Use (7.5) and (7.19)–(7.21) to show that

$$\alpha^5\beta^4 + \beta^5\gamma^4 + \gamma^5\alpha^4 = (\alpha^6\beta^2 + \beta^6\gamma^2 + \gamma^6\alpha^2) + (\alpha^3\beta + \beta^3\gamma + \gamma^3\alpha),$$

$$\alpha^6\beta^2 + \beta^6\gamma^2 + \gamma^6\alpha^2 = (\alpha^7 + \beta^7 + \gamma^7) - (\alpha^5\gamma + \beta^5\alpha + \gamma^5\beta)$$

and

$$\alpha^5\gamma + \beta^5\alpha + \gamma^5\beta = (\alpha^3\gamma^2 + \beta^3\alpha^2 + \gamma^3\beta^2) - (\alpha^3\beta + \beta^3\gamma + \gamma^3\alpha)$$

and hence deduce (7.74).

3. Use Theorems 7.18 and 7.21 to prove (7.63).

4. a. Use (3.3) to show that

$$\sum_{\substack{j>0 \\ j \text{ odd}}} q^{j^2} = q\psi(q^8).$$

b. Use (1.38) from Example 1.25 to show that

$$q\psi(q)\psi(q^7) = \sum_{j=1}^{\infty} \left(\frac{j}{7}\right) \frac{q^j}{1-q^j}.$$

c. Use Part b to show that

$$q\psi(q)\psi(q^7) - q\psi(-q)\psi(-q^7) = 2q^2\psi(q^2)\psi(q^{14}).$$

d. Use the result of Part c to show that

$$\varphi(-q^2)\varphi(-q^{14}) - \varphi(-q)\varphi(-q^7) = 2q\psi(-q)\psi(-q^7).$$

e. Use the result of Part c to deduce Guetzlaff's identity, [166]:

$$\sqrt{\varphi(q)\varphi(q^7)} - \sqrt{\varphi(-q)\varphi(-q^7)} = 2q\sqrt{\psi(q^2)\psi(q^{14})}.$$

f. Deduce the following septic analogue of Jacobi's 'aequatio identica satis abstrusa':

$$\prod_{n=1}^{\infty}(1+q^{2n-1})(1+q^{14n-7})-\prod_{n=1}^{\infty}(1-q^{2n-1})(1-q^{14n-7})=2q\prod_{n=1}^{\infty}(1+q^{2n})(1+q^{14n}).$$

5. Use the last part of the previous exercise to deduce the following result of Farkas and Kra [149, 150, Chapter 7]:

   *Let $S$ be the set of positive integers such that multiples of 7 occur in two copies, say $(7k)_A$ and $(7k)_B$. Show that the number of partitions of $2N+1$ into distinct odd elements of $S$ equals the number of partitions of $2N$ into distinct even elements of $S$.*

   For example, the partitions in the case $N = 7$ are:

$$15,\ 11+3+1,\ 9+5+1,\ 7A+7_B+1,\ 7_A+5+3,\ 7_B+5+3$$

   and

$$14_A,\ 14_B,\ 12+2,\ 10+4,\ 8+6,\ 8+4+2,$$

   with the number of partitions of each type, in this case 6, being the same.

   For a different proof, see Hirschhorn [182]. For further proofs and generalizations, see the work of Warnaar [295].

6. Let $n$ be a positive integer, and let $f(n)$ denote the number of solutions of the equation $x^2 + 7y^2 = 8n$ in odd integers.

   a. Give a bijective proof of the result $f(n) = f(2n)$.
   b. Show that

$$\sum_{n=1}^{\infty} f(n)q^n = 4q\psi(q)\psi(q^7).$$

   c. Hence, give another proof of the identity in Exercise 4 Part c.

   See work with Hirschhorn [120].

7. Let

$$z_1 = \frac{(q^3, q^4, q^7, q^7; q^7)_\infty}{(q, q, q^6, q^6; q^7)_\infty},$$

$$z_2 = -q\frac{(q, q^6, q^7, q^7; q^7)_\infty}{(q^2, q^2, q^5, q^5; q^7)_\infty}$$

and

$$z_3 = q\frac{(q^2, q^5, q^7, q^7; q^7)_\infty}{(q^3, q^3, q^4, q^4; q^7)_\infty}.$$

a. Use Theorem 7.18 to show that

$$1 + 2\sum_{j=1}^{\infty}\left(\frac{j}{7}\right)\frac{q^j}{1-q^j} = z_1 + z_2 + z_3.$$

b. Use (7.62) to show that $z_1z_2 + z_2z_3 + z_3z_1 = 0$ and deduce the "freshman's dream"

$$(z_1 + z_2 + z_3)^2 = z_1^2 + z_2^2 + z_3^2.$$

c. Use Theorem 7.20 and the result of Part b of this exercise to show that

$$\frac{1}{6}\left(7P(q^7) - P(q)\right) = z_1^2 + z_2^2 + z_3^2.$$

Further properties of $z_1$, $z_2$, and $z_3$, including a system of nonlinear differential equations, have been given by T. Huber and D. Lara [188].

8. a. Use Theorems 7.12 and 7.21 to prove that

$$z_7^3(q) = 1 - \frac{7}{8}\sum_{j=1}^{\infty}\left(\frac{j}{7}\right)\frac{j^2q^j}{(1-q^j)} + \frac{49}{8}\sum_{j=1}^{\infty}\left(\frac{j}{7}\right)\frac{(1+q^j)}{(1-q^j)^3} + \frac{3}{4}\eta_1^3\eta_7^3.$$

b. Prove that

$$z_7^4(q) = \frac{1}{50}\left(Q(q) + 49Q(q^7)\right) + \frac{16}{5}w_7(q)z_7(q)y_7(q).$$

The analogous formulas for $z_7(q)$ and $z_7^2(q)$ were given in Theorems 7.16 and 7.20, respectively. For a generalization that gives a formula for $z_7^k(q)$ for any positive integer $k$, see Theorem 11.2 and the identity (11.20).

9. a. Let $D$ be the differential operator defined by

$$D = X\sqrt{1 - 26X - 27X^2}\,\frac{\mathrm{d}}{\mathrm{d}X}.$$

Show that the function $Z = Z_7$ satisfies the nonlinear second order differential equation

$$D^2Z - \frac{1}{2Z}(DZ)^2 = 4X(1 + 3X)Z$$

with respect to $X = X_7$.

b. Let $Z$ and $X$ be any of the pairs of functions in Table 6.3 and let $(\alpha, \beta, \gamma)$ be the corresponding parameters in the first column of that table. Let $D$ be the differential operator defined by

$$D = X\sqrt{1 - 4\alpha X - 16\gamma X^2}\,\frac{\mathrm{d}}{\mathrm{d}X}.$$

Show that

$$D^2Z - \frac{1}{2Z}(DZ)^2 = 2X(\beta + 3\gamma X)Z.$$

Hint: Refer to the identity (5.113).

c. Conclude that the differential equation in Part a cannot be obtained by specializing the values $(\alpha, \beta, \gamma)$ in Part b.

This gives another sense in which the level 7 theory is different from the theories for the levels in Table 6.3.

10. Use (0.46) to prove that

$$q\prod_{j=1}^{\infty}(1-q^j)^3(1-q^{7j})^3$$
$$= \frac{1}{2}\sum_{m=-\infty}^{\infty}\sum_{n=-\infty}^{\infty}\left(m + n\left\{\frac{1+i\sqrt{7}}{2}\right\}\right)^2 q^{m^2+mn+2n^2}.$$

See work with H. H. Chan and W.-C. Liaw [82, Th. 3.1].

11. Prove that

$$q\frac{\mathrm{d}}{\mathrm{d}q}\left(P(q) - P(q^7)\right)$$
$$= \frac{7}{72}\left(P(q) - P(q^7)\right)^2 - \frac{7}{75}\left(Q(q) - Q(q^7)\right) - \frac{8}{5}w_7(q)z_7(q)y_7(q).$$

See work with P. C. Toh [124, Cor. 3.7].

12. Let $w = \exp(2\pi i/7)$ and define $A$, $B$, $\Gamma$, and $V$ by

$$A = A(q) = \frac{\sin\frac{2\pi}{7}}{\sin\frac{\pi}{7}} \prod_{j=1}^{\infty} \frac{(1-q^j w^2)(1-q^j w^5)}{(1-q^j w)(1-q^j w^6)},$$

$$B = B(q) = -\frac{\sin\frac{4\pi}{7}}{\sin\frac{2\pi}{7}} \prod_{j=1}^{\infty} \frac{(1-q^j w^3)(1-q^j w^4)}{(1-q^j w^2)(1-q^j w^5)}$$

$$\Gamma = \Gamma(q) = \frac{\sin\frac{\pi}{7}}{\sin\frac{4\pi}{7}} \prod_{j=1}^{\infty} \frac{(1-q^j w)(1-q^j w^6)}{(1-q^j w^3)(1-q^j w^4)}$$

and

$$V = V(q) = \frac{\eta_{49}}{\eta_1}.$$

Suppose $\mathrm{Im}(\tau) > 0$ and put $q = \exp(2\pi i\tau)$ and $p = \exp(-2\pi i/7\tau)$.

a. Show that

$$A(q) = \alpha(p), \qquad B(q) = \beta(p) \qquad \text{and} \qquad \Gamma(q) = \gamma(p),$$

where $\alpha$, $\beta$, and $\gamma$ are defined by (7.1)–(7.3).

b. Deduce that the following identities hold:

$$A + B + \Gamma = 1 + 7V,$$

$$AB + B\Gamma + \Gamma A = -\frac{1}{2}\left(21V + 4 + 7\sqrt{4V + 21V^2 + 28V^3}\right),$$

$$AB\Gamma = -1.$$

Hint: Apply the result of Part a and the modular transformation for the Dedekind eta function in Theorem 2.8 to the corresponding identities involving $\alpha$, $\beta$, and $\gamma$ in Theorems 7.1 and 7.3.

c. Show that the following identities hold:

$$AB^2 - A^2 + \Gamma = 0,$$

$$B\Gamma^2 - B^2 + A = 0,$$

$$\Gamma A^2 - \Gamma^2 + B = 0.$$

d. Show that the following identities hold:

$$A^3B + B^3\Gamma + \Gamma^3A = -8 - 49\,\frac{\eta_7^4}{\eta_1^4},$$

$$A^3\Gamma^2 + B^3A^2 + \Gamma^3B^2 = -5 - 49\,\frac{\eta_7^4}{\eta_1^4},$$

$$A^7 + B^7 + \Gamma^7 = 57 + 2\times 7^3\,\frac{\eta_7^4}{\eta_1^4} + 7^4\,\frac{\eta_7^8}{\eta_1^8},$$

$$A^7B^7 + B^7\Gamma^7 + \Gamma^7A^7 = 289 + 18\times 7^3\,\frac{\eta_7^4}{\eta_1^4} + 19\times 7^4\,\frac{\eta_7^8}{\eta_1^8} + 7^6\,\frac{\eta_7^{12}}{\eta_1^{12}}.$$

13. Prove that

$$\frac{\eta_1^7}{\eta_7} = \sum_{n=0}^{\infty} u_7(n)\left(\frac{\eta_7}{\eta_1}\right)^{4n}$$

where the sequence $\{u_7(n)\}$ satisfies the five-term quartic recurrence relation

$$\begin{aligned}(n+1)^4u_7(n+1) &+ \left(26\left(n+\tfrac{1}{2}\right)^4 + 19\left(n+\tfrac{1}{2}\right)^2 + \tfrac{5}{8}\right)u_7(n)\\ &+ (267n^4 + 268n^2 + 18)u_7(n-1)\\ &+ 49\left(26\left(n-\tfrac{1}{2}\right)^4 + 19\left(n-\tfrac{1}{2}\right)^2 + \tfrac{5}{8}\right)u_7(n-2)\\ &+ 2401(n-1)^4u_7(n-3)\\ &= 0.\end{aligned}$$

This is the analogue of the expansion of $y_5$ in powers of $w_5$ in Theorem 5.32. See O'Brien's thesis [239].

# Chapter 8
# Level 8: The Ramanujan–Göllnitz–Gordon Continued Fraction

## 8.1 Introduction

There are two primitive characters modulo 8, which we denote by $\chi_2$ and $\chi_{-2}$. They are defined by

$$\chi_2(j) = \begin{cases} 1 & \text{if } j \equiv 1 \text{ or } 7 \pmod 8, \\ -1 & \text{if } j \equiv 3 \text{ or } 5 \pmod 8, \\ 0 & \text{otherwise} \end{cases}$$

and

$$\chi_{-2}(j) = \begin{cases} 1 & \text{if } j \equiv 1 \text{ or } 3 \pmod 8, \\ -1 & \text{if } j \equiv 5 \text{ or } 7 \pmod 8, \\ 0 & \text{otherwise.} \end{cases}$$

Both characters will occur in this chapter. For example, the character $\chi_{-2}$ occurs in Theorem 8.13 in the formula

$$\begin{aligned}\varphi(q)\varphi(q^2) &= 1 + 2\sum_{j=1}^{\infty} \chi_{-2}(j)\frac{q^j}{1-q^j} \\ &= 1 + 2\sum_{j=1}^{\infty}\left(\frac{q^{8j-7}}{1-q^{8j-7}} + \frac{q^{8j-5}}{1-q^{8j-5}} - \frac{q^{8j-3}}{1-q^{8j-3}} - \frac{q^{8j-1}}{1-q^{8j-1}}\right)\end{aligned}$$

while the other character $\chi_2$ appears in the function $r_8$ defined by

$$r_8 = r_8(q) = q^{1/2}\prod_{j=1}^{\infty}(1-q^j)^{\chi_2(j)} = q^{1/2}\prod_{j=1}^{\infty}\frac{(1-q^{8j-7})(1-q^{8j-1})}{(1-q^{8j-5})(1-q^{8j-3})}. \tag{8.1}$$



S. Cooper, *Ramanujan's Theta Functions*, DOI 10.1007/978-3-319-56172-1_9

The function $r_8$ can be expanded as a continued fraction. One such expansion is called the Ramanujan–Göllnitz–Gordon continued fraction and given by

$$r_8(q) = \cfrac{q^{1/2}}{1+q+\cfrac{q^2}{1+q^3+\cfrac{q^4}{1+q^5+\cfrac{q^6}{1+q^7+\cdots}}}}. \tag{8.2}$$

We will prove this result and show that the function $r_8$ has several properties that are analogous to those of the Rogers–Ramanujan continued fraction studied in Chapter 5. Another, different continued fraction expansion for $r_8$ will be given in the exercises at the end of the chapter.

It will be useful to define the allied function $\xi_8$ by

$$\xi_8 = \xi_8(q) = \frac{\sin\frac{\pi}{8}}{\sin\frac{3\pi}{8}} \prod_{j=1}^{\infty} \frac{(1-\omega q^j)(1-\omega^7 q^j)}{(1-\omega^3 q^j)(1-\omega^5 q^j)} \tag{8.3}$$

where $\omega = \exp(i\pi/4)$. Note that

$$\xi_8(0) = \frac{\sin\frac{\pi}{8}}{\sin\frac{3\pi}{8}} = \sqrt{2}-1.$$

By (2.27), the functions $r_8$ and $\xi_8$ are related by

$$\xi_8(e^{-\pi t/\sqrt{2}}) = r_8(e^{-\pi/t\sqrt{2}}), \qquad \text{for } \operatorname{Re}(t) > 0. \tag{8.4}$$

## 8.2 Proof of the continued fraction expansion

In this section we prove that the function $r_8$ defined by the infinite product in (8.1) has the continued fraction expansion given by (8.2). We will first establish a recurrence relation in Lemma 8.1 and then use it to prove a more general continued fraction in Lemma 8.2. The continued fraction in (8.2) will then be obtained by specialization.

**Lemma 8.1.** *Suppose $|a| < 1$ and $|q| < 1$. For any nonnegative integer $s$, let $P_s$ be defined by*

$$P_s = \sum_{n=0}^{\infty} \frac{(bq^s/a;q)_{2n}}{(q;q)_{2n+2s}} a^{2n} \times (q^{2n+2};q^2)_s.$$

*Then for any nonnegative integer s,*

$$P_s = (1-q^{2s+1})P_{s+1} + q^s(a-bq^{s+1})(aq^{s+1}-b)P_{s+2}.$$

*Proof.* Consider $P_s-(1-q^{2s+1})P_{s+1}$. The terms in the series corresponding to $n=0$ cancel, so

$$\begin{aligned}
&P_s-(1-q^{2s+1})P_{s+1}\\
&=\sum_{n=1}^{\infty}\frac{(bq^s/a;q)_{2n}}{(q;q)_{2n+2s}}a^{2n}\times(q^{2n+2};q^2)_s\\
&\qquad-(1-q^{2s+1})\sum_{n=1}^{\infty}\frac{(bq^{s+1}/a;q)_{2n}}{(q;q)_{2n+2s+2}}a^{2n}\times(q^{2n+2};q^2)_{s+1}\\
&=\sum_{n=1}^{\infty}\frac{(bq^{s+1}/a;q)_{2n-1}}{(q;q)_{2n+2s+2}}a^{2n}\times(q^{2n+2};q^2)_{s+1}\\
&\qquad\times\left\{(1-bq^s/a)(1-q^{2n+2s+1})-(1-q^{2s+1})(1-bq^{2n+s}/a)\right\}.
\end{aligned}$$

On simplifying the terms in braces and then rearranging the factors, we obtain

$$\begin{aligned}
&P_s-(1-q^{2s+1})P_{s+1}\\
&=\sum_{n=1}^{\infty}\frac{(bq^{s+1}/a;q)_{2n-1}}{(q;q)_{2n+2s+2}}a^{2n}\times(q^{2n+2};q^2)_{s+1}\times q^s(1-q^{2n})(q^{s+1}-b/s)\\
&=q^sa^2(1-bq^{s+1}/a)(q^{s+1}-b/a)\sum_{n=1}^{\infty}\frac{(bq^{s+2}/a;q)_{2n-2}}{(q;q)_{2n+2s+2}}a^{2n-2}\times(q^{2n};q^2)_{s+2}.
\end{aligned}$$

Now shift the summation index by 1 to obtain

$$\begin{aligned}
&P_s-(1-q^{2s+1})P_{s+1}\\
&=q^s(a-bq^{s+1})(aq^{s+1}-b)\sum_{n=0}^{\infty}\frac{(bq^{s+2}/a;q)_{2n}}{(q;q)_{2n+2s+4}}a^{2n}\times(q^{2n+2};q^2)_{s+2}\\
&=q^s(a-bq^{s+1})(aq^{s+1}-b)P_{s+2}.
\end{aligned}$$

□

Iteration of the recurrence relation in Lemma 8.1 leads to a continued fraction.

**Lemma 8.2.** *The following continued fraction expansion holds*:

$$\frac{(-a;q)_\infty(b;q)_\infty-(a;q)_\infty(-b;q)_\infty}{(-a;q)_\infty(b;q)_\infty+(a;q)_\infty(-b;q)_\infty}$$
$$=\cfrac{a-b}{1-q+\cfrac{(a-bq)(aq-b)}{1-q^3+\cfrac{q(a-bq^2)(aq^2-b)}{1-q^5+\cfrac{q^2(a-bq^3)(aq^3-b)}{\cfrac{\cdots}{1-q^{2s-1}+\cfrac{q^{s-1}(a-bq^s)(aq^s-b)}{1-q^{2s+1}+\cdots}}}}}}.$$

*Proof.* First, consider the case $|a|<1$. On the one hand, repeated application of the recurrence relation in Lemma 8.1 gives

$$\begin{aligned}\frac{a-b}{P_0/P_1}&=\cfrac{a-b}{1-q+\cfrac{(a-bq)(aq-b)}{P_1/P_2}}\\&=\cfrac{a-b}{1-q+\cfrac{(a-bq)(aq-b)}{1-q^3+\cfrac{q(a-bq^2)(aq^2-b)}{P_2/P_3}}}\\&=\cdots\\&=\cfrac{a-b}{1-q+\cfrac{(a-bq)(aq-b)}{1-q^3+\cfrac{q(a-bq^2)(aq^2-b)}{1-q^5+\cfrac{q^2(a-bq^3)(aq^3-b)}{\cfrac{\cdots}{1-q^{2s-1}+\cfrac{q^{s-1}(a-bq^s)(aq^s-b)}{P_s/P_{s+1}}}}}}}.\end{aligned}\tag{8.5}$$

On the other hand, by the definitions of $P_0$ and $P_1$,

$$\frac{a-b}{P_0/P_1} = \frac{(a-b)\sum_{n=0}^{\infty}\frac{(bq/a;q)_{2n}}{(q;q)_{2n+2}}a^{2n}(1-q^{2n+2})}{\sum_{n=0}^{\infty}\frac{(b/a;q)_{2n}}{(q;q)_{2n}}a^{2n}}$$

$$= \frac{\sum_{n=0}^{\infty}\frac{(b/a;q)_{2n+1}}{(q;q)_{2n+1}}a^{2n+1}}{\sum_{n=0}^{\infty}\frac{(b/a;q)_{2n}}{(q;q)_{2n}}a^{2n}} = \frac{g(a,b)-g(-a,-b)}{g(a,b)+g(-a,-b)}$$

where

$$g(a,b) = \sum_{n=0}^{\infty}\frac{(b/a;q)_n}{(q;q)_n}a^n.$$

The series for $g(a,b)$ can be converted to a product by the $q$-binomial theorem (Theorem 0.29) to give

$$g(a,b) = \frac{(b;q)_\infty}{(a;q)_\infty},$$

and it follows that

$$\frac{a-b}{P_0/P_1} = \frac{\frac{(b;q)_\infty}{(a;q)_\infty} - \frac{(-b;q)_\infty}{(-a;q)_\infty}}{\frac{(b;q)_\infty}{(a;q)_\infty} + \frac{(-b;q)_\infty}{(-a;q)_\infty}} = \frac{(-a;q)_\infty(b;q)_\infty - (a;q)_\infty(-b;q)_\infty}{(-a;q)_\infty(b;q)_\infty + (a;q)_\infty(-b;q)_\infty}. \tag{8.6}$$

The required result may be obtained by combining (8.5) and (8.6) and letting $s\to\infty$. The restriction $|a| < 1$ can be removed by analytic continuation. □

We are now ready to prove the continued fraction expansion for $r_8$.

**Theorem 8.3.** *The functions $r_8(q)$ given by* (8.1) *and* (8.2) *are the same, that is,*

$$q^{1/2}\prod_{j=1}^{\infty}\frac{(1-q^{8j-7})(1-q^{8j-1})}{(1-q^{8j-5})(1-q^{8j-3})} = \cfrac{q^{1/2}}{1+q+\cfrac{q^2}{1+q^3+\cfrac{q^4}{1+q^5+\cfrac{q^6}{1+q^7+\cdots}}}}.$$

*Proof.* Replace $q$ with $-q$ in the result of Lemma 8.2, then put $a = iq^{1/2}$ and $b = 0$ to obtain

$$\frac{(-iq^{1/2};-q)_\infty - (iq^{1/2};-q)_\infty}{(-iq^{1/2};-q)_\infty - (iq^{1/2};-q)_\infty} = \cfrac{iq^{1/2}}{1+q+\cfrac{q^2}{1+q^3+\cfrac{q^4}{1+q^5+\cfrac{q^6}{1+q^7+\cdots}}}}. \tag{8.7}$$

It remains to simplify the left-hand side. Using the rearrangement

$$(x;q)_\infty = (x;q^2)_\infty (qx;q^2)_\infty$$

we find that

$$(-iq^{1/2};-q)_\infty = (-iq^{1/2};q^2)_\infty (iq^{3/2};q^2)_\infty = \frac{f(iq^{1/2},-iq^{3/2})}{(q^2;q^2)_\infty}$$

and

$$(iq^{1/2};-q)_\infty = (iq^{1/2};q^2)_\infty (-iq^{3/2};q^2)_\infty = \frac{f(-iq^{1/2},iq^{3/2})}{(q^2;q^2)_\infty}$$

where $f(a,b)$ is Ramanujan's theta function given in Theorem 0.4. Moreover, by Theorem 0.5,

$$f(iq^{1/2},-iq^{3/2}) = f(-q^3,-q^5) + iq^{1/2} f(-q,-q^7)$$

and

$$f(-iq^{1/2},iq^{3/2}) = f(-q^3,-q^5) - iq^{1/2} f(-q,-q^7).$$

It follows that

$$\begin{aligned}\frac{(-iq^{1/2};-q)_\infty - (iq^{1/2};-q)_\infty}{(-iq^{1/2};-q)_\infty - (iq^{1/2};-q)_\infty} &= \frac{f(iq^{1/2},-iq^{3/2}) - f(-iq^{1/2},iq^{3/2})}{f(iq^{1/2},-iq^{3/2}) + f(-iq^{1/2},iq^{3/2})} \\ &= \frac{iq^{1/2} f(-q,-q^7)}{f(-q^3,-q^5)} \\ &= iq^{1/2} \prod_{j=1}^{\infty} \frac{(1-q^{8j-7})(1-q^{8j-1})}{(1-q^{8j-5})(1-q^{8j-3})},\end{aligned} \tag{8.8}$$

where the last equality holds by Theorem 0.4. The proof is completed by substituting the result of (8.8) into (8.7) and dividing by $i$. □

## 8.3 Basic properties

In this section, we will establish infinite product formulas for the eight functions $1 + c\, r(q^2)$, where $c \in \left\{\pm 1, \pm i, \pm 1 \pm \sqrt{2}\right\}$. See Lemmas 8.4 and 8.5 below. Many of the other results in this chapter are consequences of these identities.

Recall from Theorem 0.4 that Ramanujan's theta function $f(a, b)$ may be expressed either as a series or as an infinite product by the formulas

$$f(a,b) = \sum_{j=-\infty}^{\infty} a^{j(j+1)/2} b^{j(j-1)/2} = \prod_{n=1}^{\infty} (1 + a^n b^{n-1})(1 + a^{n-1} b^n)(1 - a^n b^n).$$

In this notation, the function $r_8$ can be written as

$$r_8(q) = q^{1/2} \prod_{j=1}^{\infty} \frac{(1 - q^{8j-7})(1 - q^{8j-1})}{(1 - q^{8j-5})(1 - q^{8j-3})} = q^{1/2} \frac{f(-q, -q^7)}{f(-q^3, -q^5)}. \tag{8.9}$$

Let us also recall that by Theorem 0.5, Ramanujan's theta function satisfies the identity

$$f(a,b) = f(a^3 b, ab^3) + af\left(a^5 b^3, \frac{b}{a}\right). \tag{8.10}$$

**Lemma 8.4.** *Let $\epsilon \in \{1, -1, i, -i\}$ and let $f(a, b)$ be Ramanujan's theta function. The following identity holds:*

$$1 + \epsilon\, r_8(q^2) = \frac{f(\epsilon q, -\epsilon q^3)}{f(-q^6, -q^{10})}.$$

*Proof.* By (8.9),

$$\begin{aligned} 1 + \epsilon\, r_8(q^2) &= 1 + \epsilon q \frac{f(-q^2, -q^{14})}{f(-q^6, -q^{10})} \\ &= \frac{1}{f(\ q^6,\ q^{10})} \left\{ f(-q^6, -q^{10}) + \epsilon q f(-q^2, -q^{14}) \right\}. \end{aligned}$$

Now apply the identity (8.10) with $a = \epsilon q$ and $b = -\epsilon q^3$ to complete the proof. □

The next result provides factorizations for $1 \pm (1 \pm \sqrt{2})\, r_8$.

**Lemma 8.5.** *Let* $\omega = \exp(i\pi/4)$. *The following infinite product formulas hold*:

$$1-(1+\sqrt{2})\,r_8(q^2) = \prod_{j=1}^{\infty} \frac{(1-\omega\, q^j)\,(1-\omega^7\, q^j)\,(1-q^j)}{(1-q^{16j-10})(1-q^{16j-6})(1-q^{16j})},$$

$$1-(1-\sqrt{2})\,r_8(q^2) = \prod_{j=1}^{\infty} \frac{(1-\omega^3\, q^j)\,(1-\omega^5\, q^j)\,(1-q^j)}{(1-q^{16j-10})(1-q^{16j-6})(1-q^{16j})},$$

$$1+(1+\sqrt{2})\,r_8(q^2) = \prod_{j=1}^{\infty} \frac{(1-\omega(-q)^j)(1-\omega^7(-q)^j)\,(1-(-q)^j)}{(1-q^{16j-10})(1-q^{16j-6})(1-q^{16j})}$$

*and*

$$1+(1-\sqrt{2})\,r_8(q^2) = \prod_{j=1}^{\infty} \frac{(1-\omega^3(-q)^j)\,(1-\omega^5(-q)^j)\,(1-(-q)^j)}{(1-q^{16j-10})(1-q^{16j-6})(1-q^{16j})}.$$

*Proof.* Take $x = \omega$ in Theorem 0.3 to obtain

$$q\prod_{j=1}^{\infty}(1-\omega\, q^{8j})\,(1-\omega^7\, q^{8j})\,(1-q^{8j}) = \sum_{j=-\infty}^{\infty} q^{(4j+1)^2}\,\frac{\sin(4j+1)\pi/8}{\sin\pi/8}.$$

Since

$$\frac{\sin(4j+1)\pi/8}{\sin\pi/8} = \begin{cases} (-1)^k & \text{if } k = 2k, \\ -(-1)^k(1+\sqrt{2}) & \text{if } j = 2k-1, \end{cases}$$

this may be written as

$$q\prod_{j=1}^{\infty}(1-\omega\, q^{8j})\,(1-\omega^7\, q^{8j})\,(1-q^{8j})$$
$$= \sum_{k=-\infty}^{\infty}(-1)^k q^{(8k+1)^2} - (1+\sqrt{2})\sum_{k=-\infty}^{\infty}(-1)^k q^{(8k-3)^2}.$$

The series on the right may be summed by Jacobi's triple product identity (0.7) to give

$$q\prod_{j=1}^{\infty}(1-\omega\, q^{8j})\,(1-\omega^7\, q^{8j})\,(1-q^{8j})$$
$$= q(q^{48},q^{80},q^{128};q^{128})_\infty - (1+\sqrt{2})q^9(q^{16},q^{112},q^{128};q^{128})_\infty.$$

Divide by $q$ and replace $q^8$ with $q$ to obtain

$$\prod_{j=1}^{\infty}(1-\omega\, q^j)\,(1-\omega^7\, q^j)\,(1-q^j)$$

$$= (q^6, q^{10}, q^{16}; q^{16})_\infty \left\{1-(1+\sqrt{2})\, q\, \frac{(q^2, q^{14}; q^{16})_\infty}{(q^6, q^{10}; q^{16})_\infty}\right\}$$

$$= (q^6, q^{10}, q^{16}; q^{16})_\infty \left\{1-(1+\sqrt{2})\, r_8(q^2)\right\}.$$

This proves the first result. The second result can be proved in the same way by taking $x=\omega^3$ in Theorem 0.3 and noting that

$$\frac{\sin(4j+1)3\pi/8}{\sin 3\pi/8} = \begin{cases} (-1)^k & \text{if } k=2k, \\ -(-1)^k(1-\sqrt{2}) & \text{if } j=2k-1. \end{cases}$$

Finally, the third and fourth results follow by replacing $q$ with $-q$ in the first and second results, respectively. □

**Theorem 8.6.** *The following identities hold*:

$$\frac{1}{r_8} - r_8 = \frac{\eta_4^6}{\eta_2^2\eta_8^4}, \qquad \frac{1}{r_8} + r_8 = \frac{\eta_2^5}{\eta_1^2\eta_4\eta_8^2}, \tag{8.11}$$

$$\frac{1}{r_8} + 2 - r_8 = \frac{\eta_1^5\eta_4}{\eta_{1/2}^2\eta_2^2\eta_8^2}, \qquad \frac{1}{r_8} - 2 - r_8 = \frac{\eta_{1/2}^2\eta_4}{\eta_1\eta_8^2} \tag{8.12}$$

*and*

$$\frac{1}{r_8^2} - 6 + r_8^2 = \frac{\eta_1^4\eta_4^2}{\eta_2^2\eta_8^4}. \tag{8.13}$$

*Proof.* The proofs are by routine manipulations of infinite products, using the definition (8.1) and the results in Lemmas 8.4 and 8.5. First, by (8.9) and the cases $\epsilon=\pm 1$ of Lemma 8.4 we find that

$$\frac{1}{r_8(q^2)} - r_8(q^2) = \frac{(1+r_8(q^2))(1-r_8(q^2))}{r_8(q^2)}$$

$$= \frac{f(q,-q^3)f(-q,q^3)}{f(-q^6,-q^{10})^2} \times \frac{1}{q}\frac{f(-q^6,-q^{10})}{f(-q^2,-q^{14})}$$

$$= \frac{1}{q}\frac{f(q,-q^3)f(-q,q^3)}{f(-q^6,-q^{10})f(-q^2,-q^{14})}$$

$$= \frac{1}{q}\frac{(-q,q^3,q,-q^3,-q^4,-q^4;-q^4)_\infty}{(q^2,q^6,q^{10},q^{14},q^{16},q^{16};q^{16})_\infty}. \tag{8.14}$$

Using $(a,-a;-q)_\infty = (a^2;q^2)_\infty$, we deduce that

$$(q,-q;-q^4)_\infty (q^3,-q^3;-q^4)_\infty = (q^2;q^8)_\infty (q^6;q^8)_\infty = (q^2;q^4)_\infty.$$

It is easy to check that

$$(q^2,q^6,q^{10},q^{14};q^{16})_\infty = (q^2;q^4)_\infty.$$

Therefore, (8.14) simplifies to

$$\frac{1}{r_8(q^2)} - r_8(q^2) = \frac{1}{q}\frac{(-q^4;-q^4)_\infty^2}{(q^{16};q^{16})_\infty^2} = \frac{1}{q}\frac{E(-q^4)^2}{E(q^{16})^2}.$$

By (0.4) this simplifies further to

$$\frac{1}{r_8(q^2)} - r_8(q^2) = \frac{1}{q}\frac{E(q^8)^6}{E(q^4)^2E(q^{16})^4} = \frac{\eta_8^6}{\eta_4^2\eta_{16}^4}.$$

On replacing $q^2$ with $q$ we obtain the first result in (8.11).

The second identity in (8.11) is obtained in the same way by taking $\epsilon = \pm i$ in Lemma 8.4. We omit the details as they consist mainly of routine manipulations of infinite products.

Let us prove the identities in (8.12). By Corollary 3.4,

$$\varphi(q^4) + 2q\psi(q^8) = \varphi(q). \tag{8.15}$$

On dividing by $q\psi(q^8)$ and using Theorem 3.5 to express the result in terms of eta-quotients, we obtain

$$\frac{\eta_8^6}{\eta_4^2\eta_{16}^4} + 2 = \frac{\eta_2^5\eta_8}{\eta_1^2\eta_4^2\eta_{16}^2}.$$

By (8.11), this is equivalent to

$$\frac{1}{r_8(q^2)} - r_8(q^2) + 2 = \frac{\eta_2^5\eta_8}{\eta_1^2\eta_4^2\eta_{16}^2},$$

and replacing $q$ with $q^{1/2}$ gives the first identity in (8.12). The second identity in (8.12) can be proved by following the same procedure after first replacing $q$ with $-q$ in (8.15); we omit the details. Alternatively, the identities in (8.12) can be proved by taking the product of identities in Lemma 8.5.

The identity (8.13) is obtained by multiplying the identities in (8.12). For,

$$\frac{1}{r_8^2} - 6 + r_8^2 = \left(\frac{1}{r_8} - 2 - r_8\right)\left(\frac{1}{r_8} + 2 - r_8\right)$$
$$= \left(\frac{\eta_{1/2}^2\eta_4}{\eta_1\eta_8^2}\right) \times \left(\frac{\eta_1^5\eta_4}{\eta_{1/2}^2\eta_2^2\eta_8^2}\right) = \frac{\eta_1^4\eta_4^2}{\eta_2^2\eta_8^4}.$$

□

Properties of the allied function $\xi_8$ are inherited from the corresponding properties of $r_8$ via the transformation formula (8.4). We present just three properties in the next theorem, which will be useful later.

**Theorem 8.7.** *Let* $\alpha = \sqrt{2} - 1$. *The following identities hold*:

$$\frac{1}{\xi_8} - \xi_8 = \frac{2\,\eta_2^6}{\eta_1^4\eta_4^2}, \qquad \frac{1}{\xi_8} + \xi_8 = \frac{2\sqrt{2}\,\eta_4^5}{\eta_1^2\eta_2\eta_8^2}, \tag{8.16}$$

*and*

$$\xi_8(q) = \frac{\alpha - r_8(q^2)}{1 + \alpha r_8(q^2)}. \tag{8.17}$$

*Proof.* By (8.11) we have

$$\frac{1}{r_8(e^{2\pi i\tau})} - r_8(e^{2\pi i\tau}) = \frac{\eta^6(4\tau)}{\eta^2(2\tau)\eta^4(8\tau)}.$$

Replace $\tau$ with $-1/8\tau$ to obtain

$$\frac{1}{r_8(e^{-\pi i/4\tau})} - r_8(e^{-\pi i/4\tau}) = \frac{\eta^6(-1/2\tau)}{\eta^2(-1/4\tau)\eta^4(-1/\tau)}.$$

The terms of the left-hand side may be expressed in terms of the function $\xi_8$ by (8.4), and the right-hand side can be transformed by Theorem 2.8. We deduce that

$$\frac{1}{\xi_8(e^{2\pi i\tau})} - \xi_8(e^{2\pi i\tau}) = 2 \times \frac{\eta^6(2\tau)}{\eta^2(4\tau)\eta^4(\tau)}.$$

This gives the first identity in (8.16). The second identity in (8.16) can be proved in the same way and we omit the details as they are similar.

It remains to prove (8.17). By the first two results in Lemma 8.5 and the definition of $\xi$ in (8.3) we find that

$$\frac{1-(1+\sqrt{2})\,r_8(q^2)}{1-(1-\sqrt{2})\,r_8(q^2)} = \prod_{j=1}^{\infty}\frac{(1-\omega q^j)(1-\omega^7 q^j)}{(1-\omega^3 q^j)(1-\omega^5 q^j)} = \frac{1}{\sqrt{2}-1}\,\xi_8(q).$$

The proof may be completed by multiplying both sides by $\sqrt{2}-1$. □

**Theorem 8.8.** *As $q$ increases from 0 to 1, $r_8(q)$ strictly increases from 0 to $\sqrt{2}-1$.*

*Proof.* By Theorem 8.6 and (4.11),

$$\frac{1}{r_8} - r_8 = \frac{\eta_4^6}{\eta_2^2\eta_8^4} = \frac{2}{\sqrt[4]{x_4(q^2)}}.$$

On solving for $r_8$ and selecting the solution that is positive when $0 < q < 1$ we find that

$$r_8 = \sqrt{\frac{1}{t^2}+1} - \frac{1}{t} = \frac{t}{\sqrt{t^2+1}+1}$$

where

$$t = \sqrt[4]{x_4(q^2)}.$$

By Theorem 4.8, $t$ increases from 0 to 1 as $q$ increases from 0 to 1. Moreover, the function

$$t \to \frac{t}{\sqrt{t^2+1}+1}$$

increases from 0 to $1/(\sqrt{2}+1) = \sqrt{2}-1$ as $t$ increases from 0 to 1. The claimed result follows by composition of functions. □

The next result gives a quadratic transformation formula that is satisfied by the Ramanujan–Göllnitz–Gordon continued fraction. It is an analogue of Theorems 5.11 and 6.13 for the Rogers–Ramanujan and cubic continued fractions, respectively.

**Theorem 8.9.** *Let $u = r_8(q)$ and $v = r_8(q^2)$. Let $\alpha = \sqrt{2}-1$. Then*

$$u^2 = v\frac{(1-v)}{(1+v)}. \tag{8.18}$$

*and*

$$v = \frac{\alpha - \left(\frac{\alpha^2-u^2}{1-\alpha^2u^2}\right)^{1/2}}{1+\alpha\left(\frac{\alpha^2-u^2}{1-\alpha^2u^2}\right)^{1/2}}. \tag{8.19}$$

*Proof.* By (8.9) and taking $\epsilon = \pm 1$ in Lemma 8.4, we have

$$\begin{aligned} r_8(q^2)\,\frac{(1-r_8(q^2))}{(1+r_8(q^2))} &= q\,\frac{f(-q^2,-q^{14})}{f(-q^6,-q^{10})} \times \frac{f(-q,q^3)}{f(q,-q^3)} \\ &= q\,\frac{(q^2,q^{14};q^{16})_\infty}{(q^6,q^{10};q^{16})_\infty} \times \frac{(q,-q^3;-q^4)_\infty}{(-q,q^3;-q^4)_\infty}. \end{aligned}$$

On using the product identities

$$(a^2;q^2)_\infty = (a,-a;q)_\infty \qquad \text{and} \qquad (a;-q)_\infty = (a;q^2)_\infty(-aq;q^2)_\infty$$

this becomes

$$\begin{aligned} r_8(q^2)\,\frac{(1-r_8(q^2))}{(1+r_8(q^2))} &= q\,\frac{(q,-q,q^7,-q^7;q^8)_\infty}{(q^3,-q^3,q^5,-q^5;q^8)_\infty} \times \frac{(q,-q^3,-q^5,q^7;q^8)_\infty}{(-q,q^3,q^5,-q^7;q^8)_\infty} \\ &= q\,\frac{(q,q^7;q^8)^2_\infty}{(q^3,q^5;q^8)^2_\infty} \\ &= r_8(q)^2. \end{aligned}$$

This gives (8.18). Next, the identity (8.18) can be rearranged to give

$$\frac{\alpha^2-u^2}{1-\alpha^2u^2} = \left(\frac{\alpha-v}{1+\alpha v}\right)^2$$

and this may be solved for $v$ to give (8.19). □

The next result gives a modular transformation for $r_8$. A different transformation formula will be given in Theorem 8.12.

**Theorem 8.10.** *Let* $\alpha = \sqrt{2}-1$ *and suppose* $\mathrm{Re}(t) > 0$. *The following transformation formula holds*:

$$r_8(e^{-\pi/t}) = \frac{\alpha - r_8(e^{-\pi t})}{1+\alpha r_8(e^{-\pi t})}. \tag{8.20}$$

*Furthermore, the following evaluation holds*:

$$r_8(e^{-\pi}) = \sqrt{4+2\sqrt{2}} - \sqrt{2} - 1. \tag{8.21}$$

*Proof.* Put $q = e^{-\pi t}$ in (8.17) and apply (8.4) to obtain

$$\frac{\alpha - r_8(e^{-2\pi t})}{1 + \alpha r_8(e^{-2\pi t})} = \xi_8(e^{-\pi t}) = r_8(e^{-\pi/2t}).$$

Replace $t$ with $t/2$ to obtain (8.20). The identity (8.20) may be rearranged to give the symmetric form

$$\Big(1 + \alpha r_8(e^{-\pi t})\Big)\Big(1 + \alpha r_8(e^{-\pi/t})\Big) = 1 + \alpha^2.$$

Now take $t = 1$ to obtain

$$1 + \alpha r_8(e^{-\pi}) = \sqrt{1+\alpha^2},$$

where the positive sign of the square root is taken because $1 + \alpha r_8(e^{-\pi})$ is clearly positive. On solving for $r_8(e^{-\pi})$ we obtain the required evaluation given by (8.21). □

## 8.4 The modular function $x_8$

Of the five eta-quotients in Theorem 8.6, the function in (8.13) stands out as having a high degree of symmetry. Take its reciprocal and consider the function $x_8$ defined by

$$x_8 = x_8(q) = \frac{r_8^2(q)}{1 - 6r_8^2(q) + r_8^4(q)}. \tag{8.22}$$

Since the function

$$r \to \frac{r^2}{1 - 6r^2 + r^4}$$

increases from 0 to $+\infty$ as $r$ increases from 0 to $\sqrt{2}-1$, it follows from Theorem 8.8 that $x_8$ increases from 0 to $+\infty$ as $q$ increases from 0 to 1.

Here are some identities for $x_8$ that involve eta-quotients.

**Theorem 8.11.** *The following formulas involving infinite products hold:*

$$x_8 = \frac{\eta_2^2\eta_8^4}{\eta_1^4\eta_4^2}, \qquad 1+4x_8 = \frac{\eta_4^{10}}{\eta_1^4\eta_2^2\eta_8^4} \qquad \text{and} \qquad 1+8x_8 = \frac{\eta_2^{12}}{\eta_1^8\eta_4^4}. \tag{8.23}$$

*Hence,*

$$\frac{1}{x_8} + 12 + 32x_8 = \frac{\eta_2^8\eta_4^8}{\eta_1^8\eta_8^8}, \tag{8.24}$$

$$-x_8(-q) = \frac{x_8(q)}{1+8x_8(q)} \tag{8.25}$$

*and*

$$1+8x_8(q^2) = \frac{1+4x_8(q)}{\sqrt{1+8x_8(q)}}.$$

*Proof.* The infinite product formula for $x_8$ is immediate from the definition (8.22) and the result (8.13) in Theorem 8.6. Next, by the definition (8.22) and simple rearrangement, we find that

$$1+4x_8 = 1 + \frac{4r_8^2}{1-6r_8^2+r_8^4} = \frac{1-2r_8^2+r_8^4}{1-6r_8^2+r_8^4} = \frac{\left(\frac{1}{r_8}-r_8\right)^2}{\frac{1}{r_8^2}-6+r_8^2}.$$

Hence, by Theorem 8.6,

$$1+4x_8 = \frac{\left(\frac{1}{r_8}-r_8\right)^2}{\frac{1}{r_8^2}-6+r_8^2} = \left(\frac{\eta_4^6}{\eta_2^2\eta_8^4}\right)^2 \Big/ \left(\frac{\eta_1^4\eta_4^2}{\eta_2^2\eta_8^4}\right) = \frac{\eta_4^{10}}{\eta_1^4\eta_2^2\eta_8^4}.$$

In a similar way, we find that

$$1+8x_8 = \frac{\left(\frac{1}{r_8}+r_8\right)^2}{\frac{1}{r_8^2}-6+r_8^2} = \left(\frac{\eta_2^5}{\eta_1^2\eta_4\eta_8^2}\right)^2 \Big/ \left(\frac{\eta_1^4\eta_4^2}{\eta_2^2\eta_8^4}\right) = \frac{\eta_2^{12}}{\eta_1^8\eta_4^4}.$$

Next, by the results obtained above,

$$\begin{aligned} \frac{1}{x_8} + 12 + 32x_8 &= \frac{(1+4x_8)(1+8x_8)}{x_8} \\ &= \left(\frac{\eta_4^{10}}{\eta_1^4\eta_2^2\eta_8^4}\right) \times \left(\frac{\eta_2^{12}}{\eta_1^8\eta_4^4}\right) \Big/ \left(\frac{\eta_2^2\eta_8^4}{\eta_1^4\eta_4^2}\right) = \frac{\eta_2^8\eta_4^8}{\eta_1^8\eta_8^8} \end{aligned}$$

and this proves (8.24). To prove (8.25), write

$$x_8(q) = \frac{\eta_2^2\eta_8^4}{\eta_1^4\eta_4^2} = q\,\frac{E^2(q^2)E^4(q^8)}{E^4(q)E^2(q^4)}.$$

Now replace $q$ with $-q$. On applying (0.4) along with the results in (8.23) we find that

$$\begin{aligned}
-x_8(-q) &= q\,\frac{E^2(q^2)E^4(q^8)}{E^4(-q)E^2(q^4)}\\
&= q\,\frac{E^4(q)E^2(q^4)E^4(q^8)}{E^{10}(q^2)}\\
&= \left(\frac{\eta_2^2\eta_8^4}{\eta_1^4\eta_4^2}\right)\Big/\left(\frac{\eta_2^{12}}{\eta_1^8\eta_4^4}\right)\\
&= \frac{x_8(q)}{1+8x_8(q)}.
\end{aligned}$$

Finally, by the results in (8.23) we deduce that

$$1+8x_8(q^2) = \frac{\eta_4^{12}}{\eta_2^8\eta_8^4} = \frac{\eta_4^{10}}{\eta_1^4\eta_2^2\eta_8^4}\times\frac{\eta_1^4\eta_4^2}{\eta_2^6} = (1+4x_8(q))\times\frac{1}{\sqrt{1+8x_8(q)}}.$$

□

The function $x_8$ has a particularly simple transformation formula because of the high degree of symmetry in its eta-quotient.

**Theorem 8.12.** *The following transformation formula holds for* $\mathrm{Re}\, t > 0$:

$$x_8\left(e^{-2\pi/\sqrt{8t}}\right)\times x_8\left(e^{-2\pi\sqrt{t/8}}\right) = \frac{1}{32}. \tag{8.26}$$

*Hence,*

$$x_8\left(e^{-\pi/\sqrt{2}}\right) = \frac{1}{\sqrt{32}} \tag{8.27}$$

*and*

$$x_8\left(-e^{-\pi/\sqrt{2}}\right) = -\left(\frac{2-\sqrt{2}}{8}\right). \tag{8.28}$$

*Proof.* By (8.23),

$$x_8(e^{2\pi i\tau}) = \frac{\eta^2(2\tau)\eta^4(8\tau)}{\eta^4(\tau)\eta^2(4\tau)}.$$

Replacing $\tau$ with $-1/8\tau$ and using the modular transformation for the Dedekind eta function given by Theorem 2.8 we find that

$$x_8(e^{-\pi i/4\tau}) = \frac{\eta^2(-1/4\tau)\eta^4(-1/\tau)}{\eta^4(-1/8\tau)\eta^2(-1/2\tau)} = \frac{1}{32} \times \frac{\eta^2(4\tau)\eta^4(\tau)}{\eta^4(8\tau)\eta^2(2\tau)} = \frac{1}{32\, x_8(e^{2\pi i\tau})}.$$

The transformation formula (8.26) follows by putting $\tau = i\sqrt{t/8}$.

Next, taking $t = 1$ in (8.26) and noting that $x_8\left(e^{-\pi/\sqrt{2}}\right)$ is positive, we obtain (8.27). Finally, by (8.25) and (8.27),

$$-x_8\left(-e^{-\pi/\sqrt{2}}\right) = \frac{x_8\left(e^{-\pi/\sqrt{2}}\right)}{1 + 8x_8\left(e^{-\pi/\sqrt{2}}\right)} = \frac{1/\sqrt{32}}{1 + 8/\sqrt{32}}.$$

This gives (8.28), after simplification. □

The definition (8.22) means that any evaluation of $x_8(q)$ implies an evaluation of $r_8(q)$, and vice versa. In general, the evaluations of $x_8$ are simpler. For example, by (8.21) we have

$$r_8(e^{-\pi}) = \sqrt{4 + 2\sqrt{2}} - \sqrt{2} - 1$$

whereas (8.22) gives

$$x_8(e^{-\pi}) = \frac{\sqrt{2} - 1}{8}.$$

## 8.5 Weight one modular forms

In this section we give several results that are analogues of the sum of two squares and sum of two triangular numbers identities in Theorem 3.10. The main identities are (8.29) and (8.36) that provide series expansions for the theta functions $\varphi(q)\varphi(q^2)$ and $q^3\psi(q^8)\psi(q^{16})$.

**Theorem 8.13.** *The following identities hold*:

$$\varphi(q)\varphi(q^2) = \frac{\eta_2^3\eta_4^3}{\eta_1^2\eta_8^2} = 1 + 2\sum_{j=1}^{\infty} \chi_{-2}(j)\frac{q^j}{1-q^j}, \tag{8.29}$$

$$\varphi(-q)\varphi(q^2) = \frac{\eta_1^2\eta_4^5}{\eta_2^3\eta_8^2} = 1 - 2\sum_{j=1}^{\infty} \chi_{-2}(j)\frac{q^j}{1+q^j}, \tag{8.30}$$

$$q\psi(q^8)\varphi(q^2) = \frac{\eta_4^5\eta_{16}^2}{\eta_2^2\eta_8^3} = \sum_{j=1}^{\infty} \chi_{-2}(j)\frac{q^j}{1-q^{2j}} \tag{8.31}$$

*and*

$$q\psi(q^8)\varphi(-q^2) = \frac{\eta_2^2\eta_{16}^2}{\eta_4\eta_8} = \sum_{j=1}^{\infty} \chi_2(j)\frac{q^j}{1+q^{2j}}. \tag{8.32}$$

*Proof.* The first equality in each of the identities (8.29)–(8.32) is immediate from Theorem 3.5. It remains to prove the second equality in each identity.

The identity (8.29) has been proved in Exercise 8 at the end of Chapter 3. The results in (8.30) can be obtained from (8.29) by replacing $q$ with $-q$ and using (0.4). Let us prove (8.31). Replace $(q, x, y)$ with $(q^4, iq, q)$ in Theorem 0.26 and multiply by $q/(1+i)$ to obtain

$$q\frac{(iq^4, -iq^4, q^4, q^4; q^4)_\infty}{(q^2, q^2, iq^2, -iq^2; q^4)_\infty} = \frac{1}{1+i}\sum_{k=-\infty}^{\infty} \frac{i^k q^{2k+1}}{1-q^{4k+2}}.$$

The product simplifies to

$$\begin{aligned} q\frac{(q^4, q^4, iq^4, -iq^4; q^4)_\infty}{(q^2, q^2, iq^2, -iq^2; q^4)_\infty} &= q\frac{(q^4; q^4)_\infty^2}{(q^2; q^4)_\infty^2}\frac{(-q^8; q^8)_\infty}{(-q^4; q^8)_\infty} \\ &= q\frac{(q^4; q^4)_\infty^4}{(q^2; q^2)_\infty^2}\frac{(-q^8; q^8)_\infty^2}{(-q^4; q^4)_\infty} \\ &= \frac{\eta_4^5\eta_{16}^2}{\eta_2^2\eta_8^3} \end{aligned}$$

while the series is

$$\frac{1}{1+i}\sum_{k=-\infty}^{\infty} \frac{i^k q^{2k+1}}{1-q^{4k+2}} = \frac{1}{1+i}\left(\sum_{k=0}^{\infty} \frac{i^k q^{2k+1}}{1-q^{4k+2}} + \sum_{k=-1}^{-\infty} \frac{i^k q^{2k+1}}{1-q^{4k+2}}\right)$$

$$= \frac{1}{1+i}\left(\sum_{k=0}^{\infty}\frac{i^k q^{2k+1}}{1-q^{4k+2}} + \sum_{k=0}^{\infty}\frac{i^{-k-1}q^{-2k-1}}{1-q^{-4k-2}}\right)$$

$$= \sum_{k=0}^{\infty}\left(\frac{i^k - i^{-k-1}}{1+i}\right)\frac{q^{2k+1}}{1-q^{4k+2}}$$

$$= \sum_{j=1}^{\infty}\chi_{-2}(j)\frac{q^j}{1-q^{2j}}.$$

This completes the proof of (8.31).

The results in (8.32) can be obtained from (8.31) by replacing $q$ with $iq$ and dividing by $i$, and applying (0.4). □

**Corollary 8.14.** *The following identity holds:*

$$\varphi(q)\varphi(q^8)$$
$$= 1 + 2\sum_{n=1}^{\infty}\chi_{-2}(n)\frac{q^{4n}}{1-q^{4n}} + \sum_{n=1}^{\infty}\chi_{-2}(n)\frac{q^n}{1-q^{2n}} + \sum_{n=1}^{\infty}\chi_2(n)\frac{q^n}{1+q^{2n}}.$$

*Proof.* By the properties of theta functions in Lemma 3.3 we deduce that

$$\begin{aligned}\varphi(q)\varphi(q^8) &= \left(\varphi(q^4) + 2q\psi(q^8)\right)\varphi(q^8)\\ &= \varphi(q^4)\varphi(q^8) + 2q\psi(q^8)\varphi(q^8)\\ &= \varphi(q^4)\varphi(q^8) + q\psi(q^8)\varphi(q^2) + q\psi(q^8)\varphi(-q^2).\end{aligned}$$

Now apply the results (8.29), (8.31), and (8.32) to complete the proof. □

**Theorem 8.15.** *The following results hold:*

$$\varphi(-q^2)\varphi(-q^4) = 1 + 2\sum_{j=1}^{\infty}(-1)^{j(j+1)/2}\left(\frac{q^{2j} + (-1)^j q^{6j}}{1+q^{8j}}\right) \tag{8.33}$$

$$= 1 + \sqrt{2}\sum_{j=1}^{\infty}\sum_{k=1}^{\infty}q^{jk}\sin\frac{(j+3k)\pi}{4}. \tag{8.34}$$

*and*

$$q^3\psi(q^8)\psi(q^{16}) = \sum_{j=-\infty}^{\infty}\frac{q^{8j+3}}{1-q^{64j+24}} \tag{8.35}$$

$$= \sum_{j=0}^{\infty}\sum_{k=0}^{\infty}\left\{q^{(8j+1)(8k+3)} - q^{(8j+5)(8k+7)}\right\}. \tag{8.36}$$

*Proof.* Replace $(q,x,y)$ with $(q^2,-iq,i)$ in Theorem 0.26 and multiply by $1-i$ to deduce that

$$\frac{(q,q,q^2,q^2;q^2)_\infty}{(iq^2,-iq^2,iq,-iq;q^2)_\infty} = (1-i)\sum_{j=-\infty}^{\infty}\frac{(-iq)^j}{1-iq^{2j}}.$$

The product is, using Theorem 3.5,

$$\frac{(q;q)_\infty^2}{(iq,-iq;q)_\infty} = \frac{(q;q)_\infty^2}{(-q^2;q^2)_\infty} = \frac{(q;q)_\infty^2(q^2;q^2)_\infty}{(q^4;q^4)_\infty} = \varphi(-q)\varphi(-q^2)$$

and the series is

$$\begin{aligned}(1-i)\sum_{j=-\infty}^{\infty}\frac{(-iq)^j}{1-iq^{2j}} &= 1+(1-i)\sum_{j=1}^{\infty}\left\{\frac{(-iq)^j}{1-iq^{2j}}+\frac{i(iq)^j}{1+iq^{2j}}\right\}\\ &= 1+(1-i)\sum_{j=1}^{\infty}\frac{q^j(i^{3j}+i^{j+1})+q^{3j}(i^{3j+1}-i^{j+2})}{1+q^{4j}}\\ &= 1+2\sum_{j=1}^{\infty}(-1)^{j(j+1)/2}\left(\frac{q^j+(-1)^jq^{3j}}{1+q^{4j}}\right).\end{aligned}$$

Combining the last two results, and replacing $q$ with $q^2$ we obtain (8.33).

The identity (8.34) may be obtained by taking $x=\exp(i\pi/4)$ and $y=\exp(3\pi i/4)$ in Theorem 0.27.

Proofs for the identities for $q^3\psi(q^8)\psi(q^{16})$ were outlined in Exercise 8 at the end of Chapter 3. □

## 8.6 Weight two modular forms

This section contains results that are analogues of the sum of four squares identity in Theorem 3.26. The first theorem involves Ramanujan's Eisenstein series $P$.

**Theorem 8.16.** *The following identities hold*:

$$\varphi^2(q)\varphi^2(q^2) = \frac{\eta_2^6\eta_4^6}{\eta_1^4\eta_8^4} = \frac{1}{6}\left(-P(q)+P(q^2)-2P(q^4)+8P(q^8)\right), \qquad (8.37)$$

$$\varphi^2(q^2)\varphi^2(-q) = \frac{\eta_1^4\eta_4^{10}}{\eta_2^6\eta_8^4} = \frac{1}{6}\left(P(q)-5P(q^2)+2P(q^4)+8P(q^8)\right), \qquad (8.38)$$

$$q\varphi^2(q)\psi^2(q^4) = \frac{\eta_2^{10}\eta_8^4}{\eta_1^4\eta_4^6} = \frac{1}{24}\left(-P(q) - P(q^2) + 10P(q^4) - 8P(q^8)\right), \tag{8.39}$$

$$q\varphi^2(-q)\psi^2(q^4) = \frac{\eta_1^4\eta_8^4}{\eta_2^2\eta_4^2} = \frac{1}{24}\left(-P(q) + 7P(q^2) - 14P(q^4) + 8P(q^8)\right), \tag{8.40}$$

$$q\varphi^2(q^4)\psi^2(q^4) = \frac{\eta_8^{14}}{\eta_4^6\eta_{16}^4} = -\frac{1}{96}\left(P(q) - iP(iq) + i^2P(i^2q) - i^3P(i^3q)\right), \tag{8.41}$$

$$q^3\psi^2(q^4)\psi^2(q^8) = \frac{\eta_8^2\eta_{16}^4}{\eta_4^2} = -\frac{1}{384}\left(P(q) + iP(iq) + i^2P(i^2q) + i^3P(i^3q)\right) \tag{8.42}$$

*and*

$$\varphi^2(-q)\varphi^2(-q^2) = \frac{\eta_1^4\eta_2^2}{\eta_4^2} = 1 + 4\sum_{j=1}^{\infty}\frac{(-1)^j(2j-1)q^{2j-1}}{1+q^{4j-2}} + 8\sum_{j=1}^{\infty}\frac{(-1)^j j q^{4j}}{1+q^{4j}}, \tag{8.43}$$

*where* $i^2 = -1$.

*Proof.* The first equality in each of the identities (8.37)–(8.43) is immediate from Theorem 3.5. It remains to prove the second equality in each identity. Applying Corollary 3.4 and then Lemma 3.6 we find that

$$\varphi^2(q)\varphi^2(q^2) = \left(\varphi^2(q^2) + 4q\psi^2(q^4)\right)\varphi^2(q^2) = \varphi^4(q^2) + 4q\psi^4(q^2).$$

The terms on the right-hand side can be expanded by Theorem 3.26 to give

$$\begin{aligned}\varphi^2(q)\varphi^2(q^2) &= \frac{1}{3}\left(4P(q^8) - P(q^2)\right) + \frac{1}{6}\left(-P(q) + 3P(q^2) - 2P(q^4)\right)\\ &= \frac{1}{6}\left(-P(q) + P(q^2) - 2P(q^4) + 8P(q^8)\right).\end{aligned}$$

This proves (8.37). The identity (8.38) may be obtained from (8.37) by replacing $q$ with $-q$ and using (0.110).

Now we will prove (8.39). By Corollary 3.4 and then Lemma 3.6 we deduce that

$$q\varphi^2(q)\psi^2(q^4) = \left(\varphi^2(q^2) + 4q\psi^2(q^4)\right)q\psi^2(q^4) = q\psi^4(q^2) + 4q^2\psi^4(q^4).$$

Applying Theorem 3.26 we deduce that

$$
\begin{aligned}
&q\varphi^2(q)\psi^2(q^4)\\
&= \frac{1}{24}\left(-P(q)+3P(q^2)-2P(q^4)\right)+\frac{1}{6}\left(-P(q^2)+3P(q^4)-2P(q^8)\right)\\
&= \frac{1}{24}\left(-P(q)-P(q^2)+10P(q^4)-8P(q^8)\right).
\end{aligned}
$$

This proves (8.39). The identity (8.40) follows from (8.39) by replacing $q$ with $-q$ and using (0.110).

Let us prove (8.41) and (8.42). Applying Corollary 3.4 twice to (8.39) we find that

$$
\begin{aligned}
&\frac{1}{24}\left(-P(q)-P(q^2)+10P(q^4)-8P(q^8)\right)\\
&= q\varphi^2(q)\psi^2(q^4)\\
&= q\left(\varphi^2(q^2)+4q\psi^2(q^4)\right)\psi^2(q^4)\\
&= q\left(\varphi^2(q^4)+4q\psi^2(q^4)+4q^2\psi^2(q^8)\right)\psi^2(q^4).
\end{aligned}
$$

The right-hand side gives a 4-dissection. If we extract the terms involving powers of $q^{4j+1}$, the result is (8.41). Extracting the terms that involve powers of $q^{4j+3}$, we obtain (8.42).

Finally, to prove (8.43), take $x = i$ in the second part of Theorem 0.28 to obtain

$$
\prod_{j=1}^{\infty}\frac{(1-q^j)^4}{(1-iq^j)^2(1+iq^j)^2} = 1-2\sum_{j=1}^{\infty}\sum_{k=1}^{\infty} jq^{jk}(i^{j-k}+i^{k-j}).
$$

On simplifying both the product and the series, we deduce

$$
\begin{aligned}
\frac{\eta_1^4\eta_2^2}{\eta_4^2} &= 1-4\sum_{j \text{ odd}}\sum_{k \text{ odd}} jq^{jk}(-1)^{(j-k)/2}-4\sum_{j \text{ even}}\sum_{k \text{ even}} jq^{jk}(-1)^{(j-k)/2}\\
&= 1+4\sum_{j=1}^{\infty}\frac{(2j-1)q^{2j-1}(-1)^j}{1+q^{4j-2}}+4\sum_{j=1}^{\infty}\frac{2jq^{4j}(-1)^j}{1+q^{4j}}.
\end{aligned}
$$

This completes the proof of the Theorem 8.16. □

The series in the next two theorems involve the character $\chi_2$.

**Theorem 8.17.** *The following identities hold:*

$$
\varphi(-q)\varphi(-q^2)\varphi^2(-q^4) = 1-2\sum_{j=1}^{\infty}\chi_2(j)\frac{jq^j}{1-q^j}
$$

*and*

$$q\psi^2(q)\psi(q^2)\psi(q^4) = \sum_{j=1}^{\infty} \frac{j(q^j - q^{3j} - q^{5j} + q^{7j})}{1 - q^{8j}}.$$

*Proof.* To prove the first result, let $\omega = \exp(i\pi/4)$ and take $x = \omega$, $y = \omega^3$, in Theorem 1.17 and follow the procedure in Section 1.7. The second result may be obtained by replacing $q$ with $q^4$ in Theorem 1.18 and then letting $a = q$, $b = q^2$. □

The next set of results imply formulas, in terms of divisor sums, for the number of solutions in integers for each of

$$x^2 + y^2 + z^2 + 2z^2 = n,$$
$$x^2 + 2y^2 + 2z^2 + 2z^2 = n,$$
$$x^2 + y^2 + 2z^2 + 4z^2 = n$$

and

$$x^2 + 2y^2 + 4z^2 + 4z^2 = n.$$

**Theorem 8.18.** *Let*

$$f_1(q) = 1 - 2\sum_{j=1}^{\infty} \chi_2(j)\frac{jq^j}{1 - q^j}$$

*and*

$$f_2(q) = \sum_{j=1}^{\infty} \frac{j(q^j - q^{3j} - q^{5j} + q^{7j})}{1 - q^{8j}}.$$

*Then*

$$\varphi^3(q)\varphi(q^2) = f_1(q) + 8f_2(q),$$
$$\varphi(q)\varphi^3(q^2) = f_1(q) + 4f_2(q),$$
$$\varphi^2(q)\varphi(q^2)\varphi(q^4) = f_1(q^2) + 4f_2(q)$$

*and*

$$\varphi(q)\varphi(q^2)\varphi^2(q^4) = f_1(q^2) + 2f_2(q).$$

*Proof.* We shall make repeated use of the elementary properties of theta functions given by Lemmas 3.3 and 3.6. By (3.10) followed by two applications of each of (3.11) and (3.12), we find that

$$\begin{aligned}\varphi^3(q)\varphi(q^2) &= \varphi(q)\varphi(q^2)\left\{\varphi^2(-q) + 8q\psi^2(q^4)\right\}\\ &= \varphi(-q)\varphi^2(-q^2)\varphi(q^2) + 8q\varphi(q)\psi^2(q^2)\psi(q^4)\\ &= \varphi(-q)\varphi(-q^2)\varphi^2(-q^4) + 8q\psi^2(q)\psi(q^2)\psi(q^4). \end{aligned} \tag{8.44}$$

Similarly, by (3.9) and (3.10), followed by two applications of each of (3.11) and (3.12), we deduce

$$\begin{aligned}\varphi(q)\varphi^3(q^2) &= \varphi(q)\varphi(q^2)\left\{\varphi^2(-q) + 4q\psi^2(q^4)\right\}\\ &= \varphi(-q)\varphi^2(-q^2)\varphi(q^2) + 4q\varphi(q)\psi^2(q^2)\psi(q^4)\\ &= \varphi(-q)\varphi(-q^2)\varphi^2(-q^4) + 4q\psi^2(q)\psi(q^2)\psi(q^4). \end{aligned} \tag{8.45}$$

By (3.8) followed by three applications of each of (3.11) and (3.12), we obtain

$$\begin{aligned}\varphi^2(q)\varphi(q^2)\varphi(q^4) &= \varphi(q)\varphi(q^2)\varphi(q^4)\left\{\varphi(-q) + 4q\psi(q^8)\right\}\\ &= \varphi^2(-q^2)\varphi(q^2)\varphi(q^4) + 4q\varphi(q)\varphi(q^2)\psi^2(q^4)\\ &= \varphi(-q^2)\varphi^2(-q^4)\varphi(q^4) + 4q\varphi(q)\psi^2(q^2)\psi(q^4)\\ &= \varphi(-q^2)\varphi(-q^4)\varphi^2(-q^8) + 4q\psi^2(q)\psi(q^2)\psi(q^4). \end{aligned} \tag{8.46}$$

Similarly, by (3.7) and (3.8), followed by three applications of each of (3.11) and (3.12), we find that (omitting some details which are similar to the above)

$$\begin{aligned}\varphi(q)\varphi(q^2)\varphi^2(q^4) &= \varphi(q)\varphi(q^2)\varphi(q^4)\left\{\varphi(-q) + 2q\psi(q^8)\right\}\\ &= \varphi(-q^2)\varphi(-q^4)\varphi^2(-q^8) + 2q\psi^2(q)\psi(q^2)\psi(q^4). \end{aligned} \tag{8.47}$$

Substituting the results of Theorem 8.17 into the right-hand sides of (8.44)–(8.47), we complete the proof. □

## 8.7 Weight three modular forms

The results in this section for weight three modular forms involve the character $\chi_{-2}$.

**Theorem 8.19.** *The following identities hold:*

$$\sum_{j=1}^{\infty} \frac{j^2(q^j + q^{3j} - q^{5j} - q^{7j})}{1 - q^{8j}} = \frac{\eta_2^3\eta_8^6}{\eta_1^2\eta_4}\left(\frac{1}{r_8^2} + r_8^2\right) \tag{8.48}$$

*and*

$$1-\frac{2}{3}\sum_{j=1}^{\infty}\chi_{-2}(j)\frac{j^2q^j}{1-q^j}=\frac{1}{6}\frac{\eta_1^6\eta_4^3}{\eta_2\eta_8^2}\left(\frac{1}{\xi_8^2}+\xi_8^2\right). \tag{8.49}$$

*Proof.* Replace $q$ with $q^8$ in Theorem 1.23, then let $x=q$ and simplify the infinite product to deduce that

$$\sum_{j=1}^{\infty}\frac{j^2(q^j-q^{7j})}{1-q^{8j}}=q\frac{(q^2,q^6;q^8)_\infty(q^8;q^8)_\infty^6}{(q,q^7;q^8)_\infty^4}=q\frac{(q^2;q^2)_\infty(q^8;q^8)_\infty^6}{(q^4;q^4)_\infty(q,q^7;q^8)_\infty^4}.$$

Now multiply numerator and denominator by $(q^3,q^5;q^8)_\infty^2$ and manipulate the resulting infinite products to obtain

$$\begin{aligned}\sum_{j=1}^{\infty}\frac{j^2(q^j-q^{7j})}{1-q^{8j}}&=q\frac{(q^2;q^2)_\infty(q^8;q^8)_\infty^6}{(q^4;q^4)_\infty}\times\frac{1}{(q,q^7;q^8)_\infty^4}\times\frac{(q^3,q^5;q^8)_\infty^2}{(q^3,q^5;q^8)_\infty^2}\\&=q\frac{(q^2;q^2)_\infty(q^8;q^8)_\infty^6}{(q^4;q^4)_\infty}\times\frac{1}{(q;q^2)_\infty^2}\times\frac{(q^3,q^5;q^8)_\infty^2}{(q,q^7;q^8)_\infty^2}\\&=\frac{\eta_2^3\eta_8^6}{\eta_1^2\eta_4}\times\frac{1}{r_8^2}.\end{aligned} \tag{8.50}$$

In a similar way, if we replace $q$ with $q^8$ in Theorem 1.23 and then let $x=q^3$ we find that

$$\sum_{j=1}^{\infty}\frac{j^2(q^{3j}-q^{5j})}{1-q^{8j}}=\frac{\eta_2^3\eta_8^6}{\eta_1^2\eta_4}\times r_8^2. \tag{8.51}$$

Adding (8.50) and (8.51), we deduce (8.48).

Now we will prove (8.49). Let $\omega=\exp(i\pi/4)$ and substitute the values $x=\omega$ and $x=\omega^3$ in Theorem 1.21 to obtain

$$\frac{(qi,-qi;q)_\infty(q;q)_\infty^6}{(q\omega,q\omega^7;q)_\infty^4}=1+c_1\sum_{j=1}^{\infty}\frac{j^2q^j}{1-q^j}\left(\omega^j-\omega^{-j}\right) \tag{8.52}$$

and

$$\frac{(qi,-qi;q)_\infty(q;q)_\infty^6}{(q\omega^3,q\omega^5;q)_\infty^4}=1+c_2\sum_{j=1}^{\infty}\frac{j^2q^j}{1-q^j}\left(\omega^{3j}-\omega^{-3j}\right), \tag{8.53}$$

respectively, where

$$c_1 = \frac{(1-\omega)^3}{\omega(1+\omega)} = i\sqrt{2}(\sqrt{2}-1)^2$$

and

$$c_2 = \frac{(1-\omega^3)^3}{\omega^3(1+\omega^3)} = \frac{i\sqrt{2}}{(\sqrt{2}-1)^2}.$$

Multiply (8.52) by $1/(\sqrt{2}-1)^2$, multiply (8.53) by $(\sqrt{2}-1)^2$, and add to obtain

$$\frac{1}{(\sqrt{2}-1)^2}\frac{(qi,-qi;q)_\infty(q;q)_\infty^6}{(q\omega,q\omega^7;q)_\infty^4} + (\sqrt{2}-1)^2\frac{(qi,-qi;q)_\infty(q;q)_\infty^6}{(q\omega^3,q\omega^5;q)_\infty^4}$$
$$= 6 + i\sqrt{2}\sum_{j=1}^\infty \frac{j^2q^j}{1-q^j}\left(\omega^j+\omega^{3j}-\omega^{-j}-\omega^{-3j}\right).$$

The left-hand side simplifies to

$$\frac{\eta_1^6\eta_4^3}{\eta_2\eta_8^2}\left(\frac{1}{\xi_8^2}+\xi_8^2\right)$$

and, since

$$\omega^j+\omega^{3j}-\omega^{-j}-\omega^{-3j} = i\sqrt{8}\chi_{-2}(j),$$

the right-hand side simplifies to

$$6 - 4\sum_{j=1}^\infty \chi_{-2}(j)\frac{j^2q^j}{1-q^j}.$$

This completes the proof of (8.49). □

**Theorem 8.20.** *The following identities hold*:

$$\sum_{j=1}^\infty \frac{j^2(q^j+q^{3j}-q^{5j}-q^{7j})}{1-q^{8j}} = \frac{1}{8}\varphi(q)\varphi(q^2)\left(\varphi^4(q)-\varphi^4(q^2)\right)$$

*and*

$$1-\frac{2}{3}\sum_{j=1}^\infty \chi_{-2}(j)\frac{j^2q^j}{1-q^j} = \frac{1}{3}\varphi(q)\varphi(q^2)\left(4\varphi^4(q^2)-\varphi^4(q)\right).$$

*Hence,*

$$\varphi^5(q)\varphi(q^2) = 1 - \frac{2}{3}\sum_{j=1}^{\infty}\chi_{-2}(j)\frac{j^2q^j}{1-q^j} + \frac{32}{3}\sum_{j=1}^{\infty}\frac{j^2(q^j+q^{3j}-q^{5j}-q^{7j})}{1-q^{8j}}$$

*and*

$$\varphi(q)\varphi^5(q^2) = 1 - \frac{2}{3}\sum_{j=1}^{\infty}\chi_{-2}(j)\frac{j^2q^j}{1-q^j} + \frac{8}{3}\sum_{j=1}^{\infty}\frac{j^2(q^j+q^{3j}-q^{5j}-q^{7j})}{1-q^{8j}}.$$

*Proof.* By (8.48) and simple rearrangement we have

$$\begin{aligned}\sum_{j=1}^{\infty}\frac{j^2(q^j+q^{3j}-q^{5j}-q^{7j})}{1-q^{8j}} &= \frac{\eta_2^3\eta_8^6}{\eta_1^2\eta_4}\left(\frac{1}{r_8^2}+r_8^2\right)\\ &= \frac{1}{2}\times\frac{\eta_2^3\eta_8^6}{\eta_1^2\eta_4}\left(\left(\frac{1}{r_8}+r_8\right)^2+\left(\frac{1}{r_8}-r_8\right)^2\right).\end{aligned}$$

By Theorem 8.6 this may be expressed as

$$\begin{aligned}\sum_{j=1}^{\infty}\frac{j^2(q^j+q^{3j}-q^{5j}-q^{7j})}{1-q^{8j}} &= \frac{1}{2}\times\frac{\eta_2^3\eta_8^6}{\eta_1^2\eta_4}\left(\frac{\eta_2^{10}}{\eta_1^4\eta_4^2\eta_8^4}+\frac{\eta_4^{12}}{\eta_2^4\eta_8^8}\right)\\ &= \frac{1}{2}\times\frac{\eta_2^3\eta_4\eta_8^2}{\eta_1^2}\left(\frac{\eta_2^{10}}{\eta_1^4\eta_4^4}+\frac{\eta_4^{10}}{\eta_2^4\eta_8^4}\right).\end{aligned}$$

Theorem 3.5 may be used to express the eta-quotients in terms of Ramanujan's theta functions, giving

$$\sum_{j=1}^{\infty}\frac{j^2(q^j+q^{3j}-q^{5j}-q^{7j})}{1-q^{8j}} = \frac{\varphi(q)\varphi(q^2)}{2}\times q\,\psi^2(q^4)\times\left(\varphi^2(q)+\varphi^2(q^2)\right).$$

Now, Corollary 3.4 gives

$$q\psi^2(q^4) = \frac{1}{4}\left(\varphi^2(q)-\varphi^2(q^2)\right)$$

and hence the previous identity becomes

$$\sum_{j=1}^{\infty}\frac{j^2(q^j+q^{3j}-q^{5j}-q^{7j})}{1-q^{8j}} = \frac{1}{8}\varphi(q)\varphi(q^2)\left(\varphi^4(q)-\varphi^4(q^2)\right).$$

This proves the first result. By a similar procedure starting with (8.49) we deduce

$$1-\frac{2}{3}\sum_{j=1}^{\infty}\chi_{-2}(j)\frac{j^2q^j}{1-q^j}=\frac{1}{6}\frac{\eta_1^6\eta_4^3}{\eta_2\eta_8^2}\left(\frac{1}{\xi_8^2}+\xi_8^2\right)$$
$$=\frac{1}{3}\varphi(q)\varphi(q^2)\left(4\varphi^4(q^2)-\varphi^4(q)\right).$$

This proves the second result. The last two results are linear combinations of the first two results. □

## 8.8 The six level 8 functions

Jacobi's inversion formula in Theorem 4.1 provides an expansion of $\varphi^2(q)$ in powers of $q\psi^4(q^2)/\varphi^4(q)$. Using (4.3) and (4.8), Jacobi's result may be stated in the form

$$\varphi^2(q)=\sum_{n=0}^{\infty}\binom{2n}{n}^2\left(\frac{q\psi^4(q^2)}{\varphi^4(q)}\right)^n.$$

In this section we will prove an analogous formula for the expansion of $\varphi^2(q)$ in powers of $q\psi^2(q^4)/\varphi^2(q)$, given by

$$\varphi^2(q)=\sum_{n=0}^{\infty}\left\{\sum_{j=0}^{n}\binom{n}{j}\binom{2j}{j}\binom{2n-2j}{n-j}\right\}\left(\frac{q\psi^2(q^4)}{\varphi^2(q)}\right)^n. \tag{8.54}$$

We will also show that

$$\frac{\eta_2^6\eta_4^6}{\eta_1^4\eta_8^4}=\sum_{n=0}^{\infty}\left\{\sum_{j=0}^{n}\binom{n}{j}^2\binom{2j}{n}^2\right\}\left(\frac{\eta_1\eta_8}{\eta_2\eta_4}\right)^{8n}. \tag{8.55}$$

This may be compared with the case $\ell=4$ of Theorem 4.30, which is equivalent to

$$\frac{\eta_2^{20}}{\eta_1^8\eta_4^8}=\sum_{n=0}^{\infty}\binom{2n}{n}^3\left(\frac{\eta_1\eta_4}{\eta_2^2}\right)^{24n}.$$

To facilitate the study, we define six functions that come in three pairs and are analogous to the functions studied in Section 5.5.

1. Let $x_8 = x_8(q)$ and $z_8 = z_8(q)$ be defined by

$$x_8 = q \prod_{j=1}^{\infty} \frac{(1-q^{2j})^2(1-q^{8j})^4}{(1-q^j)^4(1-q^{4j})^2}$$

   and

$$z_8 = \prod_{j=1}^{\infty} \frac{(1-q^j)^4}{(1-q^{2j})^2}.$$

   Some basic properties of the function $x_8$ have been given in Section 8.4. By the product formulas for Ramanujan's theta functions in Theorem 3.5 we have

$$x_8(q) = q\,\frac{\psi^2(q^4)}{\varphi^2(-q)} \qquad \text{and} \qquad z_8(q) = \varphi^2(-q). \tag{8.56}$$

2. Let $w_8 = w_8(q)$ and $y_8 = y_8(q)$ be the quotients of eta functions defined by

$$w_8 = q \prod_{j=1}^{\infty} \frac{(1-q^j)^8(1-q^{8j})^8}{(1-q^{2j})^8(1-q^{4j})^8}$$

   and

$$y_8 = \prod_{j=1}^{\infty} \frac{(1-q^{2j})^6(1-q^{4j})^6}{(1-q^j)^4(1-q^{8j})^4}.$$

   Again, by Theorem 3.5,

$$w_8(q) = q\,\frac{\varphi^2(-q)\psi^2(q^4)}{\varphi^2(q^2)\varphi^2(q)} \qquad \text{and} \qquad y_8(q) = \varphi^2(q)\varphi^2(q^2).$$

   Moreover, $y_8$ is the function that occurs in (8.37) and $w_8$ is the quotient of the function in (8.40) with the function in (8.37).
3. Let $X_8$ and $Z_8$ be defined by

$$Z_8 = \frac{1}{3}\left(8P(q^8) - 4P(q^4) - 2P(q^2) + P(q)\right)$$

   and

$$X_8 = \frac{1}{Z_8^2} \times q \prod_{j=1}^{\infty} (1-q^{2j})^4(1-q^{4j})^4. \tag{8.57}$$

The functions $x_8$, $w_8$, and $X_8$ are modular functions. The function $z_8$ is a modular form of weight one, while $y_8$ and $Z_8$ are modular forms of weight two.

We now establish the basic properties of the functions $w_8$, $x_8$, $y_8$, $z_8$, $X_8$, and $Z_8$. The first result provides parameterizations for the eta function in terms of $z_8$ and $x_8$.

**Theorem 8.21.** *The following parameterizations hold*:

$$\eta_1^{24} = z_8^{12}\, x_8(1+4x_8)(1+8x_8)^2,$$
$$\eta_2^{24} = z_8^{12}\, x_8^2(1+4x_8)^2(1+8x_8)^4,$$
$$\eta_4^{24} = z_8^{12}\, x_8^4(1+4x_8)^4(1+8x_8)^2,$$
$$\eta_8^{24} = z_8^{12}\, x_8^8(1+4x_8)^2(1+8x_8).$$

*Proof.* These follow from the results in Theorem 8.11 and the definition of $z_8$ given above. □

The next result gives parameterizations of Ramanujan's Eisenstein series $Q$ in terms of $z_8$ and $x_8$.

**Theorem 8.22.** *The following parameterizations hold*:

$$Q(q) = z_8^4\left(1+256\,x_8+5120\,x_8^2+32768\,x_8^3+65536\,x_8^4\right),$$
$$Q(q^2) = z_8^4\left(1+16\,x_8+320\,x_8^2+2048\,x_8^3+4096\,x_8^4\right),$$
$$Q(q^4) = z_8^4\left(1+16\,x_8+80\,x_8^2+128\,x_8^3+256\,x_8^4\right),$$
$$Q(q^8) = z_8^4\left(1+16\,x_8+80\,x_8^2+128\,x_8^3+16\,x_8^4\right).$$

*Proof.* By Corollary 3.40,

$$Q(q) = \frac{\eta_1^{16}}{\eta_2^8} + 256\frac{\eta_2^{16}}{\eta_1^8} \qquad \text{and} \qquad Q(q^2) = \frac{\eta_1^{16}}{\eta_2^8} + 16\frac{\eta_2^{16}}{\eta_1^8}. \tag{8.58}$$

The results for $Q(q)$ and $Q(q^2)$ now follow on using Theorem 8.21; the results for $Q(q^4)$ and $Q(q^8)$ follow by replacing $q$ with $q^4$ in (8.58) and then using Theorem 8.21. □

The next result provides formulas for the derivatives of $x_8$ and $w_8$.

**Theorem 8.23.** *The following differentiation formulas hold*:

$$q\frac{\mathrm{d}}{\mathrm{d}q}\log x_8 = y_8 \qquad \textit{and} \qquad q\frac{\mathrm{d}}{\mathrm{d}q}\log w_8 = Z_8.$$

*Proof.* By the definition of $x_8$ and a calculation using (0.61) we find that

$$q\frac{\mathrm{d}}{\mathrm{d}q}\log x_8 = q\frac{\mathrm{d}}{\mathrm{d}q}\log\frac{\eta_2^2\eta_8^4}{\eta_1^4\eta_4^2} = \frac{1}{6}\left(-P(q)+P(q^2)-2P(q^4)+8P(q^8)\right). \tag{8.59}$$

Next, by (8.37) and the product formulas for Ramanujan's theta functions in Theorem 3.5 we find that

$$\frac{1}{6}\left(-P(q)+P(q^2)-2P(q^4)+8P(q^8)\right) = \varphi^2(q)\varphi^2(q^2) = \frac{\eta_2^6\eta_4^6}{\eta_1^4\eta_8^4} = y_8. \tag{8.60}$$

Combining (8.59) and (8.60), we obtain the first result.

Next, a calculation using (0.61) and the definitions of $w_8$ and $Z_8$ gives

$$\begin{aligned} q\frac{\mathrm{d}}{\mathrm{d}q}\log w_8 &= q\frac{\mathrm{d}}{\mathrm{d}q}\log\frac{\eta_1^8\eta_8^8}{\eta_2^8\eta_4^8} \\ &= \frac{1}{3}\left(8P(q^8)-4P(q^4)-2P(q^2)+P(q)\right) = Z_8. \end{aligned}$$

□

The next result relates $x_8$ and $z_8$ to $w_8$ and $y_8$.

**Theorem 8.24.** *The following identities hold:*

$$\frac{1}{w_8} = \frac{1}{x_8} + 12 + 32x_8 \tag{8.61}$$

*and*

$$x_8 z_8^2 = w_8 y_8. \tag{8.62}$$

*Hence,*

$$w_8 = \frac{x_8}{1+12x_8+32x_8^2}, \tag{8.63}$$

$$y_8 = (1+12x_8+32x_8^2)z_8^2, \tag{8.64}$$

$$\frac{1}{x_8} - 32x_8 = \frac{1}{w_8}\sqrt{1-24w_8+16w_8^2} \tag{8.65}$$

*and*

$$z_8^2(1-32x_8^2) = y_8\sqrt{1-24w_8+16w_8^2}. \tag{8.66}$$

*Proof.* The identity (8.61) is a restatement of (8.24), and the identity (8.62) follows immediately from the definitions of $w_8$, $x_8$, $y_8$, and $z_8$ as infinite products. The identities (8.63) and (8.64) follow from (8.61) and (8.62). The identity (8.65) can be obtained from (8.61), since

$$\begin{aligned}\frac{1}{x_8} - 32x_8 &= \sqrt{\left(\frac{1}{x_8} + 32x_8\right)^2 - 128}\\ &= \sqrt{\left(\frac{1}{w_8} - 12\right)^2 - 128}\\ &= \frac{1}{w_8}\sqrt{1 - 24w_8 + 16w_8^2}.\end{aligned}$$

Finally, the identity (8.66) may be obtained by taking the product of (8.62) and (8.65). □

The next result relates $X_8$ and $Z_8$ to $w_8$ and $y_8$, and hence to $x_8$ and $z_8$.

**Theorem 8.25.** *The following identities hold*:

$$Z_8 = y_8\sqrt{1 - 24w_8 + 16w_8^2} = z_8^2(1 - 32x_8^2) \tag{8.67}$$

*and*

$$X_8 = \frac{w_8}{1 - 24w_8 + 16w_8^2} = \frac{x_8(1 + 12x_8 + 32x_8^2)}{(1 - 32x_8^2)^2}. \tag{8.68}$$

*Proof.* By repeated application of Theorems 8.23 and 8.24 we obtain

$$\begin{aligned}Z_8 &= q\frac{\mathrm{d}}{\mathrm{d}q}\log w_8\\ &= -q\frac{\mathrm{d}}{\mathrm{d}q}\log\left(\frac{1}{x_8} + 12 + 32x_8\right)\\ &= \frac{\left(\frac{1}{x_8} - 32x_8\right)}{\left(\frac{1}{x_8} + 12 + 32x_8\right)}\,q\frac{\mathrm{d}}{\mathrm{d}q}\log x_8\\ &= \sqrt{1 - 24w_8 + 16w_8^2} \times y_8\\ &= z_8^2(1 - 32x_8^2).\end{aligned}$$

This proves (8.67). Next, by the definition of $X_8$ and Theorem 8.21 we have

$$X_8 = \frac{\eta_2^4\eta_4^4}{Z_8^2} = \frac{z_8^4\,x_8(1 + 4x_8)(1 + 8x_8)}{Z_8^2}.$$

By (8.67) this can be written as

$$X_8 = \frac{x_8(1+4x_8)(1+8x_8)}{(1-32x_8^2)^2}$$

and this proves part of (8.68). Finally, by (8.63) we deduce that

$$\frac{w_8}{1-24w_8+16w_8^2} = \frac{x_8(1+12x_8+32x_8^2)}{(1-32x_8^2)^2}$$

and this completes the proof of (8.68). □

## 8.9 Differential equations

The next result relates the level 8 theory to the level 4 theory of Jacobi.

**Lemma 8.26.** *Let $x_8$ and $z_8$ be as defined above, and let $x_4$ and $z_4$ be the level 4 modular forms given by Lemma 4.6, that is,*

$$x_4 = 16\,\frac{\eta_1^8\eta_4^{16}}{\eta_2^{24}} \qquad \text{and} \qquad z_4 = \frac{\eta_2^{10}}{\eta_1^4\eta_4^4}.$$

*Then*

$$x_4 = \frac{16x_8(1+4x_8)}{(1+8x_8)^2}, \qquad 1-x_4 = \frac{1}{(1+8x_8)^2}$$

*and*

$$z_4 = (1+8x_8)z_8.$$

*Proof.* By Theorem 8.11,

$$\frac{16x_8(1+4x_8)}{(1+8x_8)^2} = 16\left(\frac{\eta_2^2\eta_8^4}{\eta_1^4\eta_4^2}\right)\times\left(\frac{\eta_4^{10}}{\eta_1^4\eta_2^2\eta_8^4}\right)\Big/\left(\frac{\eta_2^{12}}{\eta_1^8\eta_4^4}\right)^2 = 16\frac{\eta_1^8\eta_4^{16}}{\eta_2^{24}} = x_4.$$

This proves the result for $x_4$. The result for $1-x_4$ follows by simple rearrangement. Finally, by Theorem 8.11 and the definition of $z_8$ given above, we find that

$$(1+8x_8)z_8 = \left(\frac{\eta_2^{12}}{\eta_1^8\eta_4^4}\right)\times\left(\frac{\eta_1^4}{\eta_2^2}\right) = \frac{\eta_2^{10}}{\eta_1^4\eta_4^4} = z_4.$$

□

A formula for $z_8$ as a function of $x_8$ in terms of the hypergeometric function can be deduced from Lemma 8.26.

**Theorem 8.27.** *The following identity holds in a neighborhood of* $q = 0$:

$$z_8 = \frac{1}{1+8x_8}\, {}_2F_1\left(\begin{matrix}\frac{1}{2}, \frac{1}{2} \\ 1\end{matrix}; \frac{16x_8(1+4x_8)}{(1+8x_8)^2}\right).$$

*Proof.* By Theorem 4.13,

$$z_4 = {}_2F_1\left(\begin{matrix}\frac{1}{2}, \frac{1}{2} \\ 1\end{matrix}; x_4\right).$$

Now use Lemma 8.26 to change variables and express $x_4$ and $z_4$ in terms of $x_8$ and $z_8$. □

The next goal is to expand $z_8$ in powers of $x_8$. This will be accomplished by deriving a differential equation for $z_8$ from the differential equation that is satisfied by $z_4$.

**Theorem 8.28.** *The following differential equation holds*:

$$\frac{\mathrm{d}}{\mathrm{d}x_8}\left(x_8(1+4x_8)(1+8x_8)\frac{\mathrm{d}z_8}{\mathrm{d}x_8}\right) + 4(1+8x_8)z_8 = 0.$$

*Proof.* From Theorem 4.11,

$$\frac{\mathrm{d}}{\mathrm{d}x_4}\left(x_4(1-x_4)\frac{\mathrm{d}z_4}{\mathrm{d}x_4}\right) = \frac{z_4}{4}. \tag{8.69}$$

We will use the results of Lemma 8.26 to change variables to produce a differential equation for $z_8$ in terms of $x_8$. Since

$$1 - x_4 = \frac{1}{(1+8x_8)^2}$$

it follows that

$$\frac{\mathrm{d}}{\mathrm{d}x_8} = \frac{\mathrm{d}x_4}{\mathrm{d}x_8}\frac{\mathrm{d}}{\mathrm{d}x_4} = \frac{16}{(1+8x_8)^3}\frac{\mathrm{d}}{\mathrm{d}x_4}.$$

Multiplying by $x_8(1+4x_8)/(1+8x_8)$ and applying Lemma 8.26 to the right-hand side, we find that

$$\frac{x_8(1+4x_8)}{1+8x_8}\frac{\mathrm{d}}{\mathrm{d}x_8} = x_4(1-x_4)\frac{\mathrm{d}}{\mathrm{d}x_4}. \tag{8.70}$$

On multiplying the differential equation (8.69) by $x_4(1-x_4)$ and applying the change of variables given by Lemma 8.26 and (8.70) we find that

$$\frac{x_8(1+4x_8)}{1+8x_8}\frac{\mathrm{d}}{\mathrm{d}x_8}\left\{\frac{x_8(1+4x_8)}{1+8x_8}\frac{\mathrm{d}}{\mathrm{d}x_8}((1+8x_8)z_8)\right\}$$
$$= \frac{4x_8(1+4x_8)}{(1+8x_8)^4}\times(1+8x_8)z_8.$$

This simplifies to the required result. □

**Theorem 8.29.** *Let $t_8(n)$ and $s_8(n)$ be the sequences defined by the recurrence relations*

$$(n+1)^2 t_8(n+1) = -4(3n^2+3n+1)t_8(n) - 32n^2 t_8(n-1) \tag{8.71}$$

*and*

$$(n+1)^3 s_8(n+1) = 4(2n+1)(3n^2+3n+1)s_8(n) - 16n^3 s_8(n-1)$$

*and initial conditions*

$$t_8(0) = s_8(0) = 1,$$

*noting that one initial condition is sufficient to start each sequence. Then in a neighborhood of $q=0$, the following series expansions hold:*

$$z_8 = \sum_{n=0}^{\infty} t_8(n)x_8^n, \tag{8.72}$$

$$y_8 = \sum_{n=0}^{\infty} s_8(n)w_8^n \tag{8.73}$$

*and*

$$Z_8 = \sum_{n=0}^{\infty}\binom{2n}{n}t_8(n)X_8^n. \tag{8.74}$$

*Proof.* If we substitute the series expansion

$$z_8 = \sum_{n=0}^{\infty} t_8(n)x_8^n$$

into the differential equation in Theorem 8.28 and equate coefficients of $x_8^n$, we deduce the recurrence relation (8.71). This proves (8.72). To prove the other results (8.73) and (8.74), apply Theorem 5.43 with $(\alpha, \beta, \gamma) = (-12, -4, -32)$ and $x = x_8$ to deduce

$$(1 + 12x_8 + 32x_8^2) \left( \sum_{n=0}^{\infty} t_8(n) x_8^n \right)^2 = \sum_{n=0}^{\infty} s_8(n) \left( \frac{x_8}{1 + 12x_8 + 32x_8^2} \right)^n$$

and

$$(1 - 32x_8^2) \left( \sum_{n=0}^{\infty} t_8(n) x_8^n \right)^2 = \sum_{n=0}^{\infty} \binom{2n}{n} t_8(n) \left( \frac{x_8(1 + 12x_8 + 32x_8^2)}{(1 - 32x_8^2)^2} \right)^n .$$

By (8.63), (8.64), and Theorem 8.25, these are equivalent to (8.73) and (8.74), respectively. □

Explicit formulas for the coefficients are given by

$$t_8(n) = (-1)^n \sum_j 4^{n-2j} \binom{n}{2j} \binom{2j}{j}^2 = (-1)^n \sum_j \binom{n}{j} \binom{2j}{j} \binom{2n-2j}{n-j}$$

and

$$s_8(n) = \sum_j \binom{n}{j}^2 \binom{2j}{n}^2 .$$

These formulas may be verified by the method of creative telescoping described in Section 5.9.

The identity (8.72), written in full using (8.56), is

$$\varphi^2(-q) = \sum_{n=0}^{\infty} \left\{ (-1)^n \sum_{j=0}^{n} \binom{n}{j} \binom{2j}{j} \binom{2n-2j}{n-j} \right\} \left( \frac{q\psi^2(q^4)}{\varphi^2(-q)} \right)^n .$$

Replacing $q$ with $-q$, we obtain (8.54). The identity (8.73) is equivalent to (8.55). Finally, the identity (8.74), written in full, is

$$\begin{aligned} &\frac{1}{3} \left( 8P(q^8) - 4P(q^4) - 2P(q^2) + P(q) \right) \\ &= \sum_{n=0}^{\infty} \binom{2n}{n} \left\{ (-1)^n \sum_{j=0}^{n} \binom{n}{j} \binom{2j}{j} \binom{2n-2j}{n-j} \right\} \\ &\qquad \times \left( \frac{3\eta_2^4 \eta_4^4}{8P(q^8) - 4P(q^4) - 2P(q^2) + P(q)} \right)^n . \end{aligned}$$

## 8.10 Notes

The continued fraction in Lemma 8.2 was given by Ramanujan in his second notebook [253, Ch. 16, Entry 11]. The proof we have given is due to Berndt [46, pp. 22–23], which is a simplification of a proof by Adiga, Berndt, Bhargava, and Watson [1]. The continued fraction in Theorem 8.3 appears in Ramanujan's second notebook [253, Chapter 19, Entry 1, (ii)]. It was independently rediscovered by Göllnitz [164] and Gordon [165], and has come to be known as the Ramanujan–Göllnitz–Gordon continued fraction. Another continued fraction, also originally due to Ramanujan, that converges to the same infinite product, will be given in the exercises. See the work of Andrews and Berndt [25, Chapter 6] for an analysis of this and other continued fractions in Ramanujan's lost notebook.

The results in Lemmas 8.4 and 8.5 are based on work of Yuttanan [305] and Hirschhorn [184]. The identity (8.13) has been studied, along with other identities of a similar type, by Horie and Kanou [186, Theorem 2.2]. The quadratic transformation formula (8.18) was studied by Chan and Huang [87]. The evaluation (8.21) was given in [164]. Further evaluations have been given in [87].

The identity (8.29) is equivalent to the result

$$\sum 1 \left(n = x^2 + 2y^2\right) = 2 \sum_{d|n} \chi_{-2}(d),$$

which was known to Dirichlet [137, p. 463]. The identity (8.29) was given by Ramanujan in his second notebook [253, Chapter 17, Entry 8 (iii)]. A proof using Ramanujan's summation formula (0.89) has been given by Askey [32]. A different proof has been given by Berndt [49, p. 73].

The results in Theorem 8.18 were known to Liouville [210]– [212]. Proofs have been given by many authors, including Nazimoff [238, p. 17], Walfisz [289, 290, 291], Fine [151, p. 74], A. Alaca, Ş. Alaca, M. Lemire, and K. Williams [6], and Williams [301]. All four of the results in Theorem 8.18 are proved in [107].

The results in Theorem 8.20 were also known to Liouville [215, 216]. Further results have been given by Alaca, Alaca, Uygul, and Williams [11].

## 8.11 Exercises

1. Prove the following identities:

   a. $\varphi(q) - \varphi(q^2) = 2r_8 \dfrac{\eta_8^2}{\eta_4}$

   b. $\varphi(q) + \varphi(q^2) = \dfrac{2}{r_8} \dfrac{\eta_8^2}{\eta_4}$

c. $\varphi^2(q) - \varphi^2(q^2) = 4\,\dfrac{\eta_8^4}{\eta_4^2}$

d. $\varphi^2(q) + \varphi^2(q^2) = 2\left(\dfrac{1}{r_8^2} + r_8^2\right)\dfrac{\eta_8^4}{\eta_4^2}$

e. $\varphi^4(q) - \varphi^4(q^2) = 8\left(\dfrac{1}{r_8^2} + r_8^2\right)\dfrac{\eta_8^8}{\eta_4^4}$

f. $\sqrt{2}\,\varphi(q^2) - \varphi(q) = \xi_8\,\dfrac{\eta_1^2}{\eta_2}$

g. $\sqrt{2}\,\varphi(q^2) + \varphi(q) = \dfrac{1}{\xi_8}\,\dfrac{\eta_1^2}{\eta_2}$

h. $2\varphi^2(q^2) - \varphi^2(q) = \dfrac{\eta_1^4}{\eta_2^2}$

i. $2\varphi^2(q^2) + \varphi^2(q) = \dfrac{1}{2}\left(\dfrac{1}{\xi_8^2} + \xi_8^2\right)\dfrac{\eta_1^4}{\eta_2^2}$

j. $4\varphi^4(q^2) - \varphi^4(q) = \dfrac{1}{2}\left(\dfrac{1}{\xi_8^2} + \xi_8^2\right)\dfrac{\eta_1^8}{\eta_2^4}$

2. Let $L_1 = P(q)$, $L_2 = P(q^2)$, $L_4 = 2P(q^4)$, and $L_8 = 8P(q^8)$. Prove that

$$(L_1 + L_4)^2 + (L_2 + L_8)^2 = 6(L_1L_2 - L_2L_4 + L_4L_8).$$

3. This exercise follows A. Selberg [266]. Suppose $k$ is real and $k > -1/2$. Let

$$C_{k,n}(x) = 1 - q^n x^n + \sum_{j=1}^{\infty}(-1)^j\, q^{(2k+1)j(j+1)/2-nj}\, x^{jk} \times \frac{(qx;q)_j}{(q;q)_j} \times \left(1 - q^{(2j+1)n}x^n\right). \tag{8.75}$$

a. Show that $C_{k,0}(x) = 0$.
b. Show that $C_{k,-n}(x) = -x^{-n}q^{-n}\,C_{k,n}(x)$.
c. Show that

$$C_{k,n}(x) - C_{k,n-1}(x) = q^{n-1}x^{n-1}(1-qx)C_{k,k+1-n}(qx).$$

d. Let $f(x) = C_{3/2,1/2}(x)$ and $g(x) = C_{3/2,1}(x)$. Use the results of Parts a, b, and c to show that

$$\frac{f(x)}{1-\sqrt{qx}} = g(qx) + q^2x(1-q^2x)f(q^2x)$$

and

$$\frac{g(x)}{1-qx} = f(qx) + q\sqrt{x}(1-q^2x)g(q^2x).$$

e. Deduce that

$$\frac{g(x)}{(1-qx)f(qx)} = 1 + \cfrac{q\sqrt{x}(1+q\sqrt{x})}{1 + \cfrac{q^3x}{\cfrac{g(q^2x)}{(1-q^3x)f(q^3x)}}}.$$

Take the reciprocal and iterate to conclude that

$$\frac{(1-qx)f(qx)}{g(x)} = \cfrac{1}{1 + \cfrac{q\sqrt{x}(1+q\sqrt{x})}{1 + \cfrac{q^3x}{1 + \cfrac{q^2\sqrt{x}(1+q^2\sqrt{x})}{1 + \cfrac{q^5x}{1 + \cfrac{q^3\sqrt{x}(1+q^3\sqrt{x})}{1 + \cfrac{q^7x}{1+\cdots}}}}}}}.$$

f. Replace $q$ with $q^2$ then put $x = 1/q^2$ to deduce the result of Ramanujan [25, p. 154], [254, p. 44] and Selberg [266, (53)]:

$$\prod_{j=1}^{\infty} \frac{(1-q^{8j-7})(1-q^{8j-1})}{(1-q^{8j-5})(1-q^{8j-3})} = \cfrac{1}{1 + \cfrac{q+q^2}{1 + \cfrac{q^4}{1 + \cfrac{q^3+q^6}{1 + \cfrac{q^8}{1+\cdots}}}}}.$$

Hint: Show that

$$f(1) = \sum_{j=-\infty}^{\infty} (-1)^j q^{(4j^2+3j)/2}$$

and

$$\lim_{x\to 1/q} \frac{g(x)}{1-qx} = \sum_{j=-\infty}^{\infty} (-1)^j q^{(4j^2+j)/2}.$$

Then replace $q$ with $q^2$ and use Jacobi's triple produce identity to convert the series to products.

g. Show that

$$\prod_{j=1}^{\infty} \frac{(1+q^{2j})}{(1+q^{2j-1})} = \cfrac{1}{1+\cfrac{q}{1+\cfrac{q+q^2}{1+\cfrac{q^3}{1+\cfrac{q^2+q^4}{1+\cfrac{q^5}{1+\cfrac{q^3+q^6}{1+\cdots}}}}}}}.$$

The continued fraction on the right has come to be known as the Ramanujan–Selberg continued fraction.
Hint: Take $x = 1$ in the result of Part e.

4. Let $G(x)$ be the function defined by (5.6), that is,

$$G(x) = 1 + \sum_{j=1}^{\infty} (-1)^j q^{(5j^2-j)/2} x^{2j} (1-q^{2j}x) \frac{(qx;q)_{j-1}}{(q;q)_j},$$

and let $C_{k,n}(x)$ be as defined above in (8.75). Show that

$$G(x) = C_{2,2}(x).$$

Thus, Selberg's function $C_{k,n}(x)$ is a generalization of the function used by Ramanujan in the proofs of the Rogers–Ramanujan identities and the Rogers–Ramanujan continued fraction in Theorems 5.6 and 5.7. Further special cases of $C_{k,n}$ have been considered by Selberg [266].

5. Prove that

$$\varphi(-q)\varphi(-q^4) = \varphi(-q^8)^2 - 2q\psi(-q^4)^2$$
$$= 1 + 2\sum_{j=1}^{\infty}\sum_{k=1}^{\infty} q^{jk} \sin\frac{(j+5k)\pi}{4}.$$

6. a. Dissect the first result in Theorem 8.17 to deduce the identities:

$$q\psi(q^8)\varphi(q^8)^3 = \sum_{n=0}^{\infty}\left(\sum_{d|8n+1}\chi_2(d)d\right)q^{8n+1},$$

$$q^3\psi(q^8)^3\varphi(q^8) = -\frac{1}{2}\sum_{n=0}^{\infty}\left(\sum_{d|8n+3}\chi_2(d)d\right)q^{8n+3},$$

$$q^5\psi(q^8)^3\psi(q^{16}) = -\frac{1}{4}\sum_{n=0}^{\infty}\left(\sum_{d|8n+5}\chi_2(d)d\right)q^{8n+5},$$

$$q^7\psi(q^8)\psi(q^{16})^3 = \frac{1}{8}\sum_{n=0}^{\infty}\left(\sum_{d|8n+7}\chi_2(d)d\right)q^{8n+7}.$$

b. Hence, obtain formulas for

$$\sum 1\left(n = \frac{x(x+1)}{2} + \frac{y(y+1)}{2} + \frac{z(z+1)}{2} + w(w+1);\ x, y, z, w \geq 0\right)$$

and

$$\sum 1\left(n = \frac{x(x+1)}{2} + y(y+1) + z(z+1) + w(w+1);\ x, y, z, w \geq 0\right).$$

See [107].

# Chapter 9
# Level 9

## 9.1 Two identities

All of the results in this short chapter center around the two identities in the next theorem, which can be deduced easily from the cubic theory.

**Theorem 9.1.** *The following identities hold*:

$$\frac{\eta_1^3}{\eta_9^3} + 9 + 27\frac{\eta_9^3}{\eta_1^3} = \frac{\eta_3^{12}}{\eta_1^6\eta_9^6} \tag{9.1}$$

*and*

$$\frac{1}{8}\left(9P(q^9) - P(q)\right) = \frac{\eta_3^{10}}{\eta_1^3\eta_9^3}. \tag{9.2}$$

*Proof.* We will use properties of the cubic theta functions along with the additional result

$$3c(q^3) = a(q) - b(q) \tag{9.3}$$

which follows from Lemma 3.11. We have

$$\begin{aligned}
&\frac{\eta_1^3}{\eta_9^3} + 9 + 27\frac{\eta_9^3}{\eta_1^3} \\
&= \frac{3b(q)}{c(q^3)} + 9 + 9\frac{c(q^3)}{b(q)} && \text{by Theorem 3.15} \\
&= \frac{9b(q)}{a(q)-b(q)} + 9 + \frac{3(a(q)-b(q))}{b(q)} && \text{by (9.3)}
\end{aligned}$$



S. Cooper, *Ramanujan's Theta Functions*, DOI 10.1007/978-3-319-56172-1_10

$$= \frac{3(a^3(q) - b^3(q))}{b(q)(a(q) - b(q))^2} \qquad \text{after simplification}$$

$$= \frac{c^3(q)}{3b(q)c^2(q^3)} \qquad \text{by Theorem 3.20 and (9.3)}$$

$$= \frac{\eta_3^{12}}{\eta_1^6\eta_9^6} \qquad \text{by Theorem 3.15.}$$

This proves the first result. In a similar way,

$$\frac{1}{8}\left(9P(q^9) - P(q)\right)$$

$$= \frac{1}{8}\left(9P(q^9) - 3P(q^3) + 3P(q^3) - P(q)\right)$$

$$= \frac{1}{4}\left(3a^2(q^3) + a^2(q)\right) \qquad \text{by (3.76)}$$

$$= \frac{1}{3}\left(a^2(q) + a(q)b(q) + b^2(q)\right) \qquad \text{by (3.17)}$$

$$= \frac{1}{3} \times \frac{a^3(q) - b^3(q)}{a(q) - b(q)}$$

$$= \frac{1}{9} \times \frac{c^3(q)}{c(q^3)} \qquad \text{by Theorem 3.20 and (9.3)}$$

$$= \frac{\eta_3^{10}}{\eta_1^3\eta_9^3} \qquad \text{by Theorem 3.15.}$$

This proves the second result. □

## 9.2 The six level 9 functions

As in Sections 5.5 and 8.8, we define six functions that come in three pairs.

1. Let $x_9 = x_9(q)$ and $z_9 = z_9(q)$ be defined by

$$x_9 = q\prod_{j=1}^{\infty}\frac{(1-q^{9j})^3}{(1-q^j)^3} \qquad \text{and} \qquad z_9 = \prod_{j=1}^{\infty}\frac{(1-q^j)^3}{(1-q^{3j})}. \tag{9.4}$$

By the product formulas for the cubic theta functions in Theorem 3.15,

$$x_9(q) = \frac{c(q^3)}{3b(q)} = \frac{c^3(q)}{27b^3(q^3)} \qquad \text{and} \qquad z_9(q) = b(q). \tag{9.5}$$

2. Let $w_9 = w_9(q)$ and $y_9 = y_9(q)$ be the quotients of eta functions defined by

$$w_9 = q\prod_{j=1}^{\infty}\frac{(1-q^j)^6(1-q^{9j})^6}{(1-q^{3j})^{12}} \quad \text{and} \quad y_9 = \prod_{j=1}^{\infty}\frac{(1-q^{3j})^{10}}{(1-q^j)^3(1-q^{9j})^3}.$$

3. Let $X_9 = X_9(q)$ and $Z_9 = Z_9(q)$ be defined by

$$Z_9 = \frac{1}{4}\left(9P(q^9) - 6P(q^3) + P(q)\right) \quad \text{and} \quad X_9 = \frac{1}{Z_9^2}\times q\prod_{j=1}^{\infty}(1-q^{3j})^8. \tag{9.6}$$

The next result relates the level 9 theory to Ramanujan's level 3 theory.

**Theorem 9.2.** *Let $x_3$ and $z_3$ be defined by* (4.29), *that is,*

$$x_3(q) = \frac{c^3(q)}{a^3(q)} \quad \textit{and} \quad z_3(q) = a(q).$$

*Let $x_9 = x_9(q)$ and $z_9 = z_9(q)$. Then*

$$x_3(q) = \frac{27x_9(1+9x_9+27x_9^2)}{(1+9x_9)^3}, \tag{9.7}$$

$$1 - x_3(q) = \frac{1}{(1+9x_9)^3}, \tag{9.8}$$

$$z_3(q) = (1+9x_9)z_9, \tag{9.9}$$

$$x_3(q^3) = \frac{27x_9^3}{(1+3x_9)^3} \tag{9.10}$$

*and*

$$z_3(q^3) = (1+3x_9)z_9. \tag{9.11}$$

*Proof.* By (9.3) and (9.5),

$$x_9 = \frac{c(q^3)}{3b(q)} = \frac{a(q)-b(q)}{9b(q)}. \tag{9.12}$$

Multiplying by 9, adding 1, and taking the cube, we obtain

$$(1+9x_9)^3 = \frac{a^3(q)}{b^3(q)}.$$

Taking the reciprocal and applying the cubic identity $a^3 = b^3 + c^3$ from Theorem 3.20 we obtain

$$\frac{1}{(1+9x_9)^3} = \frac{b^3(q)}{a^3(q)} = \frac{a^3(q)-c^3(q)}{a^3(q)} = 1 - x_3(q).$$

This proves (9.8), and (9.7) follows by simple rearrangement.

The identity (9.12) may be written in the form

$$x_9 = \frac{z_3 - z_9}{9z_9}.$$

Solving for $z_3$, we obtain (9.9).

Now we will prove (9.10) and (9.11). By Lemma 3.11,

$$a(q^3) = \frac{a(q)+2b(q)}{3} \quad \text{and} \quad c(q^3) = \frac{a(q)-b(q)}{3}.$$

Hence,

$$\begin{aligned} x_3(q^3) = \frac{c^3(q^3)}{a^3(q^3)} &= \left(\frac{a(q)-b(q)}{a(q)+2b(q)}\right)^3 \\ &= \left(\frac{z_3-z_9}{z_3+2z_9}\right)^3 \\ &= \left(\frac{(1+9x_9)z_9 - z_9}{(1+9x_9)z_9+2z_9}\right)^3 \\ &= \frac{27x_9^3}{(1+3x_9)^3} \end{aligned}$$

and

$$\begin{aligned} z_3(q^3) = a(q^3) &= \frac{a(q)+2b(q)}{3} \\ &= \frac{z_3+2z_9}{3} \\ &= \frac{(1+9x_9)z_9+2z_9}{3} \\ &= (1+3x_9)z_9. \end{aligned}$$

□

The next result relates $x_9$ to the Dedekind eta function. It is an analogue of (5.43).

**Theorem 9.3.** *The following identity holds*:

$$\frac{1}{x_9} + 9 + 27x_9 = \frac{\eta_3^{12}}{\eta_1^6\eta_9^6}.$$

*Proof.* This is just a restatement of (9.1). □

The results in the next three theorems give parameterizations for Dedekind's eta function, Ramanujan's Eisenstein series $Q$, and the modular $j$-invariant in terms of $z_9$ and $x_9$.

**Theorem 9.4.** *The following parameterizations of eta functions in terms of* $z_9$ *and* $x_9$ *hold*:

$$\eta_1^{24} = z_9^{12}x_9(1 + 9x_9 + 27x_9^2),$$
$$\eta_3^{24} = z_9^{12}x_9^3(1 + 9x_9 + 27x_9^2)^3$$

*and*

$$\eta_9^{24} = z_9^{12}x_9^9(1 + 9x_9 + 27x_9^2).$$

*Proof.* By Theorem 9.3 and the definitions of $x_9$ and $z_9$,

$$\begin{aligned} z_9^{12}x_9(1 + 9x_9 + 27x_9^2) &= z_9^{12}x_9^2\left(\frac{1}{x_9} + 9 + 27x_9\right) \\ &= \left(\frac{\eta_1^3}{\eta_3}\right)^{12} \times \left(\frac{\eta_9^3}{\eta_1^3}\right)^2 \times \left(\frac{\eta_3^{12}}{\eta_1^6\eta_9^6}\right) \\ &= \eta_1^{24}. \end{aligned}$$

This proves the first result. The other two results can be proved in the same way. □

**Theorem 9.5.** *The following parameterizations hold*:

$$Q(q) = z_9^4(1 + 9x_9)(1 + 243x_9 + 2187x_9^2 + 6561x_9^3),$$
$$Q(q^3) = z_9^4(1 + 3x_9)(1 + 9x_9)(1 + 27x_9^2)$$

*and*

$$Q(q^9) = z_9^4(1 + 3x_9)(1 + 9x_9 + 27x_9^2 + 3x_9^3).$$

*Proof.* By Theorem 4.31,

$$Q(q) = z_3^4(1 + 8x_3) \qquad \text{and} \qquad Q(q^3) = z_3^4\left(1 - \frac{8}{9}x_3\right).$$

Hence by Theorem 9.2 it follows that

$$\begin{aligned} Q(q) &= (1 + 9x_9)^4 z_9^4\left(1 + \frac{216x_9(1 + 9x_9 + 27x_9^2)}{(1 + 9x_3)^3}\right) \\ &= z_9^4(1 + 9x_9)(1 + 243x_9 + 2187x_9^2 + 6561x_9^3) \end{aligned}$$

and

$$\begin{aligned} Q(q^3) &= (1 + 9x_9)^4 z_9^4\left(1 - \frac{216x_9(1 + 9x_9 + 27x_9^2)}{9(1 + 9x_3)^3}\right) \\ &= z_9^4(1 + 3x_9)(1 + 9x_9)(1 + 27x_9^2). \end{aligned}$$

Moreover,

$$\begin{aligned} Q(q^9) &= z_3^4(q^3)\left(1 - \frac{8}{9}x_3(q^3)\right) \\ &= (1 + 3x_9)^4 z_9^4\left(1 - \frac{24x_9^3}{(1 + 3x_9)^3}\right) \\ &= z_9^4(1 + 3x_9)(1 + 9x_9 + 27x_9^2 + 3x_9^3). \end{aligned}$$

□

**Theorem 9.6.** *Let $x_9$ be defined by* (9.4) *and let $j(\tau)$ be the modular $j$-invariant defined by* (2.21). *That is,*

$$x_9 = \frac{\eta_9^3}{\eta_1^3} \quad \text{and} \quad j(\tau) = \frac{1728Q^3(q)}{Q^3(q) - R^2(q)} = \frac{Q^3(q)}{\eta_1^{24}}.$$

*Then*

$$\begin{aligned} j(\tau) &= \frac{(1 + 9x_9)^3(1 + 243x_9 + 2187x_9^2 + 6561x_9^3)^3}{x_9(1 + 9x_9 + 27x_9^2)}, \\ j(3\tau) &= \frac{(1 + 3x_9)^3(1 + 9x_9)^3(1 + 27x_9^2)^3}{x_9^3(1 + 9x_9 + 27x_9^2)^3} \\ j(9\tau) &= \frac{(1 + 3x_9)^3(1 + 9x_9 + 27x_9^2 + 3x_9^3)^3}{x_9^9(1 + 9x_9 + 27x_9^2)}. \end{aligned}$$

*Proof.* This is immediate from Theorems 9.4 and 9.5. □

The next result provides formulas for the derivatives of $x_9$ and $w_9$.

**Theorem 9.7.** *The following differentiation formulas hold:*

$$q\frac{\mathrm{d}}{\mathrm{d}q}\log x_9 = y_9$$

*and*

$$q\frac{\mathrm{d}}{\mathrm{d}q}\log w_9 = Z_9.$$

*Proof.* Successively using the definition of $x_9$, (0.61), (9.2) and the definition of $y_9$, we deduce that

$$q\frac{\mathrm{d}}{\mathrm{d}q}\log x_9 = q\frac{\mathrm{d}}{\mathrm{d}q}\log\left(\frac{\eta_9^3}{\eta_1^3}\right) = \frac{1}{8}\left(9P(q^9) - P(q)\right) = \frac{\eta_3^{10}}{\eta_1^3\eta_9^3} = y_9.$$

Similarly, by the definitions of $w_9$ and $Z_9$ and (0.61), we find that

$$q\frac{\mathrm{d}}{\mathrm{d}q}\log w_9 = q\frac{\mathrm{d}}{\mathrm{d}q}\log\left(\frac{\eta_1^6\eta_9^6}{\eta_3^{12}}\right) = \frac{1}{4}\left(9P(q^9) - 6P(q^3) + P(q)\right) = Z_9.$$

□

The next result relates $x_9$ and $z_9$ to $w_9$ and $y_9$.

**Theorem 9.8.** *The following identities hold:*

$$\frac{1}{w_9} = \frac{1}{x_9} + 9 + 27x_9 \tag{9.13}$$

*and*

$$x_9 z_9^2 = w_9 y_9. \tag{9.14}$$

*Hence,*

$$w_9 = \frac{x_9}{1 + 9x_9 + 27x_9^2}, \tag{9.15}$$

$$y_9 = (1 + 9x_9 + 27x_9^2)z_9^2, \tag{9.16}$$

$$\frac{1}{x_9} - 27x_9 = \frac{1}{w_9}\sqrt{1 - 18w_9 - 27w_9^2} \tag{9.17}$$

*and*

$$z_9^2(1 - 27x_9^2) = y_9\sqrt{1 - 18w_9 - 27w_9^2}. \tag{9.18}$$

*Proof.* The identity (9.13) is a restatement of the result of (9.1), and the identity (9.14) follows immediately from the definitions of $w_9$, $x_9$, $y_9$, and $z_9$ as infinite products. The identities (9.15) and (9.16) follow from (9.13) and (9.14). The identity (9.17) can be obtained from (9.13), since

$$\begin{aligned}\frac{1}{x_9} - 27x_9 &= \sqrt{\left(\frac{1}{x_9} + 27x_9\right)^2 - 108} \\ &= \sqrt{\left(\frac{1}{w_9} - 9\right)^2 - 108} \\ &= \frac{1}{w_9}\sqrt{1 - 18w_9 - 27w_9^2}.\end{aligned}$$

Finally, the identity (9.18) may be obtained by taking the product of (9.14) and (9.17). □

The next result relates $X_9$ and $Z_9$ to $w_9$ and $y_9$, and hence to $x_9$ and $z_9$.

**Theorem 9.9.** *The following identities hold*:

$$Z_9 = y_9\sqrt{1 - 18w_9 - 27w_9^2} = z_9^2(1 - 27x_9^2) \tag{9.19}$$

*and*

$$X_9 = \frac{w_9}{1 - 18w_9 - 27w_9^2} = \frac{x_9(1 + 9x_9 + 27x_9^2)}{(1 - 27x_9^2)^2}. \tag{9.20}$$

*Proof.* By repeated application of Theorems 9.7 and 9.8 we deduce that

$$\begin{aligned}Z_9 &= q\frac{\mathrm{d}}{\mathrm{d}q}\log x_9 \\ &= -q\frac{\mathrm{d}}{\mathrm{d}q}\log\left(\frac{1}{x_9} + 9 + 27x_9\right) \\ &= \frac{\left(\frac{1}{x_9} - 27x_9\right)}{\left(\frac{1}{x_9} + 9 + 27x_9\right)}\, q\frac{\mathrm{d}}{\mathrm{d}q}\log x_9 \\ &= \sqrt{1 - 18w_9 - 27w_9^2} \times y_9 \\ &= z_9^2(1 - 27x_9^2).\end{aligned}$$

This proves (9.19). Next, by the definition of $X_9$ and Theorem 9.4,

$$X_9 = \frac{\eta_3^8}{Z_9^2} = \frac{z_9^4\, x_9(1 + 9x_9 + 27x_9^2)}{Z_9^2}.$$

By (9.19) this can be written

$$X_9 = \frac{x_9(1 + 9x_9 + 27x_9^2)}{(1 - 27x_9^2)^2}$$

and this proves part of (9.20). Finally, by (9.15) and (9.17) it follows that

$$\frac{w_9}{1 - 18w_9 - 27w_9^2} = \frac{x_9(1 + 9x_9 + 27x_9^2)}{(1 - 27x_9^2)^2}$$

and this completes the proof of (9.20). □

## 9.3 Differential equations

We are now ready to derive a differential equation for $z_9$.

**Theorem 9.10.** *The following differential equation holds*:

$$\frac{d}{dx_9}\left(x_9(1 + 9x_9 + 27x_9^2)\frac{dz_9}{dx_9}\right) = -3(1 + 9x_9)z_9.$$

*Proof.* By Theorem 4.24,

$$\frac{d}{dx_3}\left(x_3(1 - x_3)\frac{dz_3}{dx_3}\right) = \frac{2z_3}{9}. \tag{9.21}$$

By Theorem 9.2,

$$x_3(1 - x_3) = \frac{27x_9(1 + 9x_9 + 27x_9^2)}{(1 + 9x_9)^6} \quad \text{and} \quad \frac{dx_3}{dx_9} = \frac{27}{(1 + 9x_9)^4}.$$

On changing variables, the differential equation (9.21) becomes

$$\frac{(1 + 9x_9)^4}{27}\frac{d}{dx_9}\left(\frac{x_9(1 + 9x_9 + 27x_9^2)}{(1 + 9x_9)^2}\frac{d}{dx_9}(z_9(1 + 9x_9))\right) = \frac{2z_9(1 + 9x_9)}{9}.$$

On simplifying, we complete the proof. □

**Theorem 9.11.** *Let $t_9(n)$ and $s_9(n)$ be the sequences defined by the recurrence relations*

$$(n + 1)^2 t_9(n + 1) = -3(3n^2 + 3n + 1)t_9(n) - 27n^2 t_9(n - 1) \tag{9.22}$$

*and*

$$(n+1)^3 s_9(n+1) = 3(2n+1)(3n^2+3n+1)s_9(n) + 27n^3 s_9(n-1)$$

*and initial conditions*

$$t_9(0) = s_9(0) = 1,$$

*noting that one initial condition is sufficient to start each sequence. Then in a neighborhood of* $q = 0$*, the following series expansions hold*:

$$z_9 = \sum_{n=0}^{\infty} t_9(n) x_9^n, \tag{9.23}$$

$$y_9 = \sum_{n=0}^{\infty} s_9(n) w_9^n \tag{9.24}$$

*and*

$$Z_9 = \sum_{n=0}^{\infty} \binom{2n}{n} t_9(n) X_9^n. \tag{9.25}$$

*Proof.* If we substitute the series expansion

$$z_9 = \sum_{n=0}^{\infty} t_9(n) x_9^n$$

into the differential equation in Theorem 9.10 and equate coefficients of $x_9^n$, we deduce the recurrence relation (9.22). This proves (9.23). To prove the other results (9.24) and (9.25), apply Theorem 5.43 with $(\alpha, \beta, \gamma) = (-9, -3, -27)$ and $x = x_9$ to deduce

$$(1 + 9x_9 + 27x_9^2)\left(\sum_{n=0}^{\infty} t_9(n) x_9^n\right)^2 = \sum_{n=0}^{\infty} s_9(n) \left(\frac{x_9}{1 + 9x_9 + 27x_9^2}\right)^n$$

and

$$(1 - 27x_9^2)\left(\sum_{n=0}^{\infty} t_9(n) x_9^n\right)^2 = \sum_{n=0}^{\infty} \binom{2n}{n} t_9(n) \left(\frac{x_9(1 + 9x_9 + 27x_9^2)}{(1 - 27x_9^2)^2}\right)^n.$$

By (9.15), (9.16), and Theorem 9.9, these are equivalent to (9.24) and (9.25), respectively. □

Explicit formulas for the coefficients are given by

$$t_9(n) = \sum_j (-3)^{n-3j} \binom{n}{j} \binom{n-j}{j} \binom{n-2j}{j}$$

and

$$s_9(n) = \sum_j \sum_k \binom{n}{j}^2 \binom{n}{k} \binom{j}{k} \binom{j+k}{n}.$$

These formulas may be verified by the method of creative telescoping described in Section 5.9.

## 9.4 Notes

The identity in (9.1) appears in Ramanujan's second notebook [253, Chapter 20, Entry 1 (iv)] in the equivalent form

$$1 + 9\frac{\eta_9^3}{\eta_1^3} = \left(1 + 27\frac{\eta_3^{12}}{\eta_1^{12}}\right)^{1/3}.$$

A proof of this identity has been given by Berndt [46, p. 348]. Ramanujan also gave the identity [253, Chapter 21, Entry 7 (i)]

$$\frac{1}{8}\left(9P(q^9) - P(q)\right) = \frac{\eta_3^6}{\eta_1^2\eta_9^2}\left(\eta_1^6 + 9\eta_1^3\eta_9^3 + 27\eta_9^6\right)^{1/3}.$$

On rewriting in the form

$$\frac{1}{8}\left(9P(q^9) - P(q)\right) = \frac{\eta_3^6}{\eta_1\eta_9}\left(\frac{\eta_1^3}{\eta_9^3} + 9 + 27\frac{\eta_9^3}{\eta_1^3}\right)^{1/3}$$

and applying (9.1) to the right-hand side, the result is (9.2). This deduction of (9.2) from results in Ramanujan's notebooks is due to J. Borwein and F. Garvan [68]. Borwein and Garvan in fact discovered the identity in (9.2) using symbolic computation and used it to produce a ninth order iteration that converges to $1/\pi$.

Two proofs of (9.2) were given by B. C. Berndt, S. H. Chan, Z.-G. Liu and H. Yesilyurt [51]. The first is essentially the same as the proof of Borwein and Garvan in [68]. The second proof relies on another entry in Ramanujan's second notebook [253, Ch. 20 Entry 1(v)]. Another proof of (9.2), using meromorphic functions defined on Riemann surfaces, was given by H. Farkas and I. Kra [150, p. 307]. The proof of (9.2) that we have given is from [104]. Another proof is outlined in Exercise 1 at the end of the chapter.

## 9.5 Exercises

1. Give another proof of (9.2) by replacing $q$ with $q^3$ in Theorem 1.17, and then letting $x = q$, $y = \exp(2\pi i/3)$. See S. H. Chan [95, Cor. 3.3].
2. Let $\alpha$, $\beta$, and $\gamma$ be defined by

$$\alpha = -q^{-1/3} \prod_{j=1}^{\infty} \frac{(1-q^{9j-7})(1-q^{9j-2})}{(1-q^{9j-8})(1-q^{9j-1})},$$

$$\beta = \quad q^{-1/3} \prod_{j=1}^{\infty} \frac{(1-q^{9j-5})(1-q^{9j-4})}{(1-q^{9j-7})(1-q^{9j-2})},$$

$$\gamma = \quad q^{2/3} \prod_{j=1}^{\infty} \frac{(1-q^{9j-8})(1-q^{9j-1})}{(1-q^{9j-5})(1-q^{9j-4})}.$$

a. Prove that:

$$\alpha + \beta + \gamma = 0,$$

$$\alpha\beta + \beta\gamma + \gamma\alpha = -\frac{\eta_3^4}{\eta_1\eta_9^3},$$

and

$$\alpha\beta\gamma = -1.$$

b. Prove that

$$\alpha^3 + \beta^3 + \gamma^3 = -3,$$

$$\alpha^3\beta + \beta^3\gamma + \gamma^3\alpha = -\frac{\eta_3^8}{\eta_1^2\eta_9^6}$$

and

$$\alpha^2 + \frac{1}{\alpha} = \beta^2 + \frac{1}{\beta} = \gamma^2 + \frac{1}{\gamma} = \frac{\eta_3^4}{\eta_1\eta_9^3}.$$

3. Show that

$$\sum_{m=1}^{\infty}\sum_{n=1}^{\infty}\left(\frac{m}{3}\right)\left(\frac{n}{3}\right) m\, q^{mn} = \frac{\eta_1^3\eta_9^3}{\eta_3^2}.$$

See Farkas and Kra [150, p. 303].

4. a. Show that

$$\int_0^q \prod_{n=1}^{\infty}\frac{(1-s^{3n})^{10}}{(1-s^n)^6}\,\mathrm{d}s = q\prod_{n=1}^{\infty}\frac{(1-q^{9n})^3}{(1-q^n)^3}.$$

Hint: Use (9.2) to show that both sides have the same derivative, and both sides are 0 when $q=0$.

b. Take $\tau = i/3$ in the transformation formula for Dedekind's eta function in Theorem 2.8 to show that

$$\frac{\eta(3i)}{\eta(i/3)} = \frac{1}{\sqrt{3}}.$$

Hence, deduce that

$$q\prod_{n=1}^{\infty}\frac{(1-q^{9n})^3}{(1-q^n)^3}\Bigg|_{q=\exp(-2\pi/3)} = \frac{1}{3\sqrt{3}}.$$

c. Use the results of Parts a and b to show that

$$\int_0^{e^{-2\pi/3}} \prod_{n=1}^{\infty}\frac{(1-q^{3n})^{10}}{(1-q^n)^6}\mathrm{d}q = \frac{1}{3\sqrt{3}}.$$

5. a. Show that

$$\int_0^q \prod_{n=1}^{\infty}\frac{(1-s^{2n})^{20}}{(1-s^n)^{16}}\,\mathrm{d}s = q\prod_{n=1}^{\infty}\frac{(1-q^{4n})^8}{(1-q^n)^8},$$

$$\int_0^q \prod_{n=1}^{\infty}\frac{(1-s^{2n})^8(1-s^{4n})^4}{(1-s^n)^8}\,\mathrm{d}s = q\prod_{n=1}^{\infty}\frac{(1-q^{2n})^2(1-q^{8n})^4}{(1-q^n)^4(1-q^{4n})^2},$$

and

$$\int_0^q \prod_{n=1}^{\infty}\frac{(1-s^{2n})^{14}(1-s^{6n})^6}{(1-s^n)^8(1-s^{4n})^8}\,\mathrm{d}s = q\prod_{n=1}^{\infty}\frac{(1-q^{2n})^4(1-q^{3n})^4(1-q^{12n})^4}{(1-q^n)^4(1-q^{4n})^4(1-q^{6n})^4}.$$

Hint: Use the identities (3.67), (8.37), and (3.70), respectively.

b. Deduce that

$$\int_0^{e^{-\pi}} \prod_{n=1}^{\infty} \frac{(1-q^{2n})^{20}}{(1-q^n)^{16}}\, dq = \frac{1}{16},$$

$$\int_0^{e^{-\pi/\sqrt{2}}} \prod_{n=1}^{\infty} \frac{(1-q^{2n})^{8}(1-q^{4n})^{4}}{(1-q^n)^{8}}\, dq = \frac{1}{\sqrt{32}},$$

and

$$\int_0^{e^{-\pi/\sqrt{3}}} \prod_{n=1}^{\infty} \frac{(1-q^{2n})^{14}(1-q^{6n})^{6}}{(1-q^n)^{8}(1-q^{4n})^{8}}\, dq = \frac{1}{3}.$$

Hint: Use (5.24) and (5.33).

The results in Exercise 5 are due to N. Fine [151, pp. 86–91]. For Exercise 4 and further examples, see [104].

# Chapter 10
# Level 10: Ramanujan's Function $k$

## 10.1 Ramanujan's function $k(q) = r(q)r^2(q^2)$

Consider the function $k = k(q)$ defined by

$$k(q) = r(q)\, r^2(q^2) \tag{10.1}$$

where $r(q)$ is the Rogers–Ramanujan continued fraction. Since

$$r(q) = q^{1/5} \prod_{j=1}^{\infty} \frac{(1-q^{5j-4})(1-q^{5j-1})}{(1-q^{5j-3})(1-q^{5j-2})}, \tag{10.2}$$

a straightforward calculation gives

$$k(q) = q \prod_{j=1}^{\infty} \frac{(1-q^{10j-9})(1-q^{10j-8})(1-q^{10j-2})(1-q^{10j-1})}{(1-q^{10j-7})(1-q^{10j-6})(1-q^{10j-4})(1-q^{10j-3})}. \tag{10.3}$$

The study of $k$ was initiated by Ramanujan. We shall embark on a detailed and systematic study of its properties. Among the beautiful formulas to be unearthed is the magnificent identity

$$\begin{aligned} &q\frac{\mathrm{d}}{\mathrm{d}q} \log \frac{k(1-k^2)}{(1+k-k^2)(1-4k-k^2)} \\ &= \sum_{n=0}^{\infty} \left\{ \sum_{j=0}^{n} \binom{n}{j}^4 \right\} \left( \frac{k(1-k^2)(1+k-k^2)(1-4k-k^2)}{(1+k^2)^4} \right)^n. \end{aligned}$$



S. Cooper, *Ramanujan's Theta Functions*, DOI 10.1007/978-3-319-56172-1_11

Before getting to business, observe that $k$ increases from 0 to $\sqrt{5}-2$ as $q$ increases from 0 to 1. This follows from Corollary 5.9, which shows that $r(q)$ strictly increases from 0 to $(\sqrt{5}-1)/2$ as $q$ increases from 0 to 1, and the simple fact $((\sqrt{5}-1)/2)^3 = \sqrt{5}-2$.

## 10.2 Basic properties

The first result gives infinite product formulas for $1+k$ and $1-k$.

**Theorem 10.1.** *The following identities hold:*

$$1+k=\prod_{j=1}^{\infty}\frac{(1-q^{10j-8})^2(1-q^{10j-5})^2(1-q^{10j-2})^2}{(1-q^{10j-9})(1-q^{10j-6})^2(1-q^{10j-4})^2(1-q^{10j-1})}$$

*and*

$$1-k=\prod_{j=1}^{\infty}\frac{(1-q^{10j-9})(1-q^{10j-6})(1-q^{10j-5})^2(1-q^{10j-4})(1-q^{10j-1})}{(1-q^{10j-8})(1-q^{10j-7})^2(1-q^{10j-3})^2(1-q^{10j-2})}.$$

*Proof.* The infinite product for $k$ in (10.3) may be expressed in terms of Ramanujan's theta function by Theorem 0.4, to give

$$k(q)=q\,\frac{f(-q,-q^9)f(-q^2,-q^8)}{f(-q^3,-q^7)f(-q^4,-q^6)}.$$

Hence, using Theorem 0.6 with $(a,b,c,d)=(q,q^4,-q^2,-q^3)$, we find that

$$\begin{aligned}1+k&=\frac{f(-q^3,-q^7)f(-q^4,-q^6)+qf(-q,-q^9)f(-q^2,-q^8)}{f(-q^3,-q^7)f(-q^4,-q^6)}\\&=\frac{f(q,q^4)f(-q^2,-q^3)}{f(-q^3,-q^7)f(-q^4,-q^6)}.\end{aligned}$$

This simplifies to the required result. The infinite product for $1-k$ can be proved by a similar procedure using Theorem 0.6 with $(a,b,c,d)=(-q,-q^4,q^2,q^3)$. □

The next result gives a quadratic transformation formula for the Rogers–Ramanujan continued fraction.

**Theorem 10.2.** *Let $k$ and $r$ be given by* (10.1) *and* (10.2). *Then*

$$r^5(q)=k\left(\frac{1-k}{1+k}\right)^2 \qquad \text{and} \qquad r^5(q^2)=k^2\left(\frac{1+k}{1-k}\right).$$

*Proof.* Because of the large number of infinite products that occur, we use the compact notation of (1.25). By the infinite products in (10.3) and Theorem 10.1 we find that

$$k\left(\frac{1-k}{1+k}\right)^2 = q\,\frac{(q,q^2,q^8,q^9;q^{10})_\infty}{(q^3,q^4,q^6,q^7;q^{10})_\infty} \times \frac{(q,q^4,q^5,q^5,q^6,q^9;q^{10})^2_\infty}{(q^2,q^3,q^3,q^7,q^7,q^8;q^{10})^2_\infty}$$
$$\div \frac{(q^2,q^2,q^5,q^5,q^8,q^8;q^{10})^2_\infty}{(q,q^4,q^4,q^6,q^6,q^9;q^{10})^2_\infty}.$$

This simplifies to

$$k\left(\frac{1-k}{1+k}\right)^2 = q\,\frac{(q,q^4,q^6,q^9;q^{10})^5_\infty}{(q^2,q^3,q^7,q^8;q^{10})^5_\infty} = q\,\frac{(q,q^4;q^5)^5_\infty}{(q^2,q^3;q^5)^5_\infty} = r^5(q).$$

This proves the first result. The second result can be proved in the same way. □

The following identities will be useful to facilitate further progress. They are analogues of the results in Theorem 3.25.

**Lemma 10.3.** *Let $\eta_k$ be defined by* (0.39) *and let $\varphi(q)$ and $\psi(q)$ be Ramanujan's theta functions given by Definition 3.1. Then*

$$\varphi^2(-q^5) - \varphi^2(-q) = \frac{\eta_5^4}{\eta_{10}^2} - \frac{\eta_1^4}{\eta_2^2} = 4\frac{\eta_1\eta_{10}^3}{\eta_2\eta_5}, \tag{10.4}$$

$$5\varphi^2(-q^5) - \varphi^2(-q) = 5\frac{\eta_5^4}{\eta_{10}^2} - \frac{\eta_1^4}{\eta_2^2} = 4\frac{\eta_2^3\eta_5}{\eta_1\eta_{10}}, \tag{10.5}$$

$$q^{1/4}\psi^2(q) - q^{5/4}\psi^2(q^5) = \frac{\eta_2^4}{\eta_1^2} - \frac{\eta_{10}^4}{\eta_5^2} = \frac{\eta_2\eta_5^3}{\eta_1\eta_{10}} \tag{10.6}$$

*and*

$$q^{1/4}\psi^2(q) - 5q^{5/4}\psi^2(q^5) = \frac{\eta_2^4}{\eta_1^2} - 5\frac{\eta_{10}^4}{\eta_5^2} = \frac{\eta_1^3\eta_{10}}{\eta_2\eta_5}. \tag{10.7}$$

*Proof.* The first equality in each of (10.4)–(10.7) is a trivial consequence of the infinite product identities for $\varphi(q)$ and $\psi(q)$ in Lemma 3.5. It remains to establish the second equality in each identity.

Replace $q$ with $q^{10}$ in Theorem 1.15, then let $(a,b,c) = (q,q^3,q^5)$. The product can be expressed in terms of eta-functions and the resulting identity is

$$\begin{aligned}\frac{\eta_2^2\eta_5\eta_{20}}{\eta_1\eta_4} &= r(q^{-1};q^{10}) + r(q^3;q^{10}) + r(q^7;q^{10}) - r(q^9;q^{10}) \\ &= \sum_{k=1}^{\infty}\left(\frac{q^{20k-11}}{1-q^{20k-11}} - \frac{q^{20k-9}}{1-q^{20k-9}} + \frac{q^{20k-7}}{1-q^{20k-7}} - \frac{q^{20k-13}}{1-q^{20k-13}}\right. \\ &\qquad \left. + \frac{q^{20k-3}}{1-q^{20k-3}} - \frac{q^{20k-17}}{1-q^{20k-17}} - \frac{q^{20k-1}}{1-q^{20k-1}} + \frac{q^{20k-19}}{1-q^{20k-19}}\right).\end{aligned}$$

This can be rewritten in the form

$$q\prod_{j=1}^{\infty}\frac{(1-q^{2j})^2(1-q^{5j})(1-q^{20j})}{(1-q^j)(1-q^{4j})} = \sum_{k=1}^{\infty}\chi_4(k)\frac{q^k}{1-q^k} - \sum_{k=1}^{\infty}\chi_4(k)\frac{q^{5k}}{1-q^{5k}},$$

where $\chi_4(n) = \sin(n\pi/2)$ is the character defined in (3.13). The series on the right may be summed using (3.14) to give

$$q\prod_{j=1}^{\infty}\frac{(1-q^{2j})^2(1-q^{5j})(1-q^{20j})}{(1-q^j)(1-q^{4j})} = \frac{1}{4}\left(\varphi^2(q) - \varphi^2(q^5)\right).$$

Finally, on replacing $q$ with $-q$ and making use of (0.2), we complete the proof of (10.4).

Next, let $\omega = \exp(i\pi/10)$ and put $(x,y,z) = (\omega^3,\omega^7,\omega^{11})$ in Theorem 1.12 and simplify, to obtain

$$4\frac{\eta_1\eta_4\eta_{10}^2}{\eta_5\eta_{20}} = 5\varphi^2(q^5) - \varphi^2(q).$$

The details have been omitted as they are similar to those given in Example 1.25. On replacing $q$ with $-q$ we complete the proof of (10.5).

Now we will prove (10.6) and (10.7). If we take the following identities from (10.4) and (10.5), namely

$$\frac{\eta_5^4}{\eta_{10}^2} - \frac{\eta_1^4}{\eta_2^2} = 4\frac{\eta_1\eta_{10}^3}{\eta_2\eta_5}$$

and

$$5\frac{\eta_5^4}{\eta_{10}^2} - \frac{\eta_1^4}{\eta_2^2} = 4\frac{\eta_2^3\eta_5}{\eta_1\eta_{10}},$$

and multiply each by $\eta_2\eta_{10}/\eta_1\eta_5$ we obtain

$$\frac{\eta_2\eta_5^3}{\eta_1\eta_{10}} - \frac{\eta_1^3\eta_{10}}{\eta_2\eta_5} = 4\frac{\eta_{10}^4}{\eta_5^2} \tag{10.8}$$

and

$$5\frac{\eta_2\eta_5^3}{\eta_1\eta_{10}} - \frac{\eta_1^3\eta_{10}}{\eta_2\eta_5} = 4\frac{\eta_2^4}{\eta_1^2}, \tag{10.9}$$

respectively. On taking linear combinations of (10.8) and (10.9) we deduce

$$\frac{\eta_2^4}{\eta_1^2} - \frac{\eta_{10}^4}{\eta_5^2} = \frac{\eta_2\eta_5^3}{\eta_1\eta_{10}}$$

and

$$\frac{\eta_2^4}{\eta_1^2} - 5\frac{\eta_{10}^4}{\eta_5^2} = \frac{\eta_1^3\eta_{10}}{\eta_2\eta_5}.$$

This completes the proofs of (10.6) and (10.7).

□

The next result relates $k$ to the Dedekind eta function. It is an analogue of Theorem 5.8.

**Theorem 10.4.** *Let $\eta_m$ be defined by* (0.39) *and let k be defined by* (10.1). *Then*

$$\frac{1}{k} - k = \frac{\eta_2\eta_5^5}{\eta_1\eta_{10}^5},$$

$$\frac{1}{k} + 1 - k = \frac{\eta_2^4\eta_5^2}{\eta_1^2\eta_{10}^4}$$

*and*

$$\frac{1}{k} - 4 - k = \frac{\eta_1^3\eta_5}{\eta_2\eta_{10}^3}.$$

*Proof.* The first result follows from the infinite products for $k$, $1+k$ and $1-k$ given by (10.3) and Theorem 10.1 after simplification. We omit the details. To prove the second and third results, observe that by Lemma 10.3 we have

$$\frac{1}{k} + 1 - k = \left(\frac{1}{k} - k\right) + 1 = \frac{\eta_2\eta_5^5}{\eta_1\eta_{10}^5} + 1 = \left(\frac{\eta_2\eta_5^3}{\eta_1\eta_{10}} + \frac{\eta_{10}^4}{\eta_5^2}\right)\frac{\eta_5^2}{\eta_{10}^4} = \frac{\eta_2^4\eta_5^2}{\eta_1^2\eta_{10}^4}$$

and

$$\frac{1}{k} - 4 - k = \left(\frac{1}{k} - k\right) - 4 = \frac{\eta_2\eta_5^5}{\eta_1\eta_{10}^5} - 4 = \left(\frac{\eta_5^4}{\eta_{10}^2} - 4\frac{\eta_1\eta_{10}^3}{\eta_2\eta_5}\right)\frac{\eta_2\eta_5}{\eta_1\eta_{10}^3} = \frac{\eta_1^3\eta_5}{\eta_2\eta_{10}^3}.$$

□

The results in the next theorem can be grouped naturally in pairs.

**Theorem 10.5.** *The following identities hold*:

$$\frac{\eta_2^{24}}{\eta_1^{24}} = \left(\frac{k}{1-k^2}\right)\left(\frac{1+k-k^2}{1-4k-k^2}\right)^5,$$
$$\frac{\eta_{10}^{24}}{\eta_5^{24}} = \left(\frac{k}{1-k^2}\right)^5\left(\frac{1+k-k^2}{1-4k-k^2}\right),$$

$$\frac{\eta_5^{6}}{\eta_1^{6}} = \left(\frac{k}{1+k-k^2}\right)\left(\frac{1-k^2}{1-4k-k^2}\right)^2,$$
$$\frac{\eta_{10}^{6}}{\eta_2^{6}} = \left(\frac{k}{1+k-k^2}\right)^2\left(\frac{1-k^2}{1-4k-k^2}\right),$$

$$\frac{\eta_5^{8}}{\eta_2^{8}} = \left(\frac{k}{1-4k-k^2}\right)\left(\frac{1-k^2}{1+k-k^2}\right)^3$$

*and*

$$\frac{\eta_{10}^{8}}{\eta_1^{8}} = \left(\frac{k}{1-4k-k^2}\right)^3\left(\frac{1-k^2}{1+k-k^2}\right).$$

*Proof.* This is immediate from Theorem 10.4.

□

We will now introduce three modular functions that are analogues of $w_a$, $w_b$, and $w_c$, which were studied in Chapter 6. Let $w_A$, $w_B$, and $w_C$ be the modular functions defined by

$$w_A = \left(\frac{\eta_5\eta_{10}}{\eta_1\eta_2}\right)^2, \qquad w_B = \left(\frac{\eta_2\eta_{10}}{\eta_1\eta_5}\right)^4 \quad \text{and} \quad w_C = \left(\frac{\eta_1\eta_{10}}{\eta_2\eta_5}\right)^6. \tag{10.10}$$

**Theorem 10.6.** *The functions $w_A$, $w_B$, and $w_C$ can be expressed in terms of $k$ by the formulas*

$$w_A = \frac{k(1-k^2)}{(1+k-k^2)(1-4k-k^2)}, \tag{10.11}$$
$$w_B = \frac{k(1+k-k^2)}{(1-k^2)(1-4k-k^2)} \tag{10.12}$$

*and*

$$w_C = \frac{k(1-4k-k^2)}{(1-k^2)(1+k-k^2)}. \tag{10.13}$$

*The functions $w_A$, $w_B$, and $w_C$ are algebraically related. In fact,*

$$\frac{w_B}{(1+4w_B)^2} = \frac{w_C}{(1-w_C)^2} = \frac{k(1-k^2)(1+k-k^2)(1-4k-k^2)}{(1+k^2)^4}, \tag{10.14}$$

$$\frac{w_A}{(1+5w_A)^2} = \frac{w_C}{(1+w_C)^2} = \frac{k(1-k^2)(1+k-k^2)(1-4k-k^2)}{(1+2k-6k^2-2k^3+k^4)^2} \tag{10.15}$$

*and*

$$\frac{w_A}{(1-5w_A)^2} = \frac{w_B}{(1-4w_B)^2} = \frac{k(1-k^2)(1+k-k^2)(1-4k-k^2)}{(1-8k-6k^2+8k^3+k^4)^2}. \tag{10.16}$$

*Proof.* The identities (10.11)–(10.13) are immediate from the formulas in Theorem 10.4. The results in (10.14)–(10.16) follow by direct calculation using (10.11)–(10.13).

□

## 10.3 The weight two modular form $y_{10}$

Let $y_{10}$ be the logarithmic derivative of $k$, that is,

$$y_{10} = y_{10}(q) = q\frac{\mathrm{d}}{\mathrm{d}q}\log k. \tag{10.17}$$

It may be expressed as a Lambert series, and also as a quotient of eta functions.

**Theorem 10.7.** *The following formulas hold:*

$$y_{10} = 1 + \sum_{j=1}^{\infty}(-1)^j\left(\frac{j}{5}\right)\frac{jq^j}{1-q^j}$$

*and*

$$y_{10} = \prod_{j=1}^{\infty}\frac{(1-q^j)(1-q^{2j})^2(1-q^{5j})^3}{(1-q^{10j})^2}.$$

*Proof.* By the definition of $y_{10}$ and a direct calculation we find that

$$\begin{aligned} y_{10} &= q\frac{\mathrm{d}}{\mathrm{d}q}\log k \\ &= q\frac{\mathrm{d}}{\mathrm{d}q}\log\left(q\prod_{j=1}^{\infty}\frac{(1-q^{10j-9})(1-q^{10j-8})(1-q^{10j-2})(1-q^{10j-1})}{(1-q^{10j-7})(1-q^{10j-6})(1-q^{10j-4})(1-q^{10j-3})}\right) \\ &= 1+\sum_{j=1}^{\infty}(-1)^j\left(\frac{j}{5}\right)\frac{jq^j}{1-q^j}. \end{aligned}$$

This proves the first result. To prove the other result, recall from Lemma 5.17 that

$$q\frac{\mathrm{d}}{\mathrm{d}q}\log r(q) = \frac{1}{5}q\frac{\mathrm{d}}{\mathrm{d}q}\log w_5 = \frac{1}{5}\frac{\eta_1^5}{\eta_5}.$$

It follows that

$$y_{10} = q\frac{\mathrm{d}}{\mathrm{d}q}\log k = q\frac{\mathrm{d}}{\mathrm{d}q}\left(\log r(q)+2\log r(q^2)\right) = \frac{1}{5}\left(\frac{\eta_1^5}{\eta_5}+4\frac{\eta_2^5}{\eta_{10}}\right).$$

On rearranging and then applying (10.5), this simplifies to

$$y_{10} = \frac{1}{5}\left(\frac{\eta_1^4}{\eta_2^2}+4\frac{\eta_2^3\eta_5}{\eta_1\eta_{10}}\right)\frac{\eta_1\eta_2^2}{\eta_5} = \frac{1}{5}\times\left(5\frac{\eta_5^4}{\eta_{10}^2}\right)\times\frac{\eta_1\eta_2^2}{\eta_5} = \frac{\eta_1\eta_2^2\eta_5^3}{\eta_{10}^2}.$$

□

The next result provides parameterizations of the eta function in terms of $k$ and $y_{10}$. It is an analogue of Theorem 4.21.

**Theorem 10.8.** *The following identities hold:*

$$\begin{aligned} \eta^{24}(\tau) &= y_{10}^6\,\frac{k(1-4k-k^2)^4}{(1-k^2)^4(1+k-k^2)}, \\ \eta^{24}(2\tau) &= y_{10}^6\,\frac{k^2(1+k-k^2)^4}{(1-k^2)^5(1-4k-k^2)}, \\ \eta^{24}(5\tau) &= y_{10}^6\,\frac{k^5(1-k^2)^4}{(1+k-k^2)^5(1-4k-k^2)^4}, \\ \eta^{24}(10\tau) &= y_{10}^6\,\frac{k^{10}}{(1-k^2)(1+k-k^2)^4(1-4k-k^2)^5}. \end{aligned}$$

*Proof.* This is immediate from the results in Theorem 10.4 together with the infinite product for $y_{10}$ in Theorem 10.7. □

Let us consider the four sets of functions

$$\{a_1, a_2, a_3, a_4\}, \quad \{b_1, b_2, b_3, b_4\}, \quad \{c_1, c_2, c_3, c_4\} \quad \text{and} \quad \{d_1, d_2, d_3, d_4\},$$

that are defined by

$$a_1 = \frac{\eta_1\eta_2^2\eta_5^3}{\eta_{10}^2}, \quad a_2 = \frac{\eta_1^2\eta_2\eta_{10}^3}{\eta_5^2}, \quad a_3 = \frac{\eta_1^3\eta_5\eta_{10}^2}{\eta_2^2}, \quad \frac{\eta_2^3\eta_5^2\eta_{10}}{\eta_1^2};$$

$$b_1 = \frac{\eta_1^5}{\eta_5}, \quad b_2 = \frac{\eta_2^5}{\eta_{10}}, \quad b_3 = \frac{\eta_5^5}{\eta_1}, \quad b_4 = \frac{\eta_{10}^5}{\eta_2};$$

$$c_1 = \frac{\eta_1^6\eta_5^2}{\eta_2^3\eta_{10}}, \quad c_2 = \frac{\eta_2^6\eta_{10}^2}{\eta_1^3\eta_5}, \quad c_3 = \frac{\eta_1^2\eta_5^6}{\eta_2\eta_{10}^3}, \quad c_4 = \frac{\eta_2^2\eta_{10}^6}{\eta_1\eta_5^3};$$

and

$$d_1 = \frac{\eta_1^{10}\eta_{10}}{\eta_2^5\eta_5^2}, \quad d_2 = \frac{\eta_2^{10}\eta_5}{\eta_1^5\eta_{10}^2}, \quad d_3 = \frac{\eta_2\eta_5^{10}}{\eta_1^2\eta_{10}^5}, \quad d_4 = \frac{\eta_1\eta_{10}^{10}}{\eta_2^2\eta_5^5}.$$

We will show below in Theorem 10.9 that each set is linearly independent and spans the same vector space. Before getting to that, observe that some of the functions have been encountered before. For example, from Theorem 10.7, $a_1 = y_{10}$. The functions $b_1$ and $b_3$ were expanded in Lambert series in Theorem 5.13, while $b_2$ and $b_4$ are the respective functions with $q$ replaced with $q^2$. By the infinite products for theta functions in Theorem 3.5 we find that

$$c_1 = \varphi^3(-q)\varphi(-q^5), \quad c_2 = q\psi^3(q)\psi(q^5),$$
$$c_3 = \varphi(-q)\varphi^3(-q^5), \quad c_4 = q^2\psi(q)\psi^3(q^5)$$

and

$$d_1 = \frac{\varphi^5(-q)}{\varphi(-q^5)}, \quad d_2 = \frac{\psi^5(q)}{\psi(q^5)}, \quad d_3 = \frac{\varphi^5(-q^5)}{\varphi(-q)}, \quad d_4 = q^3\frac{\psi^5(q^5)}{\psi(q)}.$$

Observe that all sixteen of the functions $a_j, b_j, c_j, d_j$, for $j = 1, 2, 3, 4$, are of the form

$$\eta_1^{e_1}\eta_2^{e_2}\eta_5^{e_5}\eta_{10}^{e_{10}} \quad \text{where} \quad e_1 + e_2 + e_5 + e_{10} = 4.$$

Theorem 10.8 may be used to show that if $e_1 + e_2 + e_5 + e_{10} = 4$ then

$$\eta_1^{e_1}\eta_2^{e_2}\eta_5^{e_5}\eta_{10}^{e_{10}} = y_{10} \times k^{h_1}(1-k^2)^{h_2}(1+k-k^2)^{h_3}(1-4k-k^2)^{h_4},$$

for some exponents $h_1$, $h_2$, $h_3$, and $h_4$ with

$$h_1 = \frac{1}{24}(e_1 + 2e_2 + 5e_5 + 10e_{10}),$$

$$h_2 = \frac{1}{24}(-4e_1 - 5e_2 + 4e_5 - e_{10}),$$

$$h_3 = \frac{1}{24}(-e_1 + 4e_2 - 5e_5 - 4e_{10}),$$

$$h_4 = \frac{1}{24}(4e_1 - e_2 - 4e_5 - 5e_{10}).$$

As a check, it may be noted that

$$h_1 + h_2 + h_3 + h_4 = 0.$$

For each of the 16 functions $a_j$, $b_j$, $c_j$, $d_j$, for $j = 1,2,3,4$, the exponents $h_1$, $h_2$, $h_3$, and $h_4$ turn out to be integers. They are given in Table 10.1.

By considering the first few terms in the $q$-expansions, it follows that each set $\{a_1, a_2, a_3, a_4\}$, $\{b_1, b_2, b_3, b_4\}$, $\{c_1, c_2, c_3, c_4\}$, and $\{d_1, d_2, d_3, d_4\}$ is linearly independent. We will now show that each set span the same vector space.

**Table 10.1**
$f = \eta_1^{e_1}\eta_2^{e_2}\eta_5^{e_5}\eta_{10}^{e_{10}}$
$= y_{10}\, k^{h_1}(1-k^2)^{h_2}$
$\times (1+k-k^2)^{h_3}$
$\times (1-4k-k^2)^{h_4}$

| $f$ | $e_1$ | $e_2$ | $e_5$ | $e_{10}$ | $h_1$ | $h_2$ | $h_3$ | $h_4$ |
|---|---|---|---|---|---|---|---|---|
| $a_1$ | 1 | 2 | 3 | −2 | 0 | 0 | 0 | 0 |
| $a_2$ | 2 | 1 | −2 | 3 | 1 | −1 | 0 | 0 |
| $a_3$ | 3 | −2 | 1 | 2 | 1 | 0 | −1 | 0 |
| $a_4$ | −2 | 3 | 2 | 1 | 1 | 0 | 0 | −1 |
| $b_1$ | 5 | 0 | −1 | 0 | 0 | −1 | 0 | 1 |
| $b_2$ | 0 | 5 | 0 | −1 | 0 | −1 | 1 | 0 |
| $b_3$ | −1 | 0 | 5 | 0 | 1 | 1 | −1 | −1 |
| $b_4$ | 0 | −1 | 0 | 5 | 2 | 0 | −1 | −1 |
| $c_1$ | 6 | −3 | 2 | −1 | 0 | 0 | −1 | 1 |
| $c_2$ | −3 | 6 | −1 | 2 | 1 | −1 | 1 | −1 |
| $c_3$ | 2 | −1 | 6 | −3 | 0 | 1 | −1 | 0 |
| $c_4$ | −1 | 2 | −3 | 6 | 2 | −1 | 0 | −1 |
| $d_1$ | 10 | −5 | −2 | 1 | 0 | −1 | −1 | 2 |
| $d_2$ | −5 | 10 | 1 | −2 | 0 | −1 | 2 | −1 |
| $d_3$ | −2 | 1 | 10 | −5 | 0 | 2 | −1 | −1 |
| $d_4$ | 1 | −2 | −5 | 10 | 3 | −1 | −1 | −1 |

**Theorem 10.9.** *The following identities, expressed as matrix products, hold:*

$$\begin{pmatrix} a_1 \\ a_2 \\ a_3 \\ a_4 \end{pmatrix} = \begin{pmatrix} \frac{1}{5} & \frac{4}{5} & 0 & 0 \\ -\frac{1}{5} & \frac{1}{5} & 0 & 0 \\ 0 & 0 & 1 & -4 \\ 0 & 0 & 1 & 1 \end{pmatrix} \begin{pmatrix} b_1 \\ b_2 \\ b_3 \\ b_4 \end{pmatrix} = \begin{pmatrix} -\frac{1}{4} & 0 & \frac{5}{4} & 0 \\ 0 & 1 & 0 & -5 \\ -\frac{1}{4} & 0 & \frac{1}{4} & 0 \\ 0 & 1 & 0 & -1 \end{pmatrix} \begin{pmatrix} c_1 \\ c_2 \\ c_3 \\ c_4 \end{pmatrix} = \begin{pmatrix} -\frac{1}{60} & -\frac{16}{15} & \frac{25}{12} & 0 \\ -\frac{1}{15} & \frac{1}{15} & 0 & -\frac{25}{3} \\ -\frac{1}{12} & 0 & \frac{1}{12} & -\frac{16}{3} \\ 0 & \frac{1}{3} & -\frac{1}{3} & -\frac{1}{3} \end{pmatrix} \begin{pmatrix} d_1 \\ d_2 \\ d_3 \\ d_4 \end{pmatrix}$$

$$\begin{pmatrix} b_1 \\ b_2 \\ b_3 \\ b_4 \end{pmatrix} = \begin{pmatrix} 1 & -4 & 0 & 0 \\ 1 & 1 & 0 & 0 \\ 0 & 0 & \frac{1}{5} & \frac{4}{5} \\ 0 & 0 & -\frac{1}{5} & \frac{1}{5} \end{pmatrix} \begin{pmatrix} a_1 \\ a_2 \\ a_3 \\ a_4 \end{pmatrix} = \begin{pmatrix} -\frac{1}{4} & -4 & \frac{5}{4} & 20 \\ -\frac{1}{4} & 1 & \frac{5}{4} & -5 \\ -\frac{1}{20} & \frac{4}{5} & \frac{1}{20} & -\frac{4}{5} \\ \frac{1}{20} & \frac{1}{5} & -\frac{1}{20} & -\frac{1}{5} \end{pmatrix} \begin{pmatrix} c_1 \\ c_2 \\ c_3 \\ c_4 \end{pmatrix} = \begin{pmatrix} \frac{1}{4} & -\frac{4}{3} & \frac{25}{12} & \frac{100}{3} \\ -\frac{1}{12} & -1 & \frac{25}{12} & -\frac{25}{3} \\ -\frac{1}{60} & \frac{4}{15} & -\frac{1}{4} & -\frac{4}{3} \\ \frac{1}{60} & \frac{1}{15} & -\frac{1}{12} & 1 \end{pmatrix} \begin{pmatrix} d_1 \\ d_2 \\ d_3 \\ d_4 \end{pmatrix}$$

$$\begin{pmatrix} c_1 \\ c_2 \\ c_3 \\ c_4 \end{pmatrix} = \begin{pmatrix} 1 & 0 & -5 & 0 \\ 0 & -\frac{1}{4} & 0 & \frac{5}{4} \\ 1 & 0 & -1 & 0 \\ 0 & -\frac{1}{4} & 0 & \frac{1}{4} \end{pmatrix} \begin{pmatrix} a_1 \\ a_2 \\ a_3 \\ a_4 \end{pmatrix} = \begin{pmatrix} \frac{1}{5} & \frac{4}{5} & -5 & 20 \\ \frac{1}{20} & -\frac{1}{20} & \frac{5}{4} & \frac{5}{4} \\ \frac{1}{5} & \frac{4}{5} & -1 & 4 \\ \frac{1}{20} & -\frac{1}{20} & \frac{1}{4} & \frac{1}{4} \end{pmatrix} \begin{pmatrix} b_1 \\ b_2 \\ b_3 \\ b_4 \end{pmatrix} = \begin{pmatrix} \frac{2}{5} & -\frac{16}{15} & \frac{5}{3} & \frac{80}{3} \\ \frac{1}{60} & \frac{2}{5} & -\frac{5}{12} & \frac{5}{3} \\ \frac{1}{15} & -\frac{16}{15} & 2 & \frac{16}{3} \\ \frac{1}{60} & \frac{1}{15} & -\frac{1}{12} & 2 \end{pmatrix} \begin{pmatrix} d_1 \\ d_2 \\ d_3 \\ d_4 \end{pmatrix}$$

*and*

$$\begin{pmatrix} d_1 \\ d_2 \\ d_3 \\ d_4 \end{pmatrix} = \begin{pmatrix} 1 & 16 & -25 & 0 \\ 1 & -\frac{1}{4} & 0 & \frac{25}{4} \\ 1 & 0 & -\frac{1}{5} & \frac{16}{5} \\ 0 & -\frac{1}{4} & \frac{1}{5} & \frac{1}{20} \end{pmatrix} \begin{pmatrix} a_1 \\ a_2 \\ a_3 \\ a_4 \end{pmatrix} = \begin{pmatrix} -3 & 4 & -25 & 100 \\ \frac{1}{4} & \frac{3}{4} & \frac{25}{4} & \frac{25}{4} \\ \frac{1}{5} & \frac{4}{5} & 3 & 4 \\ \frac{1}{20} & -\frac{1}{20} & \frac{1}{4} & -\frac{3}{4} \end{pmatrix} \begin{pmatrix} b_1 \\ b_2 \\ b_3 \\ b_4 \end{pmatrix} = \begin{pmatrix} 6 & 16 & -5 & -80 \\ -\frac{1}{4} & 6 & \frac{5}{4} & -5 \\ -\frac{1}{5} & \frac{16}{5} & \frac{6}{5} & -\frac{16}{5} \\ -\frac{1}{20} & -\frac{1}{5} & \frac{1}{20} & \frac{6}{5} \end{pmatrix} \begin{pmatrix} c_1 \\ c_2 \\ c_3 \\ c_4 \end{pmatrix}.$$

*Hence the sets* $\{a_1, a_2, a_3, a_4\}$*,* $\{b_1, b_2, b_3, b_4\}$*,* $\{c_1, c_2, c_3, c_4\}$*, and* $\{d_1, d_2, d_3, d_4\}$ *span the same vector space.*

*Proof.* All of these identities can be proved by referring to the data in Table 10.1. For example, consider the last matrix product

$$\begin{pmatrix} d_1 \\ d_2 \\ d_3 \\ d_4 \end{pmatrix} = \begin{pmatrix} 6 & 16 & -5 & -80 \\ -\frac{1}{4} & 6 & \frac{5}{4} & -5 \\ -\frac{1}{5} & \frac{16}{5} & \frac{6}{5} & -\frac{16}{5} \\ -\frac{1}{20} & -\frac{1}{5} & \frac{1}{20} & \frac{6}{5} \end{pmatrix} \begin{pmatrix} c_1 \\ c_2 \\ c_3 \\ c_4 \end{pmatrix}.$$

From Table 10.1, we find that

$$\begin{aligned} &6c_1 + 16c_2 - 5c_3 - 80c_4 \\ &= y_{10} \times \left( \frac{6(1-4k-k^2)}{1+k-k^2} + \frac{16k(1+k-k^2)}{(1-k^2)(1-4k-k^2)} \right. \\ &\qquad \left. - \frac{5(1-k^2)}{(1+k-k^2)} - \frac{80k^2}{(1-k^2)(1-4k-k^2)} \right) \end{aligned}$$

$$= y_{10} \times \frac{(1-4k-k^2)^2}{(1-k^2)(1+k-k^2)}$$

$$= d_1.$$

This proves the identity implied by the first row of the matrix product. The remaining parts of the matrix product can be computed similarly.

□

Since each function $b_1$, $b_2$, $b_3$, and $b_4$ can be expanded as a Lambert series by Theorem 5.13, Lambert series for the other functions $a_j$, $c_j$ and $d_j$, for $1 \le j \le 4$, can be obtained using Theorem 10.9. We illustrate with the functions $a_1$, $a_2$, $a_3$, and $a_4$.

**Theorem 10.10.** *The following Lambert series expansions hold*:

$$a_1 = 1 + \sum_{j=1}^{\infty} (-1)^j \left(\frac{j}{5}\right) \frac{jq^j}{1-q^j}, \tag{10.18}$$

$$a_2 = \sum_{j=1}^{\infty} \left(\frac{j}{5}\right) \frac{jq^j}{1-q^{2j}}, \tag{10.19}$$

$$a_3 = \sum_{j=1}^{\infty} (-1)^{j-1} \frac{jq^j(1-q^j)(1-q^{2j})}{1-q^{5j}} \tag{10.20}$$

*and*

$$a_4 = \sum_{j=1}^{\infty} \frac{jq^j(1-q^{2j})(1-q^{6j})}{1-q^{10j}}. \tag{10.21}$$

*Proof.* The identity (10.18) is true by Theorem 10.7.

Let us prove (10.19). Applying Theorem 10.9 and then Theorem 5.13, we obtain

$$\begin{aligned} a_2 &= \frac{1}{5}(b_2 - b_1) = \frac{1}{5}\left(\frac{\eta_2^5}{\eta_{10}} - \frac{\eta_1^5}{\eta_5}\right) \\ &= \sum_{j=1}^{\infty} \left(\frac{j}{5}\right) \left(\frac{jq^j}{1-q^j} - \frac{jq^{2j}}{1-q^{2j}}\right) \\ &= \sum_{j=1}^{\infty} \left(\frac{j}{5}\right) \frac{jq^j}{1-q^{2j}}. \end{aligned}$$

Now we will prove (10.20). Again, by Theorem 10.9 and then Theorem 5.13 we find that

$$\begin{aligned}
a_3 = b_3 - 4b_4 &= \frac{\eta_5^5}{\eta_1} - 4\,\frac{\eta_{10}^5}{\eta_2} \\
&= \sum_{k=1}^{\infty} \left(\frac{k}{5}\right) \frac{q^k}{(1-q^k)^2} - 4 \sum_{k=1}^{\infty} \left(\frac{k}{5}\right) \frac{q^{2k}}{(1-q^{2k})^2} \\
&= \sum_{j=1}^{\infty}\sum_{k=1}^{\infty} \left(\frac{k}{5}\right) j q^{jk} - 2 \sum_{j=1}^{\infty}\sum_{k=1}^{\infty} \left(\frac{k}{5}\right) (2j) q^{2jk}.
\end{aligned}$$

Since $\sum_j f(j) - 2\sum_j f(2j) = \sum_j (-1)^{j-1} f(j)$, the above can be simplified to give

$$\begin{aligned}
a_3 &= \sum_{j=1}^{\infty}\sum_{k=1}^{\infty} (-1)^{j-1} \left(\frac{k}{5}\right) j q^{jk} \\
&= \sum_{j=1}^{\infty} (-1)^{j-1} \frac{j(q^j - q^{2j} - q^{3j} + q^{4j})}{1-q^{5j}} \\
&= \sum_{j=1}^{\infty} (-1)^{j-1} \frac{j q^j (1-q^j)(1-q^{2j})}{1-q^{5j}}.
\end{aligned}$$

The proof of (10.21) is similar:

$$\begin{aligned}
a_4 = b_3 + b_4 &= \frac{\eta_5^5}{\eta_1} + \frac{\eta_{10}^5}{\eta_2} \\
&= \sum_{k=1}^{\infty} \left(\frac{k}{5}\right) \frac{q^k}{(1-q^k)^2} + \sum_{k=1}^{\infty} \left(\frac{k}{5}\right) \frac{q^{2k}}{(1-q^{2k})^2} \\
&= \sum_{k=1}^{\infty} \left(\frac{k}{5}\right) \frac{q^k}{(1-q^k)^2} - \sum_{k=1}^{\infty} \left(\frac{2k}{5}\right) \frac{q^{2k}}{(1-q^{2k})^2} \\
&= \sum_{j=1}^{\infty}\sum_{k \text{ odd}} \left(\frac{k}{5}\right) j q^{jk} \\
&= \sum_{j=1}^{\infty} \frac{j(q^j - q^{3j} - q^{7j} + q^{9j})}{1-q^{10j}} \\
&= \sum_{j=1}^{\infty} \frac{j q^j (1-q^{2j})(1-q^{6j})}{1-q^{10j}}.
\end{aligned}$$

□

Other examples of Lambert series will be given in the exercises.

The remaining theorems in this section give parameterizations for Ramanujan's Eisenstein series $P$, $Q$, and $R$ in terms of $k$ and $y_{10}$.

**Theorem 10.11.**

$$\begin{aligned}
\frac{(1+k^2)}{(1-k^2)}\, y_{10} &= \frac{1}{24}\left(P(q) - 2P(q^2) - 25P(q^5) + 50P(q^{10})\right),\\
\frac{(1+k^2)}{(1+k-k^2)}\, y_{10} &= \frac{1}{12}\left(P(q) - 4P(q^2) - 5P(q^5) + 20P(q^{10})\right),\\
\frac{(1+k^2)}{(1-4k-k^2)}\, y_{10} &= \frac{1}{24}\left(-3P(q) + 2P(q^2) - 5P(q^5) + 30P(q^{10})\right),\\
k\frac{\mathrm{d}y_{10}}{\mathrm{d}k} &= \frac{1}{24}\left(P(q) + 4P(q^2) + 15P(q^5) - 20P(q^{10})\right).
\end{aligned}$$

*Proof.* These results follow on logarithmically differentiating the results in Theorems 10.4 and 10.7.

□

The results in Theorem 10.11 may be rearranged to give $P(q)$, $P(q^2)$, $P(q^5)$, and $P(q^{10})$ in terms of $k$ and $y_{10}$.

**Theorem 10.12.** *The following identities, expressed as matrix multiplication, hold:*

$$\begin{pmatrix} P(q) \\ P(q^2) \\ P(q^5) \\ P(q^{10}) \end{pmatrix} = \begin{pmatrix} 4 & 1 & -4 & 6 \\ \frac{5}{2} & -2 & \frac{1}{2} & 3 \\ -\frac{4}{5} & 1 & \frac{4}{5} & \frac{6}{5} \\ \frac{1}{10} & \frac{2}{5} & \frac{1}{2} & \frac{3}{5} \end{pmatrix} \begin{pmatrix} \frac{(1+k^2)}{(1-k^2)} y_{10} \\ \frac{(1+k^2)}{(1+k-k^2)} y_{10} \\ \frac{(1+k^2)}{(1-4k-k^2)} y_{10} \\ k\frac{\mathrm{d}y_{10}}{\mathrm{d}k} \end{pmatrix}.$$

*Proof.* This is immediate from Theorem 10.11.

□

The next result expresses the Eisenstein series $Q$ in terms of $k$ and $y_{10}$.

**Theorem 10.13.** *The following parameterizations hold.*

$$Q(q) = y_{10}^2\left(\frac{(1-4k-k^2)^3}{(1-k^2)(1+k-k^2)^2} + \frac{256k(1+k-k^2)^3}{(1-k^2)^2(1-4k-k^2)^2}\right),$$

$$Q(q^2) = y_{10}^2\left(\frac{(1-4k-k^2)^3}{(1-k^2)(1+k-k^2)^2} + \frac{16k(1+k-k^2)^3}{(1-k^2)^2(1-4k-k^2)^2}\right),$$

$$Q(q^5) = y_{10}^2\left(\frac{(1-k^2)^3}{(1+k-k^2)^2(1-4k-k^2)} + \frac{256k^5}{(1+k-k^2)(1-k^2)^2(1-4k-k^2)^2}\right)$$

*and*

$$Q(q^{10}) = y_{10}^2\left(\frac{(1-k^2)^3}{(1+k-k^2)^2(1-4k-k^2)} + \frac{16k^5}{(1+k-k^2)(1-k^2)^2(1-4k-k^2)^2}\right).$$

*Proof.* By Corollary 3.40,

$$Q(q) = \frac{\eta_1^{16}}{\eta_2^8} + 256\frac{\eta_2^{16}}{\eta_1^8} \quad \text{and} \quad Q(q^2) = \frac{\eta_1^{16}}{\eta_2^8} + 16\frac{\eta_2^{16}}{\eta_1^8},$$

and hence, on replacing $q$ with $q^5$,

$$Q(q^5) = \frac{\eta_5^{16}}{\eta_{10}^8} + 256\frac{\eta_{10}^{16}}{\eta_5^8} \quad \text{and} \quad Q(q^{10}) = \frac{\eta_5^{16}}{\eta_{10}^8} + 16\frac{\eta_{10}^{16}}{\eta_5^8}.$$

The required results now follow by Theorem 10.8.

□

The analogous results for the Eisenstein series $R$ are as follows. We do not record explicit results as they are more complicated and will not be required.

**Theorem 10.14.** *For $r \in \{1, 2, 5, 10\}$,*

$$R(q^r) = y_{10}^3 \times (\textit{rational function of } k).$$

*Proof.* By Theorems 10.8 and 10.13, each of $\eta^{24}(r\tau)$ and $Q^3(q^r)$ is expressible as $y_{10}^6$ times a rational function of $k$. By Corollary 0.23,

$$R^2(q^r) = Q^3(q^r) - 1728\eta^{24}(r\tau),$$

and it follows that $R^2(q^r)$ is expressible as $y_{10}^6$ times a rational function of $k$. By case-by-case calculations, these rational functions turn out to be squares. On taking square roots we obtain the required results.

□

## 10.4 The modular function $X_{10}$ and weight two modular form $Z_{10}$

In this section we study the modular function $X_{10}$ defined by

$$X_{10} = \frac{k(1-k^2)(1+k-k^2)(1-4k-k^2)}{(1+k^2)^4} \tag{10.22}$$

and the weight two modular form $Z_{10}$ defined by

$$Z_{10} = \frac{1}{12}\left(10P(q^{10}) + 5P(q^5) - 2P(q^2) - P(q)\right). \tag{10.23}$$

They are related to the modular functions $w_A$, $w_B$, and $w_C$, as follows.

**Theorem 10.15.** *Let $w_A$, $w_B$, and $w_C$ be the eta-quotients defined by* (10.10). *Then*

$$X_{10} = \frac{w_B}{(1+4w_B)^2} = \frac{w_C}{(1-w_C)^2}$$

*and*

$$Z_{10} = q\frac{\mathrm{d}}{\mathrm{d}q}\log w_A.$$

*Proof.* The first result is just a restatement of (10.14). To prove the second result, a calculation using the definitions of $w_A$ and $Z_{10}$ together with (0.61) gives

$$\begin{aligned} q\frac{\mathrm{d}}{\mathrm{d}q}\log w_A &= q\frac{\mathrm{d}}{\mathrm{d}q}\log\left(\frac{\eta_5\eta_{10}}{\eta_1\eta_2}\right)^2 \\ &= \frac{1}{12}\left(10P(q^{10}) + 5P(q^5) - 2P(q^2) - P(q)\right) = Z_{10}. \end{aligned}$$

□

The next result shows that $Z_{10}$ may be expressed in terms of $k$ and $y_{10}$.

**Theorem 10.16.** *The following identity holds*:

$$Z_{10} = y_{10} \times \frac{(1+k^2)^3}{(1-k^2)(1+k-k^2)(1-4k-k^2)}.$$

*Hence,*

$$X_{10}Z_{10} = \frac{k\,y_{10}}{1+k^2}.$$

*Proof.* The formula defining $Z_{10}$ in (10.23) may be rewritten in the form

$$\begin{aligned} Z_{10} &= -\frac{1}{24}\left(50P(q^{10}) - 25P(q^5) - 2P(q^2) + P(q)\right) \\ &+ \frac{1}{12}\left(20P(q^{10}) - 5P(q^5) - 4P(q^2) + P(q)\right) \\ &+ \frac{1}{24}\left(30P(q^{10}) - 5P(q^5) + 2P(q^2) - 3P(q)\right). \end{aligned}$$

By Theorem 10.11, this is

$$\begin{aligned} Z_{10} &= y_{10} \times \left( -\left(\frac{1+k^2}{1-k^2}\right) + \frac{1+k^2}{1+k-k^2} + \frac{1+k^2}{1-4k-k^2} \right) \\ &= y_{10} \times \frac{(1+k^2)^3}{(1-k^2)(1+k-k^2)(1-4k-k^2)}, \end{aligned}$$

and this proves the first result. The second result can be obtained from the first by multiplying by $X_{10}$ and using (10.22).

□

The next result may be compared with the formulas for $X_1, \ldots, X_4$ in (4.56), $X_5$ in (5.39), $X_a$, $X_b$, and $X_c$ in (6.32), and $X_7$, $X_8$, and $X_9$ in (7.70), (8.57), and (9.6), respectively.

**Theorem 10.17.** *The following identity holds*:

$$X_{10} = \left(\frac{\eta_1\eta_2\eta_5\eta_{10}}{Z_{10}}\right)^{4/3}.$$

*Proof.* By Theorem 10.8 we find that

$$\eta_1\eta_2\eta_5\eta_{10} = y_{10} \times \left(\frac{k^3}{(1-k^2)(1+k-k^2)(1-4k-k^2)}\right)^{1/4}$$

and by Theorem 10.16 we have

$$Z_{10} = y_{10} \times \frac{(1+k^2)^3}{(1-k^2)(1+k-k^2)(1-4k-k^2)}.$$

It follows that

$$\left(\frac{\eta_1\eta_2\eta_5\eta_{10}}{Z_{10}}\right)^{4/3} = \frac{k(1-k^2)(1+k-k^2)(1-4k-k^2)}{(1+k^2)^4} = X_{10},$$

where the last equality follows from the definition of $X_{10}$ in (10.22).

□

The remaining results in this section are technical identities that will be required in the next section.

**Theorem 10.18.** *The following identities hold*:

$$1 + 4X_{10} = \frac{(1 + 2k - 6k^2 - 2k^3 + k^4)^2}{(1 + k^2)^4},$$

$$1 - 16X_{10} = \frac{(1 - 8k - 6k^2 + 8k^3 + k^4)^2}{(1 + k^2)^4}$$

*and*

$$\frac{\mathrm{d}X_{10}}{\mathrm{d}k} = \frac{(1 + 2k - 6k^2 - 2k^3 + k^4)(1 - 8k - 6k^2 + 8k^3 + k^4)}{(1 + k^2)^5}.$$

*Proof.* These all follow by basic algebra and calculus using the definition (10.22). □

**Theorem 10.19.** *The following differentiation formula holds*:

$$q\frac{\mathrm{d}}{\mathrm{d}q}\log X_{10} = Z_{10} \times \sqrt{1 - 12X_{10} - 64X_{10}^2}.$$

*Proof.* By the chain rule,

$$q\frac{\mathrm{d}}{\mathrm{d}q}\log X_{10} = \frac{1}{X_{10}}\, q\,\frac{\mathrm{d}X_{10}}{\mathrm{d}q} = \frac{1}{X_{10}}\left(q\,\frac{\mathrm{d}k}{\mathrm{d}q}\right)\left(\frac{\mathrm{d}X_{10}}{\mathrm{d}k}\right).$$

The derivative $q\,\mathrm{d}k/\mathrm{d}q$ can be obtained from the definition of $y_{10}$ in (10.17), and the derivative $\mathrm{d}X_{10}/\mathrm{d}k$ is given above in Theorem 10.18. It follows that

$$q\frac{\mathrm{d}}{\mathrm{d}q}\log X_{10} = \frac{1}{X_{10}} \times k y_{10} \times \frac{(1 + 2k - 6k^2 - 2k^3 + k^4)(1 - 8k - 6k^2 + 8k^3 + k^4)}{(1 + k^2)^5}.$$

The terms on the right-hand side may be expressed in terms of $X_{10}$ and $Z_{10}$ by Theorems 10.16 and 10.18 to give

$$q\frac{\mathrm{d}}{\mathrm{d}q}\log X_{10} = \frac{1}{X_{10}} \times (1 + k^2)X_{10}Z_{10} \times \frac{\sqrt{(1 + 4X_{10})(1 - 16X_{10})}}{1 + k^2}.$$

This simplifies to the required result.

□

## 10.5 Differential equations

The goal of this section is to derive a third order linear differential equation for $Z_{10}$ in terms of $X_{10}$. We will proceed in three steps: first we shall obtain a second order nonlinear differential equation for $y_{10}$ with respect to $k$ that is an analogue of Theorem 5.28. Second, we will deduce a third order linear differential equation satisfied by $y_{10}$ with respect to $k$. Finally, we obtain a third order linear differential equation that is satisfied by $Z_{10}$ with respect to $X_{10}$.

Here is the second order nonlinear differential equation satisfied by $y_{10}$.

**Theorem 10.20.** *Let $k$ and $y_{10}$ be defined by* (10.1) *and* (10.17). *Recall from* (10.22) *that*

$$X_{10} = \frac{k(1-k^2)(1+k-k^2)(1-4k-k^2)}{(1+k^2)^4},$$

*and let $p = p(k)$ be defined by*

$$p = \frac{k^2}{(1+k^2)^2} \times \frac{(1-4X_{10})}{X_{10}^2}. \tag{10.24}$$

*Then*

$$k\frac{\mathrm{d}}{\mathrm{d}k}\left(k\frac{\mathrm{d}y_{10}}{\mathrm{d}k}\right) = \frac{1}{2y_{10}}\left(k\frac{\mathrm{d}y_{10}}{\mathrm{d}k}\right)^2 + \frac{(1-p)}{2}y_{10}.$$

*Proof.* By the last part of Theorem 10.12,

$$k\frac{\mathrm{d}y_{10}}{\mathrm{d}k} = \frac{1}{24}\left(P(q) + 4P(q^2) + 15P(q^5) - 20P(q^{10})\right). \tag{10.25}$$

By the chain rule and the definition of $y_{10}$, we find that

$$k\frac{\mathrm{d}}{\mathrm{d}k} = kq\frac{\mathrm{d}}{\mathrm{d}q} \Big/ \left(q\frac{\mathrm{d}k}{\mathrm{d}q}\right) = \frac{1}{y_{10}}q\frac{\mathrm{d}}{\mathrm{d}q}. \tag{10.26}$$

Applying (10.26) to (10.25) and using Ramanujan's differential equation (0.78), we obtain

$$\begin{aligned}
k\frac{\mathrm{d}}{\mathrm{d}k}\left(k\frac{\mathrm{d}y_{10}}{\mathrm{d}k}\right) &= \frac{1}{24y_{10}}q\frac{\mathrm{d}}{\mathrm{d}q}\left(P(q) + 4P(q^2) + 15P(q^5) - 20P(q^{10})\right) \\
&= \frac{1}{288y_{10}}\left((P^2(q) - Q(q)) + 8(P^2(q^2) - Q(q^2))\right. \\
&\qquad \left. +75(P^2(q^5) - Q(q^5)) - 200(P^2(q^{10}) - Q(q^{10}))\right).
\end{aligned}$$

By Theorems 10.12 and 10.13, each of $P^2(q^r)$ and $Q(q^r)$, $r = 1, 2, 5$, and 10, may be expressed in terms of $k$, $y_{10}$, and $k\frac{dy_{10}}{dk}$. The claimed result follows after simplification.

□

Here is the third order linear differential equation satisfied by $y_{10}$.

**Theorem 10.21.** *Let p be as defined in* (10.24). *Then*

$$k\frac{d}{dk}\left(k\frac{d}{dk}\left(k\frac{dy_{10}}{dk}\right)\right) + k(p-1)\frac{dy_{10}}{dk} + \frac{y_{10}}{2}k\frac{dp}{dk} = 0.$$

*Proof.* Applying the operator $k\frac{d}{dk}$ to both sides of the differential equation in Theorem 10.20, we obtain

$$\begin{aligned} & k\frac{d}{dk}\left(k\frac{d}{dk}\left(k\frac{dy_{10}}{dk}\right)\right) \\ &= \frac{1}{y_{10}}k\frac{dy_{10}}{dk}\left\{k\frac{d}{dk}\left(k\frac{dy_{10}}{dk}\right) - \frac{1}{2y_{10}}\left(k\frac{dy_{10}}{dk}\right)^2\right\} + \frac{k(1-p)}{2}\frac{dy_{10}}{dk} - \frac{y_{10}}{2}k\frac{dp}{dk}. \end{aligned}$$

Theorem 10.20 can be applied to the terms in braces. The claimed result follows on simplification.

□

Here is the third order linear differential equation satisfied by $Z_{10}$ with respect to $X_{10}$.

**Theorem 10.22.** *Let* $X_{10}$ *and* $Z_{10}$ *be as defined by* (10.22) *and* (10.23), *respectively. Then*

$$\begin{aligned} & X_{10}^2(1-16X_{10})(1+4X_{10})\frac{d^3Z_{10}}{dX_{10}^3} + 3X_{10}(1-18X_{10}-128X_{10}^2)\frac{d^2Z_{10}}{dX_{10}^2} \\ & + (1-40X_{10}-444X_{10}^2)\frac{dZ_{10}}{dX_{10}} - 2(1+30X_{10})Z_{10} = 0. \end{aligned} \tag{10.27}$$

*Proof.* Let $D$ be the differential operator defined by

$$D = (1+k^2)\frac{d}{dk}. \tag{10.28}$$

By Theorem 10.18 and the chain rule, it follows that

$$D = \sqrt{(1+4X_{10})(1-16X_{10})}\frac{d}{dX_{10}}. \tag{10.29}$$

By Theorem 10.16,

$$z = \frac{(1+k^2)}{k} X_{10} Z_{10}.$$

Differentiating with respect to $k$, we obtain

$$\begin{aligned}
k\frac{dz}{dk} &= k\frac{d}{dk}\left(\frac{(1+k^2)}{k} X_{10} Z_{10}\right) \\
&= \left(k - \frac{1}{k}\right) X_{10} Z_{10} + (1+k^2)\frac{d}{dk}(X_{10} Z_{10}) \\
&= \left(k - \frac{1}{k}\right) X_{10} Z_{10} + D(X_{10} Z_{10}).
\end{aligned}$$

Therefore

$$D(X_{10} Z_{10}) = \frac{(1-k^2)}{k} X_{10} Z_{10} + k\frac{dz}{dk}. \tag{10.30}$$

Apply the operator $D$ to (10.30) and make use of (10.28) to obtain

$$\begin{aligned}
D^2(X_{10} Z_{10}) &= (1+k^2)\frac{d}{dk}\left(\frac{1-k^2}{k}\right) \times X_{10} Z_{10} \\
&\quad + \left(\frac{1-k^2}{k}\right) D(X_{10} Z_{10}) + \frac{(1+k^2)}{k} k\frac{d}{dk}\left(k\frac{dz}{dk}\right).
\end{aligned} \tag{10.31}$$

The identity (10.30) may be used to eliminate the $D(X_{10} Z_{10})$ term from the right-hand side of (10.31), and the result simplifies to

$$D^2(X_{10} Z_{10}) + 4X_{10} Z_{10} = \frac{(1-k^2)}{k} k\frac{dz}{dk} + \frac{(1+k^2)}{k} k\frac{d}{dk}\left(k\frac{dz}{dk}\right).$$

Apply the operator $D$ again and make use of (10.28) to obtain

$$D^3(X_{10} Z_{10}) + 4D(X_{10} Z_{10}) = -\frac{(1+k^2)^2}{k^2} k\frac{dz}{dk} + \frac{(1+k^2)^2}{k^2} k\frac{d}{dk}\left(k\frac{d}{dk}\left(k\frac{dz}{dk}\right)\right).$$

This may be simplified by applying Theorem 10.21 to give

$$\begin{aligned}
&D^3(X_{10} Z_{10}) + 4D(X_{10} Z_{10}) \\
&= -\frac{(1-4X_{10})}{X_{10}^2} D(X_{10} Z_{10}) + \frac{(1+k^2)^2}{k^3}\left((1-k^2)p - \frac{(1+k^2)}{2} k\frac{dp}{dk}\right) X_{10} Z_{10}.
\end{aligned} \tag{10.32}$$

A simple but lengthy algebraic calculation, using (10.22), (10.24), and Theorem 10.18, shows that

$$\frac{(1+k^2)^2}{k^3}\left((1-k^2)p-\frac{k(1+k^2)}{2}\frac{\mathrm{d}p}{\mathrm{d}k}\right)=\frac{(1-2X_{10})\sqrt{(1+4X_{10})(1-16X_{10})}}{X_{10}^3}.$$

Therefore, (10.32) simplifies further to

$$D^3(X_{10}Z_{10})+\frac{(1-2X_{10})^2}{X_{10}^2}D(X_{10}Z_{10})=\sqrt{(1+4X_{10})(1-16X_{10})}\,\frac{(1-2X_{10})}{X_{10}^2}Z_{10},$$

and by (10.29) this is equivalent to

$$X_{10}^2\frac{\mathrm{d}}{\mathrm{d}X_{10}}(D^2(X_{10}Z_{10}))+(1-2X_{10})^2\frac{\mathrm{d}}{\mathrm{d}X_{10}}(X_{10}Z_{10})=(1-2X_{10})Z_{10}.$$

The differential equation (10.27) now follows by expanding and canceling the common factor of $X_{10}$ that occurs.

□

If we consider the power series solutions of (10.27), we readily obtain:

**Theorem 10.23.** *Let $T_{10}(n)$ be the sequence defined by the recurrence relation*

$$(n+1)^3T_{10}(n+1)=2(2n+1)(3n^2+3n+1)T_{10}(n)+4n(16n^2-1)T_{10}(n-1), \tag{10.33}$$

*and initial condition $T_{10}(0)=1$. Then*

$$Z_{10}=\sum_{n=0}^{\infty}T_{10}(n)X_{10}^n. \tag{10.34}$$

*The series in* (10.34) *converges provided $|X_{10}|<1/16$.*

*Proof.* The recurrence relation is an immediate consequence of the theory of series solutions for differential equations. The initial conditions may be verified by comparing the first few terms in the $q$-expansions of the identities in (10.34). In this way, the values of $T_{10}(0)$ and $T_{10}(1)$ may be determined, and the value of $T_{10}(1)$ is consistent with the result of putting $n=0$ in (10.33). The radius of convergence may be deduced from the standard theory of differential equations by considering the regular singular points.

□

It is a classical result that the sum of the fourth powers of the binomial coefficients satisfies the recurrence relation and initial conditions given by Theorem 10.23. That leads to:

**Theorem 10.24.** *The terms $T_{10}(n)$ in the sequence defined by the recurrence relation* (10.33) *and initial condition $T_{10}(0) = 1$ are given by*

$$T_{10}(n) = \sum_{j=0}^{n} \binom{n}{j}^4.$$

*Hence,*

$$Z_{10} = \sum_{n=0}^{\infty} \left\{ \sum_{j=0}^{n} \binom{n}{j}^4 \right\} X_{10}^n.$$

*Proof.* Use a computer, as described in Section 5.9, to show that the sequence $\sum_{j=0}^{n} \binom{n}{k}^4$ satisfies the recurrence relation (10.33), and check that values of the sequence in the cases $n = 0$ and $n = 1$ are $T_{10}(0)$ and $T_{10}(1)$, respectively. □

Expressions for the series

$$\sum_{n=0}^{\infty} \left\{ \sum_{j=0}^{n} \binom{n}{j}^4 \right\} X^n$$

in terms of hypergeometric functions will be obtained in Chapter 13—see Theorem 13.1.

## 10.6 Notes

The function $k$ was introduced by Ramanujan in his second notebook [253, p. 326] where it is denoted by $n$. Further properties are given in Ramanujan's lost notebook [254, pp. 53, 56, 208]. Ramanujan introduced $k$ in a statement that is equivalent to:

$$\text{if} \quad r^5(q) = k\left(\frac{1-k}{1+k}\right)^2 \quad \text{then} \quad r^5(q^2) = k^2\left(\frac{1+k}{1-k}\right).$$

If we multiply the first identity by the square of the second identity and then take fifth roots, we obtain

$$k = r(q)r^2(q^2),$$

which we have taken as the definition of $k$ in (10.1). We have proved Ramanujan's assertion relating $r^5(q)$ and $r^5(q^2)$ in Theorem 10.2.

Several of Ramanujan's identities for $k$ were discussed by S. Raghavan and S. S. Rangachari [245]. Ramanujan's results were analyzed further by S.-Y. Kang [192]. Kang's proofs rely on modular equations in Ramanujan's notebooks [253] that have been proved by B. C. Berndt [46]. Her work has been reproduced in the book by G. E. Andrews and B. C. Berndt [25, pp. 33–44, 81–84]. For different proofs, see the paper by C. Gugg [167]. The approach in this chapter follows [111] and [113].

The identities (10.4) and (10.6) are in Ramanujan's second notebook [253, Chapter 19, Entry 10] and the other two identities (10.5) and (10.7) were proved by Andrews and Berndt [25, p. 27]. The recurrence relation satisfied by the sum of the fourth powers of the binomial coefficients in Theorem 10.23 was known to Franel [154]. See also the works of Barrucand [37], Elsner [146], and Stanley [273, pp. 245, 278]. For more information about the origin of Theorem 10.24, see the work of Chan, Tanigawa, Yang, and Zudilin [90].

## 10.7 Exercises

1. Use Theorems 10.9 and 10.10 to prove the following identities:

a. $$\frac{1}{4}\left(5\varphi(-q)\varphi^3(-q^5) - \varphi^3(-q)\varphi(-q^5)\right) = 1 + \sum_{j=1}^{\infty}(-1)^j\left(\frac{j}{5}\right)\frac{jq^j}{1-q^j},$$

b. $$q\psi^3(q)\psi(q^5) - 5q^2\psi(q)\psi^3(q^5) = \sum_{j=1}^{\infty}\left(\frac{j}{5}\right)\frac{jq^j}{1-q^{2j}},$$

c. $$\frac{1}{4}\left(\varphi(-q)\varphi^3(-q^5) - \varphi^3(-q)\varphi(-q^5)\right) = \sum_{j=1}^{\infty}(-1)^{j-1}\frac{jq^j(1-q^j)(1-q^{2j})}{1-q^{5j}},$$

d. $$q\psi^3(q)\psi(q^5) - q^2\psi(q)\psi^3(q^5) = \sum_{j=1}^{\infty}\frac{jq^j(1-q^{2j})(1-q^{6j})}{1-q^{10j}}.$$

e. Hence, deduce Lambert series expansions for each of $\varphi^3(q)\varphi(q^5)$, $\varphi(q)\varphi^3(q^5)$, $q\psi^3(q)\psi(q^5)$, and $q^2\psi(q)\psi^3(q^5)$.

Hint for Parts a–d: The left-hand sides are $\frac{1}{4}(5c_3 - c_1)$, $c_2 - 5c_4$, $\frac{1}{4}(c_3 - c_1)$, and $c_2 - c_4$, respectively. By Theorem 10.9, these are equal to $a_1$, $a_2$, $a_3$, and $a_4$, respectively. Now apply Theorem 10.10.

Parts a and b of this exercise are from Ramanujan's second notebook [253, Chapter 19, Entry 8, (i), (ii)]. See Berndt [46, pp. 249–254] for another proof. The Lambert series for $\varphi^3(q)\varphi(q^5)$ and $\varphi(q)\varphi^3(q^5)$ in Part e go back to Liouville. See A. Alaca, Ş. Alaca and K. S. Williams [14] for references and some historical background.

2. Use Theorems 10.9 and 10.10 to prove the following identities:

   a. $25\varphi(-q)\varphi^3(-q^5) - \dfrac{\varphi^5(-q)}{\varphi(-q^5)} = 24 - 40\sum_{j \text{ odd}} \left(\dfrac{j}{5}\right) \dfrac{jq^j}{1-q^j}$,

   b. $\dfrac{\psi^5(q)}{\psi(q^5)} - 25q^2\psi(q)\psi^3(q^5) = 1 + 5\sum_{j=1}^{\infty} \left(\dfrac{j}{5}\right) \dfrac{jq^j}{1+q^j}$,

   c. $q\psi^3(q)\psi(q^5) - q^3\dfrac{\psi^5(q^5)}{\psi(q)} = \sum_{j=1}^{\infty}(-1)^{j+1}\left(\dfrac{j}{5}\right)\dfrac{q^j}{(1-q^j)^2}$,

   d. $\dfrac{\varphi^5(-q^5)}{\varphi(-q)} - \varphi^3(-q)\varphi(-q^5) = 8\sum_{j=1}^{\infty}\left(\dfrac{j}{5}\right)\dfrac{q^j(1+q^{2j})}{(1-q^{2j})^2}$.

   e. Deduce Lambert series expansions for $\varphi^5(-q)/\varphi(-q^5)$, $\varphi^5(-q^5)/\varphi(-q)$, $\psi^5(q)/\psi(q^5)$ and $q^3\psi^5(q^5)/\psi(q)$.

   Hint for Parts a–d: The left-hand sides are $25c_3 - d_1$, $d_2 - 25c_4$, $c_2 - d_4$, and $d_3 - c_1$, respectively. By Theorem 10.9, these are equal to $8b_1 + 16b_2$, $2b_2 - b_1$, $b_3 + 2b_4$, and $8b_3 - 16b_4$, respectively, which may be expanded by Theorem 5.13 to produce the required Lambert series.

   Parts a and b of this exercise are from Ramanujan's second notebook [253, Chapter 19, Entry 8, (iii), (iv)]. See Berndt [46, pp. 255–257] for another proof.

3. Show that

$$\varphi^2(q)\varphi^2(q^5) = \frac{1}{6}\varphi^4(q) + \frac{5}{6}\varphi^4(q^5) + \frac{8}{3}q\prod_{j=1}^{\infty}(1-q^{2j})^2(1-q^{10j})^2$$

and

$$q^3\psi^2(q^2)\psi^2(q^{10}) = \frac{1}{6}q\psi^4(q^2) + \frac{5}{6}q^5\psi^4(q^{10}) - \frac{1}{6}q\prod_{j=1}^{\infty}(1-q^{2j})^2(1-q^{10j})^2.$$

   See A. Alaca, Ş. Alaca and K. S. Williams [14].

4. Let $u = r(q)$, $v = -r(-q)$, and $w = r(q^2)$, where $r$ is the Rogers–Ramanujan continued fraction.

   a. Use Theorem 10.2 to show that $\dfrac{w-u^2}{w+u^2} = uw^2$.

   b. Replace $q$ with $-q$ in the result of Part a to deduce that $\dfrac{w-v^2}{w+v^2} = -vw^2$.

   c. Eliminate $w$ from the results of Parts a and b to deduce that

$$uv(u-v)^4 - u^2v^2(u-v)^2 + 2u^3v^3 + (u-v)(1+u^5v^5) = 0.$$

   See Ramanujan's second notebook [253, p. 321] and Berndt [48, p. 16].

5. Let $\zeta = \exp(2\pi i/5)$. Prove the following factorizations of the identities in Theorem 10.4:

a.

$$\frac{1}{\sqrt{k}} + \sqrt{k} = \frac{1}{q^{1/5}r^{3/2}(q)} \times \frac{(q^5, q^5; q^{10})_\infty}{(q^3, q^7; q^{10})_\infty},$$

$$\frac{1}{\sqrt{k}} - \sqrt{k} = \frac{r^{3/2}(q)}{q^{4/5}} \times \frac{(q^5, q^5; q^{10})_\infty}{(q, q^9; q^{10})_\infty}.$$

b.

$$\frac{1}{\sqrt{k}} + \frac{(1-\sqrt{5})}{2}\sqrt{k} = q^{1/12}\frac{\eta^2(2\tau)\eta^{1/2}(5\tau)}{\eta^{1/2}(\tau)\eta^2(10\tau)} \times (\zeta q, \zeta^2 q^2, \zeta^3 q^2, \zeta^4 q; q^2)_\infty,$$

$$\frac{1}{\sqrt{k}} + \frac{(1+\sqrt{5})}{2}\sqrt{k} = q^{1/12}\frac{\eta^2(2\tau)\eta^{1/2}(5\tau)}{\eta^{1/2}(\tau)\eta^2(10\tau)} \times (\zeta q^2, \zeta^2 q, \zeta^3 q, \zeta^4 q^2; q^2)_\infty.$$

c.

$$\frac{1}{\sqrt{k}} - (2-\sqrt{5})\sqrt{k} = q^{1/6}\frac{\eta^{3/2}(\tau)\eta^{1/2}(5\tau)}{\eta^2(10\tau)} \times (-\zeta q, \zeta^2 q, \zeta^3 q, -\zeta^4 q; q)_\infty,$$

$$\frac{1}{\sqrt{k}} - (2+\sqrt{5})\sqrt{k} = q^{1/6}\frac{\eta^{3/2}(\tau)\eta^{1/2}(5\tau)}{\eta^2(10\tau)} \times (\zeta q, -\zeta^2 q, -\zeta^3 q, \zeta^4 q; q)_\infty.$$

Further details can be found in work with M. Hirschhorn [121].

6. Prove the following factorizations of (10.4) and (10.6):

$$\varphi(q) + \varphi(q^5) = 2\frac{(q^2, q^8, q^{10}, q^{12}, q^{18}, q^{20}; q^{20})_\infty}{(q, q^4, q^9, q^{11}, q^{16}, q^{19}; q^{20})_\infty},$$

$$\varphi(q) - \varphi(q^5) = 2q\frac{(q^4, q^6, q^{10}, q^{14}, q^{16}, q^{20}; q^{20})_\infty}{(q^3, q^7, q^8, q^{12}, q^{13}, q^{17}; q^{20})_\infty},$$

$$\psi(q^2) + q\psi(q^{10}) = 2\frac{(q^3, q^7, q^{10}, q^{13}, q^{17}, q^{20}; q^{20})_\infty}{(q, q^6, q^9, q^{11}, q^{14}, q^{19}; q^{20})_\infty}$$

and

$$\psi(q^2) - q\psi(q^{10}) = 2\frac{(q, q^9, q^{10}, q^{11}, q^{19}, q^{20}; q^{20})_\infty}{(q^2, q^3, q^7, q^{13}, q^{17}, q^{18}; q^{20})_\infty}.$$

These results appear in Ramanujan's lost notebook [254, p. 56] and were rediscovered by R. Blecksmith, J. Brillhart, and I. Gerst [58]. See also the work of Andrews and Berndt [25, p. 28] and work with Hirschhorn [117] and [118].

7. Let $\alpha = (1+\sqrt{5})/2$ and $\beta = (1-\sqrt{5})/2$. Prove the following factorizations of (10.5) and (10.7).

$$\varphi(q) + \sqrt{5}\,\varphi(q^5) = \frac{2\alpha \prod_{j=1}^{\infty}(1-q^{2j})}{\prod_{j \text{ even}}(1-\alpha q^j + q^{2j}) \prod_{j \text{ odd}}(1+\beta q^j + q^{2j})},$$

$$\varphi(q) - \sqrt{5}\,\varphi(q^5) = \frac{2\beta \prod_{j=1}^{\infty}(1-q^{2j})}{\prod_{j \text{ odd}}(1+\alpha q^j + q^{2j}) \prod_{j \text{ even}}(1-\beta q^j + q^{2j})},$$

$$\psi(q^2) + q\sqrt{5}\,\psi(q^{10}) = \frac{\prod_{j=1}^{\infty}(1-q^{2j})}{\prod_{j \text{ odd}}(1-\alpha q^j + q^{2j}) \prod_{j \text{ odd}}(1+\beta q^j + q^{2j})}$$

and

$$\psi(q^2) - q\sqrt{5}\,\psi(q^{10}) = \frac{\prod_{j=1}^{\infty}(1-q^{2j})}{\prod_{j \text{ odd}}(1+\alpha q^j + q^{2j}) \prod_{j \text{ odd}}(1-\beta q^j + q^{2j})}.$$

See Ramanujan's lost notebook [254, p. 56] and Andrews and Berndt [25, p. 30]. The values of $\alpha$ and $\beta$ above are switched from how Ramanujan had them.

8. Show that the result of Theorem 10.23 can be expressed in the form

$$q\frac{\mathrm{d}}{\mathrm{d}q}\log\left(\frac{k(1-k^2)}{(1+k-k^2)(1-4k-k^2)}\right)$$
$$= \sum_{n=0}^{\infty}\left\{\sum_{j=0}^{n}\binom{n}{j}^4\right\}\left(\frac{k(1-k^2)(1+k-k^2)(1-4k-k^2)}{(1+k^2)^4}\right)^n$$

where $k$ is given by (10.1). See [113, Theorem 4.2].

9. Let $X_A, X_B, X_C, Z_A, Z_B$, and $Z_C$ be defined by

$$X_A = \frac{w_B}{(1+4w_B)^2} = \frac{w_C}{(1-w_C)^2},$$

$$X_B = \frac{w_A}{(1+5w_A)^2} = \frac{w_C}{(1+w_C)^2},$$

$$X_C = \frac{w_A}{(1-5w_A)^2} = \frac{w_B}{(1-4w_B)^2},$$

$$Z_A = q\frac{\mathrm{d}}{\mathrm{d}q}\log w_A, \quad Z_B = q\frac{\mathrm{d}}{\mathrm{d}q}\log w_B, \quad Z_C = q\frac{\mathrm{d}}{\mathrm{d}q}\log w_C.$$

By Theorem 10.15, $X_A = X_{10}$ and $Z_A = Z_{10}$. This exercise develops some properties of the other functions $X_B$, $X_C$, $Z_B$, and $Z_C$.

a. Show that

$$\frac{1}{X_B} - \frac{1}{X_A} = 4, \quad \frac{1}{X_B} - \frac{1}{X_C} = 20 \quad \text{and} \quad \frac{1}{X_A} - \frac{1}{X_C} = 16.$$

b. Show that

$$Z_A = y_{10} \frac{(1+k^2)^3}{(1-k^2)(1+k-k^2)(1-4k-k^2)},$$

$$Z_B = y_{10} \frac{(1+k^2)(1+2k-6k^2-2k^3+k^4)}{(1-k^2)(1+k-k^2)(1-4k-k^2)}$$

and

$$Z_C = y_{10} \frac{(1+k^2)(1-8k-6k^2+8k^3+k^4)}{(1-k^2)(1+k-k^2)(1-4k-k^2)}.$$

c. Show that

$$X_A^{1/2} Z_A = X_B^{1/2} Z_B = X_C^{1/2} Z_C$$

$$= y_{10} \frac{k^{1/2}(1+k^2)^{1/2}}{(1-k^2)^{1/2}(1+k-k^2)^{1/2}(1-4k-k^2)^{1/2}}.$$

d. Let $\{T_A(j)\}$, $\{T_B(j)\}$, and $\{T_C(j)\}$ be defined by

$$Z_A = \sum_{j=0}^{\infty} T_A(j) X_A^j, \quad Z_B = \sum_{j=0}^{\infty} T_B(j) X_B^j \quad \text{and} \quad Z_C = \sum_{j=0}^{\infty} T_C(j) X_C^j.$$

Show that $\sum_{j=0}^{\infty} T_A(j) X_A^{j+1/2} = \sum_{j=0}^{\infty} T_B(j) X_B^{j+1/2} = \sum_{j=0}^{\infty} T_C(j) X_C^{j+1/2}$ and

$$\left(\frac{1}{X_A} - 12 - 64X_A\right)^{1/2} \sum_{j=0}^{\infty} T_A(j) \left(j + \frac{1}{2}\right) X_A^{j+1/2}$$

$$= \left(\frac{1}{X_B} - 24 + 80X_B\right)^{1/2} \sum_{j=0}^{\infty} T_B(j) \left(j + \frac{1}{2}\right) X_B^{j+1/2}$$

$$= \left(\frac{1}{X_C} + 36 + 320X_C\right)^{1/2} \sum_{j=0}^{\infty} T_C(j) \left(j + \frac{1}{2}\right) X_C^{j+1/2}.$$

The sequences $\{T_B(j)\}$ and $\{T_C(j)\}$ satisfy five-term recurrence relations, whereas $T_A(j) = T_{10}(j)$ and so the sequence $\{T_A(j)\}$ satisfies a three-term recurrence relation. For further details, see [113].

10. Let $w_B = \eta_2^4\eta_{10}^4/\eta_1^4\eta_5^4$ and $y_B = \eta_1^4\eta_5^4/\eta_2^2\eta_{10}^2$. Show that $y_B = \sum_{n=0}^{\infty} b(n)w_B^n$ where the coefficients $b(n)$ are integers that satisfy the four-term recurrence relation

$$\begin{aligned}(n+1)^3b(n+1) &+ 4(2n+1)(2n^2+2n+1)b(n)\\ &+16n(4n^2+1)b(n-1) + 8(2n-1)^3b(n-2) = 0.\end{aligned}$$

# Chapter 11
# Levels 11 and 23

## 11.1 Introduction

In this chapter we investigate the theory for level 11. As the same techniques can be used to develop the theories for levels 3, 7, and 23, we shall study all four levels simultaneously. The theories for levels 3 and 7 have been analyzed already in Chapters 4 and 7, respectively, so this chapter provides an alternative approach for those levels. The other levels 11 and 23 have not been considered before in this book.

As a rule, we shall give full proofs for the level 11 case and only a brief outline of the proofs for the other levels, as the method in each case is similar. We shall rely on two results, Theorems 11.1 and 11.2, which are stated without proof. They are analogues of the sum of two squares and sum of $4k$ squares identities, and are most naturally proved using the theory of modular forms and ideas that are outside the scope of this book.

## 11.2 Basic properties

For $p = 3, 7, 11$, or $23$, let $z_p$ be the theta series defined by

$$z_p = z_p(q) = \sum_{m=-\infty}^{\infty} \sum_{n=-\infty}^{\infty} q^{m^2+mn+(p+1)n^2/4},$$

so that

$$z_3 = \sum_{m=-\infty}^{\infty} \sum_{n=-\infty}^{\infty} q^{m^2+mn+n^2},$$



S. Cooper, *Ramanujan's Theta Functions*, DOI 10.1007/978-3-319-56172-1_12

$$z_7 = \sum_{m=-\infty}^{\infty} \sum_{n=-\infty}^{\infty} q^{m^2+mn+2n^2},$$

$$z_{11} = \sum_{m=-\infty}^{\infty} \sum_{n=-\infty}^{\infty} q^{m^2+mn+3n^2}$$

and

$$z_{23} = \sum_{m=-\infty}^{\infty} \sum_{n=-\infty}^{\infty} q^{m^2+mn+6n^2}.$$

We will also consider the theta series $z_{23}^*$ defined by

$$z_{23}^* = \sum_{m=-\infty}^{\infty} \sum_{n=-\infty}^{\infty} q^{2m^2+mn+3n^2}.$$

The theta series are related to Eisenstein series as follows.

**Theorem 11.1.** *The following identities hold*:

$$z_3 = 1 + 6\sum_{j=1}^{\infty} \left(\frac{j}{3}\right) \frac{q^j}{1-q^j},$$

$$z_7 = 1 + 2\sum_{j=1}^{\infty} \left(\frac{j}{7}\right) \frac{q^j}{1-q^j},$$

$$z_{11} = 1 + 2\sum_{j=1}^{\infty} \left(\frac{j}{11}\right) \frac{q^j}{1-q^j},$$

$$z_{23} = 1 + \frac{2}{3}\sum_{j=1}^{\infty} \left(\frac{j}{23}\right) \frac{q^j}{1-q^j} + \frac{4}{3}q\prod_{j=1}^{\infty}(1-q^j)(1-q^{23j}),$$

*and*

$$z_{23}^* = 1 + \frac{2}{3}\sum_{j=1}^{\infty} \left(\frac{j}{23}\right) \frac{q^j}{1-q^j} - \frac{2}{3}q\prod_{j=1}^{\infty}(1-q^j)(1-q^{23j}).$$

*Proof.* The identities for $z_3$ and $z_7$ have been proved in Theorems 3.19 and 7.16, respectively. The identities for $z_3$, $z_7$, and $z_{11}$ are in fact equivalent to instances of a general theorem of Dirichlet; see Landau's book [203, Theorem 204]. The identities for $z_{23}$ and $z_{23}^*$ have been proved by F. van der Blij [283]. Another proof of all of the identities above has been given in work with H. H. Chan in [80].

□

The last two identities in Theorem 11.1 may be rearranged to give

$$\sum_{m=-\infty}^{\infty}\sum_{n=-\infty}^{\infty} q^{m^2+mn+6n^2} + 2\sum_{m=-\infty}^{\infty}\sum_{n=-\infty}^{\infty} q^{2m^2+mn+3n^2} = 3 + 2\sum_{j=1}^{\infty}\left(\frac{j}{23}\right)\frac{q^j}{1-q^j}$$

and

$$\sum_{m=-\infty}^{\infty}\sum_{n=-\infty}^{\infty} q^{m^2+mn+6n^2} - \sum_{m=-\infty}^{\infty}\sum_{n=-\infty}^{\infty} q^{2m^2+mn+3n^2} = 2q\prod_{j=1}^{\infty}(1-q^j)(1-q^{23j}). \tag{11.1}$$

The first is equivalent to another instance of Dirichlet's theorem [203, Theorem 204], and the second was noted by J.-P. Serre [268, p. 242].

For $p = 3, 7, 11$ or $23$, let $Z_p$ and $X_p$ be defined

$$Z_p = z_p^2 \qquad \text{and} \qquad X_p = \left(\frac{\eta_1\eta_p}{z_p}\right)^{24/(p+1)}. \tag{11.2}$$

Let $Z_{23}^*$ and $X_{23}^*$ be defined by

$$Z_{23}^* = \left(z_{23}^*\right)^2 \qquad \text{and} \qquad X_{23}^* = \frac{\eta_1\eta_{23}}{z_{23}^*}.$$

The functions $Z_3$ and $X_3$ were studied in Theorem 4.30, while the functions $Z_7$ and $X_7$ were met in (7.65), (7.70), and Theorem 7.20. The other functions have not been introduced before now. By the above definitions and (11.1) we obtain

$$z_{23} - z_{23}^* = 2z_{23}X_{23} = 2z_{23}^*X_{23}^*. \tag{11.3}$$

Therefore, any results we obtain for $z_{23}$ and $X_{23}$ can be converted into results involving $z_{23}^*$ and $X_{23}^*$ by a change of variable. Hence, we shall no longer record any further results for $z_{23}^*$, $Z_{23}^*$, or $X_{23}^*$.

The main tool for studying $Z_p$ and $X_p$ is the following analogue of Ramanujan's identity (3.115) for sums of $4k$ squares.

**Theorem 11.2.** *Let $E_{2k}$ be the Eisenstein series defined by* (0.66) *and let* $G_{2k}(q) = -4kE_{2k}(q)/B_{2k}$ *so that*

$$G_{2k}(q) = 1 - \frac{4k}{B_{2k}}\sum_{j=1}^{\infty}\frac{j^{2k-1}q^j}{1-q^j}.$$

*For $p = 3, 7, 11,$ or 23 and for any positive integer $k$, the identity*

$$\frac{p^k G_{2k}(q^p) + (-1)^k G_{2k}(q)}{p^k + (-1)^k} = Z_p^k \sum_{0 \le j \le (p+1)k/12} c_{p,k,j} X_p^j \tag{11.4}$$

*holds, where $c_{p,k,j}$ are unique constants that depend only on $p$, $k$, and $j$, and $c_{p,k,0} = 1$.*

*Proof.* See Theorem 1.2 of [80].

□

The next result is the special case $k = 1$ of (11.4).

**Theorem 11.3.** *The following identities hold:*

$$Z_3 = \frac{1}{2}(3P(q^3) - P(q)),$$

$$Z_7 = \frac{1}{6}(7P(q^7) - P(q)),$$

$$Z_{11} = \frac{1}{10}(11P(q^{11}) - P(q)) + \frac{8}{5}Z_{11}X_{11}$$

*and*

$$Z_{23} = \frac{1}{22}(23P(q^{23}) - P(q)) + \frac{8}{11}Z_{23}X_{23}(4 - 3X_{23}).$$

Before starting the proof, it may be noted that the results for $Z_3$ and $Z_7$ are equivalent to (3.76) and Theorem 7.20, respectively.

*Proof.* Take $k = 1$ in Theorem 11.2 and rearrange to solve for $Z_p$. The constants $c_{p,1,j}$ can be determined by expanding in powers of $q$ and equating coefficients.

□

The following definitions are in preparation for the next two theorems. Let

$$B_3(x) = (1 - 108x)^{1/2},$$

$$B_7(x) = (1 + x)^{1/2}(1 - 27x)^{1/2},$$

$$B_{11}(x) = \left(1 - 20x + 56x^2 - 44x^3\right)^{1/2},$$

$$B_{23}(x) = \left(1 - 11x + 22x^2 - 19x^3\right)^{1/2}\left(1 - 3x + 2x^2 + x^3\right)^{1/2}$$

and let

$$B_p = B_p(X_p).$$

**Theorem 11.4.** *The following identities hold*:

$$Q(q) = Z_3^2\,(5 - 4\,B_3)\,,$$

$$Q(q^3) = \frac{Z_3^2}{9}\,(5 + 4\,B_3)\,,$$

$$Q(q) = Z_7^2\,(25 - 80X_7 - 24\,B_7)\,,$$

$$Q(q^7) = \frac{Z_7^2}{49}\,(25 - 80X_7 + 24\,B_7)\,,$$

$$Q(q) = Z_{11}^2\left(61 - 368X_{11} + 352X_{11}^2 - 60\,B_{11}\right),$$

$$Q(q^{11}) = \frac{Z_{11}^2}{121}\left(61 - 368X_{11} + 352X_{11}^2 + 60\,B_{11}\right),$$

$$Q(q) = Z_{23}^2\Big(265 - 2000X_{23} + 4720X_{23}^2 - 4560X_{23}^3 + 1440X_{23}^4 - 24(11 - 16X_{23})\,B_{23}\Big),$$

$$Q(q^{23}) = \frac{Z_{23}^2}{529}\Big(265 - 2000X_{23} + 4720X_{23}^2 - 4560X_{23}^3 + 1440X_{23}^4 + 24(11 - 16X_{23})\,B_{23}\Big).$$

*Proof.* We only give the details of the proof for the case $p = 11$. The details in the other cases are similar. Taking $p = 11$ and $k = 2$ in Theorem 11.2, we obtain

$$11^2Q(q^{11}) + Q(q) = Z_{11}^2(122 - 736X_{11} + 704X_{11}^2) \tag{11.5}$$

while taking $p = 11$ and $k = 4$ in Theorem 11.2 gives

$$\begin{aligned} &11^4Q(q^{11})^2 + Q^2(q) \\ &= Z_{11}^4(14642 - 233792X_{11} + 759936X_{11}^2 - 834944X_{11}^3 + 247808X_{11}^4). \end{aligned}$$

It follows that

$$\begin{aligned} \left(11^2Q(q^{11}) - Q(q)\right)^2 &= 2\left(11^4Q^2(q^{11}) + Q^2(q)\right) - \left(11^2Q(q^{11}) + Q(q)\right)^2 \\ &= 120^2Z_{11}^4(1 - 20X_{11} + 56X_{11}^2 - 44X_{11}^3), \end{aligned}$$

and so

$$11^2Q(q^{11}) - Q(q) = 120Z_{11}^2\,B_{11}. \tag{11.6}$$

On taking linear combinations of (11.5) and (11.6) we obtain the two results of the theorem that express $Q(q)$ and $Q(q^{11})$ in terms of $Z_{11}$ and $X_{11}$. The other results may be obtained in a similar way.

□

**Theorem 11.5.** *The following identities hold*:

$$R(q) = Z_3^3\,(14\,B_3 - (13 - 216X_3))\,,$$

$$R(q^3) = \frac{Z_3^3}{27}\,(14\,B_3 + (13 - 216X_3))\,,$$

$$R(q) = Z_7^3\,\big((172 - 80X_7)B_7 - (171 - 1800X_7 - 648X_7^2)\big)\,,$$

$$R(q^7) = \frac{Z_7^3}{343}\,\big((172 - 80X_7)B_7 + (171 - 1800X_7 - 648X_7^2)\big)\,,$$

$$R(q) = Z_{11}^3\Bigg((666 - 1584X_{11})B_{11} - (665 - 7728X_{11} + 14784X_{11}^2 - 6776X_{11}^3)\Bigg),$$

$$R(q^{11}) = \frac{Z_{11}^3}{1331}\Bigg((666 - 1584X_{11})B_{11} + (665 - 7728X_{11} + 14784X_{11}^2 - 6776X_{11}^3)\Bigg),$$

$$R(q) = Z_{23}^3\Bigg((6084 - 30672X_{23} + 46944X_{23}^2 - 22896X_{23}^3)B_{23} - (6083 - 72744X_{23} + 298032X_{23}^2 - 590408X_{23}^3 + 600768X_{23}^4 - 278208X_{23}^5 + 28728X_{23}^6)\Bigg),$$

$$R(q^{23}) = \frac{Z_{23}^3}{12167}\Bigg((6084 - 30672X_{23} + 46944X_{23}^2 - 22896X_{23}^3)B_{23} + (6083 - 72744X_{23} + 298032X_{23}^2 - 590408X_{23}^3 + 600768X_{23}^4 - 278208X_{23}^5 + 28728X_{23}^6)\Bigg).$$

*Proof.* Once again, we give the details of the proof for the case $p = 11$. Taking $p = 11$ and $k = 3$ in Theorem 11.2, we obtain

$$11^3R(q^{11}) - R(q) = 14Z_{11}^3(95 - 1104X_{11} + 2112X_{11}^2 - 968X_{11}^3), \tag{11.7}$$

while taking $p = 11$ and $k = 5$ in Theorem 11.2 gives

$$
\begin{aligned}
&11^5 Q(q^{11})R(q^{11}) - Q(q)R(q) \qquad (11.8)\\
&\quad = 2Z_{11}^5 (80525 - 1610368X_{11} + 8118368X_{11}^2\\
&\qquad\quad - 15654584X_{11}^3 + 11879296X_{11}^4 - 2385152X_{11}^5).
\end{aligned}
$$

Now, Theorem 11.4 can be used to replace the terms $Q(q)$ and $Q(q^{11})$ in (11.8) with expressions that involve $X_{11}$ and $Z_{11}$. The result is

$$
\begin{aligned}
&11^3 \left(61 - 368X_{11} + 352X_{11}^2 + 60\, B_{11}\right) R(q^{11}) \qquad (11.9)\\
&\quad - \left(61 - 368X_{11} + 352X_{11}^2 - 60\, B_{11}\right) R(q)\\
&= 2Z_{11}^3 (80525 - 1610368X_{11} + 8118368X_{11}^2\\
&\qquad - 15654584X_{11}^3 + 11879296X_{11}^4 - 2385152X_{11}^5).
\end{aligned}
$$

The pair of equations (11.7) and (11.9) may be solved to give either $R(q)$ or $R(q^{11})$ in terms of $X_{11}$ and $Z_{11}$. This proves the theorem in the case $p = 11$. The other results may be proved in a similar way.

□

The next result gives parameterizations for the Dedekind eta function in terms of $X_p$ and $Z_p$.

**Theorem 11.6.** *The following identities hold*:

$$
\begin{aligned}
\eta_1^{24} + 3^6\eta_3^{24} &= Z_3^6 X_3\, (1 - 54X_3),\\
\eta_1^{24} - 3^6\eta_3^{24} &= Z_3^6 X_3\, B_3,\\
\eta_1^{24} + 7^6\eta_7^{24} &= Z_7^6 X_7\, (1 - 39X_7 + 360X_7^2 - 286X_7^3),\\
\eta_1^{24} - 7^6\eta_7^{24} &= Z_7^6 X_7\, (1 - 26X_7 + 120X_7^2)\, B_7,\\
\eta_1^{24} + 11^6\eta_{11}^{24} &= Z_{11}^6 X_{11}\, (1 - 42X_{11} + 617X_{11}^2 - 3696X_{11}^3 + 7744X_{11}^4 - 2662X_{11}^5),\\
\eta_1^{24} - 11^6\eta_{11}^{24} &= Z_{11}^6 X_{11}\, (1 - 32X_{11} + 319X_{11}^2 - 968X_{11}^3)\, B_{11},\\
\eta_1^{24} + 23^6\eta_{23}^{24} &= Z_{23}^6 X_{23}\, (1 - 45X_{23} + 865X_{23}^2 - 9270X_{23}^3 + 60648X_{23}^4\\
&\qquad - 249463X_{23}^5 + 640904X_{23}^6 - 987056X_{23}^7 + 821224X_{23}^8\\
&\qquad - 249920X_{23}^9 - 71232X_{23}^{10} + 20610X_{23}^{11}),\\
\eta_1^{24} - 23^6\eta_{23}^{24} &= Z_{23}^6 X_{23}\, (1 - 38X_{23} + 595X_{23}^2 - 4928X_{23}^3 + 22960X_{23}^4\\
&\qquad - 58526X_{23}^5 + 70968X_{23}^6 - 24784X_{23}^7 - 2968X_{23}^8)\, B_{23}.
\end{aligned}
$$

*Proof.* By Corollary 0.23,

$$\eta_n^{24} = \frac{1}{1728}\left(Q^3(q^n) - R^2(q^n)\right).$$

The required results now follow from Theorems 11.4 and 11.5. □

The next result gives parameterizations of the modular $j$-invariant in terms of $X_p$.

**Theorem 11.7.** *Suppose* $p = 3, 7, 11,$ *or* $23$. *Let* $f_p$ *and* $g_p$ *be the polynomials defined in Table 11.1. Then*

$$j(\tau) = \frac{1}{2X_p^p}\left(f_p(X_p) - g_p(X_p)\,B_p(X_p)\right)$$

*and*

$$j(p\tau) = \frac{1}{2X_p^p}\left(f_p(X_p) + g_p(X_p)\,B_p(X_p)\right).$$

*Moreover,*

$$\begin{aligned} f_3^2(x) - g_3^2(x)B_3^2(x) &= 4x^2(1+192x)^3, \\ f_7^2(x) - g_7^2(x)B_7^2(x) &= 4x^6(1+224x+448x^2)^3, \\ f_{11}^2(x) - g_{11}^2(x)B_{11}^2(x) &= 4x^{10}(1+224x-192x^2-832x^3+1024x^4)^3 \end{aligned}$$

**Table 11.1** Data for Theorem 11.7

| | |
|---|---|
| $f_3(x) =$ | $1-126x+2944x^2$ |
| $g_3(x) =$ | $(1-8x)(1-64x)$ |
| $f_7(x) =$ | $(1-21x+8x^2)(1-42x+454x^2-1008x^3-1280x^4)$ |
| $g_7(x) =$ | $(1-2x)(1-8x)(1-24x)(1-16x-8x^2)$ |
| $f_{11}(x) =$ | $1-66x+1793x^2-26048x^3+221056x^4-1132670x^5+3535840x^6$ $-6683072x^7+7418752x^8-4460544x^9+1183744x^{10}-65536x^{11}$ |
| $g_{11}(x) =$ | $(1-x)(1-2x)(1-4x)(1-7x)(1-16x)(1-12x+16x^2)(1-14x+4x^2)$ |
| $f_{23}(x) =$ | $1-69x+2185x^2-42182x^3+556416x^4-5329215x^5+38458944x^6$ $-214256960x^7+936702784x^8-3249036672x^9+9000420352x^{10}$ $-19964195454x^{11}+35397850080x^{12}-49817482592x^{13}+54842552032x^{14}$ $-45954407744x^{15}+27781303296x^{16}-10639601280x^{17}+1359138816x^{18}$ $+902135808x^{19}-489646080x^{20}+68677632x^{21}+7962624x^{22}-1769472x^{23}$ |
| $g_{23}(x) =$ | $(1-x)(1-2x)(1-3x)(1-4x)(1-6x)(1-8x-2x^2)(1-10x+12x^2)$ $\times(1-10x+24x^2-18x^3)(1-10x+20x^2-16x^3)(1-8x+18x^2-16x^3+4x^4)$ |

*and*

$$f_{23}^2(x) - g_{23}^2(x)B_{23}^2(x) = 4x^{22}(1 + 224x - 864x^2 - 544x^3 + 9664x^4 - 26112x^5 + 36288x^6 - 27648x^7 + 9216x^8)^3.$$

*Proof.* The results all follow by direct calculation using the definition of the modular $j$-invariant in (2.21) and the results in Theorems 11.4 and 11.5. □

## 11.3 Differential equations

In this section we show that $Z_p$ satisfies second order nonlinear and third order linear differential equations with respect to $X_p$ and give explicit forms of the differential equations.

We begin with a formula for the derivative of $X_p$ with respect to $q$. The case $p = 3$ has already been encountered in Exercise 9 of Chapter 4 and the case $p = 7$ was studied in (7.76). Here we repeat those results, along with the results for $p = 11$ and $p = 23$ which are new.

**Lemma 11.8.** *Suppose* $p = 3, 7, 11,$ *or* $23$. *The following differentiation formula holds*:

$$q\frac{\mathrm{d}}{\mathrm{d}q}X_p = Z_p X_p B_p.$$

*Proof.* We shall give a complete proof of the result for the case $p = 11$. The proofs in the other cases are similar.

By the definition of $X_{11}$ in (11.2) and Theorem 11.3 we have

$$X_{11} = \frac{\eta_1^2\eta_{11}^2}{Z_{11}} \qquad \text{and} \qquad 11P(q^{11}) - P(q) = 2Z_{11}(5 - 8X_{11}).$$

Eliminating $Z_{11}$ gives

$$2\eta_1^2\eta_{11}^2\left(\frac{5}{X_{11}} - 8\right) = 11P(q^{11}) - P(q). \tag{11.10}$$

Apply the differential operator $q\frac{\mathrm{d}}{\mathrm{d}q}$ to both sides. The derivative on the left can be calculated by the product rule and by noting

$$q\frac{\mathrm{d}}{\mathrm{d}q}(\eta_1^2\eta_{11}^2) = \eta_1^2\eta_{11}^2\, q\frac{\mathrm{d}}{\mathrm{d}q}\log\left(\eta_1^2\eta_{11}^2\right) = \frac{\eta_1^2\eta_{11}^2}{12}\left(11P(q^{11}) + P(q)\right),$$

and the derivative on the right may be calculated by one of Ramanujan's differential equations in Theorem 0.21. This gives

$$\frac{\eta_1^2\eta_{11}^2}{6}\left(11P(q^{11})+P(q)\right)\left(\frac{5}{X_{11}}-8\right)-\frac{10\eta_1^2\eta_{11}^2}{X_{11}^2}\,q\frac{\mathrm{d}}{\mathrm{d}q}X_{11}$$
$$=\frac{121P^2(q^{11})-P^2(q)}{12}-\frac{121Q(q^{11})-Q(q)}{12}. \tag{11.11}$$

The identity (11.10) can be used to show that the first term on the left-hand side of (11.11) cancels with the first term on the right-hand side. On simplifying the remaining terms, we obtain

$$q\frac{\mathrm{d}}{\mathrm{d}q}X_{11}=\frac{X_{11}^2}{\eta_1^2\eta_{11}^2}\left(\frac{121Q(q^{11})-Q(q)}{120}\right).$$

The proof may be completed on appealing to the definition of $X_{11}$ in (11.2) again, and making use of the identities in Theorem 11.4 to express $121Q(q^{11})-Q(q)$ in terms of $Z_{11}$ and $X_{11}$.

The details of the proofs for the other cases $p=3$, 7, or 23 are very similar to the proof for the case $p=11$ given above: in place of (11.10) we have

$$2\eta_1^2\eta_3^2\times\frac{1}{X_3^{1/3}}=3P(q^3)-P(q),$$

$$6\eta_1^2\eta_7^2\times\frac{1}{X_7^{2/3}}=7P(q^7)-P(q)$$

and

$$2\eta_1^2\eta_{23}^2\times\left(\frac{11}{X_{23}^2}-\frac{32}{X_{23}}+24\right)=23P(q^{23})-P(q).$$

Now apply $q\frac{\mathrm{d}}{\mathrm{d}q}$ to each of the above identities and follow the procedure in the proof that was given in the case $p=11$.

□

Parameterizations for $pP(q^p)-P(q)$ in terms of $X_p$ and $Z_p$ can be obtained by rearranging the formulas in Theorem 11.3. We will now obtain the analogous parameterizations for $pP(q^p)+P(q)$ and $p^2P^2(q^p)+P^2(q)$; these will also involve the derivative $\mathrm{d}Z_p/\mathrm{d}X_p$.

**Lemma 11.9.** *Suppose* $p=3$, 7, 11, *or* 23. *Then*

$$pP(q^p)+P(q)=B_p\left((p+1)Z_p+12X_p\frac{\mathrm{d}Z_p}{\mathrm{d}X_p}\right)$$

*and*

$$
p^2P^2(q^p)+P^2(q)
$$

$$
=\begin{cases}
\dfrac{B_p^2}{2}\left((p+1)Z_p+12X_p\dfrac{\mathrm{d}Z_p}{\mathrm{d}X_p}\right)^2+\dfrac{(p-1)^2}{2}Z_p^2, & \textit{if } p=3 \textit{ or } 7,\\[2ex]
\dfrac{B_p^2}{2}\left((p+1)Z_p+12X_p\dfrac{\mathrm{d}Z_p}{\mathrm{d}X_p}\right)^2+\dfrac{(p-1)^2}{2}Z_p^2\left(1-\dfrac{8X_p}{5}\right)^2, & \textit{if } p=11,\\[2ex]
\dfrac{B_p^2}{2}\left((p+1)Z_p+12X_p\dfrac{\mathrm{d}Z_p}{\mathrm{d}X_p}\right)^2 & \\
\qquad+\dfrac{(p-1)^2}{2}Z_p^2\left(1-\dfrac{32X_p}{11}+\dfrac{24X_p^2}{11}\right)^2, & \textit{if } p=23.
\end{cases}
$$

*Proof.* By the definition of $X_p$ in (11.2) we have

$$
\eta_1^{24}\eta_p^{24}=X_p^{p+1}Z_p^{12}.
$$

Take logarithms, apply the differential operator $q\frac{\mathrm{d}}{\mathrm{d}q}$, and then use the chain rule to obtain

$$
\begin{aligned}
pP(q^p)+P(q)&=\frac{(p+1)}{X_p}q\frac{\mathrm{d}X_p}{\mathrm{d}q}+\frac{12}{Z_p}q\frac{\mathrm{d}Z_p}{\mathrm{d}q}\\
&=q\frac{\mathrm{d}X_p}{\mathrm{d}q}\left(\frac{p+1}{X_p}+\frac{12}{Z_p}\frac{\mathrm{d}Z_p}{\mathrm{d}X_p}\right).
\end{aligned}
$$

Now invoke Lemma 11.8 to complete the proof of the first result.

The second result can be obtained from the first result and from the results in Theorem 11.3 by using the elementary identity

$$
p^2P^2(q^p)+P^2(q)=\frac{1}{2}\left(pP(q^p)+P(q)\right)^2+\frac{1}{2}\left(pP(q^p)-P(q)\right)^2.
$$

□

We are now ready to show that $Z_p$ satisfies a second order nonlinear differential equation with respect to $X_p$.

**Theorem 11.10.** *Suppose* $p=3,\ 7,\ 11,$ *or* $23$. *The following differential equation for* $Z_p$ *in terms of* $X_p$ *holds*:

$$
B_pX_p\frac{\mathrm{d}}{\mathrm{d}X_p}\left(B_pX_p\frac{\mathrm{d}Z_p}{\mathrm{d}X_p}\right)-\frac{1}{2Z_p}\left(B_pX_p\frac{\mathrm{d}Z_p}{\mathrm{d}X_p}\right)^2=H_pZ_p
$$

*where*

$$H_3 = 12X_3, \tag{11.12}$$

$$H_7 = 4X_7(1 + 3X_7), \tag{11.13}$$

$$H_{11} = 4X_{11}(1 - 8X_{11} + 11X_{11}^2) \tag{11.14}$$

*and*

$$H_{23} = 4X_{23}(1 - 11X_{23} + 37X_{23}^2 - 48X_{23}^3 + 12X_{23}^4 + 19X_{23}^5). \tag{11.15}$$

*Proof.* Let $D_q$ denote the operator $D_q = q\,\frac{\mathrm{d}}{\mathrm{d}q}$. By Lemma 11.8,

$$D_q = q\,\frac{\mathrm{d}}{\mathrm{d}q} = Z_p\,X_p\,B_p\,\frac{\mathrm{d}}{\mathrm{d}X_p}.$$

On applying the operator $D_q$ to the first identity in Lemma 11.9 we find that

$$Z_p\,X_p\,B_p\,\frac{\mathrm{d}}{\mathrm{d}X_p}\left(B_p\left((p+1)Z_p + 12X_p\frac{\mathrm{d}Z_p}{\mathrm{d}X_p}\right)\right) = q\frac{\mathrm{d}}{\mathrm{d}q}\left(pP(q^p) + P(q)\right).$$

The derivative on the left-hand side can be evaluated by direct calculation and (0.78) can be used to calculate the derivative on the right-hand side. The result is

$$\begin{aligned} &12X_p^2\,B_p^2\,Z_p\,\frac{\mathrm{d}^2Z_p}{\mathrm{d}X_p^2} + X_p\left((p+13)B_p^2 + 6X_p\frac{\mathrm{d}\,B_p^2}{\mathrm{d}X_p}\right)Z_p\,\frac{\mathrm{d}Z_p}{\mathrm{d}X_p} \\ &+ \frac{(p+1)}{2}\,X_p\,\frac{\mathrm{d}\,B_p^2}{\mathrm{d}X_p}\,Z_p^2 = \frac{1}{12}\left(p^2P^2(q^p) + P^2(q) - p^2Q(q^p) - Q(p)\right). \end{aligned} \tag{11.16}$$

The right-hand side may be expressed in terms of $X_p$, $Z_p$ and $\frac{\mathrm{d}Z_p}{\mathrm{d}X_p}$ by Theorems 11.4 and 11.9. On dividing by $12Z_p$, the identity (11.16) simplifies to

$$B_p^2X_p^2\frac{\mathrm{d}^2Z_p}{\mathrm{d}X_p^2} + B_pX_p\frac{\mathrm{d}Z_p}{\mathrm{d}X_p}\left(B_p + X_p\,\frac{\mathrm{d}B_p}{\mathrm{d}X_p}\right) = \frac{B_p^2X_p^2}{2Z_p}\left(\frac{\mathrm{d}Z_p}{\mathrm{d}X_p}\right)^2 + H_pZ_p,$$

that is,

$$B_pX_p\,\frac{\mathrm{d}}{\mathrm{d}X_p}\left(B_pX_p\frac{\mathrm{d}Z_p}{\mathrm{d}X_p}\right) = \frac{1}{2Z_p}\left(B_pX_p\,\frac{\mathrm{d}Z_p}{\mathrm{d}X_p}\right)^2 + H_pZ_p,$$

where $H_p$ is a function of $X_p$ that may be determined from (11.16) by a case-by-case calculation according to whether $p = 3, 7, 11$, or 23. In this way, the functions $H_p$ are found to be given by (11.12)–(11.15). This completes the proof.

□

The next result shows that $Z_p$ satisfies a third order linear differential equation with respect to $X_p$.

**Theorem 11.11.** *Suppose* $p = 3, 7, 11,$ *or* 23. *Let* $B_p$ *and* $H_p$ *be as for Theorem 11.10. Then*

$$\frac{\mathrm{d}}{\mathrm{d}X_p}\left(B_pX_p\frac{\mathrm{d}}{\mathrm{d}X_p}\left(B_pX_p\frac{\mathrm{d}Z_p}{\mathrm{d}X_p}\right)\right) = 2H_p\frac{\mathrm{d}Z_p}{\mathrm{d}X_p} + \frac{\mathrm{d}H_p}{\mathrm{d}X_p}Z_p.$$

*Proof.* Apply $d/dX_p$ to the result of Theorem 11.10 to obtain

$$\begin{aligned}
&\frac{\mathrm{d}}{\mathrm{d}X_p}\left(B_pX_p\frac{\mathrm{d}}{\mathrm{d}X_p}\left(B_pX_p\frac{\mathrm{d}Z_p}{\mathrm{d}X_p}\right)\right)\\
&-\frac{1}{Z_p}\frac{\mathrm{d}Z_p}{\mathrm{d}X_p}\left\{B_pX_p\frac{\mathrm{d}}{\mathrm{d}X_p}\left(B_pX_p\frac{\mathrm{d}Z_p}{\mathrm{d}X_p}\right) - \frac{1}{2Z_p}\left(B_pX_p\frac{\mathrm{d}Z_p}{\mathrm{d}X_p}\right)^2\right\}\\
&= H_p\frac{\mathrm{d}Z_p}{\mathrm{d}X_p} + \frac{\mathrm{d}H_p}{\mathrm{d}X_p}Z_p.
\end{aligned}$$

The term in braces simplifies to $H_pZ_p$ by Theorem 11.10, and the required result follows.

□

The differential equation in Theorem 11.11 may be converted to the form

$$r_p(X_p)\frac{\mathrm{d}^3Z_p}{\mathrm{d}X_p^3} + s_p(X_p)\frac{\mathrm{d}^2Z_p}{\mathrm{d}X_p^2} + t_p(X_p)\frac{\mathrm{d}Z_p}{\mathrm{d}X_p} + u_p(X_p)Z_p = 0. \tag{11.17}$$

In the case $p = 7$, the result has been given before in Theorem 7.31. When $p = 11$ we find that

$$\begin{aligned}
&X_{11}^2(1 - 20X_{11} + 56X_{11}^2 - 44X_{11}^3)\frac{\mathrm{d}^3Z_{11}}{\mathrm{d}X_{11}^3}\\
&+ 3X_{11}(1 - 30X_{11} + 112X_{11}^2 - 110X_{11}^3)\frac{\mathrm{d}^2Z_{11}}{\mathrm{d}X_{11}^2}\\
&+ (1 - 68X_{11} + 400X_{11}^2 - 528X_{11}^3)\frac{\mathrm{d}Z_{11}}{\mathrm{d}X_{11}}\\
&= 4(1 - 16X_{11} + 33X_{11}^2)Z_{11}.
\end{aligned}$$

The differential equations in the form (11.17) can be used to expand each function $Z_p$ in powers of the corresponding $X_p$. This leads to recurrence relations for the coefficients.

**Theorem 11.12.** *For* $p = 3, 7, 11,$ *or* $23$, *let* $\{T_p(n)\}$ *be the sequences defined by the recurrence relations*

$$\begin{aligned}
(n+1)^3 T_3(n+1) &= 6(2n+1)(9n^2+9n+2)T_3(n),\\
(n+1)^3 T_7(n+1) &= (2n+1)(13n^2+13n+4)T_7(n) + 3n(9n^2-1)T_7(n-1),\\
(n+1)^3 T_{11}(n+1) &= 2(2n+1)(5n^2+5n+2)T_{11}(n) - 8n(7n^2+1)T_{11}(n-1)\\
&\quad + 22n(n-1)(2n-1)T_{11}(n-2),\\
(n+1)^3 T_{23}(n+1) &= (2n+1)(7n^2+7n+4)T_{23}(n) - n(57n^2+31)T_{23}(n-1)\\
&\quad + (2n-1)(53n^2-53n+42)T_{23}(n-2)\\
&\quad - 6(n-1)(15n^2-30n+19)T_{23}(n-3)\\
&\quad + 8(n-1)(n-2)(2n-3)T_{23}(n-4)\\
&\quad + 19(n-1)(n-2)(n-3)T_{23}(n-5)
\end{aligned}$$

*and initial conditions*

$$T_3(0) = T_7(0) = T_{11}(0) = T_{23}(0) = 1,$$
$$T_{23}(-5) = T_{23}(-4) = T_{23}(-3) = T_{23}(-2) = T_{23}(-1) = 0.$$

*Let* $F_p = F_p(x)$ *be the function defined in a neighborhood of* $x = 0$ *by*

$$F_p(x) = \sum_{n=0}^{\infty} T_p(n) x^n. \tag{11.18}$$

*Then*

$$Z_p = F_p(X_p).$$

*Proof.* This is immediate on converting the differential equations in Theorem 6.30 into the form given by (11.17) and then seeking the series solution in powers of $X_p$ that satisfies $Z_p = 1$ when $X_p = 0$.

□

It may be pointed out that the single initial condition $T_7(0) = 0$ is sufficient to start the three-term recurrence relation for the sequence $\{T_7(n)\}$, because taking $n = 0$ in the recurrence relation we see that the coefficient of $T_7(-1)$ is zero. Similarly, the single initial condition $T_{11}(0) = 0$ is enough to start the four-term recurrence relation for $\{T_{11}(n)\}$. However, the sequence $T_{23}(n)$ requires additional initial conditions to start the sequence.

It is easy to verify that

$$T_3(n) = \binom{3n}{n}\binom{2n}{n}^2$$

and by Theorem 7.32 we have

$$T_7(n) = \sum_{j=0}^{n} \binom{n}{j}^2 \binom{2j}{n}\binom{j+n}{j}.$$

Analogous formulas for the coefficients $T_{11}(n)$ or $T_{23}(n)$ are not known at the time of writing.

We now express each function $Z_p$ in terms of $X_p$ via the hypergeometric function in two ways. This in turn leads to hypergeometric transformation formulas of degrees 3, 7, 11, and 23.

**Theorem 11.13.** *Suppose $p = 3$, 7, 11, or 23. Let $f_p$ and $g_p$ be the polynomials defined in Table 11.1. Let $F_p$ be defined by* (11.18) *and let $y_p$ and $\overline{y}_p$ be defined by*

$$y_p = \frac{2x^p}{f_p(x) - g_p(x)B_p(x)}, \qquad \overline{y}_p = \frac{2x^p}{f_p(x) + g_p(x)B_p(x)}.$$

*Then*

$$y_p = x + O(x^2) \qquad \text{and} \qquad \overline{y}_p = x^p + O(x^{p+1}) \quad \text{as } x \to 0. \tag{11.19}$$

*And, the following representations of $F_3$, $F_7$, $F_{11}$, and $F_{23}$ in terms of the hypergeometric function hold*:

$$\begin{aligned}
F_3(x) &= (5 - 4\,B_3)^{-1/2}\ {}_3F_2\left(\begin{matrix}\frac{1}{6},\ \frac{1}{2},\ \frac{5}{6}\\ 1,\ 1\end{matrix};1728\,y_3\right)\\
&= 3\,(5 + 4\,B_3)^{-1/2}\ {}_3F_2\left(\begin{matrix}\frac{1}{6},\ \frac{1}{2},\ \frac{5}{6}\\ 1,\ 1\end{matrix};1728\,\overline{y}_3\right),\\
F_7(x) &= (25 - 80x - 24\,B_7)^{-1/2}\ {}_3F_2\left(\begin{matrix}\frac{1}{6},\ \frac{1}{2},\ \frac{5}{6}\\ 1,\ 1\end{matrix};1728\,y_7\right)\\
&= 7\,(25 - 80x + 24\,B_7)^{-1/2}\ {}_3F_2\left(\begin{matrix}\frac{1}{6},\ \frac{1}{2},\ \frac{5}{6}\\ 1,\ 1\end{matrix};1728\,\overline{y}_7\right),
\end{aligned}$$

$$F_{11}(x) = \left(61 - 368x + 352x^2 - 60\,B_{11}\right)^{-1/2}\ {}_3F_2\left(\begin{matrix}\frac{1}{6},\ \frac{1}{2},\ \frac{5}{6}\\ 1,\ 1\end{matrix};1728\,y_{11}\right)$$

$$= 11\left(61 - 368x + 352x^2 + 60\,B_{11}\right)^{-1/2}\ {}_3F_2\left(\begin{matrix}\frac{1}{6},\ \frac{1}{2},\ \frac{5}{6}\\ 1,\ 1\end{matrix};1728\,\overline{y}_{11}\right)$$

*and*

$$F_{23}(x) = \left(265 - 2000x + 4720x^2 - 4560x^3 + 1440x^4 - (264 + 384x)B_{23}\right)^{-1/2}$$

$$\times\ {}_3F_2\left(\begin{matrix}\frac{1}{6},\ \frac{1}{2},\ \frac{5}{6}\\ 1,\ 1\end{matrix};1728\,y_{23}\right)$$

$$= 23\left(265 - 2000x + 4720x^2 - 4560x^3 + 1440x^4 + (264 + 384x)B_{23}\right)^{-1/2}$$

$$\times\ {}_3F_2\left(\begin{matrix}\frac{1}{6},\ \frac{1}{2},\ \frac{5}{6}\\ 1,\ 1\end{matrix};1728\,\overline{y}_{23}\right).$$

*Proof.* The expansions in (11.19) follow by direct calculation in each case. We omit the details. Next, from Theorem 4.30,

$$Q^{1/2}(q) = {}_3F_2\left(\begin{matrix}\frac{1}{6},\ \frac{1}{2},\ \frac{5}{6}\\ 1,\ 1\end{matrix};\frac{1728}{j(\tau)}\right).$$

The required results now follow by using Theorem 11.4 to express $Q(q)$ and $Q(q^p)$ in terms of $X_p$ and $Z_p$, and using Theorem 11.7 to express $j(\tau)$ and $j(p\tau)$ in terms of $X_p$.

□

The expansions of $y_p$ and $\overline{y}_p$ in powers of $x$ given by (11.19) imply that the hypergeometric transformation formulas in Theorem 11.13 associated with $F_3$, $F_7$, $F_{11}$, and $F_{23}$ are of degrees 3, 7, 11, and 23, respectively.

## 11.4 Notes

The development in this chapter closely follows work with J. Ge and D. Ye [115]. See also Ge's thesis [160].

The identity in Theorem 11.2 is an analogue of Ramanujan's theorem for sums of $4k$ squares in (3.115). There is also an analogue of Ramanujan's sum of $4k+2$ squares identity (3.116) that may be stated as follows. For $p \in \{3, 7, 11, 23\}$ and for any positive integer $k$, the identity

$$1 - \frac{4k+2}{B_{2k+1,p}} \left\{ \sum_{j=1}^{\infty} \left(\frac{j}{p}\right) \frac{j^{2k} q^j}{1-q^j} + (-p)^k \sum_{j=1}^{\infty} \frac{j^{2k}}{1-q^{pj}} \sum_{i=1}^{p-1} \left(\frac{i}{p}\right) q^{ij} \right\}$$

$$= Z_p^{k+1/2} \sum_{0 \le j \le (p+1)(2k+1)/24} c_{p,k,j} X_p^j \tag{11.20}$$

holds, where $B_{2k+1,p}$ are the generalized Bernoulli numbers defined by (2.42), $c_{p,k,j}$ are unique constants that depend only on $p$, $k$, and $j$, and $c_{p,k,0} = 1$. For a proof, see Theorem 1.2 of [80]. Some examples are worked out in Exercise 2 below.

## 11.5 Exercises

1. Take $p = 3$ and let $k \in \{1, 2, 3, 4, 5, 6\}$ in Theorem 11.2 to deduce that the cubic theta function $a(q)$ satisfies the identities

$$a^2(q) = 1 + 12 \left( \sum_{j=1}^{\infty} \frac{jq^j}{1-q^j} - 3 \sum_{j=1}^{\infty} \frac{jq^{3j}}{1-q^{3j}} \right),$$

$$a^4(q) = 1 + 24 \left( \sum_{j=1}^{\infty} \frac{j^3 q^j}{1-q^j} + 9 \sum_{j=1}^{\infty} \frac{j^3 q^{3j}}{1-q^{3j}} \right),$$

$$a^6(q) = 1 + \frac{252}{13} \left( \sum_{j=1}^{\infty} \frac{j^5 q^j}{1-q^j} - 27 \sum_{j=1}^{\infty} \frac{j^5 q^{3j}}{1-q^{3j}} \right) + \frac{216}{13} \eta_1^6 \eta_3^6,$$

$$a^8(q) = 1 + \frac{240}{81} \left( \sum_{j=1}^{\infty} \frac{j^7 q^j}{1-q^j} + 81 \sum_{j=1}^{\infty} \frac{j^7 q^{3j}}{1-q^{3j}} \right) + \frac{1728}{41} a^2(q) \eta_1^6 \eta_3^6,$$

$$a^{10}(q) = 1 + \frac{12}{11} \left( \sum_{j=1}^{\infty} \frac{j^9 q^j}{1-q^j} - 243 \sum_{j=1}^{\infty} \frac{j^9 q^{3j}}{1-q^{3j}} \right) + \frac{648}{11} a^4(q) \eta_1^6 \eta_3^6,$$

$$a^{12}(q) = 1 + \frac{6552}{50443} \left( \sum_{j=1}^{\infty} \frac{j^{11} q^j}{1-q^j} + 729 \sum_{j=1}^{\infty} \frac{j^{11} q^{3j}}{1-q^{3j}} \right)$$
$$+ \frac{3625344}{50443} a^6(q) \eta_1^6 \eta_3^6 - \frac{2332800}{50443} \eta_1^{12} \eta_3^{12}.$$

These are the cubic analogues of the sum of 4, 8, 12, 16, 20, and 24 squares identities, respectively. The identities for $a^2(q)$ and $a^4(q)$ have been given already in (3.76) and Theorem 3.43, respectively, and the identity for $a^6(q)$ was proved in Exercise 2 at the end of Chapter 4. Apart from the trivial case $(c_1, c_2) = (0,0)$, the coefficients in $c_1\, a^6(q)\eta_1^6\eta_3^6 + c_2\,\eta_1^{12}\eta_3^{12}$ are multiplicative if and only if $(c_1, c_2) = (1, 48)$ or $(c_1, c_2) = (1, -54)$

2. Take $p = 3$ and set $k = 1$, 2, and 3 in (11.20) to show that

$$a^3(q) = 1 - 9\left\{\sum_{j=1}^{\infty}\left(\frac{j}{3}\right)\frac{j^2 q^j}{1-q^j} - 3\sum_{j=1}^{\infty}\frac{j^2(q^j - q^{2j})}{1-q^{3j}}\right\},$$

$$a^5(q) = 1 + 3\left\{\sum_{j=1}^{\infty}\left(\frac{j}{3}\right)\frac{j^4 q^j}{1-q^j} + 9\sum_{j=1}^{\infty}\frac{j^4(q^j - q^{2j})}{1-q^{3j}}\right\},$$

$$a^7(q) = 1 - \frac{3}{7}\left\{\sum_{j=1}^{\infty}\left(\frac{j}{3}\right)\frac{j^6 q^j}{1-q^j} - 27\sum_{j=1}^{\infty}\frac{j^6(q^j - q^{2j})}{1-q^{3j}}\right\} + \frac{216}{7}a(q)\eta_1^6\eta_3^6.$$

These are cubic analogues of the sum of 6, 10, and 14 squares identities, respectively. The identity for $a^3(q)$ has been given earlier in (3.92), and a proof of the result for $a^5(q)$ was outlined in Exercise 17 of Chapter 3.

3. Prove the identity

$$1 + 2\sum_{j=1}^{\infty}\left(\frac{j}{11}\right)\frac{q^{2j}}{1-q^{2j}} = \frac{1}{11}\left(\frac{2\psi^3(q)}{q\psi(q^{11})} - \frac{2\psi^3(-q)}{q\psi(-q^{11})} - \frac{\varphi^3(-q^2)}{\varphi(-q^{22})}\right).$$

Hint: Take $(v, w, x, y, z) = (2\pi/11, -4\pi/11, 6\pi/11, 8\pi/11, -12\pi/11)$ in (1.112).
See the work by Z.-G. Liu [224, Prop. 3.2].

4. Use (11.3) and Theorem 11.3 to show that

$$23P(q^{23}) - P(q) = 2z_{23}^2 + 8z_{23}z_{23}^* + 12\left(z_{23}^*\right)^2.$$

# Chapter 12
# Level 12

## 12.1 Introduction

This chapter is about the function $h = h(q)$ defined for $|q| < 1$ by

$$h = h(q) = q \prod_{j=1}^{\infty} \frac{(1-q^{12j-1})(1-q^{12j-11})}{(1-q^{12j-5})(1-q^{12j-7})}. \tag{12.1}$$

It is an analogue of Ramanujan's function $k$ studied in Chapter 10. In terms of characters, recall that $\chi_3$ and $\chi_4$ are defined by

$$\chi_3(n) = \begin{cases} 1 \text{ if } n \equiv 1 \pmod 3, \\ -1 \text{ if } n \equiv 2 \pmod 3, \\ 0 \text{ otherwise,} \end{cases}$$

and

$$\chi_4(n) = \begin{cases} 1 \text{ if } n \equiv 1 \pmod 4, \\ -1 \text{ if } n \equiv 3 \pmod 4, \\ 0 \text{ otherwise.} \end{cases}$$

Let $\chi_{12}$ be defined by

$$\chi_{12}(n) = \chi_3(n)\chi_4(n) = \begin{cases} 1 \text{ if } n \equiv 1 \text{ or } 11 \pmod{12}, \\ -1 \text{ if } n \equiv 5 \text{ or } 7 \pmod{12}, \\ 0 \text{ otherwise.} \end{cases}$$



S. Cooper, *Ramanujan's Theta Functions*, DOI 10.1007/978-3-319-56172-1_13

Then, the product in (12.1) may be written in the form

$$h(q) = q\prod_{n=1}^{\infty}(1-q^n)^{\chi_{12}(n)}.$$

The results of this chapter will be used in Chapter 13 to prove a large number of hypergeometric modular transformation formulas — see Theorem 13.3. Many of those transformation formulas have application in Chapter 14 in the study of Ramanujan's series for $1/\pi$.

## 12.2 Basic properties of the function *h*

We begin by establishing infinite product representations for eight lincar combinations of $\sqrt{h}$ and $1/\sqrt{h}$.

**Theorem 12.1.** *Let* $\omega = \exp(2\pi i/3)$, $\xi = \exp(\pi i/6)$, $a = 2-\sqrt{3}$, *and* $b = 2+\sqrt{3}$. *The following identities hold*:

$$\frac{1}{\sqrt{h}} + \sqrt{h} = \left(\frac{\eta_2^7\,\eta_3}{\eta_1^3\,\eta_4^2\,\eta_6\,\eta_{12}^2}\right)^{1/2}, \tag{12.2}$$

$$\frac{1}{\sqrt{h}} - \sqrt{h} = \left(\frac{\eta_1\,\eta_4^2\,\eta_6^9}{\eta_2^3\,\eta_3^3\,\eta_{12}^6}\right)^{1/2}, \tag{12.3}$$

$$\frac{1}{\sqrt{h}} + i\sqrt{h} = \left(\frac{\eta_2\,\eta_3^3}{\eta_1\,\eta_6\,\eta_{12}^2}\right)^{1/2} \times \frac{1}{q^{1/12}}\prod_{j=1}^{\infty}(1-iq^{3j-1})(1+iq^{3j-2}), \tag{12.4}$$

$$\frac{1}{\sqrt{h}} - i\sqrt{h} = \left(\frac{\eta_2\,\eta_3^3}{\eta_1\,\eta_6\,\eta_{12}^2}\right)^{1/2} \times \frac{1}{q^{1/12}}\prod_{j=1}^{\infty}(1+iq^{3j-1})(1-iq^{3j-2}), \tag{12.5}$$

$$\frac{1}{\sqrt{h}} + \omega\sqrt{h} = \left(\frac{\eta_1\,\eta_4^4\,\eta_6^3}{\eta_2^3\,\eta_3\,\eta_{12}^4}\right)^{1/2} \times \frac{1}{q^{1/24}}\prod_{j=1}^{\infty}(1-\omega q^{4j-1})(1-\omega^2 q^{4j-3}), \tag{12.6}$$

$$\frac{1}{\sqrt{h}} + \omega^2\sqrt{h} = \left(\frac{\eta_1\,\eta_4^4\,\eta_6^3}{\eta_2^3\,\eta_3\,\eta_{12}^4}\right)^{1/2} \times \frac{1}{q^{1/24}}\prod_{j=1}^{\infty}(1-\omega^2 q^{4j-1})(1-\omega q^{4j-3}), \tag{12.7}$$

$$\frac{1}{\sqrt{h}} - a\sqrt{h} = \left(\frac{\eta_1^3\,\eta_4^2\,\eta_6^3}{\eta_2^3\,\eta_3\,\eta_{12}^4}\right)^{1/2} \times q^{1/12}\prod_{j=1}^{\infty}(1-\xi^5 q^j)(1-\xi^7 q^j) \tag{12.8}$$

*and*

$$\frac{1}{\sqrt{h}} - b\sqrt{h} = \left(\frac{\eta_1^3\,\eta_4^2\,\eta_6^3}{\eta_2^3\,\eta_3\,\eta_{12}^4}\right)^{1/2} \times q^{1/12}\prod_{j=1}^{\infty}(1-\xi q^j)(1-\xi^{11}q^j). \tag{12.9}$$

*Proof.* We begin by proving (12.2)–(12.5). Replace $q$ with $-q$ in Euler's identity (0.44), and then separate the terms in the series according to whether the summation index is even or odd to obtain

$$\prod_{j=1}^{\infty}(1-(-q)^j) = \sum_{n=-\infty}^{\infty}(-1)^n q^{6n^2-n} + \sum_{n=-\infty}^{\infty}(-1)^n q^{6n^2+5n+1}.$$

Applying (0.4) to the left-hand side and Jacobi's triple product identity (0.7) to each series on the right we obtain

$$\begin{aligned}\prod_{j=1}^{\infty}\frac{(1-q^{2j})^3}{(1-q^j)(1-q^{4j})} &= \prod_{j=1}^{\infty}(1-q^{12j-5})(1-q^{12j-7})(1-q^{12j})\\ &\quad + q\prod_{j=1}^{\infty}(1-q^{12j-1})(1-q^{12j-11})(1-q^{12j}).\end{aligned}$$

Now divide both sides by

$$q^{1/2}\prod_{j=1}^{\infty}(1-q^{12j-1})^{1/2}(1-q^{12j-5})^{1/2}(1-q^{12j-7})^{1/2}(1-q^{12j-11})^{1/2}(1-q^{12j})$$

and use the fact that

$$\prod_{j=1}^{\infty}(1-q^{12j-1})(1-q^{12j-5})(1-q^{12j-7})(1-q^{12j-11}) = \prod_{j=1}^{\infty}\frac{(1-q^j)(1-q^{6j})}{(1-q^{2j})(1-q^{3j})} \tag{12.10}$$

to obtain

$$\begin{aligned}&\left(\frac{1}{q}\prod_{j=1}^{\infty}\frac{(1-q^{2j})^7(1-q^{3j})}{(1-q^j)^3(1-q^{4j})^2(1-q^{6j})(1-q^{12j})^2}\right)^{1/2}\\ &\quad = \frac{1}{\sqrt{q}}\prod_{j=1}^{\infty}\frac{(1-q^{12j-5})^{1/2}(1-q^{12j-7})^{1/2}}{(1-q^{12j-1})^{1/2}(1-q^{12j-11})^{1/2}}\\ &\qquad + \sqrt{q}\prod_{j=1}^{\infty}\frac{(1-q^{12j-1})^{1/2}(1-q^{12j-11})^{1/2}}{(1-q^{12j-5})^{1/2}(1-q^{12j-7})^{1/2}}.\end{aligned}$$

This gives (12.2). The identities (12.3)–(12.5) may be proved by a similar procedure, starting with the identities

$$\sum_{n=-\infty}^{\infty}(-q)^{n(3n-1)/2}=\sum_{n=-\infty}^{\infty}(-1)^n q^{6n^2-n}-\sum_{n=-\infty}^{\infty}(-1)^n q^{6n^2+5n+1},$$

$$\sum_{n=-\infty}^{\infty}i^n q^{n(3n-1)/2}=\sum_{n=-\infty}^{\infty}(-1)^n q^{6n^2-n}+i\sum_{n=-\infty}^{\infty}(-1)^n q^{6n^2+5n+1}$$

and

$$\sum_{n=-\infty}^{\infty}(-i)^n q^{n(3n-1)/2}=\sum_{n=-\infty}^{\infty}(-1)^n q^{6n^2-n}-i\sum_{n=-\infty}^{\infty}(-1)^n q^{6n^2+5n+1},$$

respectively, and using Jacobi's triple product identity to convert each series to a product. We omit the details as they are similar to those given above.

Now we will prove (12.6) and (12.7). Apply the series rearrangement

$$\sum_{m,n}c_{m,n}=\sum_{m,n}c_{m+2n,m-n}+\sum_{m,n}c_{m+2n+1,m-n}+\sum_{m,n}c_{m+2n-1,m-n}$$

with

$$c_{m,n}=(-1)^n q^{2m^2-m+4n^2+2n}\omega^{m+2n}$$

to obtain

$$\begin{aligned}&\left(\sum_{m=-\infty}^{\infty}q^{2m^2-m}\omega^m\right)\left(\sum_{n=-\infty}^{\infty}q^{4n^2+2n}(-\omega^2)^n\right)\\&\quad=\left(\sum_{m=-\infty}^{\infty}(-1)^m q^{6m^2+m}\right)\left(\sum_{n=-\infty}^{\infty}(-1)^n q^{12n^2-4n}\right)\\&\qquad+\omega q\left(\sum_{m=-\infty}^{\infty}(-1)^m q^{6m^2+5m}\right)\left(\sum_{n=-\infty}^{\infty}(-1)^n q^{12n^2+4n}\right)\\&\qquad+\omega^2q^3\left(\sum_{m=-\infty}^{\infty}(-1)^m q^{6m^2-3m}\right)\left(\sum_{n=-\infty}^{\infty}(-1)^n q^{12n^2-12n}\right).\end{aligned}$$

Now use Jacobi's triple product identity to convert each series to a product. The result is

$$\begin{aligned}&(-q\omega,-q^3\omega^2,q^4;q^4)_\infty(q^6\omega^2,q^2\omega,q^8;q^8)_\infty \qquad (12.11)\\&\quad=(q^5,q^7,q^{12};q^{12})_\infty(q^8;q^8)_\infty+\omega q(q,q^{11},q^{12};q^{12})_\infty(q^8;q^8)_\infty+0,\end{aligned}$$

the last term on the right being identically zero. The left-hand side of (12.11) simplifies to

$$\prod_{j=1}^{\infty} \frac{(1-q^j)(1-q^{4j})^2(1-q^{6j})^2(1-q^{8j})}{(1-q^{2j})^2(1-q^{3j})(1-q^{12j})} \times (1-\omega q^{4j-1})(1-\omega^2 q^{4j-3}). \tag{12.12}$$

The identity (12.6) may be obtained by dividing both sides of (12.11) by

$$q^{1/2}\prod_{j=1}^{\infty}(1-q^{12j-1})^{1/2}(1-q^{12j-5})^{1/2}(1-q^{12j-7})^{1/2}(1-q^{12j-11})^{1/2}(1-q^{12j})(1-q^{8j})$$

and using (12.10) and (12.12). The identity (12.7) may be deduced from (12.6) by replacing $\omega$ by its complex conjugate $\omega^2$.

Now we will prove (12.8) and (12.9). If $x = \xi = e^{\pi i/6}$ or $x = \xi^5 = e^{5\pi i/6}$, then elementary calculations show that

$$x^n + x^{n-1} + \cdots + x^{-n} = \begin{cases} 1 & \text{if } n \equiv 0, 5 \pmod{12}, \\ -1 & \text{if } n \equiv 6, 11 \pmod{12}, \\ 1 \pm \sqrt{3} & \text{if } n \equiv 1, 4 \pmod{12}, \\ -1 \mp \sqrt{3} & \text{if } n \equiv 7, 10 \pmod{12}, \\ 2 \pm \sqrt{3} & \text{if } n \equiv 2, 3 \pmod{12}, \\ -2 \mp \sqrt{3} & \text{if } n \equiv 8, 9 \pmod{12} \end{cases}$$

where the upper part of the $\pm$ or $\mp$ symbol is used if $x = \xi$ and the lower part is used if $x = \xi^5$. Substituting these values into Jacobi's triple product identity (0.16) and breaking the series into terms according to the power of $q$ modulo 12, we find that

$$\begin{aligned}(\xi q, \xi^{11} q, q; q)_\infty = &\sum_{j=-\infty}^{\infty} (-1)^j q^{18j^2+3j} - (1+\sqrt{3}) \sum_{j=-\infty}^{\infty} (-1)^j q^{18j^2+9j+1} \\ &+ (2+\sqrt{3}) \sum_{j=-\infty}^{\infty} (-1)^j q^{18j^2+15j+3}\end{aligned} \tag{12.13}$$

and

$$\begin{aligned}(\xi^5 q, \xi^7 q, q; q)_\infty = &\sum_{j=-\infty}^{\infty} (-1)^j q^{18j^2+3j} - (1-\sqrt{3}) \sum_{j=-\infty}^{\infty} (-1)^j q^{18j^2+9j+1} \\ &+ (2-\sqrt{3}) \sum_{j=-\infty}^{\infty} (-1)^j q^{18j^2+15j+3}.\end{aligned} \tag{12.14}$$

Now, replacing $q$ with $q^{12}$ and letting $x = -q^{-1}$ in the version of the quintuple product identity given by (0.35) we find that

$$\sum_{j=-\infty}^{\infty} (-1)^j \left( q^{18j^2+3j} + q^{18j^2+9j+1} \right) = \prod_{j=1}^{\infty} \frac{(1-q^{12j-2})(1-q^{12j-10})(1-q^{12j})}{(1-q^{12j-1})(1-q^{12j-11})}, \tag{12.15}$$

and replacing $q$ with $q^6$ and letting $x = -q^{-1}$ in the version of the quintuple product identity given by (0.36) we obtain

$$\sum_{j=-\infty}^{\infty} (-1)^j \left( q^{18j^2+9j+1} - q^{18j^2+15j+3} \right) = q \prod_{j=1}^{\infty} \frac{(1-q^{12j-2})(1-q^{12j-10})(1-q^{12j})}{(1-q^{12j-5})(1-q^{12j-7})}. \tag{12.16}$$

Using (12.15) and (12.16) in (12.13) we deduce

$$(\xi q, \xi^{11} q, q; q)_\infty = \frac{(q^2, q^{10}, q^{12}; q^{12})_\infty}{(q, q^{11}; q^{12})_\infty} - (2+\sqrt{3})\, q \frac{(q^2, q^{10}, q^{12}; q^{12})_\infty}{(q^5, q^7; q^{12})_\infty}.$$

Multiplying both sides by

$$\frac{(q, q^5, q^7, q^{11}; q^{12})_\infty^{1/2}}{q^{1/2}(q^2, q^{10}, q^{12}; q^{12})_\infty} \quad \left( = \frac{(q;q)_\infty^{1/2} (q^4;q^4)_\infty (q^6;q^6)_\infty^{3/2}}{q^{1/2}(q^2;q^2)_\infty^{3/2} (q^3;q^3)_\infty^{1/2} (q^{12};q^{12})_\infty^2} \right)$$

we obtain (12.9). In a similar way, using (12.15) and (12.16) in (12.14), we obtain (12.8).

□

The next result shows that certain rational functions of $h$ may be expressed as eta-quotients.

**Theorem 12.2.**

$$\frac{1}{h} - h = \frac{\eta_2^2 \eta_6^4}{\eta_1 \eta_3 \eta_{12}^4}, \tag{12.17}$$

$$\frac{1}{h} + h = \frac{\eta_3^3 \eta_4}{\eta_1 \eta_{12}^3}, \tag{12.18}$$

$$\frac{1}{h} - 1 + h = \frac{\eta_4^4 \eta_6^2}{\eta_2^2 \eta_{12}^4}, \tag{12.19}$$

$$\frac{1}{h} - 4 + h = \frac{\eta_1^3 \eta_4 \eta_6^2}{\eta_2^2 \eta_3 \eta_{12}^3}, \tag{12.20}$$

$$\frac{1}{h} - 2 + h = \frac{\eta_1 \eta_4^2 \eta_6^9}{\eta_2^3 \eta_3^3 \eta_{12}^6} \tag{12.21}$$

*and*

$$\frac{1}{h}+2+h=\frac{\eta_2^7\eta_3}{\eta_1^3\eta_4^2\eta_6\eta_{12}^2}. \tag{12.22}$$

*Proof.* The identity (12.17) follows by multiplying the first two identities in Theorem 12.1. Similarly, the identities (12.18)–(12.20) may be obtained by multiplying the other conjugate pairs of identities in Theorem 12.1. The identities (12.21) and (12.22) follow by squaring the identities (12.2) and (12.3), respectively. □

**Corollary 12.3.** *As $q$ increases from 0 to 1, the function $h(q)$ defined by* (12.1) *strictly increases from* 0 *to* $2-\sqrt{3}$.

*Proof.* For $0<q<1$ let $f(q)$ be the function defined by

$$f(q)=\frac{1}{q}\prod_{j=1}^{\infty}\frac{(1-q^j)^3(1-q^{4j})(1-q^{6j})^2}{(1-q^{2j})^2(1-q^{3j})(1-q^{12j})^3}.$$

By simple algebra,

$$f(q)=\frac{1}{q}\prod_{j=1}^{\infty}\frac{1}{(1+q^j)^2(1+q^j+q^{2j})(1+q^{4j}+q^{8j})(1+q^{6j})^2}.$$

It follows that $f$ is strictly decreasing on the interval $0<q<1$ and

$$\lim_{q\to 0+}f(q)=+\infty \qquad \text{and} \qquad \lim_{q\to 1-}f(q)=0.$$

By (12.20),

$$\frac{1}{h(q)}-4+h(q)=f(q).$$

On solving for $h(q)$ and using the fact that $h(0)=0$ to determine the sign, we find that

$$h(q)=\frac{1}{2}\left(4+f(q)-\sqrt{(4+f(q))^2-4}\right).$$

It is easy to check that as $y$ increases from 0 to $+\infty$, the function

$$\frac{1}{2}\left(4+y-\sqrt{(4+y)^2-4}\right)$$

strictly decreases from $2-\sqrt{3}$ to 0. The result follows. □

The results of Theorem 12.2 can be used to derive the following quadratic transformation formula for Ramanujan's cubic continued fraction that is an analogue of Theorem 10.2.

**Corollary 12.4.** *Let $r_c(q)$ be Ramanujan's cubic continued fraction defined by* (6.1). *Then*

$$r_c^3(q) = h\left(\frac{1-h}{1+h^2}\right)^2 \quad \text{and} \quad r_c^3(q^2) = \frac{h^2}{(1-h)^2(1+h^2)}$$

*and hence*

$$r_c(q)r_c(q^2) = \frac{h}{1+h^2}.$$

*Proof.* By (6.1),

$$r_c(q) = \frac{\eta_1\eta_6^3}{\eta_2\eta_3^3}.$$

Hence, by the results in Theorem 12.2,

$$h\left(\frac{1-h}{1+h^2}\right)^2 = \frac{\frac{1}{h}-2+h}{\left(\frac{1}{h}+h\right)^2} = \frac{\eta_1^3\eta_6^9}{\eta_2^3\eta_3^9} = r_c^3(q)$$

and

$$\frac{h^2}{(1-h)^2(1+h^2)} = \frac{1}{\left(\frac{1}{h}-2+h\right)\left(\frac{1}{h}+h\right)} = \frac{\eta_2^3\eta_{12}^9}{\eta_4^3\eta_6^9} = r_c^3(q^2).$$

□

## 12.3 Weight 2 modular forms

In this section we study the weight two modular form $y_{12}$ defined by

$$y_{12} = q\frac{\mathrm{d}}{\mathrm{d}q}\log h = 1 - \sum_{j=1}^{\infty}\chi_{12}(j)\frac{jq^j}{1-q^j}. \tag{12.23}$$

We begin by establishing its representation as an infinite product.

**Theorem 12.5.** *Let $f_1, f_2, f_3$, and $f_4$ be the infinite products defined by*

$$f_1 = \frac{\eta_2^2\eta_3^2\eta_4\eta_{12}}{\eta_1^2}, \quad f_2 = \frac{\eta_1^2\eta_4\eta_6^2\eta_{12}}{\eta_3^2}, \quad f_3 = \frac{\eta_1\eta_2^2\eta_3\eta_{12}^2}{\eta_4^2}, \quad f_4 = \frac{\eta_1\eta_3\eta_4^2\eta_6^2}{\eta_{12}^2}.$$

*Then*

$$f_1 = \sum_{j=1}^{\infty}\sum_{k=1}^{\infty} \chi_{12}(k)\, j\, q^{jk}, \qquad f_2 = \sum_{j=1}^{\infty}\sum_{k=1}^{\infty} \chi_3(j)\chi_4(k)\, j\, q^{jk},$$

$$f_3 = \sum_{j=1}^{\infty}\sum_{k=1}^{\infty} \chi_4(j)\chi_3(k)\, j\, q^{jk} \quad \text{and} \quad f_4 = 1 - \sum_{j=1}^{\infty}\sum_{k=1}^{\infty} \chi_{12}(j)\, j\, q^{jk}.$$

*In particular, the function $y_{12}$ defined by (12.23) has a representation as an infinite product given by*

$$y_{12} = \frac{\eta_1\,\eta_3\,\eta_4^2\,\eta_6^2}{\eta_{12}^2}.$$

*Proof.* Replace $(x, y, q)$ with $(q, q^5, q^{12})$, $(q\,e^{2\pi i/3}, q\,e^{4\pi i/3}, q^4)$, $(iq, -iq, q^3)$, and $(e^{i\pi/6}, e^{5\pi i/6}, q)$, respectively, in Theorem 1.17, and follow the procedure in Examples 1.25 and 1.26.

□

The next result provides parameterizations of various eta-functions in terms of $y_{12}$ and $h$.

**Theorem 12.6.** *The following identities hold:*

$$\eta^{24}(\tau) = y_{12}^6 \frac{h(1-4h+h^2)^6}{(1-h)^4(1+h^2)^2(1-h+h^2)^3},$$

$$\eta^{24}(2\tau) = y_{12}^6 \frac{h^2(1+h)^6}{(1-h)^2(1+h^2)^4},$$

$$\eta^{24}(3\tau) = y_{12}^6 \frac{h^3(1+h^2)^6}{(1-h+h^2)^5(1+h)^4(1-4h+h^2)^2},$$

$$\eta^{24}(4\tau) = y_{12}^6 \frac{h^4(1-h+h^2)^6}{(1+h^2)^5(1-h)^4(1-4h+h^2)^3},$$

$$\eta^{24}(6\tau) = y_{12}^6 \frac{h^6(1-h)^6}{(1+h)^2(1-h+h^2)^4(1-4h+h^2)^4}$$

*and*

$$\eta^{24}(12\tau) = y_{12}^6 \frac{h^{12}}{(1+h)^4(1+h^2)^3(1-h+h^2)^2(1-4h+h^2)^5}.$$

**Table 12.1**
$f = y_{12}\, h^{e_1}(1+h)^{e_2}$
$\times(1-h)^{e_3}(1-4h+h^2)^{e_4}$
$\times(1+h^2)^{e_5}(1-h+h^2)^{e_6}$

| $f$ | $e_1$ | $e_2$ | $e_3$ | $e_4$ | $e_5$ | $e_6$ |
|---|---|---|---|---|---|---|
| $f_1$ | −2 | 2 | 2 | 1 | 0 | 1 |
| $f_2$ | 2 | 0 | −2 | 1 | 2 | 1 |
| $f_3$ | 1 | 2 | 1 | −2 | 0 | 2 |
| $f_4$ | 1 | 0 | 1 | 2 | 2 | −2 |
| $\varphi^4(q)$ | 0 | 5 | 1 | −1 | −1 | −1 |
| $\varphi^4(q^3)$ | 0 | 1 | 5 | −1 | −1 | −1 |
| $\varphi^3(q)\varphi(q^3)$ | 0 | 4 | 2 | −1 | −1 | −1 |
| $\varphi(q)\varphi^3(q^3)$ | 0 | 2 | 4 | −1 | −1 | −1 |
| $\varphi^4(-q)$ | 0 | −1 | −1 | −1 | 0 | 2 |
| $\varphi^4(-q^2)$ | 0 | 2 | 0 | −1 | −1/2 | 1/2 |
| $\varphi^4(-q^3)$ | 0 | −1 | −1 | −1 | 2 | 0 |
| $\varphi^4(-q^6)$ | 0 | 0 | 2 | −1 | 1/2 | −1/2 |
| $q^{1/2}\psi^4(q)$ | 1/2 | 2 | 0 | 1/2 | −1 | −1 |
| $q\psi^4(q^2)$ | 1 | −1 | −1 | 2 | −1 | −1 |
| $q^{3/2}\psi^4(q^3)$ | 3/2 | 0 | 2 | −1/2 | −1 | −1 |
| $q^3\psi^4(q^6)$ | 3 | −1 | −1 | 0 | −1 | −1 |

*Proof.* These are immediate from the results of Theorems 12.2 and 12.5. □

The results in Theorem 12.6 can be used to produce parameterizations of various theta functions and eta-quotients in the form

$$y_{12}^{e_0}h^{e1}(1+h)^{e_2}(1-h)^{e_3}(1-4h+h^2)^{e_4}(1+h^2)^{e_5}(1-h+h^2)^{e_6}.$$

For example, the functions $f_1, f_2, f_3$, and $f_4$ have parametric representations given by

$$f_1 = \frac{y_{12}h}{1-4h+h^2}, \quad f_2 = \frac{y_{12}h}{1+h^2}, \quad f_3 = \frac{y_{12}h}{1-h+h^2} \quad \text{and} \quad f_4 = y_{12}.$$

These results, along with further examples of weight two modular forms, for which $e_0 = 1$, are listed in Table 12.1.

Since Lambert series for the functions $f_1$, $f_2$, $f_3$, and $f_4$ have been given in Theorem 12.5, the results in the next theorem provide Lambert series expansions for various products of theta functions.

**Theorem 12.7.** *The following identities hold:*

$$\varphi^3(q)\varphi(q^3) = 6f_1 - 2f_2 + 3f_3 + f_4,$$

$$\varphi(q)\varphi^3(q^3) = 2f_1 + 2f_2 - f_3 + f_4,$$

$$q\psi(q)\psi^2(q^2)\psi(q^3) = \frac{1}{4}(3f_1 + f_2),$$

$$q^2\psi(q)\psi(q^3)\psi^2(q^6) = \frac{1}{4}(f_1 - f_2).$$

*Proof.* For each identity, use Table 12.1 to express both sides as algebraic functions of $y_{12}$ and $h$, and check that the two sides are equal.

□

We now show that $P(q^\ell)$, where $\ell \in \{1, 2, 3, 4, 6, 12\}$, can be parameterized in terms of $y_{12}$, $h$ and the derivative $\mathrm{d}y_{12}/\mathrm{d}h$. We begin with

**Theorem 12.8.** *The following identities hold:*

$$h\frac{\mathrm{d}y_{12}}{\mathrm{d}h} = \frac{1}{24}\left(P(q) + 3P(q^3) + 8P(q^4) + 12P(q^6) - 24P(q^{12})\right),$$

$$\frac{(1-h^2)}{(1+h^2)}y_{12} = \frac{1}{24}\left(P(q) - 9P(q^3) - 4P(q^4) + 36P(q^{12})\right),$$

$$\frac{(1-h^2)}{(1-h+h^2)}y_{12} = \frac{1}{24}\left(4P(q^2) - 16P(q^4) - 12P(q^6) + 48P(q^{12})\right),$$

$$\frac{(1-h^2)}{(1-4h+h^2)}y_{12} = \frac{1}{24}\left(-3P(q) + 4P(q^2) + 3P(q^3) - 4P(q^4) - 12P(q^6) + 36P(q^{12})\right),$$

$$\frac{(1-h^2)}{(1-2h+h^2)}y_{12} = \frac{1}{24}\left(-P(q) + 6P(q^2) + 9P(q^3) - 8P(q^4) - 54P(q^6) + 72P(q^{12})\right),$$

$$\frac{(1-h^2)}{(1+2h+h^2)}y_{12} = \frac{1}{24}\left(3P(q) - 14P(q^2) - 3P(q^3) + 8P(q^4) + 6P(q^6) + 24P(q^{12})\right).$$

*Proof.* By the chain rule,

$$h\frac{\mathrm{d}y_{12}}{\mathrm{d}h} = q\frac{\mathrm{d}y_{12}}{\mathrm{d}q} \Big/ q\frac{\mathrm{d}}{\mathrm{d}q}\log h$$

On applying (12.23) and Theorem 12.5, this becomes

$$h\frac{\mathrm{d}y_{12}}{\mathrm{d}h} = \left(q\frac{\mathrm{d}y_{12}}{\mathrm{d}q}\right) \times \left(\frac{1}{y_{12}}\right) = q\frac{\mathrm{d}}{\mathrm{d}q}\log y_{12} = q\frac{\mathrm{d}}{\mathrm{d}q}\log\left(\frac{\eta_1\eta_3\eta_4^2\eta_6^2}{\eta_{12}^2}\right).$$

On computing the logarithmic derivative on the right-hand side, we obtain the first result in the theorem.

To prove the remaining results, take the logarithmic derivative (with respect to $q$) of each of (12.18), …, (12.22), respectively, and apply (12.23) and Theorem 12.5.

□

Here are the parameterizations of $P(q^\ell)$, for $\ell \in \{1, 2, 3, 4, 6, 12\}$, in terms of $y_{12}$, $h$ and $\mathrm{d}y_{12}/\mathrm{d}h$.

**Theorem 12.9.** *The following identity, expressed as a matrix product, holds:*

$$\begin{pmatrix} P(q) \\ P(q^2) \\ P(q^3) \\ P(q^4) \\ P(q^6) \\ P(q^{12}) \end{pmatrix} = \begin{pmatrix} 6 & 2 & 3 & -6 & 2 & 0 \\ 3 & 2 & 0 & 0 & \frac{1}{2} & -\frac{3}{2} \\ 2 & -2 & \frac{5}{3} & \frac{2}{3} & 0 & \frac{2}{3} \\ \frac{3}{2} & \frac{5}{4} & -\frac{3}{2} & \frac{3}{4} & \frac{1}{2} & 0 \\ 1 & 0 & \frac{2}{3} & \frac{2}{3} & -\frac{1}{2} & \frac{1}{6} \\ \frac{1}{2} & \frac{1}{4} & \frac{1}{6} & \frac{5}{12} & 0 & \frac{1}{6} \end{pmatrix} \begin{pmatrix} h\frac{\mathrm{d}y_{12}}{\mathrm{d}h} \\ \frac{(1-h^2)}{(1+h^2)}y_{12} \\ \frac{(1-h^2)}{(1-h+h^2)}y_{12} \\ \frac{(1-h^2)}{(1-4h+h^2)}y_{12} \\ \frac{(1-h^2)}{(1-2h+h^2)}y_{12} \\ \frac{(1-h^2)}{(1+2h+h^2)}y_{12} \end{pmatrix}.$$

*Proof.* This is immediate from Theorem 12.8. □

Here are the corresponding parameterizations for the Eisenstein series $Q$.

**Theorem 12.10.** *The following parameterizations hold.*

$$Q(q) = y_{12}^2\left(\frac{(1-4h+h^2)^4}{(1-h)^2(1+h)^2(1-h+h^2)^2} + \frac{256h(1+h)^4(1-h+h^2)}{(1+h^2)^2(1-4h+h^2)^2}\right),$$

$$Q(q^2) = y_{12}^2\left(\frac{(1-4h+h^2)^4}{(1-h)^2(1+h)^2(1-h+h^2)^2} + \frac{16h(1+h)^4(1-h+h^2)}{(1+h^2)^2(1-4h+h^2)^2}\right)$$

$$= y_{12}^2\left(\frac{(1+h)^4(1-4h+h^2)}{(1+h^2)(1-h+h^2)^2} + \frac{256h^2(1-h+h^2)^4}{(1-h)^2(1+h)^2(1+h^2)^2(1-4h+h^2)^2}\right),$$

$$Q(q^4) = y_{12}^2\left(\frac{(1+h)^4(1-4h+h^2)}{(1+h^2)(1-h+h^2)^2} + \frac{16h^2(1-h+h^2)^4}{(1-h)^2(1+h)^2(1+h^2)^2(1-4h+h^2)^2}\right),$$

$$Q(q^3) = y_{12}^2\left(\frac{(1+h^2)^4}{(1-h)^2(1+h)^2(1-h+h^2)^2} + \frac{256h^3(1-h)^4}{(1+h^2)^2(1-h+h^2)(1-4h+h^2)^2}\right),$$

$$Q(q^6) = y_{12}^2\left(\frac{(1+h^2)^4}{(1-h)^2(1+h)^2(1-h+h^2)^2} + \frac{16h^3(1-h)^4}{(1+h^2)^2(1-h+h^2)(1-4h+h^2)^2}\right)$$

$$= y_{12}^2\left(\frac{(1-h)^4(1+h^2)}{(1-h+h^2)^2(1-4h+h^2)}\right.$$

$$\left.+\frac{256h^6}{(1+h^2)^2(1-h)^2(1+h)^2(1-4h+h^2)^2}\right)$$

*and*

$$Q(q^{12}) = y_{12}^2 \left( \frac{(1-h)^4(1+h^2)}{(1-h+h^2)^2(1-4h+h^2)} + \frac{16h^6}{(1+h^2)^2(1-h)^2(1+h)^2(1-4h+h^2)^2} \right).$$

*Proof.* By Corollary 3.40,

$$Q(q) = \frac{\eta_1^{16}}{\eta_2^8} + 256\frac{\eta_2^{16}}{\eta_1^8} \quad \text{and} \quad Q(q^2) = \frac{\eta_1^{16}}{\eta_2^8} + 16\frac{\eta_2^{16}}{\eta_1^8}. \tag{12.24}$$

On using the parameterizations for the eta-function in Theorem 10.8 we immediately obtain the result for $Q(q)$ and the first result for $Q(q^3)$.

The remaining results are obtained in the same way, by first replacing $q$ with $q^2$, $q^3$, or $q^6$ in (12.24). □

Parameterizations for the Eisenstein series $R(q^\ell)$, $\ell \in \{1, 2, 3, 4, 6, 12\}$, in terms of $y_{12}$ and $h$ are described in Exercise 7 at the end of the chapter.

## 12.4 The modular function $X_{12}$ and weight two modular form $Z_{12}$

In this section we define a modular function $X_{12}$ and a weight two modular form $Z_{12}$, and find a third order differential equation that is satisfied by $Z_{12}$ with respect to $X_{12}$. We begin by considering the modular functions $w_A$, $w_B$, and $w_C$, defined by

$$w_A = \left(\frac{\eta_2\,\eta_3\,\eta_{12}}{\eta_1\,\eta_4\,\eta_6}\right)^4, \qquad w_B = \left(\frac{\eta_1\,\eta_{12}}{\eta_3\,\eta_4}\right)^4 \quad \text{and} \quad w_C = \left(\frac{\eta_4\,\eta_{12}}{\eta_1\,\eta_3}\right)^2. \tag{12.25}$$

By Theorem 12.6, the following representations in terms of $h$ hold:

$$w_A = \frac{h(1+h^2)}{(1-4h+h^2)(1-h+h^2)},$$

$$w_B = \frac{h(1-4h+h^2)}{(1-h+h^2)(1+h^2)},$$

$$w_C = \frac{h(1-h+h^2)}{(1+h^2)(1-4h+h^2)}.$$

It follows that $w_A$, $w_B$, and $w_C$ are algebraically related, and in fact

$$\frac{w_B}{(1+w_B)^2}=\frac{w_C}{(1+4w_C)^2}=\frac{h(1+h^2)(1-h+h^2)(1-4h+h^2)}{(1-h^2)^4}, \tag{12.26}$$

$$\frac{w_A}{(1-3w_A)^2}=\frac{w_C}{(1-4w_C)^2}=\frac{h(1+h^2)(1-h+h^2)(1-4h+h^2)}{(1-8h+6h^2-8h^3+h^4)^2} \tag{12.27}$$

and

$$\frac{w_A}{(1+3w_A)^2}=\frac{w_B}{(1-w_B)^2}=\frac{h(1+h^2)(1-h+h^2)(1-4h+h^2)}{(1-2h+6h^2-2h^3+h^4)^2}. \tag{12.28}$$

Let $X_{12}$ and $Z_{12}$ be defined by

$$X_{12}=\frac{w_B}{(1+w_B)^2}=\frac{w_C}{(1+4w_C)^2} \tag{12.29}$$

and

$$Z_{12}=q\frac{\mathrm{d}}{\mathrm{d}q}\log w_A. \tag{12.30}$$

The next result shows that both $X_{12}$ and $Z_{12}$ have a high degree of symmetry.

**Theorem 12.11.** *Let $X_{12}$ and $Z_{12}$ be defined by* (12.29) *and* (12.30). *Then,*

$$X_{12}=\frac{\eta_1^6\,\eta_3^6\,\eta_4^6\,\eta_{12}^6}{\eta_2^{12}\,\eta_6^{12}} \quad \text{and} \quad Z_{12}=\frac{\eta_2^{10}\,\eta_6^{10}}{\eta_1^4\,\eta_3^4\,\eta_4^4\,\eta_{12}^4}. \tag{12.31}$$

*Proof.* By (12.26),

$$X_{12}=\frac{h(1+h^2)(1-h+h^2)(1-4h+h^2)}{(1-h^2)^4}. \tag{12.32}$$

Theorem 12.2 can be used to express the right-hand side as an eta-quotient, and this gives (12.31).

Next, by direct calculation using (12.25) and (12.30), we find that

$$\begin{aligned}Z_{12}&=q\frac{\mathrm{d}}{\mathrm{d}q}\log\left(\frac{\eta_2\,\eta_3\,\eta_{12}}{\eta_1\,\eta_4\,\eta_6}\right)^4\\&=\frac{1}{6}\left(-P(q)+2P(q^2)+3P(q^3)-4P(q^4)-6P(q^6)+12P(q^{12})\right).\end{aligned}$$

On applying Theorem 12.9, it follows that

$$Z_{12}=y_{12}\frac{(1-h^2)^3}{(1+h^2)(1-h+h^2)(1-4h+h^2)}. \tag{12.33}$$

This may be converted to an eta-quotient by the formulas in Theorem 12.6, and this completes the proof.

□

In contrast with Theorem 12.11, none of the functions

$$\frac{w_A}{(1-3w_A)^2}=\frac{w_C}{(1-4w_C)^2},\qquad \frac{w_A}{(1+3w_A)^2}=\frac{w_B}{(1-w_B)^2},$$

$$q\frac{\mathrm{d}}{\mathrm{d}q}\log w_B \qquad \text{or} \qquad q\frac{\mathrm{d}}{\mathrm{d}q}\log w_C$$

have simple expressions as eta-quotients.

The next result provides a differentiation formula for $X_{12}$.

**Theorem 12.12.** *The following formula holds*:

$$q\frac{\mathrm{d}}{\mathrm{d}q}\log X_{12}=Z_{12}\sqrt{1-20X_{12}+64X_{12}^2}.$$

*Proof.* By a calculation using (12.32), we may deduce that

$$1-4X_{12}=\frac{(1-2h+6h^2-2h^3+h^4)^2}{(1-h^2)^4} \tag{12.34}$$

and

$$1-16X_{12}=\frac{(1-8h+6h^2-8h^3+h^4)^2}{(1-h^2)^4}. \tag{12.35}$$

On the other hand, by (12.23) and (12.32) and a calculation using the chain rule, we find that

$$\begin{aligned}q\frac{\mathrm{d}}{\mathrm{d}q}\log X_{12}&=q\frac{\mathrm{d}h}{\mathrm{d}q}\times\frac{\mathrm{d}}{\mathrm{d}h}\log X_{12}\\&=hy_{12}\times\frac{\mathrm{d}}{\mathrm{d}h}\log\left(\frac{h(1+h^2)(1-h+h^2)(1-4h+h^2)}{(1-h^2)^4}\right)\\&=y_{12}\times\frac{(1-2h+6h^2-2h^3+h^4)(1-8h+6h^2-8h^3+h^4)}{(1+h^2)(1-h^2)(1-h+h^2)(1-4h+h^2)}.\end{aligned}$$

On using (12.33), this may be written in the form

$$q\frac{\mathrm{d}}{\mathrm{d}q}\log X_{12}=Z_{12}\times\frac{(1-2h+6h^2-2h^3+h^4)(1-8h+6h^2-8h^3+h^4)}{(1-h^2)^4}. \tag{12.36}$$

On using (12.34) and (12.35) in (12.36) we obtain the required result.

□

The next goal will be to show that $Z_{12}$ satisfies a third order linear differential equation with respect to $X_{12}$. We will need

**Lemma 12.13.** *Ramanujan's Eisenstein series P and Q satisfy the identity*

$$9Q(q^3) + Q(q) = \frac{5}{2}\left(3P(q^3) - P(q)\right)^2.$$

*Proof.* By Theorems 3.31 and 3.43,

$$a^2(q) = \frac{1}{2}\left(3P(q^3) - P(q)\right) \qquad \text{and} \qquad a^4(q) = \frac{1}{10}\left(9Q(q^3) + Q(q)\right).$$

On eliminating $a(q)$ we obtain the required result. □

Let $D$ be the differential operator defined by

$$D = \frac{1}{Z_{12}}\, q\frac{\mathrm{d}}{\mathrm{d}q}. \tag{12.37}$$

By Theorem 12.12 and the chain rule, we also have

$$D = X_{12}\sqrt{1 - 20X_{12} + 64X_{12}^2}\,\frac{\mathrm{d}}{\mathrm{d}X_{12}}.$$

**Theorem 12.14.** *Let $X_{12}$ and $Z_{12}$ be defined by* (12.29) *and* (12.30)*. Then $Z_{12}$ satisfies the second order nonlinear differential equation with respect to $X_{12}$ given by*

$$D^2 Z_{12} - \frac{1}{2Z_{12}}\left(DZ_{12}\right)^2 = 4Z_{12}X_{12}(1 - 8X_{12}). \tag{12.38}$$

*Moreover, the following third order linear differential equation holds:*

$$D^3 Z_{12} - 8X_{12}(1 - 8X_{12})DZ_{12} = 4Z_{12}(1 - 16X_{12})DX_{12}. \tag{12.39}$$

*Proof.* Let us use the abbreviations $P_n = P(q^n)$ and $Q_n = Q(q^n)$. By a direct calculation using the eta-quotient for $Z_{12}$ in (12.31), we find that

$$q\frac{\mathrm{d}}{\mathrm{d}q}\log Z_{12} = \frac{1}{6}\left(-P_1 + 5P_2 - 3P_3 - 4P_4 + 15P_6 - 12P_{12}\right). \tag{12.40}$$

Applying the operator $q\frac{\mathrm{d}}{\mathrm{d}q}$ and using the differentiation formula (0.78) we obtain

$$q\frac{\mathrm{d}}{\mathrm{d}q}\left(q\frac{\mathrm{d}}{\mathrm{d}q}\log Z_{12}\right) = \frac{1}{72}(-P_1^2 + 10P_2^2 - 9P_3^2 - 16P_4^2 + 90P_6^2 - 144P_{12}^2)$$
$$+\frac{1}{72}(Q_1 - 10Q_2 + 9Q_3 + 16Q_4 - 90Q_6 + 144Q_{12}). \tag{12.41}$$

Lemma 12.13 may be used to eliminate $Q_1, \ldots, Q_{12}$ from (12.41). On simplifying, we obtain

$$q\frac{\mathrm{d}}{\mathrm{d}q}\left(q\frac{\mathrm{d}}{\mathrm{d}q}\log Z_{12}\right) = \frac{1}{48}(P_3 - P_1)(9P_3 - P_1) - \frac{5}{24}(P_6 - P_2)(9P_6 - P_2)$$
$$+\frac{1}{3}(P_{12} - P_4)(9P_{12} - P_4). \tag{12.42}$$

Hence, by (12.37), (12.40), and (12.42), we may deduce that

$$Z_{12} \times \left(D^2 Z_{12} - \frac{1}{2Z_{12}}\left(DZ_{12}\right)^2\right)$$
$$= \frac{1}{48}(P_3 - P_1)(9P_3 - P_1) - \frac{5}{24}(P_6 - P_2)(9P_6 - P_2) + \frac{1}{3}(P_{12} - P_4)(9P_{12} - P_4)$$
$$- \frac{1}{72}\left(-P_1 + 5P_2 - 3P_3 - 4P_4 + 15P_6 - 12P_{12}\right)^2.$$

By Theorem 12.9, the right-hand side can be expressed in terms of $y_{12}$ and $h$ and $h\,\mathrm{d}y_{12}/\mathrm{d}h$. On simplification, we find that

$$Z_{12} \times \left(D^2 Z_{12} - \frac{1}{2Z_{12}}\left(DZ_{12}\right)^2\right)$$
$$= 4y_{12}^2 \times \frac{h(1 - 8h + 36h^2 - 56h^3 + 86h^4 - 56h^5 + 36h^6 - 8h^7 + h^8)}{(1-h)^2(1+h)^2(1+h^2)(1-4h+h^2)(1-h+h^2)},$$

the terms involving $h\,\mathrm{d}y_{12}/\mathrm{d}h$ all canceling. On using (12.32) and (12.33), the result simplifies further to give

$$Z_{12} \times \left(D^2 Z_{12} - \frac{1}{2Z_{12}}\left(DZ_{12}\right)^2\right) = 4Z_{12}^2 X_{12}(1 - 8X_{12}).$$

After dividing by $Z_{12}$ we obtain (12.38). Now we will prove (12.39). Applying the differential operator $D$ to both sides of (12.38) and using the basic rules of calculus, we deduce that

$$D^3 Z_{12} + \frac{DZ_{12}}{Z_{12}} \left\{ \frac{(DZ_{12})^2}{2Z_{12}} - D^2 Z_{12} \right\} = 4X_{12}(1-8X_{12})DZ_{12} + 4Z_{12}(1-16X_{12})DX_{12}.$$

The differential equation (12.38) may be used to simplify the term in braces, and the identity (12.39) follows on further simplification.

□

**Theorem 12.15.** *The following series expansion holds*:

$$Z_{12} = \sum_{n=0}^{\infty} \left\{ \sum_{j=0}^{n} \binom{n}{j}^2 \binom{2n-2j}{n-j} \binom{2j}{j} \right\} X_{12}^n.$$

*Proof.* The differential equation (12.39) may be written in the form

$$X_{12}^2(1-4X_{12})(1-16X_{12})\frac{\mathrm{d}^3 Z_{12}}{\mathrm{d}X_{12}^3} + 3X_{12}(1-30X_{12}+128X_{12}^2)\frac{\mathrm{d}^2 Z_{12}}{\mathrm{d}X_{12}^2}$$
$$+(1-68X_{12}+448X_{12}^2)\frac{\mathrm{d}Z_{12}}{\mathrm{d}X_{12}} - 4(1-16X_{12})Z_{12} = 0.$$

On writing

$$Z_{12} = \sum_{n=0}^{\infty} T(n) X_{12}^n$$

we deduce that the coefficients $T(n)$ satisfy the recurrence relation

$$(n+1)^3 T(n+1) = 2(2n+1)(5n^2+5n+2)T(n) - 64n^3 T(n-1)$$

and initial conditions

$$T(0) = 1, \qquad T(1) = 4.$$

The method of creative telescoping may be used to show that sequence whose $n$-th term is given by

$$\sum_{j=0}^{n} \binom{n}{j}^2 \binom{2n-2j}{n-j} \binom{2j}{j}$$

satisfies the same recurrence relation and initial conditions. It follows that

$$T(n) = \sum_{j=0}^{n} \binom{n}{j}^2 \binom{2n-2j}{n-j} \binom{2j}{j},$$

and so $T(n)$ are the Domb numbers from (6.64). This completes the proof. □

The result of Theorem 12.15 may be compared with some earlier results. On replacing $q$ with $-q$ in the eta-quotients for $X_{12}$ and $Z_{12}$ in (12.31) and making use of (0.4), we find that

$$-X_{12}(-q) = \frac{\eta_2^6\,\eta_6^6}{\eta_1^6\,\eta_3^6} \qquad \text{and} \qquad Z_{12}(-q) = \frac{\eta_1^4\,\eta_3^4}{\eta_2^2\,\eta_6^2}.$$

Referring to Table 6.1 we find that $-X_{12}(-q) = w_b$ and $Z_{12}(-q) = y_b$. Hence, the expansion in Theorem 12.15 is equivalent to the identity (6.62) under the change of sign $q \to -q$.

Next, the result of Theorem 4.30 in the case $\ell = 4$ is

$$\varphi^4(q) = \sum_{n=0}^{\infty} \binom{2n}{n}^3 \left( \frac{\eta_1^4\eta_4^4}{\eta_2^4} \times \frac{1}{\varphi^4(q)} \right)^{2n}. \tag{12.43}$$

In a similar way, the result of Theorem 12.15 may be presented in the form

$$\varphi^2(q)\varphi^2(q^3) = \sum_{n=0}^{\infty} \left\{ \sum_{j=0}^{n} \binom{n}{j}^2 \binom{2n-2j}{n-j} \binom{2j}{j} \right\} \left( \frac{\eta_1^2\eta_3^2\eta_4^2\eta_{12}^2}{\eta_2^2\eta_6^2} \times \frac{1}{\varphi^2(q)\varphi^2(q^3)} \right)^n.$$

This is clearly an analogue of (12.43).

## 12.5 Notes

A continued fraction expansion for $h$ has been derived by M. S. M. Naika, B. N. Dharmendra, and K. Shivashankara [237]. Several modular equations for $h$ have been given by M. S. M. Naika, S. Chandankumar, and K. S. Bairy [236], Naika et al [237], and K. R. Vasuki, A. A. A. Kahtan, G. Sharath, and C. S. Kumar [286]. The signs of the coefficients in the $q$-series expansions of $h$ and its reciprocal, along with the 2-, 3-, 4-, 6-, and 12-dissections, have been studied by B. L. S. Lin [208].

The identities and proofs in Theorem 12.5 were given by A. Alaca, Ş. Alaca and K. S. Williams [16]. See also [109]. The first two identities in Theorem 12.7 were essentially known to Eisenstein and Liouville. Proofs have been given by

K. Petr [242], A. Alaca, Ş. Alaca, and K. S. Williams [16]. See also the series of papers [289]– [291] by A. Walfisz, for a huge number of results on representations of integers by quadratic forms. For last two identities in Theorem 12.7, see [109].

Many of the results in Section 12.3 involve parameterizations in terms of $y_{12}$ and $h$. A different parameterization given by

$$\kappa = \frac{\varphi^3(q^3)}{\varphi(q)} \quad \text{and} \quad p = \frac{\varphi^2(q) - \varphi^2(q^3)}{2\varphi^2(q^3)}.$$

was used by A. Alaca, Ş. Alaca, and K. S. Williams [16]. The parameterizations are connected by the formulas

$$\kappa = \left(y_{12} \times \frac{(1-h)^7}{(1+h)(1+h^2)(1-h+h^2)(1-4h+h^2)}\right)^{1/2} \quad \text{and} \quad p = \frac{2h}{(1-h)^2}.$$

Finally, much of the material in this chapter is based on work with D. Ye [130].

## 12.6 Exercises

For Exercises 1, 2, and 3, let $f_1, f_2, f_3$, and $f_4$ be as for Theorem 12.5.

1. Prove that

$$f_1(q) = \frac{1}{8}\left(\varphi(q)\varphi(q^3) - \varphi(-q)\varphi(-q^3)\right)\left(\varphi^2(q) + \varphi^2(q^3)\right),$$

$$f_2(q) = \frac{1}{8}\left(\varphi(q)\varphi(q^3) - \varphi(-q)\varphi(-q^3)\right)\left(3\varphi^2(q^3) - \varphi^2(q)\right),$$

$$f_3(q) = \frac{1}{4}\varphi(-q)\varphi(-q^3)\left(\varphi^2(q) - \varphi^2(q^3)\right),$$

$$f_4(q) = \frac{1}{4}\varphi(-q)\varphi(-q^3)\left(\varphi^2(q) + 3\varphi^2(q^3)\right).$$

See Petr [242].

2. Let $A(n)$, $B(n)$, $C(n)$, and $D(n)$ be defined by

$$A(n) = \sum_{d|n} d\,\chi_{12}(n/d), \qquad B(n) = \sum_{d|n} d\,\chi_3(d)\chi_4(n/d),$$

$$C(n) = \sum_{d|n} d\,\chi_3(n/d)\chi_4(d) \quad \text{and} \quad D(n) = \sum_{d|n} d\,\chi_{12}(d).$$

a. Use Theorem 12.5 to show that

$$f_1(q) = \sum_{n=1}^{\infty} A(n)q^n, \quad f_2(q) = \sum_{n=1}^{\infty} B(n)q^n, \quad f_3(q) = \sum_{n=1}^{\infty} C(n)q^n$$

and

$$f_4(q) = 1 - \sum_{n=1}^{\infty} D(n)q^n.$$

b. Let $n$ be a positive integer, and write $n = 2^\alpha 3^\beta m$, where $\alpha$ and $\beta$ are nonnegative integers and $m$ is a positive integer with $(m, 6) = 1$. Show that

$$\begin{aligned}
A(n) &= 2^\alpha 3^\beta A(m), \\
B(n) &= 2^\alpha (-1)^{\alpha+\beta} \chi_3(n/d)A(m), \\
C(n) &= 3^\beta (-1)^{\alpha+\beta} \chi_4(n/d)A(m), \\
D(n) &= \chi_{12}(n/d)A(m).
\end{aligned}$$

Thus, the coefficients in $f_1(q)$, $f_2(q)$, $f_3(q)$, and $f_4(q)$ can be expressed in terms of the numbers $A(m)$, where $(m, 6) = 1$.
See A. Alaca, Ş. Alaca, M. F. Lemire, and K. S. Williams, [8].

3. For a given function $f(q)$, let $f^o(q)$ denote the odd part of $f(q)$. That is, let

$$f^o(q) = \frac{1}{2}\left(f(q) - f(-q)\right).$$

a. Prove that the odd parts of $f_1, f_2, f_3$, and $f_4$ are given by

$$\begin{aligned}
f_1^o(q) &= \frac{1}{8}\left(\varphi(q)\varphi(q^3) - \varphi(-q)\varphi(-q^3)\right)\left(\varphi^2(q^2) + \varphi^2(q^6)\right), \\
f_2^o(q) &= \frac{1}{8}\left(\varphi(q)\varphi(q^3) - \varphi(-q)\varphi(-q^3)\right)\left(3\varphi^2(q^6) - \varphi^2(q^2)\right), \\
f_3^o(q) &= \frac{1}{8}\left(\varphi(q)\varphi(-q^3) - \varphi(-q)\varphi(q^3)\right)\left(\varphi^2(-q^2) + \varphi^2(-q^6)\right), \\
f_4^o(q) &= \frac{1}{8}\left(\varphi(q)\varphi(-q^3) - \varphi(-q)\varphi(q^3)\right)\left(\varphi^2(-q^2) - 3\varphi^2(-q^6)\right).
\end{aligned}$$

b. Prove that

$$q^3\psi^3(q^4)\psi(q^{12}) = \frac{1}{16}\left(3f_1^o(q) + f_2^o(q) - 3f_3^o(q) + f_4^o(q)\right)$$

and

$$q^5\psi(q^4)\psi^3(q^{12}) = \frac{1}{16}\left(f_1^o(q) - f_2^o(q) + f_3^o(q) + f_4^o(q)\right).$$

See [109] for further details.

4. Let $g_1$, $g_2$, and $g_3$ be defined by

$$g_1(q) = q\prod_{j=1}^{\infty}\frac{(1-q^{2j})^3(1-q^{12j})^2}{(1-q^{6j})},$$

$$g_2(q) = q\prod_{j=1}^{\infty}\frac{(1-q^{8j})^3(1-q^{12j})^2}{(1-q^{24j})},$$

$$g_3(q) = q\prod_{j=1}^{\infty}\frac{(1-q^{4j})^2(1-q^{6j})^3}{(1-q^{2j})}.$$

a. Show that $g_1(q) + 3g_3(q) = 4g_2(q)$.
b. Show that

$$q^3\psi(q^2)\psi(q^6)\psi^2(q^8) = \frac{1}{16}\{3f_1^o(q) - f_2^o(q)\} - \frac{1}{8}g_1(q),$$

$$q^5\psi^2(q^4)\psi(q^8)\psi(q^{24}) = \frac{1}{32}\{3f_1^o(q) - f_2^o(q) + 3f_3^o(q) + f_4^o(q)\} - \frac{1}{8}g_2(q),$$

$$q^7\psi(q^2)\psi(q^6)\psi^2(q^{24}) = \frac{1}{16}\{f_1^o(q) + f_2^o(q)\} - \frac{1}{8}g_3(q).$$

See [109] for further details.

5. Let $h = h(q)$ be defined by (12.1).
Use the results of Theorem 12.2 to prove:

a. If $H = h(-q)$, then $(H-1)^2 h + (1+h^2)H = 0$.
b. If $H = h(q^2)$, then $(h+H)^2 = H(1+h^2)$.
c. If $H = h(q^3)$, then $(H-h)^3 = H(H-1)(1+h)(1-4h+h^2)$.
d. If $H = h(q^4)$, then $(H-h)^4 = H\left(1-H+H^2\right)(1+h^2)(1-4h+h^2)$.
e. If $H = h(q^6)$, then $(H-h)^6 = H(1-4h+h^2)b(h,H)$, where

$$\begin{aligned} b(h,H) = {} & h^4H^4 + h^4H - h^4H^2 + 6h^3H^3 - 2h^3H^4 - 2h^3H \\ & - 4h^3H^2 - 3h^2H - 3h^2H^3 + 12h^2H^2 + 6hH - 2hH^3 \\ & - 2h - 4hH^2 - H^2 + H^3 + 1. \end{aligned}$$

Hint for Part b: By the identities in Theorem 12.2,

$$h\left(\frac{1-h}{1+h^2}\right)^2 = \frac{\eta_1^3\eta_6^9}{\eta_2^3\eta_3^9} \quad \text{and} \quad \frac{h^2}{(1-h)^2(1+h^2)} = \frac{\eta_2^3\eta_{12}^9}{\eta_4^3\eta_6^9}.$$

It follows that

$$H\left(\frac{1-H}{1+H^2}\right)^2 = \frac{h^2}{(1-h)^2(1+h^2)}.$$

This can be rearranged to give the required result.

6. Show that the result of Theorem 12.15 can be expressed in the form

$$q\frac{\mathrm{d}}{\mathrm{d}q}\log\left(\frac{h(1+h^2)}{(1-4h+h^2)(1-h+h^2)}\right)$$
$$= \sum_{n=0}^{\infty}\left\{\sum_{j=0}^{n}\binom{n}{j}^2\binom{2n-2j}{n-j}\binom{2j}{j}\right\}\left(\frac{h(1+h^2)(1-h+h^2)(1-4h+h^2)}{(1-h^2)^4}\right)^n$$

where

$$h = q\prod_{j=1}^{\infty}\frac{(1-q^{12j-1})(1-q^{12j-11})}{(1-q^{12j-5})(1-q^{12j-7})}.$$

See [130, (1.2)].

7. For each $\ell \in \{1,2,3,4,6,12\}$, show that there is a polynomial $t_\ell(h)$ of degree 24, such that

$$R(q^\ell) = \frac{y_{12}^3}{(h-1)^3(h+1)^3(h^2+1)^3(h^2-h+1)^3(h^2-4h+1)^3}\times t_\ell(h).$$

Hint: Use Corollary 0.23 in the form

$$R^2(q^\ell) = Q^3(q^\ell) - 1728\eta_\ell^{24} \tag{12.44}$$

together with the parameterizations for the eta-function and Eisenstein series $Q$ in Theorems 12.6 and 12.10, and show for each value of $\ell$ that the right-hand side of (12.44) is a square.

# Chapter 13
# Hypergeometric Modular Transformations

## 13.1 Introduction

At the end of Chapter 6 it was shown how modular parameterizations can be used to deduce transformation formulas for hypergeometric functions. For example, in Theorems 6.45 and 6.46 we obtained the transformation formulas

$$\frac{1}{1+16x}\,{}_3F_2\left(\frac{1}{3},\frac{1}{2},\frac{2}{3};\,1,1;\,\frac{108x}{(1+16x)^3}\right)=\frac{1}{1+4x}\,{}_3F_2\left(\frac{1}{3},\frac{1}{2},\frac{2}{3};\,1,1;\,\frac{108x^2}{(1+4x)^3}\right) \tag{13.1}$$

and

$$\frac{1}{1+27x}\,{}_3F_2\left(\frac{1}{4},\frac{1}{2},\frac{3}{4};\,1,1;\,\frac{256x}{(1+27x)^4}\right)=\frac{1}{1+3x}\,{}_3F_2\left(\frac{1}{4},\frac{1}{2},\frac{3}{4};\,1,1;\,\frac{256x^3}{(1+3x)^4}\right). \tag{13.2}$$

Because the identities (13.1) and (13.2) are transformation formulas for hypergeometric functions that arise from parameterizations by modular forms, we will call them hypergeometric modular transformations.

This chapter contains a systematic analysis of the transformation formulas implied by the results for modular forms in Chapters 10 and 12. The transformation formulas will be presented in Theorems 13.1 and 13.3, respectively. The theorems contain a large number of identities. In particular, Theorem 13.3 gives the equality of forty-two functions, of which (13.1) and (13.2) are special cases.

To shorten the appearance of the formulas, let $F_\ell$ be defined for $\ell = 1,2,3,4$ by

$$F_\ell(x) = {}_3F_2\left(\begin{matrix}\frac{1}{2},\ s,\ 1-s\\ 1,\ 1\end{matrix};4\alpha x\right)$$

where $\alpha = 432, 64, 27, 16$ and $s = 1/6,\ 1/4,\ 1/3,\ 1/2$, respectively. The values of the parameter $\alpha$ are the same as for Table 6.4, while the values of $s$ are as in Definition 4.23. We also have, e.g., see Theorem 4.30,



S. Cooper, *Ramanujan's Theta Functions*, DOI 10.1007/978-3-319-56172-1_14

$$F_1(x) = \sum_{j=0}^{\infty} \binom{6j}{3j}\binom{3j}{j}\binom{2j}{j} x^j, \qquad F_2(x) = \sum_{j=0}^{\infty} \binom{4j}{2j}\binom{2j}{j}^2 x^j,$$

$$F_3(x) = \sum_{j=0}^{\infty} \binom{3j}{j}\binom{2j}{j}^2 x^j \quad \text{and} \quad F_4(x) = \sum_{j=0}^{\infty} \binom{2j}{j}^3 x^j.$$

Thus, $F_1$, $F_2$, $F_3$, and $F_4$ have power series expansions with integer coefficients.

## 13.2 Transformations from the level 10 theory

The parameterizations from the level 10 theory for Ramanujan's function $k$ can be used to deduce the identities in the next theorem. The results are grouped according to the level of the underlying modular forms.

**Theorem 13.1.** *Let $F_5(x)$ and $G_5(x)$ be defined in a neighborhood of $x = 0$ by*

$$F_5(x) = \sum_{n=0}^{\infty} \binom{2n}{n} t_5(n)\, x^n \quad \text{and} \quad G_5(x) = \sum_{n=0}^{\infty} s_5(n)\, x^n$$

*where $\{t_5(n)\}$ and $\{s_5(n)\}$ are the sequences studied in Section 5.9, so that*

$$F_5(x) = \sum_{n=0}^{\infty} \binom{2n}{n} \left\{ \sum_{k=0}^{n} \binom{n}{k}^2 \binom{n+k}{k} \right\} x^n$$

*and*

$$G_5(x) = \sum_{n=0}^{\infty} \left\{ \sum_{k=0}^{n} (-1)^{n+k} \binom{n}{k}^3 \binom{4n-5k}{3n} \right\} x^n.$$

*Let $H_{10}(x)$ be defined by*

$$H_{10}(x) = \sum_{n=0}^{\infty} \left\{ \sum_{k=0}^{n} \binom{n}{k}^4 \right\} x^n.$$

*The following thirteen functions are equal, in a neighborhood of $t = 0$:*

*Level* 1:

$$\frac{1}{(1+236\,t+1440\,t^2+1920\,t^3+3840\,t^4+256\,t^5+256\,t^6)^{1/2}} \times F_1\left(\frac{t\,(1-4\,t)^{10}\,(1+t)^5}{(1+236\,t+1440\,t^2+1920\,t^3+3840\,t^4+256\,t^5+256\,t^6)^3}\right) \tag{13.3}$$

$$= \frac{1}{(1-4\,t+240\,t^2-480\,t^3+1440\,t^4-944\,t^5+16\,t^6)^{1/2}} \times F_1\left(\frac{t^2\,(1-4\,t)^5\,(1+t)^{10}}{(1-4\,t+240\,t^2-480\,t^3+1440\,t^4-944\,t^5+16\,t^6)^3}\right) \tag{13.4}$$

$$= \frac{1}{(1-4\,t+256\,t^5+256\,t^6)^{1/2}}\,F_1\left(\frac{t^5\,(1-4\,t)^2\,(1+t)}{(1-4\,t+256\,t^5+256\,t^6)^3}\right) \tag{13.5}$$

$$= \frac{1}{(1-4\,t+16\,t^5+16\,t^6)^{1/2}}\,F_1\left(\frac{t^{10}\,(1-4\,t)\,(1+t)^2}{(1-4\,t+16\,t^5+16\,t^6)^3}\right) \tag{13.6}$$

*Level* 2:

$$= \frac{1}{(1+4\,t^2)^{1/2}\,(1+22\,t-4\,t^2)}\,F_2\left(\frac{t\,(1+t)^5\,(1-4\,t)^5}{(1+4\,t^2)^2\,(1+22\,t-4\,t^2)^4}\right) \tag{13.7}$$

$$= \frac{1}{(1+4\,t^2)^{1/2}\,(1-2\,t-4\,t^2)}\,F_2\left(\frac{t^5\,(1+t)\,(1-4\,t)}{(1+4\,t^2)^2\,(1-2\,t-4\,t^2)^4}\right) \tag{13.8}$$

*Level* 4:

$$= \frac{1}{(1-4\,t)^{5/2}}\,F_4\left(-t\left(\frac{1+t}{1-4\,t}\right)^5\right) \tag{13.9}$$

$$= \frac{1}{(1-4\,t)^{1/2}}\,F_4\left(-t^5\left(\frac{1+t}{1-4\,t}\right)\right) \tag{13.10}$$

*Level* 5, *function* $F_5$:

$$= \frac{1}{(1+4t^2)^{1/2}(1+4t+8t^2)} F_5\left(\frac{t(1-4t)^2(1+t)}{(1+4t^2)(1+4t+8t^2)^2}\right) \tag{13.11}$$

$$= \frac{1}{(1+4t^2)^{1/2}(1-2t+2t^2)} F_5\left(\frac{t^2(1-4t)(1+t)^2}{(1+4t^2)(1-2t+2t^2)^2}\right) \tag{13.12}$$

*Level* 5, *function* $G_5$:

$$= \frac{1}{(1+t)(1-4t)^2} G_5\left(\left(\frac{t}{1+t}\right)\left(\frac{1}{1-4t}\right)^2\right) \tag{13.13}$$

$$= \frac{1}{(1+t)^2(1-4t)} G_5\left(\left(\frac{t}{1+t}\right)^2\left(\frac{1}{1-4t}\right)\right) \tag{13.14}$$

*Level* 10, *function* $H_{10}$:

$$= \frac{1}{(1+4t^2)^{3/2}} H_{10}\left(\frac{t(1+t)(1-4t)}{(1+4t^2)^2}\right). \tag{13.15}$$

*Proof.* By Theorem 4.30 for the case of level 1 we have

$$Q^{1/2}(q) = F_1\left(\frac{\eta^{24}(\tau)}{Q^3(q)}\right). \tag{13.16}$$

On substituting the parameterizations for $\eta(\tau)$ and $Q(q)$ from Theorems 10.8 and 10.13, we find that

$$y \times (f_1(k))^{1/2} = F_1\left(\frac{k(1-4k-k^2)^4}{(1-k^2)^4(1+k-k^2)} \Big/ (f_1(k))^3\right) \tag{13.17}$$

where

$$f_1(k) = \frac{(1-4k-k^2)^3}{(1-k^2)(1+k-k^2)^2} + \frac{256k(1+k-k^2)^3}{(1-k^2)^2(1-4k-k^2)^2}.$$

Next, replacing $\tau$ with $2\tau$, and hence $q$ with $q^2$, in (13.16) and then substituting the parameterizations for $\eta(2\tau)$ and $Q(q^2)$ from Theorems 10.8 and 10.13, we obtain

$$y \times (f_2(k))^{1/2} = F_1\left(\frac{k^2(1+k-k^2)^4}{(1-k^2)^5(1-4k-k^2)} \Big/ (f_2(k))^3\right) \tag{13.18}$$

where

$$f_2(k) = \frac{(1-4k-k^2)^3}{(1-k^2)(1+k-k^2)^2} + \frac{16k(1+k-k^2)^3}{(1-k^2)^2(1-4k-k^2)^2}.$$

On eliminating $y$ from (13.17) and (13.18) and then making the change of variable $t = k/(1-k^2)$, we obtain the expressions in (13.3) and (13.4) and deduce that they are equal. The expressions for (13.5) and (13.6) can be obtained in the same way, by replacing $\tau$ with $5\tau$ or $10\tau$, respectively, in (13.16).

The proofs of (13.7)–(13.15) are similar to the above, so we only provide a sketch with references to the underlying background formulas. The results for (13.7) and (13.8) can be obtained by using the level 2 theory in Theorem 4.30. The results for (13.9) and (13.10) are obtained by replacing $q$ with $-q$ in the level 4 results in Theorem 4.30 and then proceeding as before. To prove (13.11)–(13.14), start with the series expansions in Theorem 5.32 and use the formulas for $s_5(n)$ and $T_5(n)$ given in Theorems 5.34 and 5.35. Finally, for (13.15) use Theorem 10.23 and the definitions of $X_{10}$ and $Z_{10}$ given by (10.22) and (10.23), respectively.

□

It is worthwhile stating in explicit form the modular parameterization that underlies each function in Theorem 13.1.

**Theorem 13.2.** *Let the common power series of each expression in* (13.3)–(13.15) *be denoted by*

$$\sum_{n=0}^{\infty} b(n)t^n = 1 + 2t + 6t^2 + 20t^3 + 70t^4 + 244t^5 + 868t^6 + 3144t^7 + O\left(t^8\right).$$

*Then*

$$\frac{\eta(2\tau)\eta^{10}(5\tau)}{\eta^2(\tau)\eta^5(10\tau)} = \sum_{n=0}^{\infty} b(n)\left(\frac{\eta(\tau)\eta^5(10\tau)}{\eta(2\tau)\eta^5(5\tau)}\right)^n.$$

*Proof.* Recalling that $t = k/(1-k^2)$ and then using Theorem 10.4, we deduce that

$$t = \frac{\eta(\tau)\eta^5(10\tau)}{\eta(2\tau)\eta^5(5\tau)}. \tag{13.19}$$

Next, starting with (13.7) and using the formula for $Z_2$ in Theorem 4.30, we obtain

$$\begin{aligned}\sum_{n=0}^{\infty} b(n)t^n &= \frac{1}{(1+4t^2)^{1/2}(1+22t-4t^2)} \times F_2\left(\frac{t(1+t)^5(1-4t)^5}{(1+4t^2)^2(1+22t-4t^2)^4}\right)\\ &= \left(\frac{1}{t(1+t)^5(1-4t)^5}\right)^{1/4} \times \left(\frac{\eta^2(\tau)\eta^2(2\tau)}{Z_2(q)}\right) \times Z_2(q)\\ &= \left(\frac{1}{t(1+t)^5(1-4t)^5}\right)^{1/4} \times \eta^2(\tau)\eta^2(2\tau).\end{aligned}$$

Once again, recall that $t = k/(1-k^2)$ and then use Theorem 10.4 to express the resulting rational function of $k$ in terms of eta-functions. The result is

$$\sum_{n=0}^{\infty} b(n)t^n = \frac{\eta^{10}(5\tau)}{\eta^4(\tau)\eta(2\tau)\eta^5(10\tau)} \times \eta^2(\tau)\eta^2(2\tau) = \frac{\eta(2\tau)\eta^{10}(5\tau)}{\eta^2(\tau)\eta^5(10\tau)}. \tag{13.20}$$

The proof may be completed by substituting the result of (13.19) into (13.20). □

## 13.3 Transformations from the level 12 theory

The main result of this section is the following result that asserts that forty-two functions are equal. The identities are arranged according to the level of the underlying modular forms.

**Theorem 13.3.** *Let $\{t_a(n)\}$, $\{t_b(n)\}$, and $\{t_c(n)\}$ be the sequences analyzed in Theorems 6.35–6.37 and let $F_a$, $F_b$, and $F_c$ be the generating functions defined in a neighborhood of $x=0$ by*

$$F_a(x)=\sum_{n=0}^{\infty}\binom{2n}{n}t_a(n)x^n,\quad F_b(x)=\sum_{n=0}^{\infty}\binom{2n}{n}t_b(n)x^n \text{ and } F_c(x)=\sum_{n=0}^{\infty}\binom{2n}{n}t_c(n)x^n.$$

*Let $\{s_a(n)\}$, $\{s_b(n)\}$, and $\{s_c(n)\}$ be the sequences of Apéry, Domb, and Almkvist–Zudilin numbers, respectively, studied in Theorem 6.38. Let $G_a$, $G_b$, and $G_c$ be the generating functions defined in a neighborhood of $x=0$ by*

$$G_a(x) = \sum_{n=0}^{\infty} s_a(n)x^n,\quad G_b(x) = \sum_{n=0}^{\infty} s_b(n)x^n \text{ and } G_c(x) = \sum_{n=0}^{\infty} s_c(n)x^n.$$

*The following forty-two functions are equal, in a neighborhood of $t=0$:*

*Level* 1:

$$\frac{1}{(1+4t-8t^2)^{1/2}(1+228t-408t^2-128t^3-192t^4+768t^5-512t^6)^{1/2}}$$
$$\times F_1\left(\frac{t(1-t)^3(1-4t)^{12}(1-2t)(1+2t)^3}{(1+4t-8t^2)^3(1+228t-408t^2-128t^3-192t^4+768t^5-512t^6)^3}\right) \tag{13.21}$$

$$= \frac{1}{(1-2t+4t^2)^{1/2}(1-6t+240t^2-920t^3+960t^4-96t^5+64t^6)^{1/2}}$$
$$\times F_1\left(\frac{t^2(1-t)^6(1-4t)^6(1-2t)^2(1+2t)^6}{(1-2t+4t^2)^3(1-6t+240t^2-920t^3+960t^4-96t^5+64t^6)^3}\right) \tag{13.22}$$

$$= \frac{1}{(1+4t-8t^2)^{1/2}(1-12t+72t^2-128t^3-192t^4+768t^5-512t^6)^{1/2}}$$
$$\times F_1\left(\frac{t^3(1-t)(1-4t)^4(1-2t)^3(1+2t)}{(1+4t-8t^2)^3(1-12t+72t^2-128t^3-192t^4+768t^5-512t^6)^3}\right) \tag{13.23}$$

$$= \frac{1}{(1-2t-2t^2)^{1/2}(1-6t+6t^2+16t^3+204t^4-456t^5-8t^6)^{1/2}}$$
$$\times F_1\left(\frac{t^4(1-t)^{12}(1-4t)^3(1-2t)(1+2t)^3}{(1-2t-2t^2)^3(1-6t+6t^2+16t^3+204t^4-456t^5-8t^6)^3}\right) \tag{13.24}$$

$$= \frac{1}{(1-2t+4t^2)^{1/2}(1-6t+40t^3-96t^5+64t^6)^{1/2}}$$
$$\times F_1\left(\frac{t^6(1-t)^2(1-4t)^2(1-2t)^6(1+2t)^2}{(1-2t+4t^2)^3(1-6t+40t^3-96t^5+64t^6)^3}\right) \tag{13.25}$$

$$= \frac{1}{(1-2t-2t^2)^{1/2}(1-6t+6t^2+16t^3-36t^4+24t^5-8t^6)^{1/2}}$$
$$\times F_1\left(\frac{t^{12}(1-t)^4(1-4t)(1-2t)^3(1+2t)}{(1-2t-2t^2)^3(1-6t+6t^2+16t^3-36t^4+24t^5-8t^6)^3}\right) \tag{13.26}$$

$$= \frac{1}{(1-8t+4t^2)^{1/2}(1-240t+1932t^2-5700t^3+7728t^4-3840t^5+64t^6)^{1/2}}$$
$$\times F_1\left(\frac{-t(1-t)^3(1-4t)^3(1-2t)^4(1+2t)^{12}}{(1-8t+4t^2)^3(1-240t+1932t^2-5700t^3+7728t^4-3840t^5+64t^6)^3}\right) \tag{13.27}$$

$$= \frac{1}{(1-8t+4t^2)^{1/2}(1+12t^2-128t^3+48t^4+64t^6)^{1/2}}$$

$$\times F_1\left(\frac{-t^3(1-t)(1-4t)(1-2t)^{12}(1+2t)^4}{(1-8t+4t^2)^3(1+12t^2-128t^3+48t^4+64t^6)^3}\right) \tag{13.28}$$

*Level* 2:

$$= \frac{1}{1+20t-48t^2+32t^3-32t^4}\, F_2\left(\frac{t(1-t)^3(1-4t)^6(1-2t)(1+2t)^3}{(1+20t-48t^2+32t^3-32t^4)^4}\right) \tag{13.29}$$

$$= \frac{1}{1-4t+24t^2-40t^3-8t^4}\, F_2\left(\frac{t^2(1-t)^6(1-4t)^3(1-2t)(1+2t)^3}{(1-4t+24t^2-40t^3-8t^4)^4}\right) \tag{13.30}$$

$$= \frac{1}{1-4t+32t^3-32t^4}\, F_2\left(\frac{t^3(1-t)(1-4t)^2(1-2t)^3(1+2t)}{(1-4t+32t^3-32t^4)^4}\right) \tag{13.31}$$

$$= \frac{1}{1-4t+8t^3-8t^4}\, F_2\left(\frac{t^6(1-t)^2(1-4t)(1-2t)^3(1+2t)}{(1-4t+8t^3-8t^4)^4}\right) \tag{13.32}$$

$$= \frac{1}{1-28t+96t^2-112t^3+16t^4}\, F_2\left(-\frac{t(1-t)^3(1-4t)^3(1-2t)^2(1+2t)^6}{(1-28t+96t^2-112t^3+16t^4)^4}\right) \tag{13.33}$$

$$= \frac{1}{1-4t-16t^3+16t^4}\, F_2\left(-\frac{t^3(1-t)(1-4t)(1-2t)^6(1+2t)^2}{(1-4t-16t^3+16t^4)^4}\right) \tag{13.34}$$

*Level* 3:

$$= \frac{1}{(1+4t-8t^2)^2}\, F_3\left(\frac{t\,(1-t)\,(1-4\,t)^4\,\left(1-4\,t^2\right)}{(1+4t-8t^2)^6}\right) \tag{13.35}$$

$$= \frac{1}{(1-2\,t+4\,t^2)^2}\, F_3\left(\frac{t^2\,(1-t)^2\,(1-4\,t)^2\,\left(1-4\,t^2\right)^2}{(1-2\,t+4\,t^2)^6}\right) \tag{13.36}$$

$$= \frac{1}{(1-2\,t-2\,t^2)^2}\, F_3\left(\frac{t^4\,(1-t)^4\,(1-4\,t)\,\left(1-4\,t^2\right)}{(1-2\,t-2\,t^2)^6}\right) \tag{13.37}$$

$$= \frac{1}{(1-8\,t+4\,t^2)^2}\, F_3\left(-\frac{t\,(1-t)\,(1-4\,t)\,\left(1-4\,t^2\right)^4}{(1-8\,t+4\,t^2)^6}\right) \tag{13.38}$$

*Level* 4:

$$= \frac{1}{(1-2\,t)\,(1+2\,t)^3}\, F_4\left(\frac{t\,(1-t)^3\,(1-4\,t)^3}{(1-2\,t)^2\,(1+2\,t)^6}\right) \tag{13.39}$$

$$= \frac{1}{(1-2\,t)^3\,(1+2\,t)}\, F_4\left(\frac{t^3\,(1-t)\,(1-4\,t)}{(1-2\,t)^6\,(1+2\,t)^2}\right) \tag{13.40}$$

$$= \frac{1}{(1-4\,t)^3}\, F_4\left(-\frac{t\,(1-2\,t)\,(1+2\,t)^3\,(1-t)^3}{(1-4\,t)^6}\right) \tag{13.41}$$

$$= \frac{1}{(1-4\,t)}\, F_4\left(-\frac{t^3\,(1-2\,t)^3\,(1+2\,t)\,(1-t)}{(1-4\,t)^2}\right) \tag{13.42}$$

$$= \frac{1}{(1-2\,t)^{1/2}\,(1+2\,t)^{3/2}\,(1-4\,t)^{3/2}}\, F_4\left(\frac{-t^2\,(1-t)^6}{(1-2\,t)\,(1+2\,t)^3\,(1-4\,t)^3}\right) \tag{13.43}$$

$$= \frac{1}{(1-2\,t)^{3/2}\,(1+2\,t)^{1/2}\,(1-4\,t)^{1/2}}\, F_4\left(\frac{-t^6\,(1-t)^2}{(1-2\,t)^3\,(1+2\,t)\,(1-4\,t)}\right) \tag{13.44}$$

*Level* 6, *functions* $F$:

$$= \frac{1}{1-16t+24t^2+32t^3-32t^4}\,F_a\left(\frac{t(1-t)(1-4t)^2(1-2t)(1+2t)}{(1-16t+24t^2+32t^3-32t^4)^2}\right) \tag{13.45}$$

$$= \frac{1}{1-4t-12t^2+32t^3-8t^4}\,F_a\left(\frac{t^2(1-t)^2(1-4t)(1-2t)(1+2t)}{(1-4t-12t^2+32t^3-8t^4)^2}\right) \tag{13.46}$$

$$= \frac{1}{1+8t-48t^2+32t^3+16t^4}\,F_a\left(\frac{-t(1-t)(1-4t)(1-2t)^2(1+2t)^2}{(1+8t-48t^2+32t^3+16t^4)^2}\right) \tag{13.47}$$

$$= \frac{1}{(1-2t+4t^2)(1+4t-8t^2)}\,F_b\left(\frac{t(1-t)(1-4t)^2(1-2t)(1+2t)}{(1-2t+4t^2)^2(1+4t-8t^2)^2}\right) \tag{13.48}$$

$$= \frac{1}{(1-2t-2t^2)(1-2t+4t^2)}\,F_b\left(\frac{t^2(1-t)^2(1-4t)(1-2t)(1+2t)}{(1-2t-2t^2)^2(1-2t+4t^2)^2}\right) \tag{13.49}$$

$$= \frac{1}{(1-2t+4t^2)(1-8t+4t^2)}\,F_b\left(\frac{-t(1-t)(1-4t)(1-2t)^2(1+2t)^2}{(1-2t+4t^2)^2(1-8t+4t^2)^2}\right) \tag{13.50}$$

$$= \frac{1}{1+8t^2-32t^3+32t^4}\,F_c\left(\frac{t(1-t)(1-4t)^2(1-2t)(1+2t)}{(1+8t^2-32t^3+32t^4)^2}\right) \tag{13.51}$$

$$= \frac{1}{1-4t+4t^2+8t^4}\,F_c\left(\frac{t^2(1-t)^2(1-4t)(1-2t)(1+2t)}{(1-4t+4t^2+8t^4)^2}\right) \tag{13.52}$$

$$= \frac{1}{1-8t+32t^2-32t^3+16t^4}\,F_c\left(\frac{-t(1-t)(1-4t)(1-2t)^2(1+2t)^2}{(1-8t+32t^2-32t^3+16t^4)^2}\right) \tag{13.53}$$

*Level* 6, *functions* $G$:

$$= \frac{1}{(1+2t)(1-t)}\,G_a\left(\frac{t(1-4t)^2(1-2t)}{(1+2t)(1-t)}\right) \tag{13.54}$$

$$= \frac{1}{(1-2t)(1-t)^2}\,G_a\left(\frac{t^2(1+2t)(1-4t)}{(1-2t)(1-t)^2}\right) \tag{13.55}$$

$$= \frac{1}{(1-t)(1-4t)(1-2t)^2}\, G_a\left(\frac{-t(1+2t)^2}{(1-t)(1-4t)(1-2t)^2}\right) \tag{13.56}$$

$$= \frac{1}{(1-4t)^2}\, G_b\left(\frac{t(1+2t)(1-2t)(1-t)}{(1-4t)^2}\right) \tag{13.57}$$

$$= \frac{1}{(1-2t)(1+2t)(1-4t)}\, G_b\left(\frac{t^2(1-t)^2}{(1-2t)(1+2t)(1-4t)}\right) \tag{13.58}$$

$$= \frac{1}{(1-2t)^2(1+2t)^2}\, G_b\left(\frac{-t(1-t)(1-4t)}{(1-2t)^2(1+2t)^2}\right) \tag{13.59}$$

$$= \frac{1}{(1-t)(1+2t)(1-4t)^2}\, G_c\left(\frac{t(1-2t)}{(1-t)(1+2t)(1-4t)^2}\right) \tag{13.60}$$

$$= \frac{1}{(1+2t)(1-t)^2(1-4t)}\, G_c\left(\frac{t^2(1-2t)}{(1+2t)(1-t)^2(1-4t)}\right) \tag{13.61}$$

$$= \frac{1}{(1-t)(1-4t)(1+2t)^2}\, G_c\left(\frac{-t(1-2t)^2}{(1-t)(1-4t)(1+2t)^2}\right). \tag{13.62}$$

*Proof.* The proof is almost identical to the proof of Theorem 13.1, except that we use the level 12 parameterizations of the eta-function and the Eisenstein series $Q$ and $R$ given by Theorems 12.6, 12.9, and 12.10, respectively, and then put

$$t = \frac{h}{1+h^2}.$$

The idea is to substitute the relevant parameterizations into the results of Theorem 4.30 to obtain the identities (13.21)–(13.44) for levels $1 \le \ell \le 4$; use Theorem 6.39 to obtain (13.45)–(13.53); and use Theorem 6.38 to obtain (13.54)–(13.62).

□

Here is the modular parameterization that underlies each function in Theorem 13.3.

**Theorem 13.4.** *Let the common power series of each expression in* (13.21)–(13.62) *be denoted by*

$$\sum_{n=0}^{\infty} c(n)t^n = 1 + 4t + 16t^2 + 56t^3 + 200t^4 + 704t^5 + 2520t^6 + 9072t^7 + O\left(t^8\right).$$

*Then*

$$\frac{\eta^2(2\tau)\eta^{12}(3\tau)}{\eta^4(\tau)\eta^6(6\tau)} = \sum_{n=0}^{\infty} c(n) \left(\frac{\eta(\tau)\eta^3(12\tau)}{\eta^3(3\tau)\eta(4\tau)}\right)^n.$$

*Proof.* See Exercise 3.

□

The identities in Theorem 13.3 contain a huge number of special cases, and many identities can be simplified by making an appropriate change of variables. Some of the most interesting special cases are detailed in the following examples.

*Example 13.5.* The following identity holds in a neighborhood of $x = 0$:

$$\begin{aligned} {}_3F_2\left(\begin{matrix}\frac{1}{2}, \frac{1}{2}, \frac{1}{2} \\ 1, 1\end{matrix}; x\right) &= \frac{1}{\sqrt{1-x}}\, {}_3F_2\left(\begin{matrix}\frac{1}{4}, \frac{1}{2}, \frac{3}{4} \\ 1, 1\end{matrix}; \frac{-4x}{(1-x)^2}\right) \\ &= \frac{2}{\sqrt{4-x}}\, {}_3F_2\left(\begin{matrix}\frac{1}{6}, \frac{1}{2}, \frac{5}{6} \\ 1, 1\end{matrix}; \frac{27x^2}{(4-x)^3}\right) \\ &= \frac{1}{\sqrt{1-4x}}\, {}_3F_2\left(\begin{matrix}\frac{1}{6}, \frac{1}{2}, \frac{5}{6} \\ 1, 1\end{matrix}; \frac{-27x}{(1-4x)^3}\right). \end{aligned}$$

*Proof.* Let

$$x = 64\, \frac{t\,(1-t)^3\,(1-4\,t)^3}{(1-2\,t)^2\,(1+2\,t)^6}$$

in each of (13.39), (13.33), (13.22) and (13.27), respectively.

□

*Example 13.6.* The following identity holds in a neighborhood of $x = 0$:

$$\begin{aligned} {}_3F_2\left(\begin{matrix}\frac{1}{3}, \frac{1}{2}, \frac{2}{3} \\ 1, 1\end{matrix}; 4x(1-x)\right) &= \frac{1}{\sqrt{1+8x}}\, {}_3F_2\left(\begin{matrix}\frac{1}{6}, \frac{1}{2}, \frac{5}{6} \\ 1, 1\end{matrix}; \frac{64x(1-x)^3}{(1+8x)^3}\right) \\ &= \frac{3}{\sqrt{9-8x}}\, {}_3F_2\left(\begin{matrix}\frac{1}{6}, \frac{1}{2}, \frac{5}{6} \\ 1, 1\end{matrix}; \frac{64x^3(1-x)}{(9-8x)^3}\right). \end{aligned}$$

*Proof.* Let

$$x = \frac{27t^2(1-4t+4t^2)}{(1-2t+4t^2)^3}$$

in (13.36), (13.22), and (13.25), respectively.

□

*Example 13.7.* This example shows that Theorems 6.45 and 6.46, and hence (13.1) and (13.2), are special cases of Theorem 13.3:

$$G_b(x) = \frac{1}{1+16x}\,{}_3F_2\left(\begin{matrix}\frac{1}{3},\ \frac{1}{2},\ \frac{2}{3}\\ 1,\ 1\end{matrix};\frac{108x}{(1+16x)^3}\right)$$
$$= \frac{1}{1+4x}\,{}_3F_2\left(\begin{matrix}\frac{1}{3},\ \frac{1}{2},\ \frac{2}{3}\\ 1,\ 1\end{matrix};\frac{108x^2}{(1+4x)^3}\right)$$

and

$$G_c(x) = \frac{1}{1+27x}\,{}_3F_2\left(\begin{matrix}\frac{1}{4},\ \frac{1}{2},\ \frac{3}{4}\\ 1,\ 1\end{matrix};\frac{256x}{(1+27x)^4}\right)$$
$$= \frac{1}{1+3x}\,{}_3F_2\left(\begin{matrix}\frac{1}{4},\ \frac{1}{2},\ \frac{3}{4}\\ 1,\ 1\end{matrix};\frac{256x^3}{(1+3x)^4}\right).$$

*Proof.* The identities in the group that involves $G_b(x)$ are obtained by taking

$$x = \frac{t(1-t)(1+2t)(1-2t)}{(1-4t)^2}$$

in (13.35), (13.36), and (13.57). The identities in the other group may be proved by taking

$$x = \frac{t(1-2t)}{(1-t)(1+2t)(1-4t)^2}$$

in (13.29), (13.31), and (13.60).

□

The special cases in Examples 13.5–13.7 have been selected because of their use in the next chapter. Some more examples are listed in the exercises and further special cases are detailed in [131].

## 13.4 Notes

The material in this chapter is based on work with W. Zudilin [131]. Example 13.5 occurs in a paper by N. D. Baruah and B. C. Berndt [39] and in the book by J. M. Borwein and P. B. Borwein [60, pp. 180–181]. Example 13.6 occurs in work by J. Guillera and W. Zudilin [172, (20)]. Example 13.7 was studied by M. Rogers [257, Theorem 3.1] and by H. H. Chan and W. Zudilin [94, Theorems 3.2 and 4.2].

## 13.5 Exercises

1. Recall the function $f_5(x)$ was defined in Section 5.12 by

$$f_5(x) = \sum_{n=0}^{\infty} t_5(n)x^n = \sum_{n=0}^{\infty}\left\{\sum_{k=0}^{n}\binom{n}{k}^2\binom{n+k}{k}\right\}x^n.$$

a. Prove that

$$f_5\left(x\left(\frac{1-x}{1+x}\right)^2\right) = \frac{(1+x)^2}{\sqrt{d(x))}}\,{}_2F_1\left(\begin{matrix}\frac{1}{4}, \frac{3}{4} \\ 1\end{matrix}; \frac{64x(1+x-x^2)^5}{d(x)^2}\right)$$

where

$$d(x) = (1+x^2)\left((1+11x-x^2)^2 - 125x^2\right).$$

Hint: Take the square root of the appropriate functions in Theorem 13.1.

b. Prove that the following identity holds in a neighborhood of $x = 0$:

$$\frac{1}{\sqrt{1+22x-4x^2}}\,{}_2F_1\left(\begin{matrix}\frac{1}{4}, \frac{3}{4} \\ 1\end{matrix}; \frac{64x(1+x)^5}{(1+4x^2)(1+22x-4x^2)^2}\right)$$
$$= \frac{1}{\sqrt{1-2x-4x^2}}\,{}_2F_1\left(\begin{matrix}\frac{1}{4}, \frac{3}{4} \\ 1\end{matrix}; \frac{64x^5(1+x)}{(1+4x^2)(1-2x-4x^2)^2}\right).$$

c. Prove that the following quadratic transformation formula holds in neighborhoods of $x = 0$:

$$f_5\left(x\left(\frac{1-x}{1+x}\right)^2\right) = \frac{(1+x)^2}{1-x} f_5\left(x^2\left(\frac{1+x}{1-x}\right)\right). \tag{13.63}$$

A quintic transformation formula for $f_5$ was given in Theorem 5.46.

The results in the exercise above are originally due to M. N. Lalin and M. D. Rogers [202].

2. Show that the following twenty-three functions are equal, in a neighborhood of $t = 0$:

Level 2:

$$\frac{1}{(1+20\,t-48\,t^2+32\,t^3-32\,t^4)^{1/2}}\;{}_2F_1\left(\begin{matrix}\frac{1}{4},\ \frac{3}{4}\\ 1\end{matrix};\frac{64\,t\,(1-t)^3\,(1-2\,t)\,(1+2\,t)^3}{(1+20\,t-48\,t^2+32\,t^3-32\,t^4)^2}\right)$$

$$=\frac{1}{(1-4\,t+24\,t^2-40\,t^3-8\,t^4)^{1/2}}\;{}_2F_1\left(\begin{matrix}\frac{1}{4},\ \frac{3}{4}\\ 1\end{matrix};\frac{64\,t^2\,(1-t)^6}{(1-4\,t+24\,t^2-40\,t^3-8\,t^4)^2}\right)$$

$$=\frac{1}{(1-4\,t+32\,t^3-32\,t^4)^{1/2}}\;{}_2F_1\left(\begin{matrix}\frac{1}{4},\ \frac{3}{4}\\ 1\end{matrix};\frac{64\,t^3\,(1-t)\,(1-2\,t)^3\,(1+2\,t)}{(1-4\,t+32\,t^3-32\,t^4)^2}\right)$$

$$=\frac{1}{(1-4\,t+8\,t^3-8\,t^4)^{1/2}}\;{}_2F_1\left(\begin{matrix}\frac{1}{4},\ \frac{3}{4}\\ 1\end{matrix};\frac{64\,t^6\,(1-t)^2}{(1-4\,t+8\,t^3-8\,t^4)^2}\right)$$

$$=\frac{1}{(1-28\,t+96\,t^2-112\,t^3+16\,t^4)^{1/2}}\;{}_2F_1\left(\begin{matrix}\frac{1}{4},\ \frac{3}{4}\\ 1\end{matrix};\frac{-64\,t\,(1-t)^3\,(1-4\,t)^3}{(1-28\,t+96\,t^2-112\,t^3+16\,t^4)^2}\right)$$

$$=\frac{1}{(1-4\,t-16\,t^3+16\,t^4)^{1/2}}\;{}_2F_1\left(\begin{matrix}\frac{1}{4},\ \frac{3}{4}\\ 1\end{matrix};\frac{-64\,t^3\,(1-t)\,(1-4\,t)}{(1-4\,t-16\,t^3+16\,t^4)^2}\right)$$

Level 3:

$$=\frac{1}{(1+4t-8t^2)}\;{}_2F_1\left(\begin{matrix}\frac{1}{3},\ \frac{2}{3}\\ 1\end{matrix};\frac{27\,t\,(1-2t)}{(1+4t-8t^2)^3}\right)$$

$$=\frac{1}{(1-2\,t+4\,t^2)}\;{}_2F_1\left(\begin{matrix}\frac{1}{3},\ \frac{2}{3}\\ 1\end{matrix};\frac{27\,t^2\,(1-2t)^2}{(1-2\,t+4\,t^2)^3}\right)$$

$$=\frac{1}{(1-2\,t-2\,t^2)}\;{}_2F_1\left(\begin{matrix}\frac{1}{3},\ \frac{2}{3}\\ 1\end{matrix};\frac{27\,t^4\,(1-2t)}{(1-2\,t-2\,t^2)^3}\right)$$

$$=\frac{1}{(1-8\,t+4\,t^2)}\;{}_2F_1\left(\begin{matrix}\frac{1}{3},\ \frac{2}{3}\\ 1\end{matrix};\frac{-27\,t\,(1-2t)^4}{(1-8\,t+4\,t^2)^3}\right)$$

Level 4:

$$=\frac{1}{(1-2\,t)^{1/2}\,(1+2\,t)^{3/2}}\;{}_2F_1\left(\begin{matrix}\frac{1}{2},\ \frac{1}{2}\\ 1\end{matrix};\frac{16\,t\,(1-t)^3}{(1-2\,t)\,(1+2\,t)^3}\right)$$

$$= \frac{1}{(1-2t)^{3/2}(1+2t)^{1/2}}\, {}_2F_1\left(\begin{matrix}\frac{1}{2}, \frac{1}{2}\\ 1\end{matrix}; \frac{16t^3(1-t)}{(1-2t)^3(1+2t)}\right)$$
$$= \frac{1}{(1-4t)^{3/2}}\, {}_2F_1\left(\begin{matrix}\frac{1}{2}, \frac{1}{2}\\ 1\end{matrix}; \frac{-16t(1-t)^3}{(1-4t)^3}\right)$$
$$= \frac{1}{(1-4t)^{1/2}}\, {}_2F_1\left(\begin{matrix}\frac{1}{2}, \frac{1}{2}\\ 1\end{matrix}; \frac{-16t^3(1-t)}{(1-4t)}\right)$$

Level 6:

$$= \frac{1}{(1-4t)^2} f_a\left(\frac{t(1-2t)}{(1-4t)^2}\right)$$
$$= \frac{1}{(1-4t)(1+2t)} f_a\left(\frac{t^2}{(1-4t)(1+2t)}\right)$$
$$= \frac{1}{(1+2t)^2} f_a\left(\frac{-t}{(1+2t)^2}\right)$$
$$= \frac{1}{(1-t)(1+2t)} f_b\left(\frac{t(1-2t)}{(1-t)(1+2t)}\right)$$
$$= \frac{1}{(1-t)^2} f_b\left(\frac{t^2}{(1-t)^2}\right)$$
$$= \frac{1}{(1-t)(1-4t)} f_b\left(\frac{-t}{(1-t)(1-4t)}\right)$$
$$= f_c\left(t(1-2t)\right)$$
$$= \frac{1}{1-2t} f_c\left(\frac{t^2}{1-2t}\right)$$
$$= \frac{1}{(1-2t)^2} f_c\left(\frac{-t}{(1-2t)^2}\right),$$

where $f_a, f_b$, and $f_c$ are the functions defined in Section 6.11.

The corresponding identities for the level 1 function ${}_2F_1\left(\begin{matrix}\frac{1}{6}, \frac{5}{6}\\ 1\end{matrix}; x\right)$ are more complicated and do not involve rational functions of $t$, so they have not been included in the above list.

Hint: Take the square root of the appropriate functions in Theorem 13.3, and use Clausen's identity in the form (4.54) and its analogues in Theorem 5.43.

3. Prove Theorem 13.4. Hint: By (12.18) we have

$$\frac{h}{1+h^2} = \frac{\eta(\tau)\eta^3(12\tau)}{\eta^3(3\tau)\eta(4\tau)}.$$

Now follow the technique of the proof of Theorem 13.2.

4. Let the common power series expansion of each of the twenty-three expressions in Exercise 2 be denoted by

$$\sum_{n=0}^{\infty} a(n)t^n = 1 + 2t + 6t^2 + 16t^3 + 50t^4 + 156t^5 + 520t^6 + 1760t^7 + O\left(t^8\right).$$

a. Show that

$$\frac{\eta(2\tau)\eta^6(3\tau)}{\eta^2(\tau)\eta^3(6\tau)} = \sum_{n=0}^{\infty} a(n)\left(\frac{\eta(\tau)\eta^3(12\tau)}{\eta^3(3\tau)\eta(4\tau)}\right)^n.$$

b. Let the eta-quotients in Part a be denoted by

$$y = \frac{\eta(2\tau)\eta^6(3\tau)}{\eta^2(\tau)\eta^3(6\tau)} \quad \text{and} \quad x = \frac{\eta(\tau)\eta^3(12\tau)}{\eta^3(3\tau)\eta(4\tau)}.$$

Show that

$$\frac{\mathrm{d}}{\mathrm{d}x}\left(x(1-x)(1-4x)(1-2x)(1+2x)\frac{\mathrm{d}y}{\mathrm{d}x}\right) = 2(1-4x)(1+4x-8x^2)y.$$

c. Hence, deduce that the coefficients $a(n)$ in Part a satisfy the five-term recurrence relation

$$\begin{aligned}&(n+1)^2a(n+1)\\ &= (5n^2+5n+2)a(n) - 4(5n^2-5n+2)a(n-2) + 16(n-1)^2a(n-3),\end{aligned}$$

the coefficient of the term $a(n-1)$ being zero.

See [131].

5. Let

$$k = q\prod_{j=1}^{\infty}\frac{(1-q^{10j-9})(1-q^{10j-8})(1-q^{10j-2})(1-q^{10j-1})}{(1-q^{10j-7})(1-q^{10j-6})(1-q^{10j-4})(1-q^{10j-3})}.$$

Show that

$$q\frac{\mathrm{d}}{\mathrm{d}q}\log k$$

$$=\frac{(1+k-k^2)(1-k^2)^{1/2}}{(1-4k-k^2)^{3/2}}\;{}_3F_2\left(\begin{matrix}\frac{1}{2},\ \frac{1}{2},\ \frac{1}{2}\\ 1,\ 1\end{matrix};-64\left(\frac{k}{1-k^2}\right)\left(\frac{1+k-k^2}{1-4k-k^2}\right)^5\right)$$

$$=\frac{(1+k-k^2)(1-4k-k^2)^{1/2}}{(1-k^2)^{3/2}}\;{}_3F_2\left(\begin{matrix}\frac{1}{2},\ \frac{1}{2},\ \frac{1}{2}\\ 1,\ 1\end{matrix};-64\left(\frac{k}{1-k^2}\right)^5\left(\frac{1+k-k^2}{1-4k-k^2}\right)\right).$$

6. Let

$$h=q\prod_{j=1}^{\infty}\frac{(1-q^{12j-11})(1-q^{12j-1})}{(1-q^{12j-7})(1-q^{12j-5})}.$$

Show that

$$q\frac{\mathrm{d}}{\mathrm{d}q}\log h$$

$$=\frac{(1-h^2)(1-h+h^2)}{(1-4h+h^2)^2}\;{}_3F_2\left(\begin{matrix}\frac{1}{2},\ \frac{1}{2},\ \frac{1}{2}\\ 1,\ 1\end{matrix};-64\left(\frac{h(1-h)^2}{(1+h^2)^2}\right)\left(\frac{(1-h+h^2)(1+h)^2}{(1-4h+h^2)^2}\right)^3\right)$$

$$=\frac{(1-h^2)(1-h+h^2)}{(1+h^2)^2}\;{}_3F_2\left(\begin{matrix}\frac{1}{2},\ \frac{1}{2},\ \frac{1}{2}\\ 1,\ 1\end{matrix};-64\left(\frac{h(1-h)^2}{(1+h^2)^2}\right)^3\left(\frac{(1-h+h^2)(1+h)^2}{(1-4h+h^2)^2}\right)\right).$$

7. Show that the following identities are special cases of Theorem 13.3:

$$\frac{1}{1+x}\,G_a\left(\frac{x(1-8x)}{1+x}\right)=\frac{1}{1-8x}\,G_b\left(\frac{x(1+x)}{1-8x}\right),$$

$$\frac{1}{1-x}\,G_a\left(\frac{x(1-9x)}{1-x}\right)=\frac{1}{1-9x}\,G_c\left(\frac{x(1-x)}{1-9x}\right)$$

and

$$\frac{1}{1+8x}\,G_b\left(\frac{x(1+9x)}{1+8x}\right)=\frac{1}{1+9x}\,G_c\left(\frac{x(1+8x)}{1+9x}\right).$$

Compare with Exercise 11 in Chapter 6.

# Chapter 14
# Ramanujan's Series for $1/\pi$

## 14.1 Introduction

In this chapter we study some remarkable series for $1/\pi$ that were first systematically investigated by Ramanujan in his paper [248]. Ramanujan's paper contains 17 series for $1/\pi$, including

$$\sum_{n=0}^{\infty} \binom{2n}{n}^3 \left(n + \frac{5}{42}\right) \left(\frac{1}{4096}\right)^n = \frac{8}{21} \times \frac{1}{\pi} \tag{14.1}$$

and

$$\sum_{n=0}^{\infty} \binom{4n}{2n} \binom{2n}{n}^2 \left(n + \frac{1103}{26390}\right) \left(\frac{1}{396}\right)^{4n} = \frac{9801\sqrt{2}}{105560} \times \frac{1}{\pi}. \tag{14.2}$$

The ratio test can be used to analyze the convergence, and the second series converges very rapidly: each additional term adds an extra 8 decimal places of accuracy. The convergence is fast enough to have been used by R. W. Gosper in 1985 to calculate 17,526,200 decimal digits of $\pi$, a world record at the time.

We will describe a variety of techniques that can be used to prove series such (14.1) and (14.2). Although a detailed proof of (14.1) will be given, a full proof of (14.2) is difficult and we will not give one. The classification in Section 14.7 and the discussion that follows will give insight into why it is more difficult to prove (14.2) than it is to prove (14.1).

At the end of the chapter, we derive an iteration that converges quadratically to $1/\pi$ and outline a method that can be used to produce other iterations.



S. Cooper, *Ramanujan's Theta Functions*, DOI 10.1007/978-3-319-56172-1_15

## 14.2 Theta function methods

In this section we use the theory of theta functions to give complete proofs of the two series:

$$\sum_{n=0}^{\infty} \binom{2n}{n}^3 \left(n+\frac{1}{6}\right)\left(\frac{1}{256}\right)^n = \frac{2}{3} \times \frac{1}{\pi} \tag{14.3}$$

and

$$\sum_{n=0}^{\infty} \binom{2n}{n}^3 \left(n+\frac{5}{42}\right)\left(\frac{1}{4096}\right)^n = \frac{8}{21} \times \frac{1}{\pi}. \tag{14.4}$$

The proofs will rely on four lemmas.

**Lemma 14.1.** *Let $X$ be defined by*

$$X = X(q) = q\prod_{j=1}^{\infty} \frac{(1-q^j)^{24}(1-q^{4j})^{24}}{(1-q^{2j})^{48}}$$

*and let $P$ and $\varphi$ be the Eisenstein series and theta function defined by* (0.61) *and Definition 3.1, respectively, that is,*

$$P(q) = 1 - 24\sum_{j=1}^{\infty} \frac{jq^j}{1-q^j} \qquad \textit{and} \qquad \varphi(q) = \sum_{j=-\infty}^{\infty} q^{j^2}.$$

*The following identities hold in a neighborhood of $q=0$:*

$$\varphi^4(q) = \sum_{n=0}^{\infty} \binom{2n}{n}^3 X^n \tag{14.5}$$

*and*

$$P(q^2) = \sqrt{1-64X}\sum_{n=0}^{\infty} \binom{2n}{n}^3 (3n+1)\, X^n. \tag{14.6}$$

*Proof.* In Theorem 4.30 it was shown that

$$Z_4 = \left(\sum_{j=-\infty}^{\infty} q^{j^2}\right)^4 = \varphi^4(q)$$

and

$$Z_4 = \sum_{n=0}^{\infty} \binom{2n}{n}^3 X_4^n. \tag{14.7}$$

The identity (14.5) is obtained by equating the two expressions for $Z_4$ and noting that $X_4 = X$ by (4.60).

Now we will prove (14.6). By Lemma 4.9,

$$P(q^2) = (1 - 2x_4)\, z_4^2 + \frac{6}{z_4} q \frac{\mathrm{d}z_4}{\mathrm{d}q}$$

where $x_4$ and $z_4$ are given by (4.6) and (4.7). By Theorem 4.7 and the chain rule this may be expressed as

$$\begin{aligned} P(q^2) &= (1 - 2x_4)\, z_4^2 + 6x_4(1 - x_4)\, z_4 \frac{\mathrm{d}z_4}{\mathrm{d}x_4} \\ &= (1 - 2x_4)\, z_4^2 + 3x_4(1 - x_4) \frac{\mathrm{d}}{\mathrm{d}x_4}(z_4^2). \end{aligned}$$

Applying the change of variables given by (4.57) and (4.59), viz.,

$$Z_4 = z_4^2 \qquad \text{and} \qquad X_4 = \frac{x_4(1 - x_4)}{16},$$

we obtain

$$P(q^2) = \sqrt{1 - 64X_4} \left( Z_4 + 3X_4 \frac{\mathrm{d}Z_4}{\mathrm{d}X_4} \right).$$

The proof may be completed by expanding $Z_4$ and its derivative in powers of $X_4$ using (14.7), and recalling from the above that $X_4 = X$.

□

We will now evaluate each of $X(q)$, $\varphi(q)$, and $P(q^2)$ in the special cases $q = \exp(-\pi\sqrt{3})$ and $q = \exp(-\pi\sqrt{7})$. We begin with the evaluations of $X$.

**Lemma 14.2.** *Let*

$$X = X(q) = q \prod_{j=1}^{\infty} \frac{(1 - q^j)^{24}(1 - q^{4j})^{24}}{(1 - q^{2j})^{48}}.$$

*Then*

$$X\left(e^{-\pi\sqrt{3}}\right) = \frac{1}{256} \qquad \textit{and} \qquad X\left(e^{-\pi\sqrt{7}}\right) = \frac{1}{4096}.$$

*Proof.* Start with the modular equation of degree 3 (the cubic analogue of Jacobi's 'aequatio identica satis abstrusa') (3.57) in the form

$$\varphi(-q)\varphi(-q^3) + 4q\psi(q^2)\psi(q^6) = \varphi(q)\varphi(q^3).$$

Divide both sides by $\varphi(q)\varphi(q^3)$ and use Theorem 3.5 to express each theta function as an infinite product. This leads to

$$\prod_{j=1}^{\infty} \frac{(1-q^j)^4(1-q^{3j})^4(1-q^{4j})^2(1-q^{12j})^2}{(1-q^{2j})^6(1-q^{6j})^6}$$
$$+ 4q \prod_{j=1}^{\infty} \frac{(1-q^j)^2(1-q^{3j})^2(1-q^{4j})^4(1-q^{12j})^4}{(1-q^{2j})^6(1-q^{6j})^6} = 1,$$

or equivalently,

$$\frac{\eta^4(\tau)\eta^4(3\tau)\eta^2(4\tau)\eta^2(12\tau)}{\eta^6(2\tau)\eta^6(6\tau)} + 4\frac{\eta^2(\tau)\eta^2(3\tau)\eta^4(4\tau)\eta^4(12\tau)}{\eta^6(2\tau)\eta^6(6\tau)} = 1 \tag{14.8}$$

where $\eta(\tau)$ is the eta-function defined by (2.20) and $q = \exp(2\pi i\tau)$. Now let $\tau = i/\sqrt{12}$ and use Theorem 2.8 to write

$$\eta\left(\frac{i}{\sqrt{12}}\right) = \sqrt[4]{12}\,\eta(i\sqrt{12}), \qquad \eta\left(\frac{i}{\sqrt{3}}\right) = \sqrt[4]{3}\,\eta(i\sqrt{3}) \tag{14.9}$$

and

$$\eta\left(\frac{2i}{\sqrt{3}}\right) = \sqrt[4]{\frac{3}{4}}\,\eta\left(\frac{i\sqrt{3}}{2}\right). \tag{14.10}$$

Then (14.8) simplifies when $\tau = i/\sqrt{12}$ to give

$$4 \times \frac{\eta^6(i\sqrt{3}/2)\,\eta^6(i\sqrt{12})}{\eta^{12}(i\sqrt{3})} = 1.$$

On taking fourth powers and converting the eta-functions back to infinite products we deduce that $X(e^{-\pi\sqrt{3}}) = 1/256$, and this proves the first result.

The second result can be proved by the same procedure, starting with the modular equation of degree 7 (Guetzlaff's identity) from Exercise 4 in Chapter 7, in the form

$$\sqrt{\varphi(-q)\varphi(-q^7)} + 2q\sqrt{\psi(q^2)\psi(q^{14})} = \sqrt{\varphi(q)\varphi(q^7)}$$

and taking $\tau = i/\sqrt{28}$. We omit the details as they are similar to the above. □

The next result contains some technical details that will be required.

**Lemma 14.3.** *The following evaluations hold*:

$$\left.\frac{\varphi^2(-q)\varphi^2(-q^3)}{\varphi^4(q^3)}\right|_{q=e^{-\pi/\sqrt{3}}} = \frac{\sqrt{3}}{4}, \qquad \left.\frac{q\psi^2(q)\psi^2(q^3)}{\varphi^4(q^3)}\right|_{q=e^{-\pi/\sqrt{3}}} = \frac{\sqrt{3}}{8},$$

$$\left.\frac{\varphi(-q)\varphi(-q^7)}{\varphi^2(q^7)}\right|_{q=e^{-\pi/\sqrt{7}}} = \frac{\sqrt[4]{7}}{4} \quad \textit{and} \quad \left.\frac{q\psi(q)\psi(q^7)}{\varphi^2(q^7)}\right|_{q=e^{-\pi/\sqrt{7}}} = \frac{\sqrt[4]{7}}{4}.$$

*Proof.* The theta functions can be expressed as infinite products by Theorem 3.5 and then expressed in terms of the eta-function to give

$$\left.\frac{\varphi^2(-q)\varphi^2(-q^3)}{\varphi^4(q^3)}\right|_{q=e^{-\pi/\sqrt{3}}} = \frac{\eta^4(i/\sqrt{12})\eta^{12}(i\sqrt{3}/2)\eta^8(i\sqrt{12})}{\eta^2(i/\sqrt{3})\eta^{22}(i\sqrt{3})}.$$

Now use (14.9), (14.10), and Lemma 14.2 to obtain

$$\begin{aligned}\left.\frac{\varphi^2(-q)\varphi^2(-q^3)}{\varphi^4(q^3)}\right|_{q=e^{-\pi/\sqrt{3}}} &= 4\sqrt{3}\,\frac{\eta^{12}(i/\sqrt{12})\eta^{12}(i\sqrt{3}/2)}{\eta^{24}(i\sqrt{3})} \\ &= 4\sqrt{3}\,\left(X(e^{-\pi\sqrt{3}})\right)^{1/2} \\ &= \frac{\sqrt{3}}{4}.\end{aligned}$$

This proves the first result. The other results can be proved by the same procedure. □

**Lemma 14.4.** *The following identities hold.*

$$P(e^{-2\pi\sqrt{3}}) = \frac{\sqrt{3}}{\pi} + \frac{\sqrt{3}}{4}\varphi^4(e^{-\pi\sqrt{3}})$$

*and*

$$P(e^{-2\pi\sqrt{7}}) - \frac{3}{\pi\sqrt{7}} + \frac{27}{16\sqrt{7}}\varphi^4(e^{-\pi\sqrt{7}}).$$

*Proof.* By Theorems 3.28 and 3.31 respectively,

$$a^2(q) = \varphi^2(-q^{1/2})\varphi^2(-q^{3/2}) + 4q^{1/2}\psi^2(q^{1/2})\psi^2(q^{3/2})$$

and

$$a^2(q) = \frac{1}{2}\left(3P(q^3) - P(q)\right),$$

where $a(q)$ is the cubic theta function. It follows that

$$\frac{1}{2}\left(3P(q^3) - P(q)\right) = \varphi^2(-q^{1/2})\varphi^2(-q^{3/2}) + 4q^{1/2}\psi^2(q^{1/2})\psi^2(q^{3/2}). \tag{14.11}$$

We evaluate each side for $q = e^{-2\pi/\sqrt{3}}$. Using (2.19) with $\tau = i\sqrt{3}$, we deduce that

$$\frac{1}{2}\left(3P(e^{-2\pi\sqrt{3}}) - P(e^{-2\pi/\sqrt{3}})\right) = 3P(e^{-2\pi\sqrt{3}}) - \frac{3\sqrt{3}}{\pi}. \tag{14.12}$$

On the other hand, Lemma 14.3 implies

$$\varphi^2(-e^{\pi/\sqrt{3}})\varphi^2(-e^{-\pi\sqrt{3}}) + 4e^{-\pi/\sqrt{3}}\psi^2(e^{-\pi/\sqrt{3}})\psi^2(e^{-\pi\sqrt{3}}) = \frac{3\sqrt{3}}{4}\,\varphi^4(e^{-\pi\sqrt{3}}). \tag{14.13}$$

Using (14.12) and (14.13) in (14.11) we obtain

$$3P(e^{-2\pi\sqrt{3}}) - \frac{3\sqrt{3}}{\pi} = \frac{3\sqrt{3}}{4}\,\varphi^4(e^{-\pi\sqrt{3}})$$

and this proves the first result.

The second result can be proved by a similar procedure, starting with

$$z_7(q) = \varphi(-q^{1/2})\varphi(-q^{7/2}) + 2q^{1/2}\psi(q^{1/2})\psi(q^{7/2})$$

and

$$z_7^2(q) = \frac{1}{6}\left(7P(q^7) - P(q)\right)$$

from Theorems 7.17 and 7.20, respectively, and setting $q = e^{-2\pi/\sqrt{7}}$. We omit the details as they are similar to those already given above.

□

We are now ready to prove the series identities (14.3) and (14.4).

*Proof.* By Lemma 14.4,

$$P(e^{-2\pi\sqrt{3}}) = \frac{\sqrt{3}}{\pi} + \frac{\sqrt{3}}{4}\varphi^4(e^{-\pi\sqrt{3}}).$$

Lemma 14.1 can be used to expand each side in powers of $X$. The result is

$$\sqrt{1-64X}\sum_{n=0}^{\infty}\binom{2n}{n}^3 (3n+1)\,X^n\Bigg|_{q=e^{-\pi\sqrt{3}}}$$

$$= \frac{\sqrt{3}}{\pi} + \frac{\sqrt{3}}{4} \sum_{n=0}^{\infty} \binom{2n}{n}^3 X^n \Bigg|_{q=e^{-\pi\sqrt{3}}} .$$

Using the value $X(e^{-\pi\sqrt{3}}) = 1/256$ from Lemma 14.2, we deduce that

$$\frac{\sqrt{3}}{2} \sum_{n=0}^{\infty} \binom{2n}{n}^3 (3n+1) \left(\frac{1}{256}\right)^n = \frac{\sqrt{3}}{\pi} + \frac{\sqrt{3}}{4} \sum_{n=0}^{\infty} \binom{2n}{n}^3 \left(\frac{1}{256}\right)^n ,$$

which simplifies to (14.3).

The identity (14.4) can be proved in the same way. We omit the details as they are similar.

□

## 14.3 Translation methods, I: hypergeometric transformations

Since

$$\sum_{n=0}^{\infty} \binom{2n}{n}^3 X^n = {}_3F_2\left(\begin{matrix} \frac{1}{2}, \frac{1}{2}, \frac{1}{2} \\ 1, 1 \end{matrix}; 64X\right),$$

the series identities (14.3) and (14.4) can be expressed in terms of the ${}_3F_2$ hypergeometric function by the formulas

$$\left(\frac{1}{6} + X\frac{\mathrm{d}}{\mathrm{d}X}\right) {}_3F_2\left(\begin{matrix} \frac{1}{2}, \frac{1}{2}, \frac{1}{2} \\ 1, 1 \end{matrix}; X\right)\Bigg|_{X=1/4} = \frac{2}{3} \times \frac{1}{\pi} \tag{14.14}$$

and

$$\left(\frac{5}{42} + X\frac{\mathrm{d}}{\mathrm{d}X}\right) {}_3F_2\left(\begin{matrix} \frac{1}{2}, \frac{1}{2}, \frac{1}{2} \\ 1, 1 \end{matrix}; X\right)\Bigg|_{X=1/64} = \frac{8}{21} \times \frac{1}{\pi}, \tag{14.15}$$

respectively. Here we are using the notation

$$\left(a + X\frac{\mathrm{d}}{\mathrm{d}X}\right) f(X) = af(X) + X\frac{\mathrm{d}}{\mathrm{d}X} f(X).$$

Let us consider whether the transformation formulas in the previous chapter may be used to convert (14.14) and (14.15) into other series. To achieve this, appropriate values of $t$ will have to be found for the arguments in the hypergeometric functions in (13.39)–(13.44). These values will be algebraic numbers and are potentially complicated to write down. Remarkably, several instances exist for which the values can be computed explicitly and used to obtain transformation formulas with rational coefficients.

The procedure of using a transformation formula, such as one from Theorem 13.3, to obtain a series for $1/\pi$ will be called a *translation method.*

It will be convenient to translate the series (14.15) first. The other series (14.14) will be translated later, in Theorem 14.6.

**Theorem 14.5.** *The following identities hold*:

$$\sum_{n=0}^{\infty} \binom{6n}{3n}\binom{3n}{n}\binom{2n}{n}\left(n+\frac{8}{133}\right)\left(\frac{1}{255^3}\right)^n = \frac{85\sqrt{255}}{7182} \times \frac{1}{\pi}, \tag{14.16}$$

$$\sum_{n=0}^{\infty} \binom{6n}{3n}\binom{3n}{n}\binom{2n}{n}\left(n+\frac{8}{63}\right)\left(\frac{-1}{15^3}\right)^n = \frac{5\sqrt{15}}{63} \times \frac{1}{\pi} \tag{14.17}$$

*and*

$$\sum_{n=0}^{\infty} \binom{4n}{2n}\binom{2n}{n}^2\left(n+\frac{8}{65}\right)\left(\frac{-1}{63^2}\right)^n = \frac{9\sqrt{7}}{65} \times \frac{1}{\pi}. \tag{14.18}$$

*Proof.* Consider the transformation formula from Example 13.5 given by

$$ {}_3F_2\left(\begin{matrix}\frac{1}{2}, \frac{1}{2}, \frac{1}{2} \\ 1, 1\end{matrix}; x\right) = \frac{2}{\sqrt{4-x}}\, {}_3F_2\left(\begin{matrix}\frac{1}{6}, \frac{1}{2}, \frac{5}{6} \\ 1, 1\end{matrix}; \frac{27x^2}{(4-x)^3}\right). \tag{14.19}$$

Applying the operator $x\frac{\mathrm{d}}{\mathrm{d}x}$, we find that

$$\sum_{n=0}^{\infty} \frac{\left(\frac{1}{2}\right)_n^3}{n!^3}\, n\, x^n$$

$$= \frac{1}{(4-x)^{3/2}} \sum_{n=0}^{\infty} \frac{\left(\frac{1}{6}\right)_n\left(\frac{1}{2}\right)_n\left(\frac{5}{6}\right)_n}{n!^3}\,(2(x+8)n+x)\left(\frac{27x^2}{(4-x)^3}\right)^n. \tag{14.20}$$

Multiplying (14.19) by $\lambda$ and adding to (14.20) we obtain

$$\sum_{n=0}^{\infty} \frac{\left(\frac{1}{2}\right)_n^3}{n!^3}\,(n+\lambda)\, x^n = \sum_{n=0}^{\infty} \frac{\left(\frac{1}{6}\right)_n\left(\frac{1}{2}\right)_n\left(\frac{5}{6}\right)_n}{n!^3}\,(An+B)\left(\frac{27x^2}{(4-x)^3}\right)^n \tag{14.21}$$

where

$$A = \frac{2(x+8)}{(4-x)^{3/2}} \qquad \text{and} \qquad B = \frac{8\lambda+(1-2\lambda)x}{(4-x)^{3/2}}.$$

Taking $x = 1/64$, $\lambda = 5/42$, we find that

$$\sum_{n=0}^{\infty} \frac{(\frac{1}{2})_n^3}{n!^3} \left(n + \frac{5}{42}\right) \frac{1}{64^n}$$
$$= \frac{912}{7225}\sqrt{255} \sum_{n=0}^{\infty} \frac{(\frac{1}{6})_n(\frac{1}{2})_n(\frac{5}{6})_n}{n!^3} \left(n + \frac{8}{133}\right)\left(\frac{4}{85}\right)^{3n}.$$

The hypergeometric functions may be expressed in terms of sums of binomial coefficients to give

$$\sum_{n=0}^{\infty} \binom{2n}{n}^3 \left(n + \frac{5}{42}\right)\left(\frac{1}{4096}\right)^n$$
$$= \frac{912}{7225}\sqrt{255} \sum_{n=0}^{\infty} \binom{6n}{3n}\binom{3n}{n}\binom{2n}{n}\left(n + \frac{8}{133}\right)\left(\frac{1}{255^3}\right)^n.$$

The sum of the series on the left is $8/(21\pi)$ by (14.4). Hence, we obtain (14.16).

The series (14.17) and (14.18) can be deduced from (14.4) in the same way by applying to operator $(\lambda + x\frac{\mathrm{d}}{\mathrm{d}x})$ to the other transformation formulas in Example 13.5 to obtain the identities

$$\sum_{n=0}^{\infty} \frac{(\frac{1}{2})_n^3}{n!^3} (n + \lambda)\, x^n = \sum_{n=0}^{\infty} \frac{(\frac{1}{6})_n(\frac{1}{2})_n(\frac{5}{6})_n}{n!^3} (Cn + D) \left(\frac{-27x}{(1-4x)^3}\right)^n \qquad (14.22)$$

and

$$\sum_{n=0}^{\infty} \frac{(\frac{1}{2})_n^3}{n!^3} (n + \lambda)\, x^n = \sum_{n=0}^{\infty} \frac{(\frac{1}{4})_n(\frac{1}{2})_n(\frac{3}{4})_n}{n!^3} (En + F) \left(\frac{-4x}{(1-x)^2}\right)^n, \qquad (14.23)$$

where

$$C = \frac{1+8x}{(1-4x)^{3/2}}, \qquad D = \frac{\lambda + (2-4\lambda)x}{(1-4x)^{3/2}},$$
$$E = \frac{1+x}{(1-x)^{3/2}} \quad \text{and} \quad F = \frac{1}{(1-x)^{3/2}}\left((1-x)\lambda + \frac{x}{2}\right).$$

Now take $x = 1/64$, $\lambda = 5/42$ in (14.22) and (14.23) to deduce (14.17) and (14.18), respectively.

□

Let us consider what happens when we attempt to translate (14.14) by (14.21), (14.22) and (14.23). Consider the transformations one at a time, starting with (14.21).

**Theorem 14.6.** *The following identity holds:*

$$\sum_{n=0}^{\infty} \binom{6n}{3n}\binom{3n}{n}\binom{2n}{n}\left(n+\frac{1}{11}\right)\left(\frac{1}{54000}\right)^n = \frac{5\sqrt{15}}{66} \times \frac{1}{\pi}. \tag{14.24}$$

*Proof.* Substitute the data $x = 1/4$ and $\lambda = 1/6$ into the transformation formula (14.21), and make use of (14.14) to evaluate the series on the left-hand side.

□

Next, consider the transformation formula (14.22). It contains the term $(1 - 4x)$ several times in the denominators, so direct substitution of the value $x = 1/4$ is not possible. We will see in the next section how the value $x = 1/4$ can be used in (14.22) via a limiting procedure. This will give rise to a different type of translation technique.

Lastly, on substituting the data $x = 1/4$ and $\lambda = 1/6$ into (14.23) we find that

$$\sum_{n=0}^{\infty} \binom{4n}{2n}\binom{2n}{n}^2\left(n+\frac{1}{5}\right)\left(\frac{-1}{144}\right)^n \text{ “=” } \frac{\sqrt{3}}{5} \times \frac{1}{\pi}. \tag{14.25}$$

The symbol "=" has been used because the series is divergent. Divergent hypergeometric series can be interpreted in terms of analytic continuation of the ${}_3F_2$ function; we will not investigate this topic further in this book.

## 14.4 Translation methods, II: the limit at $-\infty$

The topic of this section is motivated by our attempts in the previous section to let $x = 1/4$ in the transformation formula (14.22) that is a consequence of the identity

$${}_3F_2\left(\begin{matrix}\frac{1}{2},\ \frac{1}{2},\ \frac{1}{2}\\ 1,\ 1\end{matrix};x\right) = \frac{1}{\sqrt{1-4x}}\,{}_3F_2\left(\begin{matrix}\frac{1}{6},\ \frac{1}{2},\ \frac{5}{6}\\ 1,\ 1\end{matrix};\frac{-27x}{(1-4x)^3}\right). \tag{14.26}$$

Direct substitution is not possible because some of the terms become infinite. The next lemma gives a way of handling this situation.

**Lemma 14.7.** *Suppose $0 < a < b \le c$ and $d, e > 0$. Then*

$$\lim_{x\to-\infty} (-x)^a\,{}_3F_2\left(\begin{matrix}a,\ b,\ c\\ d,\ e\end{matrix};x\right) = \frac{\Gamma(d)\Gamma(e)\Gamma(b-a)\Gamma(c-a)}{\Gamma(b)\Gamma(c)\Gamma(d-a)\Gamma(e-a)}.$$

*Proof.* A proof is outside the scope of this book. The reader is referred to work by Guillera and Zudilin [172, Theorem 2].

□

We are now ready to consider the effect of taking $x = 1/4$ in (14.26). Although we are mainly interested in the identity (14.28), we also list the auxiliary identity (14.27) as it can be obtained at almost no extra effort and is also of some interest.

**Theorem 14.8.** *The following identities hold*:

$$\sum_{n=0}^{\infty} \binom{2n}{n}^3 \left(\frac{1}{256}\right)^n = \frac{2^{4/3}}{3} \times \frac{\sqrt{\pi}}{\Gamma^3(\frac{5}{6})} \tag{14.27}$$

*and*

$$\sum_{n=0}^{\infty} \binom{2n}{n}^3 \left(n + \frac{1}{6}\right) \left(\frac{1}{256}\right)^n = \frac{2}{3} \times \frac{1}{\pi}. \tag{14.28}$$

*Proof.* Let $x \to (1/4)^-$ in (14.26) to obtain

$$\begin{aligned}
&{}_3F_2\left(\begin{matrix}\frac{1}{2},\ \frac{1}{2},\ \frac{1}{2} \\ 1,\ 1\end{matrix};\frac{1}{4}\right) \\
&= \lim_{x\to(1/4)^-} \frac{1}{\sqrt{1-4x}}\, {}_3F_2\left(\begin{matrix}\frac{1}{6},\ \frac{1}{2},\ \frac{5}{6} \\ 1,\ 1\end{matrix};\frac{-27x}{(1-4x)^3}\right) \\
&= \lim_{x\to(1/4)^-} \frac{1}{(27x)^{1/6}} \left(\frac{27x}{(1-4x)^3}\right)^{1/6} {}_3F_2\left(\begin{matrix}\frac{1}{6},\ \frac{1}{2},\ \frac{5}{6} \\ 1,\ 1\end{matrix};\frac{-27x}{(1-4x)^3}\right) \\
&= \lim_{x\to(1/4)^-} \frac{1}{(27x)^{1/6}} \times \lim_{z\to-\infty} (-z)^{1/6}\, {}_3F_2\left(\begin{matrix}\frac{1}{6},\ \frac{1}{2},\ \frac{5}{6} \\ 1,\ 1\end{matrix};z\right)
\end{aligned}$$

where we have put $z = -27x/(1-4x)^3$ in the last step. The left-hand side can be converted to a binomial sum, and the limit on the right-hand side can be evaluated by Lemma 14.7. The result is

$$\sum_{n=0}^{\infty} \binom{2n}{n}^3 \left(\frac{1}{256}\right)^n = \left(\frac{4}{27}\right)^{1/6} \frac{\Gamma(1)\,\Gamma(1)\,\Gamma(\frac{1}{3})\,\Gamma(\frac{2}{3})}{\Gamma(\frac{1}{2})\,\Gamma(\frac{5}{6})\,\Gamma(\frac{5}{6})\,\Gamma(\frac{5}{6})}.$$

Now use the well-known properties

$$\Gamma(1) = 1, \quad \Gamma\left(\frac{1}{2}\right) = \sqrt{\pi} \quad \text{and} \quad \Gamma(s)\Gamma(1-s) = \frac{\pi}{\sin \pi s}$$

to obtain (14.27).

To prove (14.28), apply the operator $\frac{1}{6} + x\frac{\mathrm{d}}{\mathrm{d}x}$ to the identity (14.26), or equivalently take $\lambda = \frac{1}{6}$ in (14.22), to obtain

$$\sum_{n=0}^{\infty} \frac{(\frac{1}{2})_n^3}{n!^3} \left(n + \frac{1}{6}\right) x^n$$
$$= \frac{1+8x}{(1-4x)^{3/2}} \sum_{n=0}^{\infty} \frac{(\frac{1}{6})_n(\frac{1}{2})_n(\frac{5}{6})_n}{n!^3} \left(n + \frac{1}{6}\right) \left(\frac{-27x}{(1-4x)^3}\right)^n.$$

Rearranging the factors in front of the sum on the right-hand side and noting that

$$\left(\frac{1}{6}\right)_n \left(n + \frac{1}{6}\right) = \frac{1}{6} \times \left(\frac{7}{6}\right)_n,$$

we find that

$$\sum_{n=0}^{\infty} \frac{(\frac{1}{2})_n^3}{n!^3} \left(n + \frac{1}{6}\right) x^n$$
$$= \frac{1}{6} \times \frac{1+8x}{\sqrt{27x}} \times \left(\frac{27x}{(1-4x)^3}\right)^{1/2} \times {}_3F_2\left(\begin{matrix} \frac{1}{2}, \frac{5}{6}, \frac{7}{6} \\ 1, 1 \end{matrix}; \frac{-27x}{(1-4x)^3}\right).$$

Now let $x \to (1/4)^-$ and use Lemma 14.7 to evaluate the limit on the right-hand side, to obtain

$$\sum_{n=0}^{\infty} \frac{(\frac{1}{2})_n^3}{n!^3} \left(n + \frac{1}{6}\right) \frac{1}{4^n} = \frac{1}{\sqrt{27}} \frac{\Gamma(1)\,\Gamma(1)\,\Gamma(\frac{1}{3})\,\Gamma(\frac{2}{3})}{\Gamma(\frac{5}{6})\,\Gamma(\frac{7}{6})\,\Gamma(\frac{1}{2})\,\Gamma(\frac{1}{2})}.$$

The proof is completed by rewriting the series on the left as a binomial sum and simplifying the gamma functions. □

The identity (14.28) is identical to (14.3) which was proved in Section 14.2 by a different method. The proof in Theorem 14.8 is short and requires no knowledge of theta functions. To show that the technique is not limited to the examples in Theorem 14.8, we provide two further examples. Only sketches of the proofs are indicated, because the ideas are similar to the proof of Theorem 14.8.

**Theorem 14.9.** *The following series identities hold*:

$$\sum_{n=0}^{\infty} \binom{3n}{n} \binom{2n}{n}^2 \left(\frac{-1}{192}\right)^n = \frac{4}{3\sqrt{3}} \times \frac{\sqrt{\pi}}{\Gamma(\frac{5}{6})^3} \tag{14.29}$$

*and*

$$\sum_{n=0}^{\infty} \binom{3n}{n} \binom{2n}{n}^2 \left(n + \frac{1}{5}\right) \left(\frac{-1}{192}\right)^n = \frac{4}{5\sqrt{3}} \times \frac{1}{\pi}. \tag{14.30}$$

*Proof.* Consider the transformation formula from Example 13.6 given by

$$\sqrt{1+8x}\ {}_3F_2\left(\begin{matrix}\frac{1}{3}, \frac{1}{2}, \frac{2}{3} \\ 1, \ 1\end{matrix}; X\right) = {}_3F_2\left(\begin{matrix}\frac{1}{6}, \frac{1}{2}, \frac{5}{6} \\ 1, \ 1\end{matrix}; Y\right) \tag{14.31}$$

where

$$X = 4x(1-x) \quad \text{and} \quad Y = \frac{64x(1-x)^3}{(1+8x)^3}.$$

The identity (14.29) may be obtained by dividing (14.31) by $\sqrt{1+8x}$ and letting $x \to (-1/8)^+$.

In order to prove (14.30), apply the differential operator

$$\left(1 + 6Y\frac{\mathrm{d}}{\mathrm{d}Y}\right)$$

to (14.31), multiply the result by $\sqrt{1-Y}$ and let $x \to (-1/8)^+$.

□

**Theorem 14.10.** *The following series identities hold*:

$$\sum_{n=0}^{\infty} \binom{6n}{3n} \binom{3n}{n} \binom{2n}{n} \left(\frac{-1}{12288000}\right)^n = \frac{4\sqrt{30}}{27} \times \frac{\sqrt{\pi}}{\Gamma(\frac{5}{6})^3}$$

*and*

$$\sum_{n=0}^{\infty} \binom{6n}{3n} \binom{3n}{n} \binom{2n}{n} \left(n + \frac{31}{506}\right) \left(\frac{-1}{12288000}\right)^n = \frac{80\sqrt{30}}{2277} \times \frac{1}{\pi}. \tag{14.32}$$

*Proof.* See Exercise 3 at the end of the chapter.

□

## 14.5 Translation methods, III: the limit at 1

Another limiting procedure that can be used to produce series for $1/\pi$ is based on the following limit at $x = 1$.

**Lemma 14.11.** *Suppose* $0 < s < 1$. *Then*

$$\lim_{x\to 1^-} \sqrt{1-x} \sum_{n=0}^{\infty} \frac{(\frac{1}{2})_n (s)_n (1-s)_n}{n!^3} n x^n = \frac{\sin \pi s}{\pi}.$$

*Proof.* See Exercise 11 at the end of the chapter. □

The utility of Lemma 14.11 will be illustrated with the following result that gives two more series for $1/\pi$.

**Theorem 14.12.** *The following series identities hold*:

$$\sum_{n=0}^{\infty} \binom{2n}{n}^3 (4n+1) \left(\frac{-1}{64}\right)^n = \frac{2}{\pi} \tag{14.33}$$

*and*

$$\sum_{n=0}^{\infty} \binom{2n}{n}^3 (6n+1) \left(\frac{-1}{512}\right)^n = \frac{2\sqrt{2}}{\pi}. \tag{14.34}$$

The identity (14.33) is called Bauer's series.

*Proof.* Start with the transformation formula in Example 13.5 given by

$${}_3F_2\left(\begin{matrix} \frac{1}{4}, \frac{1}{2}, \frac{3}{4} \\ 1, 1 \end{matrix}; X\right) = \sqrt{1-x}\; {}_3F_2\left(\begin{matrix} \frac{1}{2}, \frac{1}{2}, \frac{1}{2} \\ 1, 1 \end{matrix}; x\right) \tag{14.35}$$

where

$$X = \frac{-4x}{(1-x)^2}.$$

By the chain rule,

$$X\sqrt{1-X}\,\frac{\mathrm{d}}{\mathrm{d}X} = x\,\frac{\mathrm{d}}{\mathrm{d}x}. \tag{14.36}$$

Applying the operator in (14.36) to the transformation formula (14.35), we find that

$$\sqrt{1-X} \sum_{n=0}^{\infty} \frac{(\frac{1}{4})_n (\frac{1}{2})_n (\frac{3}{4})_n}{n!^3} n X^n$$

$$= \sqrt{1-x} \sum_{n=0}^{\infty} \frac{(\frac{1}{2})_n^3}{n!^3} n x^n - \frac{x}{2\sqrt{1-x}} \sum_{n=0}^{\infty} \frac{(\frac{1}{2})_n^3}{n!^3} x^n.$$

Now let $x \to -1^+$ and note that $X \to 1^-$. The limit on the left-hand side can be evaluated using Lemma 14.11, and the limit on the right-hand side can be obtained by direct substitution. The result is

$$\frac{\sqrt{2}}{2\pi} = \sqrt{2} \sum_{n=0}^{\infty} \frac{(\frac{1}{2})_n^3}{n!^3} \left(n + \frac{1}{4}\right) (-1)^n$$

and this is equivalent to (14.33).

The identity (14.34) can be proved using the same procedure by starting with the identity, also from Example 13.5, given by

$${}_3F_2\left(\begin{matrix}\frac{1}{6}, \frac{1}{2}, \frac{5}{6} \\ 1, 1\end{matrix}; X\right) = \sqrt{1-4x}\, {}_3F_2\left(\begin{matrix}\frac{1}{2}, \frac{1}{2}, \frac{1}{2} \\ 1, 1\end{matrix}; x\right)$$

where

$$X = \frac{-27x}{(1-4x)^3}.$$

By the chain rule,

$$X\sqrt{1-X}\,\frac{\mathrm{d}}{\mathrm{d}X} = x\sqrt{\frac{1-x}{1-4x}}\,\frac{\mathrm{d}}{\mathrm{d}x}.$$

Apply the operator to the transformation formula, then take the limit as $x \to (-1/8)^+$ so that $X \to 1^-$. This gives (14.34).

□

The series in Theorem 14.12 can be translated by the method in Section 14.3 to produce further series for $1/\pi$. Here are the results.

**Corollary 14.13.** *The following series identities hold*:

$$\sum_{n=0}^{\infty} \binom{6n}{3n}\binom{3n}{n}\binom{2n}{n}\left(n + \frac{3}{28}\right)\left(\frac{1}{8000}\right)^n = \frac{5\sqrt{5}}{28} \times \frac{1}{\pi}, \tag{14.37}$$

$$\sum_{n=0}^{\infty} \binom{6n}{3n}\binom{3n}{n}\binom{2n}{n}\left(n + \frac{5}{63}\right)\left(\frac{1}{66}\right)^{3n} = \frac{11\sqrt{33}}{252} \times \frac{1}{\pi} \tag{14.38}$$

*and*

$$\sum_{n=0}^{\infty} \binom{4n}{2n}\binom{2n}{n}^2\left(n + \frac{1}{7}\right)\left(\frac{1}{648}\right)^n = \frac{9}{14} \times \frac{1}{\pi}. \tag{14.39}$$

*Proof.* The identities (14.37) and (14.38) can be deduced from (14.33) and (14.34), respectively, by taking $(x, \lambda) = (-1, 1/4)$ (resp. $(x, \lambda) = (-1/8, 1/6)$) in (14.21). The identity (14.39) can be obtained from (14.34) by taking $(x, \lambda) = (-1/8, 1/6)$ in (14.23).

□

## 14.6 A general formula

We have now established thirteen distinct series for $1/\pi$, and they are given by (14.3), (14.4), (14.16), (14.17), (14.18), (14.24), (14.30), (14.32), (14.33), (14.34), (14.37), (14.38), and (14.39). It would be useful to classify the series and have a systematic approach for generating them. In this section we will describe how the theory of modular forms can be used to classify and give a general procedure for producing series that converge to $1/\pi$. An example will be given in detail at the end of the section. The next section will describe how further series can be obtained by numerical experimentation. Moreover, the translations in the previous sections will be examined in the light of modular forms, and the scope for further series will be discussed.

Let $N$ be an integer, called the degree, and suppose that $N \geq 2$. Let $Z = Z(q)$ be a modular form of weight two, let $X = X(q)$ be a modular function, and let $Y = X(q^N)$. There are four requirements:

**Hypothesis 1.** There is a positive integer $\ell$, called the level, such that the transformation formulas

$$tZ\left(e^{-2\pi\sqrt{t/\ell}}\right) = Z\left(e^{-2\pi/\sqrt{t\ell}}\right)$$

and

$$X\left(e^{-2\pi\sqrt{t/\ell}}\right) = X\left(e^{-2\pi/\sqrt{t\ell}}\right)$$

hold for all values of a positive real variable $t$.

**Hypothesis 2.** There is a power series expansion

$$Z = \sum_{j=0}^{\infty} h(j)X^j$$

that converges in a neighborhood of $X = 0$.

**Hypothesis 3.** There is an algebraic function $B = B(X)$ such that the differentiation formula

$$q\frac{\mathrm{d}}{\mathrm{d}q}\log X = Z\,B(X)$$

holds.

**Hypothesis 4.** The functions $X$ and $Y$ are algebraically dependent. That is, there exists an irreducible polynomial $g(X, Y)$, not identically zero, such that $g(X, Y) = 0$. The equation $g(X, Y) = 0$ is said to be a modular equation of degree $N$.

Examples of pairs of functions $X$ and $Z$ that satisfy Hypotheses 1–3 are given in Tables 14.1 and 14.2 at the end of the chapter. The last column of the tables indicates a differential equation that is satisfied by $Z$ with respect to $X$, and from this, the coefficients $h(j)$ can be determined.
With this setup, there is a general formula for producing series for $1/\pi$.

**Theorem 14.14 (Chan, Chan and Liu).** *Let $X$, $Y$, $Z$, $B$, $h(j)$, $\ell$, and $N$ be as for Hypotheses 1–4.*

*Let $\lambda = \lambda(q)$ be defined by*

$$\lambda = \frac{X}{2}\frac{\mathrm{d}}{\mathrm{d}X}\left(\frac{Y B(Y)}{X B(X)} \div \frac{\mathrm{d}Y}{\mathrm{d}X}\right),$$

*where the derivatives may be computed from the modular equation by implicit differentiation.*

*Let $X_{\ell,N}$, $B_{\ell,N}$, and $\lambda_{\ell,N}$ be defined by*

$$X_{\ell,N} = X\left(e^{-2\pi\sqrt{N/\ell}}\right), \qquad B_{\ell,N} = B\left(X_{\ell,N}\right)$$

*and*

$$\lambda_{\ell,N} = \lambda\left(e^{-2\pi/\sqrt{N\ell}}\right).$$

*Then*

$$\sum_{j=0}^{\infty} h(j)\ (j + \lambda_{\ell,N})\ (X_{\ell,N})^j = \frac{1}{2\pi} \times \frac{1}{B_{\ell,N}} \times \sqrt{\frac{\ell}{N}}. \tag{14.40}$$

Strictly speaking, the series in (14.40) should be called "Ramanujan's series for $1/2\pi$." We opt for convenience over pedantry, and will continue to refer to series of the form (14.40) as "Ramanujan's series for $1/\pi$."

*Proof.* The proof breaks naturally into four steps.

**Step 1: The first formula.** Start with the transformation formula

$$t Z\left(e^{-2\pi\sqrt{t/\ell}}\right) = Z\left(e^{-2\pi/\sqrt{t\ell}}\right)$$

from Hypothesis 1. Take the logarithmic derivative with respect to $t$, multiply by $\ell^{1/2}t^{3/2}/\pi$, and then put $t = N$, to obtain

$$\frac{\sqrt{N\ell}}{\pi} = Ne^{-2\pi\sqrt{N/\ell}} \frac{Z'(e^{-2\pi\sqrt{N/\ell}})}{Z(e^{-2\pi\sqrt{N/\ell}})} + e^{-2\pi/\sqrt{N\ell}} \frac{Z'(e^{-2\pi/\sqrt{N\ell}})}{Z(e^{-2\pi/\sqrt{N\ell}})}. \tag{14.41}$$

This is the first formula.

**Step 2: The second formula.** We will obtain a second formula which involves the same terms that are on the right-hand side of (14.41). This will allow one of the terms to be eliminated.

The multiplier $M = M(q)$ is defined by

$$M(q) = \frac{Z(q)}{Z(q^N)}. \tag{14.42}$$

Take logarithms and apply $q\frac{\mathrm{d}}{\mathrm{d}q}$ to both sides. By the chain rule, we obtain

$$q\frac{M'(q)}{M(q)} = q\frac{Z'(q)}{Z(q)} - Nq^N \frac{Z'(q^N)}{Z(q^N)}.$$

Now let $q = e^{-2\pi/\sqrt{N\ell}}$ and observe from Hypothesis 1 that

$$M(e^{-2\pi/\sqrt{N\ell}}) = \frac{Z(e^{-2\pi/\sqrt{N\ell}})}{Z(e^{-2\pi\sqrt{N/\ell}})} = N$$

to obtain

$$e^{-2\pi/\sqrt{N\ell}} \frac{M'(e^{-2\pi/\sqrt{N\ell}})}{N}$$
$$= e^{-2\pi/\sqrt{N\ell}} \frac{Z'(e^{-2\pi/\sqrt{N\ell}})}{Z(e^{-2\pi/\sqrt{N\ell}})} - Ne^{-2\pi\sqrt{N/\ell}} \frac{Z'(e^{-2\pi\sqrt{N/\ell}})}{Z(e^{-2\pi\sqrt{N/\ell}})}. \tag{14.43}$$

This is the second formula.

Subtracting (14.43) from (14.41), and dividing the result by $2N$, we obtain

$$\sqrt{\frac{\ell}{N}} \times \frac{1}{2\pi} = e^{-2\pi\sqrt{N/\ell}} \frac{Z'(e^{-2\pi\sqrt{N/\ell}})}{Z(e^{-2\pi\sqrt{N/\ell}})} + e^{-2\pi/\sqrt{N\ell}} \frac{M'(e^{-2\pi/\sqrt{N\ell}})}{2N^2}. \tag{14.44}$$

The remainder of the proof consists of expanding the two terms on the right-hand side of (14.44) as series.

**Step 3: Expansion of the first term.** We shall expand the first term on the right-hand side of (14.44) as a series. By the chain rule and Hypotheses 2 and 3 we find that

$$q\frac{Z'(q)}{Z(q)} = \frac{1}{Z} \times q\frac{\mathrm{d}Z}{\mathrm{d}q} = \frac{1}{Z} \times q\frac{\mathrm{d}X}{\mathrm{d}q} \times \frac{\mathrm{d}Z}{\mathrm{d}X} = BX\frac{\mathrm{d}Z}{\mathrm{d}X} = B\sum_{j=0}^{\infty} h(j)\, j\, X^j.$$

Now let $q = e^{-2\pi\sqrt{N/\ell}}$ to find that

$$e^{-2\pi\sqrt{N/\ell}}\frac{Z'(e^{-2\pi\sqrt{N/\ell}})}{Z(e^{-2\pi\sqrt{N/\ell}})} = B_{\ell,N}\sum_{j=0}^{\infty} h(j)\, j\, (X_{\ell,N})^j\,. \tag{14.45}$$

**Step 4: Expansion of the second term.** Now we will expand the second term on the right-hand side of (14.44) as a series. By Hypothesis 3,

$$q\frac{\mathrm{d}X}{\mathrm{d}q} = Z\,X\,B(X).$$

Replacing $q$ with $q^N$ we obtain

$$q^N\frac{\mathrm{d}Y}{\mathrm{d}q^N} = Z(q^N)\,Y\,B(Y),$$

which, by the chain rule is

$$q\frac{\mathrm{d}Y}{\mathrm{d}q} = N\,Z(q^N)\,Y\,B(Y).$$

It follows that

$$\frac{\mathrm{d}Y}{\mathrm{d}X} = N\,\frac{Z(q^N)\,Y\,B(Y)}{Z(q)\,X\,B(X)}.$$

On rearranging, and using the definition of the multiplier from (14.42), we obtain

$$M(q) = N\,\frac{Y\,B(Y)}{X\,B(x)} \div \frac{\mathrm{d}Y}{\mathrm{d}X}. \tag{14.46}$$

The derivative $dY/dX$ can be computed from the modular equation in Hypothesis 4 by implicit differentiation.

Next, the chain rule implies

$$qM'(q) = q\frac{\mathrm{d}M}{\mathrm{d}q} = q\frac{\mathrm{d}X}{\mathrm{d}q} \times \frac{\mathrm{d}M}{\mathrm{d}X} = B(X)\,Z(q)\,X(q)\,\frac{\mathrm{d}M}{\mathrm{d}X}. \tag{14.47}$$

Using (14.46) in (14.47), we find that

$$qM'(q) = N\,B(X)\,Z(q)\,X(q)\,\frac{\mathrm{d}}{\mathrm{d}X}\left(\frac{Y\,B(Y)}{X\,B(X)} \div \frac{\mathrm{d}Y}{\mathrm{d}X}\right).$$

Now let $q = e^{-2\pi/\sqrt{N\ell}}$ and use Hypothesis 1 with $t = N$ to obtain

$$\begin{aligned} &e^{-2\pi/\sqrt{N\ell}}\,M'(e^{-2\pi/\sqrt{N\ell}}) \\ &= 2N^2\,B(e^{-2\pi\sqrt{N/\ell}})\,Z(e^{-2\pi\sqrt{N/\ell}})\left[\frac{X}{2}\,\frac{\mathrm{d}}{\mathrm{d}X}\left(\frac{Y\,B(Y)}{X\,B(X)} \div \frac{\mathrm{d}Y}{\mathrm{d}X}\right)\right]_{q=e^{-2\pi/\sqrt{N\ell}}} \\ &= 2N^2\,B_{\ell,N}\,Z(e^{-2\pi\sqrt{N/\ell}})\,\lambda_{\ell,N}. \end{aligned}$$

On dividing by $2N^2$ and using Hypothesis 2 to expand $Z(e^{-2\pi\sqrt{N/\ell}})$ in a series, we obtain

$$e^{-2\pi/\sqrt{N\ell}}\frac{M'(e^{-2\pi/\sqrt{N\ell}})}{2N^2} = \lambda_{\ell,N}\,B_{\ell,N}\sum_{j=0}^{\infty} h(j)\,(X_{\ell,N})^j\,. \tag{14.48}$$

**Conclusion of proof.** The proof is completed on substituting the expansions (14.45) and (14.48) into (14.44).

□

*Example 14.15.* Consider the case $\ell = 3$ and $N = 2$. Let $Z$ and $X$ be defined by

$$Z = Z_3 = \frac{1}{2}\left(3P(q^3) - P(q)\right) \qquad \text{and} \qquad X = X_3 = \frac{\eta_1^6\eta_3^6}{Z^3},$$

where $Z_3$ and $X_3$ are the level 3 functions studied in Chapter 4. Let

$$Y = Y(q) = X(q^2).$$

Let us show that $X$, $Y$, and $Z$ satisfy Hypotheses 1–4 of Theorem 14.14.

**Hypothesis 1:** By the definition of $Z(\tau,\ell)$ in (2.30) and the transformation property (2.32) with $\tau = it$ and $\ell = 3$, we have

$$Z\left(e^{-2\pi/\sqrt{3t}}\right) = Z\left(\frac{-1}{i\sqrt{3t}}, 3\right) = t\,Z\left(\frac{i\sqrt{t}}{\sqrt{3}}, 3\right) = t\,Z\left(e^{-2\pi\sqrt{t/3}}\right).$$

Using this together with the transformation formula for the eta-function in Theorem 2.8, we obtain

$$X\left(e^{-2\pi/\sqrt{3t}}\right)$$
$$=\frac{\eta^6\left(i/\sqrt{3t}\right)\eta^6\left(i\sqrt{3/t}\right)}{Z^3\left(e^{-2\pi/\sqrt{3t}}\right)}=\frac{t^3\,\eta^6\left(i\sqrt{3t}\right)\eta^6\left(i\sqrt{t/3}\right)}{t^3\,Z^3\left(e^{-2\pi\sqrt{t/3}}\right)}=X\left(e^{-2\pi\sqrt{t/3}}\right).$$

Thus, $Z$ and $X$ satisfy the transformation formulas in Hypothesis 1 for the level $\ell=3$.

**Hypothesis 2:** By Theorem 4.30,

$$Z=\sum_{j=0}^{\infty}\binom{3j}{j}\binom{2j}{j}^2X^j,$$

which converges for $|X|<1/108$. So Hypothesis 2 holds with $h(j)=\binom{3j}{j}\binom{2j}{j}^2$.

**Hypothesis 3:** Recall the change of variables given by (4.57) and (4.58) viz.,

$$Z=z_3^2\qquad\text{and}\qquad X=\frac{x_3(1-x_3)}{27}$$

and also the differentiation formula

$$q\frac{\mathrm{d}x_3}{\mathrm{d}q}=z_3^2x_3(1-x_3)$$

given by Theorem 4.19. By the chain rule,

$$\begin{aligned}q\frac{\mathrm{d}}{\mathrm{d}q}\log X&=\frac{1}{x_3(1-x_3)}q\frac{\mathrm{d}}{\mathrm{d}q}x_3(1-x_3)\\&=\frac{1}{x_3(1-x_3)}\times(1-2x_3)\times z_3^2\,x_3(1-x_3)\\&=Z\sqrt{1-108X}.\end{aligned}$$

It follows that

$$B=\sqrt{1-108X}. \tag{14.49}$$

**Hypothesis 4:** The modular equation can be deduced from the theta function identity in Theorem 3.29, viz.,

$$a(q)a(q^2)=b(q)b(q^2)+c(q)c(q^2),$$

as follows. Let $x$ and $y$ be defined by

$$x = x_3(q) = \frac{c^3(q)}{a^3(q)} \qquad \text{and} \qquad y = x_3(q^2) = \frac{c^3(q^2)}{a^3(q^2)}$$

where $x_3$ is defined by (4.29). By the cubic identity in Theorem 3.20,

$$1 - x = \frac{b^3(q)}{a^3(q)} \qquad \text{and} \qquad 1 - y = \frac{b^3(q^2)}{a^3(q^2)}.$$

In this notation, the theta function identity becomes

$$(xy)^{1/3} + (1-x)^{1/3}(1-y)^{1/3} = 1. \tag{14.50}$$

On making the further change of variables from (4.58), namely

$$X = X_3(q) = \frac{x(1-x)}{27}, \qquad Y = X_3(q^2) = \frac{y(1-y)}{27}, \tag{14.51}$$

we obtain the modular equation of degree 2 and level 3

$$\begin{aligned} &1259712X^3Y^3 - 34992(X^3Y^2 + X^2Y^3) + 324(X^3Y + XY^3) \\ &\quad - 10287X^2Y^2 - 84(X^2Y + XY^2) - X^3 - Y^3 + XY = 0. \end{aligned} \tag{14.52}$$

A step-by-step deduction of (14.52) from (14.51) is given in Exercise 1.

**The parameter values:** Now we will determine the values of $X_{3,2}$, $B_{3,2}$, and $\lambda_{3,2}$. Taking $t = 2$ in Hypothesis 1, we obtain

$$X\left(e^{-2\pi/\sqrt{6}}\right) = X\left(e^{-2\pi\sqrt{2/3}}\right) = X\left((e^{-2\pi/\sqrt{6}})^2\right) = Y\left(e^{-2\pi/\sqrt{6}}\right),$$

since $Y(q) = X(q^2)$. Therefore, the value of $X(e^{-2\pi\sqrt{2/3}})$ can be found by setting $X = Y$ in the modular equation (14.52). The modular equation factorizes as

$$(1 - 216X)(1 - 8X)(1 + 27X)^2 = 0.$$

A numerical estimate using

$$X(q) = q - 42q^2 + 981q^3 + O(q^4)$$

with $q = e^{-2\pi\sqrt{2/3}}$ shows that the appropriate root to select is

$$X_{3,2} = X\left(e^{-2\pi\sqrt{2/3}}\right) = \frac{1}{216}.$$

Next, by (14.49)

$$B_{3,2} = \sqrt{1 - 108X_{3,2}} = \frac{1}{\sqrt{2}}.$$

It remains to determine the value of $\lambda_{3,2}$. By implicit differentiation of the modular equation (14.52) we find that

$$\frac{\mathrm{d}Y}{\mathrm{d}X} = -\frac{\begin{array}{r}3779136\,X^2Y^3 - 104976\,X^2Y^2 - 69984\,XY^3 + 972\,X^2Y \\ -20574\,XY^2 + 324\,Y^3 - 3\,X^2 - 168\,XY - 84\,Y^2 + Y\end{array}}{\begin{array}{r}3779136\,X^3Y^2 - 104976\,X^2Y^2 - 69984\,X^3Y + 972\,XY^2 \\ -20574\,X^2Y + 324\,X^3 - 3\,Y^2 - 168\,XY - 84\,X^2 + X\end{array}}. \tag{14.53}$$

Then, the value of $\lambda_{3,2}$ is given by

$$\begin{aligned}\lambda_{3,2} &= \frac{X}{2}\frac{\mathrm{d}}{\mathrm{d}X}\left(\frac{Y\sqrt{1-108Y}}{X\sqrt{1-108X}} \times \left(\frac{\mathrm{d}Y}{\mathrm{d}X}\right)^{-1}\right)\Bigg|_{q=e^{-2\pi/\sqrt{6}}} \\ &= \frac{X}{2}\frac{\mathrm{d}}{\mathrm{d}X}\left(\frac{Y\sqrt{1-108Y}}{X\sqrt{1-108X}} \times \left(\frac{\mathrm{d}Y}{\mathrm{d}X}\right)^{-1}\right)\Bigg|_{X=Y=1/216}. \end{aligned} \tag{14.54}$$

The calculation may be completed by substituting the result of (14.53) into (14.54), differentiating, and then substituting the numerical values. The most efficient way is to use computer algebra. For example, with Maple, this can be done by entering the commands:

```
> alias(y = y(x)):
> b := proc(x) sqrt(1-108*x): end:
> g := 1259712*x^3*y^3 - 34992*(x^3*y^2+x^2*y^3)
          + 324*(x^3*y+x*y^3) - 10287*x^2*y^2
          - 84*(x^2*y+x*y^2) - x^3 - y^3 + x*y = 0:
> dydx := solve(diff(g, x), diff(y, x)):
> L := (x/2)*(diff(y*b(y)/(x*b(x)*dydx), x)):
> lambda := subs(diff(y, x) = dydx, L):
> subs({x = 1/216, y = 1/216}, lambda);
```

and this produces the value $\lambda_{3,2} = 1/6$.

Substituting the data into the result of Theorem 14.14, we obtain

$$\sum_{j=0}^{\infty} \binom{3j}{j}\binom{2j}{j}^2 \left(j + \frac{1}{6}\right)\left(\frac{1}{216}\right)^j = \frac{\sqrt{3}}{2\pi}. \tag{14.55}$$

This concludes Example 14.15.

The two analogues of Clausen's identity in Theorem 5.43 can be used to produce two different types of series for $1/\pi$, as the next example illustrates.

*Example 14.16.* Let $X$ and $Z$ be given by

$$Z = Z_5 = \frac{1}{4}\left(P(q^5) - P(q)\right), \qquad X = X_5 = \frac{1}{Z^2} \times q \prod_{j=1}^{\infty} (1-q^j)^4 (1-q^{5j})^4$$

where $Z_5$ and $X_5$ are as in Section 5.5. Then the series expansion in (5.41) together with the value $q = -\exp(-\pi\sqrt{47/5})$ can be used in Theorem 14.14 to produce the series

$$\sum_{j=0}^{\infty} \binom{2j}{j} t_5(j) \left(j + \frac{71}{682}\right) \left(\frac{-1}{15228}\right)^j = \frac{81\sqrt{47}}{1705} \times \frac{1}{\pi} \tag{14.56}$$

where

$$t_5(j) = \sum_{k=0}^{j} \binom{j}{k}^2 \binom{j+k}{k}.$$

On the other hand, taking

$$Z = y_5 = \prod_{j=1}^{\infty} \frac{(1-q^j)^5}{(1-q^{5j})} \quad \text{and} \quad X = w_5 = q \prod_{j=1}^{\infty} \frac{(1-q^{5j})^6}{(1-q^j)^6}$$

in Theorem 14.14 and using the series expansion (5.42) together with the same value $q = -\exp(-\pi\sqrt{47/5})$, we obtain Sato's series [260], given by

$$\sum_{j=0}^{\infty} s_5(j) \left(j + \frac{1}{2} - \frac{25\sqrt{5}}{141}\right) (-1)^j \left\{ 5^{-3/2} \left(\frac{\sqrt{5}-1}{2}\right)^{15} \right\}^{j+1/2} = \frac{\sqrt{5}}{846} \times \frac{1}{\pi}, \tag{14.57}$$

where

$$s_5(j) = \sum_{k=0}^{j} (-1)^{j+k} \binom{j}{k}^3 \binom{4j-5k}{3j}.$$

The series (14.56) can be obtained from (14.57) and vice versa, by translation using Theorem 5.43 in the case $(\alpha, \beta, \gamma) = (11, 3, 1)$, or equivalently, by using (13.11) and (13.13).

## 14.7 Classification of series for $1/\pi$

Theorem 14.14 allows series to be classified according to level and degree. To illustrate, consider the following series from the level 2 theory:

$$\sum_{n=0}^{\infty}\binom{4n}{2n}\binom{2n}{n}^2\left(n+\frac{1}{7}\right)\left(\frac{1}{648}\right)^n = \frac{1}{2\pi}\times\frac{9}{7}, \tag{14.58}$$

$$\sum_{n=0}^{\infty}\binom{4n}{2n}\binom{2n}{n}^2\left(n+\frac{1}{8}\right)\left(\frac{1}{48}\right)^{2n} = \frac{1}{2\pi}\times\frac{\sqrt{3}}{2}, \tag{14.59}$$

$$\sum_{n=0}^{\infty}\binom{4n}{2n}\binom{2n}{n}^2\left(n+\frac{3}{40}\right)\left(\frac{1}{28}\right)^{4n} = \frac{1}{2\pi}\times\frac{49\sqrt{3}}{180}, \tag{14.60}$$

$$\sum_{n=0}^{\infty}\binom{4n}{2n}\binom{2n}{n}^2\left(n+\frac{1}{5}\right)\left(\frac{-1}{144}\right)^n \text{ “=” } \frac{1}{2\pi}\times\frac{2\sqrt{3}}{5}, \tag{14.61}$$

$$\sum_{n=0}^{\infty}\binom{4n}{2n}\binom{2n}{n}^2\left(n+\frac{8}{65}\right)\left(\frac{-1}{63^2}\right)^n = \frac{1}{2\pi}\times\frac{18\sqrt{7}}{65}, \tag{14.62}$$

$$\sum_{n=0}^{\infty}\binom{4n}{2n}\binom{2n}{n}^2\left(n+\frac{3}{28}\right)\left(\frac{-1}{12288}\right)^n = \frac{1}{2\pi}\times\frac{8\sqrt{3}}{21}. \tag{14.63}$$

The series (14.58) and (14.62) were proved earlier in (14.39) and (14.18), respectively, while (14.61) is the divergent series example from (14.25). Proofs for the other three series will be outlined in Exercises 4 and 5.

All of the series in (14.58)–(14.63) are based on the parameterization of

$$Z_2 = \sum_{n=0}^{\infty}\binom{4n}{2n}\binom{2n}{n}^2 X_2^n \tag{14.64}$$

by

$$Z_2 = 2P(q^2) - P(q) \quad \text{and} \quad X_2 = \frac{1}{Z_2^4}\times q\prod_{j=1}^{\infty}(1-q^j)^8(1-q^{2j})^8.$$

The arguments $1/648$, $1/48^2$, and $1/28^4$ in the first three power series are the values of

$$X_{2,N} = X_2(e^{-2\pi\sqrt{N/2}}) \quad \text{for } N = 2, 3 \text{ and } 9, \text{ respectively.}$$

The procedure in Example 14.15 can be used to show that $B = \sqrt{1-256X}$. It follows that the right-hand sides of (14.58)–(14.60) are given by

$$\frac{1}{2\pi} \times \frac{1}{B_{2,N}} \times \sqrt{\frac{2}{N}} = \frac{1}{2\pi} \times \frac{1}{\sqrt{1-256X_{2,N}}} \times \sqrt{\frac{2}{N}} \tag{14.65}$$

for $N = 2$, 3, and 9, respectively.

The series (14.61)–(14.63) have negative values in the power series variable. They are obtained by replacing $q$ by $-q$ in the parameterization (14.64) and using (0.2) and (0.110), to obtain

$$Z_2(-q) = 2P(q^2) - P(-q) = 4P(q^4) - 4P(q^2) + P(q)$$

and

$$\begin{aligned} X_2(-q) &= \frac{1}{Z_2^4(-q)} \times (-q) \prod_{j=1}^{\infty} (1-(-q)^j)^8 (1-q^{2j})^8 \\ &= \frac{1}{Z_2^4(-q)} \times (-q) \prod_{j=1}^{\infty} \frac{(1-q^{2j})^{32}}{(1-q^j)^8(1-q^{4j})^8}. \end{aligned}$$

As such, the underlying modular forms are now of level 4. The arguments $-1/144$, $-1/63^2$, and $-1/12288$ are the values of

$$X_{2,N} = X_2(-e^{-2\pi\sqrt{N/4}}) \quad \text{for } N = 3, 7 \text{ and } 9, \text{ respectively.}$$

The values on the right-hand sides can be determined, as for (14.65).

A computer can be used to generate additional series for $1/\pi$ by expanding $X(q)$ in powers of $q$ as far as (say) $q^{100}$, and then investigating the values of $N$ for which $X$ is rational. It appears that $X_2(e^{-2\pi\sqrt{N/2}})$ is rational for $N = 2, 3, 5, 9, 11, 29$, while $X_2(-e^{-2\pi\sqrt{N/4}})$ is rational for $N = 5, 7, 9, 13, 25, 37$. The right-hand sides of the series can be determined by numerically evaluating $\frac{1}{2\pi} \times \frac{1}{B_{\ell,N}} \times \sqrt{\frac{\ell}{N}}$. Finally, the values of $\lambda_{\ell,N}$ can be determined by summing the series. The values are set out in Table 14.4. For example, the data in the table corresponding to $N = 29$ and $q > 0$ corresponds to Ramanujan's series (14.2) in the introduction, viz.,

$$\sum_{n=0}^{\infty} \binom{4n}{2n}\binom{2n}{n}^2 \left(n + \frac{1103}{26390}\right)\left(\frac{1}{396}\right)^{4n} = \frac{1}{2\pi} \times \frac{9801\sqrt{2}}{52780}.$$

A rigorous proof, following the method of Example 14.15, would require a modular equation of degree 29. Especially challenging is the calculation of $\lambda_{2,29}$.

Tables 14.3–14.14 contain parameter values that produce series for $1/\pi$ for levels $1 \le \ell \le 10$. The series for $1/\pi$ studied in this chapter, including (14.58)–(14.63), are also indicated in the tables. Ramanujan's paper contains 17 series for $1/\pi$ and references for 16 of the series are in the tables. Ramanujan's other series, that is missing from the tables, is

$$\sum_{n=0}^{\infty} \binom{2n}{n}^3 \left(n + \frac{31}{270+48\sqrt{5}}\right) \frac{(\sqrt{5}-1)^{8n}}{2^{20n}} = \frac{1}{2\pi} \times \frac{32}{15+21\sqrt{5}}. \tag{14.66}$$

It corresponds to the parameters $\ell = 4$, $q > 0$, and $N = 15$. It is omitted from the tables because the power series variable is not rational — it is a quadratic irrational. Quadratic irrational series for $1/\pi$ have been investigated and classified in the thesis by A. Aldawoud [17].

The classification given by Theorem 14.14 can be used to explain why certain transformation formulas can be used to translate a series for $1/\pi$ into another. To illustrate, consider the series identities

$$\sum_{n=0}^{\infty} \binom{6n}{3n}\binom{3n}{n}\binom{2n}{n} \left(n + \frac{8}{133}\right) \left(\frac{1}{255^3}\right)^n = \frac{85\sqrt{255}}{7182} \times \frac{1}{\pi}, \tag{14.67}$$

$$\sum_{n=0}^{\infty} \binom{6n}{3n}\binom{3n}{n}\binom{2n}{n} \left(n + \frac{8}{63}\right) \left(\frac{-1}{15^3}\right)^n = \frac{5\sqrt{15}}{63} \times \frac{1}{\pi}, \tag{14.68}$$

$$\sum_{n=0}^{\infty} \binom{4n}{2n}\binom{2n}{n}^2 \left(n + \frac{8}{65}\right) \left(\frac{-1}{63^2}\right)^n = \frac{9\sqrt{7}}{65} \times \frac{1}{\pi} \tag{14.69}$$

and

$$\sum_{n=0}^{\infty} \binom{2n}{n}^3 \left(n + \frac{5}{42}\right) \left(\frac{1}{4096}\right)^n = \frac{8}{21} \times \frac{1}{\pi}. \tag{14.70}$$

The series (14.67), (14.68), and (14.69) were obtained from (14.70) by the method of translation, in Theorem 14.5. The series (14.67) and (14.68) originate from the level 1 theory, while (14.69) and (14.70) come from the theories for levels 2 and 4, respectively. Tables 14.3, 14.4, and 14.6 can be consulted to determine the degree $N$ and nome $q$, and we find that the relevant parameters are given by

$$(\ell, N, q) = (1, 7, e^{-2\pi\sqrt{7}}),\ (1, 7, -e^{-\pi\sqrt{7}}),\ (2, 7, -e^{-\pi\sqrt{7}}) \quad \text{and} \quad (4, 7, e^{-\pi\sqrt{7}}),$$

respectively. If the $q$-parameters are denoted by $q_1$, $q_2$, $q_3$ and $q_4$, respectively, then

$$q_1 = q_0^2, \quad q_2 = -q_0, \quad q_3 = -q_0 \quad \text{and} \quad q_4 = q_0$$

where $q_0 = e^{-\pi\sqrt{7}}$. The significant point is that $q_1$, $q_2$, $q_3$, and $q_4$ are all integral powers of a common value $q_0$, and for $q_2$ and $q_3$ there is also a sign change. The value of $q_1$ suggests that a quadratic transformation formula be used, while the values of $q_2$ and $q_3$ suggest a change of sign is involved. The corresponding hypergeometric functions in Theorem 13.3, which match with the levels, degrees and change of sign as described above, are given by (13.22), (13.27), (13.33), and (13.39), respectively. As noted in Example 13.5, those particular functions can be simplified by making a change of variable, to give

$$\begin{aligned}\frac{2}{\sqrt{4-x}}\,{}_3F_2\left(\begin{matrix}\frac{1}{6}, \frac{1}{2}, \frac{5}{6}\\ 1, 1\end{matrix}; \frac{27x^2}{(4-x)^3}\right) &= \frac{1}{\sqrt{1-4x}}\,{}_3F_2\left(\begin{matrix}\frac{1}{6}, \frac{1}{2}, \frac{5}{6}\\ 1, 1\end{matrix}; \frac{-27x}{(1-4x)^3}\right)\\ &= \frac{1}{\sqrt{1-x}}\,{}_3F_2\left(\begin{matrix}\frac{1}{4}, \frac{1}{2}, \frac{3}{4}\\ 1, 1\end{matrix}; \frac{-4x}{(1-x)^2}\right)\\ &= {}_3F_2\left(\begin{matrix}\frac{1}{2}, \frac{1}{2}, \frac{1}{2}\\ 1, 1\end{matrix}; x\right).\end{aligned}$$

That explains why the particular transformation formulas were used in the proof of Theorem 14.5.

The tables at the end of the chapter can be scoured to find other instances for which the $q$ parameters are integral powers of a common value, with changes in sign also permitted. One example is the data

$$(\ell, N, q) = (2, 2, e^{-2\pi}),\ (2, 9, -e^{-3\pi})$$

from Table 14.4, which suggests a quadratic transformation be used with a cubic transformation with a change of sign. The details are outlined in Exercise 4 at the end of the chapter. Further examples are given in Exercise 5, and this topic is examined in more detail in [131].

## 14.8 WZ proofs

In this section we give a brief indication of how the Wilf–Zeilberger method, also called WZ method for short, may be used to prove one of Ramanujan's series for $1/\pi$. At the time of writing, only a small number of Ramanujan's series can be proved by this method and the topic is very much an open field of research.

The identity we will consider is Bauer's series (14.33), which may be rewritten in the form

$$\sum_{k=0}^{\infty} \frac{\left(\frac{1}{2}\right)_k^3}{k!^3}(4k+1)(-1)^k = \frac{2}{\pi}. \tag{14.71}$$

We begin by evaluating a related finite sum.

**Lemma 14.17.** *Let n be a nonnegative integer. Then*

$$\sum_{k=0}^{\infty} \frac{(\frac{1}{2})_k^2(-n)_k}{k!^2(\frac{3}{2}+n)_k}(4k+1)(-1)^k = \frac{\Gamma(\frac{3}{2}+n)}{\Gamma(\frac{3}{2})\Gamma(n+1)}.$$

Before starting the proof, note that the series on the left-hand side contains only finitely many nonzero terms because $(-n)_k = 0$ if $k > n$.

*Proof.* Let $F(n,k)$ denote the summand on the left-hand side divided by the right-hand side, that is,

$$F(n,k) = \frac{(\frac{1}{2})_k^2(-n)_k}{k!^2(\frac{3}{2}+n)_k}(4k+1)(-1)^k \times \frac{\Gamma(\frac{3}{2})\Gamma(n+1)}{\Gamma(\frac{3}{2}+n)}.$$

We want to show that $\sum_{k=0}^{\infty} F(n,k) = 1$. Let

$$G(n,k) = \frac{(2k+1)^2}{(4k+1)(2n+2k+3)}F(n,k)$$

and observe that

$$F(n+1,k) - F(n,k) = G(n,k-1) - G(n,k).$$

Summing for $1 \le k \le n$, we obtain

$$\sum_{k=1}^{n} F(n+1,k) - \sum_{k=1}^{n} F(n,k) = G(n,0) - G(n,n).$$

Inserting extra terms on each side and using the fact that $F(n,k) = 0$ if $k > n$ leads to

$$\sum_{k=0}^{\infty} F(n+1,k) - \sum_{k=0}^{\infty} F(n,k)$$
$$= G(n,0) - G(n,n) + F(n+1,n+1) + F(n+1,0) - F(n,0).$$

The right-hand side simplifies to 0 and it follows that $\sum_{k=0}^{\infty} F(n,k)$ is a constant that is independent of $n$. On taking $n = 0$ we deduce that the constant is 1, and this implies the result. □

The identity in Lemma 14.17 has been shown to hold for all nonnegative integers $n$. If we could show that Lemma 14.17 holds for all real values of $n$, then we could set $n = -\frac{1}{2}$ to deduce series (14.71). Just because an identity holds for the nonnegative integers does not imply that it holds for all real values: a counterexample is the identity $\sin(\pi n) = 0$, which holds for all integers $n$ but not for other real values of $n$. Under certain additional conditions, an identity that holds for integers (or for nonnegative integers) must also hold for all real values. The precise details are given by Carlson's theorem, and the reader is referred to the books by Andrews, Askey, and Roy [24, p. 108] and Bailey [34, p. 39] for details of the theorem and examples.

The identity (14.71) follows from Lemma 14.17 by applying Carlson's theorem and setting $n = -\frac{1}{2}$.

## 14.9 Iterations

Up to now we have focussed on series for $1/\pi$. In this section we will describe techniques for deriving iterations that converge to $1/\pi$. We will focus on obtaining a quadratic iteration, due to J. M. Borwein and P. B. Borwein, using a recent technique of J. Guillera. Interest in the technique lies in its potential to produce an enormous number of iteration schemes. The technique consists of two steps:

1. Use a hypergeometric transformation formula to set up an iteration scheme.
2. Use a Ramanujan-type series for $1/\pi$, or a translation method, to determine starting values and to identify the limit of the iteration.

We begin with two lemmas. The first is obtained by differentiating Landen's transformation.

**Lemma 14.18.** *Let a, b, and x be real numbers and suppose* $0 < x < 1$. *Let*

$$X = \frac{1-\sqrt{1-x^2}}{1+\sqrt{1-x^2}}, \quad A = a(1+X) + bX\left(\frac{1+X}{1-X}\right) \quad \textit{and} \quad B = 2b\,\frac{(1+X)^2}{1-X}.$$

*Then*

$$\sum_{k=0}^{\infty} \frac{(\frac{1}{2})_k^2}{k!^2}\,(a+bk)\,x^{2k} = \sum_{k=0}^{\infty} \frac{(\frac{1}{2})_k^2}{k!^2}\,(A+Bk)\,X^{2k}. \tag{14.72}$$

*Proof.* Landen's transformation in Corollary 4.35 can be written in the form

$${}_2F_1\left(\begin{matrix}\frac{1}{2}, \frac{1}{2}\\ 1\end{matrix}; x^2\right) = (1+X)\,{}_2F_1\left(\begin{matrix}\frac{1}{2}, \frac{1}{2}\\ 1\end{matrix}; X^2\right) \tag{14.73}$$

where

$$x^2 = 1 - \left(\frac{1-X}{1+X}\right)^2,$$

or equivalently

$$X = \frac{1-\sqrt{1-x^2}}{1+\sqrt{1-x^2}}.$$

The chain rule implies

$$\frac{x}{2}\frac{\mathrm{d}}{\mathrm{d}x} = X\left(\frac{1+X}{1-X}\right)\frac{\mathrm{d}}{\mathrm{d}X}.$$

On applying the differential operator

$$a + \frac{bx}{2}\frac{\mathrm{d}}{\mathrm{d}x} = a + bX\left(\frac{1+X}{1-X}\right)\frac{\mathrm{d}}{\mathrm{d}X}$$

to (14.73), we obtain the required result.

□

The result of the previous lemma may be recast in the form of an iteration.

**Lemma 14.19.** *Let $a_0$, $b_0$, and $x_0$ be real numbers and suppose $0 < x_0 < 1$. Consider the sequences defined by the recurrence relations*

$$x_{n+1} = \frac{1-\sqrt{1-x_n^2}}{1+\sqrt{1-x_n^2}}, \qquad b_{n+1} = 2b_n\frac{(1+x_{n+1})^3}{1-x_{n+1}}$$

*and*

$$a_{n+1} = a_n(1+x_{n+1})^2 + \frac{b_{n+1}x_{n+1}}{2(1+x_{n+1})}.$$

*Let*

$$L_n = \left(\sum_{k=0}^{\infty}\frac{(\frac{1}{2})_k^2}{k!^2}x_n^{2k}\right)\left(\sum_{k=0}^{\infty}\frac{(\frac{1}{2})_k^2}{k!^2}(a_n+b_nk)\,x_n^{2k}\right).$$

*Then $L_n$ is a constant that is independent of $n$, and*

$$\lim_{n\to\infty} a_n = L_0.$$

*Furthermore, the rate of convergence is quadratic.*

Before starting the proof, we note that particular values of $a_0$, $b_0$, and $x_0$ for which $L_0$ can be determined explicitly will be given in Example 14.21. The limiting value of $a_n$ can therefore be identified in these cases.

*Proof.* Multiply (14.72) and (14.73) and then set $a = a_n$, $b = b_n$, and $x = x_n$ to obtain

$$\left(\sum_{k=0}^{\infty} \frac{\left(\frac{1}{2}\right)_k^2}{k!^2} x_n^{2k}\right)\left(\sum_{k=0}^{\infty} \frac{\left(\frac{1}{2}\right)_k^2}{k!^2} (a_n + b_n k)\, x_n^{2k}\right)$$
$$= \left(\sum_{k=0}^{\infty} \frac{\left(\frac{1}{2}\right)_k^2}{k!^2} x_{n+1}^{2k}\right)\left(\sum_{k=0}^{\infty} \frac{\left(\frac{1}{2}\right)_k^2}{k!^2} (a_{n+1} + b_{n+1} k)\, x_{n+1}^{2k}\right). \tag{14.74}$$

It follows that $L_n$ is a constant.

Next, it is easy to check that $\lim_{n\to\infty} x_n = 0$ and that

$$x_{n+1} \sim \frac{x_n^2}{4}, \qquad b_{n+1} \sim 2\,b_n \quad \text{and} \quad a_{n+1} \sim a_n \qquad \text{as } n \to \infty.$$

It follows that

$$\lim_{n\to\infty} a_n x_n^{2k} = 0 \quad \text{and} \quad \lim_{n\to\infty} b_n x_n^{2k} = 0$$

for all positive values of $k$. We deduce that $\lim_{n\to\infty} L_n = \lim_{n\to\infty} a_n$, and since $L_n$ is constant it follows that $\lim_{n\to\infty} a_n = L_0$, as required.

The rate of convergence can be seen to be quadratic by noting that

$$a_{n+1} - a_n = O(x_{n+1}) \qquad \text{and} \qquad x_{n+1} = O(x_n^2) \qquad \text{as } n \to \infty.$$

□

The recurrence relations in Lemma 14.19 have been deliberately designed, with Landen's transformation in mind, to force the identity (14.74) to hold. The recurrences can be simplified in two steps. First, let

$$c_n = \frac{b_n}{1 - x_n^2}$$

and note that $c_{n+1} = 2c_n$, so that $c_n = c_0\, 2^n$. Then, the recurrence for $a_n$ can be written as

$$a_{n+1} = a_n(1 + x_{n+1})^2 + c_0\, 2^n\, x_{n+1}(1 - x_{n+1})$$

and the recurrence for $b_n$ can be omitted. Second, if we put

$$a_n = r_n - c_0\, 2^{n-1}\, x_n^2,$$

then the recurrence for $a_n$ may be replaced with

$$r_{n+1} = r_n(1 + x_{n+1})^2 - c_0\, 2^n\, x_{n+1}.$$

The final iteration scheme may be summarized as follows.

**Theorem 14.20.** *Suppose that an identity of the form*

$$\left(\sum_{k=0}^{\infty} \frac{(\frac{1}{2})_k^2}{k!^2} x_0^{2k}\right)\left(\sum_{k=0}^{\infty} \frac{(\frac{1}{2})_k^2}{k!^2} (a_0 + b_0 k)\, x_0^{2k}\right) = L_0$$

*holds for certain values of* $a_0$, $b_0$, $x_0$, *and* $L_0$. *Let* $c_0$ *and* $r_0$ *be defined by*

$$c_0 = \frac{b_0}{1 - x_0^2} \qquad \text{and} \qquad r_0 = a_0 + \frac{c_0 x_0^2}{2}.$$

*Consider the sequences defined by the recurrence relations*

$$x_{n+1} = \frac{1 - \sqrt{1 - x_n^2}}{1 + \sqrt{1 - x_n^2}}$$

*and*

$$r_{n+1} = r_n(1 + x_{n+1})^2 - c_0\, 2^n\, x_{n+1}.$$

*Then*

$$\lim_{n\to\infty} r_n = L_0$$

*and the rate of convergence is quadratic.*

Here are three specific examples for which the limits can be determined explicitly.

*Example 14.21.* Consider the sequences $\{x_n\}$ and $\{r_n\}$ defined by the recurrence relations in Theorem 14.20.

1. If $x_0 = \dfrac{1}{\sqrt{2}}$, $r_0 = \dfrac{1}{2}$ and $c_0 = 2$, then $\lim\limits_{n\to\infty} r_n = \dfrac{1}{\pi}$.

2. If $x_0 = \dfrac{\sqrt{6}-\sqrt{2}}{4}$, $r_0 = \dfrac{\sqrt{3}-1}{3}$ and $c_0 = \dfrac{4\sqrt{3}}{3}$, then $\lim\limits_{n\to\infty} a_n = \dfrac{2}{3\pi}$.

3. If $x_0 = \dfrac{\sqrt{6}-\sqrt{2}}{4}$, $r_0 = 1$ and $c_0 = 0$, then $\lim\limits_{n\to\infty} a_n = \dfrac{2^{4/3}}{3} \times \dfrac{\sqrt{\pi}}{\Gamma^3(\frac{5}{6})}$.

*Proof.* Consider the special case of (4.54) given by

$$\left\{ {}_2F_1\left( \begin{matrix} \frac{1}{2}, \frac{1}{2} \\ 1 \end{matrix} ; x \right) \right\}^2 = {}_3F_2\left( \begin{matrix} \frac{1}{2}, \frac{1}{2}, \frac{1}{2} \\ 1, \ 1 \end{matrix} ; 4x(1-x) \right).$$

Apply the operator

$$\alpha + \beta \, \frac{x(1-x)}{(1-2x)} \frac{\mathrm{d}}{\mathrm{d}x}$$

to obtain

$$\left( \sum_{k=0}^{\infty} \frac{(\frac{1}{2})_k^2}{k!^2} x^k \right) \left( \sum_{k=0}^{\infty} \frac{(\frac{1}{2})_k^2}{k!^2} \left( \alpha + 2\beta \left( \frac{1-x}{1-2x} \right) k \right) x^k \right)$$
$$= \sum_{k=0}^{\infty} \frac{(\frac{1}{2})_k^3}{k!^3} (\alpha + \beta k) \, (4x(1-x))^k . \qquad (14.75)$$

Take $\alpha = 0$, $\beta = 1$ and multiply by $1-2x$, and write the result in the form

$$2(1-x) \left( \sum_{k=0}^{\infty} \frac{(\frac{1}{2})_k^2}{k!^2} x^k \right) \left( \sum_{k=0}^{\infty} \frac{(\frac{1}{2})_k^2}{k!^2} k \, x^k \right)$$
$$= \sqrt{1-4x(1-x)} \sum_{k=0}^{\infty} \frac{(\frac{1}{2})_k^3}{k!^3} k \, (4x(1-x))^k .$$

Now let $x \to 1/2$ and use Lemma 14.11 to evaluate the limit on the right-hand side. The result is

$$\left(\sum_{k=0}^{\infty} \frac{(\frac{1}{2})_k^2}{k!^2} \frac{1}{2^k}\right)\left(\sum_{k=0}^{\infty} \frac{(\frac{1}{2})_k^2}{k!^2} \frac{k}{2^k}\right) = \frac{1}{\pi}. \tag{14.76}$$

This gives rise to the data

$$x_0 = \frac{1}{2}, \quad a_0 = 0, \quad b_0 = 1 \quad \text{and} \quad L_0 = \frac{1}{\pi}$$

for the hypotheses in Theorem 14.20. On computing the values of $c_0$ and $r_0$ and applying Theorem 14.20 we complete the proof of Part 1.

Next, the series on the right-hand side of (14.75) can be evaluated when $\alpha = \frac{1}{6}$, $\beta = 1$, and $4x(1-x) = \frac{1}{4}$ by (14.28) and this leads to the data

$$x_0 = \frac{\sqrt{6}-\sqrt{2}}{4}, \quad a_0 = \frac{1}{6}, \quad b_0 = 1 + \frac{2}{\sqrt{3}} \quad \text{and} \quad L_0 = \frac{2}{3} \times \frac{1}{\pi}$$

for the hypotheses in Theorem 14.20. On computing the values of $c_0$ and $r_0$ and applying Theorem 14.20 we complete the proof of Part 2.

The values in Part 3 are obtained in the same way, by using (14.27) to sum the right-hand side of the series (14.75) when $\alpha = 0$, $\beta = 1$, and $4x(1-x) = \frac{1}{4}$.

□

A different proof of (14.76) is given in Exercise 9.

It is interesting to note that the iteration scheme in Theorem 14.20 along with the parameters for the iterations in Example 14.21 have been obtained using only the theory of hypergeometric functions. Although the proof can be written out without any reference to theta functions or modular forms, it has been shown in this chapter that the theory of modular forms is indispensable in classifying the identities and putting the results into context.

## 14.10 Tables

Tables 14.1–14.14 contain data for Theorem 14.14.

**Table 14.1** Functions satisfying Hypotheses 1–3 of Theorem 14.14

| level | $Z$ | $X$ | $B = \frac{1}{Z} q \frac{\mathrm{d}}{\mathrm{d}q} \log X$ | $BX \frac{\mathrm{d}}{\mathrm{d}X}\left(BX \frac{\mathrm{d}Z}{\mathrm{d}X}\right) - \frac{1}{2Z}\left(BX \frac{\mathrm{d}Z}{\mathrm{d}X}\right)^2$ | $h(j)$ |
|---|---|---|---|---|---|
| 1 | $Q^{1/2}$ | $\frac{\eta_1^{24}}{Z^6}$ | $(1-1728X)^{1/2}$ | $120ZX$ | $\binom{6j}{3j}\binom{3j}{j}\binom{2j}{j}$ |
| 2 | $2P_2 - P_1$ | $\frac{\eta_1^8\eta_2^8}{Z^4}$ | $(1-256X)^{1/2}$ | $24ZX$ | $\binom{4j}{2j}\binom{2j}{j}^2$ |
| 3 | $\frac{3P_3 - P_1}{2}$ | $\frac{\eta_1^6\eta_3^6}{Z^3}$ | $(1-108X)^{1/2}$ | $12ZX$ | $\binom{3j}{j}\binom{2j}{j}^2$ |
| 4 | $\frac{4P_4 - P_1}{3}$ | $\frac{\eta_1^8\eta_4^8}{\eta_2^8} \times \frac{1}{Z^2}$ | $(1-64X)^{1/2}$ | $8ZX$ | $\binom{2j}{j}^3$ |
| 5 | $\frac{5P_5 - P_1}{4}$ | $\frac{\eta_1^4\eta_5^4}{Z^2}$ | $(1-44X-16X^2)^{1/2}$ | $6ZX(1+X)$ | $\binom{2j}{j}\sum_k \binom{j}{k}^2\binom{j+k}{k}$ |
| 6 | $\frac{6P_6 - 3P_3 - 2P_2 + P_1}{2}$ | $\frac{\eta_1^2\eta_2^2\eta_3^2\eta_6^2}{Z^2}$ | $(1+68X+1152X^2)^{1/2}$ | $-12ZX(1+36X)$ | $\binom{2j}{j}\sum_{k,\ell} (-8)^{j-k}\binom{j}{k}\binom{k}{\ell}^3$ |
| 6 | $\frac{6P_6 - 3P_3 + 2P_2 - P_1}{4}$ | $\frac{\eta_1^2\eta_2^2\eta_3^2\eta_6^2}{Z^2}$ | $(1-40X+144X^2)^{1/2}$ | $6ZX(1-9X)$ | $\binom{2j}{j}\sum_k \binom{j}{k}^2\binom{2k}{k}$ |
| 6 | $\frac{6P_6 + 3P_3 - 2P_2 - P_1}{6}$ | $\frac{\eta_1^2\eta_2^2\eta_3^2\eta_6^2}{Z^2}$ | $(1-28X-128X^2)^{1/2}$ | $4ZX(1+12X)$ | $\binom{2j}{j}\sum_k \binom{j}{k}^3$ |

| | | | | | |
|---|---|---|---|---|---|
| 7 | $\frac{7P_7 - P_1}{6}$ | $\left(\frac{\eta_1^2\eta_7^2}{Z}\right)^{3/2}$ | $(1 - 26X - 27X^2)^{1/2}$ | $4ZX(1 + 3X)$ | $\sum_k \binom{j}{k}^2 \binom{2k}{j}\binom{j+k}{j}$ |
| 8 | $\frac{8P_8 - 4P_4 - 2P_2 + P_1}{3}$ | $\frac{\eta_2^4\eta_4^4}{Z^2}$ | $(1 + 48X + 512X^2)^{1/2}$ | $-8ZX(1 + 24X)$ | $(-1)^j \binom{2j}{j} \sum_k \binom{j}{k} \binom{2k}{k}\binom{2j-2k}{j-k}$ |
| 9 | $\frac{9P_9 - 6P_3 + P_1}{4}$ | $\frac{\eta_3^8}{Z^2}$ | $(1 + 36X + 432X^2)^{1/2}$ | $-6ZX(1 + 27X)$ | $\binom{2j}{j} \sum_k (-3)^{j-3k}\binom{j}{k} \binom{j-k}{k}\binom{j-2k}{k}$ |
| 10 | $\frac{10P_{10} + 5P_5 - 2F_2 - P_1}{12}$ | $\left(\frac{\eta_1\eta_2\eta_5\eta_{10}}{Z}\right)^{4/3}$ | $(1 - 12X - 64X^2)^{1/2}$ | $2ZX(1 + 15X)$ | $\sum_k \binom{j}{k}^4$ |

**Table 14.2** Functions satisfying Hypotheses 1–3 of Theorem 14.14

| level | $Z$ | $X$ | $B = \frac{1}{Z} q \frac{\mathrm{d}}{\mathrm{d}q} \log X$ | $BX \frac{\mathrm{d}}{\mathrm{d}X}\left(BX \frac{\mathrm{d}Z}{\mathrm{d}X}\right) - \frac{1}{2Z}\left(BX \frac{\mathrm{d}Z}{\mathrm{d}X}\right)^2$ | $h(j)$ |
|---|---|---|---|---|---|
| 3 | $z_3^2$ | $\left(\frac{\eta_1^2\eta_3^2}{Z}\right)^3$ | $(1-108X)^{1/2}$ | $120ZX$ | $\binom{3j}{j}\binom{2j}{j}^2$ |
| 7 | $z_7^2$ | $\left(\frac{\eta_1^2\eta_7^2}{Z}\right)^{3/2}$ | $(1-26X-27X^2)^{1/2}$ | $4ZX(1+3X)$ | $\sum_k \binom{j}{k}^2\binom{2k}{j}\binom{j+k}{j}$ |
| 11 | $z_{11}^2$ | $\frac{\eta_1^2\eta_{11}^2}{Z}$ | $(1-20X+56X^2-44X^3)^{1/2}$ | $4ZX(1-8X+11X^2)$ | $T_{11}$ in Theorem 11.12 |
| 23 | $z_{23}^2$ | $\left(\frac{\eta_1^2\eta_{23}^2}{Z}\right)^{1/2}$ | $(1-3X+2X^2+X^3)^{1/2}$ $\times(1-11X+22X^2-19X^3)^{1/2}$ | $4ZX(1-11X+37X^2$ $-48X^3+12X^4+19X^5)$ | $T_{23}$ in Theorem 11.12 |

Notation: $z_p = \sum_{j=-\infty}^{\infty} \sum_{k=-\infty}^{\infty} q^{j^2+jk+(p+1)k^2/4}$

**Table 14.3** Data for Theorem 14.14: $\ell = 1, h(j) = \binom{6j}{3j}\binom{3j}{j}\binom{2j}{j}$

| $q$ | $N$ | $X_{\ell,N}$ | $\lambda_{\ell,N}$ | Reference: Ramanujan, [248] | Reference: This book |
|---|---|---|---|---|---|
| $e^{-2\pi\sqrt{N}}$ | 2 | $\frac{1}{20^3}$ | $\frac{3}{28}$ | | (14.37) |
| | 3 | $\frac{4}{60^3}$ | $\frac{1}{11}$ | (33) | (14.24) |
| | 4 | $\frac{1}{66^3}$ | $\frac{5}{63}$ | | (14.38) |
| | 7 | $\frac{1}{255^3}$ | $\frac{8}{133}$ | (34) | (14.16) |
| $-e^{-\pi\sqrt{N}}$ | 7 | $\frac{-1}{15^3}$ | $\frac{8}{63}$ | | (14.17) |
| | 11 | $\frac{-1}{32^3}$ | $\frac{15}{154}$ | | |
| | 19 | $\frac{-1}{96^3}$ | $\frac{25}{342}$ | | |
| | 27 | $\frac{-9}{480^3}$ | $\frac{31}{506}$ | | (14.32) |
| | 43 | $\frac{-1}{960^3}$ | $\frac{263}{5418}$ | | |
| | 67 | $\frac{-1}{5280^3}$ | $\frac{10177}{261702}$ | | |
| | 163 | $\frac{-1}{640320^3}$ | $\frac{13591409}{545140134}$ | | (14.77) (not proved) |

**Table 14.4** Data for Theorem 14.14: $\ell = 2$, $h(j) = \binom{4j}{2j}\binom{2j}{j}^2$

| $q$ | $N$ | $X_{\ell,N}$ | $\lambda_{\ell,N}$ | Reference | |
|---|---|---|---|---|---|
| | | | | Ramanujan, [248] | This book |
| $e^{-2\pi\sqrt{N/2}}$ | 2 | $\frac{2}{6^4}$ | $\frac{1}{7}$ | | (14.39) |
| | 3 | $\frac{1}{48^2}$ | $\frac{1}{8}$ | (40) | (14.81) |
| | 5 | $\frac{1}{12^4}$ | $\frac{1}{10}$ | (41) | |
| | 9 | $\frac{1}{28^4}$ | $\frac{3}{40}$ | (42) | (14.82) |
| | 11 | $\frac{1}{1584^2}$ | $\frac{19}{280}$ | (43) | |
| | 29 | $\frac{1}{396^4}$ | $\frac{1103}{26390}$ | (44) | (14.2) (not proved) |
| $-e^{-\pi\sqrt{N}}$ | 5 | $\frac{-1}{2^{10}}$ | $\frac{3}{20}$ | (35) | |
| | 7 | $\frac{-1}{63^2}$ | $\frac{8}{65}$ | | (14.18) |
| | 9 | $\frac{-3}{192^2}$ | $\frac{3}{28}$ | (36) | (14.80) |
| | 13 | $\frac{-1}{324^2}$ | $\frac{23}{260}$ | (37) | |
| | 25 | $\frac{-5}{360^2}$ | $\frac{41}{644}$ | (38) | |
| | 37 | $\frac{-1}{672^2}$ | $\frac{1123}{21460}$ | (39) | |

**Table 14.5** Data for Theorem 14.14: $\ell = 3$, $h(j) = \binom{3j}{j}\binom{2j}{j}^2$

| $q$ | $N$ | $X_{\ell,N}$ | $\lambda_{\ell,N}$ | Reference: Ramanujan, [248] | Reference: This book |
|---|---|---|---|---|---|
| $e^{-2\pi\sqrt{N/3}}$ | 2 | $\frac{1}{6^3}$ | $\frac{1}{6}$ | | (14.55) |
| | 4 | $\frac{4}{18^3}$ | $\frac{2}{15}$ | (31) | (14.79) |
| | 5 | $\frac{1}{15^3}$ | $\frac{4}{33}$ | (32) | (14.84) |
| $-e^{-\pi\sqrt{N/3}}$ | 9 | $\frac{-1}{192}$ | $\frac{1}{5}$ | | (14.30) |
| | 17 | $\frac{-1}{12^3}$ | $\frac{7}{51}$ | | |
| | 25 | $\frac{-25}{60^3}$ | $\frac{1}{9}$ | | |
| | 41 | $\frac{-1}{48^3}$ | $\frac{53}{615}$ | | |
| | 49 | $\frac{-49}{252^3}$ | $\frac{13}{165}$ | | |
| | 89 | $\frac{-1}{300^3}$ | $\frac{827}{14151}$ | | |

**Table 14.6** Data for Theorem 14.14: $\ell = 4$, $h(j) = \binom{2j}{j}^3$

| $q$ | $N$ | $X_{\ell,N}$ | $\lambda_{\ell,N}$ | Reference: Ramanujan, [248] | Reference: This book |
|---|---|---|---|---|---|
| $e^{-\pi\sqrt{N}}$ | 3 | $\frac{1}{256}$ | $\frac{1}{6}$ | (28) | (14.3) |
| | 7 | $\frac{1}{4096}$ | $\frac{5}{42}$ | (29) | (14.4) |
| $-e^{-\pi\sqrt{N}}$ | 2 | $\frac{-1}{64}$ | $\frac{1}{4}$ | | (14.33) |
| | 4 | $\frac{-1}{512}$ | $\frac{1}{6}$ | | (14.34) |

**Table 14.7** Data for Theorem 14.14: $\ell = 5$, $h(j) = \binom{2j}{j}\sum_k \binom{j}{k}^2\binom{j+k}{k}$

| $q$ | $N$ | $X_{\ell,N}$ | $\lambda_{\ell,N}$ |
|---|---|---|---|
| $e^{-2\pi\sqrt{N/5}}$ | 2 | $\frac{1}{72}$ | $\frac{1}{5}$ |
| | 3 | $\frac{1}{147}$ | $\frac{2}{11}$ |
| $-e^{-\pi\sqrt{N/5}}$ | 23 | $\frac{-1}{828}$ | $\frac{29}{190}$ |
| | 47 | $\frac{-1}{15228}$ | $\frac{71}{682}$ |

**Table 14.8** Data for Theorem 14.14: $\ell = 6$, $h(j) = \binom{2j}{j} \sum_{k,\ell} (-8)^{j-k} \binom{j}{k} \binom{k}{\ell}^3$

| $q$ | $N$ | $X_{\ell,N}$ | $\lambda_{\ell,N}$ |
|---|---|---|---|
| $e^{-2\pi\sqrt{N/6}}$ | 3 | $\frac{1}{64}$ | $\frac{3}{10}$ |
| | 5 | $\frac{1}{288}$ | $\frac{1}{5}$ |
| | 7 | $\frac{1}{864}$ | $\frac{11}{70}$ |
| | 13 | $\frac{1}{10368}$ | $\frac{241}{2210}$ |
| | 17 | $\frac{1}{39168}$ | $\frac{73}{770}$ |
| $-e^{-\pi\sqrt{N/3}}$ | 3 | $\frac{-1}{48}$ | $0$ |
| | 5 | $\frac{-1}{81}$ | $\frac{4}{35}$ |
| | 7 | $\frac{-1}{144}$ | $\frac{1}{7}$ |
| | 11 | $\frac{-1}{432}$ | $\frac{8}{55}$ |
| | 19 | $\frac{-1}{2736}$ | $\frac{8}{65}$ |
| | 31 | $\frac{-1}{24336}$ | $\frac{107}{1085}$ |
| | 59 | $\frac{-1}{1123632}$ | $\frac{25808}{359605}$ |

**Table 14.9** Data for Theorem 14.14: $\ell = 6$, $h(j) = \binom{2j}{j} \sum_k \binom{j}{k}^2 \binom{2k}{k}$

| $q$ | $N$ | $X_{\ell,N}$ | $\lambda_{\ell,N}$ |
|---|---|---|---|
| $e^{-2\pi\sqrt{N/6}}$ | 2 | $\frac{1}{54}$ | $\frac{1}{5}$ |
| | 3 | $\frac{1}{100}$ | $\frac{3}{16}$ |
| | 5 | $\frac{1}{324}$ | $\frac{13}{80}$ |
| | 7 | $\frac{1}{900}$ | $\frac{1}{7}$ |
| | 13 | $\frac{1}{10404}$ | $\frac{7}{65}$ |
| | 17 | $\frac{1}{39204}$ | $\frac{899}{9520}$ |
| $-e^{-\pi\sqrt{N/3}}$ | 5 | $\frac{-1}{45}$ | $\frac{2}{7}$ |
| | 7 | $\frac{-1}{108}$ | $\frac{13}{56}$ |
| | 11 | $\frac{-1}{396}$ | $\frac{7}{40}$ |
| | 19 | $\frac{-1}{2700}$ | $\frac{253}{1976}$ |
| | 31 | $\frac{-1}{24300}$ | $\frac{2239}{22568}$ |
| | 59 | $\frac{-1}{1123596}$ | $\frac{2587}{36040}$ |

**Table 14.10** Data for Theorem 14.14: $\ell = 6$, $h(j) = \binom{2j}{j} \sum_k \binom{j}{k}^3$

| $q$ | $N$ | $X_{\ell,N}$ | $\lambda_{\ell,N}$ |
|---|---|---|---|
| $e^{-2\pi\sqrt{N/6}}$ | 2 | $\frac{1}{50}$ | $\frac{2}{9}$ |
| | 3 | $\frac{1}{96}$ | $\frac{1}{5}$ |
| | 5 | $\frac{1}{320}$ | $\frac{1}{6}$ |
| | 7 | $\frac{1}{896}$ | $\frac{13}{90}$ |
| | 13 | $\frac{1}{10400}$ | $\frac{11}{102}$ |
| | 17 | $\frac{1}{39200}$ | $\frac{53}{561}$ |
| $-e^{-\pi\sqrt{N/3}}$ | 5 | $\frac{-1}{49}$ | $\frac{4}{15}$ |
| | 7 | $\frac{-1}{112}$ | $\frac{2}{9}$ |
| | 11 | $\frac{-1}{400}$ | $\frac{17}{99}$ |
| | 19 | $\frac{-1}{2704}$ | $\frac{109}{855}$ |
| | 31 | $\frac{-1}{24304}$ | $\frac{58}{585}$ |
| | 59 | $\frac{-1}{1123600}$ | $\frac{14903}{207621}$ |

**Table 14.11** Data for Theorem 14.14: $\ell = 7$, $h(j) = \sum_k \binom{j}{k}^2 \binom{2k}{j} \binom{j+k}{j}$

| $q$ | $N$ | $X_{\ell,N}$ | $\lambda_{\ell,N}$ |
|---|---|---|---|
| $e^{-2\pi\sqrt{N/7}}$ | 4 | $\frac{1}{5^3}$ | $\frac{4}{21}$ |
| $-e^{-\pi\sqrt{N/7}}$ | 13 | $\frac{-1}{4^3}$ | $\frac{10}{39}$ |
| | 61 | $\frac{-1}{22^3}$ | $\frac{1286}{11895}$ |

**Table 14.12** Data for Theorem 14.14: $\ell = 8$, $h(j) = (-1)^j \binom{2j}{j} \sum_k \binom{j}{k} \binom{2k}{k} \binom{2j-2k}{j-k}$

| $q$ | $N$ | $X_{\ell,N}$ | $\lambda_{\ell,N}$ |
|---|---|---|---|
| $e^{-2\pi\sqrt{N/8}}$ | 3 | $\frac{1}{64}$ | $\frac{1}{6}$ |
| | 5 | $\frac{1}{160}$ | $\frac{1}{6}$ |
| | 9 | $\frac{1}{800}$ | $\frac{1}{7}$ |
| | 11 | $\frac{1}{1600}$ | $\frac{61}{462}$ |
| | 29 | $\frac{1}{156832}$ | $\frac{193}{2310}$ |
| $-e^{-2\pi\sqrt{N/8}}$ | 5 | $\frac{-1}{128}$ | $\frac{7}{30}$ |
| | 9 | $\frac{-1}{768}$ | $\frac{11}{70}$ |
| | 11 | $\frac{-1}{1568}$ | $\frac{23}{165}$ |
| | 29 | $\frac{-1}{156800}$ | $\frac{2081}{24882}$ |

**Table 14.13** Data for Theorem 14.14: $\ell = 9$, $h(j) = \binom{2j}{j} \sum_k (-3)^{j-3k} \binom{j}{k} \binom{j-k}{k} \binom{j-2k}{k}$

| $q$ | $N$ | $X_{\ell,N}$ | $\lambda_{\ell,N}$ |
|---|---|---|---|
| $e^{-2\pi\sqrt{N/9}}$ | 4 | $\frac{1}{54}$ | $\frac{2}{7}$ |
| | 7 | $\frac{1}{243}$ | $\frac{26}{133}$ |
| $-e^{-\pi\sqrt{N/9}}$ | 7 | $\frac{-1}{27}$ | $\frac{2}{7}$ |
| | 11 | $\frac{-1}{44}$ | $\frac{3}{14}$ |
| | 19 | $\frac{-1}{108}$ | $\frac{7}{38}$ |
| | 43 | $\frac{-1}{972}$ | $\frac{85}{602}$ |
| | 67 | $\frac{-1}{5292}$ | $\frac{481}{4154}$ |
| | 163 | $\frac{-1}{640332}$ | $\frac{58831}{786638}$ |

**Table 14.14** Data for Theorem 14.14: $\ell = 10$, $h(j) = \sum_k \binom{j}{k}^4$

| $q$ | $N$ | $X_{\ell,N}$ | $\lambda_{\ell,N}$ |
|---|---|---|---|
| $e^{-2\pi\sqrt{N/10}}$ | 3 | $\frac{1}{36}$ | $\frac{1}{4}$ |
| | 7 | $\frac{1}{196}$ | $\frac{11}{60}$ |
| | 13 | $\frac{1}{1296}$ | $\frac{9}{65}$ |
| | 19 | $\frac{1}{5776}$ | $\frac{47}{408}$ |
| $-e^{-\pi\sqrt{N/5}}$ | 5 | $\frac{-1}{20}$ | $\frac{1}{3}$ |
| | 9 | $\frac{-1}{64}$ | $\frac{1}{4}$ |
| | 17 | $\frac{-1}{324}$ | $\frac{3}{17}$ |

## 14.11 Notes

The series (14.33) was discovered long before Ramanujan by G. Bauer [42] in 1859 and is called Bauer's series. Bauer was interested in expanding functions in series of Legendre polynomials, and the identity (14.33) is obtained as a byproduct. Bauer's method has been analyzed and extended by P. Levrie [205].

The original work that gave rise to the topics in this chapter is Ramanujan's paper [248]. Ramanujan gave 17 series for $1/\pi$, which correspond to the levels $1 \leq \ell \leq 4$. Data for 16 of Ramanujan's series can be found in Tables 14.3–14.6 above, and the remaining series is detailed in (14.66). This chapter contains proofs of 19 series for levels $1 \leq \ell \leq 4$, including 9 of Ramanujan's original series — see the references in Tables 14.3–14.6. Despite not being one of the 17 series in [248], Bauer's series (14.33) was known to Ramanujan and occurs as Example 14 in Section 7 of Chapter 10 of Ramanujan's second notebook [253]; see also [45, pp. 23–24].

The modern-day study of series and iterations for $1/\pi$ began with the book by J. M. Borwein and P. B. Borwein [60] and several papers by the Borweins and coauthors. See their expository articles [61] and [64], their paper with Bailey [65], and the source book with Berggren [43].

The particularly impressive series

$$\sum_{n=0}^{\infty} \binom{6n}{3n}\binom{3n}{n}\binom{2n}{n}\left(n + \frac{13591409}{545140134}\right)\left(\frac{-1}{640320}\right)^{3n}$$
$$= \frac{2^6 \times 5^{3/2} \times 23^{3/2} \times 29^{3/2}}{3^{3/2} \times 7 \times 11 \times 19 \times 127 \times 163} \times \frac{1}{\pi} \qquad (14.77)$$

yields 14 decimal digits per term. It corresponds to the data for $N = 163$ in Table 14.3. It is one of the series found by D. V. Chudnovsky and G. V. Chudnovsky [98] and was used by the Chudnovskys in several world record setting calculations for $\pi$. An interesting account of the lives and work of the Chudnovskys has been given in the New Yorker article by R. Preston [244].

An historical account of Ramanujan's series for $1/\pi$ has been given by N. D. Baruah, B. C. Berndt, and H. H. Chan [40].

The first series for $1/\pi$ for a level $\ell > 4$ appears to be the series (14.57) due to Takeshi Sato in a two-page abstract for a conference [260]. Sato also indicated that there are similar series that involve the Apéry numbers.

The theory for level 6 has been developed in detail by H. H. Chan in a series of papers including works with S. H. Chan and Z.-G. Liu [79], with Y. Tanigawa, Y. Yang, and W. Zudilin [90], with H. Verrill [92] and with Zudilin [94].

The level 7 theory is worked out in [114], the details for levels 8 and 9 are given in [81], and the level 10 theory is in [113]. Series for levels 11 and 23 are in the thesis by J. Ge [160]. See also work with J. Ge and D. Ye [115]. Results for level 13 are in the honours thesis by Ye [304]. See also [127] and [128]. Series for levels 14 and 15 have been given in work with Ye [129].

The theory of Ramanujan's series for $1/\pi$ was extended significantly by the announcement of a large number of conjectures by Z.-W. Sun that are summarized in [279]. Sun's conjectures have stimulated and inspired works by W. Zudilin and coauthors, including work with H. H. Chan and J. Wan [93], the paper [125], works with J. Guillera [172], with J. Wan [294] and the paper [308]. See also the work of M. Rogers and A. Straub [258] and the works of J. Wan [292], [293].

Let us briefly discuss the contents of this chapter. The approach we have used in Section 14.2 follows Baruah and Berndt [39]. The material in Sections 14.3–14.5 is based on ideas of A. Aycock [33] that have been developed further by Zudilin [309] and by Guillera and Zudilin [172]. The general formula in Theorem 14.14 is due to H. H. Chan, S. H. Chan, and Z.-G. Liu [79]. The classification in Section 14.7 is from [81], and its use in determining which series can be translated to other series is described in more detail in [131]. The WZ proof of Bauer's series in Section 14.8 is due to D. Zeilberger and his computer Shalosh B. Ekhad [145]. A sign error in one of the identities in [145] has been corrected in (14.8). The theory of WZ proofs has been greatly extended in a series of papers by J. Guillera [168, 169, 170]. The idea of using Landen's transformation to obtain the quadratic iteration for $1/\pi$ is due to J. Guillera [171]. The method has the potential to create other iterations that converge to $1/\pi$. For example, a quintic iteration that is based on (13.9) and (13.10) is derived in [116], and a cubic iteration that is an analogue of Theorem 14.20 is outlined below in Exercise 10. Tables 14.3–14.10, 14.12, and 14.13 are from [81], and Tables 14.11 and 14.14 are reproduced from [114] and [113], respectively.

## 14.12 Exercises

1. Consider the modular equation of degree 2 and level 3 from (14.50):

$$(xy)^{1/3} + (1-x)^{1/3}(1-y)^{1/3} = 1. \tag{14.78}$$

Let $X = x(1-x)/27$ and $Y = y(1-y)/27$. The goal of this exercise is to deduce the polynomial form of the modular equation for $X$ and $Y$.

a. Take the cube of the modular equation (14.78) and simplify, to show that

$$2xy - x - y + 3\left(xy(1-x)(1-y)\right)^{1/3} = 0.$$

b. Show that the result of Part (a) is equivalent to the identity

$$\sqrt{1-108X}\sqrt{1-108Y} = 1 - 54\,(XY)^{1/3}.$$

c. Square the result of Part (b) and show that the result may be written in the form

$$\alpha - 27\alpha^2 + 108\alpha^3 = X + Y$$

where $\alpha = (XY)^{1/3}$.

d. By direct calculation, show that

$$\begin{aligned}(\alpha - 27\alpha^2 + 108\alpha^3)^3 &- 324\alpha^3(\alpha - 27\alpha^2 + 108\alpha^3)^2\\ &+(81\alpha^3 + 34992\alpha^6)(\alpha - 27\alpha^2 + 108\alpha^3)\\ &= 1259712\alpha^9 - 10935\alpha^6 + \alpha^3.\end{aligned}$$

Hence, deduce the modular equation for $X$ and $Y$ given by (14.52).

2. a. Find the polynomial form of the modular equation of degree 4 and level 3. Hint: Use computer algebra to eliminate $Y$ from the pair of modular equations

$$g(X, Y) = 0 \qquad \text{and} \qquad g(Y, W) = 0,$$

where $g(X, Y) = 0$ is the polynomial form of the modular equation of degree 2 and level 3 given by (14.52).

b. Deduce that

$$X_{3,4} = X\left(e^{-2\pi\sqrt{4/3}}\right) = X\left(e^{-2\pi/\sqrt{12}}\right) = \frac{1}{1458}$$

and

$$\lambda_{3,4} = \frac{2}{15}.$$

c. Hence obtain Ramanujan's formula [248, Eq. (31)]:

$$\sum_{j=0}^{\infty} \binom{3j}{j} \binom{2j}{j}^2 \left(j + \frac{2}{15}\right) \left(\frac{1}{1458}\right)^j = \frac{9}{20\pi}. \tag{14.79}$$

3. Use the transformation formula from Example 13.6 given by

$$\left(\frac{1+8x}{1-\frac{8x}{9}}\right)^{1/2} {}_3F_2\left(\begin{matrix}\frac{1}{6}, \frac{1}{2}, \frac{5}{6} \\ 1, 1\end{matrix}; \frac{64x^3(1-x)}{(9-8x)^3}\right)$$
$$= {}_3F_2\left(\begin{matrix}\frac{1}{6}, \frac{1}{2}, \frac{5}{6} \\ 1, 1\end{matrix}; \frac{64x(1-x)^3}{(1+8x)^3}\right)$$

together with Lemma 14.7 to prove the identities in Theorem 14.10.

4. Use the transformation formula

$$\frac{1-4t-16t^3+16t^4}{1-4t+24t^2-40t^3-8t^4}$$
$$\times {}_3F_2\left(\begin{matrix}\frac{1}{4}, \frac{1}{2}, \frac{3}{4} \\ 1, 1\end{matrix}; 256\,\frac{t^2\,(1-t)^6\,(1-4t)^3\,(1-2t)\,(1+2t)^3}{(1-4t+24t^2-40t^3-8t^4)^4}\right)$$
$$= {}_3F_2\left(\begin{matrix}\frac{1}{4}, \frac{1}{2}, \frac{3}{4} \\ 1, 1\end{matrix}; -256\,\frac{t^3\,(1-t)\,(1-4t)\,(1-2t)^6\,(1+2t)^2}{(1-4t-16t^3+16t^4)^4}\right)$$

from Theorem 13.3 and the value

$$t = \frac{1}{4}\left(7 + 3\sqrt{3} - \sqrt{72 + 42\sqrt{3}}\right)$$

to translate the identity (14.39), and hence prove the identity

$$\sum_{n=0}^{\infty} \binom{4n}{2n} \binom{2n}{n}^2 \left(n + \frac{3}{28}\right) \left(\frac{-1}{12288}\right)^n = \frac{4\sqrt{3}}{21} \times \frac{1}{\pi}. \tag{14.80}$$

5. a. Use (14.55) and a translation technique to deduce the identity

$$\sum_{n=0}^{\infty} \binom{4n}{2n} \binom{2n}{n}^2 \left(n + \frac{1}{8}\right) \left(\frac{1}{48}\right)^{2n} = \frac{\sqrt{3}}{4} \times \frac{1}{\pi}. \tag{14.81}$$

b. Use (14.33) and a translation technique to deduce the identity

$$\sum_{n=0}^{\infty} \binom{4n}{2n}\binom{2n}{n}^2 \left(n+\frac{3}{40}\right)\left(\frac{1}{28}\right)^{4n} = \frac{49\sqrt{3}}{360} \times \frac{1}{\pi} \tag{14.82}$$

6. This exercise is due to M. Rogers [257, (3.9), (3.10)]. Use the identity

$$G_b(x) = \frac{1}{1+4x}\, {}_3F_2\left( \begin{matrix} \frac{1}{3}, \frac{1}{2}, \frac{2}{3} \\ 1, 1 \end{matrix} ; \frac{108x^2}{(1+4x)^3} \right)$$

from Example 13.7 to translate the identities (14.55) and (14.79), and hence prove

$$\sum_{n=0}^{\infty} s_b(n) \left(n+\frac{1}{3}\right)\left(\frac{-1}{32}\right)^n = \frac{2}{3\pi}$$

and

$$\sum_{n=0}^{\infty} s_b(n) \left(n+\frac{1}{5}\right)\left(\frac{-1}{64}\right)^n = \frac{8\sqrt{3}}{15\pi},$$

where $s_b(n)$ are the Domb numbers studied in Section 6.9. The last identity is originally due to H. H. Chan, S. H. Chan, and Z.-G. Liu [79, (1.3)].

7. Let $Z$ and $X$ be given by

$$Z = Z_5 = \frac{1}{4}\left(5P(q^5) - P(q)\right) \qquad \text{and} \qquad X = X_5 = \frac{\eta_1^4\eta_5^4}{Z^2},$$

where $Z_5$ and $X_5$ are the level 5 functions studied in Chapter 5.

a. Follow the techniques of Example 14.15 to show that Hypotheses 1–3 of Theorem 14.14 are satisfied, with $\ell = 5$,

$$B = \sqrt{1-44X-16X^2}$$

and

$$h(j) = \binom{2j}{j} \sum_{k=0}^{j} \binom{j}{k}^2 \binom{j+k}{k}.$$

Hint: Use Theorems 5.21, 5.32, and 5.35.

b. Let $N = 2$ and put $Y = X(q^2)$. Show that

$$X^3 + Y^3 - 36(X^3Y + XY^3) + 324(X^3Y^2 + X^2Y^3) \\ + 32(X^2Y + XY^2) + 495X^2Y^2 - XY = 0. \qquad (14.83)$$

Hint: By Theorem 5.20,

$$X(q) = \frac{r^5(q)\left(1 - 11r^5(q) - r^{10}(q)\right)}{\left(1 + r^{10}(q)\right)^2}$$

where $r(q)$ is the Rogers–Ramanujan continued fraction. And, by Theorem 10.2,

$$r^5(q) = k\left(\frac{1-k}{1+k}\right)^2 \qquad \text{and} \qquad r^5(q^2) = k^2\left(\frac{1+k}{1-k}\right).$$

Use these to show that the left-hand side of (14.83) simplifies identically to zero.

c. Show that

$$X\left(e^{-2\pi\sqrt{2/5}}\right) = \frac{1}{72} \quad \text{and} \quad \lambda_{5,2} = \frac{1}{5}.$$

d. Use Theorem 14.14 to deduce that

$$\sum_{j=0}^{\infty} \binom{2j}{j} \left\{ \sum_{k=0}^{j} \binom{j}{k}^2 \binom{j+k}{k} \right\} \left(j + \frac{1}{5}\right) \left(\frac{1}{72}\right)^j = \frac{9\sqrt{2}}{10} \times \frac{1}{\pi}.$$

8. This exercise relies on the modular equation of degree 5 and level 3 given by

$$a(q)a(q^5) = b(q)b(q^5) + c(q)c(q^5) + 3\sqrt{b(q)b(q^5)c(q)c(q^5)}.$$

See Berndt [48, p. 124], work with Ye [129, Theorem 5.5] or Hirschhorn, Garvan, and Borwein [185, Section 10].

a. Let

$$x = \frac{c^3(q)}{a^3(q)} \qquad \text{and} \qquad y = \frac{c^3(q^5)}{a^3(q^5)}.$$

Show that

$$(xy)^{1/3} + (1-x)^{1/3}(1-y)^{1/3} + 3(xy)^{1/6}(1-x)^{1/6}(1-y)^{1/6} = 1.$$

b. Let

$$X = X_3(q) = \frac{x(1-x)}{27}, \qquad Y = X_3(q^5) = \frac{y(1-y)}{27}.$$

Find a polynomial form of the modular equation satisfied by $X$ and $Y$.

c. Show that $X_{3,5} = 1/3375$ and $\lambda_{3,5} = 4/33$, and hence deduce Ramanujan's formula [248, Eq. (32)]:

$$\sum_{j=0}^{\infty} \binom{3j}{j} \binom{2j}{j}^2 \left(j + \frac{4}{33}\right) \left(\frac{1}{3375}\right)^j = \frac{5\sqrt{3}}{22\pi}. \tag{14.84}$$

9. Give another proof of (14.76) as follows.

a. Consider the inversion formula in Theorem 4.15, viz.,

$$q = \exp\left(-\pi \frac{F(1-x_4)}{F(x_4)}\right) \qquad \text{where} \quad F(x) = {}_2F_1\left(\begin{matrix}\frac{1}{2}, \frac{1}{2} \\ 1\end{matrix}; x\right).$$

Take the logarithmic derivative with respect to $x_4$ and use Theorem 4.7 to show that

$$\frac{1}{z_4^2 x_4(1-x_4)} = \pi \left(\frac{F(x_4)F'(1-x_4) + F'(x_4)F(1-x_4)}{F^2(x_4)}\right).$$

b. By Theorem 4.13 we have $z_4 = F(x_4)$. Hence, deduce that

$$x_4(1-x_4)\left(F(x_4)F'(1-x_4) + F'(x_4)F(1-x_4)\right) = \frac{1}{\pi}.$$

c. Now let $x_4 = \frac{1}{2}$ to obtain

$$\left(\sum_{k=0}^{\infty} \frac{(\frac{1}{2})_k^2}{k!^2} \frac{1}{2^k}\right)\left(\sum_{k=0}^{\infty} \frac{(\frac{1}{2})_k^2}{k!^2} \frac{k}{2^k}\right) = \frac{1}{\pi}.$$

10. Let $x_0$, $r_0$, and $c_0$ be given real numbers with $0 < x_0 < 1$, and consider the iteration scheme given by the recurrence relations

$$x_{n+1} = \frac{1 - \sqrt[3]{1-x_n^3}}{1 + 2\sqrt[3]{1-x_n^3}},$$

$$r_{n+1} = r_n(1 + 2x_{n+1})^2 - c_0\, 3^n\, x_{n+1}(1 + x_{n+1}).$$

a. Show that

$$\lim_{n\to\infty} r_n = L_0$$

where

$$L_0 = \left(\sum_{k=0}^{\infty} \frac{(\frac{1}{3})_k(\frac{2}{3})_k}{k!^2} x_0^{3k}\right)\left(\sum_{k=0}^{\infty} \frac{(\frac{1}{3})_k(\frac{2}{3})_k}{k!^2} (a_0 + b_0 k)\, x_0^{3k}\right),$$

$$a_0 = r_0 - \frac{c_0\, x_0^3}{3}, \quad \text{and} \quad b_0 = (1 - x_0^3)c_0.$$

b. Show that the rate of convergence in Part a is cubic.

c. Show that if $x_0 = \frac{1}{\sqrt[3]{2}}$, $r_0 = \frac{1}{3}$ and $c_0 = 2$, then $\lim_{n\to\infty} r_n = \frac{\sqrt{3}}{2\pi}$.

Hints: Start with the cubic analogue of Landen's transformation in Exercise 10 in Chapter 4, and mimic the procedure in Lemma 14.18, Lemma 14.19, Theorem 14.20, and Example 14.21. The iterations for $a_k$, $b_k$, and $x_k$ should be self-evident. Then put

$$c_n = \frac{b_n}{1 - x_n^3} \quad \text{so that} \quad c_{n+1} = 3c_n$$

and define $r_n$ by

$$r_n = a_n + c_0\, 3^{n-1}\, x_n^3.$$

The initial conditions and limiting value in Part c can be obtained from the case $\ell = 3$ of Theorem 4.27 and following the procedure in Exercise 9.

11. This exercise follows Guillera and Zudilin [172, Theorem 1].

a. Prove the Stolz–Cesàro theorem: Let $\{a_n\}$ be an increasing sequence of real numbers with the property that $\lim_{n\to\infty} a_n = +\infty$. Let $\{b_n\}$ be another sequence and suppose that the limit

$$\lim_{n\to\infty} \frac{b_{n+1} - b_n}{a_{n+1} - a_n} = \ell$$

exists. Then $\lim_{n\to\infty} b_n/a_n$ also exists, and

$$\lim_{n\to\infty} \frac{b_n}{a_n} = \ell.$$

Hint (reproduced nearly verbatim from [227]): With $\ell-\epsilon < \frac{b_{n+1}-b_n}{a_{n+1}-a_n} < \ell+\epsilon$ for $n \geq N$, all the points $(x, y) = (a_n, b_n)$ for $n \geq N$ will lie in the wedge formed by the two lines through the point $(x, y) = (a_N, b_N)$ with slopes $\ell - \epsilon$ and $\ell + \epsilon$. For large enough $x$, this wedge will stay entirely within the wider wedge formed by the lines $y = (\ell - 2\epsilon)x$ and $y = (\ell + 2\epsilon)x$ through the origin. Since $a_n \to +\infty$, all points $(x, y) = (a_n, b_n)$ for $n \geq M$, say, will have large enough $x$-coordinate to lie in the part of the narrower wedge that lies inside the wider wedge.

b. Suppose $s_1, \ldots, s_m$ are real numbers with $0 < s_1, \ldots, s_m < 1$, and let

$$a_n = \sum_{k=0}^{n} \frac{(\frac{1}{2})_k}{k!} \quad \text{and} \quad b_n = \sum_{k=0}^{n} \frac{(\frac{1}{2})_k (s_1)_k (1 - s_1)_k \cdots (s_m)_k (1 - s_m)_k}{k!^{2m+1}} n^m.$$

Show that

$$\lim_{n\to\infty} \frac{b_{n+1} - b_n}{a_{n+1} - a_n} = \prod_{j=1}^{m} \left( \frac{\sin \pi s_j}{\pi} \right).$$

Hint: This only requires Euler's definition of the gamma function

$$\Gamma(s) = \lim_{n\to\infty} \frac{n! n^{s-1}}{(s)_n}$$

and the reflection formula

$$\Gamma(s)\Gamma(1 - s) = \frac{\pi}{\sin \pi s}.$$

See Chapter 1 of the book by Andrews, Askey, and Roy [24] for properties of the gamma function.

c. Use the results of Parts a and b, and the binomial series (0.109) with $a = 1/2$, to deduce that

$$\lim_{x\to 1^-} \sqrt{1 - x} \sum_{n=0}^{\infty} \frac{(\frac{1}{2})_n (s_1)_n (1 - s_1)_n \cdots (s_m)_n (1 - s_m)_n}{n!^{2m+1}} n^m x^n = \prod_{j=1}^{m} \left( \frac{\sin \pi s_j}{\pi} \right).$$

In particular, the value $m = 1$ gives Lemma 14.11.

# References


1. C. Adiga, B.C. Berndt, S. Bhargava, G.N. Watson, Chapter 16 of Ramanujan's second notebook: theta-functions and $q$-series. Mem. Am. Math. Soc. **53** (315), v + 85 pp. (1985)
2. C. Adiga, S. Cooper, J.H. Han, A general relation between sums of squares and sums of triangular numbers. Int. J. Number Theory **1**, 175–182 (2005)
3. C. Adiga, T. Kim, M.S. Mahadeva Naika, H.S. Madhusudhan, On Ramanujan's cubic continued fraction and explicit evaluations of theta-functions. Indian J. Pure Appl. Math. **35**, 1047–1062 (2004)
4. C. Adiga, K.R. Vasuki, On sums of triangular numbers. Math. Stud. **70**, 185–190 (2001)
5. M. Aigner, G. Ziegler, *Proofs from the Book*, 2nd edn. (Springer, Berlin, 2001)
6. A. Alaca, Ş. Alaca, M.F. Lemire, K.S. Williams, Jacobi's identity and representations of integers by certain quaternary quadratic forms. Int. J. Modern Math. **2**, 143–176 (2007)
7. A. Alaca, Ş. Alaca, M.F. Lemire, K.S. Williams, Nineteen quaternary quadratic forms. Acta Arith. **130**, 277–310 (2007)
8. A. Alaca, Ş. Alaca, M.F. Lemire, K.S. Williams, Theta function identities and representations by certain quaternary quadratic forms. Int. J. Number Theory **4**, 219–239 (2008)
9. A. Alaca, Ş. Alaca, M.F. Lemire, K.S. Williams, Theta function identities and representations by certain quaternary quadratic forms II. Int. Math. Forum **3**, 539–579 (2008)
10. A. Alaca, Ş. Alaca, M.F. Lemire, K.S. Williams, The number of representations of a positive integer by certain quaternary quadratic forms. Int. J. Number Theory **5**, 13–40 (2009)
11. A. Alaca, Ş. Alaca, F. Uygul, K.S. Williams, Representations by sextenary quadratic forms whose coefficients are 1, 2 and 4. Acta Arith. **141**, 289–309 (2010)
12. A. Alaca, Ş. Alaca, K.S. Williams, On the two-dimensional theta functions of the Borweins. Acta Arith. **124**, 177–195 (2006)
13. A. Alaca, Ş. Alaca, K.S. Williams, Lambert series and Liouville's identities. Diss. Math. (Rozprawy Mat.) **445**, 1–72 (2007)
14. A. Alaca, Ş. Alaca, K.S. Williams, On the quaternary forms $x^2 + y^2 + z^2 + 5t^2$, $x^2 + y^2 + 5z^2 + 5t^2$ and $x^2 + 5y^2 + 5z^2 + 5t^2$. JP J. Algebra Number Theory Appl. **9**, 37–53 (2007)
15. A. Alaca, Ş. Alaca, K.S. Williams, Liouville's sextenary quadratic forms $x^2 + y^2 + z^2 + t^2 + 2u^2 + 2v^2$, $x^2 + y^2 + 2z^2 + 2t^2 + 2u^2 + 2v^2$ and $x^2 + 2y^2 + 2z^2 + 2t^2 + 2u^2 + 4v^2$. Far East J. Math. Sci. **30**, 547–556 (2008)
16. A. Alaca, Ş. Alaca, K.S. Williams, Some infinite products of Ramanujan type. Can. Math. Bull. **52**, 481–492 (2009)
17. A.M. Aldawoud, Ramanujan-type series for $1/\pi$ with quadratic irrationals, Master of Science Thesis, Massey University, New Zealand, 2012


 
S. Cooper, *Ramanujan's Theta Functions*, DOI 10.1007/978-3-319-56172-1

18. G. Almkvist, D. van Straten, W. Zudilin, Generalizations of Clausen's formula and algebraic transformations of Calabi–Yau differential equations. Proc. Edinb. Math. Soc. **54**, 273–295 (2011)
19. G. Almkvist, W. Zudilin, Differential equations, mirror maps and zeta values, in *Mirror Symmetry V*, ed. by N. Yui, S.-T. Yau, J.D. Lewis. AMS/IP Studies in Advanced Mathematics, vol. 38 (International Press and American Mathematical Society, Providence, RI, 2007), pp. 481–515
20. G.E. Andrews, On $q$-difference equations for certain well-poised basic hypergeometric series. Q. J. Math. Oxf. Ser. (2) **19**, 433–447 (1968)
21. G.E. Andrews, An introduction to Ramanujan's "lost" notebook. Am. Math. Mon. **86**, 89–108 (1979)
22. G.E. Andrews, *The Theory of Partitions* (Cambridge University Press, Cambridge, 1998)
23. G.E. Andrews, The well-poised thread: an organized chronicle of some amazing summations and their implications. Ramanujan J. **1**, 7–23 (1997)
24. G.E. Andrews, R. Askey, R. Roy, *Special Functions* (Cambridge University Press, Cambridge, 1999)
25. G.E. Andrews, B.C. Berndt, *Ramanujan's Lost Notebook, Part I* (Springer, New York, 2005)
26. G.E. Andrews, B.C. Berndt, *Ramanujan's Lost Notebook, Part II* (Springer, New York, 2009)
27. G.E. Andrews, B.C. Berndt, *Ramanujan's Lost Notebook, Part III* (Springer, New York, 2012)
28. G.E. Andrews, R.P. Lewis, Z.-G. Liu, An identity relating a theta function to a sum of Lambert series. Bull. Lond. Math. Soc. **33**, 25–31 (2001)
29. T.M. Apostol, *Modular Functions and Dirichlet Series in Number Theory*, 2nd edn. (Springer, New York, 1990)
30. R. Askey, Ramanujan's extensions of the gamma and beta functions. Am. Math. Mon. **87**, 346–359 (1980)
31. R. Askey, Math 805 course notes. Hand-written notes reproduced by spirit duplicator (ditto machine), University of Wisconsin, Madison, WI, 1989
32. R. Askey, The number of representations of an integer as the sum of two squares. Indian J. Math. **32**, 187–192 (1990)
33. A. Aycock, On proving some of Ramanujan's formulas for $1/\pi$ with an elementary method. `arXiv:1309.1140v2 [math.NT]` 5 September 2013
34. W.N. Bailey, *Generalized Hypergeometric Series* (Cambridge University Press, Cambridge, 1935)
35. W.N. Bailey, A note on two of Ramanujan's formulae. Q. J. Math. Oxf. Ser. (2) **3**, 29–31 (1952)
36. W.N. Bailey, A further note on two of Ramanujan's formulae. Q. J. Math. Oxf. Ser. (2) **3**, 158–160 (1952)
37. P. Barrucand, Sur la somme des puissances des coefficients multinomiaux et les puissances successives d'une fonction de Bessel. C. R. Acad. Sci. Paris **258**, 5318–5320 (1964)
38. P. Barrucand, S. Cooper, M. Hirschhorn, Relations between squares and triangles. Discr. Math. **248**, 245–247 (2002)
39. N.D. Baruah, B.C. Berndt, Eisenstein series and Ramanujan-type series for $1/\pi$. Ramanujan J. **23**, 17–44 (2010)
40. N.D. Baruah, B.C. Berndt, H.H. Chan, Ramanujan's series for $1/\pi$: a survey. Am. Math. Mon. **116**, 567–587 (2009)
41. N.D. Baruah, S. Cooper, M. Hirschhorn, Sums of squares and sums of triangular numbers induced by partitions of 8. Int. J. Number Theory **4**, 525–538 (2008)
42. G. Bauer, Von den Coefficienten der Reihen von Kugelfunctionen einer Variablen. J. Reine Angew. Math. **56**, 101–121 (1859)
43. L. Berggren, J.M. Borwein, P.B. Borwein, *Pi: A Source Book* (Springer, New York, 1997)
44. A. Berkovich, H. Yesilyurt, Ramanujan's identities and representation of integers by certain binary and quaternary quadratic forms. Ramanujan J. **20**, 375–408 (2009)
45. B.C. Berndt, *Ramanujan's Notebooks, Part II* (Springer, New York, 1989)
46. B.C. Berndt, *Ramanujan's Notebooks, Part III* (Springer, New York, 1991)

47. B.C. Berndt, *Ramanujan's Notebooks, Part IV* (Springer, New York, 1994)
48. B.C. Berndt, *Ramanujan's Notebooks, Part V* (Springer, New York, 1998)
49. B.C. Berndt, *Number Theory in the Spirit of Ramanujan* (American Mathematical Society, Providence, RI, 2006)
50. B.C. Berndt, S. Bhargava, F.G. Garvan, Ramanujan's theories of elliptic functions to alternative bases. Trans. Am. Math. Soc. **347**, 4163–4244 (1995)
51. B.C. Berndt, S.H. Chan, Z.-G. Liu, H. Yesilyurt, A new identity for $(q;q)_\infty^{10}$ with an application to Ramanujan's partition congruence modulo 11. Q. J. Math. **55**, 13–30 (2004)
52. B.C. Berndt, R.A. Rankin, *Ramanujan. Letters and Commentary* (American Mathematical Society, Providence, RI, 1995)
53. B.C. Berndt, L.-C. Zhang, A new class of theta-function identities originating in Ramanujan's notebooks. J. Number Theory **48**, 224–242 (1994)
54. F. Beukers, Irrationality of $\pi^2$, periods of an elliptic curve and $\Gamma_1(5)$, in *Diophantine Approximations and Transcendental Numbers (Luminy, 1982)*. Progress in Mathematics, vol. 31 (Birkhäuser, Boston, MA, 1983), pp. 47–66
55. S. Bhargava, Unification of the cubic analogues of the Jacobian theta-function. J. Math. Anal. Appl. 193, 543–558 (1995)
56. S. Bhargava, D. Somashekara, Some eta-function identities deducible from Ramanujan's ${}_1\psi_1$ summation formula. J. Math. Anal. Appl. **176**, 554–560 (1993)
57. G. Bhatnagar, How to prove Ramanujan's $q$-continued fractions, in *Ramanujan 125*. Contemporary Mathematics, vol. 627 (American Mathematical Society, Providence, RI, 2014), pp. 49–68
58. R. Blecksmith, J. Brillhart, I. Gerst, Some infinite product identities. Math. Comput. **51**, 301–314 (1988)
59. J.M. Borwein, D. Bailey, R. Girgensohn, *Experimentation in Mathematics. Computational Paths to Discovery* (A. K. Peters, Natick, MA, 2004)
60. J.M. Borwein, P.B. Borwein, *Pi and the AGM. A Study in Analytic Number Theory and Computational Complexity* (Wiley, New York, 1987)
61. J.M. Borwein, P.B. Borwein, Ramanujan and pi. Sci. Am. **256**, 112–117 (1988). Reprinted in [43, pp. 588–595]
62. J.M. Borwein, P.B. Borwein, A remarkable cubic mean iteration, in *Computational Methods and Function Theory (Valparaíso, 1989)*. Lecture Notes in Mathematics, vol. 1435 (Springer, Berlin, 1990), pp. 27–31
63. J.M. Borwein, P.B. Borwein, A cubic counterpart of Jacobi's identity and the AGM. Trans. Am. Math. Soc. **323**, 691–701 (1991)
64. J.M. Borwein, P.B. Borwein, Some observations on computer aided analysis. Not. Am. Math. Soc. **39**, 825–829 (1992)
65. J.M. Borwein, P.B. Borwein, D.H. Bailey, Ramanujan, modular equations, and approximations to pi or how to compute one billion digits of pi. Am. Math. Mon. **96**, 201–219 (1989)
66. J.M. Borwein, P.B. Borwein, F.G. Garvan, Hypergeometric analogues of the arithmetic-geometric mean iteration. Constr. Approx. **9**, 509–523 (1993)
67. J.M. Borwein, P.B. Borwein, F.G. Garvan, Some cubic modular identities of Ramanujan. Trans. Am. Math. Soc. **343**, 35–47 (1994)
68. J.M. Borwein, F.G. Garvan, Approximations to $\pi$ via the Dedekind eta function, in *Organic Mathematics (Burnaby, BC, 1995)*. CMS Conference Proceedings, vol. 20 (American Mathematical Society, Providence, RI, 1997), pp. 89–115
69. L. Carlitz, Note on some partition formulae. Q. J. Math. Oxf. Ser. (2) **4**, 168–172 (1953)
70. L. Carlitz, Note on sums of four and six squares. Proc. Am. Math. Soc. **8**, 120–124 (1957)
71. A.-L. Cauchy, Second Mémoire sur les fonctions dont plusieurs valeurs sont liées entre elles par une équation linéaire. C. R. Acad. Sci. Paris **17**, 567 (1843). Reprinted in Oeuvres de Cauchy [1893], Series 1, vol. 8, 50–55
72. A.-L. Cauchy, Mémoire sur les fonctions complémentaires. C. R. Acad. Sci. Paris **19**, 1377–1384 (1844). Reprinted in Oeuvres de Cauchy [1893], Series 1, vol. 8, 378–385

73. H.-C. Chan, Another simple proof of the quintuple product identity. Int. J. Math. Math. Sci. **15**, 2511–2515 (2005)
74. H.-C. Chan, Ramanujan's cubic continued fraction and an analog of his "most beautiful identity". Int. J. Number Theory **6**, 673–680 (2010)
75. H.-C. Chan, Ramanujan's cubic continued fraction and Ramanujan type congruences for a certain partition function. Int. J. Number Theory **6**, 819–834 (2010)
76. H.-C. Chan, S. Cooper, Congruences modulo powers of 2 for a certain partition function. Ramanujan J. **22**, 101–117 (2010)
77. H.H. Chan, On Ramanujan's cubic continued fraction. Acta Arith. **73**, 343–355 (1995)
78. H.H. Chan, Triple product identity, quintuple product identity and Ramanujan's differential equations for the classical Eisenstein series. Proc. Am. Math. Soc. **135**, 1987–1992 (2007)
79. H.H. Chan, S.H. Chan, Z.-G. Liu, Domb's numbers and Ramanujan–Sato type series for $1/\pi$. Adv. Math. **186**, 396–410 (2004)
80. H.H. Chan, S. Cooper, Powers of theta functions. Pac. J. Math. **235**, 1–14 (2008)
81. H.H. Chan, S. Cooper, Rational analogues of Ramanujan's series for $1/\pi$. Math. Proc. Camb. Philos. Soc. **153**, 361–383 (2012)
82. H.H. Chan, S. Cooper, W.-C. Liaw, On $\eta^3(a\tau)\eta^3(b\tau)$ with $a+b=8$. J. Aust. Math. Soc. **84**, 301–313 (2008)
83. H.H. Chan, S. Cooper, F. Sica, Congruences satisfied by Apéry-like numbers. Int. J. Number Theory **6**, 89–97 (2010)
84. H.H. Chan, S. Cooper, P.C. Toh, The 26th power of Dedekind's $\eta$-function. Adv. Math. **207**, 532–543 (2006)
85. H.H. Chan, S. Cooper, P.C. Toh, Ramanujan's Eisenstein series and powers of Dedekind's eta-function. J. Lond. Math. Soc. (2) **75**, 225–242 (2007)
86. H.H. Chan, T.G. Chua, An alternative transformation formula for the Dedekind $\eta$-function via the Chinese Remainder Theorem. Int. J. Number Theory **12**, 513–526 (2016)
87. H.H. Chan, S.-S. Huang, On the Ramanujan–Göllnitz–Gordon continued fraction. Ramanujan J. **1**, 75–90 (1997)
88. H.H. Chan, Z.-G. Liu, Analogues of Jacobi's inversion formula for the incomplete elliptic integral of the first kind. Adv. Math. **174**, 69–88 (2003)
89. H.H. Chan, Z.-G. Liu, S.T. Ng, Circular summation of theta functions in Ramanujan's lost notebook. J. Math. Anal. Appl. **316**, 628–641 (2006)
90. H.H. Chan, Y. Tanigawa, Y. Yang, W. Zudilin, New analogues of Clausen's identities arising from the theory of modular forms. Adv. Math. **228**, 1294–1314 (2011)
91. H.H. Chan, P.C. Toh, New analogues of Ramanujan's partition identities J. Number Theory **130**, 1898–1913 (2010)
92. H.H. Chan, H. Verrill, The Apéry numbers, the Almkvist–Zudilin numbers and new series for $1/\pi$. Math. Res. Lett. **16**, 405–420 (2009)
93. H.H. Chan, J.G. Wan, W. Zudilin, Legendre polynomials and Ramanujan-type series for $1/\pi$. Isr. J. Math. **194**, 183–207 (2013)
94. H.H. Chan, W. Zudilin, New representations for Apéry-like sequences. Mathematika **56**, 107–117 (2010)
95. S.H. Chan, Generalized Lambert series identities. Proc. Lond. Math. Soc. (3) **91**, 598–622 (2005)
96. R. Chapman, Cubic identities for theta series in three variables. Ramanujan J. **8**, 459–465 (2004)
97. W. Chu, L. Di Claudio, Classical partition identites and basic hypergeometric series, Dipartimento di Matematica, Università di Lecce, Lecce, 2004
98. D.V. Chudnovsky, G.V. Chudnovsky, Approximations and complex multiplication according to Ramanujan, in *Ramanujan Revisited*, Urbana-Champaign, IL, 1987 (Academic, Boston, MA, 1988), pp. 375–472
99. E. van Fossen Conrad, Some continued fraction expansions of Laplace transforms of elliptic functions, PhD Thesis, The Ohio State University, 2002

100. E. van Fossen Conrad, P. Flajolet, The Fermat cubic, elliptic functions, continued fractions, and a combinatorial excursion, Sém. Lothar. Combin. **54**, 44pp. (2005). Article B54g
101. S. Cooper, Multiplying Macdonald identities, in *Special Functions and Differential Equations (Madras, 1997)* (Allied Publishers, New Delhi, 1998), pp. 73–82
102. S. Cooper, On sums of an even number of squares, and an even number of triangular numbers: an elementary approach based on Ramanujan's ${}_1\psi_1$ summation formula, in *q-series with Applications to Combinatorics, Number Theory, and Physics (Urbana, IL, 2000)*. Contemporary Mathematics, vol. 291 (American Mathematical Society, Providence, RI, 2001), pp. 115–137
103. S. Cooper, Cubic theta functions. J. Comput. Appl. Math. **160**, 77–94 (2003)
104. S. Cooper, A simple proof of an expansion of an eta-quotient as a Lambert series. Bull. Aust. Math. Soc. **71**, 353–358 (2005)
105. S. Cooper, Cubic elliptic functions. Ramanujan J. **11**, 355–397 (2006)
106. S. Cooper, The quintuple product identity. Int. J. Number Theory **2**, 115–161 (2006)
107. S. Cooper, On the number of representations of integers by certain quadratic forms. Bull. Aust. Math. Soc., **78**, 129–140 (2008)
108. S. Cooper, Construction of Eisenstein series for $\Gamma_0(p)$. Int. J. Number Theory **5**, 765–778 (2009)
109. S. Cooper, On the number of representations of integers by certain quadratic forms, II. J. Combin. Number Theory. **1**, 311–328 (2009)
110. S. Cooper, Inversion formulas for elliptic functions. Proc. Lond. Math. Soc. **99**, 461–483 (2009)
111. S. Cooper, On Ramanujan's function $k(q) = r(q)r^2(q^2)$. Ramanujan J. **20**, 311–328 (2009)
112. S. Cooper, Theta function identities for level 15, in *Ramanujan Rediscovered*, ed. by N.D. Baruah, B.C. Berndt, S. Cooper, T. Huber, M. Schlosser. Ramanujan Mathematical Society Lecture Notes Series, vol. 14 (Ramanujan Mathematical Society, Mysore, 2010), pp. 79–86
113. S. Cooper, Level 10 analogues of Ramanujan's series for $1/\pi$. J. Ramanujan Math. Soc. **27**, 59–76 (2012)
114. S. Cooper, Sporadic sequences, modular forms and new series for $1/\pi$. Ramanujan J. **29**, 163–183 (2012)
115. S. Cooper, J. Ge, D. Ye, Hypergeometric transformation formulas of degrees 3, 7, 11 and 23. J. Math. Anal. Appl., **421**, 1358–1376 (2015)
116. S. Cooper, J. Guillera, A. Straub, W. Zudilin, Crouching AGM, hidden modularity. `arXiv:1604.01106v1 [math.NT]` 5 April 2016
117. S. Cooper, M.D. Hirschhorn, On some infinite product identities. Rocky Mt. J. Math. **31**, 131–139 (2001)
118. S. Cooper, M.D. Hirschhorn, On some sum-to-product identities. Bull. Aust. Math. Soc. **63**, 353–365 (2001)
119. S. Cooper, M. Hirschhorn, A combinatorial proof of a result from number theory. Integers **4**, A9, 1–4 (2004)
120. S. Cooper, M. Hirschhorn, Sums of squares and sums of triangular numbers. Georgian Math. J. **13**, 675–686 (2006)
121. S. Cooper, M.D. Hirschhorn, Factorizations that involve Ramanujan's function $k(q) = r(q)r^2(q^2)$. Acta Math. Sin. Engl. Ser. **27**, 2301–2308 (2011)
122. S. Cooper, B. Kane, D. Ye, Analogues of the Ramanujan–Mordell theorem. J. Math. Anal. Appl. **446**, 568–579 (2017)
123. S. Cooper, H.Y. Lam, Sums of two, four, six and eight squares and triangular numbers: an elementary approach. Indian J. Math. **44**, 21–40 (2002)
124. S. Cooper, P.C. Toh, Quintic and septic Eisenstein series. Ramanujan J. **19**, 163–181 (2009)
125. S. Cooper, J. Wan, W. Zudilin, Holonomic alchemy and series for $1/\pi$. `arXiv:1512.04608v2 [math.NT]` 17 August 2016
126. S. Cooper, D. Ye, Eisenstein series to the tredecic base. Bull. Aust. Math. Soc. **91**, 19–28 (2015)

127. S. Cooper, D. Ye, Explicit evaluations of a level 13 analogue of the Rogers–Ramanujan continued fraction. J. Number Theory **139**, 91–111 (2014)
128. S. Cooper, D. Ye, The Rogers–Ramanujan continued fraction and its level 13 analogue. J. Approx. Theory **193**, 99–127 (2015)
129. S. Cooper, D. Ye, Level 14 and 15 analogues of Ramanujan's elliptic functions to alternative bases. Trans. Am. Math. Soc. **368**, 7883–7910 (2016)
130. S. Cooper, D. Ye, The level 12 analogue of Ramanujan's function $k$. J. Aust. Math. Soc. **101**, 29–53 (2016)
131. S. Cooper, W. Zudilin, Hypergeometric modular equations. `arXiv:1609.07276 [math.NT]` 23 September 2016
132. D.A. Cox, The arithmetic-geometric mean of Gauss. L'Enseignement Mathématique **30**, 275–330 (1984)
133. D.A. Cox, *Primes of the Form* $x^2 + ny^2$ (Wiley, New York, 1989)
134. G. Darboux, Mémoire sur la théorie des coordonnées curvilignes, et des systèmes orthogonaux. Annales scientifiques de l'École Normale Supérieure, Sér. 2 **7**, 101–150 (1878)
135. L.E. Dickson, *History of the Theory of Numbers*, vol. III (Carnegie Institute of Washington, 1923). Republished by AMS Chelsea, Providence, RI, 1999
136. L.E. Dickson, *Introduction to the Theory of Numbers* (University of Chicago Press, Chicago, 1929). Republished by Dover, New York, 1957
137. P.G.L. Dirichlet, *Werke*, vol. 1 (G. Reimer, Berlin, 1889)
138. A.C. Dixon, On the doubly periodic functions arising out of the curve $x^3 + y^3 - 3\alpha xy = 1$. Q. J. Pure Appl. Math. **24**, 167–233 (1890)
139. A.C. Dixon, *The Elementary Properties of the Elliptic Functions* (Macmillan, London, 1894)
140. J.M. Dobbie, A simple proof of some partition formulae of Ramanujan's. Q. J. Math. Oxf. Ser. (2) **6**, 193–196 (1955)
141. C. Domb, On the theory of cooperative phenomena in crystals. Adv. Phys. **9**, 149–361 (1960)
142. W. Duke, Continued fractions and modular functions. Bull. Am. Math. Soc. (N.S.) **42**, 137–162 (2005)
143. W.F. Eberlein, On Euler's infinite product for the sine. J. Math. Anal. Appl. **58**, 147–151 (1977)
144. G. Eisenstein, Neue Theoreme der höheren Arithmetik. J. Reine Angew. Math. **35**, 117–136 (1847). Reprinted in *Mathematische Werke*, vol. 1 (Chelsea, New York, 1989), pp. 483–502
145. S.B. Ekhad, D. Zeilberger, A WZ proof of Ramanujan's formula for $\pi$, in *Geometry, Analysis and Mechanics* (World Scientific Publishing, River Edge, NJ, 1994), pp. 107–108
146. C. Elsner, On recurrence formulae for sums involving binomial coefficients. Fibonacci Q. **43**, 31–45 (2005)
147. A. Erdélyi, W. Magnus, F. Oberhettinger, F.G. Tricomi, *Higher Transcendental Functions*, vol. III (Robert E. Krieger Publishing Co., Inc., Malabar, FL, 1981). Reprint of the 1955 original
148. R. Evans, Theta function identities. J. Math. Anal. Appl. **147**, 97–121 (1990)
149. H.M. Farkas, I. Kra, Partitions and theta constant identities, in *Analysis, Geometry, Number Theory: The Mathematics of Leon Ehrenpreis (Philadelphia, PA, 1998)*. Contemporary Mathematics, vol. 251 (American Mathematical Society, Providence, RI, 2000), pp. 197–203
150. H.M. Farkas, I. Kra, *Theta Constants, Riemann Surfaces and the Modular Group*. Graduate Studies in Mathematics, vol. 37 (American Mathematical Society, Providence, RI, 2001)
151. N. Fine, *Basic Hypergeometric Series and Applications* (American Mathematical Society, Providence, RI, 1988)
152. K.B. Ford, Some infinite series identities. Proc. Am. Math. Soc. **119**, 1019–1020 (1993)
153. J. Franel, Solution to a question of Laisant. L'intermédiaire des mathématiciens **1**, 45–47 (1894)
154. J. Franel, Solution to a question of Laisant. L'intermédiaire des mathématiciens **2**, 33–35 (1895)
155. R. Fricke, *Die Elliptischen Funktionen und ihre Anwendungen, Erster Teil* (B. G. Teubner, Leipzig, 1916)

156. G. Frobenius, L. Stickelberger, Ueber die Addition und Multiplication der elliptischen Functionen. J. Reine Angew. Math. **88**, 146–184 (1879)
157. F.G. Garvan, A simple proof of Watson's partition congruences for powers of 7. J. Aust. Math. Soc. Series A **36**, 316–334 (1984)
158. F.G. Garvan, Cubic modular identities of Ramanujan, hypergeometric functions and analogues of the arithmetic-geometric mean iteration, in *The Rademacher Legacy to Mathematics (University Park, PA, 1992)*. Contemporary Mathematics, vol. 166 (American Mathematical Society, Providence, RI, 1994), pp. 245–264
159. C.F. Gauss, Hundert Theoreme über die neuen Transzendenten. Werke, vol. 3 (1973), pp. 461–469
160. J. Ge, Level 11 and level 23 analogues of Ramanujan's series for $1/\pi$, Master of Philosophy Thesis, Massey University, New Zealand, 2015
161. I. Gessel, Some congruences for Apéry numbers. J. Number Theory **14**, 362–368 (1982)
162. J.W.L. Glaisher, On the numbers of representations of a number as a sum of $2r$ squares, where $2r$ does not exceed eighteen. Proc. Lond. Math. Soc. (2) **5**, 479–490 (1907)
163. J.W.L. Glaisher, On the representations of a number as the sum of two, four, six, eight, ten and twelve squares. Q. J. Pure Appl. Math. **38**, 1–62 (1907)
164. H. Göllnitz, Partitionen mit Differenzenbedingungen. J. Reine Angew. Math. **225**, 154–190 (1967)
165. B. Gordon, Some continued fractions of the Rogers–Ramanujan type. Duke Math. J. **32**, 741–748 (1965)
166. C. Guetzlaff, Aequatio modularis pro transformatione functionum ellipticarum septimi ordinis. J. Reine Angew. Math. **12**, 173–177 (1834)
167. C. Gugg, Two modular equations for squares of the Rogers–Ramanujan functions with applications. Ramanujan J. **18**, 183–207 (2009)
168. J. Guillera, Generators of some Ramanujan formulas. Ramanujan J. **11**, 41–48 (2006)
169. J. Guillera, On WZ-pairs which prove Ramanujan series. Ramanujan J. **22**, 249–259 (2010)
170. J. Guillera, WZ-Proofs of "divergent" Ramanujan-Type Series. Advances in Combinatorics (Springer, Heidelberg, 2013), pp. 187–195
171. J. Guillera, New proofs of Borwein-type algorithms for pi. Integr. Transf. Spec. Funct. **27**, 775–782 (2016)
172. J. Guillera, W. Zudilin, Ramanujan-type formulae for $1/\pi$: the art of translation, in *The Legacy of Srinivasa Ramanujan*, ed. by B.C. Berndt, D. Prasad. Ramanujan Mathematical Society Lecture Notes Series, vol. 20 (Ramanujan Mathematical Society, Mysore, 2013), pp. 181–195
173. G.-H. Halphen, *Traité des Fonctions Elliptiques et de leurs Applications, Part 1* (Gauthier-Villars, Paris, 1886)
174. G.H. Hardy, *Ramanujan. Twelve Lectures on Subjects Suggested by his Life and Work*, 3rd edn. (AMS, Providence, RI, 1999)
175. G.H. Hardy, E.M. Wright, *An Introduction to the Theory of Numbers*, 5th edn. (The Clarendon Press, Oxford University Press, New York, 1979)
176. M.D. Hirschhorn, A continued fraction of Ramanujan. J. Aust. Math. Soc. Ser. A **29**, 80–86 (1980)
177. M.D. Hirschhorn, A simple proof of an identity of Ramanujan. J. Aust. Math. Soc. Ser. A **34**, 31–35 (1983)
178. M.D. Hirschhorn, Three classical results on representations of a number. Sém. Lothar. Combinatoire **42**, 8 pp. (1999). Article B42f
179. M.D. Hirschhorn, An identity of Ramanujan, and applications, in *q-Series from a Contemporary Perspective*. Contemporary Mathematics, vol. 254 (American Mathematical Society, Providence, RI, 2000), pp. 229–234
180. M.D. Hirschhorn, Partial fractions and four classical theorems of number theory. Am. Math. Mon. **107**, 260–264 (2000)
181. M.D. Hirschhorn, Ramanujan's "most beautiful identity". Aust. Math. Soc. Gazette **32**, 259–262 (2005)

182. M.D. Hirschhorn, The case of the mysterious sevens. Int. J. Number Theory **2**, 213–216 (2006)
183. M.D. Hirschhorn, A sequence considered by Shaun Cooper. N. Z. J. Math. **43**, 37–42 (2013)
184. M.D. Hirschhorn, Factorizations of certain $q$-series identities of Ramanujan and others. Ramanujan J. **31**, 15–22 (2013)
185. M. Hirschhorn, F. Garvan, J. Borwein, Cubic analogues of the Jacobian theta functions $\theta(z, q)$. Can. J. Math. **45**, 673–694 (1993)
186. T. Horie, N. Kanou, Certain modular functions similar to the Dedekind eta function. Abh. Math. Sem. Univ. Hamburg **72**, 89–117 (2002)
187. J.G. Huard, P. Kaplan, K.S. Williams, The Chowla-Selberg formula for genera. Acta Arith. **73**, 271–301 (1995)
188. T. Huber, D. Lara, Differential equations for septic theta functions. Ramanujan J. **38**, 211–226 (2015)
189. C.G.J. Jacobi, *Fundamenta Nova Theoriae Functionum Ellipticarum*. Reprinted in Gesammelte Werke, vol. I (Chelsea, New York, 1969), pp. 49–239
190. W.P. Johnson, How Cauchy missed Ramanujan's ${}_1\psi_1$ summation. Amer. Math. Mon. **111**, 791–800 (2004)
191. C. Jordan, *Course d'Analyse,* vol. II, 3rd edn. (Gauthier-Villars, Paris, 1913). Reprinted by Jacques Gabay, Sceaux (1991)
192. S.Y. Kang, Some theorems on the Rogers–Ramanujan continued fraction and associated theta function identities in Ramanujan's lost notebook. Ramanujan J. **3**, 91–111 (1999)
193. L. Kiepert, Zur Transformationstheorie der elliptischen Functionen. J. Reine Angew. Math. **87**, 199–216 (1879)
194. S. Kim, A bijective proof of the quintuple product identity. Int. J. Number Theory **6**, 247–256 (2010)
195. D.E. Knuth, *The Art of Computer Programming. Fundamental Algorithms*, vol. 1 (1968), 3rd edn. (Addison-Wesley, Reading, MA, 1997)
196. G. Köhler, *Eta Products and Theta Series Identities* (Springer, Heidelberg, 2011)
197. O. Kolberg, Some identities involving the partition function. Math. Scand. **5**, 77–92 (1957)
198. O. Kolberg, Note on the Eisenstein series of $\Gamma_0(p)$, Årbok for Universitetet i Bergen. Matematisk-Naturvidenskapelig Serie, no. 6, 20 pp. Printed October 1968
199. T.H. Koornwinder, On the equivalence of two fundamental theta identities. Anal. Appl. **12**, 711–725 (2014)
200. L. Kronecker, Zur Theorie der elliptischen Functionen, Monatsberichte der Königlich Preussischen Akademie der Wissenschaften zu Berlin vom Jahre 1881, pp. 1165–1172. Reprinted in Leopold Kronecker's Werke, vol. 4, pp. 311–318, Teubner, Leipzig, 1929. Reprinted by Chelsea, New York, 1968
201. G. Lachaud, Ramanujan modular forms and the Klein quartic. Mosc. Math. J. **5**, 829–856 (2005)
202. M.N. Lalin, M.D. Rogers, Functional equations for Mahler measures of genus-one curves. Algebra Number Theory **1**, 87–117 (2007)
203. E. Landau, *Elementary Number Theory* (AMS Chelsea, Providence, RI, 1999)
204. P.J. Larcombe, D.R. French, On the "other" Catalan numbers: a historical formulation re-examined. Congr. Numer. **143**, 33–64 (2000)
205. P. Levrie, Using Fourier-Legendre expansions to derive series for $1/\pi$ and $1/\pi^2$. Ramanujan J. **22**, 221–230 (2010)
206. S. Levy (ed.), *The Eightfold Way. The Beauty of Klein's Quartic Curve.* Mathematical Sciences Research Institute Publications, vol. 35 (Cambridge University Press, Cambridge, 1999)
207. R.P. Lewis, Z.-G. Liu, The Borweins' cubic theta functions and $q$-elliptic functions, in *Symbolic Computation, Number Theory, Special Functions, Physics and Combinatorics*, Gainesville, FL (1999). Developments in Mathematics, vol. 4 (Kluwer Academic Publishers, Dordrecht, 2001), pp. 133–145
208. B.L.S. Lin, On the expansion of a continued fraction of order 12. Int. J. Number Theory **9**, 2019–2031 (2013)

209. J. Liouville, Sur la forme $x^2 + y^2 + 3(z^2 + t^2)$. J. Math. Pures Appl. Ser. 2 **5**, 147–152 (1860)
210. J. Liouville, Sur les deux formes quadratiques $x^2 + y^2 + z^2 + 2t^2$, $x^2 + 2(y^2 + z^2 + t^2)$. J. Math. Pures Appl. Ser. 2 **6**, 225–230 (1861)
211. J. Liouville, Sur la forme $x^2 + 2y^2 + 4z^2 + 4t^2$. J. Math. Pures Appl. Ser. 2 **7**, 62–64 (1862)
212. J. Liouville, Sur la forme $x^2 + y^2 + 2z^2 + 4t^2$. J. Math. Pures Appl. Ser. 2 **7**, 99–102 (1862)
213. J. Liouville, Remarque nouvelle sur la forme $x^2 + y^2 + 3(z^2 + t^2)$. J. Math. Pures Appl. Ser. 2 **8**, 296 (1863)
214. J. Liouville, Sua la forme $x^2 + xy + y^2 + 2z^2 + 2zt + 2t^2$. J. Math. Pures Appl. Ser. 2 **8**, 308–310 (1863)
215. J. Liouville, Sur la forme $x^2 + y^2 + z^2 + t^2 + u^2 + 2v^2$. J. Math. Pures Appl. Ser. 2 **9**, 161–174 (1864)
216. J. Liouville, Sur la forme $x^2 + 2(y^2 + z^2 + t^2 + u^2 + v^2)$. J. Math. Pures Appl. Ser. 2 **9**, 175–180 (1864)
217. J. Liouville, Extrait d'une lettre adressée à M. Besge. J. Math. Pures Appl. Ser. 2 **9**, 296–298 (1864)
218. J. Liouville, Nombre des représentations d'un entier quelconque sous la forme d'une somme de dix carrés. J. Math. Pures Appl. Ser. 2 **11**, 1–8 (1866)
219. Z.-G. Liu, On certain identities of Ramanujan. J. Number Theory **83**, 59–75 (2000)
220. Z.-G. Liu, The Borweins' cubic theta function identity and some cubic modular identities of Ramanujan. Ramanujan J. **4**, 43–50 (2000)
221. Z.-G. Liu, Some Eisenstein series associated with the Borwein functions, in *Symbolic Computation, Number Theory, Special Functions, Physics and Combinatorics*, Gainesville, FL (1999). Developments in Mathematics, vol. 4 (Kluwer Academic Publishers, Dordrecht, 2001), pp. 147–169
222. Z.-G. Liu, Some Eisenstein series identities related to modular equations of the seventh order. Pac. J. Math. **209**, 103–130 (2003)
223. Z.-G. Liu, A theta function identity and the Eisenstein series on $\Gamma_0(5)$. J. Ramanujan Math. Soc. **22**, 283–298 (2007)
224. Z.-G. Liu, A theta function identity and applications, in *Ramanujan Rediscovered*, ed. by N.D. Baruah, B.C. Berndt, S. Cooper, T. Huber, M. Schlosser, Ramanujan Mathematical Society Lecture Notes Series, vol. **14** (Ramanujan Mathematical Society, Mysore, 2010), pp. 165–183
225. L. Lorentzen, H. Waadeland, *Continued Fractions with Applications*. Studies in Computational Mathematics, vol. 3 (North-Holland Publishing Co., Amsterdam, 1992)
226. L. Lorenz, Bidrag til tallenes theori. Tidsskrift for Mathematik (3) **1**, 97–114 (1871)
227. H. Lundmark, http://math.stackexchange.com/questions/109069/stolz-cesàro-theorem (webpage accessed: July 2016)
228. J. Lützen, *Joseph Liouville 1809–1882. Master of Pure and Applied Mathematics* (Springer, New York, 1990)
229. I. Macdonald, Affine root systems and Dedekind's $\eta$-function. Inv. Math. **15**, 91–143 (1972)
230. R.S. Maier, The 192 solutions of the Heun equation. Mathematics of Computation, vol. 76 (2007), pp. 811–843
231. A. Malik, A. Straub, Divisibility properties of sporadic Apéry-like numbers. Res. Number Theory **2**, 1–26 (2016). Article 5
232. T. Miyake, *Modular Forms* (Springer, Berlin, 2006)
233. L.J. Mordell, On the representation of numbers as the sum of $2r$ squares. Q. J. Pure Appl. Math. Oxf. **48**, 93–104 (1917)
234. L.J. Mordell, An identity in combinatory analysis. Proc. Glasgow Math. Assoc. **5**, 197–200 (1962)
235. C. Moreau, Solution to a question of Laisant. L'intermédiaire des mathématiciens **1**, 47 (1894)
236. M.S. Mahadeva Naika, S. Chandankumar, K. Sushan Bairy, Some new identities for a continued fraction of order 12. South East Asian J. Math. Math. Sci. **10**, 129–140 (2012)
237. M.S. Mahadeva Naika, B.N. Dharmendra, K. Shivashankara, A continued fraction of order twelve. Centr. Eur. J. Math. **6**, 393–404 (2008)

238. P.S. Nazimoff, *Applications of the Theory of Elliptic Functions to the Theory of Numbers*, Moscow, 1884. Translated from Russian into English by A.E. Ross, Chicago, 1928
239. L. O'Brien, Modular forms and two new integer sequences at level 7, Master of Science Thesis, Massey University, New Zealand, 2016
240. Y. Ohyama, Differential equations for modular forms with level three. Funkcialaj Ekvacioj **44**, 377–389 (2001)
241. H. Petersson, *Modulfunktionen und quadratische Formen*. Ergebnisse der Mathematik und ihrer Grenzgebiete (Springer, Berlin, 1982)
242. K. Petr, O počtu tříd forem kvadratických záporného diskriminantu. Rozpravy Ceské Akademie Císare Frantiska Josefa I **10**, 1–22 (1901). JFM 32.0212.02
243. G. Pólya, G.L. Alexanderson, Gaussian binomial coefficients. Elem. Math. **26**, 102–109 (1971)
244. R. Preston, Profiles: The mountains of pi, The New Yorker, March 2, 1992, pp. 36–67
245. S. Raghavan, S.S. Rangachari, Ramanujan's elliptic integrals and modular identities, in *Number Theory and Related Topics* (Oxford University Press, Bombay, 1989), pp. 119–149
246. V. Ramamani, Some identities conjectured by Srinivasa Ramanujan found in his lithographed notes connected with partition theory and elliptic modular functions – their proofs – interconnection with various other topics in the theory of numbers and some generalisations thereon, PhD Thesis, University of Mysore, India, 1971
247. K.G. Ramanathan, Generalisations of some theorems of Ramanujan. J. Number Theory **29**, 118–137 (1988)
248. S. Ramanujan, Modular equations and approximations to $\pi$. Q. J. Math. (Oxf.) **45**, 350–372 (1914). Reprinted in [255, 23–39]
249. S. Ramanujan, On certain arithmetical functions. Trans. Camb. Philos. Soc. **22**, 159–184 (1916). Reprinted in [255, pp. 136–162]
250. S. Ramanujan, On certain trigonometrical sums. Trans. Camb. Philos. Soc. **22**, 259–276 (1918). Reprinted in [255, pp. 179–199]
251. S. Ramanujan, Some properties of $p(n)$, the number of partitions of $n$. Proc. Camb. Philos. Soc. **19**, 207–210 (1919). Reprinted in [255, pp. 210–213]
252. S. Ramanujan, Proof of certain identities in combinatory analysis. Proc. Camb. Philos. Soc. **19**, 214–216 (1919) Reprinted in [255, 214–215]
253. S. Ramanujan, *Notebooks*, 2 vols. (Tata Institute of Fundamental Research, Bombay, 1957)
254. S. Ramanujan, *The Lost Notebook and Other Unpublished Papers* (Narosa, New Delhi, 1988)
255. S. Ramanujan, *Collected Papers*, Third printing (AMS Chelsea, Providence, RI, 2000)
256. L.J. Rogers, Second memoir on the expansion of certain infinite products. Proc. Lond. Math. Soc. **25**, 318–343 (1894)
257. M. Rogers, New ${}_5F_4$ hypergeometric transformations, three-variable Mahler measures, and formulas for $1/\pi$. Ramanujan J. **18**, 327–340 (2009)
258. M. Rogers, A. Straub, A solution of Sun's \$520 challenge concerning $520/\pi$. Int. J. Number Theory **9**, 1273–1288 (2013)
259. R. Roy, *Sources in the Development of Mathematics* (Cambridge University Press, Cambridge, 2011)
260. T. Sato (Takeshi Sato), Apéry numbers and Ramanujan's series for $1/\pi$. Abstract of a talk presented the Annual Meeting of the Mathematical Society of Japan, 28–31 March, 2002
261. B. Schoeneberg, *Elliptic Modular Functions: An Introduction* (Springer, New York-Heidelberg, 1974)
262. M. Schlosser, Elliptic enumeration of nonintersecting lattice paths. J. Combin. Theory Ser. A **114**, 505–521 (2007)
263. D. Schultz, Cubic theta functions. Adv. Math. **248**, 618–697 (2013)
264. D. Schultz, Identities involving theta functions and analogues of theta functions, PhD Thesis, University of Illinois, 2013
265. H.A. Schwarz, *Formeln und Lehrsätze zum Gebrauche der elliptischen Funktionen, Nach Vorlesungen und Aufzeichnungen des Herrn K. Weierstrass* (Julius Springer, Berlin, 1893)

266. A. Selberg, Über einige arithmetische Identitäten. Avhandlinger utgitt av Det Norske Videnskaps-Akademi i Oslo I. Mat.-Naturv. Klasse, no. 8 (1936), pp. 1–23. Reprinted in Collected Papers, vol. 1 (Springer, Berlin, 1989)
267. J.-P. Serre, *A course in Arithmetic* (Springer, New York, 1973)
268. J.-P. Serre, Modular forms of weight one and Galois representations, in *Algebraic Number Fields: L-Functions and Galois Properties* (Academic, London, 1977), pp. 193–268. Reprinted in Œuvres, vol. III (Springer, Berlin, 1986)
269. L.-C. Shen, On the products of three theta functions. Ramanujan J. **3**, 343–357 (1999)
270. N.-P. Skoruppa, A quick combinatorial proof of Eisenstein series identities. J. Number Theory **43**, 68–73 (1993)
271. N.J.A. Sloane, The on-line encyclopedia of integer sequences. http://www.research.att.com/njas/sequences/ (webpage accessed: February 2017)
272. H.J.S. Smith, Report on the theory of numbers. Part VI. Report of the British Association for 1865, pp. 322–375. Reprinted in Collected Mathematical Papers, vol. 1, 1894. Reprinted by Chelsea, New York, 1979
273. R.P. Stanley, *Enumerative Combinatorics*, vol. 2 (Cambridge University Press, Cambridge, 1999)
274. A. Straub, W. Zudilin, Positivity of rational functions and their diagonals. J. Approx. Theory **195**, 57–69 (2015)
275. V. Strehl, Binomial identities—combinatorial and algorithmic aspects. Discr. Math. **136**, 309–346 (1994)
276. Z.-H. Sun, On the number of representations of $n$ by $ax(x-1)/2 + by(y-1)/2$. J. Number Theory **129**, 971–989 (2009)
277. Z.-H. Sun, K.S. Williams, On the number of representations of $n$ by $ax^2 + bxy + cy^2$. Acta Arith. **122**, 101–171 (2006)
278. Z.-H. Sun, K.S. Williams, Ramanujan identities and Euler products for a type of Dirichlet series. Acta Arith. **122**, 349–393 (2006)
279. Z.-W. Sun, List of conjectural series for powers of $\pi$ and other constants, `arXiv:1102.5649v47 [math.CA]`. 29 December 2014
280. J. Tannery, J. Molk, *Elements de la Theorie des Fonctions Elliptiques*, 4 vols. (Gauthier-Villars, Paris, 1893–1902). Reprinted by Chelsea, New York, 1972
281. P.C. Toh, Representations of certain binary quadratic forms as Lambert series. Acta Arith. **143**, 227–237 (2010)
282. P.C. Toh, Differential equations satisfied by Eisenstein series of level 2. Ramanujan J. **25**, 179–194 (2011)
283. F. van der Blij, Binary quadratic forms of discriminant −23. Nederl. Akad. Wetensch. Proc. Ser. A. **55**; Indagationes Math. **14**, 498–503 (1952)
284. B. van der Pol, On a non-linear partial differential equation satisfied by the logarithm of the Jacobian theta-functions, with arithmetical applications. I, II. Nederl. Akad. Wetensch. Proc. Ser. A **54**; Indagationes Math. **13**, 261–271, 272–284 (1951)
285. A. van der Poorten, A proof that Euler missed ⋯ Apéry's proof of the irrationality of $\zeta(3)$. An informal report. Math. Intell. **1**, 195–203 (1978/1979)
286. K.R. Vasuki, A.A.A. Kahtan, G. Sharath, C. Sathish Kumar, On a continued fraction of order 12. Ukr. Math. J. **62**, 1866–1878 (2011)
287. K. Venkatachaliengar, Elementary proofs of the infinite product formula for $\sin z$ and allied formulae. Am. Math. Mon. **69**, 541–545 (1962)
288. K. Venkatachaliengar, *Development of Elliptic Functions According to Ramanujan,* Department of Mathematics, Madurai Kamaraj University, Technical Report 2, 1988. Edited and revised by S. Cooper, World Scientific, Singapore, 2012
289. A. Walfisz, Zur additiven Zahlentheorie. VI. Trav. Inst. Math. Tbilissi [Trudy Tbiliss. Mat. Inst.] **5**, 197–254 (1938)
290. A. Walfisz, Zur additiven Zahlentheorie. VIII. Trav. Inst. Math. Tbilissi [Trudy Tbiliss. Mat. Inst.] **8**, 69–107 (1940)
291. A. Walfisz, Zur additiven Zahlentheorie. IX. Trav. Inst. Math. Tbilissi [Trudy Tbiliss. Mat. Inst.] **9**, 75–96 (1941)

292. J.G. Wan, Random walks, elliptic integrals and related constants, PhD Thesis, The University of Newcastle, Australia, 2013
293. J.G. Wan, Series for $1/\pi$ using Legendre's relation. Integr. Transf. Spec. Funct. **25**, 1–14 (2014)
294. J.G. Wan, W. Zudilin, Generating functions of Legendre polynomials: a tribute to Fred Brafman. J. Approx. Theory **164**, 488–503 (2012)
295. S.O. Warnaar, A generalization of the Farkas and Kra partition theorem for modulus 7. J. Combin. Theory Ser. A **110**, 43–52 (2005)
296. G.N. Watson, Theorems stated by Ramanujan (IX): two continued fractions. J. Lond. Math. Soc. **4**, 231–237 (1929)
297. G.N. Watson, Ramanujan's note books. J. Lond. Math. Soc. **6**, 137–153 (1931)
298. G.N. Watson, Ramanujans Vermutung über Zerfällungsanzahlen. J. Reine Angew. Math. **179**, 97–128 (1938)
299. E.T. Whittaker, G.N. Watson, *A Course of Modern Analysis*, 4th edn. (Cambridge University Press, Cambridge, 1927)
300. K.S. Williams, On Liouville's twelve squares theorem. Far East J. Math. Sci. **29**, 239–242 (2008)
301. K.S. Williams, On the representations of a positive integer by the forms $x^2 + y^2 + z^2 + 2t^2$ and $x^2 + 2y^2 + 2z^2 + 2t^2$. Int. J. Mod. Math. **3**, 225–230 (2008)
302. K.S. Williams, *Number Theory in the Spirit of Liouville*. London Mathematical Society Student Texts, vol. 76 (Cambridge University Press, Cambridge, 2011)
303. L. Winquist, An elementary proof of $p(11m+6) \equiv 0 \pmod{11}$. J. Combin. Theory **6**, 56–59 (1969)
304. D. Ye, Level 13 modular forms and their applications. Bachelor of Science Honours Report, Massey University, New Zealand, 2013
305. B. Yuttanan, New properties for the Ramanujan–Göllnitz–Gordon continued fraction. Acta Arith. **151**, 293–310 (2012)
306. D. Zagier, Elliptic modular forms and their applications, in *The 1-2-3 of Modular Forms*, 1–123 (Springer, Berlin, 2008)
307. D. Zagier, Integral solutions of Apéry-like recurrence equations, in *Groups and Symmetries*. CRM Proceedings of Lecture Notes, vol. 47 (American Mathematical Society, Providence, RI, 2009), pp. 349–366
308. W. Zudilin, A generating function of the squares of Legendre polynomials. Bull. Aust. Math. Soc. **89**, 125–131 (2014)
309. W. Zudilin, Lost in translation. Advances in Combinatorics (Springer, Heidelberg, 2013), pp. 287–293

# Notation Index




S. Cooper, *Ramanujan's Theta Functions*, DOI 10.1007/978-3-319-56172-1

# Author Index





S. Cooper, *Ramanujan's Theta Functions*, DOI 10.1007/978-3-319-56172-1

# Index





S. Cooper, *Ramanujan's Theta Functions*, DOI 10.1007/978-3-319-56172-1

Printed by Printforce, the Netherlands